Handbook on Anaerobic Fermentations

Bioprocess Technology

Series Editor

W. Courtney McGregor

Xoma Corporation
Berkeley, California

Volume 1 Membrane Separations in Biotechnology, *edited by W. Courtney McGregor*

Volume 2 Commercial Production of Monoclonal Antibodies: A Guide for Scale-Up, *edited by Sally S. Seaver*

Volume 3 Handbook on Anaerobic Fermentations, *edited by Larry E. Erickson and Daniel Yee-Chak Fung*

Additional Volumes in Preparation

Handbook on Anaerobic Fermentations

edited by

LARRY E. ERICKSON and DANIEL YEE-CHAK FUNG

Kansas State University
Manhattan, Kansas

MARCEL DEKKER, INC. New York • Basel

Library of Congress Cataloging-in-Publication Data

Handbook on anaerobic fermentations.

(Bioprocess technology ; v. 3)
Includes bibliographies and index.
1. Fermentation--Congresses. I. Erickson, L. E.
(Larry Eugene) . II. Fung, Daniel Yee-Chak
. III. Series.
TP156.F4H36 1989 660'.62 88-18960
ISBN 0-8247-7974-6

MARCEL DEKKER, INC.
270 Madison Avenue, New York, New York 10016

Current printing (last digit):
10 9 8 7 6 5 4 3 2 1

PRINTED IN THE UNITED STATES OF AMERICA

Series Introduction

The revolutionary developments in recombinant DNA and hybridoma technologies that began in the mid-1970s have helped to spawn several hundred new business enterprises. Not all these companies are aimed at producing gene products or cell products, as such. Many are supportive in nature; that is, they provide contract research, processing equipment, and various other services in support of companies that actually produce cell products. With time, some small companies will probably drop out or be absorbed by larger, more established firms. Others will mature and manufacture their own product lines. As this evolution takes place, an explosive synergism among the various industries and the universities will result in the conversion of laboratory science into industrial processing. Such a movement, necessarily profit driven, will result in many benefits to humanity.

New bioprocessing techniques will be developed and more conventional ones will be revised because of the influence of the new biotechnology. As bioprocess technology evolves, there will be a need to provide substantive documentation of the developments for those who follow the field. It is expected that the technologies will continue to develop rapidly, just as the life sciences have developed rapidly over the past 10-15 years. No single book could cover all of these developments adequately. Indeed, any single book will be in need of replacement or revision every few years. Therefore, our continuing series in this rapidly moving field will document the growth of bioprocess technology as it happens.

The numerous cell products already in the marketplace, and the others expected to arrive, in most cases come from three types of bioreactors: (1) classical fermentation; (2) cell culture technology; and (3) enzyme bioreactors. Common to the production of all cell products or cell product analogs will be bioprocess control, downstream processing (recovery and purification), and bioproduct finishing and formulation. These major branches of bioprocess technology will be represented by cornerstone books, even though they may not appear first. Other subbranches will appear, and over time, the bioprocess technology "tree" will take shape and continue growing by natural selection.

W. Courtney McGregor

Preface

During the last 20 years, there has been significant progress in the development of biotechnology and biochemical engineering. Much of the progress has been in anaerobic fermentations. The industrial importance of anaerobic fermentations in the production of food and beverages is well known and continuing. The expected future importance of anaerobic production of industrial chemicals and alcohol fuels has provided the incentive for continued research. The growth of the ethanol fermentation industry has been the motivation for many studies.

The purpose of this handbook is to bring together in one place much of the current knowledge on anaerobic fermentations. The emphasis is on the biotechnology and biochemical engineering associated with anaerobic fermentations. The contents have been ordered so that important topics such as kinetics, redox potential, membrane transport, bioenergetics, and reactor design are emphasized. The rationale for this approach is that descriptive treatments of each of the important anaerobic fermentations are available elsewhere.

The objective is to provide a book that is a useful reference for applied microbiologists, applied biochemists, biochemical engineers, and environmental engineers. The handbook is intended to be a reference book for professors and academic classes, as well as those involved in research, development, and design.

Each chapter is written to include references to the most important literature; however, a concise account of the important basic principles, methods, models, analyses, and results is included.

The intent is to provide a handbook that contains both important basic information and references where additional literature reviews and experimental and theoretical results are presented. The handbook should be an important addition to the personal library of those working in biotechnology and bioengineering.

Each chapter is written such that it can be understood by a scientist or engineer with a B.S. degree and a basic reading knowledge in biochemistry, microbiology, and biochemical engineering (equivalent to a good course in each).

The idea for a handbook on anaerobic fermentations should be credited to Dr. Maurits Dekker. The invitation to edit this work and his encouragement in the early stages of developing the scope and contents are appreciated. The early work to identify authors, develop chapter titles, and invite others to contribute was done while Dr. Erickson was on sabbatical leave at the University of California, Berkeley. The excellent library facilities and constructive suggestions received from many professional colleagues were very helpful.

The willingness of the authors to contribute and their efforts to help keep preparations for the book on schedule are appreciated. One of the goals of the editors was to have each chapter reviewed by at least one reviewer selected by the author of the chapter. We extend our thanks to all who contributed by helping with the review process.

Kansas State University provides an excellent environment for scholarly work. The encouragement of colleagues and administrators, the fine secretarial services, and other supporting services are appreciated. Financial support for our research from the National Science Foundation, Halliburton Education Foundation, U.S. Department of Agriculture, and the Kansas Agricultural and Engineering Experiment Stations is gratefully acknowledged.

Larry E. Erickson
Daniel Yee-Chak Fung

Contents

Contributors

Graham F. Andrews Biotechnology Unit, Idaho National Engineering Laboratory, Idaho Falls, Idaho

Mohammad M. Ataai* Department of Chemical Engineering, Polytechnic Institute, Brooklyn, New York

Hubert Bahl Institut für Mikrobiologie, Georg-August-Universität, Göttingen, Federal Republic of Germany

Rakesh Kumar Bajpai Department of Chemical Engineering, University of Missouri-Columbia, Columbia, Missouri

John Patrick Barford Department of Chemical Engineering, University of Sydney, Sydney, New South Wales, Australia

Mark Anthony Buono Department of Chemical Engineering, Kansas State University, Manhattan, Kansas

---

**Present affiliation*: Chemical Engineering Department, University of Pittsburgh, Pittsburgh, Pennsylvania.

Edgar C. Clausen Department of Chemical Engineering, University of Arkansas, Fayetteville, Arkansas

Rathin Datta* Moffett Technical Center, CPC International, Inc., Summit-Argo, Illinois

Peter Dürre Institut für Mikrobiologie, Georg-August-Universität, Göttingen, Federal Republic of Germany

Larry E. Erickson Department of Chemical Engineering, Kansas State University, Manhattan, Kansas

L. T. Fan Department of Chemical Engineering, Kansas State University, Manhattan, Kansas

Daniel Yee-Chak Fung Department of Animal Sciences and Industry, Kansas State University, Manhattan, Kansas

James L. Gaddy Department of Chemical Engineering, University of Arkansas, Fayetteville, Arkansas

Gerhard Gottschalk Institut für Mikrobiologie, Georg-August-Universität, Göttingen, Federal Republic of Germany

Eugene L. Iannotti Department of Agricultural Engineering, University of Missouri-Columbia, Columbia, Missouri

James C. Linden Department of Microbiology and Department of Agricultural and Chemical Engineering, Colorado State University, Fort Collins, Colorado

Charles L. Meyer† Department of Chemical Engineering, Rice University, Houston, Texas

R. Mutharasan Department of Chemical Engineering, Drexel University, Philadelphia, Pennsylvania

---

Present affiliations:
*Michigan Biotechnology Institute, Lansing, Michigan.
†Department of Environmental Science, Shell Development Company, Houston, Texas.

Mehmet D. Öner Department of Food Engineering, University of Gaziantep, Gaziantep, Turkey

E. Terry Papoutsakis* Department of Chemical Engineering, Rice University, Houston, Texas

Govind Rao† Department of Chemical Engineering, Drexel University, Philadelphia, Pennsylvania

William M. Reed‡ Department of Biotechnology, CPC International, Inc., Summit-Argo, Illinois

Ka-Yiu San Department of Chemical Engineering, Rice University, Houston, Texas

Michael L. Shuler School of Chemical Engineering, Cornell University, Ithaca, New York

S. P. Srinivas§ Department of Chemical Engineering, Drexel University, Philadelphia, Pennsylvania

Gregory Stephanopoulos Department of Chemical Engineering, Massachusetts Institute of Technology, Cambridge, Massachusetts

Adriaan H. Stouthamer Biological Laboratory, Vrije Universiteit, Amsterdam, The Netherlands

Jürgen H. Thiele ecoTech—Environmental Biotechnology, Tübingen, Federal Republic of Germany, and Michigan Biotechnology Institute, Lansing, Michigan

---

Present affiliations:
*Department of Chemical Engineering, Northwestern University, Evanston, Illinois.
†Department of Chemical Engineering, University of Maryland, Baltimore County, Catonsville, Maryland.
‡Agricultural Department, American Cyanamid Company, Hannibal, Missouri.
§Department of Ophthalmology, Stanford University Medical Center, Stanford, California.

Ebenezer R. Vedamuthu Department of Science-Research, Microlife Technics, Sarasota, Florida

Jose L. Vega Department of Chemical Engineering, University of Arkansas, Fayetteville, Arkansas

Shie-Shien Yang Department of Statistics, Kansas State University, Manhattan, Kansas

J. Gregory Zeikus Departments of Biochemistry and Microbiology, Michigan Biotechnology Institute, Lansing, Michigan, and Michigan State University, East Lansing, Michigan

Introduction

Anaerobic fermentations are important industrial operations in the food and pharmaceutical industries. The food industry has employed the art and science of biotechnology for thousands of years in making bread and fermented beverages. Many of the important food fermentations are anaerobic (Crueger and Crueger, 1982; Gilliland, 1985; Peppler and Perlman, 1979; Reed, 1982).

Industrial chemicals such as acetone, butanol, ethanol, and acetic acid can be produced by anaerobic fermentation. The ability to compete in the marketplace with other processes has influenced the extent to which these products have been made by fermentation during the past 100 years. Recently, ethanol production has increased significantly because of government incentives to produce liquid fuels from renewable resources and surplus agricultural commodities (Murtagh, 1986; Rosillo-Calle, 1986). For anaerobic fermentation to be competitive in the industrial and fuels markets, higher petroleum prices are needed; however, technological developments in the fermentation processes can also contribute significantly. The size of the markets, the long-term estimates of petroleum prices, and the progress in biotechnology provide strong incentives for further research (Busche, 1985).

Anaerobic digestion has been used in waste treatment for many years. Methane is produced as a useful product in some cases. Although the process has been extensively used, our understanding of this complex mixed culture fermentation is still far from complete. Several chapters in this book consider various aspects of anaerobic digestion.

The anaerobic biodegradation of toxic and hazardous chemicals is being investigated in many laboratories; one chapter is devoted to this topic. Anaerobic biodegradation is an important part of environmental microbiology. There is strong interest in understanding natural anaerobic ecosystems as well as developing technology to clean up spills and treat waste streams (Gibson, 1984; Green and Kramer, 1979; Rochkind-Dubinsky et al., 1987).

Anaerobic fermentation is important in agriculture. The greatest commercial success to date in anaerobic cellulose utilization is grazing cattle and sheep on pasture. Rumen microbiology and biochemistry have been reviewed by Clarke and Bauchop (1977), Hobson (1979), and Hungate (1966). The silage fermentation is used to preserve some agricultural products (Woolford, 1984). Nitrogen fixation in plants also involves anaerobic microbiology (Bergersen, 1982).

Anaerobic fermentation is interpreted broadly in developing the coverage for this book. Anaerobic conditions are considered to mean that oxygen is absent. Anaerobic fermentation refers to the growth of microorganisms under anaerobic conditions. The bioenergetics of growth with nitrate and sulfate as electron acceptors is considered in the chapter by Stouthamer, for example.

The rapid advances in microbiology, biochemistry, and biochemical engineering have resulted in many new publications and renewed interest in anaerobic fermentations. This book provides a summary of current knowledge with emphasis on phenomenological topics such as membrane transport, bioenergetics, kinetics, and product inhibition, rather than considering specific fermentations. A number of publications have considered the science and technology associated with many of the important fermentations (Atkinson and Mavituna, 1983; Bu'Lock and Bu'Lock, 1983; Grayson, 1978, 1984; Jones and Woods, 1986; Marlatt and Datta, 1986; McKetta, 1976; Moo Young, 1985; Reed, 1982; Rehm and Reed, 1984; Steinkraus, 1983). Because of this, the chapters do not focus on individual fermentations in the present work.

This work is intended to complement the textbooks (Aiba et al., 1973; Bailey and Ollis, 1986; Pirt, 1975; Roels, 1982; Stanbury and Whitaker, 1984; Wang et al., 1979) and other handbooks that are already available (Atkinson and Mavituna, 1983; Fasman, 1976; Laskin and Lechevalier, 1984; Steinkraus, 1983).

REFERENCES

Aiba, S., Humphrey, A. E., and Millis, N. F. (1973). *Biochemical Engineering*, 2nd ed. Academic Press, New York.

Atkinson, B., and Mavituna, F. (1983). *Biochemical Engineering and Biotechnology Handbook*. Nature Press, New York.

Bailey, J. E., and Ollis, D. F. (1986). *Biochemical Engineering Fundamentals*. McGraw-Hill, New York.

Bergersen, F. J. (1982). *Root Nodules of Legumes: Structure and Function*. Research Studies Press, Letchworth, Hertfordshire, England.

Bu'Lock, J. D., and Bu'Lock, A. J. (1983). *The Acetone-Butanol Fermentation and Related Topics* 1980-1983. Science and Technology Letters, Kew, Surrey, England.

Busche, R. M. (1985). The business of biomass. *Biotech. Progress* *1*:165-180.

Clarke, R. T. J., and Bauchop, T. (1977). *Microbial Ecology of the Gut*. Academic Press, New York.

Crueger, W., and Crueger, A. (1982). *Biotechnology: A Textbook of Industrial Microbiology*. Science Tech., Madison, Wisconsin.

Fasman, G. D. (Ed.) (1976). *Handbook of Biochemistry and Molecular Biology*, 3rd ed. CRC Press, Boca Raton, Florida, 8 vol.

Gibson, D. T. (Ed.) (1984). *Microbial Degradation of Organic Compounds*. Marcel Dekker, New York.

Gilliland, S. E. (1985). *Bacterial Starter Cultures for Foods*. CRC Press, Boca Raton, Florida.

Grayson, Martin (Ed.) (1978). *Kirk-Othmer Encyclopedia of Chemical Technology*, 3rd ed. Wiley, New York, 24 vol.

Grayson, Martin (Ed.) (1984). *Encyclopedia of Chemical Technology*, Supplemental Volume. Wiley, New York.

Green, J. H., and Kramer, A. (1979). *Food Processing Waste Management*. AVI Publishing Co., Westport, Connecticut.

Hobson, P. N. (1979). Polysaccharide degradation in the rumen. In *Microbial Polysaccharides and Polysaccharases*, R. C. W. Berkeley, G. W. Gooday, and D. C. Ellwood (Eds.). Academic Press, New York, pp. 377-397.

Hungate, R. E. (1966). *The Rumen and Its Microbes*. Academic Press, New York.

Jones, D. T., and Woods, D. R. (1986). Acetone-butanol fermentation revisited. *Microbiol. Rev.* *50*:484-524.

Laskin, A. I., and Lechevalier, H. (Eds.) (1984). *CRC Handbook of Microbiology*, 2nd ed. CRC Press, Boca Raton, Florida, 8 vol.

Marlatt, J. A., and Datta, R. (1986). Acetone-butanol fermentation process development and economic evaluation. *Biotechnol. Progress* *2*:23-28.

McKetta, J. J. (Ed.) (1976). *Encyclopedia of Chemical Processing and Design*. Marcel Dekker, New York, 25 vol.

Moo Young, M. (Ed.) (1985). *Comprehensive Biotechnology*. Pergamon Press, New York, 4 vol.

Murtagh, J. E. (1986). Fuel ethanol production—The U.S. experience. *Process Biochem.* *21*(2):61-65.

Pirt, S. J. (1975). *Principles of Microbe and Cell Cultivation*. Wiley, New York.

Peppler, H. J., and Perlman, D. (1979). *Microbial Technology: Fermentation Technology*. Academic Press, New York.

Reed, Gerald (Ed.) (1982). *Prescott and Dunn's Industrial Microbiology*, 4th ed. AVI, Westport, Connecticut.

Rehm, H. J., and Reed, G. (Eds.) (1984). *Biotechnology—A Comprehensive Treatise in 8 Volumes*. Verlag Chemie, Deerfield Beach, Florida.

Rochkind-Dubinsky, M. L., Sayler, G. S., and Blackburn, J. W. (1987). *Microbiological Decomposition of Chlorinated Aromatic Compounds*. Marcel Dekker, New York.

Roels, J. A. (1982). *Energetics and Kinetics in Biotechnology*. Elsevier, New York.

Rosillo-Calle, F. (1986). The Brazilian ethanol chemistry industry. *Biomass 11*:19–38.

Stanbury, P. F., and Whitaker, A. (1984). *Principles of Fermentation Technology*. Pergamon Press, Elmsford, New York.

Steinkraus, K. H. (Ed.) (1983). *Handbook of Indigenous Fermented Foods*. Marcel Dekker, New York.

Wang, D. I. C., Cooney, C. L. Demain, A. L., Dunnill, P., Humphrey, A. E., and Lilly, M. D. (1979). *Fermentation and Enzyme Technology*. Wiley, New York.

Woolford, M. K. (1984). *The Silage Fermentation*. Marcel Dekker, New York.

Handbook on Anaerobic Fermentations

I

MICROBIOLOGY OF ANAEROBES

1

Methodology of Anaerobic Cultivation

DANIEL YEE-CHAK FUNG / *Kansas State University, Manhattan, Kansas*

I. INTRODUCTION

The microbial world consists of aerobic and anaerobic organisms, defined by the ability to tolerate molecular oxygen. Sonnenwirth (1972) traced the evolution of anaerobic methodology and recorded that Leeuwenhoek demonstrated that some life forms could exist in the presence of gases other than oxygen and that Pasteur discovered anaerobiosis in 1861 and coined the terms "aerobes" and "anaerobes." Subsequent observations by later scientists showed the existence of aerotolerant anaerobes as well as facultative anaerobes. The separation between these groups is not clearly defined. A convenient way to categorize these organisms is by observing their ability to grow at different oxidation-reduction potentials E_h. E_h can be measured and expressed as + or - millivolts (mV). Thus, aerobes can grow between +300 and -50 mV, facultative anaerobes, +300 to -420 mV, aerotolerant anaerobes, +180 to -350 mV, and obligate anaerobes, -150 to -420 mV (American Society for Microbiology, 1980).

This handbook deals mainly with facultative anaerobes and obligate anaerobes, although various aerotolerant organisms and aerobes are mentioned as well. The classic definition of fermentation is by Pasteur: "La fermentation est la vie sans air." In recent years the word included such aerobic processes as acetic acid fermentation.

The purpose of this chapter is to describe various methods for isolating and growing anaerobes from food, clinical specimens, and industrial and environmental samples. This information is helpful in appreciating the roles of these organisms in many fermentation practices described in subsequent chapters.

Excellent publications related to this topic include the *Anaerobe Laboratory Manual* (Holdeman et al., 1977), *Anaerobic Bacteriology Manual* (Sutter et al., 1980), *Recent Advances in Anaerobic Bacteriology* (Borriello and Hardie, 1987), and *Isolation of Anaerobes* (Shapton and Board, 1971).

II. ISOLATION OF ANAEROBES

The microbiologic procedures for studying obligate anaerobes are similar to those used for the more familiar aerobic organisms with the exception of the need to exclude oxygen from the working environment during manipulation, inoculation, and incubation of cultures and samples. The following books contain the necessary information on media preparation, aseptic techniques, and maintenance and preservation of cultures, as well as quantitative and qualitative evaluation of microbes in general: *Methods in Microbiology*, Vol. 3A (Norris and Ribbons, 1970), *Manual of Methods for General Bacteriology* (Gerhardt, 1981), *Manual of Clinical Microbiology* (Lennette et al., 1985), and *Bacteriological Analytical Manual* (Food and Drug Administration, 1984). Several excellent chapters on growth and maintenance of anaerobes are included in *Methods in Microbiology*, Vol. 3B (Norris and Ribbons, 1969).

As a class of organisms, anaerobes are very diversified in their behavior in terms of oxygen tolerance and nutritional requirements for growth. Thus there is no single set of procedures for isolation of all anaerobes. Clinically important anaerobes constitute the most studied class. Specific methods were developed for isolating anaerobes from various sites in patients. For example, procedures are described for specimens from pulmonary, pleural, urinary tract, abscesses, female genital tract, and sinus tracts or draining wounds (Sutter et al., 1980).

A detailed treatment of clinically important anaerobic organisms is recorded in the *Manual for the Determination of the Clinical Role of Anaerobic Microbiology* by Gall and Riely (1981). This manual contains a definition of clinically important anaerobic bacteria and methods for the collection and transportation of anaerobic specimens, culturing of anaerobic specimens, and identification of anaerobes. Information on commonly used techniques for the primary culturing and isolation of clinically important anaerobic bacteria in the book is of particular relevance to topics discussed in this chapter.

The book, *Isolation of Anaerobes*, by Shapton and Board (1971) contains chapters describing the isolation of clostridia from animal tissues, feces, soil, and foods. Another series of chapters in the same book describes the isolation of anaerobes from the intestinal tracts of humans and animals. Other chapters deal with the isolation of anaerobes from rumen as well as sulfate-reducing bacteria and photosynthetic bacteria. Zeikus (1977) described the biology of methanogenic bacteria and methods of studying these organisms.

In the area of food microbiology very little attention has been given to the anaerobic microbiology. *The Compendium of Methods for the Microbiological Examination of Foods* (Speck, 1984) contains some chapters on anaerobes, mainly on the spore formers and the detection of toxins in foods. Anderson and Fung (1983) reviewed the status of anaerobic methods, techniques, and principles for food bacteriology upon which this present chapter is developed.

The general conclusion on isolation procedures is that samples must be free from contact with air during collection time. The best method of collecting samples is by aspiration with needle and syringe. Samples should be placed in media that are designed to handle anaerobes. A variety of liquid media are marketed for anaerobic cultivation by such companies as DIFCO, BBL, Oxoid, and GIBCO. Solid media for the enumeration of anaerobes are also available from those companies. Many researchers have developed their own methods and procedures to satisfy their experimental protocol. The *Journal of Applied and Environmental Microbiology*, *Journal of Bacteriology*, *Journal of Applied Bacteriology*, *Journal of General Microbiology*, *Canadian Journal of Microbiology*, *Journal of Microbiological Methods*, and other related journals regularly record new developments in anaerobic microbiology.

III. DEVELOPMENT OF ANAEROBIC CULTIVATION SYSTEMS

A. Physical and Mechanical Reduction of Oxygen

Pasteur removed oxygen by boiling the cultivation solution (Hall, 1929). This method is not adequate for strict anaerobes. Pasteur and Roux used vacuum to remove oxygen through constricted tubes (Figs. 1 and 2) (Hunziker, 1902). Novy developed a glass cylinder with a rubber ring on a broad rim tip (Fig. 2). Petri dishes were placed in the cylinder upon which a bell jar was placed. Vacuum was drawn to remove air (Hunzinker, 1902). Later, glass stopcocks were replaced with brass covers with small metal pinchcocks (Gage and Spurr, 1913). Vacuum alone was not effective because it was difficult to construct airtight chambers (Hall, 1929) for the effective removal of air by mercury aspirator, water aspirator, or air pumps. Since strict obligate anaerobes are inhibited by less than 0.5% O_2, residual O_2 due to ineffective seals will kill some obligate anaerobes

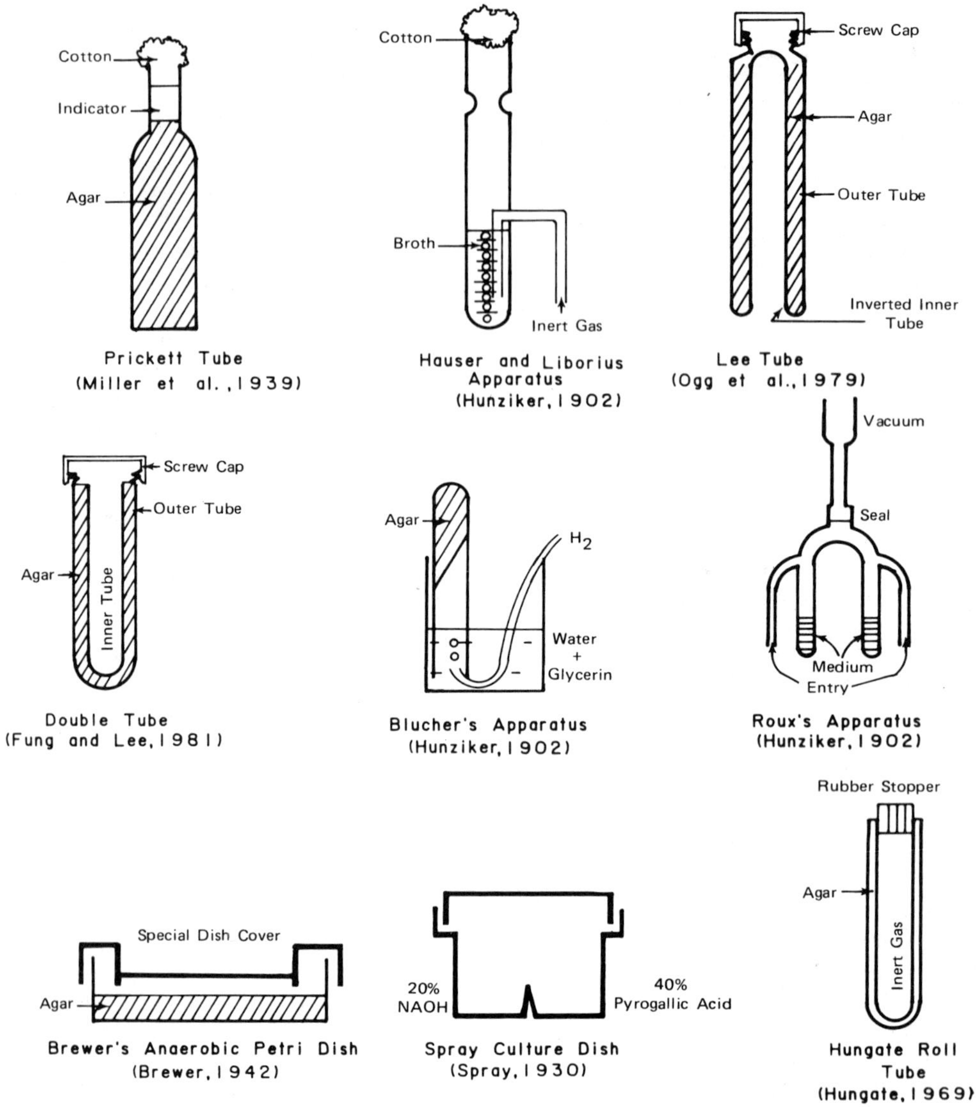

Figure 1 Anaerobic systems: tubes and dishes. (From Anderson and Fung, 1983. Used with permission from *J. Food Protec.*, Ames, Iowa.)

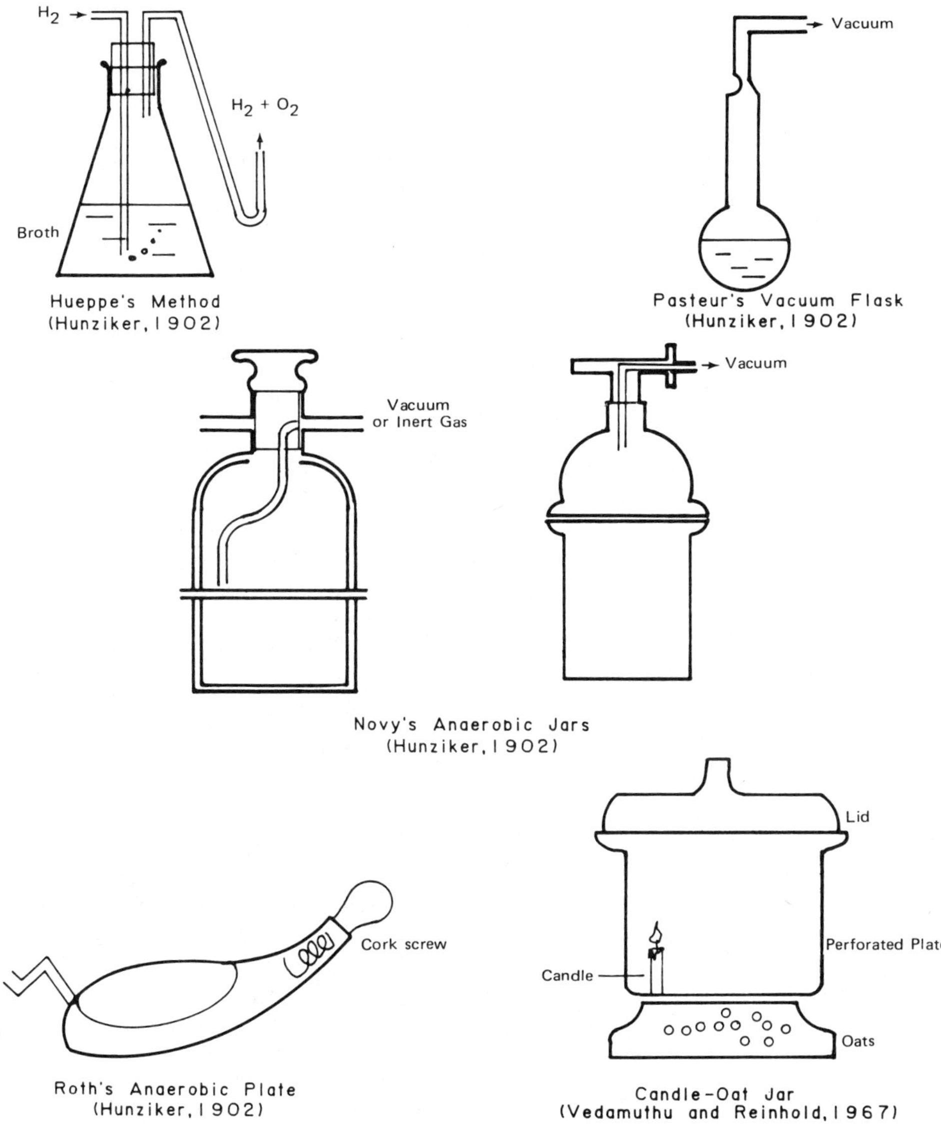

Figure 2 Anaerobic systems: flasks and jars. (From Anderson and Fung, 1983. Used with permission from *J. Food Protect.*, Ames, Iowa.)

(Loesche, 1969). Prolonged incubation under vacuum may result in the dehydration of cultivating medium (Hall, 1929).

The next approach developed was the application of inert gas (CO_2 or H_2). Pasteur was accredited with achieving anaerobiosis by this method (Hunziker, 1902). Hauser, Liborius, Hueppe, and Franenke devised various apparatuses for passing gas into or over inoculated liquid and solid media to displace O_2 (Figs. 1 and 2) (Hunziker, 1902). Blucher placed an inverted tube of solidified agar into a beaker of water and glycerol and passed gas through a curved tube in the inverted tube (Fig. 1) (Hunziker, 1902). Kitasato and Roth used a flask tube with flattened sides and two openings for gassing (Hall, 1929). Jones (1916) described a procedure in which a petri dish is placed on a trough, two openings are used to flush out oxygen by hydrogen, and the container is sealed by paraffin. Spray (1930) used a deep glass dish with a ridge in the middle. On one side, 40% pyrogallic acid was placed and the other side was filled with 20% NaOH. A petri dish with agar and culture was inverted and sealed to the top of the dish (Fig. 1). The chemicals in the bottom dish were mixed, and the reducing gas was produced to achieve anaerobiosis. McLeod (1913) developed a similar system for anaerobic cultivation of agar plates. Rosenthal (1937) used chromium-sulfuric acid to achieve anaerobiosis in a sealed jar. The mixture produced H_2 to bind with O_2. Mueller and Miller (1941) modified Rosenthal's chromium-sulfuric acid by adding sodium carbonate to generate CO_2 as well as H_2. Miles (1940) indicated that this method of producing H_2 to bind with O_2 also produced toxic gas for anaerobes.

A variety of anaerobic jars were developed in the early studies of anaerobes. Liborius, in 1886, combined evacuation and hydrogen gassing in a bell jar for anaerobic cultivation on several plates (Hall, 1929). Novy's jar was adapted for use with an inert gas (Fig. 2) (Hall, 1929). Kedrowski, in 1885, used a container with perforations to permit the flow of gas before sealing the container (Hunziker, 1902). Gage and Spurr (1913) used brass lids on large jars to accommodate more samples.

A new approach utilizing the Laidlaw principle (1915) of removal of oxygen by the oxidation of hydrogen through a catalyst was developed by McIntosh and Fildes (1916). Several workers (Eggerth and Gagnon, 1933; Zinsser and Bayne-Jones, 1934; Fildes and McIntosh, 1921; Brown, 1921, 1922) used platinized or palladinized asbestos or nichrome wrappings and activated the catalysts by heat or electric current to achieve the binding of H_2 with O_2 in a closed environment.

Boez (1927) evacuated the jar before catalytic activity took place, thus reducing the chances of explosion and increasing the speed of reduction. Weiss and Spaulding (1937) redesigned the jar lid structure to make the jar more sturdy and versatile. Brewer developed a series of modifications of the Brown jar and developed what was subsequently named Brewer's anaerobic jar (Fig. 3) (Brewer, 1939; Brewer and Brown, 1938). U.S. Army researchers modified the Brewer anaerobe jar (Evans et al., 1948) for field use for cultivating anaerobes by using an aspirator and natural gas or CO_2.

The use of deep agar tubes and overlay to exclude oxygen from the cultures started more than a century ago (Hunziker, 1902). Esmarch, in 1866, developed a method in which warm agar with organisms could be "rolled" so that the agar solidified on the inside wall of the tube. To expel O_2 gelatin was poured into the cavity of the rolled tube and formed an anaerobic environment (Hall, 1929). Schill and Marpmann introduced a small sterile tube into the cavity before agar solidification (Hall, 1929) and developed the forerunner of the double-tube system of Fung and Lee (1981) (Fig. 1). The current Prickett tube was developed by Miller et al. (1939), who made use of oval tubes with broad sides so that colonies formed in the oval tubes could be easily observed. The agar was first prereduced by heat and then poured into the Prickett tube without addition of reducing agents. After solidification of agar with culture, a layer of agar containing methylene blue was used to fill the neck of the tube to provide an anaerobic seal (Fig. 1).

Oil, mineral oil, paraffin, lanolin, and petrolatum have been used to act as overlay on solid and liquid medium to physically displace O_2 (Hall, 1915, 1929; Hunziker, 1902). Koch placed sterile mica blades on top of inoculated medium in a petri dish (Hunziker, 1902). Sanfelice placed the face of a petri dish onto the medium contained in the inverted lid (Hall, 1929) to create an anaerobic environment. Brewer (1942) designed a petri dish lid specifically for this purpose (Fig. 1) and introduced sodium thioglycolate into the solid medium.

B. Chemical and Biologic Reduction of Oxygen

As early as 1886, Liborius observed that glucose stimulated anaerobic growth of organisms (Hunziker, 1902). Smith showed that glucose has nutritive and reducing properties (Hall, 1929). The reducing power of glucose has attributed to the formation of reductones, which are strong reducing agents (Hodges and Osman, 1976). Novy determined that litmus, gelatin, and glucose have useful reducing ability (Hall, 1929). Kitasato and Wiel, in 1890, proposed that sodium formate is a stronger reducing agent than glucose (Hall, 1929). They also found that pyrogallic acid was a good reducing agent.

Although at 0.5% and higher pyrogallic acid was found to be inhibitory to bacterial growth, this compound has been used for a long time. It was used as a component of the growth medium as well as a compound placed in a closed container along with culture plates. After Buchner popularized the use of alkaline pyrogallic acid, much of the research on anaerobes was done with this compound as reducing agent (Hall, 1929; Hunziker, 1902).

Hammerl, in 1901, and Rivas, in 1902, proposed sulfur-containing compounds (Na_2S, K_2S, and NH_4SH) as reducing agents and showed that using a reducing agent alone was not adequate to promote the surface growth of obligate anaerobes. By 1925, Hosoya determined that L-cysteine-HCl was better than any other reducing agents used up to that time, such as thioglycolic acid, L-cystine, and sodium sulfite. Hosoya and Kishino, in 1925, found that cysteine in combination with sulfuretted hydrogen or sodium sulfide was an even more effective reducing agent (Hall, 1929). Quastel and Stephenson (1926) confirmed Hosoya's results with L-cysteine and considered the -SH group the source of reducing power.

The use of sodium thioglycolate in culture medium was first reported by Brewer (1940). The natural reducing ability of cooked meat medium was described by Robertson (1915). Iron filings added to cooked meat medium further reduced the E_h. These reducing agents (cooked meat and sodium thioglycolate) are still widely used today.

The use of biologic activity as a means of removal of oxygen was first proposed by Pasteur, who claimed that aerobic bacteria were able to use up the free O_2 in the medium, thereby creating an oxygen-free environment for anaerobes to grow. Kedrowski considered that the by-products of aerobes created a suitable environment in which to grow anaerobes. *Bacillus subtilis* and *Escherichia coli* were used in conjunction with anaerobes so that free oxygen was first consumed by these organisms to create an anaerobic environment for anaerobes to proliferate (Hunziker, 1902; Smith and Hungate, 1958). Porter (1948) was able to achieve an E_h of -440 mV by using *E. coli* to remove O_2.

Animal tissues were used in the medium to help create a more anaerobic environment (Smith et al., 1905), but the effect was thought to be the result of better nutrient contents in the medium rather than the direct reducing ability of tissues (McClung, 1944). Other ways of achieving anaerobiosis include the use of the respiratory activities of living plants, such as potato (Carrol and Hastings, 1925; Novy, 1925) or grain (McClung et al., 1935). In general this approach of achieving anaerobiosis was questioned by McClung (1935). Turner et al. (1963) and Vedamuthu and Reinbold (1967) burned a candle along with moist oats sealed in a jar to create anaerobiosis for cultivating anaerobes. This is the "candle-oat jar" method (Fig. 2). An ingenious plastic pouch method for isolation and enumeration of propionibacteria was developed by Hettinga et al. (1968), who used the candle-oat jar method as a reference comparison method. The sample and suitable melted agar are introduced into a plastic pouch (the size of a standard petri dish) with a small elongated opening "neck." The pouch is then inserted between two metal plates. Excess agar is forced to fill the elongated opening, thereby creating an airtight plug for anaerobiosis. The pouch method has the advantages of freedom from unwieldy equipment, ease of colony counting and retrieval, reduced incubation time, increased colony counts, and need for less incubator space.

A novel method using biologic materials to achieve anaerobiosis was recently developed by Adler and Crow (1981). Sterile suspensions of membrane fragments derived from *E. coli* were able to remove oxygen rapidly and efficiently as a result of the presence of a cytochrome-based electron transport system that transfers hydrogen from suitable donors in the bacteriologic media to oxygen and produces water as an end product. The membranes are stable, nontoxic, and active in both liquid and solid media. Applications of the membrane technique include (1) growth of a large variety of anaerobes (*Clostridium* spp., *Desulfovibrio*, *Peptostreptococcus* spp., *Propionibacterium*, *Eubacterium*, *Fusobacterium*, *Bacteriodes*, and *Viellionella*) (Adler and Brow, 1981); (2) growth studies of *Clostridium butyricum* and *Clostridium acetobutylicum* (Adler et al., 1983); (3) isolation of acetone-butanol-producing organisms (Crow et al., 1985); (4) growth studies of thermophilic anaerobic bacteria (Reed and Adler, 1986); and (5) growth of *Clostridium perfringens* under aerobic conditions (Hoskins et al., 1986).

Indicators of oxidation and reduction in media were developed when workers realized that some compounds, although possessing color in the oxidized state, became colorless in the reduced state. Kitasato and Wiel found that sodium sulfonate (indigo blue) was a good indicator of the oxidation and reduction potential of a medium. Cahen is credited to be the first to recommend litmus as an E_h indicator (Hall, 1921). Smith (1893) was the first to use methylene blue as an indicator of anaerobiosis. Brewer put methylene blue in paper strip to indicate reduction in a closed jar, similar to the procedure described by van Reimsdijk in 1922 (Brewer et al., 1966). Other indicators used include resazurin (Hungate, 1950; Pittman, 1946), phenosafranine (Bryant, 1963), and benzyl viologen (Hungate, 1969).

Electrochemical methods (similar to a pH meter) are now used to measure the redox potential (E_h) of anaerobic systems (Jacob, 1970; Tabatabai and Walker, 1970). Some form of platinum electrode and reference electrode is attached to the potentiometer and inserted into the test system for measurement of the E_h as volts or millivolts. Positive millivolts indicate the aerobic state and negative millivolts indicate the anaerobic state. There are some theoretical questions about the accuracy of the millivolt electrode in atmospheres containing H_2. Redox potential and its measurement are the subject of Chapter 6.

C. Current Methods

Current anaerobic systems include anaerobic jars, anaerobic glove boxes, roll tubes, and anaerobic tube systems.

Anaerobic jars that utilize both H_2 and CO_2 were developed by Brewer and Allgeier (1966). The chemical generator consisted of a sodium borohydride tablet and a citric acid-sodium bicarbonate tablet in a convenient package. These two tablets, when activated by the addition of water to the package, product H_2 and CO_2 gas, respectively (Brewer et al., 1955, 1966). The system required no vacuum or electric current since the $H_2 + O_2$ reaction was catalyzed by a "cold" catalyst (alumina pellets coated with 0.5% palladium) (Heller, 1954). CO_2 is used in the system for the promotion of growth of some anaerobic bacteria. Researchers have used this system with success for clinical specimens (Watt, 1972; Collee et al., 1961, 1972). Some companies are marketing these "anaerobic jars" (e.g., BBL, B.T.L., and Oxoid). Seip and Evans (1980) determined that in the BBL anaerobic jar, after properly applying the GasPak, the E_h dropped to -100 mV in 1 h and -300 mV in 3 h. The medium inside the jar achieved -200 mV in 6 h. The advantages of the anaerobic jar include (1) ease of operation, especially for the small laboratory, (2) economy of space, (3) convenience of prepackaged materials for the generation of gases, and (4) safety of the units. The disadvantages include the following: (1) the culture plates are inoculated under aerobic conditions, (2) the jar remains aerobic for 15-20 minutes even while the GasPaks are activated, which may affect some strict anaerobes, and (3) occasionally the jar fails to achieve anaerobiosis.

To eliminate the chance of exposure of anaerobes to the natural environment during bacteriologic manipulation, the anaerobic chamber or glove box was developed. Socransky et al. (1959) reported a glove box for anaerobic inoculation of cultures. After inoculation the cultures were incubated in a Brewer jar. They demonstrated the procedure was better than the aerobic or "bench" inoculation. The system consisted of an inoculation chamber with a 12 ft meteorologic balloon inside filled with O_2-free gas.

Rosenbury and Reynolds (1964) constructed a similar chamber using metal and an air lock. Oxygen was removed by evacuation of air followed by admission of 10% H_2 in N_2 passed over a palladium catalyst. Drasar (1967) constructed a clear plastic cabinet with glove port and an air lock. Oxygen was consumed by the burning of an alcohol lamp inside the cabinet, after which incoming gas (5% CO_2 in N_2) is applied through alkaline pyrogallic acid. Other similar structures were reported by Lee et al. (1968) and Gordon and Dubos (1970). Aranki and Freter (1972) and Aranki et al. (1969) designed a clear flexible vinyl plastic anaerobic incubator with the addition of a palladium chloride catalyst in the chamber and palladium black as a reducing agent in the culture medium. They used phenosafranine as indicator of reduced state. With these additional reducing steps they obtained an E_h of -280 mV in the broth. Anaerobic bacterial counts in this system were comparable to the direct microscopic count and data from the Hungate roll tube (to be discussed). Leach et al. (1971) designed a rigid anaerobic chamber with glove ports and a clear plastic viewing window (Fig. 3) for ease of operation and visualization of the experiment in progress.

Dowell (1972) and Killgore et al. (1973) compared glove boxes, GasPak jar, and roll rubes for cultivating clinically important anaerobes and found that they are equally reliable. Edwards and McBride (1975) and Cox and Herbert (1978) developed an Ultra Low Oxygen Chamber (ULCO) to grow methanogenic bacteria, one of the strictest anaerobes (Morris, 1975). The ULCO, made of an anaerobic jar with palladium catalysts and silica gel inside, was wrapped with rubber tubing through which warm water was circulated to keep the chamber at a desired temperature. Inoculated petri dishes were placed inside the ULCO, where the O_2 concentration was 0.0001%, 10 times lower than the inhibitory level of 0.001% for methanogens.

The basic principles for achieving anaerobiosis in the glove box, as well as promoting binding of residual O_2 with H_2 in the presence of a catalyst (Laidlaw principle), are to remove air first and then flush the chamber with inert gas. Reducing agents in the medium also help to reduce the E_h to a desired low level. The advantages of the glove box include the following: (1) the bacteriologic procedure can be operated under anaerobic conditions from the sampling stage to the incubation stage; (2) the area for manipulation is larger; and (3) in combination with temperature control the culture can be incubated in the chamber without further disturbance of the cultural environment as far as atmospheric conditions are concerned. The disadvantages are (1) high initial cost, (2) high operating expenses (gas and electricity), (3) high maintenance cost, (4) requirement for operator training, (5) oxygen contamination through leakage, operator error, or mechanical malfunction, (6) occasional high labor requirements, and (7) moisture buildup in the box.

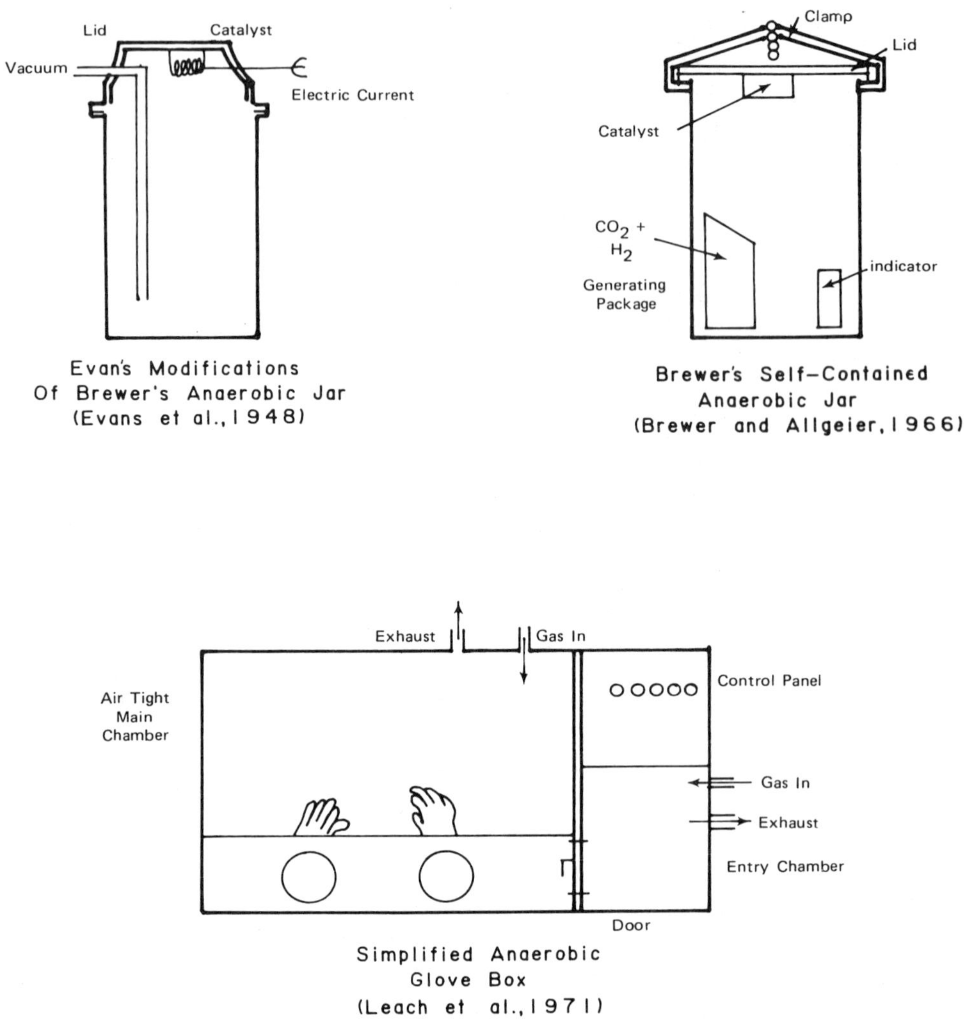

Figure 3 Anaerobic units and instruments. (From Anderson and Fung, 1983. Used with permission from *J. Food Protect.*, Ames, Iowa.)

One of the most widely used anaerobic cultivation systems is the Hungate roll tube, developed by Hungate (1947, 1950). Sterile tubes were flushed with CO_2 while the medium was aseptically added to the tubes to prevent air contamination. The tubes were then inoculated, sealed with rubber stoppers, cooled (to 48°C), and hand spun so that the agar with culture forms a thin film in the inner wall of the test tube. After incubation the colonies could be examined with ease. Isolation of colonies was also feasible with minimum effort. This method was similar to Esmarch's roll tube except that CO_2 flushing allowed the center cavity to remain reduced, making Esmarch's addition of gelatin unnecessary. Hungate was not aware of Esmarch's method when he worked on the roll tube technique (Holdeman et al., 1977). This method and its subsequent modifications allow the growth of a wide variety of strict anaerobes (Bryant and Burkey, 1953; Eller et al., 1971; Holdeman and Moore, 1972; Hungate et al., 1966; Kistner, 1960; Moore, 1966), including the extremely strict anaerobes—methanogenic bacteria (Balch and Wolfe, 1976; Paynter and Hungate, 1968; Smith, 1966).

Technical operation of the roll tube procedures was discussed in detail by Hungate (1969), Bryant (1972), and Holdeman et al. (1977). Anderson and Fung (1983) provided some observations on the operation of the procedures. Roll tubes are gassed by means of a hollow needle (cannula) that fits over the lip of the tube, and O_2 free gas is passed into the tube while bacteriologic manipulations are done. If the medium is prepared in the tube initially, the cannula need only extend past the lip of the tube to keep O_2 out. Since the roll tube does not utilize the Laidlaw principle, the E_h depends on the effectiveness of keeping O_2 out during the inoculation process or flushing the O_2 out before placing the medium into the tube.

Anaerobiosis also depends on media preparation. The medium is boiled initially and cooled while an O_2-free gas (usually CO_2) is bubbled into it. At this point the reducing agent should be added and the pH adjusted. The exposure of reducing agent to O_2 should be minimized because excessive amount of oxidized reducing agent may be toxic to some fastidious anaerobes (Holdeman et al., 1977). The tubes are then sealed airtight (usually with a rubber stopper) and autoclaved. Agar may be added before the initial boiling or dispensed into each tube before autoclaving. A popular version of the roll tube system is the VPI Anaerobic Culture System developed by Moore (1966). This apparatus has a platform so that the tubes can stand upright with cannulas positioned above each tube. The system also allows for streaking cultures into the roll tube anaerobically. The system is continuously flushed by CO_2 to exclude O_2 from the system. New Models of the system utilize electric spinners and foot pedals to facilitate operations.

The advantages of the Hungate roll tube system include (1) good anaerobiosis with minimum investment, (2) unit operation for convenience of small sample numbers, and (3) ease of observation of colonies and isolation of cultures for further studies. Disadvantages of the system include (1) time requirement for large number of samples (varies with technicians), (2) need for manual dexterity, and (3) occasional failure of some tubes to achieve anaerobiosis.

A few other anaerobic tube systems were also developed. Ogg et al. (1979) designed a tube with double walls so that agar with culture can be introduced into the tube and form a thin film (Fig. 1). Anaerobiosis is initially achieved by boiling and autoclaving of the medium. By the action of reducing agent in the medium and minimum reabsorption of O_2 due to the design of the tube (commercially called the "Lee Tube"), good anaerobiosis is maintained in the system. The glass cylinder forming the inner wall (the inverted smaller tube) makes it easy for enumeration and observation of colony morphologies. Lee et al. (1981) reported good growth of *Pectinatus cerevisiae*, an organism that is difficult to grow on solid medium in ordinary anaerobic systems. The Lee Tube has the advantage of ease of operation and low operating cost (no gas or special equipment needed). The disadvantages include (1) difficulty in picking isolates from the agar, (2) difficulty in cleaning the tubes, and (3) fragility of the system.

Fung and Lee (1981) developed a double-tube system for anaerobic cultivation of bacteria from foods. It consists of a glass tube (15 × 1.0 cm outer diameter) placed into a larger glass tube (15 × 1.5 cm outer diameter) with anaerobic agar (DIFCO) and culture sandwiched between the two tubes. The opening of a smaller tube is securely fastened into the rubber-stoppered cap of the larger tube. This design is similar to Marpmann's tubes developed 90 years earlier, but the Fung and Lee double tube (Fig. 1) uses a reducing agent and is more tightly sealed than Marpmann's design [Fung was not aware of Marpmann's work (Hall, 1929) at the time of initial experiments]. Diluted food samples can first be introduced into the cavity of the sterilized double tube while the agar is still warm (48°C). The sample is then distributed evenly by the action of insertion of the small inner tube. After the tubes are sealed the assemblage is inverted five times before being placed in the appropriate temperature for incubation. With this method a comparative evaluation of anaerobic counts of ground pork, fish, and mushrooms was made using the anaerobic petri dish method of Brewer (1942). The data (Fung and Lee, 1981) indicated that the double-tube method provided higher anaerobic counts than Brewer's method.

Subsequently, Anderson and Fung (1982) modified the double tube by the use of screw caps (more anaerobic than the original double tube) and improved reducing agents in the medium. They were successful in cultivating strict anaerobic rumen microbes, such as *Megasphaera elsdenii* and *Butyrivibrio fibrisolvens*, as well as *C. perfringens*, *Eubacterium limosum*, and anaerobes from beef and poultry. A detailed discussion of the theoretical concepts contributing to the oxygen sensitivity of anaerobes, the roles of hydrogen peroxide, superoxide dismutases, redox potential measurements, and effectiveness of various reducing agents, and indicator dyes can be found in the work by Anderson (1982).

IV. IDENTIFICATION OF ANAEROBES

The identification of anaerobes is a concern for applied microbiologists. Classic methods for identification can be found in *Bergey's Manual of Determinative Bacteriology*, 8th Edition (Buchanan and Gibbons, 1974) and *Bergey's Manual of Systematic Bacteriology*, Volume 1 (Krieg and Holt, 1984). A variety of rapid and automated methods in microbiology have been developed for the isolation, enumeration, and identification of microorganisms (Fung et al., 1988). Many of these new methods can be adapted to anaerobic bacteriology. For example, Fung and Hartman (1975) developed a variety of miniaturized microbiologic methods for aerobic bacteria. The entire concept and system were adapted for miniaturized anaerobic bacteriology simply by operating the procedures in an anaerobic chamber (Wilkins and Walker, 1975; Wilkins et al., 1975). Recently, at Kansas State University, Guy Miller and T. G. Nagaraja (personal communication, 1987) found that miniaturized techniques could be used for studying rumen bacteria. Minitek, a commercial test kit, offers miniaturized plates for the cultivation of anaerobes in the BBL anaerobic jar. Diagnostic test kits using enzymes of anaerobes are also available. For example, the API-ZYM and IDS systems are marketed to identify anaerobic bacteria by the enzyme activities in an aerobic environment in as short as 4 h. The Vitek AMS system, probably the most automated and comprehensive microbiologic system, can detect a variety of aerobes and anaerobes in 4 h automatically. The BACTEC system utilizes the radiometric measurement of radioactive gas generated by microorganisms (aerobic and anaerobic) during the metabolism of radioactive substrates.

"Fingerprinting" techniques have been explored to identify anaerobes by matching profiles from unknown cultures with known profiles in the data base. One such approach is the gas-liquid chromatography method popularized by the VPI Anaerobic Laboratory (Holdeman et al., 1977). Mitruka (1976) provided a detailed discussion on this method. The Hewlett-Packard Company developed a procedure to fingerprint the fatty acid profiles of microorganisms using a high-performance liquid chromatographing technique. The AMBIS system involves computer-integrated fingerprinting of electrophoretically separated labeled proteins of bacteria to match profiles of unknowns versus knowns. The field of rapid techniques and automation in microbiology will no doubt continue to develop many very useful methods for the rapid identification of anaerobes from food, clinical specimens, and industrial and environmental samples.

V. CONCLUSIONS

The emphasis in this chapter is on laboratory microbiologic methods for the isolation and growth of strict anaerobes. For facultative anaerobe cultivation, most conventional aerobic procedures are sufficient. Cultivation of these organisms in larger fermenters is relatively simple because of the availability of modern industrial gases, such as nitrogen, hydrogen, and carbon dioxide. Nitrogen can be sparged into large fermenters to purge the system of oxygen before inoculation. Gas sparging can be used during fermentation to maintain a positive pressure and to remove product gases from the fermenter. The concentration of carbon dioxide may need to be controlled or kept below a certain level in some applications. Ho and Shanahan (1986) have reviewed the effects of carbon dioxide concentration on a number of fermentations; this subject is also considered by Erickson in Chapter 5. Gas sparging into the fermenter and agitation can be used to remove carbon dioxide from the fermentation broth. Recently, Ho and Shanahan (1986) have reviewed carbon dioxide transfer in bioreactors.

The removal of samples from anaerobic fermenters is relatively easy provided a positive pressure is maintained within the fermenter by gas sparging. Microbial samples are best handled with a needle and syringe, as discussed earlier. General information on measurement, data analysis, and control are included in Chapter 14.

Elsewhere in this handbook, the methodology of anaerobic biodegradation of toxic chemicals is included in Chapter 20 by Erickson and Fan. The last chapter considers the methodology associated with anaerobic digestion. Anaerobic waste treatment is the most widely used technology in which anaerobes are cultivated industrially.

In summary, methodologies for anaerobic cultivation are constantly being developed and refined. This chapter describes some of the key classic methods and touches upon some of the modern approaches in anaerobic cultivation methodologies. This field certainly deserves more attention by applied microbiologists with innovative ideas and ingenious approaches.

ACKNOWLEDGMENT

Contribution no. 88-18-B, Kansas Agricultural Experiment Station, Manhattan, Kansas 66506.

REFERENCES

Adler, H. I., and Crow, W. D. (1981). A novel approach to the growth of anaerobic microorganisms. *Biotechnol. Bioeng. Symp. 11*:533-540.

Adler, H. I., Crow, W. D., Hadden, C. T., Hall, J., and Machanoff, R. (1983). New technique for growing anaerobic bacteria: Experiments with *Clostridium butyricum* and *Clostridium acetobutylicum*. *Biotechnol. Bioeng. Symp. 13*:153-161.

American Society for Microbiology (1980). *Fundamentals of Anaerobic Bacteriology as Related to the Clinical Laboratory*. American Society for Microbiology, Washington, D.C.

Anderson, K. L. (1982). Development of the double-tube system for the cultivation of anaerobic bacteria from foods. M.S. Thesis, Kansas State University, Manhattan, Kansas.

Anderson, K. L., and Fung, D. Y. C. (1982). Proceedings of the 42nd Annual Meeting of the Institute of Food Technologists, Abstract 163.

Anderson, K. L., and Fung, D. Y. C. (1983). Anaerobic methods, technologies and principles for food bacteriology: A review. *J. Food Protect. 40*:811-827.

Aranki, A., and Freter, R. (1972). Use of anaerobic glove boxes for the cultivation of strictly anaerobic bacteria. *Am. J. Clin. Nutr. 25*:1329-1334.

Aranki, A., Seyed, S. A., Kenny, E. B. and Freter, R. (1969). Isolation of anaerobic bacteria from human gingiva and mouse cecum by means of a simplified glove box procedure. *Appl. Microbiol. 17*:568-576.

Balch, W. E., and Wolfe, R. S. (1976). New approach to the cultivation of methanogenic bacteria: 2-Mercaptoethanesulfonic acid (HS-CoM)-dependent growth of *Methanobacterium ruminantium* in a pressurized atmosphere. *Appl. Environ. Microbiol. 32*:781-791.

Boez, L. (1927). Anaerobiosis produced by means of a central catalyzing station. *J. Bacteriol. 13*:227-233.

Borriello, S. P., and Hardi, J. M. (Eds.) (1987). *Recent Advances in Anaerobic Bacteriology*. Martinus Nijhoff, Boston.

Brewer, J. H. (1939). A modification of the Brown anaerobe jar. *J. Lab. Clin. Med. 24*:1190-1192.

Brewer, J. H. (1940). A clear liquid medium for the aerobic cultivation of anaerobes. *J. Bacteriol. 39*:10.

Brewer, J. H. (1942). A new petri dish cover and technique for the use in the cultivation of anaerobes and microaerophiles. *Science 95*:587.

Brewer, J. H., and Allgeier, D. L. (1966). Safe self-contained carbon dioxide-hydrogen anaerobic system. *Appl. Microbiol. 14*:985-988.

Brewer, J. H., and Brown, J. H. (1938). A method for utilizing illuminating gas in the Brown, Fildes and McIntosh or other anaerobic jars of the Laidlaw principle. *J. Lab. Clin. Med. 23*: 870-874.

Brewer, J. H., Heer, A. A., and McLaughlin, C. B. (1955). The use of sodium borohydride for producing hydrogen in an anaerobe jar. *Appl. Microbiol.* *3*:136.

Brewer, J. H., Allgeier, D. L., and McLaughlin, C. B. (1966). Improved anaerobic indicator. *Appl. Microbiol.* *14*:135-136.

Brown, J. H. (1921). An improved anaerobe jar. *J. Exp. Med.* *33*:677-681.

Brown, J. H. (1922). Modification of an improved anaerobe jar. *J. Exp. Med.* *35*:467.

Bryant, M. P. (1963). Symposium on microbial digestion in ruminants; identification of groups of anaerobic bacteria active in the rumen. *J. Anim. Sci.* *22*:801-813.

Bryant, M. P. (1972). Commentary on the Hungate technique for culture of anaerobic bacteria. *Am. J. Clin. Nutr.* *25*:1324-1328.

Bryant, M. P., and Burkey, L. A. (1953). Cultural methods and some characteristics of some of the more numerous groups of bacteria in the bovine rumen. *J. Dairy Sci.* *36*:205-217.

Buchanan, R. E., and Gibbons, N. E. (1974). *Bergey's Manual of Determinative Bacteriology*, 8th ed. Williams & Wilkins, Baltimore.

Carrol, W. R., and Hastings, E. G. (1925). The use of living vegetable tissue for securing a favorable gaseous environment for anaerobic. *J. Lab. Clin. Med.* *11*:205-217.

Collee, J. G., Knowlde, J. A., and Hobbs, B. C. (1961). Studies on the growth, sporulation and carriage of *Clostridium welchii* with special reference to food poisoning strains. *J. Appl. Bacteriol.* *24*:326-339.

Collee, J. G., Watte, B., Fowler, E. B., and Brown, R. (1972). An evaluation of the GasPak system in the culture of anaerobic bacteria. *J. Appl. Bacteriol.* *35*:71-82.

Cox, D. J., and Herbert, S. D. (1978). An anaerobic glove box for the isolation and cultivation of methanogenic bacteria. *J. Appl. Bacteriol.* *45*:411-415.

Crow, W. D., Machanoff, R., and Adler, H. I. (1985). Isolation of anaerobes using and oxygen reducing membrane fraction: Experiments with acetone butanol producing organisms. *J. Microbiol. Methods* *4*:133-139.

Dowell, V. R. (1972). Comparison of techniques for isolation and identification of anaerobic bacteria. *Am. J. Clin. Nutr.* *25*:1335-1343.

Drasar, B. S. (1967). Cultivation of anaerobic intestinal bacteria. *J. Pathol. Bacteriol.* *94*:417-427.

Edwards, T., and McBride, B.C. (1975). New method for the isolation and identification of methanogenic bacteria. *Appl. Microbiol.* *29*:540-545.

Eggerth, A. H., and Gagnon, B. H. (1933). The *Bacteriodes* of human feces. *J. Bacteriol.* *25*:389-413.

Eller, C., Crabill, M. R., and Bryant, M. P. (1971). Anaerobic roll tube media for nonselective enumeration and isolation of bacteria in human feces. *Appl. Microbiol.* *25*:522-529.

Evans, J. M., Carguist, P. R., and Brewer, J. H. (1948). A modification of the Brewer anaerobic jar. *Am. J. Clin. Pathol.* *18*:747-749.

Fildes, P., and McIntosh, J. (1921). An improved form of McIntosh and Fildes anaerobic jar. *Br. J. Exp. Pathol.* *2*:153-154.

Food and Drug Administration (1984). *Bacteriological Analytical Manual.* Association of Official Analysts and Chemists, Arlington, Virginia.

Fung, D. Y. C., Cox, N. A., and Bailey, J. S. (1988). Rapid methods and automation in microbiology. *Dairy Food Sanita.* in press.

Fung, D. Y. C., and Hartman, P. A. (1975). Miniaturized microbiological techniques for rapid characterization of bacteria. In *New Approaches to the Identification of Microorganisms*, C. G. Heden and T. Illeni (Eds.). John Wiley and Sons, New York.

Fung, D. Y. C., and Lee, C. M. (1981). Double-tube anaerobic bacterial cultivation system. *Food Sci.* (Republic of China) *7*:209-213.

Gage, S. D., and Spurr, J. H. (1913). A convenient anaerobic jar for plate cultures. *Am. J. Public Health* *3*:677.

Gall, L. S., and Riely, P. E. (1981). *Manual for the Determination of the Clinical Role of Anaerobic Microbiology.* CRC Press, Boca Raton, Florida.

Gerhardt, P. (1981). *Manual of Methods for General Bacteriology.* American Society for Microbiology, Washington, D.C.

Gordon, J. H., and Dubos, R. (1970). The anaerobic bacterial flora of the mouse cecum. *J. Exp. Med.* *132*:251-260.

Hall, I. C. (1915). A new aerobe-anaerobe culture tube. *Univ. Calif. Publ. Pathol.* *2*:147.

Hall, I. C. (1921). The early history of litmus in bacteriology. *Science* *53*:388-389.

Hall, I. C. (1929). A review of the development and application of physical and chemical principles in the cultivation of obligately anaerobic bacteria. *J. Bacteriol.* *17*:255-301.

Heller, C. C. (1954). A simple method for producing anaerobiosis. *J. Appl. Bacteriol.* *17*:202.

Hettinga, D. H., Vedamuthu, E. R., and Reinbold, G. W. (1968). Pouch method for isolating and enumerating propionibacteria. *J. Dairy Sci.* *51*:1707-1709.

Ho, C. S., and Shanahan, J. F. (1986). Carbon dioxide transfer in bioreactors. *CRC Crit. Rev. Technol.* 5-252.

Hodges, J. E., and Osman, E. M. (1976). Carbohydrates. In *Food Chemistry*, O. R. Fennema (Ed.). Marcel Dekker, New York, pp. 41-138.

Holdeman, L. V., and Moore, W. E. C. (1972). Roll-tube techniques for anaerobic bacteria. *Am. J. Clin. Nutr.* *25*:1314-1317.

Holdeman, L. V., Cato, E. P., and Moore, W. E. C. (1977). *Anaerobe Laboratory Manual*, 4th Ed. Virginia Polytechnic Institute and State University, Blacksburg, Virginia.

Hoskin, C. B., Rico-Munoz, E., and Davidson, P. M. (1986). Proceedings of the 46th Annual Meeting Institute of Food Technologists, abstract 79.

Hungate, R. E. (1947). The culture and isolation of cellulose-decomposing bacteria from the rumen of cattle. *J. Bacteriol. 53*: 631-645.

Hungate, R. E. (1950). The anaerobic mesophilic cellulotytic bacteria. *Bacteriol. Rev. 14*:1-49.

Hungate, R. E. (1969). A roll tube method for cultivation of strict anaerobes. In *Methods in Microbiology*, Vol. 3B, J. R. Norris and D. W. Ribbons (Eds.). Academic Press, New York, pp. 117-132.

Hungate, R. E., Smith, W., and Clarke, R. T. J. (1966). Suitability of butyl rubber stoppers for closing anaerobic roll culture tubes. *J. Bacteriol. 91*:908-909.

Hunziker, O. F. (1902). A review of the existing methods for cultivating anaerobic bacteria. *J. Appl. Microsc. 5*:1694-1697, 1741-1758, 1800-1813, 1854-1856.

Jacob, H. E. (1970). Redox potential. In *Methods in Microbiology*, Vol. 2, J. R. Norris and D. W. Ribbons (Eds.). Academic Press, New York, pp. 91-123.

Jones, H. M. (1916). A method of plating anaerobes permitting observation of growth. *J. Bacteriol. 1*:229-341.

Killgore, G. E., Starr, S. E., DelBene, V. E., Whaley, D. N., and Dowell, V. R. (1973). Comparison of three anaerobic systems for the isolation of anaerobic bacteria from clinical specimens. *Am. J. Clin. Pathol. 59*:552-559.

Kistner, A. (1960). An improved method for viable counts of bacteria of the bovine rumen which ferment carbohydrates. *J. Gen. Microbiol. 23*:565-576.

Krieg, N. R., and Holt, J. G. (1984). *Bergey's Manual of Systematic Bacteriology*, Vol. 1. Williams & Wilkins, Baltimore.

Laidlaw, P. O. (1915). Some simple anaerobic methods. *Br. Med. J. 1*:497-498.

Leach, P. A., Bullen, J. J., and Grant, I. D. (1971). Anaerobic CO_2 cabinet for the cultivation of strict anaerobes. *Appl. Microbiol. 22*:824-827.

Lee, A., Gordon, J., and Dubos, R. (1968). Enumeration of the oxygen sensitive bacteria usually present in the intestine of healthy mice. *Nature 220*:1137-1139.

Lee, S. Y., Moore, S. E., and Mabee, M. S. (1981). Selective-differential medium for isolation and differentiation of *Pectinatus* from other brewery microorganisms. *Appl. Environ. Microbiol.* *41*:386-387.

Lennette, E. H., Balows, A., Hausler, W., and Shadomy, H. (1985). *Manual of Clinical Microbiology*, 4th ed. American Society for Microbiology, Washington, D.C.

Loesche, W. J. (1969). Oxygen sensitivity of various anaerobic bacteria. *Appl. Microbiol.* *18*:723-727.

McClung, L. S. (1935). Studies on anaerobic bacteria. III. Historical review and technique of culture of certain thermophilic anaerobes. *J. Bacteriol.* *29*:173-187.

McClung, L. S. (1944). Recent developments concerning the anaerobic bacteria and their activities, with particular reference to the tetanus and gangrene organisms. *Proc. Indiana Acad. Sci.* *53*:30-46.

McClung, L. S., McCoy, E., and Fred, E. B. (1935). Studies on anaerobic bacteria. II. Further extensive uses of the vegetable tissue anaerobic system. *Zentraebl. Bakterial.* *91*:225-227.

McIntosh, J., and Fildes, P. (1916). A new apparatus for the isolation and cultivation of anaerobic microorganisms. *Lancet* *1*:768-770.

McLeod, J. W. (1913). A method for the plate culture of anaerobic bacteria. *J. Pathol. Bacteriol.* *17*:454-457.

Miles, A. A. (1940). The Rosenthal anaerobic method. *Lancet* *2*:7.

Miller, J. J., Garrett, O. W., and Prickett, P. S. (1939). Anaerobic technique—a modified deep agar shake. *J. Food Res.* *4*:447-451.

Mitruka, B. M. (1976). *Methods of Detection and Identification of Bacteria*. CRC Press, Boca Raton, Florida.

Moore, W. E. C. (1966). Techniques for routine culture of fastidious anaerobes. *Inst. J. Syst. Bacteriol.* *16*:173-190.

Morris, J. G. (1975). The physiology of obligate anaerobiosis. *Adv. Microbiol. Physiol.* *12*:169-246.

Mueller, J. H., and Miller, P. A. (1941). A modification of Rosenthal's chromium-sulfuric acid method for anaerobic cultures. *J. Bacteriol.* *41*:301-303.

Norris, J. R., and Ribbons, D. W. (1969). *Methods in Microbiology*, Vol. 3B. Academic Press, New York.

Norris, J. R., and Ribbons, D. W. (1970). *Methods in Microbiology*, Vol. 3A. Academic Press, New York.

Novy, F. G., Jr. (1925). Microbic respiration. IV. The so-called aerobic growth of anaerobes: Potato respiration. *J. Infect. Dis.* *36*:343-382.

Ogg, J. E., Lee, S. Y., and Ogg, B. J. (1979). A modified tube method for the cultivation and enumeration of anaerobic bacteria. *Can. J. Microbiol.* *25*:987-990.

Paynter, M. J. B., and Hungate, R. E. (1968). Characterization of *Methanobacterium mobilis* sp. n., isolated from the bovine rumen. *J. Bacteriol.* *95*:1943-1951.

Pittman, M. (1946). A study of fluid thioglycolate medium for the sterility test. *J. Bacteriol. 51*:19-32.

Porter, J. R. (1948). *Bacterial Chemistry and Physiology*. Wiley, New York.

Quastal, J. H., and Stephenson, M. (1926). Experiments on "strict" anaerobes. I. The relationship of *B. sporogenes* to oxygen. *Biochemistry 20*:1125-1137.

Reed, W. M., and Adler, H. I. (1986). Application of bacterial membrane for the growth of thermophilic anaerobic bacteria. Eighth Symposium on the Biotechnology of Fuels and Chemicals. Oak Ridge National Lab. Tenn. Feb. 1, 1986.

Robertson, M. (1915). Notes upon certain anaerobes isolated from wounds. *J. Pathol. Bacteriol. 20*:327-349.

Rosenbury, T., and Reynolds, J. B. (1964). Continuous anaerobiosis for cultivation of spirochetes. *Proc. Soc. Exp. Biol. Med. 117*:813-815.

Rosenthal, L. (1937). "Chromium-sulfuric acid" method for anaerobic cultures. *J. Bacteriol. 34*:317-320.

Seip, W. F., and Evens, G. (1980). Atmospheric analysis and redox potentials of culture media in the GasPak system. *J. Clin. Microbiol. 11*:226-233.

Shapton, D. A., and Board, R. G. (1971). *Isolation of Anaerobes*. Academic Press, New York.

Smith, P. H. (1966). The microbial ecology of sludge methanogenesis. *Dev. Indust. Microbiol. 7*:156-161.

Smith, P. H., and Hungate, R. E. (1958). Isolation and characterization of *Methanobacterium ruminantium* N. sp. *J. Bacteriol. 75*:713-718.

Smith, T. (1893). The fermentation tube with special reference to anaerobiosis and gas production among bacteria. Wilder Quarter Century Book, p. 187.

Smith, T., Brown, H. R., and Walker, E. L. (1905). The fermentation tube in the study of anaerobic bacteria with special reference to gas production and the use of milk as a culture medium. *J. Med. Res. 14*:193-206.

Socransky, S., McDonald, J. B., and Sawyer, S. (1959). The cultivation of *Treponema microdentium* as surface colonies. *Arch. Oral Biol. 1*:171-172.

Sonnenwirth, A. C. (1972). Evolution of anaerobic methodology. *Am. J. Clin. Nutr. 25*:1295-1298.

Speck, M. L. (Ed.) (1984). *Compendium of Methods for the Microbiological Examination of Foods*. American Public Health Association, Washington, D.C.

Spray, R. S. (1930). An improved anaerobic culture dish. *J. Lab. Clin. Med. 16*:203-206.

Sutter, V. L., Citron, D. M., and Finegold, S. M. (1980). *Anaerobic Bacteriology Manual.* C. V. Mosby, St. Louis.

Tabatabai, L. B., and Walker, H. W. (1970). Oxidation-reduction potential growth of *Clostridium perfringens* and *Pseudomonas fluorescens. Appl. Microbiol. 20*:441-446.

Turner, N. S., Sandine, W. E., Elliker, P. R., and Day, E. A. (1963). Use of tetrazolium dyes in an agar medium for differentiation of *Streptococcus lactis* and *Streptococcus cremoris. J. Dairy Sci. 46*:380-385.

Vedamuthu, E. R., and Reinbold, G. W. (1967). The use of candle oats jar incubation for the enumeration, characterization, and taxonomic study of propionibacterium. *Milchwissenchaft. 22*:428-431.

Watt, N. (1972). The recovery of clinically important anaerobes on solid media. *J. Med. Microbiol.* 5:211-218.

Weiss, J. E., and Spaulding, E. H. (1937). A simple method for obtaining effective anaerobiosis. *J. Lab. Clin. Med. 22*:726-728.

Wilkins, T. D., and Walker, C. B. (1975). Development of a micromethod for identification of anaerobic bacteria. *Appl. Microbiol. 30*:825-830.

Wilkins, T. D., Walker, C. B., and Moore, W. E. C. (1975). Micromethod for identification of anaerobic bacteria: Design and operation of apparatus. *Appl. Microbiol. 30*:831-837.

Zeikus, J. C. (1977). The biology of methanogenic bacteria. *Bacteriol. Rev. 41*(2):514-541.

Zinsser, H., and Bayne-Jones, S. (1934). *A Textbook of Bacteriology.* D. Appleton-Century Company, New York.

2

Mutation and Genetic Engineering of Anaerobic Bacteria

WILLIAM M. REED* / *CPC International, Inc., Summit-Argo, Illinois*

I. INTRODUCTION

Genetic studies of obligate anaerobic bacteria have been hampered by the technical difficulty involved in maintaining anaerobiosis during cultural manipulation. Nevertheless, the increasingly common inclusion of anaerobic methodology in the microbiology curriculum, the availability of commercial anaerobic culturing instrumentation, and a rapidly growing interest in genetic systems in anaerobic bacteria have given rise to a dramatic increase in the number of publications concerning the genetics of anaerobic bacteria. For example, there was a 400% increase in the number of publications concerning the genetics of anaerobes between 1980 and 1985, as referenced in this review. Most of the research has focused on the involvement of anaerobes in disease or the fermentative production of useful products (solvents, organic acids, methane, and so on) from a diverse range of substrates, such as cellulose, hemicellulose, and starch. Mutation and genetic engineering of the latter class of anaerobes are reviewed here.

A discussion of the genetics of all anaerobic bacteria could not be properly presented in a single volume, let alone a single chapter. Therefore, I have restricted my discussion to three groups of strict anaerobic bacteria that have major importance from a commercial standpoint: industrially important saccharolytic clostridia, anaerobic ruminal bacteria, and methanogens.

**Present affiliation*: American Cyanamid Company, Hannibal, Missouri.

The fermentative production of solvents and organic acids will become increasingly important as the supplies of nonrenewable chemical feedstocks decrease (Lipinsky, 1981; Ng et al., 1983). Solventogenesis and acidogenesis by many clostridia have been well established (Zeikus, 1985). Also, the potential for the production of unique and useful enzymes by clostridia has been demonstrated (Hyun and Zeikus, 1985a, b). As a result of their useful products, the saccharolytic clostridia have received substantial attention. Several excellent reviews are available concerning the biochemistry, physiology, and beginning genetics of clostridia (Minton and Morris, 1983; Woods and Jones, 1984; Zeikus, 1985; Jones and Woods, 1986a, b; Rogers, 1986). Steady progress has been made toward the application of genetic engineering for cultural improvement and the elucidation of the molecular biology of this important genus. These new discoveries and achievements are included in the discussion of the mutation and genetic engineering of clostridia.

There are numerous applications of genetic engineering to the development of anaerobic ruminal bacteria and subsequently the improvement of ruminant nutrition. Areas of application include alterations in fermentation products, the addition or improvement of metabolic pathways for the degradation of feedstuffs, and the production of biologically active compounds that can directly affect ruminant physiology. These and other potential applications of genetic engineering to ruminal bacteria have been reviewed in detail by Smith and Hespell (1983). Armstrong and Gilbert (1985), Teather (1985), and Forsberg et al. (1986). The status of mutation and genetic engineering of anaerobic ruminal bacteria are discussed below. Methanogenic and clostridial rumen isolates are covered in their respective, unique sections. Some reference is made to research involving nonruminal but closely related clinical anaerobes. In general, disease-related anaerobes are not covered in this review. However, some reports on the molecular biology of these isolates are discussed to fill in the blanks in specific research areas in which similar investigations with nonclinical isolates have not been completed.

The methanogens complete the terminal step in the degradation of biopolymers to methane. Therefore, they play a critical part in the cycling of carbon and other elements in the biosphere. They also represent the key biocatalyst in the treatment of waste and generation of methane in anaerobic digesters.

As a subgroup of the archaebacteria, the methanogens are unique among the three groups of bacteria discussed in this review. They differ from other prokaryotes in many ways, including the composition and structure of the cell wall and membrane (Kandler and Konig, 1978; Makula and Singer, 1978; Tornabene and Langworthy, 1978; Kandler, 1979), 5S and 16S ribosomal RNA sequences (Fox et al., 1977, 1980), and RNA polymerase (Zillig et al., 1980).

The study of mutagenesis and genetic engineering of methanogens is still in its infancy. Nevertheless, significant research has been completed and is reviewed here.

II. CLOSTRIDIA

A. Mutagenesis in Clostridia

Intense study of mutagenesis in saprophytic clostridia was initiated only at the beginning of this decade (Gomez et al., 1980). Nevertheless, the numbers of mutagenesis procedures and isolated mutants have rapidly increased, as shown in Table 1. In general, EMS (ethylmethanesulfonate) has been the most successfully used mutagenic agent. NTG (*N*-methyl-*N'*-nitrosoguanidine) and ultraviolet (UV) irradiation have been used with mixed success.

The success in obtaining mutants with some mutagens may simply reflect the application of effective procedures that allowed the selection of spontaneous mutants after treatment with ineffective mutagens. Most of the reports cited in Table 1 describe attempts to select for mutants after exposing a culture to a mutagen. Only Gomez et al. (1980), Lemmel (1985), and Bowring and Morris (1985) actually measured mutation frequencies induced by various mutagens. Gomez et al. (1980) evaluated the mutagenesis of *Clostridium thermocellum* using UV irradiation, gamma irradiation, and NTG. UV treatments produced the highest mutation frequency (about 2×10^{-5} 5-fluorouracil-resistant cells per total viable cells), and NTG mutation frequencies were not significantly above what might be expected for spontaneous mutation (3-4 $\times 10^{-6}$ 5-fluorouracil-resistant cells per total viable cells). Lemmel (1985) was unable to obtain *Clostridium acetobutylicum* mutants using UV irradiation or NTG but increased the spontaneous mutation frequency (32.2 Em^r or 4.3 Amp^r cells $\times 10^{-8}$ per total viable cells) by about 100- to 200-fold using EMS (see Sec. VI for explanation of the nomenclature used in this chapter). Bowring and Morris (1985) also obtained their highest relative induced mutation frequency in *C. acetobutylicum* with EMS (1000-fold higher than spontaneous mutation). However, NTG was similarly mutagenic.

These results indicate that NTG is variable for mutagenesis in saccharolytic clostridia. Lemmel (1985) found UV irradiation to be ineffective for *C. acetobutylicum*, but Gomez et al. (1980) identified UV irradiation as the mutagen of choice for *C. thermocellum*. Therefore, the most effective mutagen must be determined for each species and perhaps each strain studied. Nevertheless, EMS generally appears to be a consistently effective mutagen. Also, the EMS kill rate is low enough that moderate errors in concentration or time of exposure can be tolerated.

Table 1 Mutagenesis in Clostridia

Species	Method[a]	Mutant trait[b]	Reference
Clostridium acetobutylicum	EMS	Autolysis deficient	Allcock et al. (1981)
	EMS	Rif^r, $Spor^-$	Jones et al. (1982)
	SP	$Butanol^r$	Lin and Blaschek (1983)
	SP	$Spor^-$	Meinecke et al. (1984)
	EMS	Arg^-, His^-, Met^-, Trp^-,	Jones et al. (1985)
	EMS	Amp^r, Em^r, $Butanol^r$	Lemmel (1985)
	EMS, NTG	Rif^r auxotrophs	Bowring and Morris (198
	NTG	$Butanol^r$	Hermann et al. (1985
	EMS	$Granulose^-$	Reysenbach et al. (1986)
C. saccharolyticum	NTG	$Pyruvate^-$	Murray et al. (1983)
C. thermoaceticum	EMS	$Acid^r$	Schwartz and Keller (198
	SP	$Acid^r$, high u_{max}	Keller et al. (1985)
	UV/EMS	Acetic $acid^r$	Reed et al. (1987)
C. thermocellum	UV	5-$Fluorouracil^r$ Rif^r	Gomez et al. (1980)
	SP	$Ethanol^r$	Herrero and Gomez (1980
	UV, SP	Leu^-, Ade^-	Mendez and Gomez (1982
	G	Low acid producer	Duong et al. (1983)
C. thermosaccharo-lyticum	NTG	$Fluoroacetate^r$ low acetate prod.	Rothstein (1986)
C. thermohydro-sulfuricum	NTG	GA cat. $rep.^r$	Hyun and Zeikus (1985a)
C. thermo-sulfurogenes	NTG	B-Amy $cat.^r$ B-Amy const. B-Amy hyper.	Hyun and Zeikus (1985b

[a]Abbreviations: EMS, ethylmethanesulfonate; SP, spontaneous mutation; NTG, *N*-methyl-*N*'-nitrosoguanidine; UV, ultraviolet light; G, γ irradiation.
[b]Abbreviations: GA cat. rep. r, catabolite repression resistant for glucoamylase production; B-Amy cat. rep. r, catabolite repression resistant for β-amylase production; B-Amy hyper., hyperproducer of β-amylase.

An interesting correlation between certain *C. acetobutylicum* mutants and colony morphology has been made by Adler and Crow (1987). Mutants with the trait of decreased sporulation or solventogenesis can be isolated as spontaneous outgrowths near the edges of wild-type colonies. These intermediate mutant cultures upon subculturing degenerated to genetically stable organisms that produce neither spores nor solvents. Colonies of the genetically stable, fully degenerate mutant cells do not produce outgrowths. The correlation of solventogenesis, sporulation, and colony morphology suggests that their genes may be located in close proximity to each other on the chromosome. This would explain the common development of degenerate cultures with simultaneous mutant phenotypes in these three traits. The development of the degenerate traits may be due to a deletion as a result of errors in DNA replication or packaging during rapid growth and sporulation.

The relatively low induced mutation frequency of clostridia may reflect the presence of a very efficient repair system in some strict anaerobes. This is consistent with the observation that EMS, which does not require error-prone repair processing, is the most consistently effective mutagen for anaerobes.

B. Clostridial Plasmids

Pathogenic clostridia have been shown to contain plasmids coding for the functions of antibiotic resistance in *Clostridium perfringens* (Brefort et al., 1977), bacteriocinogenic activity in *C. perfringens* (Sebald and Brefort, 1975), caseinase in *C. perfringens* (Blaschek and Solberg, 1981), tetanus toxin in *Clostridium tetani* (Laird et al., 1980), and mercury catabolism in *Clostridium cochlearium* (Pan Hou et al., 1980). Also, many industrially important saccharolytic clostridia and closely related species have been shown to contain native plasmids (Table 2). These plasmids are classified as cryptic since their functions have not been determined. Therefore, the presence of these cryptic plasmids cannot be easily determined through the loss or gain of a phenotype, such as antibiotic resistance.

Once identified, plasmids must be characterized and selectable markers identified or added to obtain a useful cloning vector. Several research groups (Table 3) have performed restriction analyses of the native plasmids. These analyses, along with further manipulation, may eventually permit the development of these cryptic plasmids into effective clostridial cloning vectors. Restriction analyses of the native plasmids are summarized in Table 4. All the small plasmids (i.e., < 15 kb) contain unique cleavage sites that permit the development of vector plasmids through the addition of selectable markers.

Hybrid plasmids have been constructed from the cryptic clostridial plasmids and one or more nonclostridial plasmids. Portions of (Luczak

Table 2 Observations of Clostridial Plasmids

Species	Plasmids	Reference
Clostridium acetobutylicum	1–2 cryptic plasmids per strain; 7.9, 10.2, 16.7, 75.8 kb present in 5/17 strains	Truffaut and Sebald (1983)
C. beijerinckii	1–4 plasmids per strain; 5.3–91 kb present in 6/14 strains	Popoff and Truffaut (1985)
C. butylicum	1 cryptic plasmid per strain; 3.9 kb present in 1/2 strains	Truffaut and Sebald (1983)
C. butyricum	1–2 cryptic plasmids per strain; 5.9, 6.5, 7.9 kb present in 3/7 strains	Minton and Morris (1981)
	3 cryptic plasmids per strain; 14.2, 48.5, 77.5 kb present in 1/1 strain	Urano et al. (1983)
	1–6 plasmids per strain; 2.3–106 kb, most common, 6.8 kb present in 43/50 strains	Popoff and Truffaut (1985)
C. saccharoper-butylacetonicum	1 plasmid per strain; 3.9 kb present in 1/2 strains	Truffaut and Sebald (1983)
C. thermoaceticum	1 cryptic plasmid per strain; 87 kb present in 2/2 strains	Chou and Ljungdahl (1987) (personal communication)

Table 3 Characterization of Clostridial Plasmids

Species/strain	Plasmid	Size (kb)	Reference
Clostridium acetobutylicum			
ATCC 8529, NRRL 6441 NRRL 6442, NRRL 6443	pCA1	11.0	Ginsburgh (1987)
C. butylicum			
NRRL B-593	pAB10	4.0	Ginsburgh (1987)
C. butyricum			
NCIB 7423	pCB101	6.0	Minton and Morris (1981) Collins et al. (1985) Luczak et al. (1985)
NCIB 7423	pCB102	7.8	Luczak et al. (1985)
NCIB 7423	pCBU1	6.4	Luczak et al. (1985
SA1, SA11 NCTC 6084	pCB103	6.5	Minton and Morris (1981)
IFO 3847	pSSK1	77.3	Urano et al. (1983)
IFO 3847	pSSK2	48.5	Urano et al. (1983)
IFO 3847	pSSK3	14.2	Urano et al. (1983)

et al., 1985) or complete native plasmids (Collins et al., 1985) of *Clostridium butyricum* NCIB 7423 have been cloned into plasmids containing the chloramphenicol acetyltransferase (CAT) gene of the *Staphylococcus aureus* plasmid pC194 and portions of the *Escherichia coli* plasmid pBR322. The CAT gene of these chimeric plasmids is expressed in *E. coli* and *Bacillus subtilis*, but the plasmids are not replicated in Gram-positive hosts. Recombinant plasmids were selected for in *E. coli* and subsequently transformed into *B. subtilis*. Positive transformants were obtained with recombinant plasmids containing a complete clostridial plasmid insert and in some cases portions of the clostridial plasmid that contained a replicon that functioned in *B. subtilis*. The recombinant plasmids were segregationally and structurally stable in the presence of chloramphenicol selection pressure for the CAT gene. However, in the absence of chloramphenicol, the *C. butyricum* DNA of a recombinant plasmid (pRB7) underwent integration into the *B. subtilis* chromosome (Collins et al., 1985).

The formation of structural rearrangements and deletions was observed with high frequency by Luczak et al. (1985) when clostridial plasmid DNA was cloned in the *B. subtilis* vector pUB110. They suggested that this instability in *B. subtilis* may be due to an effi-

Table 4 Clostridial Plasmid Restriction Analyses

Enzyme	*C. acetobutylicum* PCA1 11.0 kb	*C. butylicum* pAB10 4.0 kb	*C. butyricum* pSSK1 77.3 kb	pSSK2 48.5 kb	pSSK3 14.2 kb	pCB101 6.0 kb	pCB102 7.8 kb	pCB103 6.5 kb	pCBU1 6.4 kb
Acc I	5	3				1	3		1
Ava II			10–20	4	2				
Bam HI	1	0	10	5	0	0	0		
Bcl I						0	0		3
Bgl II	3	1				0	0		4
Cfo I						2	2		1
Cla I	0	0				1	2		3
Eco RI	0	2	10	4	1	0	2	1	2
Eco RV	2	2				2	3		3
Hae II						1	1		2
Hae III						1	1		1
Hind III	4	1				3	2	4	4
Hpa I	2	0					2		
Hpa II						2	2		
Pst I	0	1	10–20	4	1	0	1		2
Sal I	0	0	8	3	1	0	0		
Sau 3A						4	2		
Sau 96I						1	1		2
Sma I						0	0		1
Sst I						0	0		
Xba I	1	1				0	0		
Xho I	0	0				0	0		

cient recombination system that requires very little homology either within the cloned DNA or between insert and vector DNA.

The structural analysis of the cryptic plasmid pCA1 from *C. acetobutylicum* was made by Ginsburgh (1987). The complete plasmid was cloned into pBR322. The recombinant plasmid was expressed and replicated in *E. coli* with no segregational or structural instability. As a result, sufficient quantities of plasmid DNA for structural analysis were obtained (Table 4).

The work just described represents substantial progress toward the development of an efficient gene transfer system for nonpathogenic clostricia of biotechnological interest. Native clostridial plasmids have been identified and structurally analyzed. Plasmid vectors have been constructed from the clostridial plasmids and nonclostridial plasmids containing selectable markers. These plasmid vectors can be used for effectively transferring clostridial plasmid DNA into and between *E. coli* and *B. subtilis*. However, until recently, the absence of an efficient clostridial transformation system has prevented the evaluation of these cloning vectors in clostridia. A notable exception has been the development of a transformation system in the clinically important *C. perfringens*. Squires et al. (1984) first created a chimeric plasmid from a small cryptic *C. perfringens* plasmid and pBR322. Then a *C. perfringens* Tet^r gene from a larger plasmid was introduced into the recombinant plasmid. The resulting vector contained both a clostridial origin of replication and a Tet^r gene. The plasmid vector was transformed into *E. coli* cells or L-phase variants and autoplasts of *C. perfringens*. The intact transforming plasmid could be isolated from walled *C. perfringens* cells following cell wall regeneration.

C. Clostridial Genetic Transfer Systems

As mentioned, the only transformation of clostridia with chimeric plasmids derived in part from native clostridial plasmids was achieved in *C. perfringens* (Squires et al., 1984). However, several reports have been made of the transformation of nonpathogenic clostridia with chromosomal clostridial DNA or nonclostridial DNA. Reid et al. (1983) transformed bacteriophage DNA into *C. acetobutylicum* protoplasts in the absence of polyethylene glycol (PEG). Lin and Blaschek (1983) used PEG to transform heat-treated *C. acetobutylicum* protoplasts with pUB110 (K_m). The pretransformation heating of the protoplasts was necessary to inactive DNase (deoxyribonuclease). This specialized protoplast transformation procedure using pUB110 has not been repeated in other strains of *C. acetobutylicum* by other investigators (M. Sebald, C. L. Ginsburgh, and H. I. Adler, personal communications, 1986). This may be an indication of complex transformation factors that vary from one strain to another. However, Soutschek-Bauer et al. (1985) recently transformed competent *Clostridium thermohydrosulfuricum* DSM 586 whole cells with pUB110 and pGS13.

Chromosomal clostridial DNA has been introduced into clostridial protoplasts by Shirai et al. (1986). They developed a hybrid cell that hydrolyzes cellulose and produces butanol (Cel^+, But^+) by transforming *Clostridium thermosaccharolyticum* B-258 (But^+, Cel^-) protoplasts with chromosomal DNA from *C. thermocellum* C-27 (Cel^+, But^-).

An alternative method of DNA transfer that prevents the exposure of the naked DNA to extracellular DNase is protoplast fusion. Jones et al. (1985) used this technique to achieve genetic transfer in *C. acetobutylicum*. Protoplasts of two double auxotroph strains were fused through a short exposure to PEG. Hybrid cells were obtained at a frequency of 0.3-8.5% hybrid cells in the total regenerated cell population.

The most promising research into plasmid vector development for genetic transfer in clostridia has involved streptococcal plasmids, especially pAMβ1 (Clewell et al., 1974). These plasmids are conjugation proficient and in many cases are capable of mobilizing other plasmids. The transfer of pAMβ1 into clostridia was first demonstrated by Oultram and Young (1985) and further expanded by Reysset and Sebald (1985) and Yu and Pearce (1986). The results of these three reports are summarized in Tables 5 and 6. In these studies pAMβ1 was transferred between and maintained by 19 strains of four species of Gram-positive bacteria. In general, plasmids were transferred into and from *C. acetobutylicum* at average frequencies of about 10^{-5}-10^{-6} transconjugants per recipient cell. The pAMβ1 plasmid has also been transfered to *Clostridium pasteurianum* and *C. butyricum* (J. D. Oultram, personal communication, 1986).

The best transfer rates into *C. acetobutylicum* have been achieved using streptococcal donors (average of about 10^{-4}-10^{-5}). Conjugation from *C. acetobutylicum* to *C. acetobutylicum* was achieved at moderate transfer rates of 10^{-5}-10^{-6}. The lowest transfer rates (10^{-6}-10^{-8}) were observed in transfers from *B. subtilis* to *C. acetobutylicum*. This is particularly unfortunate since it would be desirable to use *B. subtilis* as a donor and use the proven techniques for genetic analysis in this organism.

Mobilizations of small nonconjugative plasmids were achieved at very low transfer rates (10^{-6}-10^{-8}). Plasmid pAM160 was mobilized using the conjugative plasmids pAMβ1 or pVA979 (Yu and Pearce, 1986). The nonconjugative plasmid pOD1 was constructed by Young et al. (1987). The Em^r determinant of pAMβ1 is contained in pOD1 and provides significant homology for cointegrate formation with

Table 5 Conjugal Plasmid Transfer from *Streptococcus* Species and *Bacillus subtilis* into *Clostridium acetobutylicum*

Donor cell	Donor plasmid	Plasmid size (kb)	Conjugation selection	Recipient cell	Transfer frequency[a]	Reference
S. lactis	pAMB1	26.5	Em^r	*C. acetobutylicum*		
MG3020	pAMB1			ATCC 8052	1.5×10^{-3}	Oultram and Young (1985)
MG3020	pAMB1			SBA53	3.6×10^{-4}	Oultram and Young (1985)
MG3020	pAMB1			SBA28	1.4×10^{-3}	Oultram and Young (1985)
SH4174	pAMB1			SBA53	4.1×10^{-5}	Oultram and Young (1985)
4558	pAMB1			P262	1.4×10^{-4}	Yu and Pearce (1986)
4874	pAMB1			P262	9.1×10^{-5}	Yu and Pearce (1986)
4737	pVA797	30.7	Cm^r	P262	2.3×10^{-4}	Yu and Pearce (1986)
4737	pAM610	9.5	Tet^r	P262	3.1×10^{-7}[b]	Yu and Pearce (1986)
4874	pAM610			P262	6.8×10^{-7}[b]	Yu and Pearce (1986)
S. faecalis						
4485	pAMB1	26.5	Em^r	P262	6.8×10^{-5}	Yu and Pearce (1986)
JH2.2	pAMB1			903	1.4×10^{-5}	Reysset and Sebald (1985)
JH2.2	pIP501	35	Em^r, Cm^r	903	3.5×10^{-5}	Reysset and Sebald (1985)
JH2.2	pJH4	39	Em^r, Sm^r, Km^r	903	1.3×10^{-6}	Reysset and Sebald (1985)
4479	pVA797:Tn917	35.7	Em^r, Cm^r	P262	1.8×10^{-4}	Yu and Pearce (1986)
4485	pAM610	9.5	Tet^r	P262	2.3×10^{-7}[b]	Yu and Pearce (1986)
S. sanguis						
4641	pVA797	30.7	Cm^r	P262	6.8×10^{-5}	Yu and Pearce (1986)
4641	pAM610	9.5	Tet^r	P262	4.5×10^{-7}[b]	Yu and Pearce (1986)
B. subtilis						
SA147	pAMB1	26.5	Em^r	ATCC 8052	4.8×10^{-6}	Oultram and Young (1985)
SA147	pAMB1			SBA28	1.2×10^{-8}	Oultram and Young (1985)

[a]Transfer frequencies are expressed as transconjugants per recipient cell.
[b]Transfer dependent on the presence of a second conjugation proficient plasmid.

Table 6 Conjugal Plasmid Transfer from *Clostridium acetobutylicum* into *Bacillus subtilis*, *Streptococcus lactis*, and *Clostridium acetobutylicum*

Donor cell	Donor plasmid	Plasmid size (kb)	Recipient	Transfer frequency[a]	Reference
C. acetobutylicum	pAMB1 Em^r	26.5	*B. subtilis*		
ET101	pAMB1		168	1.9×10^{-6}	Oultram and Young (1985)
ET102	pAMB1		168	4.0×10^{-6}	Oultram and Young (1985)
			S. lactis		
ET101	pAMB1		SH1614	1.5×10^{-5}	Oultram and Young (1985)
ET102	pAMB1		SH1614	1.4×10^{-7}	Oultram and Young (1985)
			C. acetobutylicum		
ET103	pAMB1		SBA53	9.1×10^{-6}	Oultram and Young (1985)
ET104	pAMB1		SBA28	1.3×10^{-5}	Oultram and Young (1985)
903	pAMB1		903-9	3.2×10^{-5}	Reysset and Sebald (1985)
903	pAMB1		ATCC 824-7	2.2×10^{-6}	Reysset and Sebald (1985)
ATCC 824	pAMB1		903-9	2.4×10^{-6}	Reysset and Sebald (1985)
903	pIP501 Em^r	35	903-9	1.6×10^{-5}	Reysset and Sebald (1985)
903	pJH4 Em^r	39	903-9	1.5×10^{-6}	Reysset and Sebald (1985)

[a]Transfer frequencies are expressed as transconjugants per recipient cell.

pAMβ1. The cointegrate plasmid (pAMβ1:pOD1) was transferred from *B. subtilis* to *C. acetobutylicum* and back again with frequencies comparable to those for pAMβ1 alone (10^{-7}-10^{-8} transcipients per recipient). This mobilization procedure was somewhat cumbersome and achieved at a low transfer rate. Nevertheless, it demonstrated the potential for the establishment of small and, possibly, multicopied elements in clostridia.

D. Cloning of *Clostridium* Genes

Chromosomal genes from a number of saccharolytic clostridia have been cloned into *E. coli*. There have been intensive efforts by several groups to clone clostridial genes involved in cellulose hydrolysis (*cel* genes). This is due to the importance of cellulase in chemicals from biomass research programs. Cornet et al. (1983a) cloned cel genes (*celA* and *celB*) as well as genes for biosynthesis of leucine and tryptophan from *C. thermocellum* into *E. coli* through the use of the pHC79 cosmid vector. The *celA* and *celB* genes were subsequently subcloned in pBR322 (Cornet et al., 1983b). Both genes were carried in a DNA sequence of about 2.7 kb. The expression of these genes was independent of their orientation. They were not contiguous on the *C. thermocellum* and shared no homology as shown by ^{32}P-labeled probes. The genes did show homology to DNA from various cellulolytic and noncellulolytic bacteria (Petre et al., 1985). This was not surprising since enzymes that hydrolyze soluble derivatives of cellulose are widespread in bacteria that are not able to digest native cellulose (Bisaria and Ghose, 1981).

Millet et al. (1985) cloned 10 distinct DNA fragments (including the *celA* and *celB* genes described by Cornet et al., 1983a) coding for cellulases of *C. thermocellum* into *E. coli*. Of the cloned fragments, 7 (including the *celA* and *celB* genes) coded for endoglucases, and 3 fragments coded for proteins with properties that suggested they were cellobiohydrolases.

Cellulase genes have also been cloned from *C. acetobutylicum* into *E. coli* by Zappe et al. (1986). Endoglucanase and cellobiase genes were cloned on a 4.9 kb DNA fragment in a recombinant plasmid (pHZ100). The cloned fragments were expressed from their own clostridial promoter in *E. coli*. This suggests that it may be possible to transfer the cloned cellulase genes back into *C. acetobutylicum* and study their expression in the natural solvent-producing host. Zappe et al. (1986) also observed the complementation of *E. coli* arginine and histidine mutations by cloned *C. acetobutylicum* DNA. Complementation of *E. coli* mutations (*proA2* or *leuB6*) by cloned *C. acetobutylicum* DNA was also observed by Efstathiou and Truffaut (1986). Recombinant plasmid pKN391 was isolated from a prototrophic *E. coli* clone and used to transfer *B. subtilis* CU661. Leucine prototrophs were obtained from the transformed *B. subtilis* auxotrophs (*leuC7*).

Usdin et al. (1986) have initiated studies into the regulation of nitrogen metabolism of *C. acetobutylicum*. They cloned the glutamine synthetase gene *glnA* from *C. acetobutylicum* on recombinant plasmid pHZ200. Transformation with plasmid pHZ200 into *E. coli glnA* deletion mutants allowed the utilization of $(NH_4)_2SO_4$ as the sole source of nitrogen. Expression of the cloned gene was regulated by a region contained within the cloned DNA fragment. The cloned gene and gene product functioned very efficiently in *E. coli* despite the lack of any DNA or protein homology with the native *E. coli* gene and/or product.

C. butyricum genes have also been cloned into *E. coli*. Ishii et al. (1983) cloned the β-isopropylmalate dehydrogenase gene from *C. butyricum* into *E. coli*. EcoRI DNA fragments from *C. butyricum* were cloned into pBR322 to obtain the recombinant plasmid pCE4. Competent *E. coli* HB101 (*leuB* mutant) cells were transformed with pCE4. The isolation of Leu^+ transformants demonstrated the functional expression of the *C. butyricum* gene in *E. coli*. Restriction analysis of the recombinant plasmid pCE4 showed that the β-isopropylmalate dehydrogenase gene was located in a 1.4 kb BglII/EcoRI fragment. A single BamHI site was located inside the β-isopropylmalate dehydrogenase gene. Ishii et al. (1983) suggested that inactivation of this gene by cloning into the BamHI site may make the pCE4 a good candidate as a plasmid vector for future genetics studies in *C. butyricum*.

The hydrogenase gene from *C. butyricum* was cloned into *E. coli* by Karube et al. (1983). The clostridial DNA was digested with PstI and cloned into pBR322 to obtain the chimeric plasmic, pCBH1 (11.1 kb). The mutant strain HK16 with decreased hydrogenase was derived from *E. coli* strain C600. HK16 was transformed with pCBH1. The transformants demonstrated hydrogenase activity at a level approximately 870-fold greater than the pretransformed *E. coli* HK16 and 3-fold greater than the donor *C. butyricum*.

III. ANAEROBIC RUMINAL BACTERIA

A. Mutagenesis in Anaerobic Ruminal Bacteria

There have been very few studies on mutagenesis of anaerobic ruminal bacteria and related special. Taya et al. (1983) applied NTG and UV mutagenesis and screened for increased cellulose hydrolysis by clones of *Ruminococcus albus*. Van Tassell and Wilkins (1978) obtained auxotrophic mutants of *Bacteroides fragilis* through ethylmethanesulfonate and a modified penicillin enrichment. Auxotrophs with phenotypes of Arg^-, His^-, Met^-, Thr^-, Trp^-, Ade^-, and Gua^- were obtained. Complementation of these mutations via transduction or conjugation was unsuccessful.

Abratt et al. (1985) isolated mitomycin C-sensitive mutants of *B. fragilis* after mutagenesis with EMS. Most of the mutants demonstrated UV sensitivity equivalent to the parent strain. One mutant that demonstrated only a moderate increase in mitomycin C sensitivity was markedly more sensitive to UV irradiation. These different mutant phenotypes suggest different mechanisms for repair of mitomycin C- and UV-induced DNA damage in *B. fragilis.*

B. fragilis has been shown to be more sensitive to far-UV irradiation under aerobic conditions than under anaerobic conditions (Jones et al., 1980; Jones and Woods, 1981; Slade et al., 1981, 1984). This was not due to a change in the frequency of pyrimidine dimer formation or to an increase in oxygen sensitivity. Schumann et al. (1982, 1983) demonstrated that *B. fragilis* did not produce a typical RecA protein, as defined in *E. coli* (38,000-43,000 molecular weight as observed in *E. coli*; Bagg et al., 1981), following far-UV irradiation under anaerobic conditions. However, synthesis of a new 95,000 molecular weight protein was induced. Also, the synthesis of two constituitive proteins (90,000 and 70,000 molecular weights) was increased following anaerobic UV irradiation. The lack of a typical recA protein was also expressed by the fact that DNA synthesis is unaffected by far-UV irradiation in *B. fragilis*. This is contrary to the usual inhibition of DNA synthesis by the RecA protein in *E. coli* (Trgovcevic et al., 1980). Mutation via error-prone repair is dependent on a functional RecA protein. Woods and Jones (1984) have suggested that the lack of a RecA protein and the associated error-prone repair system could help to explain the low mutation rates observed in *B. fragilis* and related species.

B. Plasmids and Genetic Transfer Systems of Anaerobic Ruminal Bacteria

Teather (1985) demonstrated the conjugal transfer of the broad host range plasmid pRP4 (Am^r) from *E. coli* to *Butyrivibrio fibrisolvens*. Transformation of vancomycin-induced spheroplasts of *B. fibrisolvens* by plasmid DNA was also reported. However, the report did not give details of the transformation procedure. Hazelwood et al. (1986) also investigated the conjugal transfer of pRP4 as well as pR1, pRK2, pRP1, and pRK404 from *E. coli* to anaerobic isolates. Transfer was unsuccessful using 10 butyrivibrio strains and 6 selenomonad strains. Successful conjugation was reported only into an unidentified ruminal isolate. Therefore, the use of conjugation as a routine means of genetic exchange to and from ruminal anaerobes is far from being well established.

The results of conjugation studies with clinical *Bacteroides* are very encouraging and suggest that under the right conditions conjugation in the related ruminal isolates will be developed. The most extensively studied genetic transfer in *Bacteroides* has involved the conjugal transfer of antibiotic resistance plasmids (R plasmids). These plasmids have been shown to transfer between *Bacteroides* species (Privitera et al., 1979; Tally et al., 1979) and from *Bacteroides* species to *E. coli* (Mancini and Behme, 1977; Welch et al., 1979; Young and Mayer, 1979; Rashtchian and Booth, 1981). The conjugation transfer frequencies for R plasmids were enhanced by pretreatment with antibiotics (Tally et al., 1981). Transfer of the self-transmissible clindamycin-resistant plasmid pBFTM10 was not enhanced by pretreatment of tetracycline-sensitive donor cells with clindamycin. However, when this plasmid was transferred from a tetracycline-resistant donor, pretreatment with clindamycin or tetracycline increased the transfer of both determinants 100-fold or 10,000-fold, respectively. The tetracycline resistance determinant appears to be located on the *B. fragilis* chromosome. Therefore, its movement from one cell to another as described earlier represents chromosomal gene transfer via conjugation. This type of genetic transfer has been observed in *Bacteroides* by several investigators (Macrina, 1982; Mays et al., 1981, 1982; Rashtchian et al. 1982; Smith et al., 1982). Cotransfer of other chromosomal genes along with the tetracycline determinant has not been demonstrated.

The enhancement of conjugal transfer by antibiotic pretreatment has also been observed for chimeric plasmids composed in part of *Bacteroides* plasmids pBFTM10 or pB8-51. Shoemaker et al. (1986) constructed *Bacteroides-E. coli* shuttle vectors from a nonmobilizable pBR322 derivative and either pBFTM10 or pB8-51. In general, antibiotic resistances that are expressed and plasmids that replicate in *E. coli* do not function in *Bacteroides* species (Guiney et al., 1984a, Shoemaker et al., 1985). Shuttle plasmids, which include mobilizing regions that are recognized by the IncP class of broad host range conjugative *E. coli* plasmids, have been constructed by Shoemaker et al. (1986). The *E. coli* mobilization regions were effective in allowing the mobilization of shuttle vectors from *E. coli* or *Bacteroides* cells. However, the *E. coli* mobilization regions were not always necessary. *E. coli-Bacteroides* shuttle vectors whose construction included *Bacteroides* plasmids pBFTM10 or pB8-51, cloned using the proper restriction sites, could be mobilized by the IncP plasmids and by the nonplasmid-associated, conjugal *Bacteroides* Tc^r element found in a clinical isolate of *B. fragilis* ERL. Pretreatment with tetracycline increased the mobilization frequencies of the vectors to both *E. coli* and *Bacteroides* recipients 50- to 100-fold.

C. Cloning of Genes from Anaerobic Ruminal Bacteria

Recombinant DNA research involving anaerobic ruminal isolates has been restricted to the cloning of cellulase genes into *E. coli*. These genes have been cloned from *Bacteroides succinogenes* (Crosby et al., 1984), *Butyrivibrio* strain A46 (Hazelwood et al., 1986), and *R. albus* (Romaniec et al., 1986). *E. coli* clones with high cellulose activity demonstrated 6.6 kb *B. succinogenes* DNA inserts. These inserts could be decreased to 2.3 kb with retention of full enzymatic activity (Crosby et al., 1984). The presence of six separate genes encoding polypeptides, each with individual endoglucanase activity, were demonstrated by restriction enzyme analysis and cross-hybridization of the inserts from various clones.

Hazelwood et al. (1986) also observed high- and low-activity clones when cellulase genes were cloned from *Butyrovibrio* strain A46 into *E. coli* using phage λgt11 or plasmid pBR322. The cellulase activity was shown to reside in a 2.3 kb fragment (Romaniec et al., 1986), which, interestingly, is the same size as the cellulase determinant for *B. succinogenes* as shown by Crosby et al. (1984). The cause of the low-activity versus high-activity clones as observed by Hazelwood et al. (1986) or Crosby et al. (1984) has not been determined. Romaniec et al. (1986) also found high- and low-activity clones derived from *R. albus*. However, the more active clone contained a large DNA insert (10.5 kb).

IV. METHANOGENS

A. Mutagenesis in Methanogens

Research on the mutagenesis of methanogens has only recently begun. As shown in Table 7, there have been very few reports of mutagenesis in methanogens. Most of the mutants derived to date have been spontaneous mutants derived at frequencies of about 10^{-7}-10^{-8}. There have been no reports on the efficiency of various chemical or physical mutagens. More basic research in this area, as well as the nature of DNA replication and repair, is needed for the effective application of genetic engineering technology for studying and utilizing methanogens.

The most commonly used mutants for genetic studies are those of antibiotic resistance. Owing to their unique cell envelopes, the methanogens are unaffected by many of the antibiotics commonly used to isolate drug-resistant mutants (Hilpert et al., 1981; Pecher

Table 7 Mutagenesis in Methanogens[a]

Species	Method	Mutant trait	Reference
Methanobacterium thermoautotrophicum	NTG	Analog^r Auxotrophic	Kiener et al. (1984) Rechsteiner et al. (1986)
	NTG	Pseudomonic acid	Kiener et al. (1986)
M. formicicum	SP	Anisomycin^r Narciclasiner Bruceantin^r Trichodermin^r Verrucarin A^r	Hummel and Bock (1985)
Methanosarcina strain 227	SP	2-BES^r	Smith and Mah (1981)
Methanosarcina strain 227	SP	FA^r, acetate$^-$	Smith and Lequerica (1985)
Methanobacterium strain FR-2	SP	Bacitracin^r Nisin^r	Harris and Pinn (1985)

[a]Abbreviations: SP, spontaneous mutation; NTG, *N*-methyl-*N*'-nitrosoguanidine; 2-BES, 2-bromoethanesulfonate; FA, fluoroacetate.

and Bock, 1981). Nevertheless, effective antibiotics have been identified and mutants obtained (Hilpert et al., 1981; Hummel and Bock, 1985; Harris and Pinn, 1985).

The unique nature of the methanogen cell envelope and the corresponding ineffectiveness of penicillin have also made it difficult to obtain auxotrophic mutants. Such mutants are commonly enriched in eubacterial populations through the use of penicillin following mutagenesis. Recently, Rechsteiner et al. (1986) developed an effective alternative enrichment method for auxotrophic methanogenic mutants. They substituted bacitracin for penicillin in auxotrophic enrichments. Bacitracin is a polypeptide antibiotic that interferes with the formation of putative lipid-activated precursors for pseudomurein biosynthesis and interferes with the synthesis of the archaebacterial isoprenylglycerol ether lipids (Hilpert et al., 1981). Therefore, this antibiotic acts preferentially on growing methanogens. Using bacitracin enrichment following nitrosoguanidine treatment, Rechsteiner et al. (1986) achieved a 50-fold enrichment and a 5% yield of auxotrophs. Auxotrophs requiring L-leucine, L-phenylalanine, L-tryptophan, adenosine, or thiamine were obtained.

B. Plasmids and Genetic Transfer Systems in Methanogens

The first report of a methanogen plasmid was made by Thomm et al. (1983). No plasmids were observed in 14 cultural collection strains. As a result, cultural isolation from nature was undertaken to obtain a plasmid-containing methanogenic strain. The plasmid, pMP1 (4.6 kb), was obtained from the coccoid isolate PL-12M. It was not observed in the cleared cell lysate, which is commonly the source for plasmid DNA using the isolation procedure developed for *E. coli* (Clewell and Helinski, 1969). Instead it was obtained from the viscous pellet along with the chromosomal DNA. pMP1 was cloned into pBR322 using Pst I, for which it contained a single restriction site. The recombinant plasmid was amplified in *E. coli* and used for detailed mapping of the pMP1 restriction sites. Thomm et al. (1983) suggested that possible plasmids in methanogens coding for additional functions during life in nature had been lost during the serial transfers in the laboratory on artificial media.

A second methanogen plasmid was isolated from the thermophilic methanogen *Methanobacterium thermoautotrophicum* by Meile et al. (1983). The plasmid, pME2001, was similar in size (4.5 kb) to pMP1 but demonstrated a different restriction map. The yield of plasmid pME2001 was independent of the growth medium utilized, the growth phase of the bacterial culture, and the colony type of the inoculum used in the mass cultivation of cells. Functions coded for by plasmids pMP1 and pME2001 have not been determined. Therefore, both plasmids are cryptic. Conjugation has not been demonstrated for methanogens.

Knox and Harris (1986) reported the observation of a bacteriophage of *Methanobrevibacter smithii*. Spontaneous lysis in late-exponential phase was demonstrated by a 6-mercaptopurine-resistant mutant of *M. smithii*. Bacteriophage particles were observed by electron microscopy. The phage, PMS1, had a hexagonal head and a flexible tail. Restriction enzyme analysis of the DNA of phage PMS1 indicated a genome size of approximately 35 kb. Transduction has not been demonstrated in methanogens.

The formation of stable, metabolically active protoplasts has been demonstrated for *Methanospirillum hungatei* (Sprott et al., 1979), *Methanobacterium bryantii* (Jarrel et al., 1982), *Methanosarcina barkeri* (Davis and Harris, 1985), *M. thermoautotrophicum* (Mountfort et al., 1986), and *Methanosarcina mazei* (Harris, 1986). However, protoplast regeneration, transformation, and fusion have not been achieved with methanogens.

These reports describe components that may ultimately be involved in genetic transfer in methanogens. However, to date, the systems of genetic transfer (conjugation, transformation, and transduction) used commonly to construct genetic maps and develop improved strains have not been demonstrated for methanogens.

C. Cloning of Methanogen Genes

The transcription and translation in *E. coli* of methanogen-derived DNA was first demonstrated using DNA for *M. barkeri* (Reeve et al., 1982) and *Methanobrevibacter arborphilus* (Bollschweiler and Klein, 1982). The functional expression in *E. coli* of DNA derived from *Methanococcus voltae* and *M. smithii* has been reported by Wood et al. (1983) and Hamilton and Reeve (1984), respectively.

Wood and Konisky (1985) observed that complementation of the *E. coli argG* mutation was strong or weak depending on the orientation of a *M. voltae* DNA fragment cloned into the plasmid vector pBR322. Some variants containing plasmids that had undergone genetic rearrangements grew faster. Some of these plasmids were shown to have acquired insertion elements (IS2 or IS5) from the *E. coli* chromosome that activated the expression of the methanogen-derived DNA. A similar positive effect was achieved by dimerization of the recombinant plasmid followed by deletion of sequences between the *tet* promoter of pBR322 and the *arg*G-complementing region.

Functional expression of methanogen-derived DNA has been observed in *B. subtilis* by Morris and Reeve (1984). Transcription of the cloned *M. barkeri* DNA at an internal site was suggested by the fact that orientation of the cloned DNA in the vector plasmids did not affect expression.

Meile et al. (1985) identified and cloned autonomously replicating sequences (ARS) from *Methanococcus vannielii*. Such sequences are thought to act as initiation sites for yeast DNA replication. Only one prokaryotic DNA sequence (pC194, the *Bacillus* or *Staphylococcus* plasmid) and no archaebacterial DNA had been previously reported capable of autonomous replication in yeast. The methanogen ARS exhibited low transformation efficiencies for yeast and promoted the slow growth of yeast transformants. Nevertheless, these replicating units may prove valuable if they can be re-established in their native methanogen hosts as extrachromosomal elements.

V. CONCLUSIONS

In general, the elucidation of the molecular biology of obligate anaerobic bacteria is well underway. Some early success has been achieved in the application of genetic engineering to culture development.

Procedures for mutagenesis and gene cloning in the clostridia have been developed. Genetic transfer via conjugation and transformation has been achieved. The functions of the cryptic plasmids observed in saccharolytic clostridia still need to be determined. It is hoped that these native plasmids or the nonclostridial broad host range plasmids, such as pAMβ1, can be developed into efficient shuttle vectors.

Cloning of cellulase genes from the anaerobic ruminal bacteria has been achieved. Genetic transfer via conjugation has been observed in closely related nonruminal bacteria. Nevertheless, much more research is needed before conjugation can be considered a routine genetic transfer mechanism in anaerobic ruminal bacteria. Also, the mechanisms of DNA repair and mutagenesis in this group of bacteria, as well as all anaerobes, need intense study.

The genetics of methanogens is the least well understood of the three groups of anaerobes. Cloning of some genes has been achieved. Very little is known about mutagenesis, however, and none of the systems of genetic transfer (conjugation, transformation, or transduction) normally used by eubacteria have been observed.

In conclusion, much more basic research is needed into the molecular biology of obligate anaerobic bacteria. However, the practical application of genetic engineering for the cultural development of anaerobes is being realized.

VI. NOMENCLATURE*

Mutant type	Phenotypic character
Acetic acid^r	Acetic acid resistant
Acid^r	Low pH resistant
Ade$^-$	Adenine auxotroph
Amp^r	Ampicillin resistant
Arg$^-$	Arginine auxotroph
Auxotroph	Unable to synthesize one or more cellular constituents
Butanol^r	Butanol resistant
Em^r	Erythromycin resistant
Ethanol^r	Ethanol resistant
Fluoroacetate^r	Fluoroacetate resistant
5-Fluorouracil^r	5-Fluorouracil resistant
Granulose^r	Unable to synthesize granulose
Gua$^-$	Guanine auxotroph

Mutant type	Phenotypic character
His^-	Histidine auxotroph
Leu^-	Leucine auxotroph
Met^-	Methionine auxotroph
$Pyruvate^-$	Unable to metabolize pyruvate
Rif^r	Rifamycin resistant
$Spor^-$	Unable to sporulate
Tet^r	Tetracycline resistant
Thr^-	Threonine auxotroph
Trp^-	Tryptophan auxotroph

*For a complete discussion of commonly used restriction endonucleases and plasmids, refer to any of several available genetic engineering texts (Bukhari et al., 1977; Maniatis et al., 1982; Old and Primrose, 1985).

ACKNOWLEDGMENTS

I would like to thank Dr. Howard I. Adler of Oak Ridge National Laboratory and Dr. Charles L. Ginsburgh of CPC International, Inc., for reviewing the manuscript. Their comments and suggestions have greatly enhanced the quality of the review.

REFERENCES

Abratt, V. R., Jones, D. T., and Woods, D. R. (1985). Isolation and physiological characterization of mitomycin C-sensitive/UV-sensitive mutants in *Bacteroides fragilis*. *J. Gen. Microbiol. 131*:2479-2483.

Adler, H. I., and Crow, W. (1987). Genetic instability in *Clostridium acetobutylicum*. *Abstracts of the Annual Meeting of the American Society for Microbiology*.

Allcock, E. R., Reid, S. J., Jones, D. T., and Woods, D. R. (1981). Autolytic activity and an autolysis-deficient mutant of *Clostridium acetobutylicum*. *Appl. Environ. Microbiol. 42*:929-935.

Armstrong, D. G., and Gilbert, H. J. (1985). Biotechnology and the rumen: A mini review. *J. Sci. Food Agric. 36*:1039-1046.

Bagg, A., Kenyon, C. J., and Walker, G. C. (1981). Inducibility of a gene product required for UV and chemical mutagenesis in *Escherichia coli*. *Proc. Natl. Acad. Sci. USA 78*:5749-5753.

Bisaria, V. S., and Ghose, T. K. (1981). Biodegradation of cellulosic materials: Substrates, microorganisms, enzymes and products. *Enzyme Microb. Technol. 3*:90-104.
Blaschek, H. P., and Solberg, M. (1981). Isolation of a plasmid responsible for caseinase activity in *Clostridium perfringens* ATCC 3626B. *J. Bacteriol. 147*:262-266.
Bollschweiler, C., and Klein, A. (1982). Polypeptide synthesis in *Escherichia coli* directed by cloned *Methanobrevibacter arboriphilus* DNA. *Zentraebl. Bakterial. Hyg. [C] 3*:101-109.
Bowring, S. N., and Morris, J. G. (1985). Mutagenesis of *Clostridium acetobutylicum. J. Appl. Bacteriol. 58*(6):577-584.
Brefort, G., Magot, M., Ionesco, H., and Sebald, M. (1977). Characterization and transferability of *Clostridium perfringens* plasmids. *Plasmid 1*:52-66.
Bukhari, A. I., Shapiro, J. A., and Adhya, S. L. (Eds.). (1977). *DNA insertion Elements, Plasmids, and Episomes.* Cold Spring Harbor Laboratory, Cold Spring Harbor, New York.
Chou, C.-F., and Ljungdahl, L. G. (1987). Isolation of a large plasmid from *Clostridium thermoaceticum.* Personal communication.
Clewell, D. B., and Helinski, D. R. (1969). Supercoiled circular DNA-protein complex in *Escherichia coli*: Purification and induced conversion to an open circular DNA form. *Proc. Natl. Acad. Sci. USA 62*:1159-1169.
Clewell, D. B., Yagi, Y., Dunny, G. M., and Schultz, S. K. (1974). Characterization of three plasmid deoxyribonucleic acid molecules in a strain of *Streptococcus faecalis*: Identification of a plasmid determining erythromycin resistance. *J. Bacteriol. 117*:283-289.
Collins, M. E., Oultram, J. D., and Young, M. (1985). Identification of restriction fragments from two cryptic *Clostridium butyricum* plasmids that promote the establishment of a replication-defective plasmid in *Bacillus subtilis. J. Gen. Microbiol. 131*:2097-2105.
Cornet, P., Tronik, D., Millet, J., and Aubert, J. P. (1983a). Cloning and expression in *Escherichia coli* of *Clostridium thermocellum* genes coding for amino acid synthesis and cellulose hydrolysis. *FEMS Microbiol. Lett. 16*:137-141.
Cornet, P., Millet, J., Beguin, P., and Aubert, J. P. (1983b). Characterization of two CEL (cellulose degradation) genes of *Clostridium thermocellum* coding for endoglucanases. *Bio/Technology 1*:589-594.
Crosby, B., Collier, B., and Thomas, D. Y. (1984). Cloning and expression in *Escherichia coli* of cellulase genes from *Bacteroides succinogenes Can.* In *Bioenergy R + D Seminar (Proceedings)*, 5th Ed., S. Hasnain (Ed.). Elsevier, London.
Davis, R. P., and Harris, J. E. (1985). Spontaneous protoplast formation by *Methanosarcina barkeri. J. Gen. Microbiol. 131*:1481-1486.

Duong, T.-V. C., Johnson, E. A., and Demain, A. L. (1983). Thermophilic, anaerobic and cellulolytic bacteria. In *Topics in Enzyme and Fermentation Biotechnology*, Vol. 7, A. Weisman (Ed.). Wiley, New York, pp. 156-195.

Efstathiou, I., and Truffaut, N. (1986). Cloning of *Clostridium acetobutylicum* genes and their expression in *Escherichia coli* and *Bacillus subtilis*. *Mol. Gen. Genet.* *204*:317-321.

Forsberg, C. W., Crosby, B., and Thomas, D. Y. (1986). Potential for manipulation of the rumen fermentation through the use of recombinant DNA techniques. *J. Anim. Sci.* *63*:310-325.

Fox, G. F., Magrum, L. J., Balch, W. E., Wolfe, R. S., Woese, C. R. (1977). Classification of methanogenic bacteria by 16S ribosomal RNA characterization. *Proc. Natl. Acad. Sci. USA* *74*: 4537-4541.

Fox, G. E., Stackebrandt, E., Hespell, R. B., Gibson, J. Maniloff, J., Dyer, T. A., Wolfe, R. S., Balch, W. E., Tanner, R. S., Magrum, L. J., Zablen, L. B., Blakemore, R., Gupta, R., Bonen, L., Lewis, B. J., Stahl, D. A., Luehrsen, K. R., Chen, K. N., and Woese, C. R. (1980). The phylogeny of prokaryotes. *Science* *209*:457-463.

Ginsburgh, C. L. (1987). The isolation of cryptic plasmids from *Clostridium butylicum* and *Clostridium acetobutylicum*. *Appl. Environ. Microbiol.* (manuscript submitted).

Gomez, R. F., Snedecor, B., and Mendez, B. (1980). Development of genetic principles in *Clostridium thermocellu*. *Dev. Ind. Microbiol.* *22*:87-96.

Guiney, D. G., Hasegawa, P., and Davis, C. E. (1984a). Plasmid transfer from *Escherichia coli* to *Bacteroides fragilis*: Differential expression of antibioltic resistance genes. *Proc. Natl. Acad. Sci. USA* *81*:7203-7206.

Guiney, D. G., Hasegawa, P., and Davis, C. E. (1984b). Expression in *E. coli* of cryptic tetracycline resistance genes from *Bacteroides R. plasmids*. *Plasmid* *11*:248-252.

Hamilton, P. T., and Reeve, J. N. (1984). Cloning and expression of archaebacterial DNA from methanogens in *Escherichia coli*. *Microbial Chemoautotrophy*, W. R. Strohl and O. H. Tuovinen (Eds.). Ohio State University Press.

Harris, J. E. (1986). Development of genetic techniques for the aggregate-forming methanogen *Methanosarcina mazei*. Abstracts, XIV International Microbiology Congress, Manchester, England, p. 240.

Harris, J. E., and Pinn, P. A. (1985). Bacitracin-resistant mutants of a mesophilic *Methanobacterium* species. *Arch. Microbiol.* *143*:1.

Hazelwood, G. P., Mann, S. P., Orpin, C. G., and Romaniac, M. P. M. (1986). Prospects for the genetic manipulation of rumen microorganisms. In *Recent Advances in Anaerobic Bacteriology*,

S. P. Borriello (Ed.). Anaerobic Discussion Group, Proceedings of the Fourth Annual Meeting, July 1985. Martinus Nijhoff, Boston.

Hermann, M., Fayolle, F., Marchal, R., Podvin, L., Sebald, M., and Vandecasteele, J.-P. (1985). Isolation and characterization of butanol-resistant mutants of *Clostridium acetobutylicum*. *Appl. Environ. Microbiol.* *50*(5):1238-1243.

Herrero, A. A., and Gomez, R. F. (1980). Development of ethanol tolerance in *Clostridium thermocellum*: Effect of growth temperature. *Appl. Environ. Microbiol.* *40*:571-577.

Hilpert, R., Winter, J., Hammes, W. P., and Kandler, O. (1981). The sensitivity of Archaebacteria to antibiotics. *Zentraebl. Bakteriol. I [C]* *2*:11-20.

Hummel, H., and Bock, A. (1985). Mutations in *Methanobacterium formicicum* conferring resistance to anti-80S ribosome-targeted antibiotics. *Mol. Gen. Genet.* *198*:529-533.

Hyun, H. H., and Zeikus, J. G. (1985a). Regulation and genetic enhancement of glucoamylase and pullulanase production in *Clostridium thermohydrosulfuricum*. *J. Bacteriol.* *164*(3):1146-1152.

Hyun, H. H., and Zeikus, J. G. (1985b). Regulation and genetic enhancement of β-amylase production in *Clostridium thermosulfurogenes*. *J. Bacteriol.* *164*(3):1162-1170.

Ishii, K., Kudo, T., Honda, H., and Horikoshi, K. (1983). Molecular cloning of B-isopropylmalate dehydrogenase gene from *Clostridium butyricum* M588. *Agric. Biol. Chem.* *47*(10):2313-2317.

Jarrel, K. F., Colvin, J. R., and Sprott, G. D. (1983). Spontaneous protoplast formation in *Methanobacterium bryantii*. *J. Bacteriol.* *149*:346-353.

Jones, D. T., and Woods, D. R. (1981). Effect of oxygen on liquid holding recovery of *Bacteroides fragilis*. *J. Bacteriol.* *145*:1-7.

Jones, D. T., and Woods, D. R. (1986a). Gene transfer, recombination and gene cloning in *Clostridium acetobutylicum*. *Microbiol. Sci.* *3*(1):19-22.

Jones, D. T., and Woods, D. R. (1986a). The acetone and butanol fermentation revisited. *Microbiol. Rev.* *50*(4):484-524.

Jones, D. T., Robb, F. T., and Woods, D. R. (1980). Effect of oxygen on *Bacteroides fragilis* survival after far-ultraviolet irradiation. *J. Bacteriol* *144*:1178-1181.

Jones, D. T., Westhuizen, A., Long, S., Allcock, E. R., Reid, S. J., and Woods, D. R. (1982). Solvent production and morphological changes in *Clostridium acetobutylicum*. *Appl. Environ. Microbiol.* *43*:1434-1439.

Jones, D. T., Jones, W. A., and Woods, D. R. (1985). Production of recombinants after protoplast fusion in *Clostridium acetobutylicum* P262. *J. Gen. Microbiol.* *131*:1213-1216.

Kandler, O. (1979). Zellwandstrukturen bei methanbakterien. *Zur Evolution der Prokaryonten. Naturwissenschaften* *66*:95-105.

Kandler, O., and Konig, H. (1978). Chemical composition of peptidoglycan-free cell walls of methanogenic bacteria. *Arch. Microbiol. 118*:141-152.

Karube, I., Urano, N., Yamada, T., Hirochika, H., and Sakaguchi, K. (1983). Cloning and expression of the hydrogenase gene from *Clostridium butyricum* in *Escherichia coli*. *FEBS Lett. 158*(1):119-122.

Keller, F. A., Ganoung, J. S., and Luenser, S. J. (1985). Mutant strain of *Clostridium thermoaceticum* useful for the preparation of acetic acid. U.S. Patent 4,513,084.

Kiener, A., Holliger, C., and Leisinger, Th. (1984). Analogue-resistant and auxotrophic mutants of *Methanobacterium thermoautotrophicum*. *Arch. Microbiol. 139*:87-90.

Kiener, A., Rechsteiner, T., and Leisinger, T. (1986). Mutation to pseudomonic acid resistance of *Methanobacterium thermoautotrophicum* leads to an altered isoleucyl-tRNA synthetase. *FEMS Microbiol. Lett. 33*:15-18.

Knox, M. R., and Harris, J. E. (1986). Isolation and characterization of a bacteriophage of *Methanobrevibacter smithii* strain PS. Abstracts, XIV International Microbiology Congress, Manchester, England, p. 240.

Laird, W. J., Aaronson, W., Silver, R. P., Habig, W. H., and Hardegree, M. C. (1980). Plasmid associated toxigenicity in *Clostridium tetani*. *J. Infect. Dis. 142*:623-624.

Lemmel, S. A. (1985). Mutagenesis in *Clostridium acetobutylicum*. *Biotechnol. Lett. 7*:711-716.

Lin, Y.-L., and Blaschek, H. P. (1983). Butanol production by a butanol-tolerant strain of *Clostridium acetobutylicum* in extruded corn broth. *Appl. Environ. Microbiol. 45*(3):966-973.

Lipinsky, E. S. (1981). Chemicals from biomass: Petrochemical substitution options. *Science 212*:1465-1471.

Luczak, H., Schwarzmoser, H., and Staudenbauer, W. L. (1985). Construction of *Clostridium butyricum* hybrid plasmids and transfer to *Bacillus subtilis*. *Appl. Microbiol. Biotechnol 23*:114-122.

Macrina, F. L. (1982). Plasmid and non-plasmid-mediated gene transmission in *Bacteroides*. *Anaerobic Genetics Meeting*, Cold Spring Harbor Laboratory, Cold Spring Harbor, New York.

Makula, R. A., and Singer, M. E. (1978). Ether-containing lipids in methanogenic bacteria. *Biochem. Biophys. Res. Commun. 82*:716-722.

Mancini, C., and Behme, R. J. (1977). Transfer of multiple resistance from *Bacteroides fragilis* to *Escherichia coli*. *J. Infect. Dis. 136*:597-600.

Maniatis, T., Fritsch, E. F., and Sambrook, J. (1982). *Molecular Cloning: A Laboratory Manual*. Cold Spring Harbor Laboratory, Cold Spring Harbor, New York.

Mays, T. D., Macrina, F. L., Welch, R. A., and Smith, C. J. (1981). Unusual conjugal transfer of antibiotic resistance in *Bacteroides*: Non-involvement of plasmids. In *Molecular Biology, Pathogenicity, and Ecology of Bacterial Plasmids*, S. B. Levy, R. C. Clowes, and E. L. Koenig (Eds.). Plenum, New York, p. 631.

Mays, T. D., Smith, C. J., Welch, R. A., Delfini, C., and Macrina, F. L. (1982). Novel antibiotic resistance transfer in *Bacteroides*. *Antimicrob. Agents Chemother.* *21*:110-118.

Meile, L., Kiener, A., and Leisinger, T. (1983). A plasmid in the archaebacterium *Methanobacterium thermoautotrophicum*. *Mol. Gen. Genet.* *191*:480-484.

Meile, L., Leisinger, T., and Reeve, J. N. (1985). Cloning of DNA sequences from *Methanococcus vannielii* capable of autonomous replication in yeast. *Arch. Microbiol.* *143*:253-255.

Meinecke, B., Bahl, H., and Gottschalk, G. (1984). Selection of an asporogenous strain of *Clostridium acetobutylicum* in continuous culture under phosphate limitation. *Appl. Environ. Microbiol.* *48*:1064-1065.

Mendez, B. S., and Gomez, R. G. (1982). Isolation of *Clostridium thermocellus* auxotrophs. *Appl. Environ. Microb iol.* *43*:495-496.

Millet, J., Petre, D., Beguin, P., Raynaud, O., and Aubert, J. P. (1985). Cloning of ten distinct DNA fragments of *Clostridium thermocellum* coding for cellulases. *FEMS Microbiol. Lett.* *29*:145-149.

Minton, N. P., and Morris, J. G. (1981). Isolation and partial characterization of three cryptic plasmids from strains of *Clostridium butyricum*. *127*:325-331.

Minton, N. P., and Morris, J. G. (1983). Regeneration of protoplasts of *Clostridium pasteurianum* ATCC 6013. *J. Bacteriol.* *155*:432-434.

Morris, C. F., and Reeve, J. N. (1984). Functional expression of an archaebacterial gene from the methanogen *Methanosarcina barkeri* in *Bacillus subtilis*. In *Microbiological Growth on* C_1 *Compounds, Proceedings of an International Symposium*, R. L. Crawford and R. S. Hanson (Eds.). American Society for Microbiology, Washington, D.C., pp. 205-209.

Mountfort, D. O., Morschel, E., Beimborn, D. B., and Schonheit, P. (1986). Methanogenesis and ATP synthesis in a protoplast system of *Methanobacterium thermoautotrophicum*. *J. Bacteriol.* *168*(2):892-900.

Murray, W. D., Wemyss, K. B., and Khan, A. W. (1983). Increased ethanol production and tolerance by a pyruvate-negative mutant of *Clostridium saccharolyticum*. *Eur. J. Appl. Microbiol. Biotechnol.* *18*:71-74.

Ng, T. K., Busche, R. M., McDonald, C. C., and Hardy, R. W. F. (1983). Production of feedstock chemicals. *Science* *219*:733-740.

Old, R. W., and Primrose, S. B. (1985). *Principles of Gene Manipulation*, 3rd Ed. Blackwell Scientific, Oxford.
Oultram, J. D., and Young, M. (1985). Conjugal transfer of plasmid pAMB1 from *Streptococcus lactis* and *Bacillus subtilis* to *Clostridium acetobutylicum*. *FEMS Microbiol. Lett.* *27*:129-134.
Pan Hou, H. S. K., Hosono, M., and Imura, N. (1980). Plasmid controlled mercury biotransformation by *Clostridium cochlearium* T-2. *Appl. Environ. Microbiol.* *40*:1007-1011.
Pecher, T., and Bock, A. (1981). In vivo susceptibility of halophilic and methanogenic organisms to protein synthesis inhibitors. *FEMS Microbiol. Lett.* *10*:295-297.
Petre, D., Beguin, P., Millet, J., and Aubert, J.-P. (1985). Heterologous hybridization of bacterial DNA to the endoglucanases A and B structural genes celA and celB of *Clostridium thermocellum*. *Ann. Inst. Pasteur (Microbiol.)* *136B(2)*:113-124.
Popoff, M. R., and Truffaut, N. (1985). Survey of plasmids in *Clostridium butyricum* and *Clostridium beijerinckii* strains from different origins and different phenotypes. *Curr. Microbiol.* *12*:151-156.
Privitera, G. A., Dublanchet, A., and Sebald, M. (1979). Transfer of multiple antibiotic resistance between subspecies of *Bacteroides fragilis*. *J. Infect. Dis.* *139*:97-101.
Rashtchian, A., and Booth, S. J. (1981). Stability in *Escherichia coli* of an antibiotic resistance plasmid from *Bacteroides fragilis*. *J. Bacteriol.* *146*:212-227.
Rashtchian, A., Dubes, G. R., and Booth, S. J. (1982). Tetracycline-inducible transfer of tetracycline resistance in *Bacteroides fragilis* in the absence of detectable plasmid DNA. *J. Bacteriol.* *150*:141-147.
Rechsteiner, T., Kiener, A., and Leisinger, T. (1986). Mutants of *Methanobacterium thermoautotrophicum*. *System. Appl. Microbiol.* *7*:1-4.
Reed, W. M., Keller, F. A., Kite, F. E., Bogdan, M. E., and Ganoung, J. S. (1987). Development of increased acetic acid tolerance in anaerobic homoacetogens through induced mutagenesis and continuous selection. *Enzyme Microb. Technol.* *9*:117-120.
Reeve, J. N., Trun, N. J., and Hamilton, P. T. (1982). Beginning genetics with methanogens. In *Genetic Engineering of Microorganisms for Chemicals*, A. Hollaender, D. R. DeMoss, S. Kaplin, J. Konisky, D. Savage, and R. Wolfe (Eds.). Plenum Press, New York, pp. 233-244.
Reid, S. J., Allcock, E. R., Jones, D. T., and Woods, D. R. (1983). Transformation of *Clostridium acetobutylicum* with bacteriophage DNA. *Appl. Environ. Microbiol.* *45*(1):305-307.

Reysenbach, A. L., Ravenscroft, N., Long, S., Jones, D. T., and Woods, D. R. (1986). Characterization, biosynthesis, and regulation of granulose in *Clostridium acetobutylicum*. *Appl. Environ. Microbiol.* *52*(1):185-190.

Reysset, G., and Sebald, M. (1985). Conjugal transfer of plasmid-mediated antibiotic resistance from streptococci to *Clostridium acetobutylicum*. *Ann. Inst. Pasteur (Microbiol.)* *136 B*:275-282.

Rogers, P. (1986). Genetics and biochemistry of *Clostridium* relevant to development of fermentation processes. *Adv. Appl. Microbiol.* *31*:1-60.

Romaniec, M. P. M., Clarke, N. G. E., Orpin, C. G., and Hazlewood, G. P. (1986). Cloning and expression in *Escherichia coli* of cellulase genes from rumen anaerobic bacteria, *Ruminococcus albus* and *Butyrivibrio* sp. Abstracts, XIV International Congress of Microbiology, Manchester, England, p. 101.

Rothstein, D. M. (1986). *Clostridium thermosaccharolyticum* strain deficient in acetate production. *J. Bacteriol.* *165*(1):319-320.

Schumann, J. P., Jones, D. T., and Woods, D. R. (1982). UV light induction of proteins in *Bacteroides fragilis* under anaerobic conditions. *J. Bacteriol.* *151*:44-47.

Schumann, J. P., Jones, D. T., and Woods, D. R. (1983). Effect of oxygen and UV irradiation on nucleic acid and protein synthesis in *Bacteroides fragilis*. *J. Bacteriol.* *156*:1366-1368.

Schwartz, R. D., and Keller, F. A. (1982). Isolation of a strain of *Clostridium thermoaceticum* capable of growth and acetic acid production at pH 4.5. *Appl. Environ. Microbiol.* *43*(1):117-123.

Sebald, M., and Brefort, G. (1975). Transfer du plasmide tetracycline chloraphenicol chez *Clostridium perfringens*. *C.R. Acad. Sci. [D]* *281*:317-319.

Shirai, K., Miyama, T., and Ito, T. (1986). Breeding of cellulose-decomposing, butanol-producing microorganisms. Abstracts of the 1986 Annual Meeting of the Japan Agricultural Chemical Society.

Shoemaker, N. B., Guthrie, E. P., Salyers, A. A., and Gardner, J. F. (1985). Evidence that the clindamycin-erythromycin resistance gene of *Bacteroides* plasmid pBF4 is on a transposable element. *J. Bacteriol.* *162*:626-632.

Shoemaker, N. B., Getty, C., Gardner, J. F., and Salyers, A. A. (1986). Tn4351 transposes in *Bacteroides* spp. and mediates the integration of plasmid R751 into the *Bacteroides* chromosome. *J. Bacteriol.* *165*:929-936.

Slade, H. J. K., Jones, D. T., and Woods, D. R. (1981). Effect of oxygen radicals and peroxide on UV survival and liquid holding recovery in *Bacteroides fragilis*. *J. Bacteriol.* *147*:685-687.

Slade, H. J. K., Jones, D. T., and Woods, D. R. (1984). Effect of oxygen and peroxide on *Bacteroides fragilis* cell and phage

survival after treatment with DNA damaging agents. *FEMS Microbiol. Lett. 24*:159-163.
Smith, C. J., and Hespell, R. B. (1983). Symposium: Application of molecular genetics in ruminants. *J. Dairy Sci. 66*:1536-1546.
Smith, M. R., and Lequerica, J. L. (1985). *Methanosarcina* mutant unable to produce methane or assimilate carbon from acetate. *J. Bacteriol. 164*:618-625.
Smith, M. R., and Mah, R. A. (1981). 2-Bromoethanesulfonate: A selective agent for isolating resistant *Methanosarcina* mutants. *Curr. Microbiol. 6*:321-326.
Smith, C. J., Welch, R. A., and Macrina, F. L. (1982). Two independent conjugal transfer systems operating in *Bacteroides fragilis* V479-1. *J. Bacteriol. 151*:281-287.
Soutschek-Bauer, E., Hartl, L., and Staudenbauer, L. (1985). Transformation of *Clostridium thermohydrosulfuricum* DSM 568 with plasmid DNA. *Biotechnol. Lett.* 7(10):705-710.
Sprott, G. D., Colvin, J. R., and McKellar, R. C. (1979). Sphaeroplasts of *Methanosprillum hungatei* formed upon treatment with dithiothreitol. *J. Microbiol. 25*:730-738.
Squires, C. H., Heefner, D. L., Evans, R. J., Kopp, B. J., and Yarus, M. J. (1984). Shuttle plasmids for *Escherichia coli* and *Clostridium perfringens*. *J. Bacteriol. 159*:465-471.
Tally, F. P., Snydman, D. R., Gorbach, S. L., and Malamy, M. H. (1979). Plasmid mediated transferable resistance to clindamycin and erythromycin in *Bacteroides fragilis*. *J. Infect. Dis. 139*:83-88.
Tally, F. P., Shemell, M. J., Carson, G. R., and Malamy, M. H. (1981). Chromosomal and plasmid-mediated transfer of clindamycin resistance in *Bacteroides fragilis*. In *Molecular Biology, Pathogenecity, and Ecology of Bacterial Plasmids*, S. B. Levy, R. C. Clowes, and E. L. Koenig (Eds.). Plenum, New York, pp. 51-59.
Taya, M., Ohmiya, K., Kobayashi, T., and Shoichi, S. (1983). Enhancement of cellulose digestion by mutants from an anaerobe, *Ruminococcus albus*. *J. Ferment. Technol. 61*(2):197-199.
Teather, R. M. (1985). Application of gene manipulation to rumen microflora. *Can. J. Anim. Sci. 65*:563-574.
Thomm, M., Altenbuchner, J., and Stetter, K. O. (1983). Evidence for a plasmid in a methanogenic bacterium. *J. Bacteriol. 153*:1060-1062.
Tornabene, T. G., and Langworthy, T. A. (1978). Diphytanyl and dibiphytanyl glycerol ether lipids in methanogenic archaebacteria. *Science 203*:51-53.
Trgovcevic, Z., Petranovic, D., Petranovic, M. and Salajsmic, E. (1980). recA gene product is responsible for inhibition of deoyribonucleic acid synthesis after ultraviolet irradiation. *J. Bacteriol. 143*:1506-1508.

Truffaut, N., and Sebald, M. (1983). Plasmid detection and isolation in strains of *Clostridium acetobutylicum* and related species. *Mol. Gen. Genet. 189*:178-180.
Urano, N., Karube, I., Suzuki, S., Yamada, T., Hirochika, H., and Sakaguchi, K. (1983). Isolation and partial characterization of large plasmids in hydrogen-evolving bacterium *Clostridium butyricum. Eur. J. Appl. Microbiol. Biotechnol. 17*:349-354.
Usdin, K. P., Zappe, H., Jones, D. T., and Woods, D. R. (1986). Cloning, expression, and purification of glutamine synthetase from *Clostridium acetobutylicum. Appl. Environ. Microbiol. 52*(3): 413-419.
van Tassell, R. L., and Wilkins, T. D. (1978). Isolation of auxotrophs of *Bacteroides fragilis. Can. J. Microbiol. 24*:1619-1621.
Welch, R., Jones, K. R., and Macrina, F. L. (1979). Transferable lincosamide-macrolide resistance in *Bacteroides. Plasmid 2*:261-268.
Wood, A. G., and Konisky, J. (1985). Activation of expression of a cloned archaebacterial gene in *Escherichia coli* by IS2, IS5, or deletions. *Mol. Gen. Genet. 198*:309-314.
Wood, A. G., Redborg, A. H., Cue, D. R., Whitman, W. B., and Konisky, J. (1983). Complementation of argG and hisA mutations of *Escherichia coli* by DNA cloned from the archaebacterium *Methanococcus voltae. J. Bacteriol. 156*:19-29.
Woods, D. R., and Jones, D. T. (1984). Genetics of anaerobic bacteria. In *Herbivore Nutrition in the Subtropics and Tropics*, F. M. C. Gilchrist and R. I. Mackie (Eds.). Science Press, Craighall, South Africa, pp. 294-312.
Young, F. E., and Mayer, L. (1979). Genetic determinants of microbial resistance to antibiotics. *Rev. Infect. Dis. 1*:55-62.
Young, M., Oultram, J. D., Pennock, A., and Richards, D. R. (1987). Gene transfer in clostridia. In *Genetics of Industrial Microorganisms*, American Society for Microbiology, Washington, D.C. (In press).
Yu, P.-L., and Pearce, L. E. (1986). Conjugal transfer of streptococcal antibiotic resistance plasmids into *Clostridium acetobutylicum. Biotechnol. Lett. 8*(7):469-474.
Zappe, H., Jones, D. T., and Woods, D. R. (1986). Cloning and expression of *Clostridium acetobutylicum* endoglucanase, cellobiase and amino acid biosynthesis genes in *Escherichia coli. J. Gen. Microbiol. 132*:1367-1372.
Zeikus, J. G. (1985). Biology of spore-forming anaerobes. In *Biology of Industrial Microorganisms*, A. L. Demain and N. A. Solomon (Eds.). Benjamin/Cummigs Publishing, Menlo Park, California, pp. 79-114.
Zillig, W., Stetteer, K. O., Schulz, W., and Janekovic, D. (1980). Comparative studies of structure and function of DNA-dependent

RNA polymerase from eubacteria and archaebacteria. In *Enzyme Regulation and Mechanism of Action*, P. Mildner and B. Ries (Eds.). Pergamon Press, Oxford.

3

Industrially Important Strains and Pathways

JAMES C. LINDEN / *Colorado State University, Fort Collins, Colorado*

I. INTRODUCTION

The material presented in this chapter is organized according to the major products of anaerobic fermentation. These in turn are arranged from the simplest, most reduced fermentation product to more complex and oxidized products.

In the traditional sense, the term "fermentation" is descriptive of anaerobic processes. Methane, acetic acid, ethanol, lactic acid, propionic acid, acrylic acid, acetone, acetoin, butanediol, butyric acid, and butanol encompass the traditional fermentation products discussed in this chapter. To complete the collection of industrially important products from anaerobic bacteria, enzymes, which represent nontraditional fermentation products, are presented briefly in the concluding section.

II. METHANE

Methanogens play a key role in the carbon cycle. This group of bacteria are part of the most ancient of microorganisms, the archaebacteria, and are capable of methane production from carbon dioxide and hydrogen in the absence of oxygen. Not only do these autotrophic bacteria reduce CO_2 to methane, but some are unicarbonotrophic and also fix one-carbon substrates to cellular carbon for growth. The application of methanogens in biotechnology is receiving increasing attention, not only for the production of methane, but also for waste disposal and the production of single-cell protein, vitamins, and novel biochemicals, such as lipids, coenzymes, and metabolites. An excellent review of methanogenic bacteria by Whitman (1985) should be consulted for additional information.

Methanobacteria appear widely as mesophilic and thermophilic species. Representative species are *Methanobacterium* (rods), *Methanococcus* (cocci), *Methanosarcina* (packets), *Methanothrix* (filaments), *Methanospirillum* (spirals), and *Methanobrevibacter* (Zeikus, 1983). The methanogens are strict anaerobes and have been considered difficult organisms to cultivate in the laboratory. Since 1977 the biochemical literature on methanogens has increased rapidly. Much of the biochemical research concerns the mechanisms involved in CO_2 fixation, hydrogenase activity, and methane formation. Although progress is being achieved in these areas, no complete pathways are known. These organisms apparently lack cytochromes, the Calvin cycle, the pentose phosphate pathway, and portions of the citric acid cycle (Daniels and Zeikus, 1978; Fuchs and Stupperich, 1978). Regeneration of ATP in this organism involves a membrane-bound ATPase coupled to a transmembrane proton gradient, which is established by the action of hydrogenase on H_2 (Sauer et al., 1981; Doddema et al., 1979). The function of a novel P,P-diphosphodiester, cyclic 2,3-diphosphoglycerate (cyclic DPG), from *Methanobacterium thermoautotrophicum*, has received considerable attention (Seely and Fahrney, 1986). There is recent evidence that cyclic DPG may also play a role in energy transfer in this organism (Fahrney et al., 1987).

Zeikus (1977) has stated that the rather diverse species of methanogenic bacteria can be related by the common capacity to couple oxidation of hydrogen with the concomitant reduction of carbon dioxide:

$$CO_2 + 4H_2 \rightarrow CH_4 + 2H_2O \qquad (1)$$

The possible reactions of methane formation, which were outlined by Barker (1956), have been found to involve 2,2'-dithiodiethane, a cofactor called coenzyme M (Zeikus, 1977):

$$CO_2 + HS\text{-}CH_2\text{-}CH_2SH \rightarrow HS\text{-}CH_2\text{-}CH_2\text{-}S\text{-}COOH \qquad (2)$$

$$HS\text{-}CH_2\text{-}CH_2\text{-}S\text{-}COOH + 2[H] \rightarrow HS\text{-}CH_2\text{-}CH_2\text{-}S\text{-}CHO + H_2O \qquad (3)$$

$$HS\text{-}CH_2\text{-}CH_2\text{-}S\text{-}CHO + 2[H] \rightarrow HS\text{-}CH_2\text{-}CH_2\text{-}S\text{-}CH_2OH \qquad (4)$$

$$HS\text{-}CH_2\text{-}CH_2\text{-}S\text{-}CH_2OH + 2[H] \rightarrow HS\text{-}CH_2\text{-}CH_2\text{-}S\text{-}CH_3 + H_2O \qquad (5)$$

$$HS\text{-}CH_2\text{-}CH_2\text{-}S\text{-}CH_3 + 2[H] \rightarrow CH_4 + HS\text{-}CH_2\text{-}CH_2\text{-}SH \qquad (6)$$

The reducing equivalents, described by [H] in these equations, involve another cofactor, F_{420}, which is unique to but also ubiquitous among methanogens (Zeikus, 1977).

Growth physiology has been studied primarily with the thermophilic *M. thermoautotrophicum* and the mesophilic *Methanosarcina barkeri*. *M. thermoautotrophicum* is the most prolific autotrophic methanogen known. Growing at 65°C in a medium containing only phosphate, ammonium, and sulfide with CO_2 and H_2 as gaseous substrates, specific growth rates of 0.61 h^{-1} have been achieved in continuous culture (Krueger et al., 1986). The maximum specific growth rate for this organism has not yet been measured since methods for achieving a hydrogen-unlimited medium have not been developed. *M. thermoautotrophicum* is not an obligate autotroph since a significant portion of cell carbon can be derived from acetate with cysteine serving as a source of sulfur. In such situations it would be classified as a chemolithotroph. The maximum specific growth rate for *M. thermoautotrophicum* on acetate is given as 0.18 h^{-1} (Krueger et al., 1986).

Growth in the presence of acetate induces acetate thiokinase, with assimilation of acetate into cellular biomass (Oberlies et al., 1980). The usual role of acetate thiokinase is in the synthesis of acetyl phosphate from acetyl-coenzyme A (acetyl-CoA), which is derived from the decarboxylation of pyruvate:

$$\text{Pyruvate} + \text{CoASH} + \text{NAD} \rightarrow CO_2 + \text{acetyl-CoA} + NADH_2 \qquad (7)$$

The subsequent generation of ATP via acetylphosphate is possible, although in the chemolithotrophic sense, acetate thiokinase may be operating in the direction of acetyl-CoA synthesis:

$$\text{Acetyl-CoA} + PO_4 \rightarrow \text{acetyl-}PO_4 + \text{CoASH} \qquad (8)$$

$$\text{Acetyl-}PO_4 + \text{ADP} \rightarrow \text{acetate} + \text{ATP} \qquad (9)$$

The most common acetate utilizing methanogen, *M. barkeri*, can utilize CO_2 and H_2, acetate, methanol, and methylamine as sole carbon substrates. As above, the growth parameters of this mesophile on the gaseous substrates are greater than on acetate. Acetate utilization can be characterized by slow growth and long lag periods because of the poor energy yield of this substrate. If the gaseous substrates are provided in sufficiently high concentration, a high biomass concentration should be attained within a shorter time than when grown on acetate alone. By retaining acetate in the nutrient solution, the adapted bacteria can easily switch back to acetate utilization after depletion of the gaseous substrates (Boone et al., 1987).

From the bioenergetic point of view (Daniels et al., 1984), the methanogenic system is advantageous for treatment of high organic content wastewater compared with aerobic and other anaerobic systems. By using immobilized cell reactors, biomass and volumetric productivity can be increased to high levels. The only disadvantage of the process is the lengthy period of start-up. A period of 6-8 months is required for intensive immobilized cell reactors to come to steady state. The slow growth (μ_{max} = 0.16 h^{-1}) on acetate makes not only industrial application, but also research and development work on such systems difficult. The steady-state operation of an immobilized cell reactor could be attained within a short time if a preliminary state were provided in which the biomass were grown on gaseous substrates. One-carbon compounds are available in large quantities from renewable resources in the form of CO and CO_2 as synthetic gases from pyrolysis processes. The necessary reductant, H_2, is also available from syngases.

III. ACETATE

In addition to unicarbonotrophic synthesis of methane, which was discussed earlier, another group of anaerobic bacteria, commonly called homoacetogens or acetogens, can grow on one-carbon compounds to synthesize acetic acid as the sole fermentation end product. These would be industrially useful fermentations. Syngas produced from pyrolysis reactions could be a suitable substrate mixture of CO and H_2 in equimolar concentrations. The one-carbon substrates, CO_2 (with H_2), CO, HCOOH, and CH_3OH, participate in C_1-C_1 condensation reactions to produce the two-carbon building block for cellular synthesis. Hexoses, lactate, and other substrates may also serve as substrates for these organisms, but in this heterotrophic condition, acetate is not necessarily the only fermentation product. Formic acid, propionic acid, butyric acid, and CO_2 also appear as end products of certain species under specified conditions. Acetic acid can be formed from growth on CO_2 and H_2, as follows:

$$2CO_2 + 4H_2 \rightarrow \text{acetic acid} + 2H_2O \quad (10)$$

The most widely studied homoacetogenic anaerobic bacterium is *Clostridium thermoaceticum* (Ljungdahl et al., 1978). This spore-forming microorganism grows optimally between 55 and 60°C at pH 7-8 under 100% CO_2. Breakdown of sugars by the Embden-Meyerhof-Parnas (EMP) pathway yields pyruvic acid, which undergoes oxidative decarboxylation to yield acetate and an activated methoxy, "CO." The CO is held metabolically, not released as CO_2 as in less efficient anaerobic fermentations.

$$\text{Pyruvate} \rightarrow \text{acetate} + \text{activated methoxy} \qquad (11)$$

These chemical transformations occur with a nearly maximal theoretical yield of products and with minimal energy loss from the system (Zeikus, 1983).

The mechanism of formation of acetate from sugars has been described (Ljungdahl, 1983) as a two-step process according to reactions (12) and (13):

$$\text{Glucose} + 2H_2O \rightarrow 2 \text{ acetic acid} + 2CO_2 + 8\ H^+ + 8\ e^- \qquad (12)$$

$$2CO_2 + 8H^+ + 8e^- \rightarrow \text{acetic acid} + 2H_2O \qquad (13)$$

Net reaction:

$$\text{Glucose} \rightarrow 3 \text{ acetic acid} \qquad (14)$$

Diauxic growth has been exhibited on a mixture of xylose, fructose, and glucose (Andreesen et al., 1973). It is interesting that xylose is the first of the three sugars to be utilized; fructose utilization precedes glucose uptake in this situation.

Acetogenic bacteria also synthesize acetate from carbon monoxide, formate, and methanol according to reactions (15), (16), and (17):

$$4CO + 2H_2O \rightarrow \text{acetic acid} + 2CO_2 \qquad (15)$$

$$4 \text{ formic acid} \rightarrow \text{acetic acid} + 2CO_2 + 2H_2O \qquad (16)$$

$$4 \text{ methanol} + 2CO_2 \rightarrow 3 \text{ acetic acid} + 2H_2O \qquad (17)$$

Acetate inhibition of growth is a problem with this fermentation. An acetate-tolerant strain was derived from ethylmethanesulfonate mutagenesis using low pH and high acetate selective pressures. This mutant grew at pH 4.5 and produced 4.5 g/L of acetate (Schwartz and Keller, 1982).

Clostridium aceticum, the first acetogen isolated, grows chemolithotrophically on CO_2/H_2 and organotrophically on sugars at 30°C at alkaline pH values. Braun and Gottschalk (1981) have studied the hydrogen chemistry of this bacterium.

Clostridium thermoautotrophicum is an anaerobic spore-forming bacterium that grows autotrophically at 70°C. The synthesis of acetate follows reactions (11) through (14) at lower pH values (pH 5.7) than does *C. aceticum*. Chemo-organotrophic growth on fructose, glucose, galactose, and glycerate have been recorded. Wiegel et al. (1981) also noted two strains that grew on xylose.

Acetobacterium woodii is an anaerobic non-spore-forming rod that grows at neutral pH at 30°C. The pathway for synthesis of acetate from CO_2 is similar to that of *C. thermoaceticum* (Tanner et al., 1978), shown in reactions (11) through (14). The production of hydrogen during heterotrophic growth indicates the activity of a hydrogenase enzyme. In addition to fermenting sugars and growth on one-carbon compounds, acetogenic bacteria also utilize methoxyl groups of methoxylated aromatic compounds, such as those found in lignin. Bache and Pfennig (1981) demonstrated that *A. woodii* would demethoxylate vanillic acid, syringic acid, ferulic acid, and sinapic acid and convert the resulting methanol to acetate. For instance,

$$\text{2 syringic acid} + 4H_2 \rightarrow \text{2 } p\text{-hydroxybenzoic acid} + \text{4 methanol} \quad (18)$$

$$\text{4 methanol} + 2CO_2 \rightarrow \text{3 acetic acid} + 2H_2 \quad (19)$$

Other species of acetogenic bacteria exist that synthesize acetate; however, these may also form other products in addition to acetate, utilize other types of substrates, and use different pathways (Gottschalk, 1979).

In the process by which vinegar or food-grade acetic acid is made, ethanol is oxidized to acetic acid by *Acetobacter suboxydans*.

$$\text{Ethanol} + O_2 \rightarrow \text{acetic acid} \quad (20)$$

Trickle-bed reactors, which have traditionally been used for vinegar production, are being replaced by modern process-controlled fermenters.

Lipids, proteins, and carbohydrates found in wastewater as organic pollutants are predominantly degraded to acetate by a mixed population of acetogenic bacteria to acetic acid. Acetate serves as an intermediate substrate for methanogenesis, as discussed in the previous section.

Acetogenic bacteria will play a significant role in potential technology for producing liquid fuels from coal-derived gases. The coal-derived gases of interest consist of blends of hydrogen and carbon oxides obtainable from modern coal gasifiers. The biologic synthesis of liquid fuels from syngases offers the possibility of a highly selective process that could operate efficiently at low pressure and temperature. Alcohol fuels are probably the most feasible final products obtainable from the bioprocessing of such gas streams.

IV. ETHANOL

Ethanol produced by yeasts under anaerobic conditions is the basis of the alcohol fermentation industry. *Saccharomyces* (especially

S. cerevisiae, *S. uvarum*, and *S. carlsbergiensis*) fermentations represent the bulk of the starch- and molasses-based processes. The pathway for the fermentation of glucose to ethanol and CO_2 involves enzymes of the Embden-Meyerhof-Parnas pathway and the glycolytic pathway. Enzymes of the latter include glyceraldehyde-3-phosphate dehydrogenase (23), 3-phosphoglycerate kinase (24), phosphoglycerate mutase (25), enolase (26), pyruvate kinase (27), pyruvate decarboxylase (28), and alcohol dehydrogenase (29). The enzyme-catalyzed reactions, beginning with the overall EMP conversion of glucose to pyruvate, are as follows:

$$\text{Glucose} + 2\text{ATP} \rightarrow \text{fructose-1,6-bisphosphate} + 2\text{ADP} \tag{21}$$

$$\text{Fructose-1,6-bisphosphate} \rightarrow 2\ \text{glyceraldehyde-3-phosphate} \tag{22}$$

$$\text{Glyceraldehyde-3-phosphate} + P_i + NAD^+ \rightarrow \text{1,3-bisphosphoglycerate} + NADH + H^+ \tag{23}$$

$$\text{1,3-bisphosphoglycerate} + \text{ADP} \rightarrow \text{3-phosphoglycerate} + \text{ATP} \tag{24}$$

$$\text{3-phosphoglycerate} \rightarrow \text{2-phosphoglycerate} \tag{25}$$

$$\text{2-phosphoglycerate} \rightarrow \text{2-phosphoenolpyruvate} + H_2O \tag{26}$$

$$\text{2-phosphoenolpyruvate} + \text{ADP} \rightarrow \text{pyruvate} + \text{ATP} \tag{27}$$

$$\text{Pyruvate} \rightarrow CO_2 + \text{acetaldehyde} \tag{28}$$

$$\text{Acetaldehyde} + NADH + H^+ \rightarrow \text{ethanol} + NAD^+ \tag{29}$$

The Pasteur effect, the observed increase in glucose consumption rates under anaerobic conditions compared with aerobic conditions, hinges on the control of the first enzyme of the EMP pathway, phosphofructokinase. Greater concentrations of ADP under anaerobic conditions stimulate phosphofructokinase activity and provide sufficient substrate for the substrate-level phosphorylation reactions (24) and (27). The hydrogen balance in these anaerobic fermentations is maintained by the coupled reactions (23) and (29). Yeast growth is complicated by the requirement for the addition of controlled amounts of oxygen for sterol formation, which is necessary for membrane integrity (Beavan et al., 1982).

The conversion of xylose to ethanol by yeast has been a particular problem when considering the fermentation of lignocellulose hydrolysates. Yeasts that are xylose utilizers have the capability of converting xylose to xylitol and the xylitol to xylulose, which is further metabolized in the pentose phosphate shunt [see reactions (43) through (45)]. Yeasts that cannot utilize xylose produce ethanol when provided with xylulose. The best of the xylulose-fermenting yeasts are *Schizosaccharomyces pombe*, *Kluyveromyces lactis*, *S. cerevisiae*, and *Candida tropicalis* (Gong et al., 1981a, b; Ueng

Table 1 Fermentation Parameters

	Z. mobilis	*S. uvarum*
Specific glucose uptake rate (g/g hr^{-1})	5.45	2.08
Specific ethanol productivity (g/g hr^{-1})	2.53	0.87
Theoretical ethanol yield (%)	92.5	85.9
Volumetric ethanol productivity (g/L hr^{-1})	120	82

et al., 1981). Two yeasts that directly convert xylose to ethanol under aerobic or microaerobic conditions are *Pachysolen tannophilus* (Jeffries, 1981) and *C. tropicalis* (Wang et al., 1980).

Bacterial ethanol producers are of interest because of more favorable fermentation parameters. Rogers et al. (1982) have compared the fermentation parameters for *Zymomonas mobilis* and *S. uvarum*. These data are given in Table 1.

The *Z. mobilis* fermentation was found to be of choice in an industrial waste starch fermentation in West Germany (Bringer et al., 1984). This bacterium grows rapidly under totally anaerobic conditions. Fermentation can be conducted at low pH values (pH 4), which reduces the risk of contamination, and some strains can grow at higher temperatures (37°C) than yeasts.

These bacteria are as alcohol tolerant as are yeast and exhibit tolerance to high glucose concentrations (400 g/L) as well. *Z. mobilis* ferments glucose and fructose to ethanol and CO_2 by a different pathway compared with yeasts. The glucose to pyruvate conversion is via the Entner-Doudoroff (ED) pathway:

$$\text{Glucose} + \text{ATP} \rightarrow \text{glucose-6-PO}_4 + \text{ADP} \tag{30}$$

$$\text{Glucose-6-PO}_4 + H_2O + \text{NADP}^+ \rightarrow \text{6-phosphogluconate} + \text{NADPH} + H^+ \tag{31}$$

$$\text{6-phosphogluconate} \rightarrow \text{KDPG} + H_2O \tag{32}$$

$$\text{KDPG} \rightarrow \text{pyruvate} + \text{glyceraldehyde-3-phosphate} \tag{33}$$

Reaction (33) is catalyzed by 2-keto-3-deoxy-6-phosphogluconate (KDPG) aldolase. This metabolism allows the formation of only 1 mol of ATP per mole of glucose during the subsequent glycolysis steps (24). The more limited amount of ATP restricts cell growth in the fermentation.

Thermophilic bacteria are also being developed for ethanol fermentation (Wang et al., 1979). *Clostridium thermohydrosulfuricum*, *Thermoanaerobium brockii*, and *Thermoanaerobacter ethanolicus* are representative of the species that have been investigated. In commercial production of ethanol, the use of thermotolerant strains would be advantageous since fermentation rates would be greater and cooling costs would be reduced. Product recovery costs and risk of contamination may also be lowered, although Sonnleitner (1983) contends that the latter is not alleviated using thermophilic microorganisms. Pyruvate decarboxylase (7) is missing from these bacteria, and ethanol is formed from acetyl-CoA:

$$\text{Acetyl-CoA} + 2\text{NADH} + \text{H}^+ \rightarrow \text{ethanol} + 2\text{NAD}^+ + \text{CoASH} \qquad (34)$$

In addition to ethanol, acetate and lactate are formed by these organisms.

Recently developed strains of *Clostridium thermocellum*, an anaerobic thermophilic bacterium, hydrolyzes cellulose and ferments cellobiose as well as glucose to ethanol and acetic acid in the ratio of 3:1 (Wang et al., 1979). High specific growth rates are another advantage of thermophilic organisms; that of *C. thermocellum*, for instance, has been reported as high as 0.5 h^{-1} (Ng et al., 1977). Considerable improvement in the yields of cell mass and cellulase has been realized by optimization of medium components.

V. LACTIC ACID

The lactic acid bacteria are represented in the genera *Lactobacillus*, *Streptococcus*, *Leuconostoc*, and *Pediococcus*. The organism of choice depends on the substrate (glucose, fructose, or lactose) and on the desired product (D-, D,L-, or L-lactic acid). Lactic acid bacteria are classified according to the type of fermentation. In natural mixed population fermentations, homofermentative lactic acid bacteria, such as *Lactobacillus curvatus* and *Lactobacillus plantarum*, initiate acidification of the medium.

$$\text{Glucose} \rightarrow 2 \text{ lactate} \qquad (35)$$

The homofermentative pathway involves the Embden-Meyerhof-Parnas conversion of glucose to pyruvate. Pyruvate is reduced to lactic acid utilizing the reducing equivalents generated in the EMP oxidation of glyceraldehyde-3-phosphate:

$$\text{Glyceraldehyde-3-phosphate} + \text{NAD}^+ + \text{P}_\text{i} \rightarrow \text{1,3-bisphosphoglycerate} + \text{NADH} + \text{H}^+ \qquad (36)$$

$$\text{1,3-bisphosphoglycerate} + 2\text{ADP} \rightarrow \text{pyruvate} + 2 \text{ ATP} \qquad (37)$$

$$\text{Pyruvate} + \text{NADH} + \text{H}^+ \rightarrow \text{lactate} + \text{NAD}^+ \qquad (38)$$

Heterofermentative species, such as *Lactobacillus buchneri* and *Lactobacillus brevis*, succeed in dominating the microbial population in later stages of mixed population processes.

$$\text{Glucose} \rightarrow \text{lactate} + \text{ethanol} + CO_2 \tag{39}$$

The heterofermentative pathway involves the pentose phosphate shunt, in which the following reactions occur:

$$\text{Glucose} + \text{ATP} \rightarrow \text{glucose-6-phosphate} + \text{ADP} \tag{40}$$

$$\text{Glucose-6-phosphate} + \text{NAD}^+ \rightarrow \text{6-phosphogluconate} + \text{NADH} + \text{H}^+ \tag{41}$$

$$\text{6-phosphogluconate} + \text{NAD}^+ \rightarrow \text{ribulose-5-phosphate} + \text{NADH} + \text{H}^+ + CO_2 \tag{42}$$

$$\text{Ribulose-5-phosphate} \rightarrow \text{xylulose-5-phosphate} \tag{43}$$

$$\text{Xylulose-5-phosphate} + P_i \rightarrow \text{glyceraldehyde-3-phosphate} + \text{acetylphosphate} + H_2O \tag{44}$$

$$\text{Acetylphosphate} + \text{CoASH} \rightarrow \text{acetyl-CoA} + P_i \tag{45}$$

$$\text{Acetyl-CoA} + \text{NADH} + \text{H}^+ \rightarrow \text{acetaldehyde} + \text{CoASH} + \text{NAD}^+ \tag{46}$$

$$\text{Acetaldehyde} + \text{NADH} + \text{H}^+ \rightarrow \text{ethanol} + \text{NAD}^+ \tag{47}$$

Glyceraldehyde-3-phosphate from (44) is also reduced as in the homofermentative pathway, according to reactions (36) through (38) to lactic acid.

Heterofermentative lactic acid bacteria, of course, ferment pentose sugars via the pentose shunt pathway, reactions (42) through (44). Pentoses are also fermented by some homofermentative lactobacilli (e.g., *Lactobacillus casei*) when phosphoketolase (44), the key enzyme in the heterofermentative pathway, is part of the enzyme complement (Gottschalk, 1979). Heterofermentative bacteria also convert fructose to mannitol. Fructose may be a direct electron acceptor in the heterofermentative pathway, just as pyruvate is the electron acceptor in homofermentation. Fructose reduction to mannitol maintains the reducing equivalent balance necessary for glycolysis.

$$\text{Fructose} + \text{NADH}_2 + \text{H}^+ \rightarrow \text{mannitol} + \text{NAD}^+ \tag{48}$$

The malolactic acid fermentation of importance in wine production is carried out by *L. plantarum*, *L. casei*, *Streptococcus faecalis*,

Streptococcus faecium, *Pediococcus*, and *Leuconostoc mesenteroides* (Gottschalk, 1979). The malic enzyme induced in these bacteria in the presence of malate and a fermentable sugar converts malate to lactate and reduces wine acidity:

$$\text{Malate} + NAD^+ \rightarrow \text{oxalacetate} + CO_2 + NADH + H^+ \tag{49}$$

$$\text{Oxalacetate} \rightarrow \text{pyruvate} + CO_2 \tag{50}$$

$$\text{Pyruvate} + NADH + H^+ \rightarrow \text{lactate} + NAD^+ \tag{51}$$

The species of lactic acid bacteria used most for production of lactic acid is *Lactobacillus delbruckii*. The fermentation is very efficient, with yields of 90%, depending on the substrate. This bacterium is also important in the production of fermented milk products. In the United States, annual production represents $9 billion.

L. delbruckii is also vital to the agricultural forage preservation process called ensiling. *Streptococcus lactis* has been isolated from sugarcane top silage. Most silages are dominated by *L. plantarum*, *L. brevis*, and *Pediococcus* sp. (MacDonald, 1981). It has therefore been found beneficial to inoculate freshly chopped herbage with bacteria capable of anaerobic growth and acidification of the plant material. Lactic acid and acetic acid are desirable primary fermentation products. This process inhibits the growth of filamentous fungi and yeasts and, under proper conditions of dry matter content and buffering capacity, the growth of clostridial species as well.

Some of the ensilage inocula on the market contain the desirable homofermentative lactobacilli, streptococci, and pediococci. Although homofermentative lactobacilli convert glucose and fructose quantitatively to lactic acid, the latter two genera develop lactate and acetate from fermentation of organic acids, such as malate and citrate, from the plant material, reactions (49) through (51). Heterofermentative lactic acid bacteria yield acetate and ethanol, reactions (44) through (47), as well as lactate from the hexose sugars and also utilize fructose as an electron sink to generate significant quantities of mannitol, reaction (48).

VI. PROPIONATE AND ACRYLATE

Propionic acid is formed by some anaerobic bacteria fermenting either glucose or lactic acid:

$$1.5 \text{ glucose} \rightarrow 2 \text{ propionate} + \text{acetate} + CO_2 \tag{52}$$

$$3 \text{ lactate} \rightarrow 2 \text{ propionate} + \text{acetate} + CO_2 \tag{53}$$

The fermentation of lactate has been studied in depth, and two complex pathways have been elucidated, at least in part. The first, known as the acrylate pathway, which involves transformations of lactyl-CoA, has been studied in *Clostridium propionicum* and in *Peptostreptococcus elsdenii*:

$$\text{Lactate} + \text{CoASH} \rightarrow \text{lactyl-CoA} + H_2O \tag{54}$$

$$\text{Lactyl-CoA} \rightarrow \text{acrylyl-CoA} + H_2O \tag{55}$$

$$\text{Acrylyl-CoA} + \text{ETFP}\cdot H_2 \rightarrow \text{propionyl-CoA} + \text{ETFP} \tag{56}$$

$$\text{Propionyl-CoA} + H_2O \rightarrow \text{propionate} + \text{CoASH} \tag{57}$$

The electron-transferring flavoprotein (ETFP) in reaction (56) is regenerated by oxidation of the third mole of lactate [see reaction (38)] first to pyruvate and that to acetate and carbon dioxide (Gottschalk, 1979). Since two reducing equivalents are formed in this sequence of reactions, two lactate are converted simultaneously to two propionate. *C. propionicum* also converts alanine to acrylate for the production of propionate:

$$\text{Alanine} \rightarrow \text{acrylate} + NH_3 \tag{58}$$

A more complex series of reactions by which lactate is converted to propionate involves a carbon fixation step in what is known as the succinate-propionate pathway (Gottschalk, 1979). The overall reaction is

$$\text{Lactate} + \text{NADH} + H^+ + \text{ADP} + P_i \rightarrow \text{propionate} + \text{NAD}^+ + \text{ATP} + 2H_2O \tag{59}$$

This reaction has been studied in *Clostridium tetanomorphum*. Greater biochemical detail of these reactions can be obtained from Gottschalk (1979).

VII. ACETONE

Acetone formation is observed primarily in *Clostridium acetobutylicum*, the bacterium responsible for the commercial process known as acetone-butanol fermentation. Under certain conditions, other clostridia have been forced, primarily by oxidation-reduction constraints, to produce acetone in small quantities. Papoutsakis (1984) has calculated, based on carbon, hydrogen, ATP, and oxidation-reduction balances, that a neutral product mix of 95% acetone and 5% acetate would theoretically be possible using genetically designed strains of clostridia.

Hartmanis and Gatenbeck (1984) provided evidence for the molar equivalence of acetone production and butyrate uptake by *C. acetobutylicum*. Their results point to directed enzymatic conversion of butyrate to butanol through the involvement of acetoacetate CoA:butyrate-(acetate)-transferase and two dehydrogenase reactions [see reaction (75)]. The relationship between the butyrate uptake and acetone formation was reinforced by thermodynamic data cited by Hartmanis and Gatenbeck (1984) and by the apparent induction of acetoacetate decarboxylase (60) coincident with the acid-solvent transition, which typifies acetone-butanol fermentations. The acetoacetate decarboxylation yields acetone directly:

$$\text{Acetoacetate} \rightarrow \text{acetone} + CO_2 \qquad (60)$$

Some clostridial species (e.g., *Clostridium beijerinckii*) have the capability to further reduce acetone to isopropyl alcohol:

$$\text{Acetone} + \text{NADH} + H^+ \rightarrow \text{isopropanol} + \text{NAD}^+ \qquad (61)$$

VIII. ACETOIN AND BUTANEDIOL

An inverse relationship between the final concentrations of acetone and acetoin was found in *C. acetobutylicum* fermentations by Doremus et. al. (1985). Acetoin (3-hydroxy-2-butanone) is a structural analog of acetoacetate that cannot be decarboxylated. Relatively high concentrations of acetoin thereby impede acetone production by inhibition of acetoacetate decarboxylase. Nonpressurized fermentation exhibited relatively high final concentrations of acetone; fermentation conducted at 1 bar gage pressure produced relatively high acetoin levels. On a weight basis, acetone and acetoin collectively comprised 31 ± 4% of the neutral solvents, regardless of the fluctuations in the relative productivities of the two compounds imposed by increased pressure and reduced agitation rates. Accordingly, the difference in fermentation conditions may result in negative regulation of pyruvate decarboxylase and the accumulation of pyruvate.

In *Aerobacter* sp., 2 mol of pyruvate are condensed by action of acetolactate synthase,

$$2 \text{ pyruvate} \rightarrow \text{acetolactate} \qquad (62)$$

and acetoin is formed by decarboxylation of the resulting acetolactate:

$$\text{Acetolactate} \rightarrow \text{acetoin} + CO_2 \qquad (63)$$

Acetoin production stops at the time acetone formation begins in the *C. acetobutylicum* fermentation. Bahl and Gottschalk (1985) reported the accumulation of lactic acid by *C. acetobutylicum* constrained by sulfate limitation above pH 5.0. The formation of acetoin from the reaction of lactic acid with either acetyl-CoA or the activated acetyl moiety, which was involved in acetolactate formation (62), would provide an outlet not only for pyruvate but also reducing equivalents:

$$\text{Lactate + acetyl-CoA + NADH + H}^+ \rightarrow \text{acetoin + NAD}^+ + \text{CO}_2 \qquad (64)$$

The conversion of acetoin to diacetyl in lactic acid bacteria,

$$\text{Acetoin + NAD}^+ \rightarrow \text{diacetyl + NADH + H}^+ \qquad (65)$$

is supplied with a flow of substrates from citric acid (Gottschalk, 1979):

$$\text{Citrate} \rightarrow \text{acetate + oxalacetate} \qquad (66)$$

$$\text{Oxalacetate} \rightarrow \text{CO}_2 + \text{pyruvate} \qquad (67)$$

$$\text{2 pyruvate + CoASH} \rightarrow 2\text{CO}_2 + \text{acetyl-CoA + "activated aldehyde"} \qquad (68)$$

$$\text{Acetyl-CoA + "activated aldehyde"} \rightarrow \text{diacetyl + CoASH} \qquad (69)$$

Butanediol formation is observed primarily in *Enterobacter aerogenes*, *Bacillus* sp., *Serratia* sp., and *Erwinia* sp. bacteria. These could be used for commercial processes because of the potential for the production of 1,3-butadiene from 2,3-butanediol by chemical catalysis. Butadiene is important as a feedstock for polymer production.

One route of butanediol biosynthesis is via α-acetolactate synthase exhibited in *Aerobacter* (62). Decarboxylation yields acetoin [see reaction (63)], which may then be reduced to 2,3-butanediol:

$$\text{Acetoin + NADH + H}^+ \rightarrow \text{2,3-butanediol + NAD}^+ \qquad (70)$$

Another pathway involves pyruvate-formate lyase:

$$\text{Pyruvate + CoASH} \rightarrow \text{formate + acetyl-CoA} \qquad (71)$$

The acetyl-CoA (2 mol) is condensed to form acetoacetyl-CoA,

$$\text{2 acetyl-CoA} \rightarrow \text{acetoacetyl-CoA + CoASH} \qquad (72)$$

which is rearranged and reduced to yield butanediol:

$$\text{Acetoacetyl-CoA} + 2\text{NADH} + 2\text{H}^+ \rightarrow \text{butanediol} + 2\text{NAD}^+ + \text{CoASH} \qquad (73)$$

Some microbial processes, as exhibited in *Escherichia coli* and *E. aerogenes*, when driven by oxidation-reduction constraints, split formate into CO_2 and H_2 to dump electrons in order that glycolysis and other metabolic functions might proceed:

$$\text{Formate} + \text{H}^+ + \text{e}^- \rightarrow \text{CO}_2 + \text{H}_2 \qquad (74)$$

IX. BUTYRATE

Butyric acid is an end product of the anaerobic metabolism of many of the *Clostridium* species bacteria, known as the butyric clostridia. Bacteria that exemplify the following reactions include *Clostridium butyricum, Clostridium tyrobutyricum,* and *Clostridium lactoacetophilum*. The metabolic pathway of glucose catabolism to pyruvate is via the Embden-Meyerhof pathway, yielding 2 mol of ATP, $2NADH + H^+$, and 2 pyruvate per mole of glucose. Pyruvate then undergoes oxidative decarboxylation by the action of ferredoxin oxidoreductase to yield acetyl-CoA, CO_2, and reduced ferredoxin. More will be said of ferredoxin in Section X. The role of the three key fermentation intermediates, acetyl-CoA, acetoacetyl-CoA, and butyryl-CoA, is important to final fermentation product distribution.

The pools of these three fermentation intermediates are in intimate flux because of reactions catalyzed by the enzyme, acetoacetyl:acetate/butyrate:CoA transferase. This enzyme converts acetoacetyl-CoA to acetoacetate using either acetate or butyrate as the CoA acceptor:

$$\text{Acetoacetyl-CoA} + \text{acetate (or butyrate)} \rightarrow \text{acetyl-CoA (or butyryl-CoA)} + \text{acetoacetate} \qquad (75)$$

A second route for butyryl-CoA formation occurs through a four-step process beginning with acetyl-CoA. First, 2 mol of acetyl-CoA are condensed by action of acetyl-CoA acetyltransferase to form acetyl-CoA:

$$2\ \text{acetyl-CoA} \rightarrow \text{acetoacetyl-CoA} + \text{CoASH} \qquad (76)$$

Second, reduction catalyzed by 3-hydroxybutyryl-CoA reductase yields NAD^+ and 3-hydroxybutyryl-CoA,

$$\text{Acetoacetyl-CoA} + \text{NADH} + H^+ \rightarrow \text{3-hydroxybutyryl-CoA} + \text{NAD}^+ \quad (77)$$

which undergoes dehydration in the third step to form crotonyl-CoA by action of the enzyme crotonase:

$$\text{3-hydroxybutyryl-CoA} \rightarrow \text{crotonyl-CoA} + H_2O \quad (78)$$

Finally, butyryl-CoA reductase reacts NADH + H^+ with crotonyl-CoA to form butyryl-CoA:

$$\text{Crotonyl-CoA} + \text{NADH} + H^+ \rightarrow \text{butyryl-CoA} + \text{NAD}^+ \quad (79)$$

This process has the advantage over the transferase reaction in that the reoxidation of 2 mol of NADH + H^+ occurs.

Acetate and butyrate are also produced from acetyl-CoA and butyryl-CoA, respectively, by analogous two-step sequences. The first step involves the production of the corresponding acyl-phosphate with the concomitant regeneration of CoASH:

$$\text{Butyryl-CoA} + P_i \rightarrow \text{butyryl-phosphate} + \text{CoASH} \quad (80)$$

The generation of ATP and the acid end product occurs in the second step:

$$\text{Butyryl-phosphate} + H_2O \rightarrow \text{butyrate} + P_i \quad (81)$$

The formation of acetate in this two-step sequence is mediated by the enzymes phosphate acetyltransferase and acetate kinase. The formation of butyrate is catalyzed by the distinct, but analogous, enzymes phosphate butyryltransferase and butyrate kinase. The overall effect of these pathways is the production of an extra mole of ATP per mole of pyruvate. The acids are excreted from the cell; their accumulation results in a decrease in the pH of the medium.

The butyric clostridia also ferment lactic acid when acetate is present as electron acceptor:

$$\text{Lactate} + \text{NAD}^+ \rightarrow \text{pyruvate} + \text{NADH} + H^+ \quad (82)$$

Butyric acid is the undesired product of such reactions in spoiled ensilage (MacDonald, 1981).

Another type of butyric acid fermentation in which lactic acid is a major product and caproic acid a minor product is carried out by *Butyribacterium rettgeri*. Since lactate formation is increased with iron deficiency, the critical role of ferredoxin in the electron balance of the cell is again implicated. Without the ability to dump reducing equivalents to the medium in the form of molecular hydrogen, this organism produces lactate directly from pyruvate as an electron sink (82). Similarly, the clostridia under conditions of metal ion deficiency (Bahl and Gottschalk, 1985) or in the presence of carbon monoxide or cyanide (Datta and Zeikus, 1985) have been shown to produce lactate.

Caproic acid is also produced in an interesting fermentation of *Clostridium kluyveri*. Butyrate, caproate, and hydrogen are formed from ethanol and acetate or propionate; the latter acids serve as electron acceptors. As earlier, more detail may be obtained from Gottschalk (1979).

X. BUTANOL

Clostridial fermentations in which butanol is produced are termed butylic fermentations. *C. acetobutylicum*, *C. beijerinckii*, and *Clostridium tetranomorphum* are representative organisms; acetone-butanol, isopropanol-butanol, and ethanol-butanol, respectively, are unique end products of the anaerobic energy metabolism of the three organisms. An excellent review of the acetone-butanol literature by Jones and Woods (1986) is available.

There are two distinct metabolic phases observed in batch culture of the butylic clostridia. The initial phase is known as acetogenesis and the second as solventogenesis. The basic reactions during acetogenesis are those described for butyric clostridia in Section IX. During this phase hydrogen gas evolution predominates over that of carbon dioxide. Electron flow through ferredoxin to molecular hydrogen is important during acetogenesis. Acid production yields additional ATP, but only a fraction of the reducing equivalents produced in glycolysis are consumed. No NADH + H^+ is consumed in the conversion of pyruvate to acetate, and in only one of the three routes leading to butyrate is NADH + H^+ utilized. Production of the acid products results in a net generation of NADH + H^+, which must be oxidized to allow continued fermentation. Ferredoxin oxidoreductase can transfer electrons from NADH + H^+ to either NADPH + H^+, which is required for biosynthetic reactions, or to the hydrogenase enzyme, which catalyzes electron transfer to protons for production of hydrogen gas.

Specific rates of hydrogen production and hydrogenase activity have been shown to be high during acetogenesis and to decrease during solventogenesis (Doremus et al., 1985). During solventogenesis, a portion of the electron flow is shifted away from the production of hydrogen toward production of NADH + H^+ or NADPH + H^+, which in turn are used in the formation of butanol and ethanol. Butyric acid and acetic acid, which are produced during the initial phase of batch fermentations, are reassimilated by the cells and converted to the more reduced end products, butanol and ethanol. Sugar consumption continues at a similar rate during both phases of the fermentation.

Many of the enzymes that participate in neutral solvent production must be synthesized or activated with the shift to solventogenesis. Acetoacetyl-CoA:acetate/butyrate:CoA-transferase (75), the enzyme involved in the reassimilation of acetate and butyrate from the medium into the cells, exhibits high levels of activity in solvent-producing cells but is virtually undetectable in acid-producing cells (Hartmanis and Gatenback, 1984). Similarly, the enzyme catalyzing the final step in acetone formation in *C. acetobutylicum*, acetoacetate decarboxylase (60), has 40-fold higher activity in solvent-producing cells than in acid-producing cells (Andersch et al., 1983). This contributes to the greater ratio of CO_2 to H_2 during solventogenesis than during acetogenesis (Doremus et al., 1985). Not only is the specific H_2 production rate less during the second phase of the fermentation relative to the first phase, the specific CO_2 formation rate is greater. Finally, both the dehydrogenase enzymes involved in the formation of butanol from butyryl-CoA, (83) and (84), exhibit much higher activity in solvent-producing cells (Andersch et al., 1983):

$$\text{Butyryl-CoA} + \text{NADH} + H^+ \rightarrow \text{butyraldehyde} + \text{CoASH} + \text{NAD}^+ \quad (83)$$

$$\text{Butyraldehyde} + \text{NADH} + H^+ \rightarrow \text{butanol} + \text{NAD}^+ \quad (84)$$

XI. ENZYMES

Products of commerce, which might not necessarily be commodity materials, may also be derived from anaerobic fermentation. Enzymes, in particular cellulase enzymes, may one day be industrially important. The cellulase system from *C. thermocellum* has been studied extensively (Hon-Nami et al., 1986; Ng and Zeikus, 1981; Ng et al. 1977; and Johnson et al., 1982). *T. brockii* (Ljungdahl et al., 1981) and *Acetivibrio cellulolyticus* (MacKenzie et al., 1987) are other cellulolytic anaerobic bacteria that have received recent attention. *Caldocellum saccharolyticum*, isolated from New Zealand thermal sites, produces a cellulase capable of degrading cellulose at 75°C (Sissons et al., 1987).

REFERENCES

Andersch, W., Bahl, H., and Gottschalk, G. (1983). Level of enzymes involved in acetate, acetone and butanol formation by *Clostridium acetobutylicum*. *Eur. J. Appl. Microbiol. Biotechnol.* *18*:327-332.

Andreesen, J. R., Schaupp, A. Neurauter, C., Brown, A., and Ljungdahl, L. G. (1973). Fermentation of glucose, fructose and xylose by *Clostridium thermoaceticum*: Effects of metals on growth yield, enzymes and the synthesis of acetate from CO_2. *J. Bacteriol.* *114*:743-751.

Bache, R., and Pfennig, N. (1981). Selective isolation of *Acetobacterium woodii* on methoxylated aromatic acids and determination of growth yields. *Arch. Microbiol.* *130*:255-261.

Bahl, H., and Gottschalk, G. (1985). Parameters affecting solvent production by *Clostridium acetobutylicum* in continuous culture. *Biotechnol. Bioeng. Symp.* *14*:217-223.

Barker, H. A. (1956). *Bacterial Fermentations*. John Wiley and Sons, New York.

Beaven, M. J., Charpentier, C., and Rose, A. H. (1982). Production and tolerance of ethanol in relation to phospholipid fatty-acyl composition in *Saccharomyces cerevisiae* NCYC431. *J. Gen. Microbiol.* *128*:1447-1455.

Boone, D. R., Menaia, J. A. G. F., Boone, J. E., and Mah, R. A. (1987). Effects of hydrogen pressure during growth and effects of pregrowth with hydrogen on acetate degradation by *Methanosarcina* species. *Appl. Environ. Microbiol.* *53*:83-87.

Braun, K., and Gottschalk, G. (1981). Effect of molecular hydrogen and carbon dioxide on chemo-organotropic growth of *Acetobacterium woodii* and *Clostidium aceticum*. *Arch. Microbiol.* *128*:294-298.

Bringer, S., Sahm, H., and Swyzen, W. (1984). Ethanol production by *Zymomonas mobilis* and its application on an industrial scale. *Biotechnol. Bioeng. Symp.* *14*:311-319.

Daniels, L., and Zeikus, J. G. (1978). One-carbon metabolism in methanogenic bacteria: Analysis of short-term fixation products of $^{14}CO_2$ and $^{14}CH_3OH$ incorporated into whole cells. *J. Bacteriol.* *136*:75-84.

Daniels, L., Sparling, R., and Sprott, G. D. (1984). The bioenergetics of methanogenesis. *Biochim. Biophys. Acta* *768*:113-163.

Datta, R., and Zeikus, J. G. (1985). Modulation of acetone-butanol-ethanol fermentation by carbon monoxide and organic acids. *Appl. Environ. Microbiol.* *49*:522-529.

Doddema, H. J., van der Drift, C., Vogels, G. D., and Veenhuis, R. (1979). Chemiosmotic coupling in *Methanobacterium thermoautotrophicum*: Hydrogen-dependent adenosine-5-triphosphate synthesis by subcellular particles. *J. Bacteriol.* *140*:1081-1089.

Doremus, M. D., Linden, J. C., and Moreira, A. R. (1985). Agitation and pressure effects on the acetone-butanol fermentation. *Biotechnol. Bioeng. 27*:852-860.

Fahrney, D. E., Harper, S. H., and Krueger, R. D. (1987). The cyclic-2,3-diphosphoglycerate cycle in methanogenic bacteria. Abstract for the American Society of Biological Sciences Meeting, Philadelphia, June 7-11.

Fuchs, G., and Stupperich, E. (1978). Evidence for an incomplete reductive carboxylic acid cycle in *Methanobacterium thermoautotrophicum. Arch. Microbiol. 118*:121-125.

Gong, C. S., Chen, L. F., Flickinger, M. C., Chiang, L. C., and Tsao, G. T. (1981a). Production of ethanol from D-xylose by using D-xylose isomerase and yeasts. *Appl. Environ. Microbiol. 41*:430-436.

Gong, C. S., McCracken, L. D., and Tsao, G. T. (1981b). Direct fermentation of D-xylose to ethanol by a xylose-fermenting yeast mutant, *Candida* sp. XF217. *Biotechnol. Lett. 3*:245-250.

Gottschalk, G. (1979). *Bacterial Metabolism*, 2nd ed. Springer Verlag, Berlin, pp. 214-249.

Hartmanis, M. G. N., and Gatenbeck, S. (1984). Intermediary metabolism of *Clostridium acetobutylicum*: Levels of enzymes involved in the formation of acetate and butyrate. *Appl. Environ. Microbiol. 47*:1277-1283.

Hon-Nami, K., Coughlan, M. P., Hon-Nami, H., and Lungdahl, L. G. (1986). Separation and characterization of the complexes constituting the cellulolytic enzyme system of *Clostridium thermocellum. Arch. Microbiol. 145*:13-19.

Jeffries, T. W. (1981). Fermentation of xylulose to ethanol using xylose isomerase and yeasts. *Biotechnol. Bioeng. Symp. 11*:315-324.

Johnson, E. A., Sakajoh, M., Halliwell, G., Madia, A., and Demain, A. L. (1982). Saccharification of complex cellulosic substrates by the cellulase system from *Clostridium thermocellum. Appl. Environ. Microbiol. 43*:1125-1132.

Jones, D. T., and Woods, D. R. (1986). Acetone-butanol fermentation revisited. *Microbiol. Rev. 50*:484-524.

Krueger, R. D., Harper, S. H., Campbell, J. W., and Farhney, D. E. (1986). Phosphate-limited continuous culture of *Methanobacterium thermoautotrophicum*: Kinetics of phosphate uptake, growth and accumulation of cyclic diphosphoglycerate. *J. Bacteriol. 167*:49-56.

Ljungdahl, L. G. (1983). Formation of acetate using homoacetate fermenting anaerobic bacteria. In *Organic Chemicals from Biomass*, D. L. Wise (Ed.). Benjamin Cummings Publishing, Menlo Park, California, pp. 220-247.

Ljungdahl, L. G., Yang, S. S., Liu, M. T., and Wiegel, J. (1978). On the physiology and thermophilic properties of *Clostridium thermoaceticum* and some other thermophilic anaerobes. In *Biochemistry of Thermophily*, S. M. Friedman (Ed.). Academic Press, New York, pp. 385-400.

Ljungdahl, L. G., Bryant, F., Carreira, L., Saiki, T., and Wiegel, J. (1981). Some aspects of thermophilic and extreme thermophilic anaerobic microorganisms. *Basic Life Sci. 18*:397-419.

MacDonald, P. (1981). *The Biochemistry of Silage*. John Wiley and Sons, Chichester, pp. 63-76.

MacKenzie, C. R., Patel, G. B., and Bilous, D. (1987). Factors involved in hydrolysis of microcrystalline cellulose by *Acetivibrio cellulolyticus*. *Appl. Environ. Microbiol. 53*:304-308.

Ng, T. K., and Zeikus, J. G. (1981). Comparison of extracellular cellulase activities of *Clostridium thermocellum* LQRI and *Trichoderma reesei* QM 9414. *Appl. Environ. Microbiol. 42*:231-240.

Ng, T. K., Weimer, P. J., and Zeikus, J. G. (1977). Cellulolytic and physiological properties of *Clostridium thermocellum*. *Arch. Microbiol. 114*:1-7.

Oberlies, G., Fuchs, G., and Thauer, R. K. (1980). Acetate thiokinase and the assimulation of acetate in *Methanobacterium thermoautotrophicum*. *Arch. Microbiol. 128*:248-252.

Papoutsakis, E. T. (1984). Equations and calculations for fermentations of butyric bacteria. *Biotechnol. Bioeng. 26*:174-187.

Rogers, P. L., Lee, K. J., Skotnicki, M. L., and Tribe, D. E. (1982). Ethanol production by *Zymomonas mobilis*. *Adv. Chem. Biochem. Eng. 23*:37-84.

Sauer, F. D., Erfle, J. D., and Mahadevan, S. (1981). Evidence for an internal electrochemical proton gradient in *Methanobacterium thermoautotrophicum*. *J. Biol. Chem. 256*:9843-9848.

Schwartz, R. D., and Keller, F. A., Jr. (1982). Isolation of a strain of *Clostridium thermoaceticum* capable of growth and acetic acid production at pH 4.5. *Appl. Environ. Microbiol. 43*:117-123.

Seely, R. J., and Farhney, D. E. (1986). Levels of cyclic-2,3-diphosphoglycerate in *Methanobacterium thermoautotrophicum* during phosphate limitation. *J. Bacteriol. 160*:50-54.

Sissons, C. H., Sharrock, K. R., Daniel, R. M., and Morgan, H. W. (1987). Isolation of cellulolytic anaerobic thermophiles from New Zealand thermal sites. *Appl. Environ. Microbiol. 54*:832-838.

Sonnleitner, B. (1983). Biotechnology of thermophilic bacteria—growth, products and application. *Adv. Biochem. Eng. Biotechnol. 28*:69-138.

Tanner, R. S., Wolfe, R. S., and Ljungdahl, L. G. (1978). Tetrahydrofolate enzyme levels in *Acetobacterium woodii* and their implication in the synthesis of acetate from CO_2. *J. Bacteriol. 134*:668-670.

Ueng, P. P., Hunter, C. A., Gong, C. S., and Tsao, G. T. (1981). D-Xylulose fermentation in yeasts. *Biotechnol. Lett.* *3*:315-320.

Wang, D. I. C., Cooney, C. L., Demain, A. L., Gomez, R. F., and Sinskey, A. J. (1979). Degradation of cellulosic biomass and its subsequent utilization for the production of chemical feedstocks. Progress Report COO-4198-9 for the USDOE under contract EG-77-S-02-4198:25-44.

Wang, P. Y., Johnson, B. F., and Schneider, H. (1980). Fermentation of D-xylose by yeasts using glucose isomerase in the medium to convert D-xylose to D-xylulose. *Biotechnol. Lett.* *2*:273-278.

Whitman, W. B. (1985). Methanogenic bacteria. *Bacteria* *8*:3-84.

Zeikus, J. G. (1977). The biology of methanogenic bacteria. *Bacteriol. Rev.* *41*:514-541.

Zeikus, J. G. (1983). Chemical and fuel production from one-carbon fermentations: A microbiological assessment. In *Organic Chemicals from Biomass*, D. L. Wise (Ed.). Benjamin Cummings Publishing, Menlo Park, California, pp. 361-366.

II

BIOCHEMISTRY, KINETICS, AND TRANSPORT IN ANAEROBIC FERMENTATION

4

Detailed Stoichiometry and Process Analysis

CHARLES L. MEYER* and E. TERRY PAPOUTSAKIS† / *Rice University, Houston, Texas*

I. INTRODUCTION AND SCOPE

In anaerobic fermentation without external electron acceptors, the energy for biosynthesis and growth is derived from substrate-level phosphorylation and the use of an incomplete electron transport system by converting part of the carbon (and possibly the complex nitrogen) substrate into products. If there is more than one nongaseous product, in most cases the same amount of biosynthetic energy can be derived by producing product mixtures of different compositions. The composition of such product mixtures can be of infinite variety. In industrial applications, anaerobic fermentation is employed in order to convert carbon substrates (mostly sugars) into desirable oxychemicals (carboxylic acids, alcohols, and ketones). Since usually one of the fermentation products is the desirable one, the optimal utilization of the substrates and the minimization of product separation and purification costs dictate the maximization of the yield and selectivity for the desired product. The ideal fermentation would then convert the sugar units to the desired product to the highest allowable degree and thus produce very little or none of the other fermentation products. The maximal allowable yield of the desired product is determined by both thermodynamic

**Present affiliation*: Shell Development Company, Houston, Texas.
†*Present affiliation*: Northwestern University, Evanston, Illinois.

constraints and the biochemical topology, that is, the structure of the network of biochemical reactions used in product formation. Establishment of the thermodynamic and biochemical constraints that determine the theoretically highest yield for each product and calculation of these maximal yields would be of both fundamental and practical importance. They would allow us to rationally establish the upper bounds for the productivity of fermentation, which in turn can be used as a guide in feasibility studies and experimentation for genetic and bioreactor productivity improvements. Similarly, it would be desirable to establish any possible interrelations among the amounts and rates of appearance or disappearance of the various fermentation products, biomass, and substrates. This would allow us to check the consistency of the experimental data and possibly identify sources of experimental error.

Earlier work has employed carbon, nitrogen, and, often, the so-called available-electron balances to achieve the aforementioned objectives, as we discuss first in this chapter. Then, we discuss the derivation, validation, and use of "advanced" stoichiometric equations (the so-called fermentation equations) for a variety of anaerobic fermentations. The equations describe the interrelations among the fermentation products, biomass, and consumed substrates and can be tested using data from batch and continuous fermentations. These apparently valid equations are shown to obey the constraints imposed on growth and product formation by thermodynamics and the biochemical topology. On that basis, it is demonstrated that the equations are useful for checking the consistency of experimental data, for calculating maximal product yields and selectivities, and for predicting experimental data. The latter, coupled with the validity of the equation under both steady-state and transient conditions, render the equation a useful tool as a gateway (Rolf and Lim, 1983; Papoutsakis, 1984) or software sensor for monitoring of fermentation. A gateway sensor allows the calculation of a number of fermentation parameters using a combination of available sensors. It is shown, in fact, that the equation can be used for calculating fermentation parameters that typically cannot be measured directly, such as rates of production of ATP and of reduction energy (NADH and FdH_2, among others) and the true intracellular rates of important reactions. The equations can also be used to calculate measurable parameters, such as product concentrations, using a combination of available sensors for other fermentation products.

A number of anaerobic fermentations are discussed in this context, including butyric acid, mixed acid, butanediol, and propionic acid fermentations with various hexoses and pentoses as carbon substrates.

II. MACROSCOPIC BALANCES OF GROWTH AND PRODUCT FORMATION

For an anaerobic fermentation, the stoichiometric balance equation for carbon, oxygen, nitrogen, and hydrogen, assuming NH_3 as the nitrogen source for growth, can be written as

$$CH_mO_l + aNH_3 = y_cCH_pO_nN_q + zCH_rO_sN_t + bH_2O + cCO_2 \tag{1}$$

CH_mO_l, $CH_pO_nN_q$ and $CH_rO_sN_t$ represent the elemental compositions of substrate, dry biomass, and product(s). A *carbon mole* of dry biomass, substrate, or product(s) is the amount containing 1 mol of carbon. The values of the subscripts for these elemental formulas are either known or can be readily determined. Elemental carbon, nitrogen, oxygen, and hydrogen balances on Equation (1) yield, respectively,

$$y_c + z + c = 1 \tag{2}$$

$$y_cq + zt = a \tag{3}$$

$$y_cn + zs + b + 2c = 1 \tag{4}$$

$$y_cp + zr + 2b = m + 3a \tag{5}$$

If we eliminate b, the coefficient of H_2O in Equation (1) (which cannot be easily determined), from Equations (4) and (5) and substitute a from Equation (3) we obtain

$$2l - m = y_c(2n + 3q - p) + z(2s + 3t - r) + 4c \tag{6a}$$

If we eliminate c between Equations (2) and (6a), we obtain

$$(4 + m - 2l) = y_c(4 + p - 2n - 3q) + z(4 + r - 2s - 3t) \tag{6b}$$

Thus, Equations (2), (3), and (6a) [or (6b)] are threee linearly independent equations that represent balances of carbon, nitrogen, and the combined oxygen and hydrogen balances, respectively, for Equation (1). Equation (6a), or its equivalent, Equation (6b), has been known for quite some time as an *oxidation-reduction* (O/R) balance (Doelle, 1975; Gottschalk, 1979), balance of the *number of available hydrogens* (Gottschalk, 1979), or an *available electron balance* (Minkevich and Eroshin, 1973; Erickson et al., 1978; Roels, 1980). In the oxidation-reduction balance, the O/R value of formaldehyde and its multiples are arbitrarily taken as zero. Each H in excess contributes −0.5 to the O/R value of the stoichiometric formula; a lack of H contributes +0.5, and each excess N contributes

+1.5 to the O/R value of the formula. Units of CH_2O and H_2O can be added or subtracted as necessary. Thus, the O/R values of CH_mO_l, $CH_pO_nN_q$, $CH_rO_sN_t$, and CO_2 are (1 - 0.5m), (n + 1.5q - 0.5p), (s + 1.5t - 0.5r), and 2, respectively. The O/R values of NH_3 and H_2O are zero, and the O/R balance for Equation (1) gives (after multiplication by two) Equation (6a). The number of available hydrogens is obtained from oxidation of the "compound" to CO_2 and NH_3 with water [$CH_mO_l + (2 - l)H_2O \rightarrow (4 + m - 2l)H + CO_2$; $CH_rO_sN_t + (2 - s)H_2O \rightarrow (4 + r - 2s - 2t)H + CO_2 + tNH_3$]. Thus, the balance of the number of available hydrogens for Equation (1) is given by Equation (6b). The available electrons (e^-) of a formula are the electrons that would be transferred to oxygen upon oxidation of the "compound" represented by the formula to CO_2, H_2O, and NH_3. The balance of available e^- is most conveniently written in terms of the degrees of reduction γ of the compounds, defined as the number of available e^- equivalents per carbon mole of the compound. Using the valences of 4, 1, -2, and -3 for C, H, O, and N, respectively, the reduction degrees of CH_mO_l, $CH_pO_nN_q$, and $CH_rO_sN_t$ are $\gamma_s = 4 + m - 2l$, $\gamma_x = 4 + p - 2n - 3q$, and $\gamma_p = 4 + r - 2s - 3t$, respectively. The available e^- balance for Equation (1) is then given by Equation (6b), which can also be written as

$$\gamma_s = y_c \gamma_x + z\,\gamma_p \tag{6c}$$

If, instead of NH_3, complex organic compounds are used as nitrogen sources for growth, there are two equivalent approaches for writing a balance for the degree of reduction. In the first approach, the definition of the degree of reduction can be extended, as has been discussed by Roels (1980, 1983). By adopting Roels's notation, the more general form of Equation (1) can be written as

$$\begin{aligned} CH_{a_2}O_{b_2}N_{c_2} + aC_{d_4}H_{a_4}O_{b_4}N_{c_4} &= y_c CH_{a_1}O_{b_1}N_{c_1} \\ &+ z\, CH_{a_3}O_{b_3}N_{c_3} + bH_2O + cCO_2 \end{aligned} \tag{7}$$

The first term on the left-hand side of Equation (7) represents the carbon substrate and the second term the nitrogen substrate. On the right-hand side, the first term is for dry biomass and the second for the product(s). In order to derive the generalized form of Equation (6b) that corresponds to Equation (7), we first write the C, N, O, and H balances for Equation (7), which are analogous to Equations (2) through (5). Then, we eliminate b between the O and H balance equations, then we eliminate a using the N balance equation, and finally we eliminate c using the C balance equation. The resulting equation is

$$\left[4 + a_2 - 2b_2 - \frac{c_2}{c_4}(4d_4 + a_4 - 2b_4)\right]$$

$$= y_c\left[4 + a_1 - 2b_1 - \frac{c_1}{c_4}(4d_4 + a_4 - 2b_4)\right]$$

$$+ z\left[4 + a_3 - 2b_3 - \frac{c_3}{c_4}(4d_4 + a_4 - 2b_4)\right] \quad (8a)$$

If we define the quantities in brackets of Equation (8a) to be the *generalized degrees of reduction* of substrate γ_S', biomass γ_X', and product(s) γ_P' , respectively, we can write Equation (8a) in the form

$$\gamma_S' = y_c\gamma_X' + z\gamma_P' \quad (8b)$$

which is the balance of the *generalized degrees of reduction*. If the original meaning of the degree of reduction based on NH_3 is extended to the general case in which $C_{d_4}H_{a_4}O_{b_4}N_{c_4}$ is the N source, we observe that the generalized "valence" of N must be taken equal to $-(4d_4 + a_4 - 2b_4)/c_4$ in order to satisfy the requirement that the degree of reduction of the N source is equal to zero.

Alternatively, since the complex nitrogen source is also a source of available electrons (and also carbon), we can write an available e^- balance for Equation (7) based on the original definition of the degree of reduction, namely,

$$\gamma_{cs} + a\gamma_{ns} = y_c\gamma_X + z\gamma_P \quad (8c)$$

where γ_{cs} and γ_{ns} represent the degrees of reduction of the first and second term, respectively, of the left-hand side of Equation (7). It can be readily verified that if the nitrogen balance equation for Equation (7) is used, Equation (8b) can be derived from Equation (8c), and vice versa. We note, incidentally, that the -3 valence used for N in the definition of the degree of reduction in Equations (6c) and (8c) is the valence of N not only in NH_3 but also in the thermodynamically dead state for nitrogen in biologic systems (for example in amino acids and nucleic acids).

The available e^- or degree of reduction balance is frequently written after normalization by γ_S (or γ_S') as follows, for Equation (6c) for example,

$$\eta + \xi_P = 1 \quad (9)$$

where

$$\eta \equiv z \frac{y_c \gamma_b}{\gamma_s} \tag{10}$$

represents the fraction of available electrons [or of energy (Erickson et al., 1978)] of the substrate contained in biomass, and

$$\xi_p \equiv z \frac{\gamma_p}{\gamma_s} \tag{11}$$

is the corresponding fraction contained in the product(s).

The equations that represent the balances for carbon [Equation (2)], nitrogen [Equation (3)], and available e^- [Equations (6c) or (9)] [or, for the more general case, the equation that represents the reduction degree balance, Equation (8b)] can be used both as gateway sensors and for checking the consistency of data. For example, for a fermentation with only one product other than CO_2 (e.g., ethanol) and with NH_3 as an N source, the three balance equations can be used to calculate observable but difficult to measure (on-line or off-line) quantities, such as substrate consumption and biomass, product, or CO_2 formation, depending on the sensors available. Such applications are discussed, for example, by Roels (1983). Alternatively, complete sets of data can be checked for consistency by checking how well they satisfy one, two, or all three of the balance equations. Such applications are discussed for a variety of fermentations by Erickson and Oner (1983) and Oner et al. (1984).

Since most anaerobic fermentations are carried out with complex N sources, the nitrogen-balance equation (3) is usually of little practical use. The remaining two (carbon and available e^-) balance equations are indeed useful, but in multiproduct anaerobic fermentations they can provide no substantial information regarding the ratios or relative rates of production for the various products. Are there rules governing these ratios and relative rates? Are there additional relationships that the various parameters of the fermentation must obey? Can extracellular measurements provide information on physiological intracellular events? All these questions are addressed in Sections IV and V with the development of advanced stoichiometric equations. Before that, however, we shall address briefly the linear rule of substrate consumption in anaerobic fermentation.

III. THE LINEAR EQUATION OF SUBSTRATE CONSUMPTION

Most of the material in this section is based on Roels's book (1983, Chap. 4). The linear rule of substrate consumption is based on the premise that substrate is consumed for biomass formation and product formation and to generate energy for cell maintenance. This premise is essentially a straightforward extension of the ideas proposed by Herbert and Pirt (see, e.g., Roels, 1983). In its rate form, and assuming constant conversion (yield) coefficients, this postulate can be written as

$$r_s = \frac{r_x}{Y_{sx}} + \frac{r_p}{Y_{sp}} + m_s C_x \tag{12}$$

where r_s, r_x, and r_p are the rates of substrate consumption, biomass production, and product formation, respectively. Y_{sx} and Y_{sp} are the corresponding and assumed constant yield coefficients, m_s is the maintenance coefficient, and C_x is the biomass concentration. We consider for simplicity the case of Equation (1). Since $y_c \equiv r_x/r_s$ and $z \equiv r_p/r_s$, Equation (6b) can be written as

$$\gamma_s - \gamma_x \frac{r_x}{r_s} - \gamma_p \frac{r_p}{r_s} = 0 \tag{13}$$

If we eliminate r_p between Equations (12) and (13), we obtain

$$r_s = \frac{r_x}{Y_{sx}^a} + m_s^a C_x \tag{14}$$

where we have defined

$$Y_{sx}^a \equiv \frac{Y_{sx}(\gamma_p Y_{sp} - \gamma_s)}{\gamma_p Y_{sp} - \gamma_x Y_{sx}} \tag{15}$$

$$m_x^a \equiv \frac{m_s \gamma_p Y_{sp}}{\gamma_p Y_{sp} - \gamma_s} \tag{16}$$

where Y_{sx}^a and m_s^a are the true *anaerobic* yield and maintenance coefficients. Since in anaerobic fermentation energy generation is directly coupled to product formation, Equation (14) can be viewed as the proper form of the balance represented by Equation (12), namely, that substrate is consumed to generate energy for biomass synthesis and maintenance.

If r_s is eliminated between Equations (13) and (14), we obtain

$$r_p = \alpha r_x + \beta C_x \tag{17}$$

where

$$\alpha \equiv \frac{\gamma_s}{\gamma_p Y_{sx}^a} - \frac{\gamma_x}{\gamma_p} \tag{18}$$

and

$$\beta \equiv m_s^a \frac{\gamma_s}{\gamma_p} \tag{19}$$

Equation (17) states that the rate of product formation is the sum of a growth and a non-growth-associated term and is of the same form as the Leudeking-Piret equation (Roels, 1983). Equations (14) and (17) can be expressed in terms of the yields η and ξ_p of Equations (10) and (11). We rewrite the last two equations as

$$\eta = \frac{r_x \gamma_x}{r_s \gamma_s} \tag{20}$$

$$\xi_p = \frac{r_p \gamma_p}{r_s \gamma_x} \tag{21}$$

and use Equation (20) to rewrite Equation (14) as

$$\frac{\mu}{\eta} = \frac{\mu}{\eta_{max}} + m_s' \tag{22}$$

where

$$\eta_{max} \equiv Y_{sx}^a \frac{\gamma_x}{\gamma_s} \tag{23}$$

$$m_s' \equiv m_s^a \frac{\gamma_s}{\gamma_x} \tag{24}$$

Similarly, Equation (17) can be rewritten using Equations (20) and (21) to obtain

$$\frac{\mu \xi_p}{\eta} = \alpha' \mu + \beta' \tag{25}$$

where

$$\alpha' \equiv \alpha \frac{\gamma_p}{\gamma_x} \tag{26}$$

$$\beta' \equiv \beta \frac{\gamma_p}{\gamma_x} \tag{27}$$

The various yields (Y_{sx}^{a} and η_{max}), maintenance coefficients (m_s^a and $m_s^{\prime}$) and product formation coefficients (α, α', β, and β') can be determined from Equations (14), (17), (22), and (25), provided that these equations are valid. The validity of these equations and physiological significance of the parameters have been much debated (see, e.g., Roels, 1983). It is however safe to assume that these equations are valid under steady-state (or balanced growth) conditions for fermentations of one product or of constant-ratio product mixtures. Yield, maintenance, and product formation coefficients have been calculated for a number of fermentations (Roels, 1983; Erickson and Oner, 1983; Oner et al., 1984, 1986). A more detailed discussion of the significance and measurements of such coefficients is given in Chapters 5 and 11.

For multiproduct fermentations, the constant-ratio hypothesis for the products is rarely, if ever, satisfied. Also, in our experience, the linear equations of substrate consumption with constant coefficients are rarely valid for *batch* anaerobic fermentation, in general (Papoutsakis, 1984; Papoutsakis and Meyer, 1985a, b; McLaughlin et al., 1985). In the development of "fermentation" equations in Section IV, the emphasis is on multiproduct fermentation whereby the constant-ratio hypothesis for products is not valid. As a result, the linear equations of product formation and substrate consumption are not considered.

IV. ADVANCED STOICHIOMETRIC ("FERMENTATION") EQUATIONS

A. Derivations and Validation

The elemental balances presented earlier assume no knowledge of the biochemistry of the fermentation. In many fermentations, the sequence of reactions leading to the formation of products is known, and this information can be utilized to provide additional relationships among substrate utilization, production of the various extracellular products, and biomass production. From knowledge of the biochemical pathways of the fermentation, one can identify a set of independent chemical reactions that account for production or consumption of all substrates and products. These reactions can then be combined to form a comprehensive stoichiometric equation [the so-called fermentation equation (Papoutsakis, 1984)]. In the following discussion, the butyric acid (acetone-butanol, A/B) fermentation is used to illustrate the development and applications of a fermentation equation.

One must first identify the independent reactions leading to the formation of all products and biomass from the substrate. Each reaction is balanced only for carbon, available e^-, and ATP. The oxidized form of the electron carriers NAD and ferredoxin (Fd) and the free form of coenzyme A (CoA) are omitted for simplicity. Figure 1 shows the primary catabolic pathways for the A/B fermentation. Each reaction or linear pathway is represented by a single equation. The production of pyruvate from glucose through the Embden-Meyerhof-Parnas (EMP) pathway and the phosphoenolpyruvate phosphotransferase system can be represented as

$$\text{Glucose} = 2\ \text{pyruvate} + 2\text{ATP} + 2\text{NADH} \tag{28}$$

For the production of acetyl-CoA (AcCoA) from pyruvate, one can write

$$\text{Pyruvate} = \text{AcCoA} + \text{FdH}_2 + \text{CO}_2 \tag{29}$$

Hydrogen production from FdH_2 is given by

$$\text{FdH}_2 = \text{H}_2 \tag{30}$$

and reduction of NAD by FdH_2 can be represented as

$$\text{FdH}_2 = \text{NADH} \tag{31}$$

Note that reactions (30) and (31) are both readily reversible (Papoutsakis, 1984). The production of acetate from acetyl-CoA through phosphotransacetylase and acetate kinase can be represented by

Figure 1 Biochemistry of glucose fermentation by butyric acid bacteria (Papoutsakis, 1984). The pathways or enzyme systems for the reactions shown are as follows: (1) phosphoenolpyruvate phosphotransferase system and the Embden-Meyerhof-Parnas pathway; (2) pyruvate-ferredoxin oxidoreductase; (3) hydrogenase; (4) NADH-ferredoxin oxidoreductase; (5) acetyl-CoA-acetyltransferase; (6) L(+)-β-hydroxybutyryl-CoA dehydrogenase; (7) L-3-hydroxyacyl-CoA hydrolase; (8) butyryl-CoA dehydrogenase; (9) CoA transferase; (10) butyryl-CoA synthetase (ATP/AMP); (11) phosphotransbutyrylase; (12) butyrate kinase; (13) phosphotransacetylase; (14) acetate kinase; (15) CoA transferase; (16) acetoacetate decarboxylase; (17) *i*-propanol dehydrogenase; (18) butyraldehyde dehydrogenase; (19) butanol dehydrogenase; (20) acetaldehyde dehydrogenase; and (21) ethanol dehydrogenase.

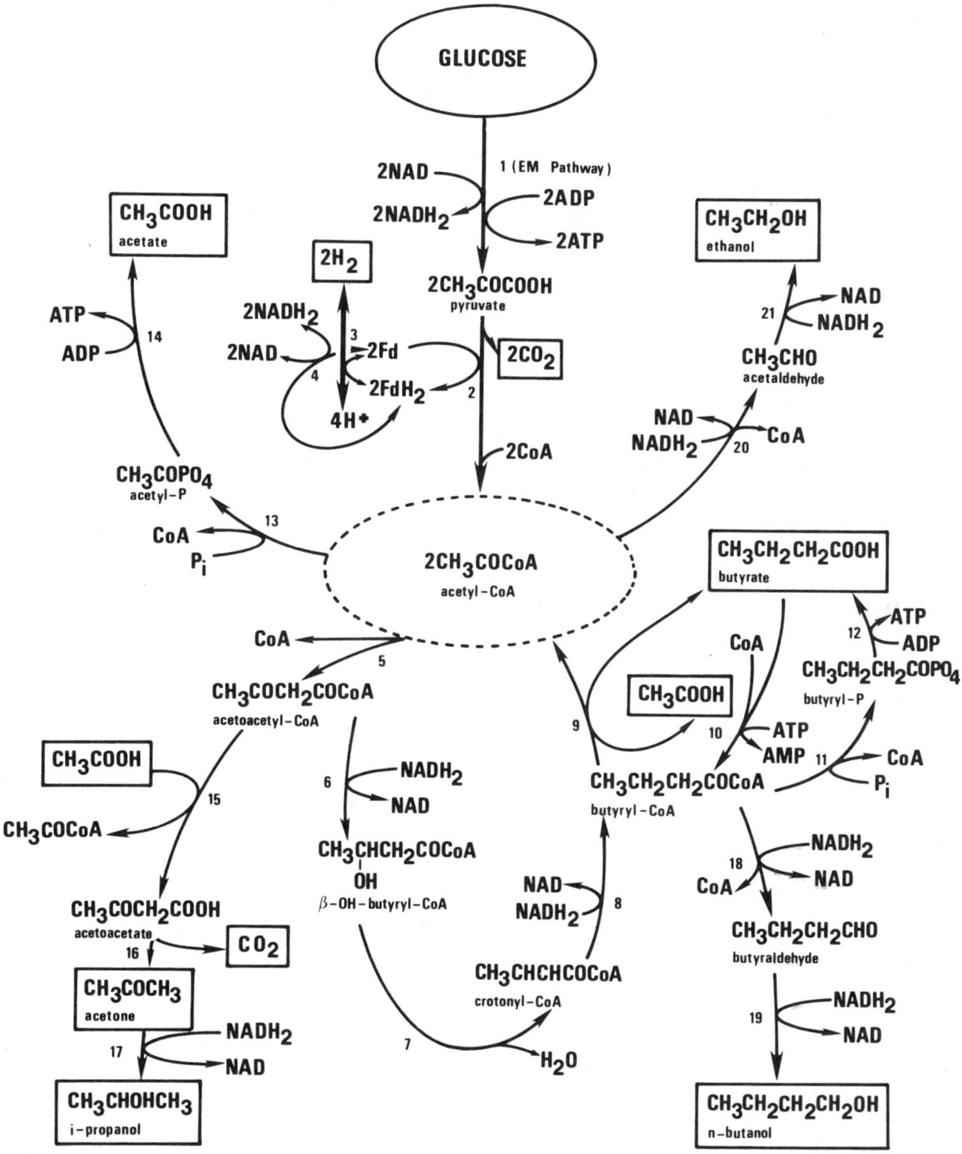
GLUCOSE
1 (EM Pathway)
2NAD
2NADH2
2ADP
2ATP
2CH3COCOOH
pyruvate
2H2
2NADH2
2NAD
3
4
2Fd
2FdH2
4H+
2
2CO2
2CoA
CH3COOH
acetate
ATP
ADP
14
CH3COPO4
acetyl-P
13
CoA
Pi
CH3CH2OH
ethanol
21
NAD
NADH2
CH3CHO
acetaldehyde
NAD
NADH2
20
CoA
2CH3COCoA
acetyl-CoA
CoA
5
CH3COCH2COCoA
acetoacetyl-CoA
CH3COOH
15
CH3COCoA
CH3COCH2COOH
acetoacetate
16
CO2
CH3COCH3
acetone
17
NADH2
NAD
CH3CHOHCH3
i-propanol
6
NADH2
NAD
CH3CHCH2COCoA
OH
β-OH-butyryl-CoA
7
H2O
CH3CHCHCOCoA
crotonyl-CoA
NAD
NADH2
8
CH3CH2CH2COCoA
butyryl-CoA
9
CH3COOH
10
CoA
ATP
AMP
CH3CH2CH2COOH
butyrate
12
ATP
ADP
CH3CH2CH2COPO4
butyryl-P
11
CoA
Pi
18
NADH2
CoA
NAD
CH3CH2CH2CHO
butyraldehyde
19
NADH2
NAD
CH3CH2CH2CH2OH
n-butanol

$$\text{AcCoA} = \text{acetate} + \text{ATP} \tag{32}$$

The sequence of reactions leading from acetyl-CoA to the formation of butyryl-CoA can be represented as

$$2\text{AcCoA} + 2\text{NADH} = \text{butyryl-CoA} \tag{33}$$

There are two reactions shown in Figure 1 for the production of butyrate from butyryl-CoA. These are given by

$$\text{Butyryl-CoA} + \text{acetate} = \text{AcCoA} + \text{butyrate} \tag{34a}$$

$$\text{Butyryl-CoA} = \text{ATP} + \text{butyrate} \tag{34b}$$

In view of Equation (32), one can see that Equations (34a) and (34b) are equivalent, and Equation (34b) will be used to represent butyrate production. For butanol production from butyryl-CoA, one can write

$$\text{Butyryl-CoA} + 2\text{NADH} = \text{butanol} \tag{35}$$

Acetone production can be represented by

$$\text{Acetate} + \text{AcCoA} = \text{acetone} + CO_2 \tag{36}$$

Ethanol production from acetyl-CoA is given by

$$\text{AcCoA} + 2\text{NADH} = \text{ethanol} \tag{37}$$

Acetoin production, although not shown in Figure 1, probably proceeds via two possible mechanisms that have been demonstrated in other bacteria (Doelle, 1975; Gottschalk, 1979). One mechanism is through acetolactate, which is represented by the reaction

$$2\ \text{pyruvate} = \text{acetoin} + 2CO_2 \tag{38a}$$

The other reaction is through diacetyl and is given by

$$\text{Pyruvate} + \text{AcCoA} + \text{NADH} = \text{acetoin} + CO_2 \tag{38b}$$

By considering Equations (29) and (31), one can see that Equations (38a) and (38b) are equivalent. For this derivation, Equation (38a) is used. The formation of isopropanol can be represented as

$$\text{Acetone} + \text{NADH} = \text{isopropanol} \tag{39}$$

For production of biomass, it would be impractical to consider a detailed biosynthetic pathway. For the purposes already outlined, it suffices to represent biomass as a single chemical species and biomass production by a single chemical reaction. Let $CH_pO_nN_q$ represent the elemental composition of biomass. From previous studies on biomass composition (Minkevich and Eroshin, 1973; Erickson, 1980) of a large variety of microorganisms, it is known that the weight fraction of carbon in biomass σ_b and the reduction degree of biomass γ_b are relatively constant, namely,

$$\sigma_b = 0.462 \pm 0.023 \tag{40}$$

$$\gamma_b = 4 + p - 2n - 3q = 4.291 \pm 0.172 \tag{41}$$

Using an available e^- balance and Equation (41), one can calculate that 0.873 mol of NADH is required for 1 mol of glucose used in biomass synthesis. The ATP requirements for biomass synthesis can be estimated from Y_{ATP}, defined as the amount of dry biomass produced in grams per mole ATP used for biomass synthesis. Using the value for σ_b given in Equation (40), the amount of ATP necessary for incorporation of 1 mol glucose into biomass is given by

$$\frac{\text{mol ATP}}{\text{mol glucose}} = \frac{(12)(6)}{Y_{ATP}\sigma_b} \tag{42}$$

The value for Y_{ATP} is not constant, however, and may vary not only from fermentation to fermentation, but even within a given fermentation. Thus, the amount of ATP required for biomass synthesis must appear as a variable in the equation representing biomass synthesis from glucose:

$$\text{Glucose} + 0.873\text{NADH} + \frac{156\ \text{ATP}}{Y_{ATP}} = 6CH_pO_nN_q \tag{43}$$

Y_{ATP} is therefore an output rather than an input in the calculations presented later. The nitrogen source for biomass synthesis has been neglected in Equation (43) but can be added if more specific information is available for the biomass composition. Small deviations in the values of σ_b and γ_b have little effect on the final calculations since only a small percentage (usually less than 10%) of substrate carbon is incorporated into biomass. Although NADPH may be used for biosynthesis and in some of the reactions of Figure 1, it can be assumed equivalent to NADH (Papoutsakis, 1984). General decarboxylation and CO_2 fixation are not expected to affect the validity and accuracy of the equation (Papoutsakis, 1984).

One now has a system of 13 independent chemical equations [Equations (28) through (39) and (43)] with 17 chemical species (ATP, acetyl-CoA, butyryl-CoA, NADH, FdH_2, pyruvate, ethanol, acetone, acetate, butanol, acetoin, butyrate, isopropanol, H_2, CO_2, biomass, and glucose) representing catabolism and biomass synthesis for this fermentation. A balance of each chemical species yields the following set of equations:

$$x_{ATP} = 2r_1 + r_5 + r_7 - \frac{156r_{13}}{Y_{ATP}} \tag{44}$$

$$x_{NADH} = 2r_1 + r_4 - 2r_6 - 2r_8 - 2r_{10} - r_{12} - 0.873r_{13} \tag{45}$$

$$x_{pyruvate} = 2r_1 - r_2 - 2r_{11} \tag{46}$$

$$x_{FdH_2} = r_2 - r_3 - r_4 \tag{47}$$

$$x_{AcCoA} = r_2 - r_5 - 2r_6 - r_9 - r_{10} \tag{48}$$

$$x_{butyryl\text{-}CoA} = r_6 - r_7 - r_8 \tag{49}$$

$$x_{H_2} = r_3 \tag{50}$$

$$x_{CO_2} = r_2 + r_9 + 2r_{11} \tag{51}$$

$$x_{acetate} = r_5 - r_9 \tag{52}$$

$$x_{butyrate} = r_7 \tag{53}$$

$$x_{butanol} = r_8 \tag{54}$$

$$x_{acetone} = r_9 - r_{12} \tag{55}$$

$$x_{ethanol} = r_{10} \tag{56}$$

$$x_{acetoin} = r_{11} \tag{57}$$

$$x_{isopropanol} = r_{12} \tag{58}$$

$$x_{biomass} = 6r_{13} \tag{59}$$

$$x_{glucose} = -r_1 - r_{13} \tag{60}$$

where r_j is the rate or extent of the jth reaction and x_i is the net rate of production of component i, or the net amount produced, depending on whether the fermentation is run in continuous or batch mode. Equations (44) through (60) can be summarized by the following overall stoichiometric reaction:

$$\begin{aligned}(r_1 + r_{13})\ \text{glucose} &= 6r_{13}\,CH_pO_nN_q + (r_2 - r_3 - r_4)FdH_2\\ &+ (2r_1 + r_4 - 2r_6 - 2r_8 - 2r_{10} - r_{12} - 0.873r_{13})NADH\\ &+ (r_2 - r_5 - 2r_6 - r_9 - r_{10})AcCoA + (r_6 - r_7 - r_8)\text{butyryl-CoA}\end{aligned}$$

$$+ (2r_1 - r_2 - 2r_{11})\text{pyruvate} + \left(2r_1 + r_5 + r_7 - \frac{156r_{13}}{Y_{ATP}}\right)\text{ATP} + r_3H_2$$
$$+ (r_2 + r_9 + 2r_{11})CO_2 + (r_5 - r_9)\text{acetate} + r_7 \text{ butyrate} + r_8 \text{ butanol}$$
$$+ (r_9 - r_{12})\text{acetone} + r_{10} \text{ ethanol} + r_{11} \text{ acetoin} + r_{12} \text{ isopropanol} \quad (61)$$

Equation (61) is termed the fermentation equation for A/B fermentation. It obeys all constraints imposed on growth and product formation by thermodynamics and the biochemical topology. The thermodynamic constraints are contained in Equations (28) through (43), which express the production or requirements of ATP and NADH for the enzymatic reactions of product and biomass formation.

Of the 16 components in these reactions, acetyl-CoA, butyryl-CoA, NADH, FdH_2, and pyruvate are intermediates and are assumed not to be produced in any net quantity. Their extent or rates of production can thus be set equal to zero:

$$r_2 - r_5 - 2r_6 - r_9 - r_{10} = 0 \quad (62)$$
$$r_6 - r_7 - r_8 = 0 \quad (63)$$
$$2r_1 + r_4 - 2r_6 - 2r_8 - 2r_{10} - r_{12} - 0.873r_{13} = 0 \quad (64)$$
$$r_2 - r_3 - r_4 = 0 \quad (65)$$
$$2r_1 - r_2 - 2r_{11} = 0 \quad (66)$$

This assumption is valid whenever the pseudo-steady-state hypothesis holds for each of the five chemical species, that is, whenever the *net* rate of production of each species is very small compared with the *total* rate of production or consumption of each species. This hypothesis should be valid for these five species during continuous steady-state and completed batch fermentation. Although levels of these intermediates may vary under transient conditions, their rate of accumulation is still very small compared with total rates of production and consumption, and Equations (62) through (66) should therefore be valid under transient conditions as well.

Of the remaining 12 species, the rates or extents of production of all but ATP can be directly measured experimentally. Thus, of the 17 chemical species, the rates or extents of production of 16 are known or can be determined directly. If one omits the equation for the rate or extent of ATP production [Equation (44)], then Equations (45) through (60) represent a system of 16 equations in 13 unknowns. One can choose a set of 13 equations to exactly solve the system and calculate the rates or extents of the 13 independent reactions. One would use Equations (62) through (66) for the five intermediate species plus a set of eight equations selected from Equations (45) through (60) and the corresponding experimental measurements. For example, one might use the measured amounts of biomass, ethanol, acetone, acetate, butanol, acetoin, butyrate, and isopropanol. Alternatively, one can choose all 16 of the equations to solve the system in the least-squares sense (Noble, 1969, p. 142) and determine the best values for r_j based upon all available data. It should be noted that the amount or rate of biomass production is typically measured on a mass basis and must be converted to a molar basis before Equation (60) can be used. The molecular weight of biomass based on one carbon atom per mole and a carbon mass fraction σ_b of 0.462 is 12/0.462 = 26.0 g/mol. Once the extents of these intracellular reactions are known, additional quantities can be determined. Values for ATP production, NADH production from FdH_2 (NfF) via reaction (31), and other "unobservable" parameters can also be calculated. By setting x_{ATP} equal to zero in Equation (44), one has for Y_{ATP}

$$Y_{ATP} = \frac{156r_{13}}{2r_1 + r_5 + r_7} \tag{67}$$

If only 13 measurements are chosen to solve the system exactly, values for the three remaining variables can be determined. In the example discussed here, total CO_2 and H_2 production and glucose utilization can be calculated from Equations (50), (51), and (60).

Equations (44) through (60) and (62) through (69) can be reduced to a set of three independent equations involving only the measurable variables. One such set of equations can be written as

$$x_{H_2} - 2x_{acetoin} - 3x_{isopropanol} - 2x_{butyrate} - 2x_{acetate} - 4x_{acetone} + \frac{0.873x_{biomass}}{6} = 0 \tag{68}$$

$$2x_{glucose} - x_{CO_2} + x_{acetone} + x_{isopropanol} - \frac{x_{biomass}}{3} = 0 \tag{69}$$

$$2x_{glucose} - 2x_{acetoin} - x_{acetate} - 2x_{acetone} - x_{isopropanol} - 2x_{butyrate} - 2x_{butanol} - x_{ethanol} - \frac{x_{biomass}}{3} = 0 \tag{70}$$

Using Equations (68) through (70), one can directly calculate 3 of the 11 measurable variables from the remaining 8. Because r_j

has been eliminated from Equations (68) through (70), however, no information concerning the various reaction rates or extents or other unobservable variables can be determined using these three equations. Note that Equations (68) through (70) represent one more relationship than could be obtained using carbon and available e^- balances alone. Because carbon and available e^- balances for each reaction were used in the derivation procedure, these balances are contained in Equations (68) through (70) and can be obtained by taking linear combinations of these three equations.

Some computational examples are shown in Table 1. The first column contains measured amounts of products from a batch fermentation of *Clostridium acetobutylicum*. The second column contains the amounts of the products calculated from a best set of r_j obtained from a least-squares calculation using all the data from column 1. Note that agreement between columns 1 and 2 is very good for all measured values. Columns 3-6 show the results of calculations using just enough measured values to allow an exact solution. In each case, all but three of the measured values were used to compute the r_j from which the remaining three values were calculated. Table 2 shows a similar set of calculations but for a continuous culture of *C. acetobutylicum* at steady state. As in Table 1, agreement between measured and calculated values is quite good. Calculations have been performed on several additional sets of data, from the authors' laboratory (Meyer et al., 1985; Roos et al., 1985; McLaughlin et al., 1985; Meyer and Papoutsakis, 1985; Meyer et al., 1986a; Papoutsakis et al., 1986) as well as from the literature (Papoutsakis, 1984). Except in a few cases, which are discussed later, the agreement between measured and calculated values is quite good.

Hydrolysis of hemicellulose, a potential source of sugar substrates, typically yields, in addition to glucose, the pentoses xylose and arabinose, the hexoses mannose and galactose, and the disaccharide cellobiose (Papoutsakis and Meyer, 1985b). It would therefore be useful to develop equations for fermentation with these additional sugars. To accommodate substrates other than glucose, the previously developed equation needs to be modified only slightly. As was shown previously (Papoutsakis and Meyer, 1985b), the equations would be unchanged for fermentation of the two hexoses galactose and mannose. For the disaccharide cellobiose, Equations (28) and (43) would be replaced with the following two equations (Papoutsakis and Meyer, 1985b):

$$\text{Cellobiose} = 4\ \text{pyruvate} + 5\text{ATP} + 4\text{NADH} \tag{71}$$

$$\text{Cellobiose} + 1.75\ \text{NADH} + \frac{312\text{ATP}}{Y_{\text{ATP}}} = 12\text{CH}_p\text{O}_n\text{N}_q \tag{72}$$

Table 1 Batch Fermentation of Glucose with *Clostridium acetobutylicum*

	Amount (mol per 100 mol glucose fermented)					
	Experimental	Calculated				
Column no.	1	2	3	4	5	6
Ethanol	2.8	3.2	2.8[a]	2.8[a]	2.8[a]	2.8[a]
Acetone	9.0	7.7	9.0[a]	9.0[a]	9.3	9.0[a]
Acetate	43.7	43.1	40.1	44.0	43.7[a]	43.7[a]
Butanol	18.8	19.6	20.3	18.8[a]	18.8[a]	18.6
Acetoin	1.8	1.6	1.8[a]	1.8[a]	1.8[a]	1.8[a]
Butyrate	36.4	36.2	36.4[a]	32.5	36.4[a]	32.8
CO_2	180	183	187	180[a]	187	180[a]
H_2	183	180	183[a]	183[a]	191	183[a]
Glucose	100	98.8	100[a]	96.5	100[a]	96.5
Biomass	66.0	66.3	66.0[a]	66.0[a]	66.0[a]	66.0[a]
NADH from FdH_2	—	-8.68	-8.60	-15.6	-16.8	-15.6
Y_{ATP} (g mol^{-1})	—	6.56	6.51	6.69	6.42	6.69

[a]Data used in calculations.
Source: McLaughlin et al. (1985).

Table 2 Continuous, Steady-State Fermentation of Glucose with *Clostridium acetobutylicum*

	Rate (mmol per g dry biomass per hour)				
	Experimental	Calculated			
Column	1	2	3	4	5
Ethanol	0.21	0.18	0.21[a]	0.21[a]	0.210[a]
Acetone	0	0.14	0[a]	0[a]	0[a]
Acetate	3.43	3.46	4.03	3.43[a]	3.69
Butanol	0.31	0.24	0.31[a]	0.31[a]	0.31[a]
Acetoin	0	-0.01	0[a]	0[a]	0[a]
Butyrate	4.69	4.68	4.41	4.69[a]	4.74
CO_2	14.0	13.9	13.7	13.6	14.0[a]
H_2	15.2	15.2	15.2[a]	14.6	15.2[a]
Glucose	8.51	8.75	8.51[a]	8.73	8.92
Biomass	11.5	11.5	11.5[a]	11.5[a]	11.5[a]
NADH from FdH_2	—	-1.4	-1.5	-0.9	-1.2
Y_{ATP} (g mol^{-1})	—	13.6	13.5	13.8	13.3

[a]Data used in calculations.
Source: Meyer et al. (1986).

For fermentations with the pentoses xylose, arabinose, or ribose, Equations (28) and (43) would be replaced by (Papoutsakis and Meyer, 1985b):

$$3 \text{ pentose} = 5 \text{ pyruvate} + 5\text{NADH} + 5\text{ATP} \tag{73}$$

$$\text{Pentose} + 0.728\text{NADH} + \frac{130\text{ATP}}{Y_{\text{ATP}}} = 5\text{CH}_p\text{O}_n\text{N}_q \tag{74}$$

There are only limited sets of complete data available for fermentation with substrates other than glucose, but calculations (not shown) give good results (Papoutsakis and Meyer, 1985b).

The procedure described here can be applied to several other fermentations as well. Fermentation equations have already been derived and tested for butanediol, mixed-acid, and propionic acid fermentation (Papoutsakis and Meyer, 1985a, b). The final form of the fermentation equation for mixed-acid and butanediol fermentation (Papoutsakis and Meyer, 1985a) is

$$\begin{aligned}
&(r_1 + r_{11} + 3r_{13} + r_{14})\text{glucose} = 6r_{14}\ \text{CH}_p\text{O}_n\text{N}_q \\
&\quad + \left(2r_1 + Br_2 + r_{10} + 5r_{13} - \frac{156r_{14}}{Y_{\text{ATP}}}\right)\text{ATP} \\
&\quad + (2r_1 - r_2 - r_6 - r_8 - 2r_9 - 2r_{11} - r_{12} + 11r_{13} - 0.873r_{14})\text{NADH} + \\
&\quad (2r_1 - r_2 - r_3 - r_4 - r_6 - 2r_7 + 5r_{13})\text{pyruvate} \\
&\quad + (-r_2 + r_3 + r_{12})\text{XH}_2 + (r_3 + r_4 - r_9 - r_{10})\text{AcCoA} \\
&\quad + (-r_2 + r_3 + r_5 + 2r_7 + 3r_{13})\text{CO}_2 + r_5\text{H}_2 + r_2\ \text{succinate} \\
&\quad + (r_7 - r_8)\text{acetoin} + (r_4 - r_5)\text{formate} + r_6\ \text{lactate} \\
&\quad + r_8\ \text{butanediol} + r_9\ \text{ethanol} + r_{10}\ \text{acetate} + 2r_{11}\ \text{glycerol}
\end{aligned} \tag{75}$$

This equation was developed based on the biochemical pathways shown in Figure 2. XH_2 represents either or both $FADH_2$ and $FMNH_2$ (reduced flavin mononucleotide). Parameter B in the coefficient for ATP has a value of 1 if succinate is produced through fumarate or a value of -1 if succinate is produced through succinyl-CoA (pathway not shown in Fig. 2). Because B is not known and can vary anywhere from -1 to 1 depending on the relative importance of these two pathways, only a lower and upper bound on the amount of ATP produced can be calculated. For the calculations presented here, a value of 1 is used for B.

The first calculations done with this fermentation used only the Embden-Meyerhof-Parnas pathway and did not consider the

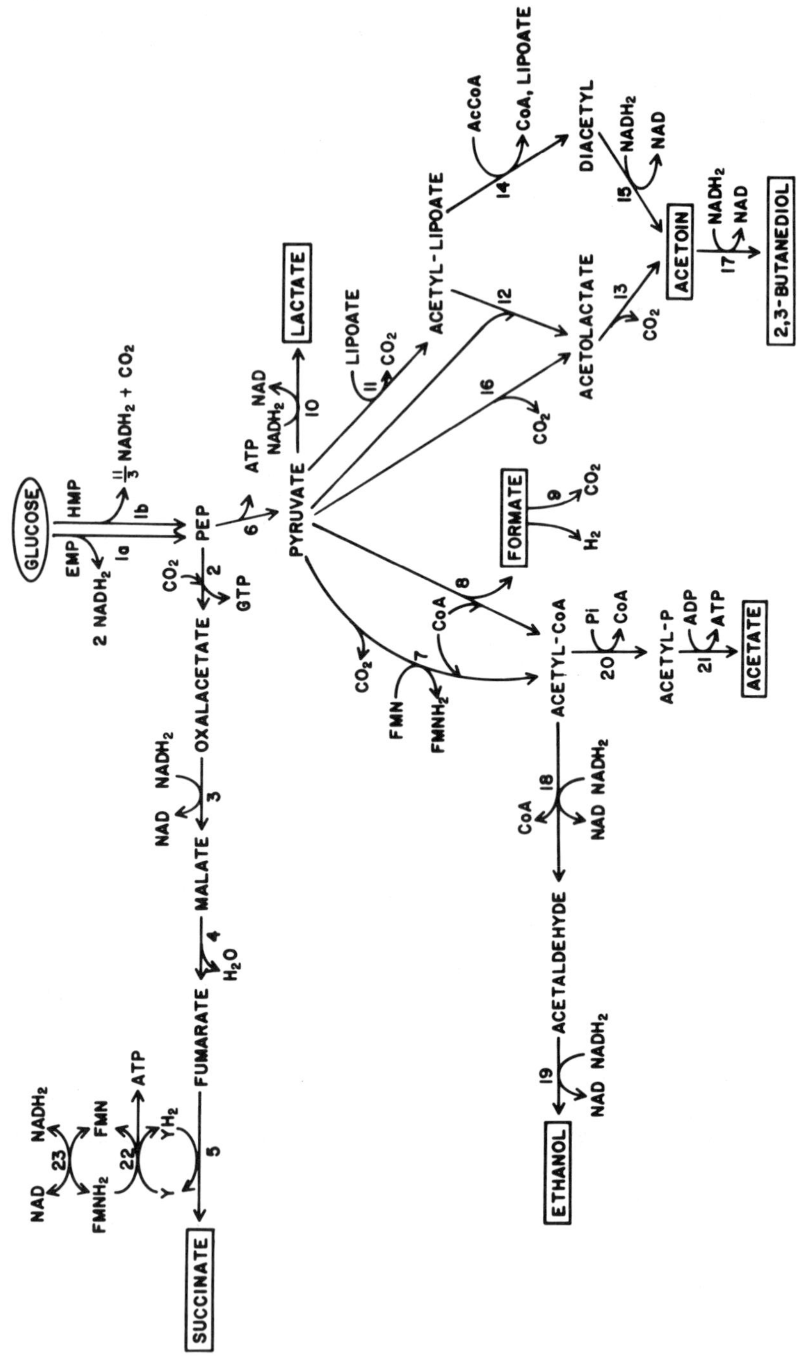
GLUCOSE
EMP
HMP
2 NADH₂
1a
1b
11/3 NADH₂ + CO₂
PEP
CO₂
2
GTP
6
ATP
PYRUVATE
NADH₂
NAD
10
LACTATE
LIPOATE
11
CO₂
ACETYL-LIPOATE
12
16
CO₂
ACETOLACTATE
13
CO₂
AcCoA
14
CoA, LIPOATE
DIACETYL
NADH₂
NAD
15
ACETOIN
NADH₂
NAD
17
2,3-BUTANEDIOL
FORMATE
9
CO₂
H₂
8
CoA
7
CO₂
FMN
FMNH₂
ACETYL-CoA
Pi
20
CoA
ACETYL-P
ADP
21
ATP
ACETATE
18
CoA
NAD NADH₂
ACETALDEHYDE
19
NAD NADH₂
ETHANOL
OXALACETATE
NAD NADH₂
3
MALATE
4
H₂O
FUMARATE
5
YH₂
Y
22
ATP
FMN
FMNH₂
23
NADH₂
NAD
SUCCINATE

Figure 2 Biochemistry of glucose fermentation by butanediol and mixed-acid bacteria (Papoutsakis and Meyer, 1985a). The pathways or enzyme systems for the reactions shown are as follows: (1a) phosphoenolpyruvate phosphotransferase system and the Embden-Meyerhof-Parnas pathway; (1b) phosphoenolpyruvate phosphotransferase system and the hexose monophosphate pathway; (2) phosphoenolpyruvate carboxylase (GDP specific); (3) NAD(P)H malate dehydrogenase; (4) fumarate hydratase; (5) succinate dehydrogenase; (6) ATP-pyruvate phosphotransferase; (7) pyruvate-flavin oxidoreductase; (8) pyruvate-formate lyase; (9) formate-hydrogen lyase; (10) lactate dehydrogenase; (11) pyruvate-lipoate oxidoreductase; (12) acetolactate synthase (pyruvate/acetyl lipoate); (13) acetolactate decarboxylase; (14) diacetylsynthase (acetyl lipoate/acetyl-CoA); (15) acetoin dehydrogenase; (16) acetolactate synthase (pyruvate/pyruvate); (17) butanediol dehydrogenase; (18) acetaldehyde dehydrogenase; (19) ethanol dehydrogenase; (20) phosphotransacetylase; (21) acetate kinase; (22) YH_2-FMN oxidoreductase (ATP producing); and (23) NADH-FMN oxidoreductase.

hexose monophosphate (HMP) pathway for glucose utilization. Many of these calculations gave very poor results. When the HMP pathway was included, however, excellent results were obtained. The HMP pathway is included in the form of the equation given here. Some results of calculations using Equation (75) are shown in Tables 3 and 4. Table 3 shows results from a fermentation of glucose with *Enterobacter cloaceae*. Results of calculations with and without the HMP pathway are shown to give good agreement with experimental values. The amount of CO_2 produced via the HMP pathway can be estimated from Equation (75) and is also presented in Table 3. Note that these values are very low, indicating very little involvement of the HMP pathway in this fermentation. Thus, one would expect good results from the calculations without the HMP pathway. Table 4 shows results from a fermentation of glucose with *Escherichia coli*. Calculations without the HMP pathway give poor results, but calculations with the HMP pathway give excellent results and show a significant proportion (>10%) of CO_2 produced via this pathway. These results demonstrate the potential of the fermentation equation for use in the investigation of the relative extents of involvement of various alternative pathways for a particular fermentation.

The Y_{ATP} is an output quantity from the fermentation equation [e.g., Equation (61) or (75)], and when pertinent, a set of Y_{ATP} data can be used to calculate the maximal or true ATP yield, Y_{ATP}^{max}, according to the equation (Roos et al., 1985)

$$\frac{1}{Y_{ATP}} = \frac{1}{Y_{ATP}^{max}} + \frac{m_e}{\mu} \qquad (76)$$

Equation (76) can be considered analogous to Equation (14) and is usually valid for steady-state continuous cultures or under conditions of balanced growth (Roos et al., 1985). m_e is the maintenance coefficient in units of mole ATP per gram dry biomass per hour.

A reduced set of equations relating only the measurable variables in Equation (75) can be derived as was done for the A/B fermentation and can be written as

$$\begin{aligned} &6x_{glucose} - x_{CO_2} - 4x_{succinate} - 2x_{ethanol} - 2x_{acetate} \\ &\quad - x_{formate} - 4x_{acetoin} - 4x_{butanediol} - 3x_{glycerol} \\ &\quad -3x_{lactate} - x_{biomass} = 0 \end{aligned} \qquad (77)$$

$$\begin{aligned} &12x_{glucose} - 7x_{succinate} - x_{H_2} - 6x_{ethanol} - 4x_{acetate} \\ &\quad - x_{formate} - 10x_{acetoin} - 11x_{butanediol} - 7x_{glycerol} \\ &\quad - 6x_{lactate} - \frac{12.873x_{biomass}}{6} = 0 \end{aligned} \qquad (78)$$

Table 3 Fermentation of Glucose with *Enterobacter cloaceae*

	Amount (mol per 100 mol glucose fermented)						
		Calculated					
	Experimental	Without HMPP			With HMPP		
CO_2	173.0	173.0[a]	172.7	173.0[a]	174.2	173.3	173.0[a]
Formate	8.5	8.5[a]	8.5[a]	8.2	8.5[a]	8.5[a]	8.5[a]
Acetoin	0.5	0.5[a]	0.5[a]	0.5[a]	0.5[a]	0.5[a]	0.5[a]
Ethanol	65.0	65.0[a]	65.0[a]	65.0[a]	65.0[a]	65.0[a]	65.0[a]
Lactate	3.2	3.1	3.4	3.4	3.2[a]	3.2[a]	3.2[a]
Succinate	0.0	0.0[a]	0.0[a]	0.0[a]	0.0[a]	0.0[a]	0.0[a]
Acetate	0.0	0.0[a]	0.0[a]	0.0[a]	0.0[a]	0.0[a]	0.0[a]
Butanediol	57.6	57.8	57.6[a]	57.6[a]	57.4	57.6[a]	57.7
H_2	47.0	43.5	43.4	43.7	47.0[a]	44.6	43.7
Glycerol	0.0	0.0[a]	0.0[a]	0.0[a]	0.0[a]	0.0[a]	0.0[a]
Carbon recovery	92.3	92.3[a]	92.3[a]	92.3[a]	92.3[a]	92.3[a]	92.3[a]
$NADH_2$ from XH_2	—	13.0	13.1	13.1	9.5	11.9	12.8
CO_2 from HMPP	—	—	—	—	1.9	0.6	0.2
Y_{ATP} (g mol^{-1})	—	6.5	6.5	6.5	6.52	6.51	6.50

[a]Data used in the calculations.
Source: Papoutsakis and Meyer (1985a).

Table 4 Fermentation of Glucose with *Escherichia coli*

	Amount (mol per 100 mol glucose fermented)					
		Calculated				
	Experimental	Without HMPP		With HMPP		
CO_2	88.0	88.0[a]	78.7	85.6	86.4	88.0[a]
Formate	2.4	2.4[a]	2.4[a]	2.4[a]	2.4[a]	2.4[a]
Acetoin	0.1	0.1[a]	0.1[a]	0.1[a]	0.1[a]	0.1[a]
Ethanol	49.8	49.8[a]	42.9	49.8[a]	50.7	49.9
Lactate	79.5	79.8	79.5[a]	80.4	79.5[a]	79.5[a]
Succinate	10.7	5.9	10.7[a]	10.7[a]	10.7[a]	10.7[a]
Acetate	36.5	45.8	48.2	36.5[a]	36.5[a]	36.5[a]
Butanediol	0.3	0.3[a]	0.3[a]	0.3[a]	0.3[a]	0.3[a]
H_2	75.0	75.0[a]	75.0[a]	75.0[a]	75.0[a]	79.7
Glycerol	1.4	1.4[a]	1.4[a]	1.4[a]	1.4[a]	1.4[a]
Carbon recovery	91.7	91.7[a]	91.7[a]	91.7[a]	91.7[a]	91.7[a]
$NADH_2$ from XH_2	—	12.3	2.9	-1.8	-1.0	-6.4
CO_2 from HMPP	—	—	—	11.7	11.7	14.0
Y_{ATP} (g mol^{-1})	—	5.54	5.37	5.74	5.74	5.76

[a]Data used in the calculations.
Source: Papoutsakis and Meyer (1985a).

Note that because of the additional metabolic pathway (HMP pathway) used to derive Equation (75), only two such relationships are possible. Also note that Equations (77) and (78) can be derived directly from a carbon balance and an available e^- balance.

B. Applications

Use of the equation as a gateway sensor to calculate various observable as well as "unobservable" parameters from measured values or other variables has already been mentioned. In addition, these types of calculations can be used to check the consistency of experimental data, calculate theoretically maximal yields, and investigate various aspects of the biochemistry of the fermentation. These applications are illustrated here with some examples.

From Tables 1 and 2, one can see that some of the observable variables can be used to calculate others. For the A/B fermentation, of the 11 measured variables (substrate, biomass, ethanol, acetone, acetate, butanol, acetoin, butyrate, isopropanol, H_2, and CO_2), a combination of 8 can be used to calculate values for the remaining 3.

This would be particularly useful if certain of the observable variables were difficult to measure. This approach has been used for the on-line monitoring of batch and continuous cultures of *C. acetobutylicum* (McLaughlin et al., 1985; Meyer et al., 1986a, b). Concentrations of liquid products in the fermentation broth were measured using on-line gas chromatography (GC). These measurements, coupled with measurements of biomass concentration, were used to calculate glucose concentration and total CO_2 and H_2 production. The results of these calculations from a batch fermentation are shown in Figures 3 through 5 and compared to experimental values.

Table 5 illustrates the use of the fermentation equation to determine possible sources of experimental error. Assuming that there is a significant error in the value of one of the measured variables, the use of this value to calculate other measureable values should lead to poor agreement between the calculated and measured values. On the other hand, using only measured values for which accurate data are available would lead to good agreement between calculated and measured values, except for the value with the experimental error. Such a situation is shown in Table 5 in which the value for H_2 seems to be in error. Using this value in the calculations produces unacceptable results for all calculated values, as seen in columns 6 and 7. If the H_2 value is not used in the calculation,

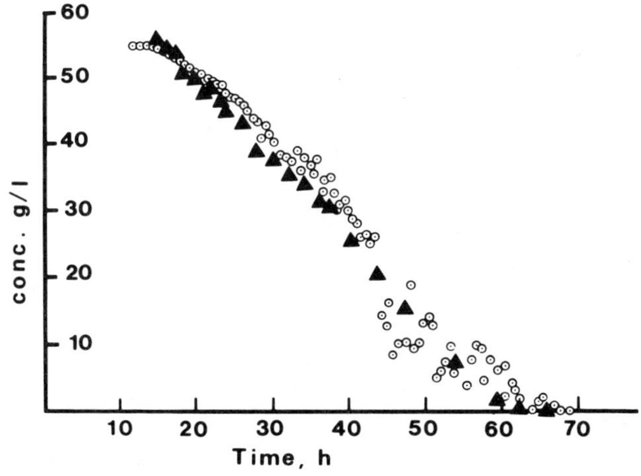

Figure 3 Glucose concentration during a batch fermentation of *Clostridium acetobutylicum* (Meyer et al., 1986) showing off-line measurements (▲) compared with calculated values using the fermentation equation with on-line liquid measurements (⊙).

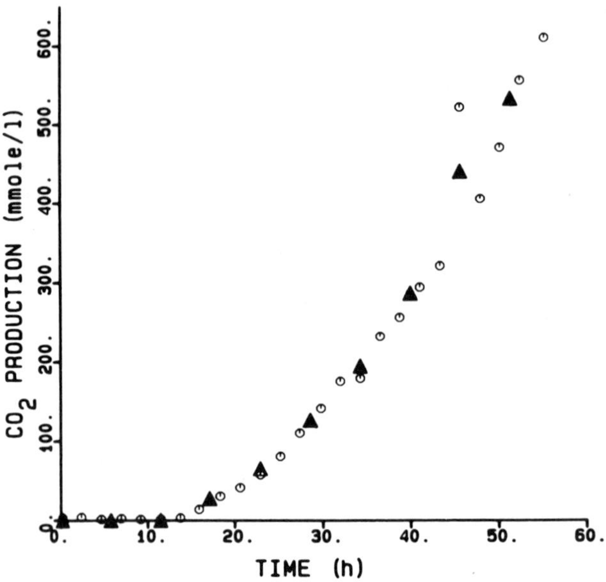

Figure 4 Time profile of total CO_2 produced during a batch fermentation of *Clostridium acetobutylicum* (Meyer et al., 1986) showing the values from on-line measurements of the gas phase (▲) compared with values calculated from the fermentation equation with on-line liquid measurements (⊙).

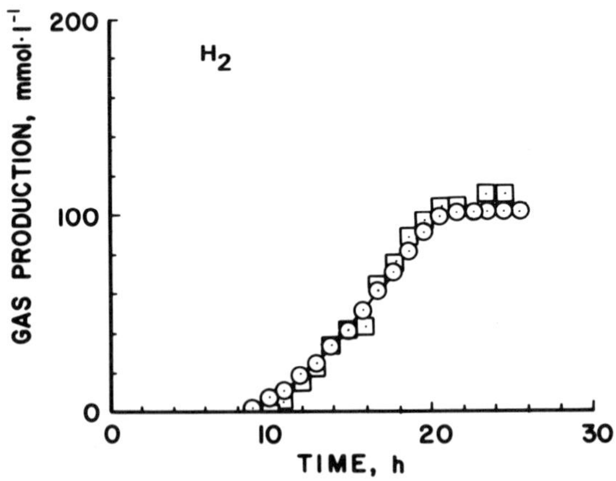

Figure 5 Time profile of total H_2 produced during a batch fermentation of *Clostridium acetobutylicum* (McLaughlin et al., 1985) showing the values from on-line measurements of the gas phase (⊙) compared with values calculated from the fermentation equation with on-line liquid measurements (⊡).

Table 5 Fermentation of Glucose with *Clostridium butylicum*

	Amount (mol per 100 mol glucose fermented)						
	Experimental	Calculated					
Column	1	2	3	4	5	6	7
Butyrate	18.2	18.2[a]	16.3	18.4	18.2[a]	18.2[a]	31.8
Acetate	12.1	8.3	12.1[a]	12.1[a]	12.1[a]	25.7	12.1[a]
H_2	121.2	93	96	94	100	121.2[a]	121.2[a]
CO_2	200	202	202	200[a]	202	200[a]	200[a]
Butanol	54.6	54.6[a]	54.6[a]	54.6[a]	52.7	48	41
Acetone	0	0[a]	0[a]	0[a]	0[a]	0[a]	0[a]
Ethanol	2.9	2.9[a]	2.9[a]	2.9[a]	2.9[a]	2.9[a]	2.9[a]
Acetoin	0	0[a]	0[a]	0[a]	0[a]	0[a]	0[a]
Isopropanol	15.1	15.1[a]	15.1[a]	13	15.1[a]	13	13
Carbon recovered (as glucose)	93.5	93.5[a]	93.5[a]	93.5[a]	93.5[a]	93.5[a]	93.5[a]
$NADH_2$ from FdH_2	—	94	91	93	87	66	66
Y_{ATP}	—	4.4	4.4	4.4	4.4	4.16	4.16

[a]Data used in calculations.
Source: Papoutsakis (1984).

all calculated values except the H_2 value agree quite well with the measured values, as shown in columns 2-5.

Using Equations (44) through (60) and (62) through (66), one can calculate the maximal possible yield for each of the main fermentation products for an assumed carbon recovery in extracellular products. The estimated maximal yields for five of the major fermentation products (butanol, acetone, butyrate, acetate, and isopropanol) are shown in Table 6. Acetone, butyrate, acetate, and isopropanol can all be produced as the sole liquid product of the fermentation, but a small amount of acetate must be produced to provide the reduction energy for biomass synthesis when butanol is the product of maximal yield. Also shown in Table 6 is the Y_{ATP} for each set of maximal yield data. Note that in each case the Y_{ATP} is well below the value of 10.5 (Papoutsakis, 1984) that is widely accepted for most bacteria, indicating that all the combinations of products in Table 6 would provide more than sufficient biosynthetic energy for growth and would therefore be thermodynamically feasible. The results of Table 6 indicate the potential of fermentation of butyric acid bacteria for industrial applications.

Since Equation (61) was developed based upon the assumed biochemistry of the fermentation, its validity is dependent on and is indicative of the accuracy of the assumed biochemical pathways. Significant and systematic disagreement between experimental and calculated values based on the equation would indicate either flaws in the experimental procedure or that the assumed biochemistry is incorrect or incomplete. Table 7 shows results from a continuous, steady-state culture of *C. acetobutylicum* in which considerable excess substrate (glucose) was present. An analysis similar to that used for the data of Table 5 would indicate that there is a systematic error in the value for glucose utilization. The calculated values of the glucose utilization based on the other measured values are significantly lower than what was actually measured (columns 2-5). The measurements have been carefully checked, however, and are accurate. This indicates that glucose is utilized by reactions other than those used to derive Equation (61). Production of an extracellular polymer, most likely a polysaccharide, in cultures of *C. acetobutylicum* under certain conditions has been reported by Häggström and Forberg (1986). This hypothesis is supported by the discrepancy between calculated and measured values in Table 7. Thus, one can see the usefulness of the equation in the analysis of the biochemical pathways of a complex fermentation. A similar but more involved case involving the HMP pathway in the butanediol fermentation has already been discussed.

The fermentation equation concept appears to have fairly broad applications. Bailey et al. (1987), Reardon et al. (1987), and Rao

Table 6 Theoretically Computed Maximal Yields for Primary Products of Glucose Fermentation with Butyric Acid Bacteria[a]

	Amount[b] (mol per 100 mol glucose fermented)				
	Butanol	Acetone	Butyrate	Acetate	Isopropanol
Butyrate	0	0	95	0	0
Acetate	2.2	0	0	190	0
H_2	0	376	186	376	281
CO_2	190	285	190	190	285
Butanol	93.9	0	0	0	0
Acetone	0	95	0	0	0
Ethanol	0	0	0	0	0
Acetoin	0	0	0	0	0
Isopropanol	0	0	0	0	95
Carbon recovery (as glucose)	95	95	95	95	95
NADH from FdH_2	190	-186	4.4	-186	-91
Y_{ATP}	4.05	2.73	2.73	2.05	2.73

[a]A 95% carbon recovery in products was assumed.
[b]Product of maximal yield.
Source: Papoutsakis (1984).

and Mutharasan (1986) have used the fermentation equation in their investigations of the metabolism of *C. acetobutylicum*. Extension of the fermentation equation concept to animal cell cultures has been addressed by Stephanopoulos et al. (1985), and Vallino and Stephanopoulos (1987) have developed and used a fermentation equation coupled with on-line high-performance liquid chromatography for state estimation of fermentations of *Corynebacterium glutamicum*.

C. Generalized Formulation

For any given fermentation, there are *N measurable macroscopic quantities* that describe the fermentation. These quantities represent the extent or rate of production or consumption of a product or substrate. In a batch fermentation, for example, these quantities

Table 7 Continuous, Steady-State Fermentation of Glucose with *Clostridium acetobutylicum*

	Rate (mmol/g dry biomass per hour)						
	Experimental	Calculated					
Column	1	2	3	4	5	6	7
Ethanol	0.28	0.28[a]	0.28[a]	0.28[a]	0.28[a]	0.28[a]	0.28[a]
Acetone	0.93	0.93[a]	0.93[a]	0.93[a]	0.93[a]	0.93[a]	0.93[a]
Acetate	5.92	5.77	5.92[a]	5.92[a]	5.92[a]	3.06	5.92[a]
Butanol	2.23	1.20	1.27	2.23[a]	2.23[a]	2.23[a]	6.57
Acetoin	0.03	0.03[a]	0.03[a]	0.03[a]	0.03[a]	0.03[a]	0.03[a]
Butyrate	6.21	6.21[a]	6.06	6.06	6.21[a]	15.0	6.21[a]
CO_2	23.7	23.7[a]	23.7[a]	25.6	25.9	34.6	34.6
H_2	25.3	25.3[a]	25.3[a]	25.3[a]	25.6	25.3[a]	25.6
Glucose	19.6	14.2	14.2	15.1	15.3	19.6[a]	19.6[a]
Biomass	16.6	16.6[a]	16.6[a]	16.6[a]	16.6[a]	16.6[a]	16.6[a]
NfF	—	-2.6	-2.6	-0.67	-0.67	8.32	8.01
Y_{ATP} (g mol^{-1})	—	12.1	12.1	11.5	11.3	9.3	9.2

[a]Data used in calculations.
Source: Unpublished data from the authors' laboratory.

are biomass, substrate, and extracellular product concentrations, or total amounts. For continuous operation, these quantities are overall or specific rates of production or consumption of biomass, extracellular products, and substrates. These rates of production can be calculated from steady-state or nonsteady-state mass balances, depending on the conditions of interest.

From knowledge of the biochemical pathways of the fermentation, one should be able to find a set of *M linearly independent chemical reactions* that account for the production or consumption of all the measured species. Additionally, other species involved in these reactions may be either intermediates that are neither produced nor consumed in any net amount (pseudo-steady-state hypothesis) or chemical species that either cannot be measured or may be consumed or produced in reactions that are not explicitly known (i.e., ATP, which may be consumed by futile cycles or substrate transport, for example). Reduced electron carriers are considered intermediates in this discussion and represent available electrons, which are conserved. Let *L represent the number of intermediates* and *P the number of nonmeasurable quantities*. This gives us a set of M linearly independent chemical reactions in N + L + P chemical species.

For A/B fermentation, one has N = 11 measurable quantities. These are biomass, substrate (glucose), and extracellular products: ethanol, acetone, acetic acid, butanol, acetoin, butyric acid, isopropanol, H_2, and CO_2. There are five intermediates (L = 5)—acetyl-CoA, butyryl-CoA, pyruvate, NADH, and FdH_2—and one nonmeasurable quantity, ATP.

A balance of each chemical species (excluding the nonmeasurable species) for the set of M chemical reactions yields the following set of equations:

$$x_i = \sum_{j=1}^{M} a_{ij} r_j \qquad i = 1, \quad N + L \tag{79}$$

where a_{ij} is the stoichiometric coefficient for the ith species in the jth reaction, r_j is the rate or extent of the jth reaction, and x_i is the net amount produced, or the net rate of production of species i. The set of Equations (79) can be written in matrix form as

$$\underset{\sim}{A}\underset{\sim}{r} = \underset{\sim}{x} \tag{80}$$

where $\underset{\sim}{A}$ is the N + L by M matrix of stoichiometric coefficients, whose element in the ith row and jth column, a_{ij}, is the coefficient for the ith component in the jth reaction. $\underset{\sim}{r}$ is a column vector composed of the rates or extents (r_j) of the M reactions, and $\underset{\sim}{x}$ is a column vector composed of the net rates or amounts of the N + L measurable quantities plus intermediates. The L elements of $\underset{\sim}{x}$ corresponding to the intermediates have a value of zero.

Since the M reactions are linearly independent, it must be true that $M < N + L + P$, since the number of independent reactions must be less than the number of species involved. P, the number of unmeasurable quantities, is usually 1 (for ATP), and thus $M \leq N + L$ (actually, this is a strong inequality, as we see later). In other words, matrix $\underset{\sim}{A}$ has more rows than columns. The rank of the matrix $\underset{\sim}{A}$ equals the number of independent reactions M. Thus, Equation (80) represents a set of N + L equations in M unknowns, where $N + L \geq M$. Such a system is an overdetermined system and can be solved in the least-squares sense (Noble, 1969, p. 142) to determine the best-fit values for the reaction rates. Alternatively, various combinations of measurable quantities can be chosen to give equations in M unknowns which can be solved exactly to yield the reaction rates. These rates can then be used to calculate the remaining measurable quantities, which were omitted from the linear problem, as well as for calculating important "unobservable" quantities, as was mentioned earlier. For example, M − L measurable quantities plus L intermediates would provide M equations, enough to solve the system exactly. The remaining $N - (M - L) = N + L - M$ measurable quantities could then be calculated.

Another way of looking at the problem is that with N + L equations in M unknowns, there are N + L − M linear relationships (equations) that the N + L measurable quantities plus intermediates must satisfy in order for the system to be consistent. Since the rate of production of intermediates is always zero, this means there are N + L − M linear equations that the N measurable quantities must satisfy. Since the N quantities must satisfy at least the carbon balance [such as Equation (2)] and the available e^- balance [such as Eq. (6c)], we have that $N + L - M \geq 2$ and thus $M < N + L$. For the A/B fermentations, there would be $11 + 5 - 13 = 3$ equations relating the 11 measurable variables. This is one more equation than was obtained using the elemental balances (Sec. II).

These N + L − M equations relating the N measurable quantities can be obtained by performing gaussian elimination on the N + L by M matrix $\underset{\sim}{A}$. From linear algebra, it is known that there exists an N + L by N + L matrix $\underset{\sim}{D}$ such that the product $\underset{\sim}{D}\underset{\sim}{A}$ is an upper triangular matrix (Noble 1969, pp. 86-87). The matrix $\underset{\sim}{D}$ is formed by performing the same row operations on the identity matrix that would be used to convert $\underset{\sim}{A}$ to upper triangular form by gaussian elimination. Since the rank of $\underset{\sim}{A}$ is M, the lower N + L − M rows of $\underset{\sim}{D}\underset{\sim}{A}$ must all be zeros (Noble, 1969, p. 89). Equation (80) can therefore be written in the following form (using notation of partitioned matrices):

$$\begin{bmatrix} \tilde{D}\tilde{A}_u \\ 0 \end{bmatrix} \tilde{r} = \begin{bmatrix} \tilde{D}_u \\ \tilde{D}_l \end{bmatrix} \tilde{x} \tag{81}$$

where $\tilde{D}\tilde{A}_u$ is the submatrix composed of the upper M rows of $\tilde{D}\tilde{A}$, and $\tilde{D}_u$ and $\tilde{D}_l$ are the submatrices composed of the upper M rows and the lower N + L - M rows of $\tilde{D}$, respectively. From this, we must have

$$\tilde{D}_l\tilde{x} = \tilde{0} \tag{82}$$

or, since the lower L components of x are zero,

$$\tilde{D}_{ll}\tilde{x}_u = \tilde{0} \tag{83}$$

where $\tilde{x}_u$ is composed of the upper N elements of $\tilde{x}$ (the N measurable variables) and $\tilde{D}_{ll}$ is the submatrix formed from the lower N + L - M rows and the left N columns of $\tilde{D}$. This gives the N + L - M equations relating the N measurable variables. These equations [which are given by Equations (68) through (70) for the A/B fermentation] can be used to calculate up to N + L - M of the variables from the measured values of the remaining variables, or it can be used to check the consistency of measured values for all measurable variables, as was discussed in Section IV.B. It should be noted that submatrix $\tilde{D}_{ll}$ is one of an infinite number of matrices that express this relationship among the measurable variables. Other matrices can be formed by performing elementary row operations on $\tilde{D}_{ll}$. The elemental balance equations (Sec. II) can be formed from the matrix $\tilde{D}_{ll}$ by taking appropriate linear combinations of the N + L - M rows of $\tilde{D}_{ll}$. In other words, the space spanned by the rows of the matrix representing the elemental balances is a subspace of the space spanned by the rows of the matrix $\tilde{D}_{ll}$.

V. SUMMARY

Using the available information on the sequence of biochemical reactions responsible for product formation in anaerobic fermentation, it is possible to derive an advanced stoichiometric equation that describes the interrelationships among the fermentation products, biomass, and consumed substrates. This equation, coupled with equations that derive from the application of the pseudo-steady-state hypothesis to a number of key intermediates, allows one to extract useful stoichiometric relationships among the aforementioned parameters. Such relationships are valuable as gateway sensors and for calculating on-line and off-line a number of important physiological parameters and the in vivo rates or extents of reaction of important intracellular enzymatic reactions. In addition, such relationships are useful as investigative tools of new or known metabolic pathways and their regulation in response to observable changes in the bioreactor environment and also useful for calculating maximal product yields and selectivities for process analysis and design.

ACKNOWLEDGMENTS

Papoutsakis acknowledges the support of the National Science Foundation through a series of grants related to the subject of this chapter, and in particular support through grants ECE-8613077 and CBT-8518959 for support during the preparation of this chapter.

REFERENCES

Bailey, J. E., Axe, D. D., Doran, P. M., Galazzo, J. L., Reardon, K. F., Seressiotis, A., and Shanks, J. V. (1987). Redirection of cellular metabolism: Analysis and synthesis. In *Biochemical Engineering V, Ann. N.Y. Acad. Sci. 506*:1-23.

Doelle, H. W. (1975). *Bacterial Metabolism.* Academic Press, New York.

Erickson, L. E. (1980). Biomass elemental composition and energy content. *Biotechnol. Bioeng. 22*:451-456.

Erickson, L. E., and Oner, M. D. (1983). Available electron and energetic yields in fermentation processes. In *Biochemical Engineering III, Ann. N.Y. Acad. Sci. 413*:99-113.

Erickson, L. E., Minkevich, I. G., and Eroshin, V. K. (1978). Application of mass and energy balance regularities in fermentation. *Biotechnol. Bioeng. 20*:1595-1621.

Gottschalk, G. (1979). *Bacterial Metabolism.* Springer-Verlag, New York

Häggström, L., and Förberg, C. (1986). Significance of an extracellular polymer for the energy metabolism in *Clostridium acetobutylicum*: A hypothesis. *Appl. Microbiol. Biotechnol 23*:234-239.

McLaughlin, J. K., Meyer, C. L., and Papoutsakis, E. T. (1985). Gas chromatography and gateway sensors for on-line state estimation of complex fermentations (butanol-acetone fermentation). *Biotechnol. Bioeng. 27*:1246-1257.

Meyer, C. L., and Papoutsakis, E. T. (1985). The interplay of electron flow, ATP balances, bioreactor conditions and product formation by *C. acetobutylicum.* Presented at the 1985 Annual AIChE Meeting, Chicago.

Meyer, C. L., McLaughlin, J. K., and Papoutsakis, E. T. (1985). The effect of CO on growth and product formation in batch cultures of *Clostridium acetobutylicum*. *Biotechnol. Lett.* *7*:37–42.

Meyer, C. L., McLaughlin, J. K., and Papoutsakis, E. T. (1986a). On-line chromatographic analysis and fermenter state characterization of butanol/acetone fermentations. In *Biochemical Engineering IV*, *Ann. N.Y. Acad. Sci.* *469*:350–363.

Meyer, C. L., Roos, J. W., and Papoutsakis, E. T. (1986b). Carbon monoxide gasing leads to alcohol production and butyrate uptake without acetone formation in continuous cultures of *Clostridium acetobutylicum*. *Appl. Microbiol. Biotechnol.* *24*:159–167.

Minkevich, I. G., and Eroshin, V. K. (1973). Productivity and heat generation of fermentation under oxygen limitation. *Folia Microbiol.* *18*:376–385.

Noble, B. (1969). *Applied Linear Algebra*. Prentice-Hall, Englewood Cliffs, New Jersey, p. 142.

Oner, M. D., Erickson, L. E., and Yang, S. S. (1984). Estimation of yield, maintenance, and product formation kinetic parameters in anaerobic fermentations. *Biotechnol. Bioeng.* *26*:1436–1444.

Oner, M. D., Erickson, L. E., and Yang, S. S. (1986). Estimation of the true growth yield and maintenance coefficient for yoghurt cultures. *Biotechnol. Bioeng.* *28*:919–926.

Papoutsakis, E. T. (1984). Equations and calculations for fermentations of butyric acid bacteria. *Biotechnol. Bioeng.* *26*:174–187.

Papoutsakis, E. T., and Meyer, C. L. (1985a). Equations and calculations of product yields and preferred pathways for butanediol and mixed-acid fermentations. *Biotechnol. Bioeng.* *27*:50–66.

Papoutsakis, E. T., and Meyer, C. L. (1985b). Fermentation equations for propionic-acid bacteria and for production of assorted oxychemicals from various sugars. *Biotechnol. Bioeng.* *27*:67–80.

Papoutsakis, E. T., Huesemann, M. H. W., Meyer, C. L., Welch, R. W., Wiesenborn, D. P., and Rudolph, F. B. (1986). The enzymatic basis of regulation of product yields and selectivities in *Clostridium acetobutylicum*. Presented at the 1986 Annual AIChE Meeting, Miami Beach.

Rao, G., and Mutharasan, R. (1986). Manipulation of catabolic pathways in *Clostridium acetobutylicum* by modifying culture reduction state. Presented at the 1986 Annual AIChE Meeting, Miami Beach.

Reardon, K. F., Scheper, T.-H., and Bailey, J. E. (1987). Metabolic pathway rates and culture fluorescence in batch fermentations of *Clostridium acetobutylicum*. *Biotechnol. Prog.* *3*:153–167.

Roels, J. A. (1980). Application of macroscopic principles to microbial metabolism. *Biotechnol. Bioeng.* *22*:2457–2514.

Roels, J. A. (1983). *Energetics and Kinetics in Biotechnology*. Elsevier Biomedical, Amsterdam.

Rolf, M. J., and Lim, H. C. (1983). Computer control of fermentation processes. *Enzyme Microb. Technol. 4*:370-380.
Roos, J. W., McLaughlin, J. K., and Papoutsakis, E. T. (1985). The effect of pH on nitrogen supply, cell lysis and solvent production in fermentations of *Clostridium acetobutylicum*. *Biotechnol. Bioeng. 27*:681-694.
Stephanopoulos, G., Cooney, C., Sinskey, A., Glacken, M., and Adema, E. (1985). The use of a fermentation equation for determining carbon and energy metabolism in hybridoma cell cultures. Presented at the 1985 Annual AIChE Meeting, Chicago.
Vallino, J. J., and Stephanopoulos, G. N. (1987). Intelligent sensors in biotechnology: Applications for the monitoring of fermentations and cellular metabolism. In *Biochemical Engineering V, Ann. N.Y. Acad. Sci. 506*:415-430.

5

Kinetics of Growth and Product Formation

LARRY E. ERICKSON / *Kansas State University, Manhattan, Kansas*

I. INTRODUCTION

This chapter introduces the subject of the kinetics of growth and product formation in anaerobic fermentation. Since much of the research with anaerobic microbial growth is related to kinetics and rate processes, this is a broad topic with many aspects. Chapter 6 on redox potential considers the effects of redox potential on anaerobic microbial processes, including kinetic aspects. Membrane transport and reaction kinetics are very intimately related at the cellular level; membrane transport is considered in Chapter 7. Substrate and product inhibition are important kinetic considerations; product inhibition is considered in Chapter 8. Structured kinetic models are treated in Chapter 9. Chapter 10 will consider some of the kinetic aspects of product formation in the acetone-butanol fermentation. Mixed culture interactions and kinetics are treated in Chapters 16, 17, and 18.

The kinetics of anaerobic fermentation has been treated together with aerobic processes in many review articles and books (Roels, 1983; Bailey and Ollis, 1986; Blanch et al., 1983). This chapter considers some basic concepts and provides an overview of the kinetics of anaerobic microbial processes. Models for cell growth may be structured or unstructured. Multicomponent cellular models are structured models; in an unstructured model cell mass is represented as a single variable and single component. Segregated cell models can consider the differences within the population; unsegregated models consider the average properties of the population.

Table 1 Categories of Cell Population Kinetic Models

	Unstructured	Structured
Unsegregated	Most idealized case: cell population treated as one component solute in balanced growth	Multicomponent average cell description
Segregated	Single component, heterogeneous individual cells	Multicomponent description of cell-to-cell heterogeneity; actual case

Source: From Bailey and Ollis (1986).

Table 1 shows that four classes of models can be treated, with unstructured and unsegregated models the simplest case. The most realistic case is the structured and segregated model; however, it is also the most complex. One needs to select a model appropriate for the task. Models help one to understand science and technology; decisions can be made based on results predicted using models. All models are approximations to reality. The principle of optimum sloppiness should be used in selecting a model; that is, select the simplest model that will provide the needed information within the error limits that can be tolerated.

Growth and product formation kinetics are considered for a number of processes. In some anaerobic fermentations, the kinetics and bioenergetics are intimately related. Thus, some bioenergetic concepts are introduced in this chapter as well. We start with an analysis of fermentation kinetics for processes in which substrate phosphorylation provides the energy for growth as products of interest are formed.

II. FERMENTATION KINETICS

Anaerobic growth and product formation in which the adenosine triphosphate (ATP) formed is the result of substrate phosphorylation has been extensively investigated. One of the unique aspects of this fermentation is that biomass growth, substrate consumption, and product formation are very closely coupled (Roels, 1983; Oner et al., 1984). When no electron acceptors are present, the available electrons initially present in the substrate are distributed to the product and the biomass, with the distribution controlled by the bioenergetics of the process. Consider the anaerobic production of a simple product, such as ethanol or lactic acid, by anaerobic fermentation. The chemical balance equation is

$$CH_mO_l + aNH_3 = y_cCH_pO_nN_q + zCH_rO_sN_t + cH_2O + dCO_2 \tag{1}$$

where CH_mO_l, $CH_pO_nN_q$, and $CH_rO_sN_t$ give the elemental compositions of the carbon, hydrogen, oxygen, and nitrogen in the substrate, biomass, and extracellular product, respectively. Using the valences C = 4, H = 1, O = -2, and N = -3, the available electron balance is (Minkevich and Eroshin, 1973; Erickson et al., 1978)

$$y_c \frac{\gamma_b}{\gamma_s} + z \frac{\gamma_p}{\gamma_s} = 1.0 \tag{2}$$

or

$$\eta + \xi_p = 1.0 \tag{3}$$

where γ is the reductance degree (Minkevich and Eroshin, 1973; Erickson et al., 1978). The subscripts s, b, and p refer to substrate, biomass, and product, respectively.

Luedeking and Piret (1959a, b) reported that product formation kinetics and growth kinetics were closely related; product formation was modeled with two parameters, a growth-associated term and a maintenance-associated term; that is,

$$\frac{dP}{dt} = \alpha \frac{dX}{dt} + \beta X \tag{4}$$

where X is biomass concentration and P is product concentration. Growth and product formation are dependent upon the availability of substrate; the Monod model

$$\frac{1}{X}\frac{dX}{dt} = \mu = \frac{\mu_{max} S}{K_s + S} \tag{5}$$

is commonly used to relate growth rate to substrate concentration in unstructured models. Here S is substrate concentration, μ is specific growth rate, μ_{max} is maximum specific growth rate, and K_s is the saturation constant. The organism uses substrate for growth and maintenance according to the linear model for substrate consumption (Pirt, 1975; Roels, 1983; Oner et al., 1984); that is,

$$-\frac{dS}{dt} = \frac{1}{Y_{x/s}^{max}} \frac{dX}{dt} + m_s X \tag{6}$$

where $Y_{x/s}^{max}$ is the true growth yield corrected for maintenance and m_s is the maintenance coefficient. In this process, product formation occurs. Thus, the product formation kinetic parameters and the bioenergetic parameters are related. Oner et al. (1984) show that Equation (6) may be written as

$$\frac{\mu}{\eta} = \frac{\mu}{\eta_{max}} + m_e \tag{7}$$

and Equation (4) may be written in the form

$$\frac{\mu}{\eta} \xi_p = \frac{\mu}{\eta} (1 - \eta) = \alpha_e \mu + \beta_e \tag{8}$$

where

$$m_e = \beta_e = m_s \frac{\sigma_s \gamma_s}{\sigma_b \gamma_b} = \beta \frac{\sigma_p \gamma_p}{\sigma_b \gamma_b} \tag{9}$$

$$\frac{1}{\eta_{max}} = \alpha_e + 1 = \frac{\sigma_s \gamma_s}{\sigma_b \gamma_b Y_{x/s}^{max}} = \alpha \frac{\sigma_p \gamma_p}{\sigma_b \gamma_b} + 1 \tag{10}$$

The true growth yield must satisfy the relationship

$$\eta_{max} = \frac{(\sigma_b \gamma_b/12) Y_{ATP}^{max}}{(\sigma_b \gamma_b/12) Y_{ATP}^{max} + \delta} \tag{11}$$

where Y_{ATP}^{max} is the true growth yield based on ATP, δ is the equivalents of available electrons transferred to products to produce 1 mol of ATP from ADP, σ_b is the weight fraction carbon in biomass, and γ_b is the reductance degree of the biomass. Knowledge of the theoretical maximum yield with respect to ATP and the value of δ for a particular fermentation allows one to estimate the acceptable range of values of the growth-associated kinetic term in the product formation model. For $\delta = 12$ equivalents of available electrons per mole of ATP generated, $Y_{ATP}^{max} = 28.8$ g cells per mole ATP, $\sigma_b = 0.462$, and $\gamma_b = 4.291$; $\alpha_e \geq 2.52$.

In summary, there are four independent kinetic parameters (μ_{max}, K_s, η_{max}, and m_e) needed to describe growth and product formation in fermentations with one chemical balance equation and one product that contains available electrons. The estimation of the parameters for such a system is considered in more detail in later chapters (see Chaps. 11 and 15, for example).

III. EFFECT OF TEMPERATURE

Temperature is an important variable that must be considered in the design of an anaerobic fermentation process. Van Uden (1984) has reviewed the literature and presented an extensive treatment of the effect of temperature on yeasts. Both growth and death processes are considered. The effect of temperature is quantitatively evaluated using an Arrhenius plot such as the one shown in Figure 1. Kinetic parameters, such as the maximum specific growth rate, may vary with absolute temperature according to the relationship

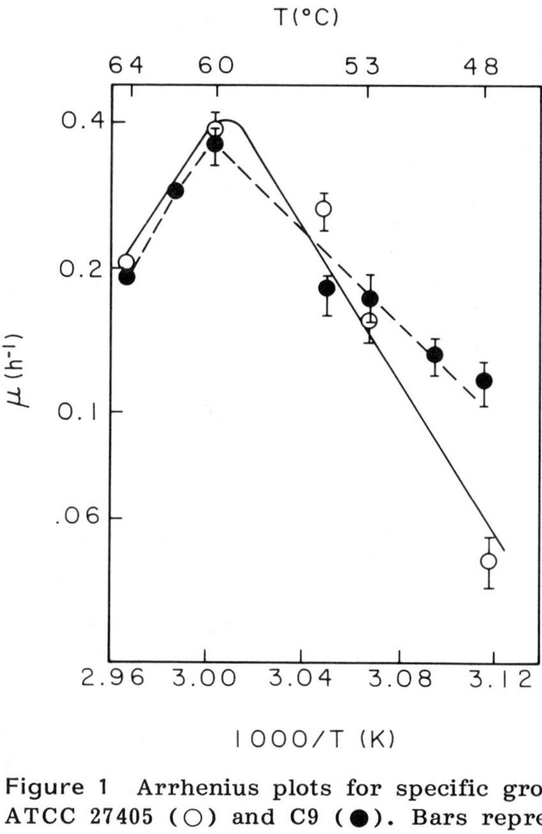

Figure 1 Arrhenius plots for specific growth rate of *C. thermocellum* ATCC 27405 (○) and C9 (●). Bars represent sample standard deviations (eight replicates). (From Herrero and Gomez, 1980.)

$$\mu_{max} = k_0 e^{-E/RT} \tag{12}$$

A plot of ln μ_{max} versus 1/T should have a slope of -E/R because

$$\ln \mu_{max} = \ln k_0 - \frac{E}{RT} \tag{13}$$

The activation energy E provides a measure of the effect of temperature on kinetic rate. Activation energies for the thermal death of microorganisms are much larger than those associated with growth processes (Van Uden, 1984). Activation energies for the thermal inactivation of microorganisms range from 50 to 100 kcal/g mol (Bailey and Ollis, 1986) or about 200-400 kJ/g mol. Activation energies for microbial growth are often in the range from 10 to 20 kcal/g mol or about 40-80 kJ/g mol (Leao and Van Uden, 1985; Herrero and Gomez, 1980; Van Uden, 1984) when exponential growth is evaluated at temperatures below the optimum growth temperature. Sometimes the slope changes with temperature; this may happen because the rate-limiting factor changes or because of phase changes in membrane lipids. When activation energies are of the order of 5 kcal/g mol or 20 kJ/g mol, diffusion processes may be rate limiting.

The effect of temperature on growth may depend on the concentration of other chemicals in the fermentation broth (Kuhn and Linden, 1986; Herrero and Gomez, 1980; Leao and Van Uden, 1985; Van Uden, 1984; Lovitt et al., 1984; Sa-Correia and Van Uden, 1983). Such solvents as ethanol and butanol have an effect on membrane transport that depends on temperature. Figures 2 and 3 show the effect of temperature on growth rate for two different strains of *Clostridium thermocellum* in the presence of several different concentrations of ethanol. In Figure 2, the optimum temperature for growth decreases as ethanol concentration is increased; however, the same effect is not observed in Figure 3. Figure 4 shows how the maximum specific growth rate varies with temperature and butanol concentration for *Clostridium acetobutylicum*.

Temperature also affects the product formation kinetics (Fieschko and Humphrey, 1983). In two separate studies, Fieschko and Humphrey (1983) found that the maintenance coefficient or non-growth-associated product formation coefficient increased with temperature and also with ethanol concentration for *Zymomonas mobilis*. The true growth yield or growth-associated product formation parameter did not change significantly with temperature but it decreased as ethanol concentration was increased.

Information on the minimum, maximum, and optimum temperature for growth of many bacteria and yeasts is tabulated by Atkinson and Mavituna (1983).

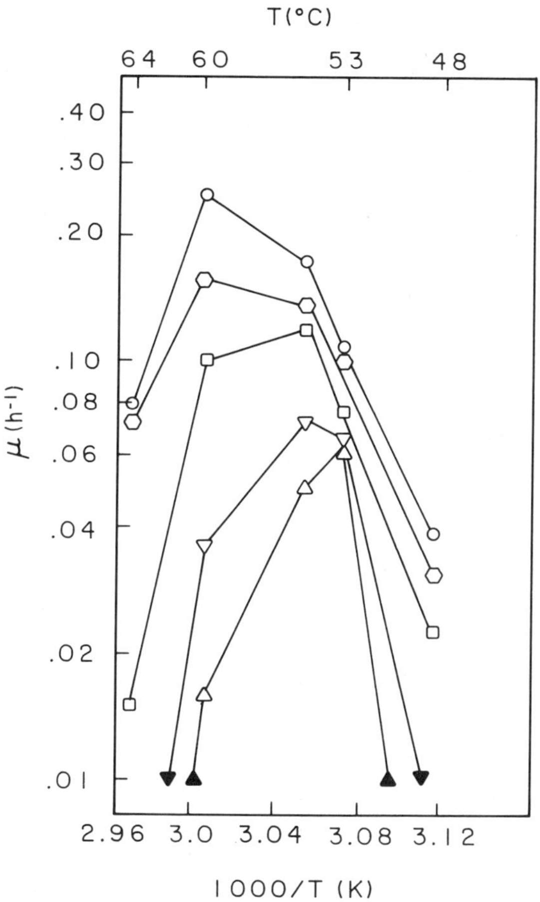

Figure 2 Effect of ethanol concentration on Arrhenius plot for *C. thermocellum* ATCC 27405: ○, 2.5 g/L; ⬡, 5 g/L; □, 10 g/L; ▽, 15 g/L; △, 20 g/L. Closed symbols represent values obtained from curves of μ versus temperature when lysis rather than growth was observed. (From Herrero and Gomez, 1980.)

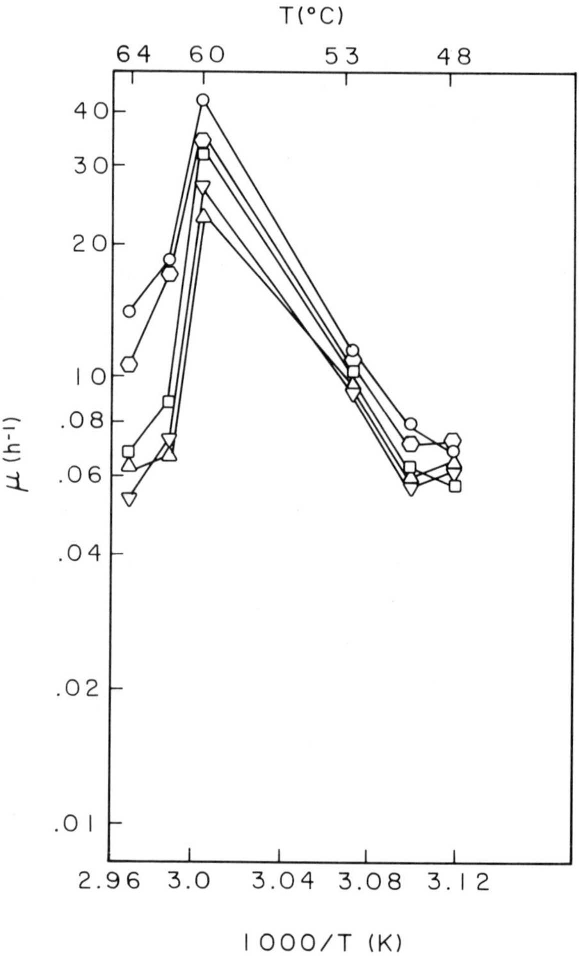

Figure 3 Effect of ethanol concentration on Arrhenius plot for *C. thermocellum* C/9: ○ , 2.5 g/L; ⬡ , 5 g/L; □ , 10 g/L; ▽, 15 g/L; △, 20 g/L. (Herrero and Gomez, 1980.)

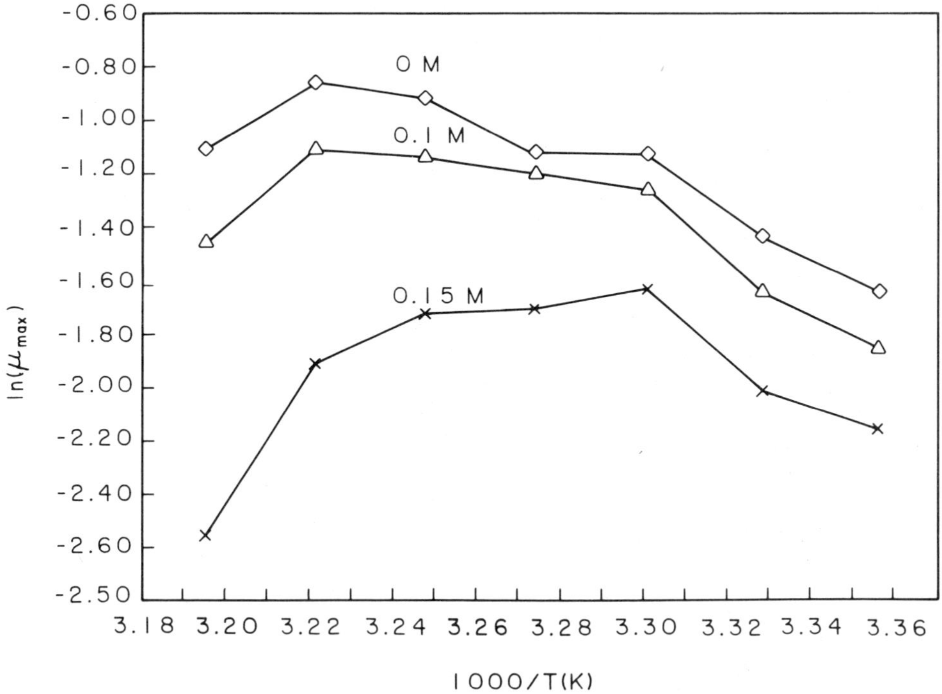

Figure 4 Effect of butanol concentration on Arrhenius plot for *Clostridium acetobutylicum* ATCC 824: ◇ , no butanol; △, 0.1 m butanol; X, 0.15M butanol. (From Kuhn and Linden, 1986.) Copyright © 1986; reprinted from *Biotechnology and Bioengineering Symposium 17* by permission of John Wiley and Sons, Inc.

IV. EFFECT OF CHEMICAL ENVIRONMENT

The chemical environment affects microbial growth and product formation kinetics. Although this is well known and many studies of the effects of media composition on kinetics have been reported, no generally useful models have been developed. Rather, the kinetics that are developed tend to be limited to specific applications under relatively specific environmental conditions. This is because many chemicals can affect growth and product formation. Water activity, pH, osmolality, carbon dioxide concentration, substrate concentration, product concentration, and inorganic nutrient concentrations are examples of important variables associated with the chemical environment in anaerobic fermentation. In order to model cell growth and product formation in complex chemical environments, it may be necessary to develop structured models that allow important aspects of cellular function, such as membrane transport, to be considered.

A. Water Activity and Osmolality

The concentration of water or water activity can be an important variable in fermentations in which high product concentrations are desired or high inorganic nutrient concentrations are encountered. In food technology, food preservation is achieved by keeping the water activity below the level at which microbial growth occurs. Sugars and salts are commonly used to lower the water activity of foods. The water activity is commonly found by measuring the equilibrium relative humidity (ERH) (Erickson, 1982; Hahn-Hagerdal, 1986). The water activity may be defined

$$a_w = \frac{f_w}{f_0} = \frac{P_w}{P_0} = \text{ERH} \qquad (14)$$

where a_w is water activity, f_w is fugacity at the conditions encountered, f_0 is the fugacity at the standard-state condition, which is usually defined as the vapor pressure of pure water P_0 at the temperature of the system. At the pressures normally encountered in fermentations, the fugacity may be set equal to the vapor pressure of the solution P_w or the equilibrium partial pressure of water. The osmolality ε_m and the water activity are related (Jones and Greenfield, 1986):

$$\varepsilon_m = -\frac{1000}{M_w} \ln a_w = -55.51 \ln a_w \qquad (15)$$

where M_w is the molecular weight of the solvent, which in this case is water. Figure 5, from Jones and Greenfield (1986), shows that the specific growth rate decreases linearly as the osmolality increases for values above 1 Osmol/kg of water. Figure 6 shows the variation in osmolality, ethanol concentration, biomass concentration, and substrate concentration with time for a batch fermentation of *Saccharomyces cerevisiae* UQM7OY growing on sucrose (Jones and Greenfield, 1986). The solution osmolality starts to increase at about the same time that exponential growth stops. Ethanol and the substrate sugars jointly contribute to the osmolality of the solution and the inhibition observed in Figure 6. However, ethanol also appears to affect the cell membranes and their function (Jones and Greenfield, 1986; Leao and Van Uden, 1985).

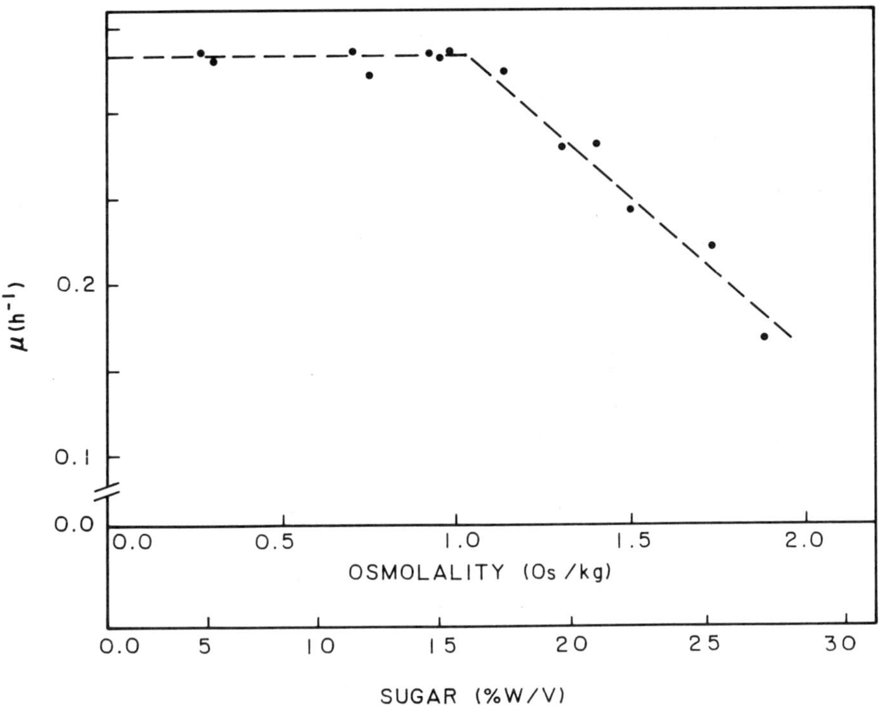

Figure 5 Variation of initial specific growth rate with osmolality for *S. cerevisiae* UQM7OY. Osmolality was adjusted by variation in initial concentration of either glucose, fructose or glucose plus fructose. (From Jones and Greenfield, 1986.) Copyright © 1986; reprinted from *Biotechnology and Bioengineering* by permission of John Wiley and Sons, Inc.

B. Hydrogen Ion Concentration or pH

The pH is commonly measured and controlled in most fermentations. It is an important variable that can affect the dominant form of ionizable biochemicals, which are important to the fermentation. Figure 7 shows the effect of pH on product yield, maximum specific growth rate, maximum specific product formation rate, and maximum volumetric rate of ethanol production for *Candida shehatae* and *Pichia stipitis* growing on D-xylose. The highest values of maximum specific growth rate and the highest values of specific product formation rate occur at about the same pH for each of these organisms. This is to be expected when the growth-associated contribution to product formation is a relatively large fraction of the total.

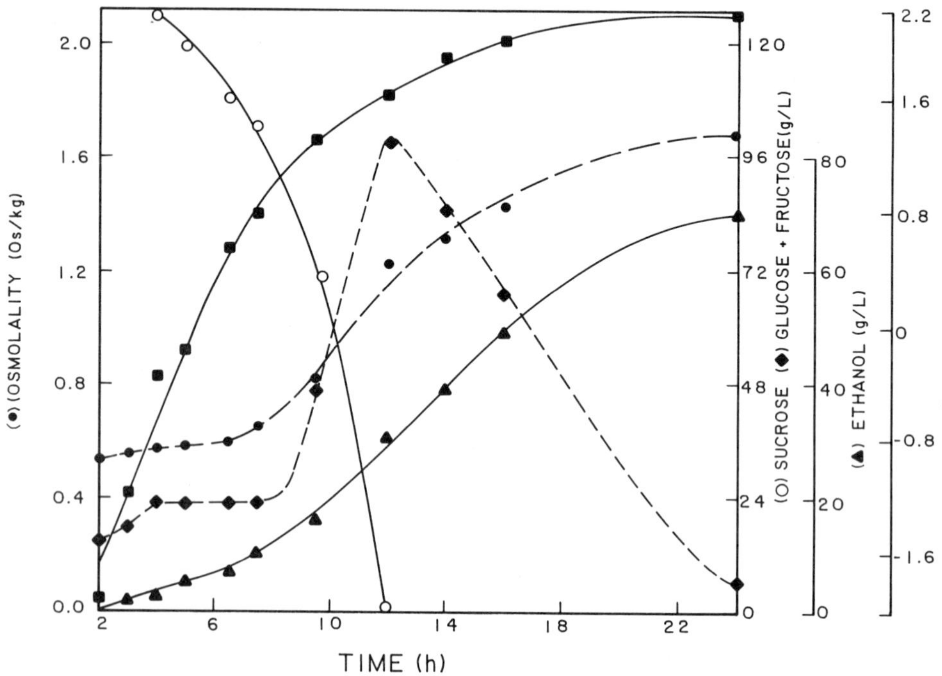

Figure 6 Batch fermentation with *S. cerevisiae* UQMTOY at 30°C, pH 4.5. (From Jones and Greenfield, 1986.) Copyright © 1986; reprinted from *Biotechnology and Bioengineering* by permission of John Wiley and Sons, Inc.

C. Carbon Dioxide

Carbon dioxide in the vapor phase is related to dissolved carbon dioxide in the liquid through the Henry's law phase equilibrium relationship

$$p_c = H_c C_c^* \tag{16}$$

where p_c is the partial pressure in the vapor phase, C_c^* is the equilibrium concentration in the liquid phase, and H_c is the Henry's law constant (Stumm and Morgan, 1970; Perry and Chilton, 1973). In the liquid phase, carbon dioxide is present as dissolved CO_2, H_2CO_3, HCO_3^-, and CO_3^{-2}. Figure 8, which is from Stumm and Morgan (1970), shows how the quantity of carbon in the liquid phase varies with pH for a fixed partial pressure of carbon dioxide in the gas phase. Increasing the pH increases the amount of carbon present

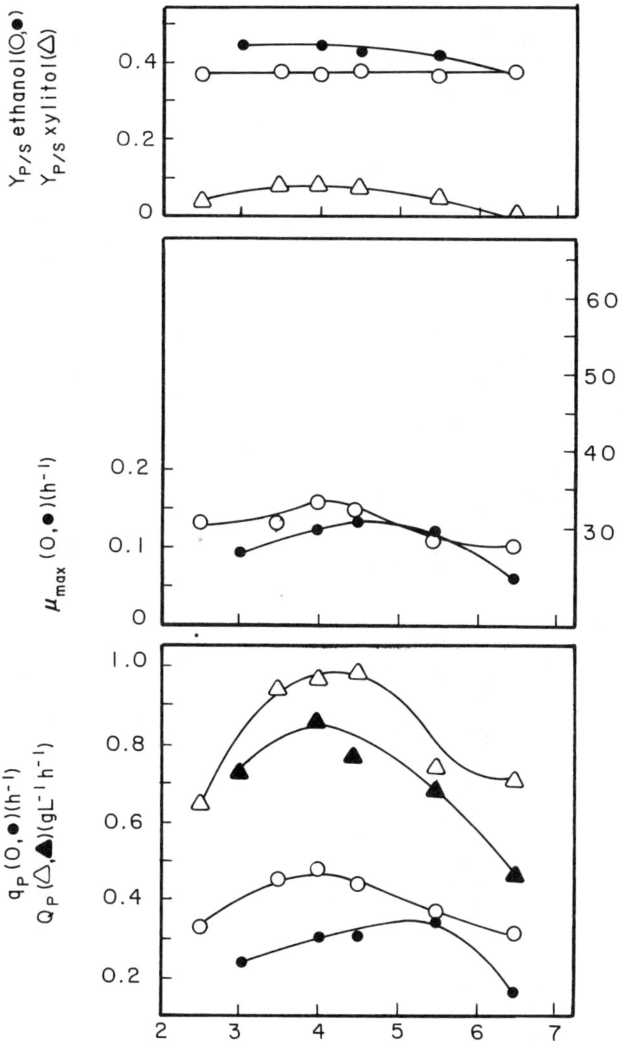

Figure 7 Effect of pH on product yields ($Y_{P/S}$), maximum specific rates of growth (μ_{max}) and ethanol production (q_p), and maximum volumetric rate of ethanol production (Q_p). Open symbols are for *Candida shehatae*; solids symbols are for *Pichia stipitis*. (From du Preez et al., 1986.)

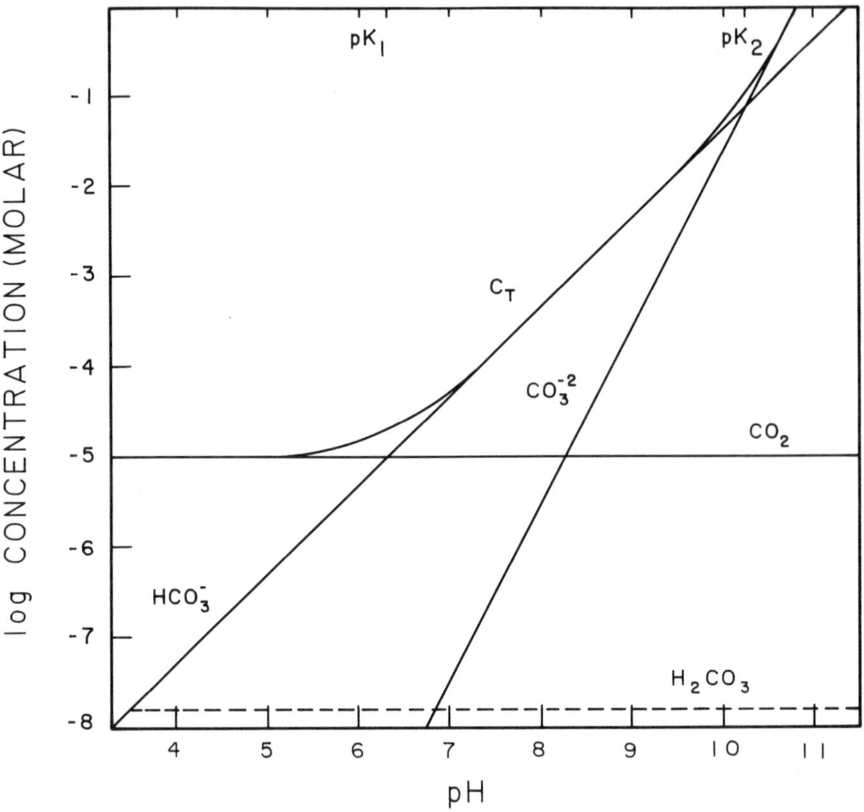

Figure 8 Aqueous carbonate equilibrium for constant CO_2 partial pressure of $10^{-3.5}$ atm. The pH is adjusted with strong base or acid. (From Stumm and Morgan, 1970, p. 127.)

as bicarbonate and carbonate. The kinetics and mass transfer rates are such that one should not automatically assume that equilibrium conditions exist (Jones and Greenfield, 1982; Erickson and Stephanopoulos, 1986).

The effects of carbon dioxide on growth and product formation have been reviewed recently by Jones and Greenfield (1982). Based on the results of their review, they conclude that at carbon dioxide partial pressures below 0.15-0.2 atm, the growth of most industrially important microorganisms is either relatively unaffected or improved by the presence of carbon dioxide. Above 0.2 atm partial pressure, carbon dioxide can adversely influence membrane function and metabolic pathways. Ethanol and carbon dioxide can exhibit additive inhibitory effects that may be due to common membrane disruptive functions. In yeasts, carbon dioxide appears to reduce cell yield and disrupt membrane function.

Figure 9 shows the variation in specific growth rate and biomass yield with carbon dioxide concentration for *Escherichia coli* growing

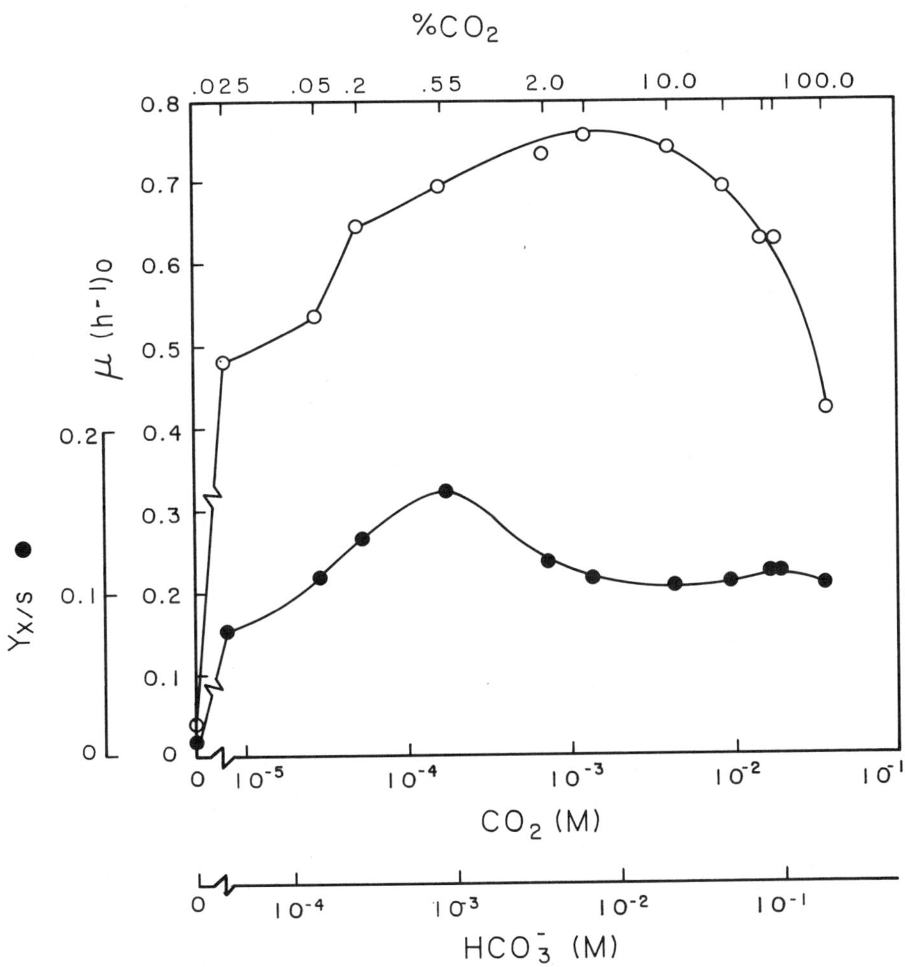

Figure 9 Effect of dissolved CO_2 and bicarbonate on specific growth rate (○) and biomass yield (●) for *E. coli* growing anaerobically on glucose at pH 6.8. (From Lacoursiere et al., 1986.)

on glucose under anaerobic conditions. These results indicate that an optimum carbon dioxide concentration exists, with the maximum specific growth rate observed at 5% carbon dioxide in the gas phase (1.3 mM dissolved carbon dioxide) and the maximum yield observed at 0.175 mM dissolved carbon dioxide in the broth. The fermentation was conducted at 1 atm total pressure and pH 6.8.

D. Inorganic Nutrients

The effects of inorganic ions on anaerobic fermentation have been reviewed by Maiorella et al. (1984). The water activity decreases as salts and inorganic ions are added to the media. Table 2, from Maiorella et al. (1984), shows the molar concentrations and corresponding water activities required to reduce the cell mass production by 80%. These authors also point out that the cell must work to maintain the desired internal ion concentration. The free energy required ΔG for expulsion of 1 mol or cations is approximately (Maiorella et al., 1984)

$$\Delta G = RT \ln \frac{I_m}{I_c} + ZF\Delta\psi \tag{17}$$

where I_c and I_m are the ion concentration in the cell and media, respectively, R is the gas constant, T is absolute temperature,

Table 2 Molar Solute Concentration, Water Activity, and Osmotic Pressure Required for 80% Inhibition of Cell Mass Production[a]

Solute	Molar concentration of solute	Water activity a_w	Osmotic pressure (atm)
$(NH_4)_2SO_4$	0.24	0.991	13.0
$CaCl_2$	0.23	0.989	15.0
NaCl	0.45	0.985	20.9
NH_4Cl	0.46	0.985	21.2
Xylose	0.92	0.982	26.0
$MgSO_4$	0.97	0.981	26.8
KH_2PO_4	0.73	0.980	28.6
$MgCl_2$	0.92	0.946	77.3
KCl	1.74	0.944	80.1

[a]This is an 80% decline in continuous culture cell mass production.
Source: From Maiorella et al. (1984). Copyright © 1984; reprinted from *Biotechnology and Bioengineering* by permission of John Wiley and Sons, Inc.

Z is the charge on the ion, F is the Faraday constant, and $\Delta\psi$ is the membrane electrical potential. Figures 10 and 11 show the variation in cell concentration, ethanol concentration, specific product formation rate, product yield, and cell yield with inorganic nutrient concentration; sodium chloride concentration is varied in Figure 10, and ammonium sulfate concentration is varied in Figure 11. These figures show that the specific rate of product formation increases as the salt concentration increases because more maintenance energy is required. Cell yields decrease because of the stress and increased maintenance requirement.

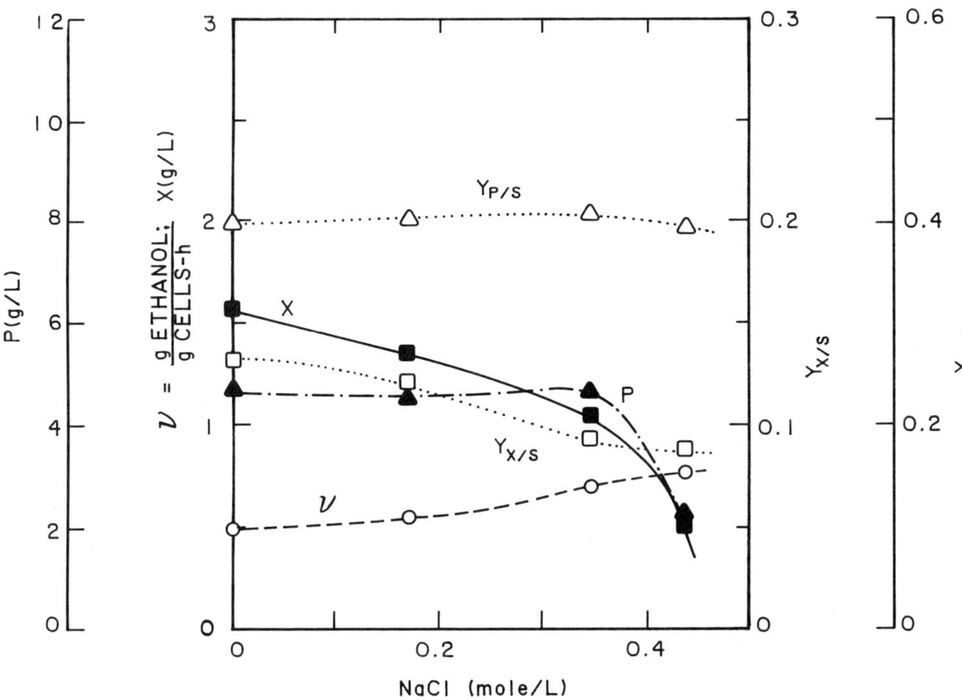

Figure 10 Effect of sodium chloride concentration on cell concentration (X), ethanol concentration (P), biomass yield ($Y_{X/S}$), ethanol yield ($Y_{P/S}$), and specific ethanol productivity (ν) for *Saccharomyces cerevisiae*; dilution rate = 0.163 h^{-1}. (From Maiorella et al. 1984.) Copyright © 1984; reprinted from *Biotechnology and Bioengineering* by permission of John Wiley and Sons, Inc.

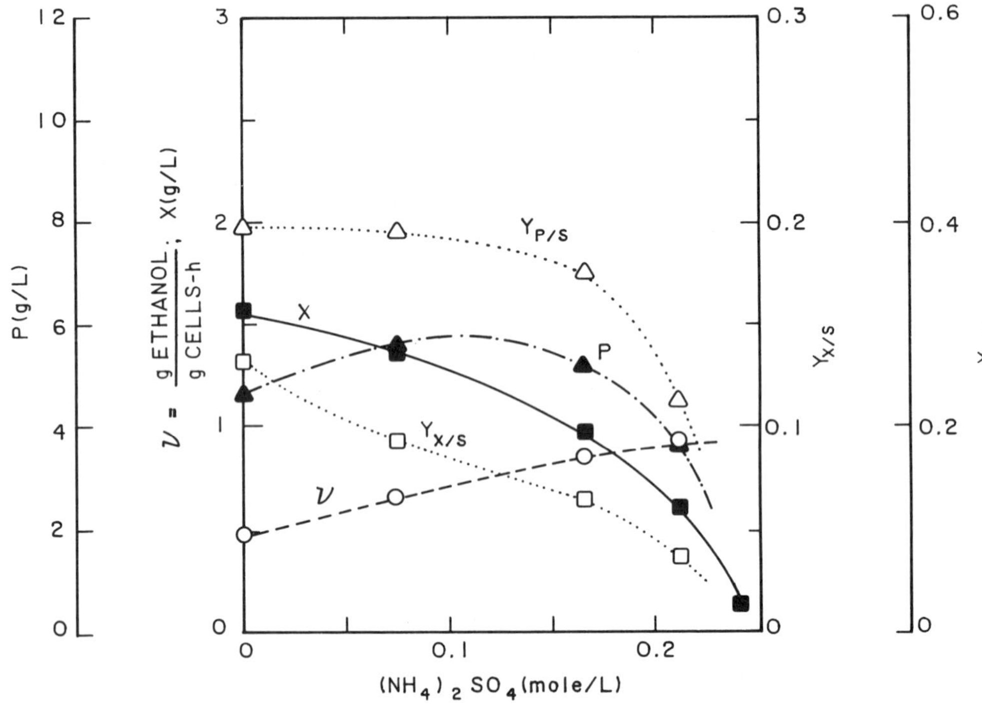

Figure 11 Effect of $(NH_4)_2SO_4$ concentration on cell concentration (X), ethanol concentration (P), biomass yield ($Y_{X/S}$), ethanol yield ($Y_{P/S}$), and specific ethanol productivity (ν) for *Saccharomyces cerevisiae*; dilution rate = 0.160 h^{-1}. (From Maiorella et al. 1984.) Copyright © 1984; reprinted from *Biotechnology and Bioengineering* by permission of John Wiley and Sons, Inc.

E. Organic Substrates and Products

Organic substrates and products have been widely studied because they are known to influence the kinetics of growth and product formation in anaerobic fermentation. Substrate-limited growth was considered in Section II. Substrate inhibition and product inhibition are considered in this section. In the development of structured models, knowledge of the mechanism of inhibition is important. Many of the models for inhibition that have been developed to date are unstructured. In these models inhibition is modeled without incorporating information on the mechanism of action into the model.

Edwards (1970) has reviewed the data on substrate inhibition and models to treat the data. He concludes that the Haldane Model,

$$\mu = \frac{\mu_{max}S}{(K_S + S)(1 + S/K_i)} \tag{18}$$

is a good model relative to the other models he considered.

Converti et al. (1985) have investigated substrate inhibition for two different strains of *Saccharomyces*. They found that *Saccharomyces oviformis* exhibited less inhibition than *S. cerevisiae*; however, they did not relate their results to other variables, such as water activity or osmolality.

Product inhibition has been considered for ethanol by a number of investigators because of its industrial importance. Aiba et al. (1968) modeled ethanol inhibition with the model

$$\mu = \mu_0 e^{-kP} \tag{19}$$

where μ_0 is the specific growth rate when P = 0 and k is the kinetic parameter associated with ethanol inhibition. Figure 12 illustrates the fit obtained with this model. In later work, Leao and Van Uden (1982, 1983, 1984a, b, 1985) showed that the modified model

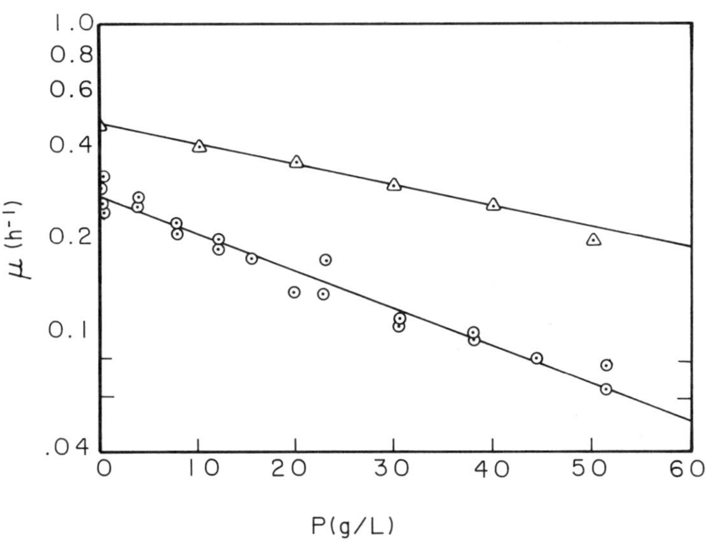

Figure 12 Variation of specific growth (μ) with ethanol concentration (P) when no substrates are limiting; ⊙, shaken flask data; △, continuous culture data using intercepts of Lineweaver-Burk plot; baker's yeast growing on glucose. (From Aiba et al., 1968.)

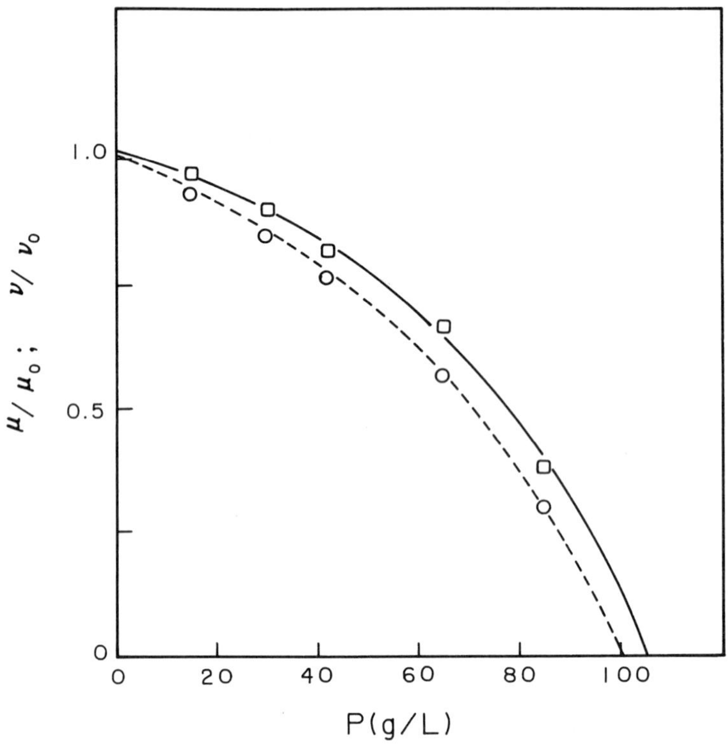

Figure 13 Effect of ethanol concentration on yeast growth (○), μ/μ_0 and product formation (□), ν/ν_0 for *Saccharomyces cerevisiae* ATCC 4126. (From Luong, 1985.)

$$\mu = \mu_0 e^{-k(P - P_{min})} \tag{20}$$

could be used to model inhibition for several different alcohols. In this equation, P_{min} is the minimum product concentration above which the toxic effect is measurable. Luong (1985) has investigated ethanol inhibition and found that the models

$$\frac{\mu}{\mu_0} = 1 - \left(\frac{P}{P_{max}}\right)^a \tag{21}$$

$$\frac{\nu}{\nu_0} = 1 - \left(\frac{P}{P'_{max}}\right)^b \tag{22}$$

fit their data very well, where ν_0 is the specific rate of product formation corresponding to $\mu = \mu_0$ when there is no product inhibition. Here P_{max} and P'_{max} are the product concentrations at which growth and product formation, respectively, stop. The specific rate of product formation ν is slightly different from the specific rate of growth, as shown in Figure 13. Another model that has been proposed and used is (Bovee et al., 1984)

$$\frac{dS}{dt} = kS^a P^b \tag{23}$$

Product inhibition is considered in detail in Chapter 8. The effects of alcohol on microorganisms have been considered in other reviews (Ingram and Buttke, 1984; Van Uden, 1985). Structured models have also been used to treat product inhibition; these are considered in the next section.

V. STRUCTURED MODELS

Structured models are considered in detail in Chapter 9. There are many potential applications of structured models in anaerobic fermentation; however, their use depends upon the experimental ability to make measurements of the variables used in the model. Progress in the development of sensors and experimental methods has progressed to the point that structured models now receive considerable attention (Bailey and Ollis, 1986; Roels, 1983; Blanch et al., 1983; Shuler, 1985; Lee and Bailey, 1984a, b). Recently, Jobses et al. (1985, 1986) have used a structured two-compartment model to consider growth, substrate utilization, and product formation with *Z. mobilis*. They employed a model consisting of a K compartment including RNA, carbohydrates, and monomers of macromolecules, and a G compartment consisting of protein, DNA, and lipids. The following kinetic model was developed and used (Jobses et al., 1985). For substrate consumption,

$$r_s = \frac{\mu_{max} SX}{Y_{sk}(K_s + S)} \tag{24}$$

For formation and conversion of K-compartment substances,

$$r_k = \frac{\mu_{max} SX}{K_s + S} - k_K KGX + k_g GX \tag{25}$$

For the G compartment,

$$r_G = Y_{Kg}k_K KGX - k_g GX \tag{26}$$

where

$$r_X = r_K + r_G \tag{27}$$

and the K and G compartments sum to give the total biomass X. This model is useful in that it provides a rational explanation of the experimental data (Jobses et al., 1985, 1986). The models provide a quantitative testing of the conceptual ideas.

Lee and Bailey (1984a, b) have developed genetically structured models for the *lac* promoter-operator function in the *E. coli* chromosome and in multicopy plasmids. The model for gene expression includes mass balances for messenger RNA and the protein of interest. Transcription efficiency is considered by including the binding of three regulatory species to *lac* region DNA in the model. This work is important in that it shows the potential and power of structured models. It also shows how the measurements and data accumulated by geneticists and molecular biologists can be effectively used to understand genetic control in a microbial population. One of the problems of this approach is the need for chemical understanding of the biochemical events that occur in the *lac* region of the DNA. Considerable experimental work is needed to generate this information.

Lee et al. (1985) have developed a model for product formation in unstable recombinant populations. They also compare their model to the Leudeking and Piret model presented earlier in Section II.

VI. SEGREGATED MODELS

Segregated models allow differences in the population to be considered. Segregation can include other aspects of the process in addition to the microbial population. When particles are used in the fluidized bed anaerobic digestion process, the differences among the particles can be treated with a segregated model. Segregation has been used to treat differences among drops in a two-liquid phase fermentation. Although there are many potential applications of segregated models for growth and product formation in anaerobic fermentation, very few prior studies have involved anaerobic processes.

VII. MIXED CULTURE MODELS

Modeling growth and product formation in mixed cultures has received some attention. Chapters 16 and 18 consider mixed culture fermentations and anaerobic digestion, respectively. The yogurt fermentation is a mixed culture anaerobic fermentation that has been considered in several studies (Oner and Erickson, 1986; Driessen, 1981). Anaerobic digestion has been extensively investigated as a mixed culture process, and several models have been presented that consider the major populations and their interactions (see also Chaps. 17 and 18 (Mosey, 1983; Bryers and Hamer, 1985).

VIII. CONCLUSIONS

The kinetics of anaerobic growth and product formation have been investigated in a large number of studies. Some of the literature is reviewed in this chapter; however, many of the chapters that follow also include kinetic considerations.

IX. NOMENCLATURE

a	Parameter in product inhibition models; dimensionless
a_w	Water activity; see Equation (14); dimensionless
b	Parameter in product inhibition models; dimensionless
C_c^*	Equilibrium CO_2 concentration in liquid phase; g mol/L
d	Moles of carbon dioxide per quantity of organic substrate containing one gram atom carbon; g mol/g atom carbon
E	Activation energy; kJ/g mol
F	Faraday constant; 96,485 C/g mol
f_0	Fugacity at standard state; atm
f_w	Fugacity; atm
G	Fraction of biomass; dimensionless
ΔG	Free energy; J/g mol
H_c	Henry's law constant; L atm/g mol
I_c	Ion concentration within cell; g mol/L
I_m	Ion concentration in media; g mol/L
K	Fraction of biomass; dimensionless
K_i	Inhibition kinetic constant; g/L
K_s	Saturation kinetic constant; g/L
k	Kinetic constant associated with product inhibition; L/g
k_g	Kinetic constant in Equation (25); h^{-1}
k_0	Kinetic constant; h^{-1}
k_K	Kinetic constant in Equation (25); h^{-1}
m_e	Rate of organic substrate consumption for maintenance; g equivalent of available electron per g equivalent of available electrons in biomass (h)
m_s	Grams of substrate expended per gram of cells per hour for maintenance; h^{-1}
N	Number of data points

P	Product concentration; g/L
P_{max}	Product concentration at which growth stops
P'_{max}	Product concentration at which product formation stops
P_{min}	Product concentration at which inhibition is first observed
P_0	Vapor pressure of pure water; atm
P_w	Vapor pressure of water in solution; atm
P_c	Partial pressure of CO_2 in vapor phase; atm
Q_{CO_2}	Specific molar production rate of CO_2; g mol/g biomass (h)
R	Gas constant
R_i	Rate of conversion of compound i; g/L (h)
S	Substrate concentration; g/L
T	Temperature; K
t	Time; h
X	Biomass concentration; g/L
Y_{ATP}	Biomass yield based on ATP; g dry biomass per g mol ATP
Y_{ATP}^{max}	True biomass yield based on ATP; g dry biomass per g mol ATP
Y_{kg}	Yield of G biomass from K biomass; dimensionless
$Y_{P/S}$	Product yield; g product per g substrate
Y_{sk}	Yield of K biomass from substrate; dimensionless
$Y_{x/s}$	Growth yield; g biomass per g substrate
$Y_{x/s}^{max}$	True growth yield corrected for maintenance; g biomass per g substrate
y_c	Biomass carbon yield (fraction of organic substrate carbon in biomass); dimensionless
z	Fraction of organic substrate carbon in products; dimensionless
Z	Charge on ion
α	Growth-associated kinetic parameter; dimensionless
β	Maintenance-associated kinetic parameter; h^{-1}
γ_b	Reductance degree of biomass; equivalents of available electrons per g atom carbon
γ_p	Reductance degree of products; equivalents of available electrons per g atom carbon
γ_s	Reductance degree of substrate; equivalents of available electrons per g atom carbon
δ	Number of equivalents of available electrons transferred to product to produce 1 g mol of ATP from ADP
ε_m	Osmolality; see Equation (15); Osmol/kg
η	Fraction of available electrons in organic substrate converted to biomass; dimensionless
η_{max}	"True" biomass available electron yield; dimensionless
μ	Specific growth rate; h^{-1}
μ_{max}	Maximum specific growth rate; h^{-1}

μ_0	Kinetic parameter in Equation (19); h^{-1}
ν	Specific rate of product formation; h^{-1}
ν_0	Specific rate of product formation when there is no inhibition; h^{-1}
ξ_p	Fraction of available electrons in organic substrate converted to products; dimensionless
σ_b	Weight fraction carbon in dry biomass; dimensionless
σ_p	Weight fraction carbon in products; dimensionless
σ_s	Weight fraction carbon in organic substrate; dimensionless
$\Delta\psi$	Electrical potential; V

REFERENCES

Aiba, S., Shoda, M., and Nagatani, M. (1968). Kinetics of product inhibition in alcohol fermentation. *Biotechnol. Bioeng.* *10*:845-864.

Atkinson, B., and Mavituna, F. (1983). *Biochemical Engineering and Biotechnology Handbook.* Nature Press, New York.

Bailey, J. E., and Ollis, D. F. (1986). *Biochemical Engineering Fundamentals*, 2nd Ed. McGraw-Hill, New York.

Blanch, H. W., Papoutsakis, E. T., and Stephanopoulos, G. (1983). *Foundations of Biochemical Engineering.* ACS Symposium Series 207, American Chemical Society, Washington, D.C.

Bovee, J. P., Strehaiano, P., Goma, G., and Sevely, Y. (1984). Alcohol fermentation: Modeling based on sole substrate and product measurement. *Biotechnol. Bioeng.* *26*:328-334.

Bryers, J. D., and Hamer, G. (1985). Structured modeling of the anaerobic digestion of biomass particulates. *Biotechnol. Bioeng.* *27*:638-649.

Converti, A., Perego, P., Lodi, A., Parisi, F., and del Borghi, M. (1985). A kinetic study of *Saccharomyces* strains: Performance at high sugar concentrations. *Biotechnol. Bioeng.* *27*:1108-1114.

Driessen, F. M. (1981). Protocooperation of yogurt bacteria in continuous culture. In *Mixed Culture Fermentations*, M. E. Bushell and J. H. Slater (Eds.). Academic Press, New York, pp. 99-120.

du Preez, J. C., Bosch, M., and Prior, B. A. (1986). Xylose fermentation by *Candida shehatae* and *Pichia stipitis*: Effects of pH, temperature, and substrate concentration. *Enzyme Microb. Technol.* *8*:360-364.

Edwards, V. H. (1970). The influence of high substrate concentrations on microbial kinetics. *Biotechnol. Bioeng.* *12*:679-712.

Erickson, L. E. (1982). Recent developments in intermediate moisture foods. *J. Food Protect.* *45*:484-491.

Erickson, L. E., and Stephanopoulos, G. (1986). Biological reactors. In *Chemical Reaction and Reactor Engineering*, J. J. Carberry and A. Varma (Eds.). Marcel Dekker, New York, pp. 779-838.

Erickson, L. E., Minkevich, I. G., and Eroshin, V. K. (1978). Application of mass and energy balance regularities in fermentation. *Biotechnol. Bioeng. 20*:1595-1621.

Fieschko, J., and Humphrey, A. E. (1983). Effects of temperature and ethanol concentration on the maintenance and yield coefficient of *Zymomonas mobilis. Biotechnol. Bioeng. 25*:1655-1660.

Hahn-Hagerdal, B. (1986). Water activity: A possible external regulator in biotechnical processes. *Enzyme Microb. Technol. 8*:322-327.

Herrero, A. A., and Gomez, R. F. (1980). Development of ethanol tolerance in *Clostridium thermocellum*: Effect of growth temperature. *Appl. Environ. Microbiol. 40*:571-577.

Ingram, L. O., and Buttke, T. M. (1984). Effects of alcohols on microorganisms. *Adv. Microb. Physiol. 25*:253-300.

Jobses, I. M. L., Egberts, G. T. C., van Baalen, A., and Roels, J. A. (1985). Mathematical modelling of growth and substrate conversion of *Zymomonas mobilis* at 30 and 35 C. *Biotechnol. Bioeng. 27*:984-995.

Jobses, I. M. L., Egberts, G. T. C., Luyben, K. C. A. M., and Roels, J. A. (1986). Fermentation kinetics of *Zymomonas mobilis* at high ethanol concentrations: Oscillations in continuous culture. *Biotechnol. Bioeng. 28*:868-877.

Jones, R. P., and Greenfield, P. F. (1982). Effect of carbon dioxide on yeast growth and fermentation. *Enzyme Microb. Technol. 4*:210-223.

Jones, R. P., and Greenfield, P. F. (1986). Role of water activity in ethanol fermentations. *Biotechnol. Bioeng. 28*:29-40.

Kuhn, R. H., and Linden, J. C. (1986). Effects of temperature and membrane fatty acid composition on butanol tolerance of *Clostridium acetobutylicum. Eighth Symposium on Biotechnology for Fuels and Chemicals, Biotechnol. Bioeng. Symp. 17*:197-207.

Lacoursiere, A., Thompson, B. G., Kole, M. M., Ward, D., and Gerson, D. F. (1986). Effects of carbon dioxide concentration on anaerobic fermentations of *Escherichia coli. Appl. Microbiol. Biotechnol. 23*:404-406.

Leao, C., and Van Uden, N. (1982). Effects of ethanol and other alkanols on the glucose transport system of *Saccharomyces cerevisiae. Biotechnol. Bioeng. 24*:2601-2604.

Leao, C., and Van Uden, N. (1983). Effects of ethanol and other alkanols on the ammonium transport system of *Saccharomyces cerevisiae. Biotechnol. Bioeng. 25*:2085-2090.

Leao, C., and Van Uden, N. (1984a). Effects of ethanol and other alkanols on the general amino acid permease of *Saccharomyces cerevisiae. Biotechnol. Bioeng. 26*:403-405.

Leao, C., and Van Uden, N. (1984b). Effects of ethanol and other alkanols on passive proton influx in the yeast *Saccharomyces cerevisiae. Biochim. Biophys. Acta 774*:43-48.

Leao, C., and Van Uden, N. (1985). Effects of ethanol and other alkanols on the temperature relations of glucose transport and fermentation in *Saccharomyces cerevisiae*. *Appl. Microbiol. Biotechnol.* *22*:359-363.

Lee, S. B., and Bailey, J. E. (1984a). Genetically structured models for *lac* promoter-operator function in the *Escherichia coli* chromosome and in multicopy plasmids: *lac* operator function. *Biotechnol. Bioeng.* *26*:1372-1382.

Lee, S. B., and Bailey, J. E. (1984b). Genetically structured models for *lac* promoter-operator function in the chromosome and in multicopy plasmids: *lac* promoter function. *Biotechnol. Bioeng.* *26*:1383-1389.

Lee, S. B., Seressiotis, A., and Bailey, J. E. (1985). A kinetic model for product formation in unstable recombinant populations. *Biotechnol. Bioeng.* *27*:1699-1709.

Lovitt, R. W., Longin, R., and Zeikus, J. G. (1984). Ethanol production by thermophilic bacteria: Physiological comparison of solvent effects on parent and alcohol-tolerant strains of *Clostridium thermohydrosulfuricum*. *Appl. Environ. Microbiol.* *48*:171-177.

Luedeking, R., and Piret, E. L. (1959a). A kinetic study of the lactic acid fermentation. Batch process at controlled pH. *Biotechnol. Bioeng.* *1*:393-412.

Luedeking, R., and Piret, E. L. (1959b). Transient and steady states in continuous fermentation. Theory and experiment. *Biotechnol. Bioeng.* *1*:431-459.

Luong, J. H. T. (1985). Kinetics of ethanol inhibition in alcohol kinetics. *Biotechnol. Bioeng.* *27*:280-285.

Maiorella, B. L., Blanch, H. W., and Wilke, C. R. (1984). Feed component inhibition in ethanolic fermentation by *Saccharomyces cerevisiae*. *Biotechnol. Bioeng.* *26*:1155-1166.

Minkevich, I. G., and Eroshin, V. K. (1973). Productivity and heat generation of fermentation under oxygen limitation. *Folia Microbiol. (Praha)* *18*:376-385.

Mosey, F. E. (1983). Mathematical modelling of the anaerobic digestion process. Regulatory mechanisms for the formation of short chain volatile acids from glucose. *Water Sci. Technol.* *15*:209-232.

Oner, M. D., and Erickson, L. E. (1986). Anaerobic fermentation of *Lactobacillus bulgaricus* and *Streptococcus thermophilus* on 3% nonfat dry milk with pure and mixed cultures. *Biotechnol. Bioeng.* *28*:883-894.

Oner, M. D., Erickson, L. E., and Yang, S. S. (1984). Estimation of yield, maintenance, and product formation kinetic parameters in anaerobic fermentations. *Biotechnol. Bioeng.* *26*:1436-1444.

Perry, R. H., and Chilton, C. H. (1973). *Chemical Engineers' Handbook*, 5th Ed. McGraw-Hill, New York.

Pirt, S. J. (1975). *Principles of Microbe and Cell Cultivation.* Wiley, New York.

Roels, J. A. (1983). *Energetics and Kinetics in Biotechnology.* Elsevier Biomedical Press, New York.

Sa-Correia, I., and Van Uden, N. (1983). Temperature profiles of ethanol tolerance; Effects of ethanol on the minimum and the maximum temperatures of growth of the yeasts *Saccharomyces cerevisiae* and *Kluyveromyces fragilis*. *Biotechnol. Bioeng. 25*: 1665-1667.

Shuler, M. L. (1985). Invited review on the use of chemically structured models for bioreactors. *Chem. Eng. Commun. 36*:161-189.

Stumm, W., and Morgan, J. J. (1970). *Aquatic Chemistry.* Wiley, New York, p. 148.

Van Uden, N. (1984). Temperature profiles of yeast. *Adv. Microb. Physiol. 25*:195-251.

Van Uden, N. (1985). Ethanol toxicity and ethanol tolerance in yeast. *Annu. Rep. Ferment. Proc. 8*:11-58.

6

Redox Potential in Anaerobic and Microaerobic Fermentation

S. P. SRINIVAS,* GOVIND RAO,† and R. MUTHARASAN / *Drexel University, Philadelphia, Pennsylvania*

I. INTRODUCTION

Historically, redox potential measurement of bacterial cultures has been motivated by the need to monitor oxygen-free conditions while cultivating obligate anaerobes (Hentges and Maier, 1972; O'Brien and Morris, 1971; Morris, 1974). The redox potential response of a culture medium due to the addition of reducing agents, such as titanium(III) citrate, cysteine, sodium sulfide, and sodium thioglycolate, is an indication of the effectiveness of these agents in removing trace levels of oxygen (Zehnder and Wuhrmann, 1976). On the other hand, redox potential measurements in aerobic and facultative bacterial cultures (and also in fungal and mammalian cultures) are motivated by an implicit assumption that the intracellular redox environment is a function of the measured culture redox potential. Another contributing factor is the ease with which redox electrodes can be used for dissolved oxygen sensing under microaerobic conditions (Wimpenny and Necklen, 1971; Shibai et al., 1974; Radjai et al., 1984; Daniels et al., 1970a, b; Noparatnaraporn et al., 1986; Akashi et al., 1978).

**Present affiliation*: Stanford University Medical Center, Stanford, California.
†*Present affiliation*: University of Maryland, Baltimore County, Catonsville, Maryland.

Redox potential measurement and its implications in biological systems can be considered from two perspectives: extracellular and intracellular. In extracellular studies, redox potential is usually obtained by a platinum electrode immersed in the culture broth and is referred to as culture redox potential (CRP) (*Note*: E_h and CRP are both used in this review to denote cultural redox potential. Unless stated otherwise, all measurements are with a platinum probe.) In intracellular studies, redox measurements of reactions involving pyridine nucleotide cofactors or electron transport chains are made (Chance et al., 1979; Bucher et al., 1972; Caswell, 1968; Dutton and Wilson, 1974; Dutton, 1971). In some of these studies, the term "redox state" is often used to connote the redox potential of intracellular reactions. The scope of this review is limited to discussions of culture redox potential in the context of anaerobic and microaerobic metabolism.

Although culture redox potential is easily measured, its mechanistic interpretation is difficult. Many factors contribute to this dilemma. A large number of redox couples present in the broth interact with the platinum electrode. Consequently, the measured CRP cannot easily be assigned to a specific intracellular (or extracellular) oxidation-reduction reaction. Furthermore, the intracellular reactions are compartmentalized and may exist under nonequilibrium conditions. In addition, uncertainty exists about the identity of the redox mediators responsible for the measured CRP. Thus, quantitative interpretation of CRP with respect to cellular metabolism is not feasible with the current state of knowledge, and further research is needed. In spite of these difficulties, many investigators have derived several useful empirical correlations between CRP and cellular metabolism, which are discussed in this chapter. A brief review of the principles and methods used in the measurement and control of CRP is also presented.

II. PRINCIPLES

A thermodynamic treatment of oxidation-reduction processes is provided to emphasize the nature of information that can be obtained from redox potential measurements. Reversible reactions in a single phase are considered here as an illustration. For excellent reviews on this and related topics, the reader is referred to Clark (1960), Dutton and Wilson (1974), and Walz (1979). The following treatment is primarily adapted from Dutton (1978).

A. Redox Reactions

A typical oxidation-reduction reaction involves an electron donor A_{red} and an electron acceptor B_{ox}. An electron donor undergoes oxidation and is represented as

$$A_{red} = A_{ox} + ne \tag{1}$$

where n is the number of electrons involved in the reaction. Similarly, an electron acceptor undergoes reduction according to the equation

$$B_{ox} + ne = B_{red} \tag{2}$$

The value of n is usually 1 or 2, but it can be greater. A reduced-oxidized pair such as A_{red}/A_{ox} is called a redox couple. Since reactions (1) and (2) occur simultaneously and are coupled, one may write

$$A_{red} + B_{ox} = A_{ox} + B_{red} \tag{3}$$

The reaction in this form is known as a redox reaction. Reactions represented by Equations (1) and (2) are referred to as half-cells or half-reactions in the parlance of electrochemistry.

At equilibrium, the equilibrium constant K_{Eq} for reaction (3) is given by

$$K_{Eq} = \frac{[A_{ox}][B_{red}]}{[B_{ox}][A_{red}]} \tag{4}$$

where the terms in brackets denote the activities of the reaction components at equilibrium. The free energy change for reaction (3) at constant temperature, not necessarily at equilibrium, is given by

$$\Delta G = \Delta G^0 + RT \ln \frac{[A_{ox}][B_{red}]}{[B_{ox}][A_{red}]} \tag{5}$$

where R is the gas constant, T is absolute temperature, and $\Delta G^0 = -RT \ln K_{Eq}$ is the standard free energy of the reaction.

B. Redox Potential

Electrochemical parameters are often used to express the free energy changes of redox reactions. An advantage afforded by this approach is that we can avoid listing K_{Eq} or ΔG^0 of the various possible redox reactions between the innumberable combinations of half-reactions, and instead, we can consider electron affinities with respect to a standard half-cell. The standard half-cell is the hydrogen half-cell consisting of H_2 at 1 atm in equilibrium with a solution of H^+ of unit activity. The reaction that takes place at the electrode is

$$H_2 = 2H^+ + 2e^- \tag{6}$$

The work done in moving n electrons across a potential difference ΔE is equal to nF ΔE, where F is Faraday's constant (=96,495 C). Hence, the free energy change of reaction (5) is

$$\Delta G = -nF\Delta E \tag{7}$$

Substituting (7) into (5) gives

$$\Delta E = \Delta E^0 - \frac{RT}{nF} \ln \frac{[A_{ox}][B_{red}]}{[A_{red}][B_{ox}]} \tag{8}$$

where ΔE^0 is defined as

$$\Delta E^0 = -\frac{\Delta G^0}{nF} \tag{9}$$

Rearranging (8),

$$\Delta E = \Delta E^0 + \frac{RT}{nF} \ln \frac{[B_{ox}]}{[B_{red}]} - \frac{RT}{nF} \ln \frac{[A_{ox}]}{[A_{red}]} \tag{10}$$

or

$$E_{(B)} - E_{(A)} = E^0_{(B)} - E^0_{(A)} + \frac{RT}{nF} \ln \frac{[B_{ox}]}{[B_{red}]} - \frac{RT}{nF} \ln \frac{[A_{ox}]}{[A_{red}]} \tag{11}$$

where $E_{(A)}$ and $E_{(B)}$ are the half-electrode potentials of redox couples A and B, respectively, and the superscript 0 refers to standard conditions. By convention, the standard hydrogen half-cell E^0 is given a value of 0 V. When A is the standard hydrogen half-cell, this reduces to the Nernst equation:

$$E_{(B)} = E^0_{(B)} + \frac{RT}{nF} \ln \frac{[B_{ox}]}{[B_{red}]} \tag{12}$$

where $E_{(B)}$ is the redox potential of the half-cell B. The subscript h is often used to denote that the reference half-cell is the standard hydrogen half-cell. Thus, the redox potential of a half-cell is written as

$$E_h = E^0 + \frac{RT}{nF} \ln \frac{[Ox]}{[red]} \tag{13}$$

C. Effect of pH in Redox Potential Measurements

Although the redox couple just discussed involves the transfer only of electrons, several biologic and inorganic redox reactions also involve protons. Oxidized and reduced forms of a redox couple may have different proton affinity. In general, the pK of the oxidized form of a couple is lower than its reduced form (Dutton, 1978). For example, consider the redox reaction

$$A_{redH} = A_{ox} + ne + mH^+ \qquad (14)$$

Application of Equation (13) yields

$$E_h = E^0 + \frac{RT}{nF} \ln \frac{[A_{ox}]}{[A_{redH}]} - 2.303 \frac{mRT}{nF} pH \qquad (15)$$

Table 1 Standard Redox Potentials of Some Half-Reactions[a]

Reaction	Half-reaction	E_m (mV)
1	$1/2O_2 + 2H^+ + 2e^- = H_2O$	816
2	Cytochrome a_3-Fe^{3+} + $1e^-$ = cytochrome a_3-Fe^{2+}	550
3	Crotonyl-S-CoA + $2H^+$ + $2e^-$ = butyryl-S-CoA	190
4	Ubiquinone + $2H^+$ + $2e^-$ = ubiquinone-H_2	100
5	Fumarate + $2H^+$ + $2e^-$ = succinate	30
6	Methylene blue(ox) + $2H^+$ + $2e^-$ = methylene blue(red)	11
7	Ketoglutarate + NH_3 + $2H^+$ + $2e^-$ = glutamate + H_2O	-14
8	Acetaldehyde + $2H^+$ + $2e^-$ = ethanol	-163
9	Oxalacetate + $2H^+$ + $2e^-$ = malate	-175
10	FAD + $2H^+$ + $2e^-$ = $FDAH_2$	-180
11	Pyruvate + $2H^+$ + $2e^-$ = lactate	-190
12	Cystine + $2H^+$ + $2e^-$ = cysteine	-220
13	Acetoacetate + $2H^+$ + $2e^-$ = β-hydroxybutyrate	-290
14	NAD^+ + $2H^+$ + $2e^-$ = NADH + H^+	-320
15	$NADP^+$ + $2H^+$ + $2e^-$ = NADPH + H^+	-320
16	Acetyl-S-CoA + $2H^+$ + $2e^-$ = acetaldehyde + CoA	-410
17	$2H^+ + 2e^- = H_2$	-414
18	Ferredoxin-Fe^{3+} + $1e^-$ = ferredoxin-Fe^{2+}	-432
19	Methyl viologen(ox) + $2H^+$ + $2e^-$ = methyl viologen(red)	-440
20	Acetate + $2H^+$ + $2e^-$ = acetaldehyde	-600
21	Acetate + CO_2 + $2H^+$ + $2e^-$ = pyruvate	-700

[a]Standard state: 25°C and pH 7.0.
Source: Segel (1976).

When the pH is zero, this equation reduces to the form of Equation (13). Thus, the redox potential of a half-cell at a pH other than zero becomes more negative at a rate of 2.3 mRT/nF per pH unit. This corresponds to -60 mV per pH unit when m = n and -30 mV when m = 1 and n = 2. In the biological literature, half-call electrode potentials are usually reported at pH 7, with the activity of the oxidant equal to that of the reductant, and are called standard midpoint potentials, denoted E_m (Dutton, 1978; Nichols, 1982). The midpoint potentials of important redox reactions are given in Table 1. The direction of electron transfer can be determined from the E_m data; the reaction with more negative E_m is capable of reducing a reaction of higher E_m.

III. MEASUREMENT AND CONTROL OF REDOX POTENTIAL

We consider here the methods available for the measurement and control of redox potential. Two distinct approaches have been used, namely, indicator dyes and platinum redox electrodes. The latter are reported as more reliable (Jacob, 1970; Kjaergaard, 1977). For controlling CRP at a constant value, the most common technique reported in the literature is the controlled addition of reducing agents. Complementary to CRP measurement, several investigators have examined intracellular redox potential. The motivation for these studies is that the rate and direction of several intracellular reactions are dependent on the intracellular redox environment. Among the various intracellular redox reactions, redox reactions involving dehydrogenases with NAD(P)H as cofactors have been studied extensively (Chance et al., 1979; Bucher et al., 1972; Sies, 1982).

A. Measurement

1. *Redox Potential of Nicotinamide Nucleotide Systems*

The ratio of [NADH] to [NADH + NAD] is an important parameter in the regulation of a number of catabolic pathways and has often been referred to as the intracellular reduction state (ICRS) or catabolic reduction charge (Andersen and Meyenburg, 1977; Sies, 1982). The redox potential of intracellular free pyridine nucleotides has been measured using two methods, namely, the metabolite indicator method and fluorescence spectroscopy (Bucher et al., 1972). Although these techniques are not applicable in a process situation, they are briefly reviewed here as they are the only means of determining the intracellular redox state.

Metabolite indicator method. In this technique, the redox potential of the free NAD/NADH system in a cellular compartment is estimated by measuring the potential of an exogenous redox metabolite couple (called the indicator metabolite couple) in equilibrium with the intracellular NAD/NADH system. For example, the redox couples lactate-pyruvate for the cytosolic compartment (involving lactate dehydrogenase) and β-hydroxybutyrate-acetoacetate for the mitochondrial compartment (involving β-hydroxybutyrate dehydrogenase) have been used as indicator metabolites by Sies (1982) in studying rat liver cells. He has reviewed this technique for the investigation of redox compartmentation (cytosolic versus mitochondrial) in alcohol and aldehyde metabolizing systems of rat liver cells in aerobic and anaerobic conditions. The success of this technique requires that equilibrium between the intracellular pyridine nucleotide pool and the indicator metabolite couple be established by a specific and unilocated dehydrogenase. Furthermore, the indicator metabolite couple should be able to permeate the cell membrane. Under these conditions, the redox potential of the pyridine nucleotide system at the site of the intervening dehydrogenase can be obtained by measurements of the concentrations of the indicator metabolite couple in the extracellular fluid.

Fluorescence spectroscopy. The use of fluorescence phenomena for measuring the intracellular reduction state has been used by a number of investigators (Chance et al., 1979; Bucher et al., 1972). Reduced pyridine nucleotides are fluorescent but their oxidized counterparts are not. Similarly, oxidized flavoprotein is fluorescent, unlike its reduced counterpart. Thus, the redox state of flavoproteins and pyridine nucleotides can be measured by fluorescence spectroscopy. NADH and oxidized flavoproteins are the major fluorophores in the mitochondrial space, and NADH is the major fluorophore in the cytosol. Since flavoproteins and pyridine nucleotides are known to be at near equilibrium, the ratio of their respective fluorescence intensities is a qualitative measure of the intracellular redox state. Simultaneous fluorescence measurement of the two fluorophores can be used to determine the ratio of their fluorescence, which is considered a more accurate measure of intracellular reduction state than their absolute values in the presence of various measurement interferences (Chance et al., 1979).

2. *Indicator Dye Method*

A number of organic dyes exhibit changes in color depending on the degree of oxidation or reduction of the system. This is analogous to the indicator method used for pH determination. In Table 2, commonly used redox dyes are listed with their standard midpoint

Table 2 Midpoint Potentials of Some Redox Indicator Dyes

Dye	E_m (mV)
Phenol-*m*-sulfonate indo-2,6-dibromophenol	273
Phenol-*m*-sulfonate indophenol	250
m-Bromophenol indophenol	248
Phenol indophenol	227
Phenol blue	224
2,6-Dichlorophenol indo-*o*-chlorophenol	219
m-Cresol indophenol	208
o-Cresol indophenol	191
Thymol indophenol	174
Prune	56
Cresyl blue	47
Azure I	11
Methylene blue	11
Ciba scarlet sulfonate	-36
Resorufin	-51
Indigo trisulfonate	-81
Nile blue	-142
Cresyl violet	-167
Neutral blue	-192
Phenosafranine	-252
Safranine-T	-289
Neutral red	-325
Benzyl viologen	-359
Methyl viologen	-440

Source: After Jacob (1970).

potentials. The useful range of a dye is about ±50 mV about its midpoint potential. Consequently, to determine the E_h of a system, a number of dyes of different E_m values would be needed. Although the indicator dye method is easily implemented, a number of practical problems are associated with their use (Hewitt, 1936). The dye should be intensely colored so that it can be added in low concentrations to avoid poising (analogous to the term "buffering" in pH studies) of the system being examined. The color change should be independent of pH. The redox reaction of the dye should be reversible. Perhaps most important of all, the selected dyes should not catalyze biologic reactions or have any toxic effect.

3. *Electrochemical Method*

Principles. The exchange of electrons between a chemically inert but electrically conductive surface (such as platinum) and a redox reaction system manifests as electrochemical potential. This potential can be measured with respect to a reference electrode, such as a calomel or an Ag/AgCl electrode. The measured potential can be converted to the redox potential with respect to the standard hydrogen electrode by

$$E_h = E_{measured} + E_{ref} \tag{16}$$

where $E_{measured}$ and E_{ref} are the measured and reference electrode redox potentials, respectively. Commercially available redox electrodes usually use a platinum surface and frequently have Ag/AgCl or calomel reference electrodes. The potential across the electrode is measured potentiometrically so that chemical composition changes in the solution are minimized and polarization of the electrode surface is reduced (Buhler and Galster, 1980).

The potential measured by a redox probe is affected by the rate at which the platinum surface is able to exchange electrons with the medium. Slow exchange results in inaccurate redox potential measurements. Electron migration may be represented by currents i+ and i-, denoting flow to and from the probe, respectively. When equilibrium is reached, the two currents have the same magnitude. The absolute value of the migration currents per unit electrode surface area is referred to as the exchange current density (Buhler and Galster, 1980). This parameter depends on the electrode material, the redox system, and the concentration of the redox components. In redox systems exhibiting a sluggish response, an exogenous electron carrier (known as a redox mediator) is often added to increase the exchange current density. The mediator accelerates the attainment of equilibrium with the platinum surface and does not affect the thermodynamic relationships of the redox couples being studied (Dutton, 1978; Walz, 1979; Caswell, 1968; Wurmser and Banerjee, 1964). Some commonly used redox mediators in biologic systems are listed in Table 3.

In vivo oxidation-reduction systems are sequestered from the medium by semipermeable membranes and hence cannot directly interact with the electrode surface. Therefore, the deliberate addition of a redox mediator (at low concentration) can facilitate the exchange of electrons from intracellular redox couples to the platinum electrode surface. As long as the extent of reaction of the primary couple is not disturbed by the addition of the redox mediator, the electrode registers the correct potential of the primary couple. The mediator

Table 3 Commonly Used Redox Mediators and Their Midpoint Potential

Mediator	E_m (mV)	n value
Potassium ferro/ferricyanide	430	1
2,3,5,6-Tetramethylphenylenediamine (Diaminodurol, DAD)	260	2
N,*N*,*N*',*N*-tetramethylphenylenediamine (TMPD)	260	1
N-methylphenazonium methosulfate (PMS)	80	2
N-ethylphenazonium methosulfate (PES)	55	2
N-methyl 1-hydroxy phenazonium methosulfate (Procyanine)	-34	2
2-Hydroxyl-1,4-naphthaquinone	-145	2
Anthraquinone-2-disulfonate	-185	2
N,*N*'-dimethyl-4,4-bipyridium chloride (benzyl viologen)	-311	1
N,*N*'-dibenzyl-4,4-bipyridium chloride (methyl viologen)	-430	1

Source: After Dutton (1978).

should be selected such that the midpoint potential, under the conditions used, is equal to that of the intracellular redox couple being examined. The use of multiple mediators is practiced in bioenergetic studies and redox potentiometry of membrane fractions (Dutton, 1978).

At constant pH, a redox electrode can be used to measure dissolved oxygen concentration (Pirt, 1975; Ishizaki et al., 1974). In the presence of oxygen, the reaction at the electrode surface can be represented by

$$O_2 \text{ (dissolved)} + 2H_2O + 4e^- = 4OH^- \qquad (17)$$

Application of Equation (13) to Equation (17) yields

$$E_h = E_m + \frac{RT}{nF} \ln \frac{[O_2 \text{ (dissolved)}]}{[OH^-]^4} \qquad (18)$$

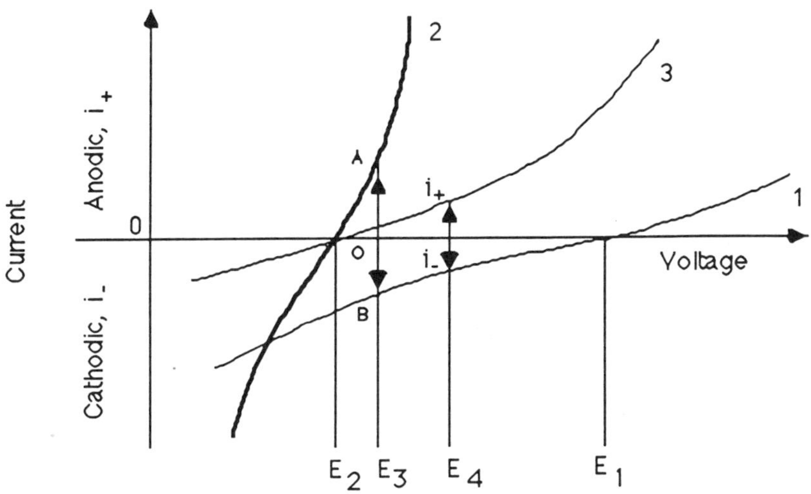

Figure 1 Polarograms showing the effect of oxygen on indicated redox potential. (From Buhler and Galster, 1980.)

The redox potential therefore depends on the dissolved oxygen concentration. The presence of oxygen significantly reduces the exchange current density (Buhler and Galster, 1980). The reason for this is illustrated by the polarograms shown in Figure 1. Water saturated with oxygen has a curve of low slope due to low exchange current density (curve 1), and its equilibrium potential is given by E_1. A system with a high exchange current density has a steep polarogram (curve 2) with an equilibrium potential of E_2. When oxygen is present in a redox system with a high exchange current, the potential established at the electrode can be determined from the data of curves 1 and 2. The resulting potential E_3 should satisfy the condition $i_+ = i_-$. The redox potential E_3 is found by requiring OA = OB, and it can be seen that the resultant E_h is not far from the actual equilibrium value E_2. However, when the redox system itself is of a low exchange current density (curve 3), then dissolved oxygen introduces significant errors in E_h measurement. The resultant potential E_4 is far from the correct equilibrium value E_2. Considerable caution must therefore be exercised in measuring the E_h of microaerobic systems. Care must be taken to use growth media containing suitable redox mediators so that exchange current density is high.

Instrumentation. Redox robes are available as combination probes; that is, the reference electrode is combined with the indicator electrode. Redox electrodes, like pH electrodes, have a high output

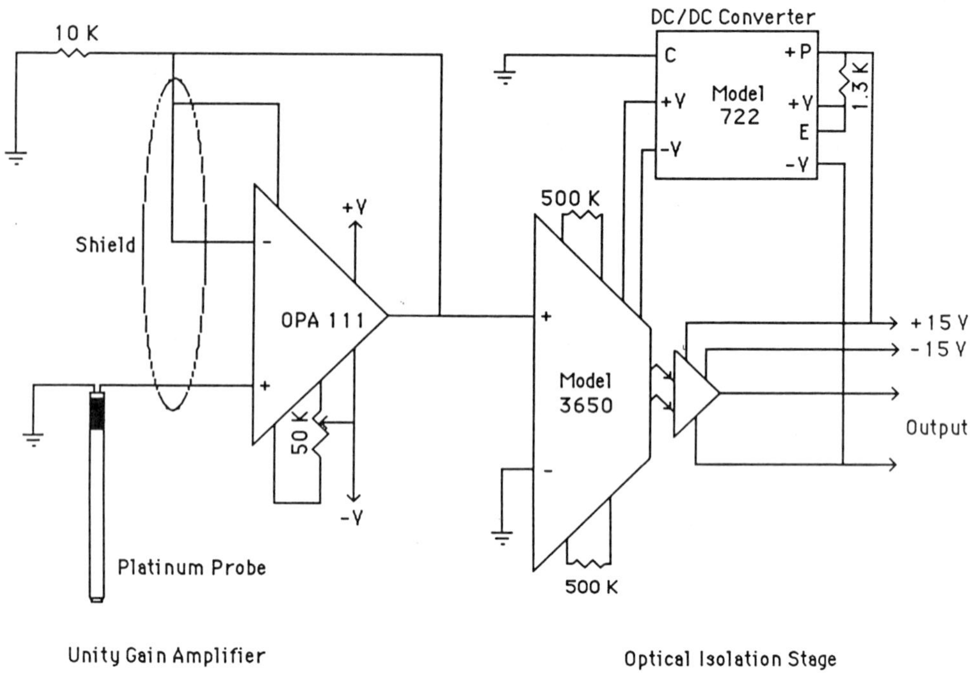

Figure 2 Optical isolation interface for pH and redox potential measurements. (From Rao, 1987.)

impedence (typically 50-100 M Ω). Commercially available pH meters are suitable for use with redox electrodes. In principle, it is possible to measure redox potential with only a platinum probe and use the reference electrode of a pH probe. The use of isolation amplifiers is strongly recommended when making simultaneous pH and E_h measurements. An isolation amplifier breaks the ohmic continuity between the input and output signals of the amplifier by providing coupling using thermal, magnetic, or optical means. It is our experience that this approach eliminates probe interactions and ground loop problems and provides stable signals. The circuitry employed in our laboratory with an optical isolation amplifier is detailed in Figure 2. This circuit is used between the probe and pH-millivolt meter. It can also be used to amplify the probe output, and the signal can be directly interfaced to a computer. It is important to note that many commercially available pH-millivolt meters do not feature isolation amplifiers.

Calibration and maintenance. The platinum redox probe can be calibrated using the method suggested by Kjaergaard (1977).

This requires E_h measurement of two saturated quinhydrone solutions at two pH values. The solution should be prepared just prior to calibration by adding excess quinhydrone to two pH buffers, typically 3 and 9. At 25°C the potential registered is given by

$$E_h \text{ (in mV)} = 699 - 59.1 \text{ pH} \tag{19}$$

This method is reproducible and convenient. Electrodes should be stored with the tip dipped in the reference electrolyte. When the probes are to be used in strongly reducing environments, probe life may be shortened. In solutions with high sulfide concentrations, diaphragm contamination can lead to erroneous measurements. In such cases, the use of a double-junction probe is recommended.

B. Poising and Control of Redox Potential

The motive for poising the redox potential of the growth medium arises from the need to examine the effects of CRP on cell metabolism. The E_h of the medium can be poised or controlled in a fashion analogous to pH buffering and control. For E_h control, reducing and oxidizing agents are used in contrast to acid and base pH control. Commonly used reducing and oxidizing agents in fermentation systems, their preparation, and recommended usage are briefly reviewed in this section. E_h can also be controlled by air sparging or by electrochemical means. The former introduces ambiguities due to a direct contribution to probe response, oxygen toxicity to the culture, and measurement interferences. The latter avoids the use of oxygen or chemical agents.

1. Reducing and Oxidizing Agents

Titanium (III) citrate is a nontoxic redox poising reagent useful in cultivating strict anaerobes (Zehnder and Wuhrmann, 1976). It has a standard midpoint potential of -480 mV and rapidly reacts with oxygen in solution with a rate constant of $1.14 \pm 0.08 \times 10^{-4}$ s^{-1}. Owing to this high affinity for oxygen, titanium (III) citrate is useful for achieving totally anoxic conditions. At the same time, this affinity requires that its preparation be done in an oxygen-free atmosphere. Titanium (III) citrate is prepared by adding 5 mL of 15% titanium (III) chloride to 50 mL of 0.2 M sodium citrate, followed by neutralization with saturated sodium carbonate solution. It is advisable to thoroughly degas all solutions with oxygen-free nitrogen before use. The titanium (III) citrate solution is blue to violet in color in the reduced form and is decolorized upon oxidation. If microbial uptake of citrate is of concern, other suggested ligands are tartarate, oxalate, and rodanide (Zehnder and Wuhrmann, 1976).

Ferrous sulfide, besides being capable of rapidly removing oxygen, is sparingly soluble. This offers two important advantages. In the first case, unlike other reducing agents ferrous sulfide addition does not alter medium pH. Second, the amount of ferrous sulfide that can be added is not critical as long as it is in excess of the solubility limit. It is prepared by mixing equimolar boiling solutions of ferrous ammonium sulfate and sodium sulfide (Brock and O'Dea, 1977). The black precipitate of ferrous sulfide is recovered, washed four to five times with boiling distilled water, and transferred while hot to screw-capped bottles and sparged with oxygen-free nitrogen prior to capping. Any oxygen leak will lead to the formation of a reddish product. The ferrous sulfide preparation is stable for several weeks.

Ascorbic acid, a relatively mild reducing agent, has been used for cultivating strict anaerobes by Kligler and Guggenheim (1937). *Clostridium welchii*, a strict anaerobe, was found to grow in the presence of air only when a minimum concentration of ascorbic acid (0.02%) was present in the medium. The addition of 0.4% agar to the medium caused a decrease in the required ascorbic acid from 0.02 to 0.003% to support growth in presence of air.

Sodium thioglycolate is frequently used as a poising agent in growth media. At 0.1% it gives a well-poised system with an initial E_h between -175 and -200 mV (Reed and Orr, 1942). Concentrations lower than this give a more positive initial E_h value and a less effective poising of the medium. This was demonstrated by bubbling 1 mL of air rapidly through 20 mL of peptone medium poised with 0.1% thioglycolate. As seen from Figure 3, the E_h rises briefly but rapidly returns to its preaeration value. In contrast, the response of a medium containing 0.01% sodium thioglycolate to 1 mL air shows a rapid increase in E_h initially followed by a gradual rise in E_h. The buffering capacity of the medium is improved by the addition of agar at 0.1% to the growth medium. The effect of agar on redox buffering is illustrated in Figure 4. The response of the E_h of the medium to continuous air exposure under three different conditions is shown. The results show that small amounts of agar along with a reducing agent significantly increase the redox buffering capacity.

Sodium hydrosulfite can lower the E_h of a typical growth medium to -420 mV. Belkin and Padan (1983) have incubated *Oscillatoria limnetica* cells in the presence of hydrosulfite at a concentration range from 0.05 to 10 mM. It should be noted that the oxidation products of hydrosulfite are strongly acidic. The use of freshly prepared hydrosulfite at a typical concentration of 0.5 M is recommended. Small volumes of this solution may be added to the culture medium to obtain the desired change in E_h.

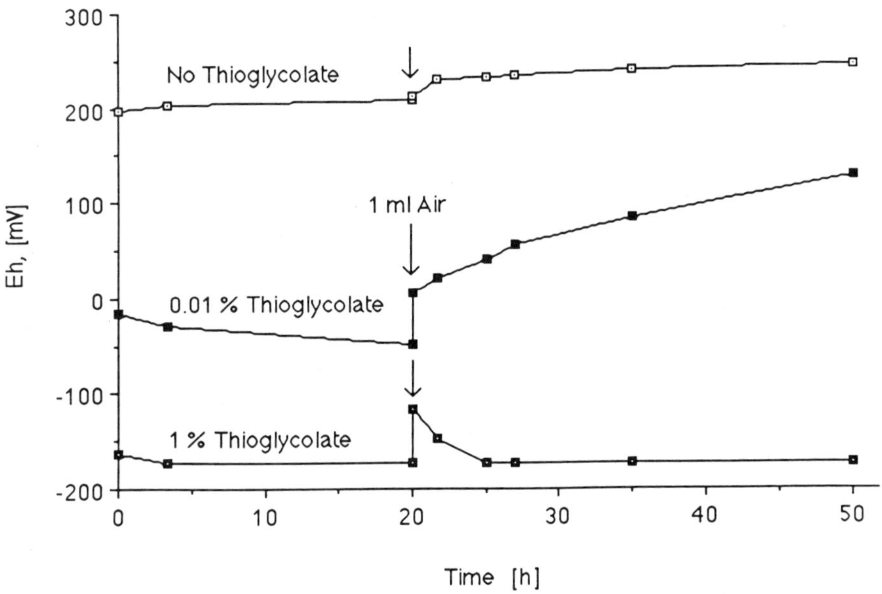

Figure 3 Effect of sodium thioglycolate concentration on redox poising of the culture medium. Arrow indicates the time of addition of 1 mL of air. (From Reed and Orr, 1942.)

Other commonly used reducing agents include cysteine, sodium sulfide, sodium formaldehyde sulfoxylate, and dithiothreitol. The compatability of a reducing agent in a cultural system of interest should first be evaluated by carrying out several small-scale (20 mL) batch experiments at various concentrations to ascertain toxic limits and to examine any undesirable metabolic effects.

Partially purified membrane fractions of *Escherichia coli* have been used as a novel "reducing agent" to remove oxygen from the culture broth enzymatically (Adler et al., 1983). Oxygen removal was found to be rapid and complete. The membrane fractions can be freeze-dried and stored over an extended period of time. The use of *E. coli* membrane fractions can simplify the handling of strict anaerobes.

Very few oxidizing agents have been used to poise culture media. The two commonly used agents are potassium ferricyanide and oxygen. Ferricyanide can increase the E_h up to +500 mV and has been used to distinguish between E_h and oxygen effects in anaerobic metabolism (O'Brien and Morris, 1971; Onderdonk et al.,

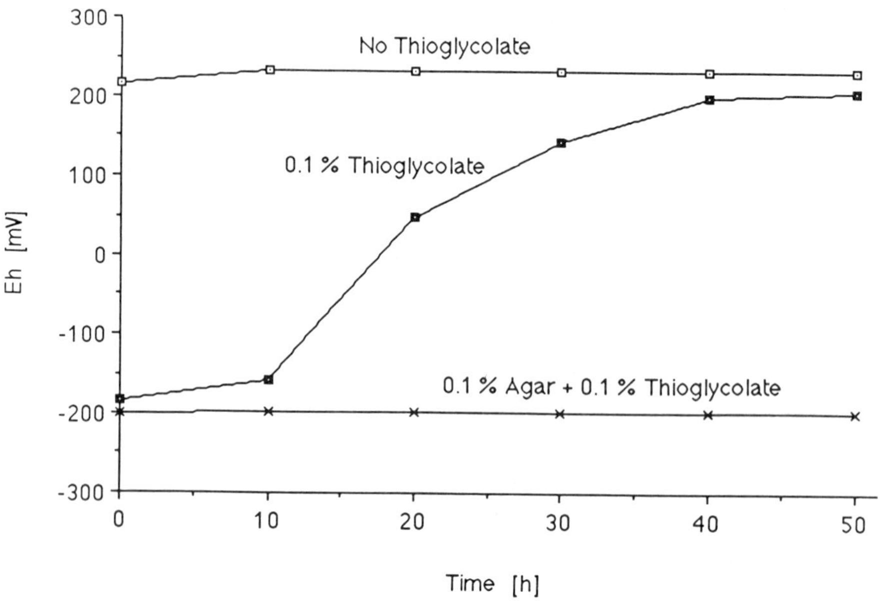

Figure 4 Effect of agar on redox poising of culture medium. (From Reed and Orr, 1942.)

1976). Positive E_h limits on ruminal bacterial growth have been determined using ferricyanide (Marounek and Wallace, 1984). The addition of 0.1-0.2 M potassium ferricyanide in small amounts (10-20 mL/L) is recommended for poising or controlling E_h.

2. Oxygen-Air Sparging

Dilute streams of oxygen in nitrogen have been employed by a number of investigators for controlling culture redox potential at a desired value. As mentioned earlier, the CRP value in the presence of oxygen is a composite signal consisting of both dissolved oxygen concentration and the culture metabolic activity. The former is the dominant component of the measured potential, and therefore the contribution due to the metabolic activity of the culture cannot easily be distinguished. The presence of oxygen also decreases the exchange current density, resulting in measurement errors. Thus the control of redox potential using oxygen is essentially equivalent to control of the dissolved oxygen level. Unfortunately, the use of oxygen for E_h control has the drawback of oxygen toxicity in strict anaerobes, and the analysis of CRP response should be viewed with caution.

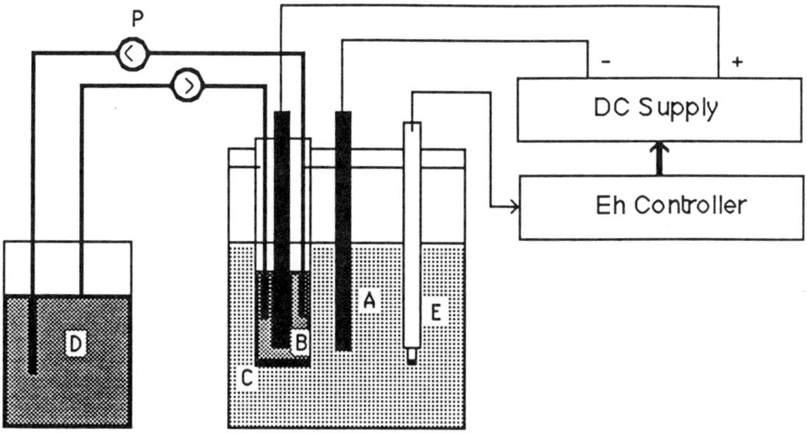

Figure 5 Electrolytic control of cultural redox potential. Abbreviations: (A) platinum electrode (cathode); (B) recirculating electrolyte electrode (anode); (C) salt bridge; (D) anode electrolyte reservoir; (E) redox potential electrode; (P) pumps for electrolyte flows to and from recirculating electrolyte electrode. (From Thompson and Gerson, 1985.)

3. *Electrolytic Control of Redox Potential*

Controlled direct current electrolysis can be used to adjust E_h (Hanke and Katz, 1943; Thompson and Gerson, 1985). A schematic of the apparatus making use of this technique is shown in Figure 5. The platinum electrode (labeled A), which is in contact with the culture medium, is connected to the negative pole of a dc power supply. The positive pole of the supply is connected to a platinum electrode (labeled B) and is immersed in a recirculating electrolyte. The platinum electrodes A and B are for E_h control and should not be confused with electrode E, which is meant for E_h measurement. When a suitable potential is applied across A and B, electrode A (in contact with the medium) reduces the broth components by electrolysis and this results in a decreased culture redox potential. Thus the E_h of the broth can be controlled at a value more negative than is obtainable without applying the external potential. If E_h control at a more positive value is desired, the electrode connections to the dc power supply should be reversed. The main advantage afforded by the electrolytic control technique is the total avoidance of oxidizing or reducing agents.

The electrolytic technique appears to offer the only means of achieving two apparently contradictory goals: lowering E_h in the continuous presence of oxygen. Hanke and Katz (1943) demonstrated the feasibility of this approach by controlling E_h at 130-150 mV in the presence of oxygen. In a more recent study, Thompson and Gerson (1985) used the same technique for aerobically cultivating *E. coli* at 380 mV, a potential 200 mV lower than that obtained under air saturation without E_h control. They found that the addition of 0.25 mg/L neutral red as a mediator was necessary to control E_h.

IV. METABOLIC SIGNIFICANCE OF REDOX POTENTIAL MEASUREMENTS

The redox potentials of whole-cell cultures have been measured in a variety of biologic situations. In spite of the ease of measurement of CRP, it is not widely reported in the context of microbial metabolism. The available literature on redox potential effects may be classified into the following four major categories: (1) lag phase and initiation of growth in strict anaerobes, (2) oxygen toxicity in strict anaerobes, (3) microaerobic metabolism, and (4) metabolic activity in terms of growth rate and end-product distribution. Prior to the discussion of these topics, let us consider the problems inherent in interpreting CRP from a metabolic perspective.

A. Interpretation of Culture Redox Potential

A fundamental question that arises when interpreting the metabolic significance of CRP is its relation to the intracellular redox potential and the mechanism of electron exchange with the platinum probe. Intracellular redox potential is the result of various intracellular reactions. These reactions are spatially segregated by semipermeable membranes, and some of the redox couples may be bound to specific ligands. At present, it is not clear how these intracellular redox reactions interact with and influence the observed CRP. Clearly, for such an interaction to occur, a redox mediator is essential. If indeed such a mediator couple exists in the broth, the question then arises of which intracellular redox couple it is in equilibrium with. A possible solution to this problem is to use an exogenous redox mediator in measuring the redox potential of a targeted intracellular redox couple.

The presence of oxygen strongly interferes with the measurement of CRP response due to metabolic activities. This interference cannot be corrected for because low DO tension cannot be measured reliably with available membrane electrodes. Besides oxygen, dissolved hydrogen can also bring about changes in the probe response and thus interfere in interpreting CRP changes. Changes in bacterial metabolism alter broth hydrogen concentration, resulting in changes in measured CRP. Experimental evidence of the effect of dissolved gases on CRP has been obtained by using gas-permeable membranes covering the platinum surface (Balakireva et al., 1974). This method eliminates problems due to direct contact with the broth and helps in isolating the response due to changes in the dissolved gas concentration from those due to metabolic effects.

Yet another difficulty arises when interpreting CRP results when redox mediators are used. Redox mediators are often added to culture media to improve exchange current densities. A few redox mediators (for example, methyl and benzyl viologen) significantly alter catabolic pathways, thus rendering the interpretation of CRP effects even more difficult (Rao and Mutharasan, 1986, 1987).

B. Microaerobic Growth

In spite of the known interference of dissolved oxygen in the measurement and interpretation of redox potential, a few investigators have been successful in eliciting and isolating CRP response due to metabolic shifts. For example, Shibai et al. (1974) suggested that in the presence of oxygen, CRP is not only proportional to oxygen but also dependent on "cellular redox components." Elegant experiments involving a deliberate change in the metabolic state of the cells using dinitrophenol, an uncoupler, have confirmed their hypothesis.

The dependency of redox potential on dissolved oxygen has been used by several investigators for controlling DO, particularly at very low levels. In this section we consider the advantages of CRP measurement as an oxygen sensor, in light of the metabolic significance of the microaerobic cultivation of facultative and aerobic bacteria.

1. Use of Redox Potential in DO Sensing

Oxygen, an avid electron acceptor, plays a significant role in microbial metabolism. Oxygen is toxic to obligate anaerobes even in trace quantities. In facultative and aerobic bacteria, fundamental changes in metabolism occur at various dissolved oxygen levels. For example, respiration rate is a function of DO tension. It is well known that expression and activity of various respiratory enzymes are strongly dependent on oxygen availability. Adaptive changes in carbon flow patterns in response to DO tension in *Klebsiella aerogenes* have been observed (Harrison, 1972). Oxygen serves at least three roles in metabolism: as an electron acceptor and as a nutrient and enzyme regulator. These metabolic characteristics are particularly discernible at such low oxygen tension as 1 mm Hg. For example, the apparent Michaelis constant for oxygen for a major-

Table 4 Relationship Between CRP and DO Tension

Condition of oxygen supply[a]			
Agitation speed (RPM)	Air flow rate (mL/min)	DO[b] (atm)	CRP (mV)
1500	300	0.163	-50
1000	150	0.067	-68
950	150	0.031	-80
900	150	0.007	-105
870	150	0	-157
850	150	0	-160
800	150	0	-175
700	150	0	-200
600	150	0	-230

[a]Inosine fermentation culture during stationary phase.
[b]As measured by a Teflon-coated oxygen electrode.
Source: After Shibai et al. (1974).

ity of cytochrome oxidase enzymes is between 10^{-7} and 10^{-8} M, well below the sensitivity limit of commercially available membrane oxygen probes. For investigating the metabolic response of various aerobic and facultative organisms at these low levels, the redox potential offers a possible solution (Squires and Hosler, 1958). For example, Shibai et al. (1974) showed that at DO levels below 0.007 atm, the CRP is a function of oxygen concentration. Typical results, given in Table 4, show large changes in redox potential when the oxygen probe is essentially insensitive. Shibai et al. (1974) used this technique in cultivating *Bacillus subtilis* at controlled CRP levels. Lactic acid was found to be the main product when cultivated at -220 mV, and 2,3-butanediol and acetoin were the products at CRP values of -195 and -160 mV, respectively. In batch fermentation of xylose, *Pacholsen tannophilus* showed a maximum ethanol yield at a potential of 180 mV (Chung and Lee, 1986).

Amino acid fermentations are normally carried out under oxygen limitation. In these fermentations, too low an oxygen concentration causes anaerobiosis; too high a value (greater than 0.02 atm) causes complete oxidation of the substrate, resulting in low amino acid yield. For example, Akashi et al. (1978) found that L-leucine production was a function of redox potential. At a CRP of -100 mV the concentration of L-leucine was 26.5 mg/L. It increased to 29.5 mg/L as the redox potential was decreased to -200 mV by decreasing the oxygen supply. As the redox potential was further lowered to

-280 mV, L-leucine concentration dropped to 12.5 mg/L. In another study, *Corynebacterium glutamicum* was cultivated at varying degrees of dissolved oxygen tension to examine its effect on relative yields of alanine, homoserine, lysine, and valine (Radjai et al., 1984). DO tension was varied using redox potential as the measured variable, with agitation rate manipulated based on a self-tuning feedback control algorithm. The minimum redox potential of the culture became less negative as the agitation rate was increased from 400, 500, and 580 RPM and is shown in Figure 6. The increase in the observed minimum CRP is due to an increase in oxygen transfer rate and a consequent increase in DO concentration. The various amino acids show different optimum oxygen tensions. The optimum for lysine, homoserine, and valine is in the range -220 to -260 mV, which corresponds to DO concentrations well below the sensitivity range of a DO probe.

C. Significance in Anaerobic Metabolism

Earlier studies of CRP measurement and control were motivated by the need to monitor oxygen in growth media used for cultivating anaerobes. This led to studies on anaerobic growth and metabolism under various reducing conditions. The degree of reducing environment was monitored by culture redox potential (O'Brien and Morris, 1971; Morris, 1974). As pointed out earlier, CRP is a more accurate measure of reducing conditions in anoxic environments due to the total absence of oxygen. Consequently, the CRP response of anaerobic cultures can perhaps be more meaningfully interpreted.

1. Redox Potential Variation During Batch Growth

When an oxygen-free medium is inoculated with a strict anaerobe, the culture redox potential is usually between +50 and -150 mV, depending upon the inoculum size and medium composition. The culture redox potential decreases rapidly during emergence from the lag phase. During this phase, illustrated in Figure 7, the amount of new cells synthesized is very small and is not easily detectable by optical density changes. The redox potential of the medium reaches a characteristically low value (-200 to -450 mV, depending on the culture) during the late exponential phase. As the cells enter the stationary phase, the culture redox potential increases, suggesting that active metabolism is responsible for the maintenance of low CRP during growth. Similar correlations have been noted by many investigators in other anaerobic facultative, and aerobic cultures (Jacob, 1970, 1974; Brosseau and Zajic, 1982). Hence, CRP can be used to determine the initiation of growth, which is useful, for example, in the start-up of open loop control algorithms for

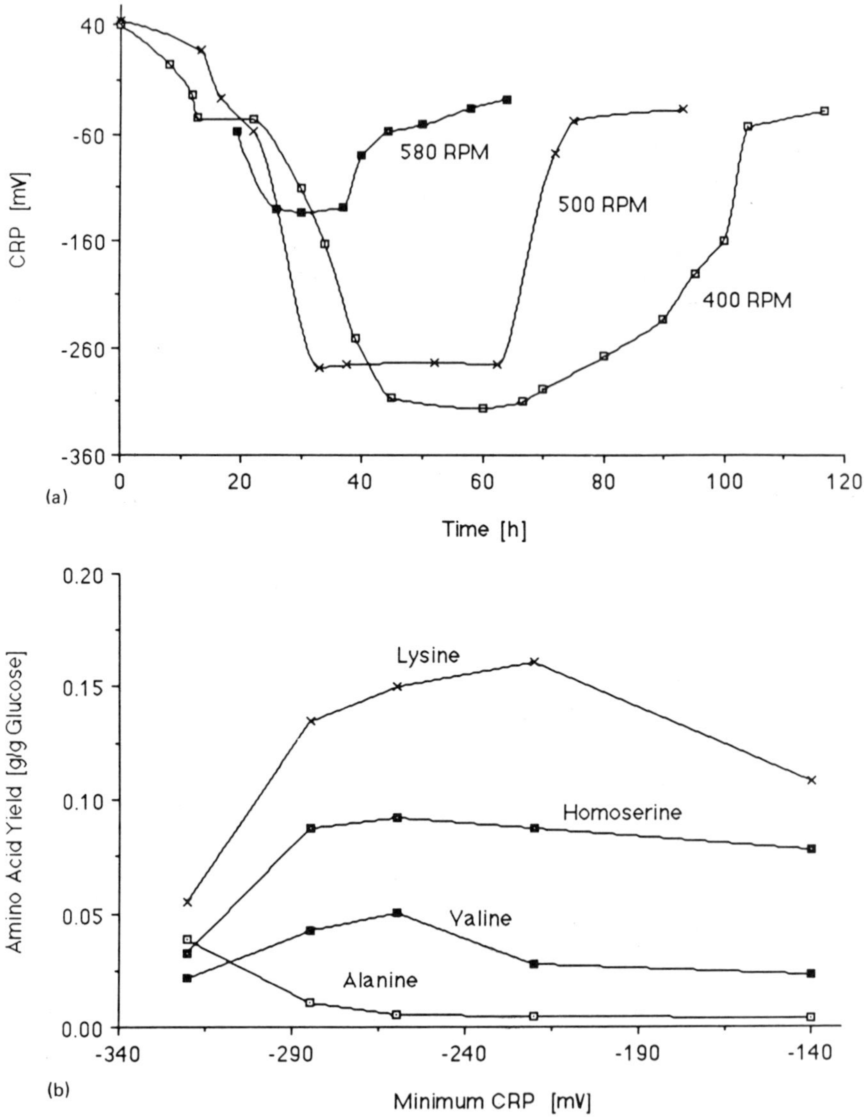

Figure 6 (a) CRP profile during batch fermentation with *Corynebacterium glutamicum* at various agitation rates but constant aeration rate. (b) Variation of product yield with CRP. (From Radjai et al., 1984.)

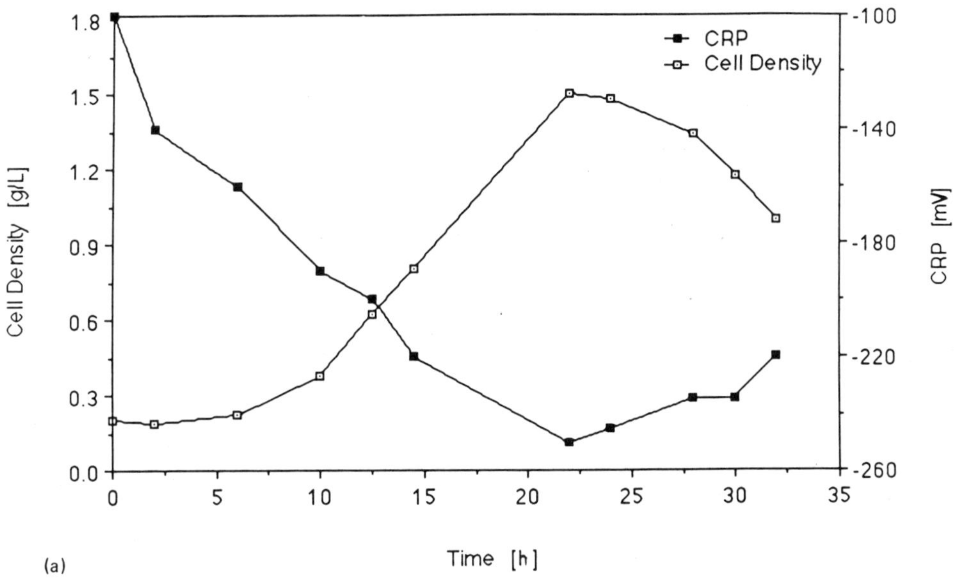

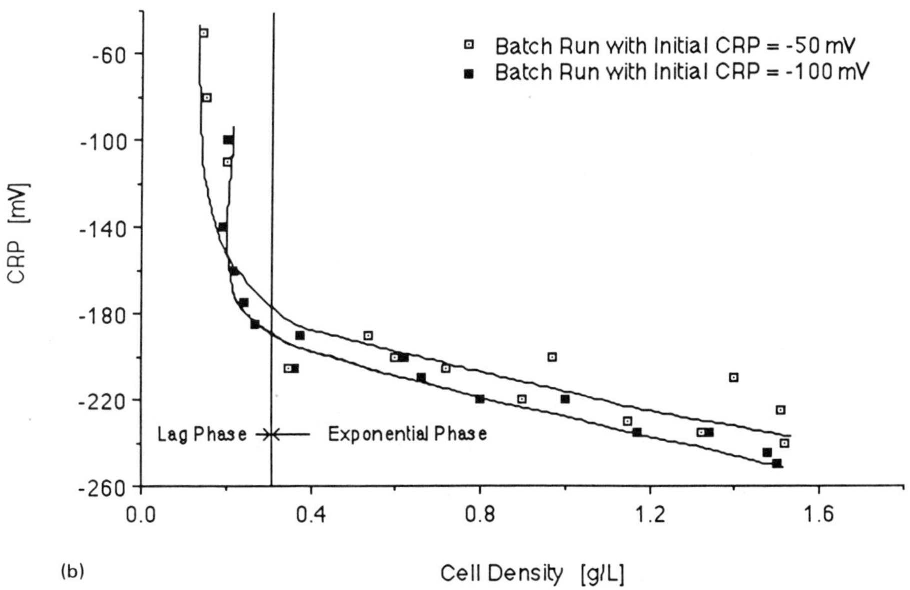

Figure 7 (a) Variation of culture redox potential during batch growth of *Thermoanaerobacter ethanolicus* at a controlled pH of 6.0. CRP was initially poised at -100 mV with a equimolar mixture of sodium sulfide and cysteine. (From Kannan and Mutharasan, 1983.) (b) Culture redox potential as a function of cell concentration. (From Kannan and Mutharasan, 1983.)

bioreactor optimization. Another useful application is that a steep decrease in CRP prior to inoculation can serve as an early warning of medium contamination.

A number of explanations have been proposed for the observed decrease in CRP during early growth phase. Jacob (1974) attributes the sharp decrease to exhaustion of trace levels of oxygen from the medium. This seems unlikely because similar profiles have been obtained even under conditions in which oxygen has been removed by adding reducing agents or special care has been taken to use oxygen-free sparging gas (Ward and Mutharasan, 1985; Rao and Mutharasan, 1985). Another explanation is that hydrogen evolved by bacteria emerging from the lag phase causes the precipitous decrease. Yet another explanation is that anaerobes produce reductants for reducing the growth environment (Vennesland and Hanke, 1940). In order to test this hypothesis, Kantere (1972) measured the reductants produced by titrating a cell-free medium of *Candida utilis* against 0.001 N ferricyanide at various stages of growth. Their results, summarized in Table 5, show that reductant concentration increased by almost an order of magnitude in 48 h. The redox potential of the medium, indicated by the inflection point of the titrations, decreased from 165 to 70 mV. These results confirm the hypothesis that the cells product reductants during growth. Unfortunately, systematic studies to identify these reductants have not been undertaken.

The changes in CRP during exponential growth phase have been empirically correlated to changes in cell density, and an illustrative case is presented in Figure 8 (Kannan and Mutharasan, 1983). The three curves in Figure 8 show sensitivity to the initial poising of the medium (-50, -100, and -150 mV) brought about by the addition of sodium sulfide and cysteine. Increased redox poising of the medium increases the slope of the correlation. Hence, CRP may be useful in monitoring cell concentration.

Table 5 Reductant Concentration in Cell-Free Medium for *Candida utilis* Cultures Determined by Titrating against 0.001 N Potassium Ferricyanide

Culture condition	Inflexion point (mV)	Concentration of reductant (M)
Culture medium	165	6.5×10^{-4}
24 h culture	130	20.0×10^{-4}
48 h culture	70	66.0×10^{-4}

Source: After Kantere (1972).

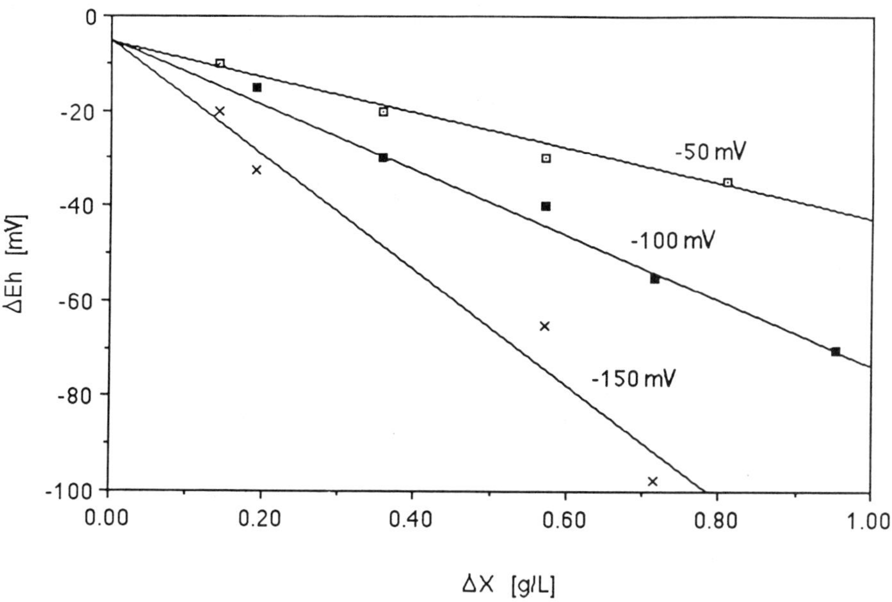

Figure 8 Correlation of change in CRP to change in cell density for *Thermoanaerobacter ethanolicus* at pH 6.5. (From Kannan and Mutharasan, 1983.)

2. Correlation with Lag Phase

Since strict anaerobes require a reducing environment, it is believed that their lag phase may be decreased by lowering the CRP. Three approaches have been used to examine the validity of this hypothesis. Increasing or decreasing of CRP can be brought about by varying the inoculum size or by changing the reducing agent concentration or oxygen sparging rate.

The addition of such reducing agents as cysteine, thioglycolate, or hydrosulfite to the medium usually decreases lag phase and facilitates growth, particularly when small inocula are used. If a large inoculum is used, the culture redox potential can be decreased to values similar to those obtained by using reducing agents (Kannan and Mutharasan, 1985). The upper limit of CRP value for growth is a function of various factors, such as the medium, the size of the inoculum, and the reagents responsible for the redox potential (Marounek and Wallace, 1984). *Clostridium thermoaceticum* requires a CRP lower than -300 and -360 mV at pH 6 and 7, respectively, for growth to occur (Schwartz and Keller, 1982). Growth of *Thermo-*

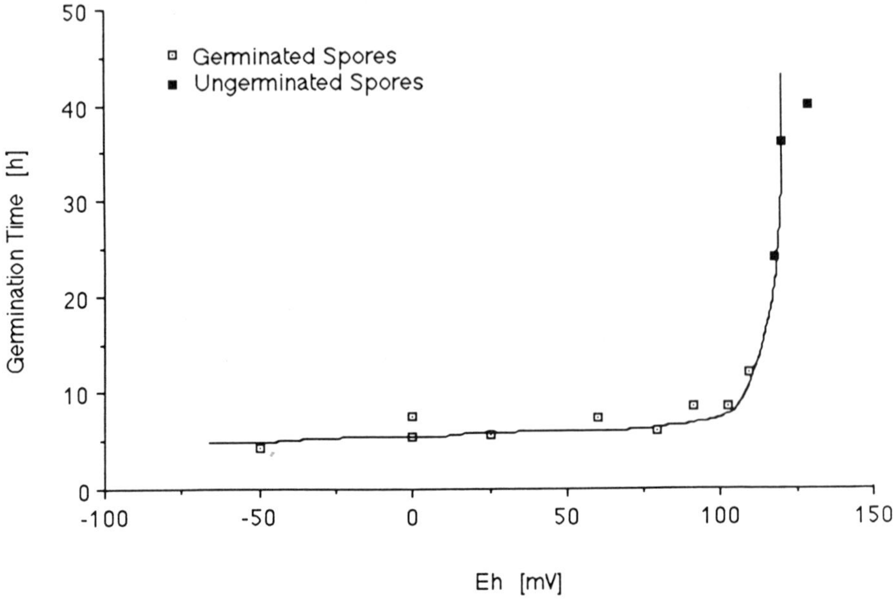

Figure 9 Effect of CRP on spore germination of *Clostridium tetanii*. CRP was adjusted using a mixture of oxygen in nitrogen. (From Knight and Fildes, 1930.)

anaerobacter ethanolicus at pH 7.5 is possible only when the medium is poised at -100 mV or lower with sodium sulfide and cysteine (Kannan and Mutharasan, 1985). The use of thioglycolate for reducing the inoculum size was investigated by Reed and Orr (1942) for 15 clostridial species. Their study showed that the inoculum required to initiate growth was a function of CRP. Lowering the CRP of the medium using sodium thioglycolate reduces the required inoculum size to initiate growth.

An upper positive limit of CRP for spore germination was reported by Knight and Fildes (1930) for *Clostridium tetanii*. The CRP was adjusted by sparging the medium with a dilute mixture of oxygen in nitrogen (Knight, 1930a, b). Their results, given in Figure 9, show that *C. tetanii* does not germinate at CRP greater than +110 mV at pH 7. On the other hand, the culture always germinated when CRP was below +110 mV even when oxygen was present. Since oxygen is toxic to *C. tetanii*, the maximum CRP for germination gives a measure of oxygen tolerance. Redox potential measurements for examining oxygen toxicity has been employed by several other investigators. For example, Knaysi and Dutky (1936) found a butyric

Clostridium to grow at +335 mV maintained by ferricyanide. However, at the same redox value obtained by oxygen, the culture did not grow. This suggested that growth inhibition was due to oxygen rather than the high CRP. In a similar study, O'Brien and Morris (1971) found that a high redox value maintained by ferricyanide provided a condition conducive for the growth of *Clostridium acetobutylicum* but the use of oxygen resulted in cell death.

Interestingly, there is another aspect to oxygen toxicity and the CRP. Hanke and Katz (1943) showed that in the presence of oxygen, growth of anaerobes is possible provided the CRP is significantly lowered. Kligler and Guggenheim (1937) showed that growth of the strict anaerobe *C. welchii* in the presence of air was made possible by adding ascorbic acid (a mild reducing agent and nonreactive to oxygen) to the culture broth at concentrations greater than 50 ppm. Under these conditions the CRP was found to be -125 mV. Successful cultivation of *Clostridium sporogenes* and *Bacteroides vulgatus* in air-sparged reactors was possible when the CRP was lowered electrolytically to 150 mV or less (Hanke and Katz, 1943). Thus it is seen that lowering the CRP usually facilitates growth and shortens the lag phase in strict anaerobes.

3. *Metabolic Effects of Reducing-Oxidizing Agents*

Although reducing agents are used for oxygen removal, recent studies indicate that some reducing agents can alter growth rate and electron flow pathways. The mode of action of the reducing agents and redox mediators on the metabolic reactions is not fully understood. One often attributes the observed metabolic changes to the apparent lack of specificity of some biologic electron transfer reactions. It is therefore possible that an exogenous reducing agent capable of donating electrons can competitively reduce a metabolic intermediate and thus modulate intracellular redox reactions. This could manifest itself macroscopically as changes in both cell growth rate and metabolic end-product distribution. Measurement of CRP, in this context, is an indicator of the effective reducing power of the culture broth. The concentration of the reducing agent employed is also of significance as it directly affects the poising capability of the medium. In this section, the effects on growth rate and end-product distribution associated with the use of reducing agents, as well as the redox mediators benzyl and methyl viologen, are discussed.

Titanium (III) citrate was found to increase the growth rate of three different anaerobes, *Clostridium formicoaceticum, Bifidobacterium bifidum,* and *Methanobacterium* strain AZ (Zehnder and Wuhrmann, 1976). A summary of these results, shown in Figure 10, indicates that as the reducing agent concentration is increased,

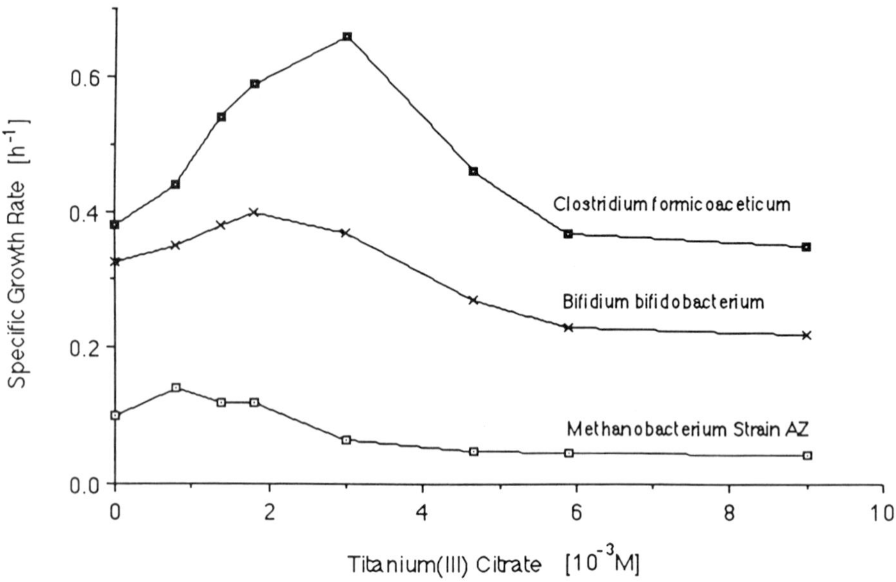

Figure 10 Effect of reductant titanium (III) citrate on specific growth rate of three strict anaerobes. (From Zehnder and Wuhrmann, 1976.)

growth rate increases. However, titanium (III) citrate is inhibitory at high concentrations. Titanium (IV) citrate, the oxidation product, was not found to have any inhibitory effect on the microorganisms studied. The authors also report that growth of the facultative anaerobes *E. coli* and *Pseudomonas denitrificans* was prevented in the presence of titanium (III) citrate. Since most contaminant microorganisms are facultative anaerobes, it may be useful to use titanium (III) citrate in anaerobic fermentations where contamination is of concern.

Similar studies on the effect of reducing agents on the growth of *Clostridium botulinum* spores in trypticase soy broth using several reducing agents were carried out by Smith and Pierson (1979). Optimal growth patterns were achieved at CRP values lower than those obtained using unpoised medium. Recently, Marounek and Wallace (1984) used titanium (III) citrate to examine the effects of CRP on the ruminal bacteria *Selenomonas ruminantium*, *Bacteroides amylophilus*, *Bacteroides succinogenes*, and *Streptococcus bovis*. Contrary to the results of Zehnder and Wuhrmann (1976) and Smith and Pierson (1979), no discernible increase in growth rate was found in these organisms. No significant change in yield or the ratio of end product was observed. It appears that the metabolic effects of titanium (III) citrate may be species specific.

In addition to affecting growth rate, reducing agents can also alter electron flow pathways. For example, Belkin and Padan (1983) found that when sodium hydrosulfite was added to a suspension of intact *O. limnetica*, the resulting low redox potential significantly improved photosynthetic sulfide-dependent hydrogen evolution. Since hydrosulfite was found to increase hydrogenase activity in cell extracts, the authors concluded that CRP exerts control over the electron flow pathway through hydrogenase and that the excess reducing power was released by hydrogen evolution.

The effect of initial CRP and poising capacity of the medium on the end products of *T. ethanolicus* were investigated by Kannan et al. (1984) and by Ward and Mutharasan (1985). The redox potential of the culture broth was poised by adding such reducing agents as sodium sulfide, sodium thioglycolate, ascorbic acid, cysteine, and methyl viologen at varying concentrations. In these studies, *T. ethanolicus* was grown in batch culture at an initial pH of 6.8 with no subsequent pH control. Their results, summarized in Table 6, show that as reducing agent concentration is increased, the initial E_h decreases and the preferred product of metabolism is ethanol. High reducing agent concentrations were growth inhibitory, a finding consistent with that of other investigators (Smith and Pierson, 1979). Since pH was found to alter the products of metabolism in *T. ethanolicus*, McCallion (1984) and Kannan et al. (1984) investigated the effects of poising CRP by sodium sulfide and cysteine under controlled pH conditions. A significant reduction in lactate formation was found at all pH values studied (Fig. 11) with decreasing initial CRP. The effects of different reducing agents on *T. ethanolicus* metabolism, shown in Table 6, clearly indicate that the reducing agents influence the end-product distribution by a mechanism other than as a nutrient. In spite of differences in chemical structure, ionization characteristics, and elemental composition of the reducing agent, it appears that a reducing environment favors a reduced end product at the expense of a relatively more oxidized one (in this case ethanol over lactate). This suggests that the degree of reducing environment, which is the common factor in all of these experiments, is the primary variable responsible for the altered metabolism.

In addition to pH, growth rate is also known to affect the end products of metabolism. In order to distinguish between growth rate and CRP effects, continuous culture experiments were performed under constant pH and growth rate conditions by Ward and Mutharasan (1985). As the steady-state CRP was lowered by increasing the sodium sulfide and cysteine concentrations in the feed, it was

Table 6 Effect of Various Reducing Agents on Product Distribution in Ethanol Fermentation by *Thermoanaerobacter ethanolicus*

Condition	Concentration of reducing agent (g/L)	CRP (mV)	Product yield (substrate; glucose) Ethanol (mol/mol)	Lactate (mol/mol)	Acetate (mol/mol)
Control	0	-100	0.524	0.939	0.163
Cysteine	0.025	-139	0.612	0.902	0.134
	0.050	-144	0.834	0.818	0.137
	0.100	-161	0.740	0.752	0.130
	0.250	-173	0.819	0.823	0.137
	0.500	-182	0.732	0.800	0.129
Ascorbic acid	0.025	-142	0.556	1.112	0.197
	0.050	-153	0.491	0.888	0.179
	0.100	-101	0.535	0.970	0.199
	0.250	-101	0.552	0.955	0.158
	0.500	-103	0.477	0.897	0.165
Sodium sulfide	0.025	-97	0.736	0.940	0.152
	0.050	-99	0.768	0.858	0.145
	0.100	-196	0.829	0.752	0.141
	0.250	-209	0.865	0.756	0.123
	0.500	-255	0.947	0.774	0.117
Sodium thio-glycolate	0.025	-100	0.721	0.800	0.160
	0.050	-120	0.647	0.757	0.134
	0.100	-147	0.814	0.783	0.149
	0.250	-153	0.783	0.787	0.167
	0.500	-155	0.610	0.724	0.186
Methyl viologen	0.010	-177	0.625	0.954	0.153
	0.025	-183	0.774	0.989	0.225
	0.050	-106	0.472	0.829	0.216
	0.100	-111	0.522	0.878	0.209

Source: After Ward and Mutharasan (1985).

found that the culture produced ethanol rather than lactate at constant glucose uptake conditions (Fig. 12).

In our laboratory, we have used the viologen dyes, methyl and benzyl viologen, as redox mediators to examine the relationship between end-product distribution and CRP. As discussed earlier, redox mediators improve the accuracy of CRP measurements. Most reducing agents are irreversibly oxidized, but redox mediators

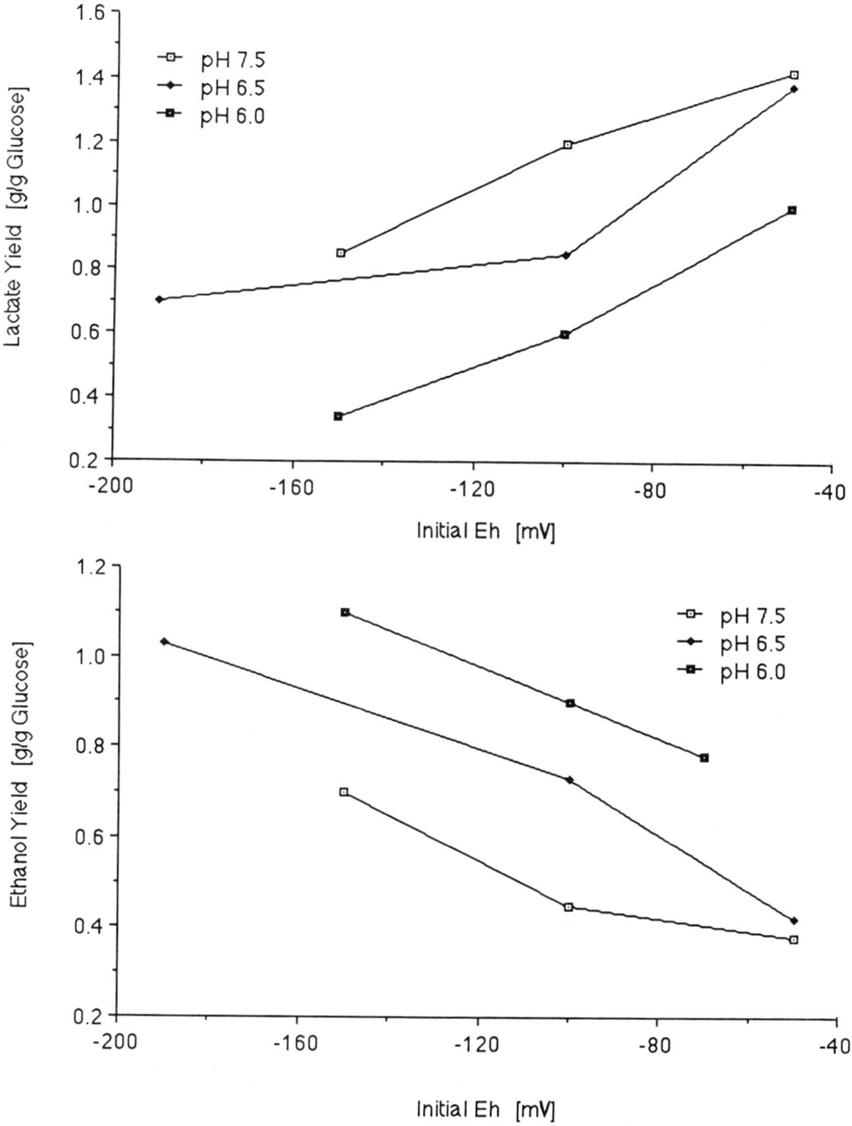

Figure 11 Effect of initial CRP on product yield at various pH values. (From Kannan et al., 1984.)

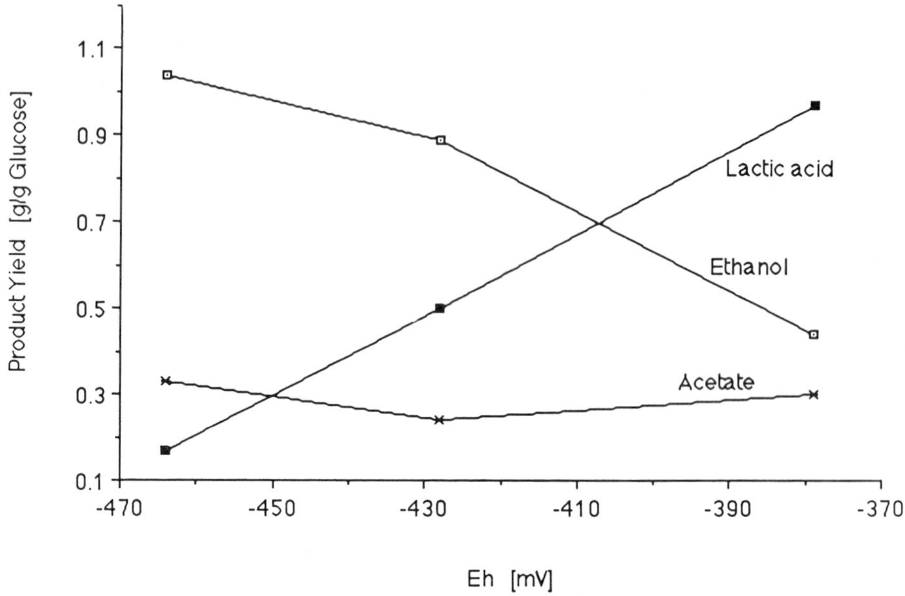

Figure 12 Steady-state product yield versus CRP at pH 6.5 for *Thermoanaerobacter ethanolicus*. (From Ward and Mutharasan, 1985.)

undergo reversible redox reactions. They behave as electron donors or acceptors depending on the ambient redox potential. Continuous culture experiments with *C. acetobutylicum* as a model system were carried out. To a glucose-limited steady-state culture a pulse of the viologen dye was added at 6 mg/L at different dilution rates and pH values. At CRP -360 mV and pH 6.4, a pulse of methyl viologen causes a sharp decrease in hydrogen, acetate, and butyrate yields with a concomitant increase in ethanol and butanol (Fig. 13a). On the other hand, at CRP -260 mV and pH 5, opposite effects (Fig. 13b), that is, sharp increases in hydrogen, butyrate, and acetate yields, are observed. The CRP was observed to fluctuate ±10 mV in both cases when the methyl viologen was added to the fermenter. Benzyl viologen, on the other hand, induces identical metabolic responses in end-product distribution under these two different CRP conditions (Fig. 14). However, the CRP response is different. At CRP -260 mV, the introduction of benzyl viologen was accompanied by a drop in CRP to -300 mV followed by a return to the pretransient value of -260 mV. At CRP -360 mV, the CRP decreased to -380 mV. The pH invariant midpoint potential of methyl viologen is -440 mV; that of benzyl viologen is -360 mV (Prince et al., 1980). At the CRP values in these experiments (namely, -360

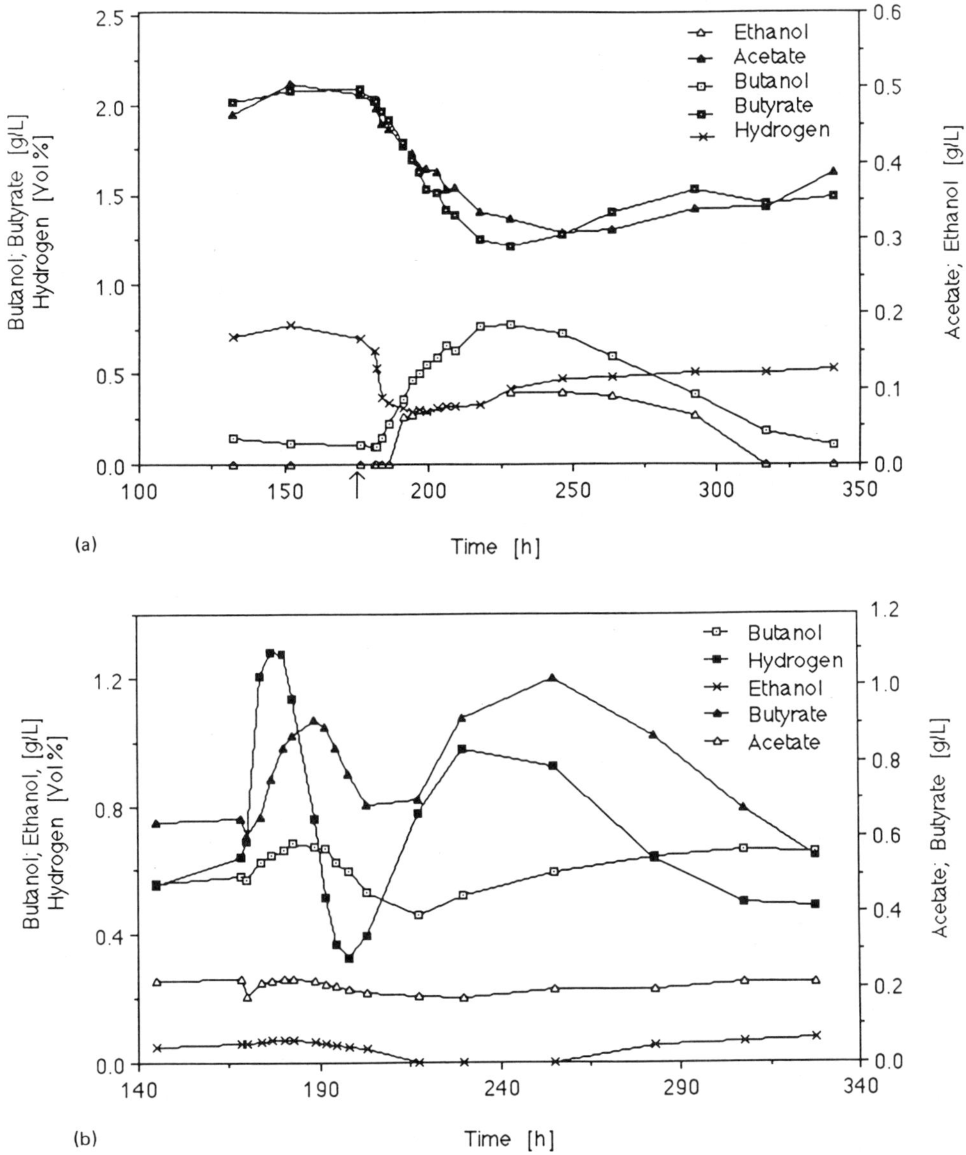

Figure 13 (a) Transient response of steady-state continuous culture of *Clostridium acetobutylicum* to a pulse of methyl viologen (6 mg/L). pH = 6.4; CRP = -360 ± 10 mV; D = 0.023 h^{-1}. Arrow denotes the time of methyl viologen addition. (From Rao, 1987.) (b) Transient response of steady-state continuous culture of *Clostridium acetobutylicum* to a pulse of methyl viologen (6 mg/L) pH = 5.3; CRP = -260 ± 10 mV; D = 0.036 h^{-1}. Arrow denotes the time of methyl viologen addition. (From Rao, 1987.)

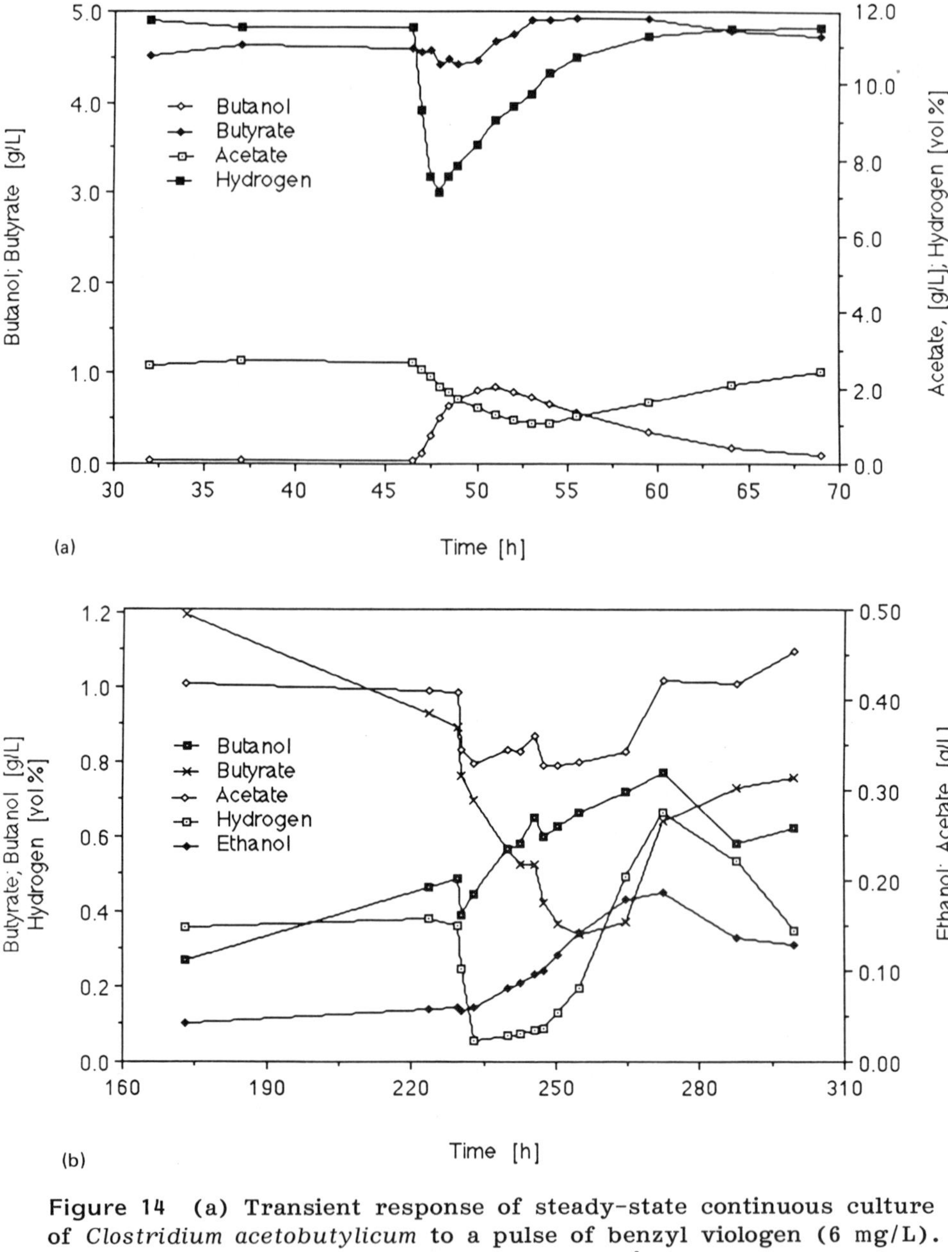

Figure 14 (a) Transient response of steady-state continuous culture of *Clostridium acetobutylicum* to a pulse of benzyl viologen (6 mg/L). pH = 6.3; CRP = -360 ± 10 mV; D = 0.228 h^{-1}. Arrow denotes the time of benzyl viologen addition. (From Rao, 1987.) (b) Transient response of steady-state continuous culture of *Clostridium acetobutylicum* to a pulse of benzyl viologen (6 mg/L). pH = 5.0; CRP = -260 ± 10 mV; D = 0.033 h^{-1}. Arrow denotes the time of benzyl viologen addition. (From Rao, 1987.)

and -260 mV), benzyl viologen is reduced but methyl viologen is reduced only at the lower of the two CRP values. Since oxidized viologens are colorless and their reduced counterparts are blue and violet (methyl and benzyl, respectively) the reduction state of the viologen dyes was visually confirmed (Rao, 1987).

Another important result from these experiments is that the CRP measured by the platinum probe differed by approximately 20-40 mV in the presence and absence of the mediators. Therefore, it is possible that the CRP measured by the probe is in fact close to a specific intracellular redox couple having a similar midpoint to that of the added viologen dyes. Further work using other redox mediators is needed to develop a better understanding of these phenomena.

V. CONCLUSIONS

We have discussed principles, methods, and applications of redox potential measurements in anaerobic and microaerobic cultures. Although the theoretical significance of the measured culture redox potential is unresolved, its practical significance is of great value. From an empirical standpoint, it is possible to use the redox potential measurements in quantifying the degree of anaerobiosis in microaerobic and anaerobic cultures. Dissolved oxygen levels can be empirically correlated using Nernst-type equations. The onset of growth or emergence from the lag phase can be detected by the precipitous drop in culture redox potential. The use of reducing agents, used to alter culture redox potential, can also bring about changes in the growth rate and metabolism of microorganisms. Similarly, the use of redox mediators for improving cultural redox potential measurements are responsible for the altered metabolic flow in *C. acetobutylicum*. These results are not generic and are dependent on several factors, such as the type of reducing agent or redox mediator, its concentration, the pH of the culture, and the species being studied. This is partly because the fundamental parameter that influences metabolic activity is the intracellular E_h. Means to measure this parameter in intact cells are now available, and use of them in future studies may enable us to better relate CRP to metabolic activity. In conclusion, the practical advantages and ease of implementation of culture redox potential measurements should stimulate further research in this field.

REFERENCES

Adler, H. I., Crow, W. D., Hadden, C. T., Hall, J., and Machanoff, R. (1983). New techniques for growing anaerobic bacteria:

Experiments with *Clostridium butyricum* and *Clostridium acetobutylicum*. *Biotechnol. Bioeng. Symp. 13*:153-161.
Akashi, K., Ikeda, S., Shibai, H., Kobayashi, K., and Hirose, Y. (1978). Determination of redox potential levels critical for cell respiration and suitable for L-leucine production. *Biotechnol. Bioeng. 20*:27-41.
Andersen, K. B., and Meyenburg, K. (1977). Charges of nicotinamide adenine nucleotides and adenylate energy charge as regulatory parameters of the metabolism in *Escherichia coli*. *J. Biol. Chem. 252*:4156-4156.
Balakireva, L. M., Kantere, V. M., and Robotnova, I. L. (1974). The redox potential in microbiological media. *Biotechnol. Bioeng. Symp. 4*:769-780.
Belkin, S., and Padan, E. (1983). Low redox potential promotes sulfide and light-dependent hydrogen evolution in *Osillatoria limnetica*. *J. Gen. Microbiol. 129*:3091-3098.
Brock, T. D., and O'Dea, K. (1977). Amorphous ferrous sulfide as a reducing agent for culture of anaerobes. *Appl. Environ. Microbiol. 33*:254-256.
Brosseau, J. D., and Zajic, J. E. (1982). Hydrogen-gas production with *Citrobacter intermedius* and *Clostridium pasteurianum*. *J. Chem. Technol. Biotechnol. 32*:496-502.
Bucher, T., Brauser, B., Conze, A., Klein, F., Langguth, O., and Sies, H. (1972). State of oxidation-reduction and state of binding in cytosolic NADH-system as disclosed by equilibration with extracellular lactate/pyruvate in hemoglobin-free perfused rat liver. *Eur. J. Biochem. 27*:301-317.
Buhler, H., and Galster, H. (1980). *Redox Measurement: Principles and Problems*. (Ingold Electrodes, Inc.),
Caswell, A. H. (1968). Potentiometric determination of interrelationships of energy conservation and ion gradients in mitochondria, *J. Biol. Chem. 243*:5827-5836.
Chance, B., Schoener, B., Oshino, R., Itshak, F., and Nakase, Y. (1979). Oxidation reduction ratio studies of mitochondria in freeze-trapped samples. *J. Biol. Chem. 254*:4764-4771.
Chung, I. S., and Lee, Y. Y. (1986). Effect of oxygen and redox potential on D-xylose fermentation by non-growing cells of *Pachysolen tannophilus*. *Enzyme Microb. Technol. 8*:503-507.
Clark, W. M. (1960). *Oxidation-Reduction Potentials of Organic Systems*. Waverley Press, Baltimore.
Daniels, W. F., Garcia, L. H., and Rosensteel, J. F. (1970a). The relationship of oxidation-reduction potential to the growth performance of tissue culture media poised prior to incubation. *Biotechnol. Bioeng. 12*:409-417.

Daniels, W. F., Garcia, L. H., and Rosensteel, J. F. (1970b). Oxidation-reduction potential and concomitant growth patterns of cultures of Earle's "L" cells in centrifuge bottle spinners. *Biotechnol. Bioeng. 12*:419-428.

Dutton, P. L. (1971). Oxidation-reduction potential dependence of the interaction of cytochromes, bacteriochlorophyll and carotenoids at 77°K in chromatophores of chromatium D and *Rhodopseudomonas gelatinosa. Biochim. Biophys. Acta 226*:63-80.

Dutton, P. L. (1978). Redox potentiometry: Determination of midpoint potentials of oxidation-reduction components of biological electron-transfer systems. *Methods Enzymol. 54*:411-435.

Dutton, P. L., and Wilson, D. F. (1974). Redox potentiometry in mitochondrial and photosynthetic bioenergetics. *Biochim. Biophys. Acta 346*:165-212.

Hanke, M. E., and Katz, Y. J. (1943). An electrolytic method for controlling oxidation-reduction potential and its application in the study of anaerobiosis. *Arch. Biochem. Biophys. 2*:183-200.

Harrison, D. E. F. (1972). Physiological effects of dissolved oxygen tension and redox potential on growing populations of microorganisms. *J. Appl. Chem. Biotechnol. 22*:417-440.

Hentges, D. J., and Maier, B. R. (1972). Theoretical basis for anaerobic methodology. *Am. J. Clin. Nutr. 25*:1299-1305.

Hewitt, L. F. (1936). *Oxidation-reduction potential in Bacteriology and Biochemistry*, 4th Ed. Livingstone, Edinburgh.

Ishizaki, A., Shibai, H., and Hirose, Y. (1974). Basic aspects of electrode potential change in submerged fermentation. *Agric. Biol. Chem. 36*:2399-2406.

Jacob, H. E. (1970). Redox potential. *Adv. Microbiol.*, 91-123.

Jacob, H. E. (1974). Reasons for the redox potential in microbial cultures. *Biotechnol. Bioeng. Symp. 4*:781-788.

Kannan, V., and Mutharasan, R. (1983). Role of redox potential on growth and product formation in *Thermoanaerobacter ethanolicus*. ACS Annual Meeting, Washington, D.C., September.

Kannan, V., and Mutharasan, R. (1985). Ethanol fermentation characteristics of *Thermoanaerobacter ethanolicus. Enzyme Microbiol. Technol. 7*:87-91.

Kannan, V., McCallion, E. J., and Mutharasan, R. (1984). Role of pH and redox potential in growth and product formation kinetics of *Thermoanaerobacter ethanolicus*. ACS Annual Meeting, Philadelphia, August.

Kantere, V. M. (1972). Redox potential measurements in microbiological media and some applications. In *Experientia (Suppl.) 18*: 355-366.

Kjaergaard, L. (1977). The redox potential: Its use and control in biotechnology. *Adv. Biochem. Eng.* *7*:131-150.

Kligler, L. J., and Guggenheim, K. (1937). The influence of vitamin C on the growth of anaerobes in the presence of air, with special reference to relative significance of E_h and oxygen in the growth of anaerobes. *J. Bacteriol.* *35*:141-156.

Knaysi, G., and Dutky, S. R. (1936). The growth of butanol *Clostridium* in relation to oxidation-reduction potential and oxygen content of the medium. *J. Bacteriol.* *31*:137-149.

Knight, B. C. J. G. (1930a). Oxidation reduction studies in relation to bacterial growth. I. Oxidation-reduction potential of sterile meat broth. *Biochem. J.* *24*:1066-1074.

Knight, B. C. J. G. (1930b). Oxidation reduction studies in relation to bacterial growth. II. A method of poising the oxidation-reduction potential in culture media. *Biochem. J.* *24*:1075-1078.

Knight, B. C. J. G., and Fildes, P. (1930). Oxidation reduction studies in relation to bacterial growth. III. The positive limit of oxidation-reduction potential required for the germination of spores. *Biochem. J.* *24*:1496-1502.

Marounek, M., and Wallace, R. J. (1984). Influence of culture Eh on the growth and metabolism of the rumen bacteria *Selenomonas ruminantium, Bacteriodes amylophilus, Bacteroides succinogenes* and *Streptococcus bovis* in batch culture. *J. Gen. Microbiol.* *130*: 223-229.

McCallion, E. J. (1984). Effect of pH on growth and product formation kinetics of *Thermoanaerobacter ethanolicus*. Master's thesis, Department of Chemical Engineering, Drexel University, Philadelphia.

Morris, J. G. (1974). The physiology of obligate anaerobiosis. *Adv. Microb. Physiol.* 169-246.

Nichols, D. G. (1982). *Bioenergetics: An Introduction to Chemisomotic Theory*. Academic Press, New York.

Noparatnaraporn, N., Sasaki, K., Nishizawa, Y., and Nagai, S. (1986). Stimulation of vitamin B12 in aerobically-grown *Rhodopseudomonas geletinosa* under microaerobic condition. *Biotech. Lett.* *8*:491-496.

O'Brien, R. W., and Morris, J. G. (1971). Oxygen and the growth and metabolism of *Clostridium acetobutylicum*. *J. Gen. Microbiol.* *68*:307-318.

Onderdonk, A.B., Johnston, J., Mayhew, J. W., and Gorbach, S. L. (1976). Effect of dissolved oxygen and E_h on *Bacteroides fragilis* during continuous culture. *Appl. Environ. Microbiol.* *31*: 168-172.

Pirt, J. S. (1975). *Principles of Microbe and Cell Cultivation*. John Wiley and Sons, New York.

Prince, R. C., Linkletter, S. J. G., and Dutton, P. L. (1980). The thermodynamic properties of some commonly used oxidation-reduction mediators, inhibitors and dyes, as determined by polarography. *Biochim. Biophys. Acta 635*:132-148.

Radjai, M. K., Hatch, R. T., and Cadman, T. W. (1984). Optimization of amino acid production by automatic self-tuning digital control of redox potential. *Biotechnol. Bioeng. Symp. 14*:656-679.

Rao, G. (1987). Directed metabolic flow in *Clostridium acetobutylicum*. Ph.D. Dissertation, Department of Chemical Engineering, Drexel University, Philadelphia.

Rao, G., and Mutharasan, R. (1985). Growth and product formation kinetics of *Clostridium* sp. cultured under different redox potentials. ACS Annual Meeting, Chicago.

Rao, G., and Mutharasan, R. (1986). Alcohol production by *Clostridium acetobutylicum* induced by methyl viologen. *Biotechnol. Lett. 8*:893-896.

Rao, G., and Mutharasan, R. (1987). Altered electron flow in continuous cultures of *Clostridium acetobutylicum* induced by viologen dyes. *Appl. Environ. Microbiol. 53*:1232-1235.

Reed, G. B., and Orr, J. H. (1942). Cultivation of anaerobes and oxidation-reduction potential. *J. Bacteriol. 43*:309-320.

Schwartz, R. D., and Keller, F. A. (1982). Acetic acid production by *Clostridium thermoaceticum* in pH controlled batch fermentations at acidic pH. *Appl. Environ. Microbiol. 43*(6):1385-1392.

Segel, I. H. (1976). *Biochemical Calculations, How to Solve Mathematical Problems in General Biochemistry*. John Wiley and Sons, New York.

Sies, H. (1982). *Metabolic Compartmentation*. Academic Press, New York.

Shibai, H., Ishizaki, A., Kobayshi, K., and Hirose, Y. (1974). Simultaneous measurement of dissolved oxygen and oxidation reduction potentials in the aerobic culture. *Agric. Biol. Chem. 38*:2407-2411.

Smith, M. V., and Pierson, M. D. (1979). Effect of reducing agents on oxidation-reduction potential and the outgrowth of *Clostridium botulinum* type E spores. *Appl. Environ. Microbiol. 37*:978-984.

Squires, R. W., and Hosler, P. (1958). Pilot plant fermenter with continuous platinum electrode measurement. *Ind. Eng. Chem. 50*:1263-1266.

Thompson, B. G., and Gerson, D. F. (1985). Electrochemical control of redox potential in batch cultures of *Escherichia coli*. *Biotechnol. Bioeng. 27*:1512-1515.

Vennesland, B., and Hanke, M. E. (1940). The oxidation-reduction potential of a non-spore forming obligate anaerobe. *J. Bacteriol. 39*:139-169.

Walz, D. (1979). Thermodynamics of oxidation-reduction reactions and its applications to bioenergetics. *Biochim. Biophys. Acta.* *505*:279-353.

Ward, P.J., and Mutharasan, R. (1985). Effect of controlled redox potential on the growth and energetics of *Thermoanaerobacter ethanolicus*. ACS Annual Meeting, Chicago, September.

Wimpenny, J. W. T., and Necklen, D. K. (1971). The redox environment and microbial physiology. I. The transition anaerobiosis to aerobiosis in continuous cultures of facultative anaerobes. *Biochim. Biophys. Acta 253*:352-359.

Wurmser, R., and Banerjee, R. (1964). Oxidation-reduction potentials. In *Comprehensive Biochemistry*, Vol. 12, M. Florkin and E. H. Stotz (Eds.). pp. 62-88.

Zehnder, A. J. B., and Wuhrmann, K. (1976). Titanium (III) citrate as a non toxic oxidation reduction buffering system for the culture of obligate anaerobes. *Science 194*:1165-1166.

7

Membrane Processes and Product Formation in Anaerobes

PETER DÜRRE, HUBERT BAHL, and GERHARD GOTTSCHALK / *Institut für Mikrobiologie, Georg-August-Universität, Göttingen, Federal Republic of Germany*

I. INTRODUCTION

The interior of unicellular organisms must be surrounded by a membrane so that the influx and efflux of solutes can be controlled and other vital functions of the cells be fulfilled. This membrane consists of a phospholipid bilayer with various proteins embedded within it and is impermeable to protons and practically all other nonlipophilic solutes except gases. The various functions of the cytoplasmic membrane make it necessary that it exists in an energized state. This means in general that a proton-motive force is operative across the membrane bilayer. This proton-motive force Δp, which is usually expressed as the electrochemical gradient of protons, $\Delta\mu_{H^+}/F$, consists of two components: the membrane potential $\Delta\psi$ and the proton gradient between outside and inside ΔpH.

$$\Delta p = \frac{\Delta\mu_{H^+}}{F} = \Delta\psi - \frac{2.3RT}{F}\,\Delta pH$$

What has been said holds true for all microorganisms, regardless of whether they live as aerobes, phototrophs, or anaerobes. In this chapter how the proton-motive force is generated in anaerobes and how the accumulation of fermentation products interferes with its functions are discussed.

II. GENERATION OF THE PROTON-MOTIVE FORCE

The proton-motive force can be generated in anaerobes by a number of mechanisms, which are summarized in Figure 1. Many anaerobic bacteria generate ATP as the primary form of metabolic energy during fermentation. Examples are organisms carrying out alcohol or lactate fermentation. They synthesize ATP by substrate-level phosphorylation during glycolysis and energize the membrane by ATP hydrolysis as catalyzed by the proton-translocating membrane-bound ATPase (Fig. 1a). A primary proton gradient may be generated in certain anaerobes, for example in sulfate-reducing bacteria, by hydrogenases facing the outer aspect of the cytoplasmic membrane (Thauer and Badziong, 1980). These hydrogenases mediate transfer of the electrons to intracellular acceptors but leave the protons outside, so-called scalar protons (Fig. 1b). At low external lactate concentrations, certain lactic acid bacteria have been shown to carry out an electrogenic lactate export (Michels et al., 1979). The driving force is the lactate gradient directed from inside to outside. Per lactate anion two protons are excreted (Fig. 1c), so that a proton gradient is established.

Like aerobes and phototrophs, a number of anaerobes have redox reactions at their disposal taking place in the cytoplasmic membrane and being coupled to proton translocation (Fig. 1d). A long-known reaction of this type is the fumarate reductase reaction (Kröger, 1980):

$$\text{Fumarate} + \text{NADH} + \text{H}^+ \ (\text{or } \text{H}_2\text{O}) \rightarrow \text{succinate} + \text{NAD}^+$$

$\Delta G_0' = -86.2$ kJ per reaction with H_2. It consists of membrane-integrated proteins, b-type cytochromes, and menaquinone. Other not as well characterized redox reactions coupled to proton translocation occur in sulfate-reducing and in methanogenic bacteria. In sulfate reducers proton translocation is accomplished by the reduction of sulfite to sulfide (Badziong and Thauer, 1978) and in methanogenic bacteria by the reduction of methyl coenzyme M to methane (Blaut and Gottschalk, 1985):

$$\text{HSO}_3^- + 3\text{H} \rightarrow \text{HS}^- + 3\text{H}_2\text{O}$$

$\Delta G_0' = -172.8$ kJ per reaction.

$$\text{CH}_3\text{-S-coenzyme M} + \text{H}_2 \rightarrow \text{CH}_4 + \text{HS-coenzyme M}$$

$\Delta G_0' = -85.0$ kJ per reaction. That the reduction of methyl coenzyme M is coupled to proton translocation was the result of a study on the conversion of methanol and H_2 to methane by *Methanosarcina*

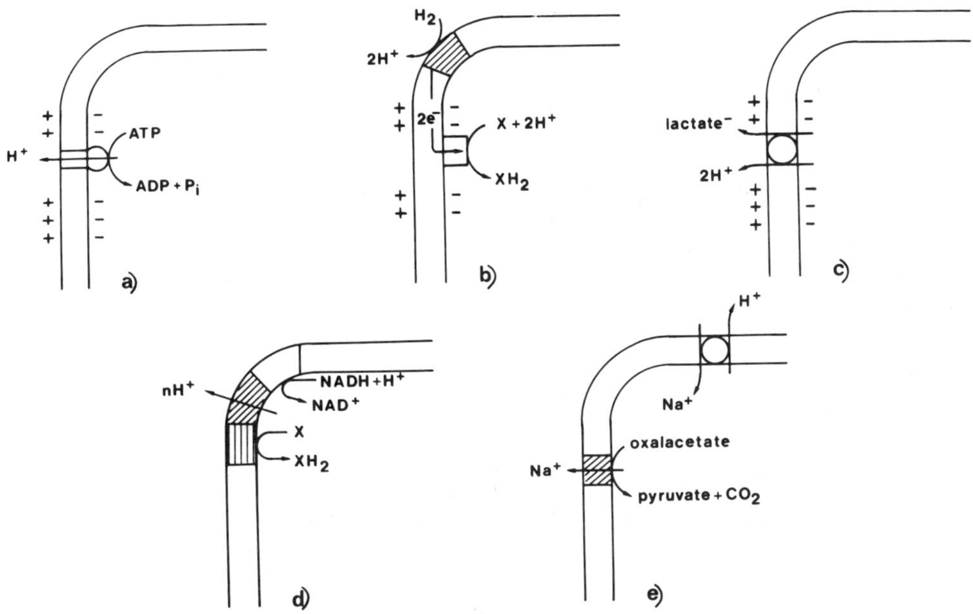

Figure 1 Mechanisms for membrane energization in anaerobes: (a) ATP hydrolysis; (b) "periplasmic" hydrogenase; (c) end-product efflux; (d) "respiratory" reactions; (e) Na^+ translocation.

barkeri (Blaut and Gottschalk, 1984). The main component of the proton-motive force generated is $\Delta\psi$. However, proton translocation is the primary event; by means of a very active H^+/Na^+ antiporter the proton gradient is converted into a sodium ion gradient, which apparently is used in *M. barkeri* and possibly in other methanogenic bacteria to drive energetically unfavorable reactions (Blaut et al., 1985, 1987; Müller et al., 1987). There is a second site of proton translocation in *M. barkeri* growing on acetate, the oxidation of CO to CO_2 as catalyzed by CO dehydrogenase (Bott et al., 1986). It is not known how the reduction of coenzyme M to methane and the oxidation of CO to CO_2 are mechanistically coupled to proton translocation. Some sort of interaction with membrane-integrated components is required. In this connection it is worth mentioning that the methyl coenzyme M methylreductase of *Methanococcus voltae*, which upon cell rupture appears in the soluble fraction (Wolfe, 1985), has been shown by immunoelectron microscopy to be membrane associated (Ossmer et al., 1986).

In a few anaerobic organisms the primary gradient formed is not a proton but a sodium gradient (Hilpert et al., 1984). This gradient is established in *Propionigenium modestum* by the oxaloacetate decarboxylase reaction and in *Acidaminococcus fermentans* by the glutaconyl-coenzyme A decarboxylase reaction. The Na^+ gradient may be used in the form of a sodium-motive force to drive certain reactions; it may also be converted in a proton gradient by a Na^+/H^+ antiporter (Fig. 1e). Anaerobes may thus employ different reactions for energization of the cytoplasmic membrane. In addition to ATP hydrolysis, a number of primary proton or sodium translocating reactions have been identified, and there may be more, as yet undiscovered reactions of this type present in certain anaerobic bacteria.

III. USE OF THE PROTON-MOTIVE FORCE

Anaerobes that generate a proton-motive force by ATP-independent reactions (organisms growing with fumarate + H_2; sulfate-reducing or methanogenic bacteria) use part of the $\Delta\mu_{H^+}$ to drive ATP synthesis via ATP synthase (Fig. 2a). Several additional processes depend on components of $\Delta\mu_{H^+}$ (Hellingwerf and Konings, 1985):

1. Transport processes. Several ions and substrates required for growth are actively taken up by symport processes (Fig. 2b). Sodium gradients are maintained with the help of Na^+/H^+ antiporters, for example in methanogenic bacteria (Müller et al., 1987).
2. Transhydrogenation reactions. The reduction of $NADP^+$ by NADH and several so-called reverse electron transport processes are driven by $\Delta\mu_{H^+}$.
3. Regulation of the intracellular pH (Booth, 1985). Aerobic bacteria growing at neutral or acidic pH maintain a constant intracellular pH (the pH homeostasis according to Padan et al., 1981). By means of the activity of the components of the proton current loop (proton extrusion by the respiratory chain and influx via the ATP synthase, respectively) intracellular pH is kept higher than extracellular pH. An interesting model of intracellular pH regulation in anaerobes has recently been described for *Streptococcus faecalis* (Kobayashi et al., 1986). Protons were extruded by the membrane-bound ATPase, and the level of this enzyme showed an inverse relationship with the cytoplasmic pH of the microorganism. With pH values below 7.5 the amount of the enzyme increased, whereas upon alkalization lower activities were found. In alkalophilic bacteria, such as *Vibrio alginolyticus*, a more acidic intracellular pH is maintained by potassium-proton or sodium-proton antiporters (Nakamura et al., 1984).

4. Motility. The flagellar motor functions with $\Delta\mu_{H^+}$ as the energy source.
5. Penetration of DNA into the cell. Such processes as conjugation or transformation depend on the existence of a $\Delta\psi$ at the membrane.
6. Protein export. Excretion of exoenzymes and translocation of periplasmic proteins require energy. This energy is provided by ATP or by $\Delta\psi$.

It is apparent from this list of functions that an energized membrane is of vital importance to anaerobes, especially in mediating the various transport processes across the membrane. In this connection it may be of interest to mention an experiment done by Harold and Van Brunt (1977) with *S. faecalis*. This organism synthesizes all its ATP in the course of glycolysis. When the membrane was de-energized by the addition of gramicidin, this anaerobe could only grow in a medium kept at a pH between 7.5 and 8.2 and containing 2 mmol/L of each amino acid and high K^+ (more than 0.3 mol/L) and low Na^+ (less than 10 mmol/L) concentrations. Such strict growth requirements that somewhat replace an energized membrane in this anaerobe are not met by conditions in nature or in a bioreactor. Therefore, growth and product formation by anaerobes only occur in conditions in which the cytoplasmic membrane exists in an energized state. Maintaining such a state becomes more of a problem for the organisms as fermentation products accumulate in the medium. This is discussed in the subsequent sections of this chapter.

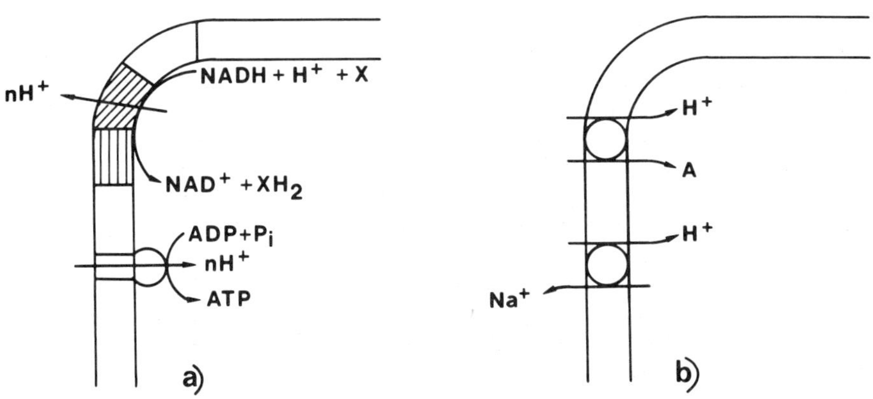

Figure 2 Use of the proton-motive force in anaerobes: (a) for ATP synthesis; (b) for solute uptake by symport processes or Na^+ extrusion by an antiport process.

IV. EFFECTS OF ORGANIC ACIDS ON MEMBRANE PROCESSES

The organic acids produced by fermentative organisms are acetic, butyric, caproic, formic, lactic, propionic, and succinic acids. Among them, acetate is probably the most important substance because it is one of the main fermentation products and is generated by a variety of genera and species. Also, it represents the predominant source of methane formation, the final step in the anaerobic food chain (Mah et al., 1976). This dominating role of acetate is not surprising since production starts from the central intermediate acetyl coenzyme A (with only few exceptions) and is coupled to energy generation in almost all anaerobes. ATP synthesis occurs via substrate-level phosphorylation during the conversion of acetyl phosphate to acetate as catalyzed by the soluble enzyme acetate kinase. Some specialized organisms able to ferment amino acids and/or purines produce acetate via glycine reductase (Stadtman, 1978; Dürre and Andreesen, 1982, 1983; Dürre et al., 1983). This membrane-associated selenoenzyme complex catalyzes ATP formation during the reductive deamination of glycine (Stadtman, 1978).

However, besides these positive effects coupled to its generation, acetate itself represents a problem for microorganisms. The production of high concentrations of this acid lowers the pH of the environment of the organisms and results in the presence of increasing amounts of its undissociated form. This form is membrane permeable (in contrast to the acetate anion) and may bring about dissipation of the proton gradient across bacterial membranes (Padan et al., 1981). This property of acetic acid has led to its use in food preservation.

Despite that many anaerobes form acetic acid as an end product and that this compound represents a major feedstock of chemical processes leading to such synthetic fibers as polyvinyl acetate and various other compounds, little attention has been given to a fermentative production of acetate. In recent years, however, the high substrate-product conversion yield of homoacetogenic bacteria (e.g., 3 mol acetate per mole glucose with *Clostridium thermoaceticum*) has stimulated research to obtain strains that could be used in an economically feasible process for the synthesis of acetic acid. It has been estimated that a price reduction of about 30% would favor the use of the fermentation-derived compound (Palsson et al., 1981). Criteria considered essential for economic viability are a fermentation down to a pH value of 4.5 and a productivity of 5 g acetic acid per liter per hour (Schwartz and Keller, 1982a). A strain growing at pH 4.5 was isolated, but the mass doubling time was too high and the final acetic acid concentration too low to meet the desired productivity (Schwartz and Keller, 1982a, b). However, at pH 7.0

a final acetate concentration of 37 g/L was reached (Wang et al., 1978), a value in the same range as observed with the aerobic *Acetobacter* strains. These aerobic acetogens are used for the commercial production of acetic acid employed as vinegar because of their higher productivity with respect to this acid. However, the process is currently not competitive with chemical synthesis and plays a role only in the food industry.

A recent investigation addresses the reasons for the different performances in acetic acid production at low pH in aerobic and anaerobic acetogens (Menzel and Gottschalk, 1985). As an anaerobic representative, *Acetobacterium wieringae* was chosen, a chemolithotroph able to grow with hydrogen plus carbon dioxide or with heterotrophic substrates, such as fructose (Braun and Gottschalk, 1982). In a weakly buffered medium, 19 mmol/L of fructose were converted to 49 mmol/L of acetate that concomitantly lowered the pH from 7.1 to 5.0. A dramatic change was observed with the internal pH; the organism was obviously unable to keep it constant. The internal values decreased in parallel with the external pH in such a way that a ΔpH of approximately 0.5 units (inside more alkaline) was the result. A similar phenomenon has been reported for other obligate anaerobic, acid-forming clostridia. *Clostridium pasteurianum* lowered its internal pH from 7.5 to 5.9 during growth in batch culture while the external pH fell from 7.1 to 5.1, thereby yielding a ΔpH of 0.4–0.8 units (Riebeling et al., 1975). The authors found the pH gradient to be essential for growth and concluded that it was formed by an ATPase-driven extrusion of protons from the cells. *C. thermoaceticum* also showed acidification of the cytoplasm during acetate production, with the result that a ΔpH of about 0.6 units (inside more alkaline) was maintained throughout most of growth in batch culture (Baronofsky et al., 1984). When the pH of the medium fell below 5.0, both the ΔpH and the $\Delta\psi$ collapsed. Another example for an internal pH becoming continuously more acidic as growth and acid production progressed in batch culture was shown to be *Clostridium acetobutylicum* (Gottwald and Gottschalk, 1985). A constant ΔpH of 0.9–1.3 units was found that could be abolished by protonovorous uncouplers and high concentrations of butanol. A slight discrepancy exists in that another group reported that the intracellular pH did not decrease further when the external pH was lowered from 5.5 to 4.5 (Huang et al., 1985). The reason could be the use of different ΔpH probes or experimental conditions.

Thus, it now seems well established that anaerobic bacteria producing large amounts of acetic and/or butyric acid are unable to maintain cytoplasmic pH homeostasis but can maintain a relatively constant transmembrane pH gradient. This finding is not too surprising, since the obligate anaerobes mostly depend on ATP hydroly-

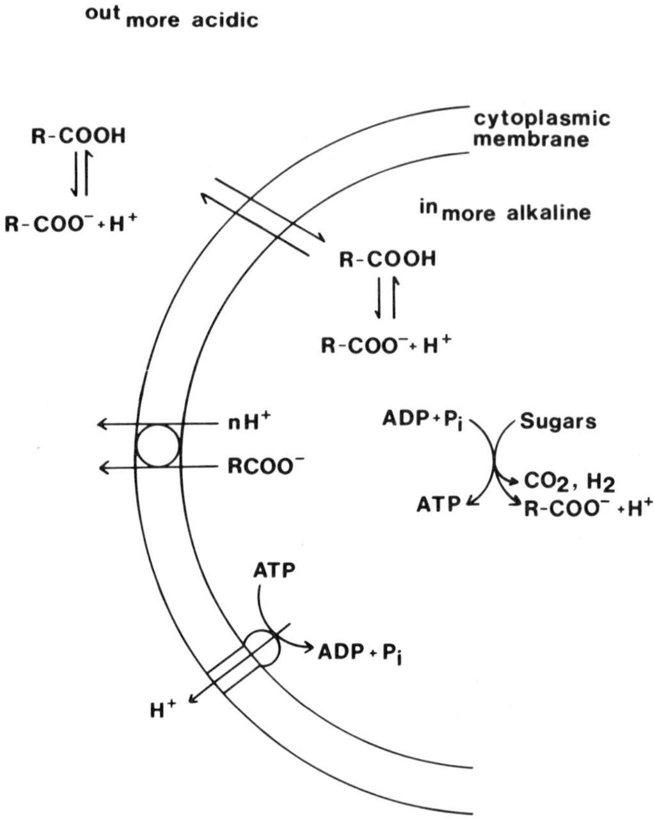

Figure 3 Hypothetical membrane reactions in obligate anaerobic acid-producing bacteria during fermentation. Undissociated acid is membrane permeable. The anion transport in symport with (perhaps variable amounts of) protons via a specific carrier is hypothesized. A proton gradient is established by hydrolysis of ATP from substrate-level phosphorylation.

sis for the generation of a proton gradient. The production via fermentation of acid anions that are in equilibrium with their membrane-permeable, undissociated forms would lead to an increasing ATP expenditure with falling external pH values, if the organisms try to maintain a constant internal pH. A generalized picture is shown in Figure 3. Substrates (e.g., sugars) are fermented to acidic and gaseous products, thereby generating ATP via substrate-level phosphorylation. Gases are membrane permeable and do not create a problem in release from the cells and the culture broth, respectively. Acids, however, are present in bacteria largely in their dissociated form, which is membrane impermeable. To excrete acids, these bacteria need a specific carrier. Such a permease has as yet only been detected for lactic acid, and it has been mentioned that the efflux of lactate in *Streptococcus cremoris* occurs in symport with protons. At high external pH values and high internal lactate concentrations, the anion is excreted together with two protons, thereby creating a proton-motive force. This could theoretically lead to an additional energy gain of 50% (Ten Brink and Konings, 1986). The process is driven by the lactate gradient. Using a coculture with a second lactate-consuming organism, an increase in the molar growth yield of *S. cremoris* of about 50% has been found. At higher external lactate concentrations the H^+/end-product stoichiometry falls to 1 or even lower. As yet, no evidence exists in favor of similar carriers for other acidic end products, but in the opinion of the authors an excretion mechanism exclusively via the undissociated, membrane-permeable forms seems very unlikely. Thus, permeases for other organic acids may exist as well. Dissociated and undissociated forms of the acids produced are in an equilibrium dictated by the pH value. The uncharged free acids can cross the membrane without carrier (Kell et al., 1981). Since all acidic end products derived by fermentation possess pK_a values between 3.5 and 5.7, an increasing percentage is converted into the undissociated form at lower pH values. Their concentration inside and outside the cell is equal, and a large ΔpH can no longer be maintained by the cells. However, the exact nature of the toxic effect of acidic end products is a matter of debate. The collapse of $\Delta\mu_{H^+}$ as soon as metabolism can no longer supply the ATP required for proton extrusion by the H^+-ATPase (Baronofsky et al., 1984), the malfunction of specific cellular enzymes at low pH values as demonstrated for cell division in *Escherichia coli* (Zilberstein et al., 1984), and a reduction of ΔpH by effective anion exclusion (Herrero et al., 1985b) are discussed. In the last case, the anion dimer HA_2^- has been proposed as the charged permeant species. Wang and Wang (1984) reported that not only the undissociated acid at low pH, but also the acetate anion at pH values above 7.0 exerts a toxic effect. The authors calculated the concentration required for total growth inhibition at about 45 mmol/L for undissociated acetic acid and approximately 800 mmol/L for the acetate anion.

The performance in acetic acid production by anaerobic and aerobic acetogens was compared (Menzel and Gottschalk, 1985). As an aerobic representative, *Acetobacter aceti* was chosen. Surprisingly, this organism was also unable to maintain a constant internal pH. However, in comparison with the obligate anaerobe,

a)

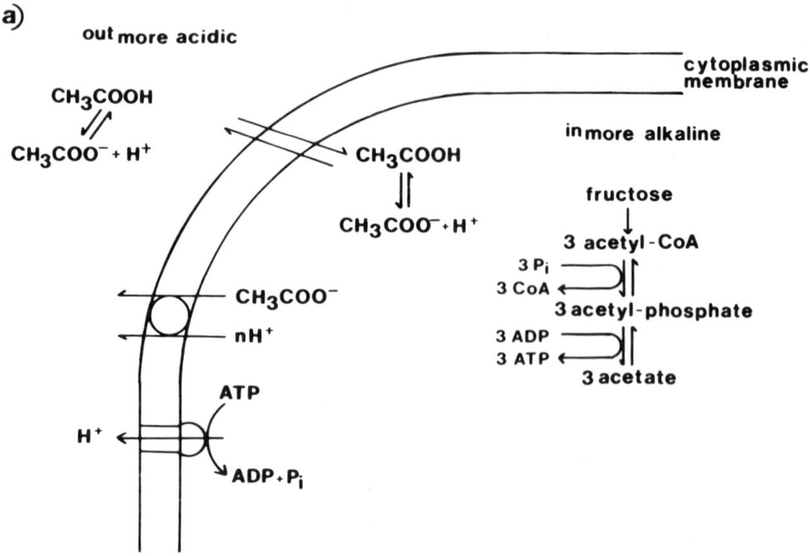

Figure 4 Mechanisms of acetic acid production by *Acetobacterium wieringae* and *Acetobacter aceti*. The existence of a special acetate carrier is hypothesized that may be shut down at high external acetate concentrations. (a) *A. wieringae*. Acetate is derived by reversible reactions. A proton gradient is established by hydrolysis of ATP from substrate-level phosphorylation. (b) *A. aceti*. Acetate is derived by irreversible reactions. A proton gradient is established by H^+ translocation from reduced PQQ. ATP is generated via electron transport phosphorylation. (From Menzel and Gottschalk, 1985.)

the external pH decreased to much lower values (pH 3.6) and much more acetic acid was produced. How was this achieved? It was suggested that the main difference in acetate production between the two species resulted from the respective terminal reactions. The anaerobic *A. wieringae* formed acetic acid via acetyl coenzyme A and acetyl phosphate in reversible reactions, whereas *A. aceti* converted ethanol irreversibly to acetate by pyrroloquinoline quinone (PQQ)-dependent dehydrogenases (Fig. 4). The cessation of acetic acid production by the obligate anaerobe is then the result of a shortage of coenzyme A or phosphate due to elevated levels of the intermediates acetyl coenzyme A and acetyl phosphate. Such product inhibition does not occur in the aerobic organism, thereby explaining the higher yield of acetic acid.

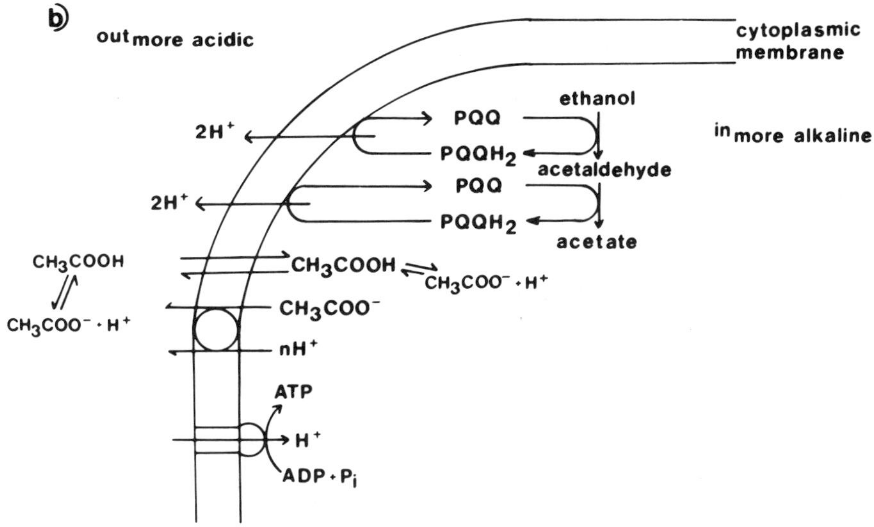

A similar phenomenon of inhibition of acid production by elevated levels of intermediates may be the cause for the shift to solvent formation in acetone-butanol fermentation. A shortage of coenzyme A could, for instance, function as a signal for the induction of the enzymes required for butanol and acetone synthesis (Gottwald and Gottschalk, 1985).

What has been said here about acetate formation may generally apply to the production of weak organic acids during fermentation. The increasing concentration of the undissociated form of the acids at low pH makes it more and more difficult for the cells to maintain a certain ΔpH across the membrane by ATP-dependent proton translocation. An additional drawback is that most acids are formed in reversible reactions (e.g., acetate kinase, butyrate kinase, and lactate dehydrogenase). An increase in intracellular acid concentration therefore results in an increase in the precursor concentrations and, at some stage, in the unfeasibility of further production.

V. EFFECTS OF NEUTRAL COMPOUNDS ON MEMBRANE PROCESSES

The neutral fermentation products of anaerobically growing microorganisms include ethanol, 2-propanol, acetone, 1-butanol, 1,2-propandiol, 2,3-butandiol, and other related compounds. Relevant for industrial processes were, are, or will be butanol fermentation by *C. acetobutylicum* and ethanol production either by yeast, *Zymomonas mobilis*, or some thermophilic bacteria.

Z. mobilis is the only known obligate ethanol producer. All other neutral compound-forming bacteria take advantage of branched catabolic pathways to control product formation (Gottschalk, 1986). Depending on the conditions, the organisms have a choice between different routes. This flexibility enables them to balance their oxidation-reduction state, a crucial problem in anaerobes, and to avoid the accumulation of high amounts of one product, which then may become toxic (see Sec. IV).

Despite their name, neutral end products exert strong effects on growth, fermentation activity, and viability of microorganisms. Several effects of alcohols on microorganisms have been reported. For a recent review about the influence of ethanol on microorganisms, see Ingram and Buttke (1984). The growth of most bacteria is inhibited at ethanol concentrations in the range of 1-10% (w/v); only a few grow at concentrations above 10% (Ingram and Buttke, 1984). The toxicity of alcohols is correlated with their chain length (up to 10 carbon atoms) and their hydrophobicity (Harold, 1970; Herrero and Gomez, 1980). Therefore, butanol is much more toxic to bacteria than ethanol. The growth rate of *C. acetobutylicum* is reduced by 50% at butanol concentrations of 1% and the maximum concentration of this alcohol produced and tolerated is about 1.8% (Moreira et al., 1981), whereas *Z. mobilis* can grow and produce ethanol at concentrations of up to 8% (Rogers et al., 1982). Fatty alcohols with more than 10 carbon atoms are relatively ineffective as inhibitors (Ingram and Buttke, 1984).

Alcohol inhibition of fermentation may be based on two mechanisms: damage to the cell membrane (Osman and Ingram, 1985) and end-product inhibition of glycolytic enzymes (Herrero et al., 1985a). It is generally accepted that the cell membrane is one target of alcohol attack (Ingram, 1986). Membrane processes affected by either ethanol or butanol are discussed in this section. The known hazardous effects of alcohols on membrane functions are described, as are the changes in membrane composition that anaerobic alcohol producers undergo to adapt to alcohol stress.

Owing to their amphipathic character, alcohols dissolve in the membrane lipids, affecting membrane fluidity (Vollherbst-Schneck et al., 1984) and increasing its permeability (Dombek and Ingram, 1984). The leakage of essential cofactors, coenzymes, and intermediary metabolites caused by the accumulation of ethanol has been shown in *Z. mobilis* (Osman and Ingram, 1985). Harmful effects of butanol on *C. acetobutylicum* have also been described. High concentrations of butanol led to a complete abolition of the pH gradient, lowered the intracellular level of ATP, caused the release of intracellular metabolites, and inhibited sugar uptake (Moreira et al., 1981; Gottwald and Gottschalk, 1985; Bowles and Ellefson, 1985; Hutkins and Kashket, 1986). On the other hand, neither the internal pH nor the intracellular ATP level was affected by ethanol in *Clostridium thermocellum* (Herrero et al., 1985a).

Why is the alcohol tolerance of microorganisms different? The growth rate of *C. thermocellum* is reduced by 50% at ethanol concentrations of 0.5% at 60°C (Herrero and Gomez, 1980), whereas *Lactobacillus homohiochii* and *Lactobacillus heterohiochii* can grow at ethanol concentrations of 20% (Demain et al., 1961). Analysis of the lipid composition of alcohol-producing or tolerating organisms revealed some answers to this question. Some bacteria are able to adapt to stress situations in which membrane damage can occur. For example, the effect of growth temperature on membrane lipid composition is well documented (Khuller and Goldfine, 1974; Russel, 1984). Thus, it was not surprising that changes in the lipid composition of anaerobic bacteria were also found when neutral compounds were added or accumulated in the medium during fermentation.

An increase in chain length of the fatty acyl component of membrane lipid seems to be a general trend connected with ethanol tolerance and was observed in such ethanol-producing microorganisms as *E. coli* (Ingram and Buttke, 1984), *Saccharomyces cerevisiae* (Beaven et al., 1982), and *Clostridium thermohydrosulfuricum* (Lovitt et al., 1984). The higher ethanol tolerance of a mutant strain of *C. thermohydrosulfuricum* is attributed to the occurrence of C-30 dicarboxylic acids in membrane lipids (Lovitt et al., 1984). Consistent with this hypothesis is that in a strain of *L. heterohiochii*, an alcoholophilic bacterium, saturated and monounsaturated fatty acids with unusually long chains (C-20 to C-30) were found (Uchida, 1974) and that in the obligate ethanol producer *Z. mobilis* over 70% of the fatty acyl chains are cis-18, 1 ($\Delta 11$) fatty acid (vaccenic acid) (Carey and Ingram, 1985). Challenge of *Z. mobilis* with ethanol did not result in a further increase in this fatty acid or other significant changes in fatty acid composition. The organism seems to be well adapted with respect to its fatty acids in the membrane lipids. Although most of the longer fatty acyl chains found in response to increased ethanol concentrations contain cis monounsaturated fatty acids, this does not seem to play an important role in ethanol tolerance. A cis double bond can result in a bend in the acyl chain that disrupts the ordered packing of the acyl chains in the hydrophobic core. Therefore, an increase in unsaturated fatty acids does not seem to be sufficient to counteract the effect of alcohols on increased membrane fluidity. However, longer hydrocarbon chains interact more strongly than do short chains, thicken the hydrophobic core, and increase hydrophobicity. Transconfiguration of monounsaturated fatty acid results in a longer mean chain length. Not surprisingly, the addition of transunsaturated 18,1 fatty acids (elaidic and *trans*-vaccenic acid) to the growth medium and incorporation in membrane lipids were more efficient in increasing ethanol tolerance in *E. coli* than the cis unsaturated oleic and vaccenic acids that occur naturally in this organism (Ingram and Buttke, 1984). A similar result was obtained in an elaidic acid-supplemented butanol fermentation by *C. acetobutylicum* (Paige and Linden, 1984).

The fatty acid changes that occur in *C. acetobutylicum* under the influence of butanol are different (Lepage et al., 1987). Whereas ethanol also resulted in an increase in 18,1 fatty acids, butanol caused a decrease in 18,1 and an increase in 16,0 and 14,0. Longer chain alcohols favored even more the synthesis of shorter chain fatty acids, such as 14,0 and 12,0, combined with a large decrease in 16,1. In general, butanol and longer chain alcohols decrease the unsaturated-saturated ratio in fatty acid composition and shorten the fatty acid chain length. The observation that a butanol-tolerant mutant of *C. acetobutylicum* did not have an increased resistance to ethanol (Hermann et al., 1985) provides further evidence for the different effects of ethanol and butanol, although both damage the cell membrane. Specific interactions of these solvents with different components of the complex membrane structure are possible.

The effects of alcohols on the lipid composition have also been examined. Two changes seem to be beneficial to the growth in the presence of ethanol: increase in the proportion of acidic phospholipids (cardiolipin and phosphatidylglycerol) as found in *E. coli* (Clark and Beard, 1979) and *Z. mobilis* (Carey and Ingram, 1983) and reduction of the phospholipid-protein ratio. The latter change would reduce the area available for ethanol action and therefore for the leakage of cytoplasmic compounds (Ingram, 1986). A substantial decrease in the phospholipid-protein ratio in the membrane of *E. coli* (Ingram, 1977) and *Z. mobilis* (Carey and Ingram, 1983) has been described.

Z. mobilis has evolved an additional way to protect itself from the deleterious effects of ethanol. It is able to synthesize hopanoids and to integrate them into the cytoplasmic membrane (Bringer et al., 1984). Hopanoids are sterol equivalents of bacteria (Poralla et al., 1980). They are able to enhance the membrane viscosity and to decrease membrane permeability (Taylor, 1984). The hopanoid content of the membrane of *Z. mobilis* reached a maximum when high concentrations of ethanol accumulated in the medium and increased sharply when ethanol was added to the medium (Schmidt et al., 1986).

Changes in lipid composition in the membrane of *C. acetobutylicum* in response to butanol have been studied only recently (Oulevey et al., 1986; Lepage et al., 1987). Butanol had no effect on the total lipid content of *C. acetobutylicum*, which amounted to 5-6% of the total dry weight in the presence or absence of butanol. The ratio of acidic to neutral phospholipids increased slightly, whereas lipids with alkenyl ether groups were reduced under solvent-producing conditions. The function of these novel ether lipids in anaerobes is not known. The possible presence of hopanoids in the membrane of *C. acetobutylicum* could not be confirmed.

The effects of alcohols on the membrane lipids of alcohol-producing bacteria clearly demonstrate the importance of the cell membrane with respect to alcohol tolerance. However, this does not exclude other sites of cellular regulation where interaction with alcohols may occur. To understand these alcohol-related phenomena in detail, more information about the mode of action of different alcohols at different concentrations in different bacteria is necessary.

VI. CONCLUDING REMARKS

Anaerobic bacteria have various mechanisms at their disposal to generate an electrochemical gradient of protons across the membrane. Accumulating fermentation products interfere with the existence of this gradient, which is of vital importance to all living cells. This sets limitations to the concentration of fermentation products that can actually be reached. It indicates that an economically feasible product recovery from dilute solutions is decisive in the production of commodity chemicals by fermentation.

REFERENCES

Badziong, W., and Thauer, R. K. (1978). Growth yields and growth rates of *Desulfovibrio vulgaris* (Marburg) growing on hydrogen plus sulfate and hydrogen plus thiosulfate as the sole energy sources. *Arch. Microbiol. 117*:209-214.

Baronofsky, J. J., Schreurs, W. J. A., and Kashket, E. R. (1984). Uncoupling by acetic acid limits growth of and acetogenesis by *Clostridium thermoaceticum. Appl. Environ. Microbiol. 48*:1134-1139.

Beaven, M. J., Charpentier, C., and Rose, A. H. (1982). Production and tolerance of ethanol in relation to phospholipid fatty-acyl composition in *Saccharomyces cerevisiae* NCYC 431. *J. Gen. Microbiol. 128*:1447-1455.

Blaut, M., and Gottschalk, G. (1984). Coupling of ATP synthesis and methane formation from methanol and molecular hydrogen in *Methanosarcina barkeri. Eur. J. Biochem. 141*:217-222.

Blaut, M., and Gottschalk, G. (1985). Evidence for a chemiosmotic mechanism of ATP synthesis in methanogenic bacteria. *Trends Biochem. Sci. 10*:486-489.

Blaut, M., Müller, V., Fiebig, K., and Gottschalk, G. (1985). Sodium ions and an energized membrane required by *Methanosarcina barkeri* for the oxidation of methanol to the level of formaldehyde. *J. Bacteriol.* *164*:95-101.

Blaut, M., Müller, V., and Gottschalk, G. (1987). Proton translocation coupled to methanogenesis from methanol plus hydrogen in *Methanosarcina barkeri*. *FEBS Lett.* *215*:53-57.

Booth, I. R. (1985). Regulation of cytoplasmic pH in bacteria. *Microbiol. Rev.* *49*:359-378.

Bott, M., Eikmanns, B., and Thauer, R. K. (1986). Coupling of carbon monoxide oxidation to CO_2 and H_2 with the phosphorylation of ADP in acetate-grown *Methanosarcina barkeri*. *Eur. J. Biochem.* *159*:393-398.

Bowles, L. K., and Ellefson, W. L. (1985). Effects of butanol on *Clostridium acetobutylicum*. *Appl. Environ. Microbiol.* *50*:1165-1170.

Braun, M., and Gottschalk, G. (1982). *Acetobacterium wieringae* sp. nov., a new species producing acetic acid from molecular hydrogen and carbon dioxide. *Zentralbl. Bakteriol. Mikrobiol. Hyg. I [C]* *3*:368-376.

Bringer, S., Härtner, T., Poralla, K., and Sahm, H. (1985). Influence of ethanol on the hopanoid content and the fatty acid pattern in batch and continuous cultures of *Zymomonas mobilis*. *Arch. Microbiol.* *140*:312-316.

Carey, V. C., and Ingram, L. O. (1983). Lipid composition of *Zymomonas mobilis*: Effects of ethanol and glucose. *J. Bacteriol.* *154*:1291-1300.

Clark, D. P., and Beard, J. P. (1979). Altered phospholipid composition in mutants of *Escherichia coli* sensitive or resistant to organic solvents. *J. Gen. Microbiol.* *113*:267-274.

Demain, A. L., Ricks, E. L., Hendlin, D., and Barnes, E. C. (1961). Nutritional studies on *Lactobacillus heterohiochii*. *J. Bacteriol.* *81*:147-153.

Dombek, K. M., and Ingram, L. O. (1984). Effects of ethanol on the *Escherichia coli* plasma membrane. *J. Bacteriol.* *157*:233-239.

Dürre, P., and Andreesen, J. R. (1982). Selenium-dependent growth and glycine fermentation by *Clostridium purinolyticum*. *J. Gen. Microbiol.* *128*:1457-1466.

Dürre, P., and Andreesen, J. R. (1983). Purine and glycine metabolism by purinolytic clostridia. *J. Bacteriol.* *154*:192-199.

Dürre, P., Spahr, R., and Andreesen, J. R. (1983). Glycine fermentation via a glycine reductase in *Peptococcus glycinophilus* and *Peptococcus magnus*. *Arch. Microbiol.* *143*:127-135.

Gottschalk, G. (1986). *Bacterial Metabolism*, 2nd Ed. Springer-Verlag, New York.

Gottwald, M., and Gottschalk, G. (1985). The internal pH of *Clostridium acetobutylicum* and its effect on the shift from acid to solvent formation. *Arch. Microbiol.* *143*:42-46.

Harold, F. M. (1970). Antimicrobial agents and membrane function. *Adv. Microb. Physiol.* *4*:45-104.

Harold, F. M., and Van Brunt, J. (1977). Circulation of H^+ and K^+ across the plasma membrane is not obligatory for bacterial growth. *Science* *197*:372-373.

Hellingwerf, K. J., and Konings, W. N. (1985). The energy flow in bacteria: The main free energy intermediates and their regulatory role. *Adv. Microb. Physiol.* *26*:125-154.

Herrero, A. A., and Gomez, R. F. (1980). Development of ethanol tolerance in *Clostridium thermocellum*: Effect of growth temperature. *Appl. Environ. Microbiol.* *40*:571-577.

Herrero, A. A., Gomez, R. F., and Roberts, M. F. (1985a). ^{31}P NMR studies of *Clostridium thermocellum*. *J. Biol. Chem.* *260*: 7442-7451.

Herrero, A. A., Gomez, R. F., Snedecor, B., Tolman, C. J., and Roberts, M. F. (1985b). Growth inhibition of *Clostridium thermocellum* by carboxylic acids: A mechanism based on uncoupling by weak acids. *Appl. Microbiol. Biotechnol.* *22*:53-62.

Hermann, M., Fayolle, F., Marchal, R., Podwin, L., Sebald, M., and Vandecasteele, J.-P. (1985). Isolation and characterization of butanol-resistant mutants of *Clostridium acetobutylicum*. *Appl. Environ. Microbiol.* *50*:1238-1243.

Hilpert, W., Schink, B., and Dimroth, P. (1984). Life by a new decarboxylation-dependent energy conservation mechanism with Na^+ as coupling ion. *EMBO J.* *3*:1665-1670.

Huang, L., Gibbins, L. N., and Forsberg, C. W. (1985). Transmembrane pH gradient and membrane potential in *Clostridium acetobutylicum* during growth under acetogenic and solventogenic conditions. *Appl. Environ. Microbiol.* *50*:1043-1047.

Hutkins, R. W., and Kashket, E. R. (1986). Phosphotransferase activity in *Clostridium acetobutylicum* from acidogenic and solventogenic phases of growth. *Appl. Environ. Microbiol.* *51*:1121-1123.

Ingram, L. O. (1977). Changes in lipid composition of *Escherichia coli* resulting from growth with organic solvents and with food additives. *Appl. Environ. Microbiol.* *33*:1233-1236.

Ingram, L. O. (1986). Microbial tolerance to alcohols: Role of the cell membrane. *Trends Biotechnol.* *4*:40-44.

Ingram, L. O., and Buttke, T. (1984). Effects of alcohols on microorganisms. *Adv. Microb. Physiol.* *25*:253-300.

Kell, D. B., Peck, M. W., Rodger, G., and Morris, J. G. (1981). On the permeability to weak acids and bases of the cytoplasmic membrane of *Clostridium pasteurianum*. *Biochem. Biophys. Res. Commun.* *99*:81-88.

Khuller, G. K., and Goldfine, H. (1974). Phospholipids of *Clostridium butyricum*. V. Effects of growth temperature on fatty acid alk-1-enyl ether group, and phospholipid composition. *J. Lipid Res. 15*: 500-507.

Kobayashi, H., Suzuki, T., and Unemoto, T. (1986). Streptococcal cytoplasmic pH is regulated by changes in amount and activity of a proton-translocating ATPase. *J. Biol. Chem. 261*:627-630.

Kröger, A. (1980). Bacterial electron transport to fumarate. In *Diversity of Bacterial Respiratory Systems*, Vol. 2, C. J. Knowles (Ed.). CRC Press, Boca Raton, Florida, pp. 1-17.

Lepage, C., Fayolle, F., Hermann, M., and Vandecasteele, J.-P. (1987). Changes in membrane lipid composition of *Clostridium acetobutylicum* during acetone-butanol fermentation: Effects of solvents, growth temperature, and pH. *J. Gen. Microbiol. 133*: 103-110.

Lovitt, R. W., Longin, R., and Zeikus, J. G. (1984). Ethanol production by thermophilic bacteria: Physiological comparison of solvent effects on parent and alcohol-tolerant strains of *Clostridium thermohydrosulfuricum. Appl. Environ. Microbiol. 48*: 171-177.

Mah, R. A., Hungate, R. E., and Ohwaki, K. (1976). Acetate, a key intermediate in methanogenesis. In *Microbial Energy Conversion*, H. G. Schlegel and J. Barnea (Eds.). E. Goltze KG, Göttingen, pp. 97-106.

Menzel, U., and Gottschalk, G. (1985). The internal pH of *Acetobacterium wieringae* and *Acetobacter aceti* during growth and production of acetic acid. *Arch. Microbiol. 143*:47-51.

Michels, P. A. M., Michels, J. P. J., Boonstra, J., and Konings, W. N. (1979). Generation of an electrochemical proton gradient in bacteria by the excretion of metabolic end products. *FEMS Microbiol. Lett. 5*:357-364.

Moreira, A. R., Ulmer, D. C., and Linden, J. C. (1981). Butanol toxicity in the butylic fermentation. *Biotechnol. Bioeng. Symp. 11*:567-579.

Müller, V., Blaut, M., and Gottschalk, G. (1987). Generation of a transmembrane gradient of Na^+ in *Methanosarcina barkeri. Eur. J. Biochem. 162*:461-466.

Nakamura, T., Tokuda, H., and Unemoto, T. (1984). K^+/H^+ antiporter functions as a regulatory of cytoplasmic pH in a marine bacterium, *Vibrio alginolyticus. Biochim. Biophys. Acta 776*:330-336.

Osman, Y. A., and Ingram, L. O. (1985). Mechanism of ethanol inhibition of fermentation in *Zymomonas mobilis* CP4. *J. Bacteriol. 164*:173-180.

Ossmer, R., Mund, T., Hartzell, P. L., Konheiser, U., Kohring, G. W., Klein, A., Wolfe, R. S., Gottschalk, G., and Mayer, F. (1986). Immunocytochemical localization of component C of the methyl-reductase system in *Methanococcus voltae* and *Methanobacterium thermoautotrophicum*. *Proc. Natl. Acad. Sci. USA 83*: 5789-5792.

Oulevey, J., Bahl, H., and Thiele, O. W. (1986). Novel alk-1-enyl ether lipids isolated from *Clostridium acetobutylicum*. *Arch. Microbiol. 144*:166-168.

Padan, E., Zilberstein, D., and Schuldiner, S. (1981). pH homeostasis in bacteria. *Biochim. Biophys. Acta 650*:151-166.

Paige, M. R., and Linden, J. C. (1984). Phosphate limitation in elaidic acid supplemented butylic fermentations. Abstracts of the Annual Meeting of the Microbial Biochemistry Technology Division, American Chemical Society, Philadelphia, August 31.

Palsson, B. O., Fathi-Afshar, S., Rudd, D. F., and Lightfoot, E. N. (1981). Biomass as a source of chemical feedstocks: An economic evaluation. *Science 213*:513-517.

Poralla, K., Kannenberg, E., and Blume, A. (1980). A glycolipid containing hopane isolated from the acidophilic, thermophilic *Bacillus acidocaldarius*, has a cholesterol-like function in membranes. *FEBS Lett. 113*:107-110.

Riebeling, V., Thauer, R. K., and Jungermann, K. (1975). The internal-alkaline pH gradient, sensitive to uncoupler and ATPase inhibitor, in growing *Clostridium pasteurianum*. *Eur. J. Biochem. 55*:445-453.

Rogers, P. L., Lee, K. J., Skotnicki, M. L., and Tribe, D. E. (1982). Ethanol production by *Zymomonas mobilis*. *Adv. Biochem. Eng. 23*:37-84.

Russel, N. J. (1984). Mechanisms of thermal adaptation in bacteria: Blueprint for survival. *Trends Biochem. Sci. 9*:108-112.

Schmidt, A., Bringer-Meyer, S., Poralla, K., and Sahm, H. (1986). Effect of alcohols and temperature on the hopanoid content of *Zymomonas mobilis*. *Appl. Microbiol. Biotechnol. 25*:32-36.

Schwartz, R. D., and Keller, F. A. (1982a). Isolation of a strain of *Clostridium thermoaceticum* capable of growth and acetic acid production at pH 4.5. *Appl. Environ. Microbiol. 43*:117-123.

Schwartz, R. D., and Keller, F. A. (1982b). Acetic acid production by *Clostridium thermoaceticum* in pH-controlled batch fermentations at acidic pH. *Appl. Environ. Microbiol. 43*:1385-1392.

Stadtman, T. C. (1978). Selenium-dependent clostridial glycine reductase. *Methods Enzymol. 53*:373-382.

Taylor, R. F. (1984). Bacterial triterpenoids. *Microbiol. Rev. 48*: 181-198.

Ten Brink, B., and Konings, W. N. (1986). Generation of a proton-motive force in anaerobic bacteria by end-product efflux. *Methods Enzymol. 125*:492-510.

Thauer, R. K., and Badziong, W. (1980). Respiration with sulfate as electron acceptor. In *Diversity of Bacterial Respiratory Systems*, Vol. 2, C. J. Knowles (Ed.). CRC Press, Boca Raton, Florida, pp. 65-85.

Uchida, K. (1974). Occurrence of saturated and mono-unsaturated fatty acids with unusually long-chains (C20-C30) in *Lactobacillus heterohiochii*, an alcoholophilic bacterium. *Biochim. Biophys. Acta 348*:86-93.

Vollherbst-Schneck, K., Sands, J. A., and Montenecourt, B. S. (1984). Effect of butanol on lipid composition and fluidity of *Clostridium acetobutylicum* ATCC 824. *Appl. Environ. Microbiol. 47*:193-194.

Wang, D. I. C., Fleischaker, R., and Wang, G. (1978). A novel route to the production of acetic acid by fermentation. *AIChE Symp. Ser. 182*:105-110.

Wang, G., and Wang, D. I. C. (1984). Elucidation of growth inhibition and acetic acid production by *Clostridium thermoaceticum. Appl. Environ. Microbiol. 47*:294-298.

Wolfe, R. S. (1985). Unusual coenzymes of methanogenesis. *Trends Biochem. Sci. 10*:396-399.

Zilberstein, D., Agmon, V., Schuldiner, S., and Padan, E. (1984). *Escherichia coli* intracellular pH, membrane potential, and cell growth. *J. Bacteriol. 158*:246-252.

8

Product Inhibition

RAKESH KUMAR BAJPAI and EUGENE L. IANNOTTI / *University of Missouri-Columbia, Columbia, Missouri*

I. INTRODUCTION

Pure culture fermentations in which microbial cells of a single species are utilized to carry out transformations are the backbone of modern biotechnology. One of the first pure culture industrial processes, the production of acetone and butanol, depended for maintenance of sterility on the toxicity of products to any contaminants (Beesch, 1952), in addition to the other sterile practices then available. Like pure culture processes, mixed-population anaerobic fermentations also began with such processes as yogurt production, which depended upon the maintenance of the desired organisms through rapid establishment of inhibitory conditions. A mixed culture is utilized in which it is not practical to maintain a sterile environment or a single organism cannot carry out the desired reactions. The study of interference (inhibition) by the products of a process to various uptake/secretion, metabolic, and growth phenomena has grown in importance because it influences the operation and yield as well as productivity in bioreactors.

This chapter focuses upon product inhibition that affects the organism's metabolism (feedback) and/or other simultaneous processes. The survival advantages of one species over others under the adverse conditions caused by inhibitory products and differences between the toxicity of microbial products are not covered (Demain et al., 1979; Katz and Demain, 1977).

A. Effect of Products on the Cell

Microbial metabolism constitutes a complex network of coupled chemical reactions that interactively result in growth and product formation; a product may affect any aspect of this metabolism. First, an inhibitory compound may interfere with the pathways involved in the formation of required enzymatic machinery. Feedback regulation by metabolites is a prime example of this type of inhibition (Demain, 1972). Second, the inhibitor may interfere with the processes involving energetic coupling in the microorganism. This results in a decreased rate of growth of cells; the product formation rate may temporarily increase if it produces energy couplers like ATP. Interactions of some organic acids and alcohols with the lipid membrane structure represent such phenomena (Finean et al., 1978). Finally, the inhibitory compound may indirectly interact with the cellular machinery by creating conditions under which the cells must exert more energy to carry out their otherwise normal functions. The net result of inhibition is decreased growth and/or product formation, formation of faulty membrane structures (by β-lactum antibiotics), or other similar phenomena (Maiorella et al., 1983). The same compound may affect the cell in different ways depending upon its concentration at the locale of action. Intracellular concentrations of compounds are more closely related to the extent of the production of biomass and the metabolites than are extracellular concentrations (Matsuo et al., 1966; Nagodawithana and Steinkraus, 1976; Nagodawithana et al., 1977). In many cases, however, the specific action of the so-called inhibitor is not clear.

In the microbial world, inhibition is often a means to regulate metabolic fluxes inside a cell. In a large number of amino acid fermentations, knowledge of the nature and location of product inhibition has been put to use in overproduction of these compounds through specific mutations that modify the sites controlling the extent of product formation (Demain, 1972). In the light of inhibition, process operations have also been modified to alleviate its effects. A case in point is the glutamic acid fermentation in which biotin deficiency is created during product formation. This causes the synthesis of defective membrane structures through which glutamic acid leaks out to the extracellular environment. The same information in several enzyme-catalyzed reactions has been utilized to develop process strategies for maximal conversions (Blotkamp et al., 1977). On a different scale, a number of molecular events in the cell are thought to be timed by product inhibition. An example is the current theory of control of initiation of chromosomal replication in *Escherichia coli* by the repressor protein produced as a burst upon initiation, preventing a second initiation until the repressor is sufficiently inactivated by the antirepressor protein (Pritchard et al., 1969). This information has been used to model cell duplication as well as copy numbers of plasmids and their control in a growing cell (Peretti and Bailey, 1986).

B. Expression of Inhibition

In pure culture systems, product inhibition is initially expressed via reduced rates of growth and/or product formation when only one primary product is formed, or as a completely new spectrum of products plus reduced rates when several products are possible. Of course, if inhibition is unchecked, it ultimately results in process failure. In continuous reactors and in cultures involving several organisms, changes in system dynamics and reactor operation are generally observed.

Product inhibition in a mixed population, such as that in anaerobic digestion, includes the fundamental aspects of failure plus unique aspects resulting from the presence of more than one organism. The process itself takes place in several steps and includes interactions between each part of the process (Bryant, 1979; Iannotti et al., 1986a; Kirsch and Sykes, 1971; McInerney and Bryant, 1981). Inhibition is usually first seen as a shift in the metabolism, especially the part that involves interactions between different members of the microbial population. The effects are seen at smaller concentrations than those associated with product inhibition in pure cultures. The inhibitor is not necessarily the desired end product; for example, acids and other intermediates are toxic (but not methane because of its low solubility). Unlike pure cultures in which the natural growth cycle results in a buildup of toxic end products, process failure in anaerobic digestion usually results from an outside cause, either a toxic compound or a change in operating conditions resulting in such products as ammonia and volatile fatty acids. Normally, the total concentration of the acids in the digester is low (below 300 mg/L at the end of the day); stressed digesters may contain up to 14,000 mg/L (Fischer et al., 1979; Georgacakis et al., 1982; Kroeker et al., 1979; McCarty and Brosseau, 1963; McCarty and McKinney, 1961). Although the cause is usually an outside perturbation, the intermediates themselves can be toxic and lead to further deterioration of the process until very high levels cause failure of the process (Buswell and Morgan, 1962; Hobson and Shaw, 1976; Iannotti and Fischer, 1984; Kroeker et al., 1979; McCarty, 1964; McCarty and McKinney, 1961; van den Berg et al., 1976). If present at low levels over a long term, inhibition can result in changes in the dominant members of the population as a result of succession to bacteria better adapted to growth in the presence of toxic end products.

C. Classification of Products

Inhibitory products can be classified not only according to the mechanism but also on the basis of their chemical nature. Chemical nature is related to the mechanism of action plus the physical conditions for optimum activity, the most toxic form, and so on. Since microbial reactions take place predominantly in the liquid phase, inhibitory products are for the most part the soluble forms even if solubility is limited. Of the gaseous products, hydrogen and carbon dioxide are prime examples distinguished by their extremely large Henry's law constants; still these are important effectors of fermentations.

The nongaseous compounds are the primary toxic compounds and are divided into dissociating and undissociating substances. The dissociating types are inhibitory mainly in the undissociated forms; hence, their inhibitory effects are strongly dependent upon ionic strength in the broth. For example, organic acids are considerably more toxic in acidic broths than under basic conditions. The inhibitory effects of nondissociating nonpolar products are strongly dependent upon their partitioning in the membrane lipids. Thus these have the potential to affect membrane functions in addition to their impact upon intracellular reactive agents and the reactions themselves.

II. EFFECTS OF GASEOUS PRODUCTS

A. Carbon Dioxide

Carbon dioxide is the most common gaseous fermentation product resulting from catabolic reactions and is often inhibitory to growth and product formation at high concentrations (Brown, 1905; Doelle et al., 1982). Ho and Shanahan (1986) and Jones and Greenfield (1982) have provided an extensive listing of carbon dioxide effects on microorganisms and their fermentation. Yet, the full nature of its effect is not firmly established. Dissolved gas probably shifts the reaction equilibrium of decarboxylation reactions that are essentially at equilibrium. This might be beneficial if a coproduct is inhibitory, or harmful if it results in the buildup of a toxic intermediate product of an irreversible reaction in the pathway (indirect effect). Carbon dioxide affects the biosynthesis of membranes (Ho and Smith, 1986).

At low partial pressures, carbon dioxide can stimulate growth (Dehority, 1971). Carbon dioxide is required for succinate production. When carboxylases are active, increases in carbon dioxide partial pressure may result in increased incorporation of the dissolved gas into cellular metabolism and thus increase product yield (Jones and Greenfield, 1982).

Different organisms have markedly different responses to carbon dioxide partial pressure. For instance, Nyiri and Lengyel (1965) reported a decrease in respiration rate by *Penicillium chrysogenum* above 4% carbon dioxide in the exit gas. No inhibition was observed below 1.5%. Emerson and Held (1969), on the other hand, found carbon dioxide inhibitory to the metabolism and to the membrane functions of *Aqualinderella fermentans* only above 20% carbon dioxide. When present with other toxic products, particularly ethanol, the carbon dioxide effects were additive.

Carbon dioxide and, to a lesser extent, methane may be inhibitory to the formation of methane in anaerobic digesters. This explains the increased methane production following sparging of digesters with nitrogen gas (Hansson, 1979; Hansson and Molin, 1981). The gases formed in a digester could also block access to substrates.

For most fermentations conducted near atmospheric pressure, the partial pressure of carbon dioxide is generally low and does not play a major inhibitory role. Burrill et al. (1983) found ethanol, not carbon dioxide, to be the primary inhibitory compound in ethanolic fermentation. One should, however, be cautious of the effect of carbon dioxide when working with systems under high pressure or when the scale of operation results in inordinately high pressures as in the ICI (Imperial Chemical Industry, England) pressure recycle reactor (Scott, 1983). The same applies to situations in which head pressure is increased to improve gas-liquid oxygen transfer or near the bottom of reactors during very large scale fermentations.

B. Hydrogen

Hydrogen is another major fermentation product in several anaerobic processes. Interspecies hydrogen transfer is one of the major forms of interaction in the microbial population (Iannotti et al., 1973; Wolin, 1975, 1982). It involves the production of hydrogen by acidogens and acetogens and subsequent utilization of hydrogen by methanogens in such a way that fermentations of individual organisms are changed and the overall fermentation becomes more efficient. In this way, compounds that would not otherwise be metabolized, are metabolized. The initial substrates metabolized include carbohydrates, lipids, proteins, and recalcitrant ring structures (Chen and Wolin, 1977; Fiebig and Gottschalk, 1983; Latham and Wolin, 1977; Nagase and Matsuo, 1982; Naumann et al., 1983; Scheifinger and Wolin, 1973; Wolin, 1975, 1982). If hydrogen levels are maintained at low levels by methanogens, acidogens are able to shunt hydrogen in the form of NADH to gaseous hydrogen, producing less reduced products, such as ethanol, lactate, propionate, and succinate. This results in the formation of more energy yielding end products, especially acetate, which is the most important intermediate in the flow of carbon to methane (Chynoweth and Mah, 1971; Iannotti et al., 1983; Jeris and McCarty, 1965; Kaspar and Wuhrmann, 1978; Mackie and Bryant, 1981; Mountfort and Asher, 1978; Smith and Mah, 1966).

Because of the thermodynamics involved in the reactions, hydrogen can be considered an inhibitor; increases in the level of hydrogen result in sequential increases in the concentration of propionate, butyrate, ethanol, and lactate (Bryant, 1979). The formation of methane from acetate and cellulolysis can be inhibited by hydrogen (Baresi et al., 1978; Ferguson and Mah, 1983; Laube and Martin, 1981; Zinder and Mah, 1979). The metabolism of propionate and butyrate to acetate, carbon dioxide, and hydrogen by acetogens is much more dependent on interspecies hydrogen transfer (Boone and Bryant, 1980; Bryant, 1979; McInerney et al., 1981).

Hydrogen is produced in almost equimolar proportions with carbon dioxide in acetone-butanol fermentations. The formation of hydrogen in the cells involves the reversible exchange of electrons between reduced nucleotides and ferredoxin and the activity of hydrogenase (Jones and Wood, 1986). Increasing the hydrogen partial pressure has been related to the H^+/hydrogen redox potential in clostridial systems (Jungermann et al., 1976), which favors the formation of butanol (Yerushalmi et al., 1985). At pH levels below 4.5, hydrogenase is present in an inactive form (Kim and Zeikus, 1985), and this results in an increased carbon flux toward solvents. Similarly, inhibition of the enzyme by carbon monoxide has the same effect (Meyer et al., 1985). Kim (1986) has observed a strong relationship between the redox potential in the broth and the flux of sugars to solvents and to acids. At redox potential values below -350 mV, mostly solvents were formed; higher values resulted in the formation primarily of acids. Yerushalmi and Volesky (1985) found that unagitated fermentations resulted in higher solvent levels than their agitated counterparts, which may also be traced back to the supersaturation of broth by released hydrogen gas.

The effects of hydrogen are a result of mass action on reactions forming the gas. There is no evidence that hydrogen is directly related to any inhibitory observations.

III. EFFECTS OF DISSOCIATING PRODUCTS

A. Organic Acids

Low-molecular-weight organic acids (formic, acetic, propionic, butyric, and lactic acids) are the primary dissociating products encountered in anaerobic fermentation. The undissociated forms are primarily responsible for the toxicity (Foster and McLaughlin, 1974; Freese, 1974; Huesemann and Papautsakis, 1986; Kashket et al., 1980; Kroeker et al., 1979; Monot et al., 1983; Sievers and Brune, 1978). Thus, the effect of organic acids upon cell growth is amplified at low pH values; for example, acetic acid is considerably less toxic for cell growth at pH 7.0 than at pH 6.0 (Schwab, 1978). This may be because only the undissociated forms of compounds are able to move across the cell membrane.

Monot et al. (1984) showed the intracellular undissociated acids are responsible for the reduction of growth and enhanced solvent production in *Clostridium acetobutylicum*. An inhibitor of membrane ATPase decreased the internal pH, thus increasing the concentration of undissociated acids within the cells.

For clostridial fermentations, the solvents and the acids behave similarly (Leung and Wang, 1981). This and several other findings indicate that the acids, like the alcohols, interfere with membrane functions (van Uden, 1985). Organic acids allow protons to enter the cell from the medium and thus counteract the effect of the proton pump (Finean et al., 1978). This also reduces the internal pH of the broth, thus increasing the concentration of undissociated acids in the cytoplasm (Herrero, 1983). In the case of *Clostridium thermocellum*, acetic, propionic, butyric, and β-hydroxy butyric acids showed similar noncompetitive inhibition of growth (Herrero, 1981). The toxicity of acids appears to be related to their solubility in the membrane lipids. Formic and acetic acids have similar solubilities in lipids (Sampson et al., 1955) and interfere with phosphate transport across the membrane at low concentrations (0.5-0.9 g/L). Lactic acid has lower solubility in membrane lipids (Conway and Downey, 1950) and is toxic only at high concentrations (10-40 g/L). Higher acids, like octanoic and decanoic acids, formed in ethanol fermentation have also been reported to be toxic to yeast (Lafon-Lafourcade, 1983; Lafon-Lafourcade et al., 1984). Maiorella et al. (1983) have reported irregular cell morphology in *Saccharomyces cerevisiae* cells in the presence of acids. Acetaldehyde also affects the active transport chain and has the same effect as formic and acetic acids.

Organic acids influence cellular metabolism in several other ways as well. Acetic acid affects the formation of RNA in *E. coli* (Schwab, 1978). Biggins and Dilworth (1968) reported that intracellular acetate inhibits acetate kinase activity in *Clostridium pasteurianum*, thus increasing the acetyl phosphate concentration, which acts as a product inhibitor of pyruvate phosphoroclastic activity. Butyric acid is associated with the induction process of acetone-butanol production; 0.2-0.8 g/L of the undissociated acids inhibits the growth of cells and induces solvent production at 0.5-2.5 g/L (Fond et al., 1985).

There are indications from both pure and mixed cultures that the acids are not equal in effect on target organisms. In digesters, propionate has been associated with reactor instability more than any other acid (Andrews, 1969; Buswell and Morgan, 1962). McCarty and Brosseau (1963) added volatile acids to digesters and found that, unlike acetate or butyrate, propionate depressed the biogas production rate. The perturbations caused by propionate also required more time for recovery. Since levels of other acids did not increase as the biogas was reduced, propionate probably affected the organisms that initiate the degradations (McCarty and Brosseau, 1963). This is substantiated in pure culture (Hobson and Shaw, 1976; Iannotti and Fischer, 1984; van den Berg et al., 1976) as acetate was the least toxic volatile acid for both acidogens and methanogens. This relative importance of butyrate and propionate is not universal, however, and it changes with the group of acidogens examined (Iannotti and Fischer, 1984). These differences were not correlated with activity of the acid or metabolic end products. The least toxic acid, acetate, was a metabolic end product for all organisms examined.

B. Ammonia

Ammonia is also a toxic end product when protein is degraded anaerobically. In a domestic waste digester, ammonia nitrogen above 1.2 g/L is toxic depending on the pH (Albertson, 1961; McCarty, 1964). The bacterial population can be acclimated to the high levels of ammonia found in animal manure digesters (Converse et al., 1979; Fischer et al., 1979; Georgacakis et al., 1982; Hobson and Shaw, 1976; Iannotti and Fischer, 1984; Kroeker et al., 1979; Rockey et al., 1978; Sievers and Brune, 1978; van Velsen, 1979). In a multi-cation system, tolerance to potentially toxic cations, such as ammonia, is developed by acclimation of the microorganism to the agent as well as by antagonism of other cations (Kugelman and Chin, 1971).

Free ammonia is the toxic form (McCarty and McKinney, 1961). Ammonia indirectly plays a dual role in digesters (Georgacakis et al., 1982). Besides being toxic, ammonia buffers volatile acids by maintaining a high level of bicarbonate.

IV. EFFECT OF NONDISSOCIATING PRODUCTS

Alcohols and sugars are the primary products of this category. Of these, the alcohols are more toxic to cells. They interact with the cellular machinery at several levels, which includes the membrane functions and cytoplasmic enzymes, as well as the genetic level.

A major portion of the inhibition is correlated with their partitioning in membrane lipids and modification of the membrane physiology (van Uden, 1985).

A. Inhibition by Alcohols

Owing to their commercial importance, a significant amount of information exists concerning ethanol and butanol. Alcohols adversely influence the specific rate of fermentation and the growth rate and viability of cells. This is done by affecting the uptake of nutrients across membranes, altering the activities of anabolic and catabolic enzymes, as well as by influencing the transcriptional and translational processes. The relative contributions of the different modes of action change with the environmental conditions and, because of this, result in a diversity of reported ethanol effects on cells (Ingram and Buttke, 1984).

Interactions of solvents with the membrane layer of cells has been elegantly discussed by Ingram and Buttke (1984). Ethanol and other alcohols cause noncompetitive inhibition of transport systems for sugars (Bowles and Ellefson, 1985; Hutkins and Kashket, 1986; Leao and van Uden, 1982; Loureiro-Dias and Peinado, 1982; Moreira et al., 1981; Ounine et al., 1985), ammonium ions (Leao and van Uden, 1983), and amino acids (Leao and van Uden, 1984; Linden and Moreira, 1982). The inhibition correlates well with the solubility of alcohols in lipids (Ingram and Buttke, 1984; Leao and van Uden, 1984). Alcohol toxicity is dependent upon its chain length and hydrophobicity (Harold, 1970; Hugo, 1967). Ethanol is inhibitory in the range of 0.5-8% (w/v). Fusel oils (propanol and butanol), on the other hand, cause membrane disruption in the concentration range of 2-12 g/L. 2,3-Butanediol has a low lipid solubility and is toxic in the range of 40-90 g/L. Those having more than 10 carbon atoms are not toxic.

Cells grown in the presence of ethanol have a decreased lipid-protein ratio (Ingram, 1982), their membranes are more fluid (Dombek and Ingram, 1983), and they have a different fatty acid composition than those grown without alcohols (Ingram, 1982). In yeasts, alcohol tolerance is related to the fat content (Gray, 1948). The importance of fatty acid composition has been exploited in designing media containing lipids, unsaturated fatty acids, and sterols for achieving high levels of ethanol in fermentation (Hayashida et al., 1975; Hayashida and Ota, 1980). As a result of membrane fluidity and ultimate disruption, potassium and ammonium ions freely leak through the cell membrane, resulting in increased energy expenditure to maintain a proper balance. A corresponding increase in maintenance requirements is also observed (Fieschko and Humphrey, 1983).

The presence of ethanol interferes with the formation of peptidoglycans in bacteria (Blumberg and Strominger, 1974). The effect is similar to that of chaotropic salts and is countered by antichaotropic salts, such as NaCl and K_2SO_4.

In *C. acetobutylicum*, butanol reduces the effectiveness of the cell membrane as a permeability barrier through an increase in membrane fluidity (Bowles and Ellefson, 1985; Vollherbst-Schneck et al., 1984) and loss of membrane-bound ATPase activity (Linden and Moreira, 1982). This causes the increased leakage of phosphorylated compounds and protons across the membrane, and the transmembrane pH gradient is destroyed (Bowles and Ellefson, 1985; Hutkins and Kashket, 1986). As a result, the alkaline interior pH, which is necessary for the cell to maintain its growth and other metabolic activities, is lost (Gottwald and Gottschalk, 1985; Huang et al., 1985). Lower temperatures reduce membrane fluidity and, thereby, counter solvent effects and increase productivity (Carnarius, 1940). Butanol affects both the transport of sugars across the cell membrane as well as their incorporation by the cell (Bowles and Ellefson, 1985; Ounine et al., 1985).

A number of glycolytic enzymes are inhibited by ethanol. Yeast hexokinase and α-glycerophosphate dehydrogenase have 45 and 64%, respectively, of their control activities in the presence of 12% (w/v) ethanol (Nagodawithana et al., 1977). Pyruvate decarboxylase and phosphoglycerate kinase are even more sensitive to alcohol inhibition (Millar et al., 1982). This inhibition is primarily noncompetitive (Bazua and Wilke, 1977; Millar et al., 1982). The allosteric enzymes phosphofructokinase and aldolase are, however, not at all influenced by ethanol (Nagodawithana et al., 1977). The irreversible denaturation of glycolytic enzymes by ethanol has also been reported (Llorente and Sols, 1969).

In bacterial (*Zymomonas mobilis*) as well as in yeast (*S. cerevisiae*) cells, the same enzymes are most strongly affected, suggesting that the enzyme inhibition is rather general. This is particularly interesting as the *Z. mobilis* utilizes the Entner-Duodoroff pathway for breakdown of hexoses, compared with Entner-Meyerhof-Parnas pathway in *S. cerevisiae*.

The toxic effect of alcohol on cell growth is dependent upon other environmental factors, such as temperature or higher concentrations of sugars (Benshoter et al., 1983; Carey and Ingram, 1983; Dombek and Ingram, 1983; Leao and van Uden, 1982; Loureiro and van Uden, 1982; van Uden and Da cruz Duarte, 1981). Higher partitioning of alcohols in membrane lipids under these conditions may be the cause (Dombek and Ingram, 1983; Janoff and Miller, 1982; Leao and van Uden, 1982).

The effect of ethanol on key metabolic enzymes is augmented at elevated temperatures; thus, less ethanol is required to reduce hexokinase activity by 50% in *S. cerevisiae* at 34°C than at 15 or 25°C (Foster and Hofmeyr, 1984). Also, enzymes of alcohol-sensitive yeasts, such as *Kloeckera apiculata*, show higher toxicity from ethanol than those from alcohol-tolerant yeasts, such as *Saccharomyces oviformis* (Llorente and Sols, 1969). Lower temperatures also adversely affect cell growth (Sa-Corriea and van Uden, 1983), suggesting alcohol interactions with the hydrophobic regions of the membrane as the more important locale of action of the compounds (Leao and van Uden, 1984).

In addition to interfering with the formation of peptidoglycans in membrane layers, ethanol inhibits the synthesis of protein and RNA as well (Ingram and Buttke, 1984).

Ethanol is more toxic to the growth of cells than to its own formation. Both these effects are more pronounced in *Z. mobilis* than in *S. cerevisiae* (Brown et al., 1981; Ghose and Tyagi, 1979; Lee and Rogers, 1983). This stronger inhibition has been suggested to stem from higher maintenance requirements in bacteria (Roels, 1983). Lactose-fermenting yeast, *Kluyveromyces fragilis*, as well as thermophilic *Saccharomyces* strains, also show considerably more toxicity (Hahn-Haegerdal, 1985).

Externally added ethanol is less inhibitory than that produced by the cells (Nagodawithana and Steinkraus, 1976; Moulin et al., 1980). This has been attributed to the accumulation of high concentrations of ethanol in the cells (Beaven et al., 1982; Burrill et al., 1983; Laudrin and Goma, 1982; Navarro and Durand, 1978; Novak et al., 1981; Thomas and Rose, 1979). However, there does not appear to be a diffusional limitation in *S. cerevisiae* or *Z. mobilis* (Dasari et al., 1983; Guijarro and Lagunas, 1984; Jobses et al., 1986; Loureiro and Ferreira, 1983). Ethanol may inhibit some intermediate activities (formation of RNA or proteins) that in turn result in delayed growth inhibition. This concept has recently been utilized by Jobses et al. (1986) to propose a structured model of ethanol fermentation systems. It may also explain why continuous cultures tend to show a stronger inhibitory effect of ethanol than that observed in batch systems.

The inhibitory effect of butanol is dependent upon the nature of the carbon source (Ounine et al., 1983). Growth on xylose stops completely at 8 g/L levels, and 12-16 g/L is required with glucose; similarly, glucose transport is reduced by 50% at 7 g/L butanol, and xylose transport is affected to the same extent at only 4.5 g/L levels.

Leung and Wang (1981) found the inhibitory effects of acids and solvents to be additive in *C. acetobutylicum*, suggesting that these are equally strong inhibitors. Operating with *C. acetobutylicum* at a constant pH of 5.0, these authors found that butanol was not inhibitory to growth at concentrations below 8 g/L but it completely stopped cell growth at 14 g/L. For butyric acid, the corresponding levels were 10 and 17 g/L, respectively. The fermentation always stopped at total acid and solvent concentrations between 18 and 20 g/L; the same results were found by Fond et al. (1985).

B. Aromatics

Compounds that are not the direct result of energy metabolism can also be inhibitory to fermentation processes. For example, a variety of benzenoid compounds have been shown to inhibit the cellulolytic activity of bacteria to varying degrees (Borneman et al., 1986; Chesson et al., 1982; Fedorak and Hrudey, 1984; Jung and Fahey, 1983; Jung et al., 1983; Jung and Sahlu, 1986; Varel and Jung, 1986). Exposure apparently results in membrane damage, lysis of bacteria, and release of cell contents (Jung and Fahey, 1983).

These aromatic compounds result from the partial metabolism of protein, lignin, and so on. Benzoic acids, hydrocinnamic acid, and phenylacetic acid are produced during the degradation of the aromatic amino acids tyrosine and phenylalanine by a wide variety of anaerobic bacteria (Allison, 1978; Bourgeau and Mayrand, 1983; Iannotti et al., 1982; Iannotti et al., 1986b; Jung, 1985; Mayrand and Bourgeau, 1982; Moss et al., 1980; van Assche, 1978). Also, phenolic acids and their esters are released from grass cell walls by cellulase action (Jung and Fahey, 1983). Phenylacetic acid can accumulate and then be degraded in swine manure digesters (Iannotti et al., 1986b). In fact, a wide variety of aromatic compounds have been found to be degraded under anaerobic conditions (Balba et al., 1979; Bouwer and McCarty, 1983; Boyd et al., 1983; Colberg and Young, 1982; Ferry and Wolfe, 1976; Grbic-Galic, 1984; Healy and Young, 1979; Nottingham and Hungate, 1969; Shelton and Tiedje, 1984; Wang et al., 1984; Zeikus, 1980). Rather complex molecules, such as lignin, can undergo a degree of slow degradation (Benner et al., 1984). This degradation is probably the result of not just a simple food chain but of microbial interaction. *Syntrophus buswellii* is a benzoate catabolizer that has a requirement for low hydrogen concentration similar to that discussed above (Mountfort et al., 1984). Thus, like the acids discussed earlier, aromatic acids are also intermediates in the anaerobic process that can become toxic.

V. MODELS OF PRODUCT INHIBITION

A. Models of Ethanol Fermentation

The economic importance of ethanol fermentation has resulted in a significant amount of published literature dealing with models of ethanol fermentation. Most of these models are unstructured in nature and are based upon the observation that inhibition is primarily noncompetitive. An excellent survey of these models is presented by van Uden (1985), and the kinetic forms of growth and product formation are summarized in Table 1. Kinetic expressions ranging from exponential to linear to a hyperbolic inhibition type have been used by different authors. It is not very clear why a particular form of kinetic expression may be favored over others. It is interesting that Taylor series expansions of all of these give similar forms; as a result, no differences are expected between these models at low ethanol concentrations. Indeed, Roman et al. (1984) failed to discriminate among the three forms, all of which fit their experimental data equally well. Aiba and Shoda (1969) also noticed that the data of Aiba et al. (1968) could be fit with hyperbolic as well as exponential functions. Levenspiel (1980) utilized data of Bazua and Wilke (1977) to establish a function of the form

$$\mu = \mu_{max} \frac{S}{K_S + S} \left(1 - \frac{P}{P_{max}}\right)^{0.45}$$

The hyperbolic function has been a favorite form for many investigators, presumably because it is akin to the noncompetitive form of enzyme inhibition. At high ethanol concentrations, however, it suggests a weaker inhibition that actually observed (Egamberdiev and Ierusalimsky, 1968). One should therefore take note that ethanol inhibition of growth and fermentation activity is a composite effect of several factors whose relative contributions may change with ethanol concentration. Of particular interest is the model proposed by Converti et al. (1986) that incorporates product inhibition through a substrate concentration-dependent parameter.

None of the models presented in Table 1 is compatible with the observation that externally added ethanol is less toxic than that produced by the cell (Nagodawithana and Steinkraus, 1976; Navarro and Durand, 1978). They also do not consider that inhibitory effects appear to be stronger in chemostats than in batch and batch-fed cultures. Are these observations only artifacts of measuring techniques, or are they real? Although it may not be possible at present to be this precise, an attempt to explain these observations has been made by Jobses et al. (1986) in the form of a structured model. This model considers an intracellular component E (perhaps RNA or protein) whose synthesis is inhibited by ethanol that participates in the formation of cell mass as well as product (ethanol). The governing equations for this structured model in continuous culture are

Table 1 Models of Growth and Product Formation

Specific growth rate μ	Specific product formation rate ν	Reference
$\mu_{max} - k_1 (P - P_{min})$ $k_1 = 0 \quad P < P_{min}$ $k_1 \neq 0 \quad P > P_{min}$	—	Holzberg et al. (1967)
$\mu_{max} \frac{S}{K_s + S} e^{-k_1 p}$	$\nu_{max} \frac{S}{K_p + S} e^{-k_2 p}$	Aiba et al. (1968)
$\mu_{max} \frac{S}{K_s + S} \frac{K_{ix}}{K_{ix} + P}$	$\nu_{max} \frac{S}{K_p + S} \frac{K_{ip}}{K_{ip} + P}$	Aiba and Shoda (1969); Kollerup and Daugulis (1985)
$\mu_{max} \frac{S}{K_s + S + S^2/K_{is}} \left(1 - \frac{P}{P_{max}}\right)$	$\nu_{max} \frac{S}{K_p + S} \left(1 - \frac{P}{P'_{max}}\right)$	Ghose and Tyagi (1979)
$\mu_{max} - \frac{aP}{b - P_{max}}$	—	Bazua and Wilke (1977)
$\mu_{max} \frac{S}{K_s + S} \frac{1}{1 + P/K_{ix_1} + P^2/K_{ix_2}}$	$\nu_{max} \frac{S}{K_p + P} \frac{K_{ip}}{K_{ip} + P}$	Jayanata (1983)

$\mu_{max} \left[1 - \left(\frac{P}{P_{max}} \right) \right]^n$	$\nu_{max} \left(1 - \frac{P}{P'_{max}} \right)^m$	Levenspiel (1980)
$\mu_{max} \left[1 - \left(\frac{P}{P_{max}} \right)^n \right]$	$\nu_{max} \left[1 - \left(\frac{P}{P'_{max}} \right)^m \right]$	Luong (1985)
—	$\nu_{max} \frac{S}{K_p + S} \left\{ 1 - \exp \left[-a' \left(a - \frac{P}{P'_{max}} \right)^m \right] \right\}$	Converti et al. (1986)
—	$a' = a_0 \left(1 - \frac{S_0}{K_3} \right)$	
$\mu_{max} \frac{S}{K_s + S} \frac{K_{ix}}{K_{ix} + Y_{p/s} (S_0 - S)}$	—	Hoppe and Hansford (1982)

$$\frac{dX}{dt} = p \frac{S}{K_s + S} E - D X$$

$$\frac{dE}{dt} = f(P) \cdot \frac{S}{K_s + S} E - D E$$

$$\frac{dS}{dt} = -\frac{1}{Y_{sx}} p \frac{S}{K_s + S} E - m_s X + D (S_{in} - S)$$

$$\frac{dP}{dt} = \frac{1}{Y_{px}} p \frac{S}{K_s + S} E + m_p X - D P$$

$$f(P) = k_1 - k_2 P - k_3 P^2$$

This type of model incorporates a significant amount of information and could be an excellent way of testing the different hypotheses. The maintenance terms have been reported to increase at increased ethanol concentrations (Fieschko and Humphrey, 1983; Lee et al., 1979), and their variation can be easily incorporated in these models.

When complex substrates are utilized as carbon sources, they must be broken down into suitable monomeric units prior to their final incorporation into cellular metabolism. This can take place in a preprocessing step (hydrolysis or saccharification) or in a simultaneous step. A number of investigators have studied the fermentation of hydrolyzates (Chung and Lee, 1985; Ghose and Tyagi, 1979; Watson et al., 1984). In most of these cases, the net effect is lumped in the values of such parameters as μ_m, ν_m, K_{is}, P_{max}, and P'_{max} (Ghose and Tyagi, 1979). The effect of by-products of the hydrolysis process has been extensively studied in acid hydrolysis (Chung and Lee, 1985; Jeffries et al., 1985). Jayanata (1983) studied the fermentation of molasses in batch and continuous cultures and accounted for the effects of the monosaccharides glucose and fructose upon each compound's metabolism. The model used by Jayanata (1983) was found to be applicable to a wide range of molasses concentrations in batch as well as continuous experiments involving different extents of cell recycling. This model exclusively involves the formation of monosaccharides by the inversion of sucrose in molasses, as follows:

Specific rate of hydrolysis of molasses = r_{inv} =

$$r_m E_{inv} \frac{1}{1 + P/K_a} \frac{1}{1 + (G + F)/k_{i1}} \frac{Sc}{K_{Sc}[1 + (G + F)/K_{i2}] + Sc}$$

where E_{inv} is the concentration of invertase in the cells, given by

$$E_{inv} = E_{max} \frac{1 + k_1 (G + F)^4}{1 + k_2 (G + F)^4}$$

The specific growth and product formation rates from the monosaccharides glucose and fructose were modeled as

$$\mu^G = \mu_m^G \frac{G}{K_G (1 + aF) + G} \frac{1}{1 + P/K_{ix1} + P^2/K_{ix2}}$$

$$\mu^F = \mu_m^F \frac{F}{K_F(1 + G/a) + F} \frac{1}{1 + P/K_{ix1} + P^2/K_{ix2}}$$

$$\nu^G = \nu_m^G \frac{G}{K'_G(1 + bF) + G} \frac{1}{1 + P/K_{ip}}$$

$$\nu^F = \nu_m^F \frac{F}{K'_F(1 + G/b) + F} \frac{1}{1 + P/K_{ip}}$$

The total specific growth and product formation rates were defined as

$$\mu = \mu^G + \mu^F$$

$$\nu = \nu^G + \nu^F$$

B. Models of Acetone-Butanol Fermentation

Leung (1982) related the concentrations of acids and solvents in clostridial fermentation to the specific growth rate of the cells and their product formation capability. Based upon his observations in batch and continuous cultures, he proposed that growth is noncompetitively inhibited by the sum total of molar concentrations of butyric acids and the butanol as follows:

$$\frac{\mu}{\mu_m} = 1 - \frac{0.10P_B}{180 - P_B}$$

where P_B is the sum of concentrations of butanol and butyric acids in millimoles per liter. The effect of acetate was modeled similarly but with an exponential function, with the net effect upon growth as the cumulative contributions of the acids:

$$\frac{\mu}{\mu_m} = e^{-0.0096P_A}$$

where P_A is the concentration of acetic acid in millimoles per liter. This model did not consider direct effect of products on their formation rates except through a growth-related substrate uptake function

$$q_S = (3.2 + 32)\mu \qquad \text{mmol/(g dry weight cells h)}$$

The product formation rates were directly related to the substrate uptake rate. The formation of butyrate was favored over that of butanol once the specific sugar uptake rate fell below 11 mmol/g DW cells per hour. The opposite was true above this critical sugar flux.

A more involved structured model has been proposed by Votruba et al. (1986) in which an intracellular marker y of physiological state, such as RNA, has been used. The only inhibitory action considered is that of butanol for the physiological marker.

$$\frac{dy}{dt} = \left[k_1 S \frac{K_i}{K_i + B} - \lambda(y - 1) \right] y$$

The relationship between the specific growth rate of cells and the intracellular marker has been defined as

$$\frac{1}{X}\frac{dX}{dt} = 0.56(y - 1) - k_2 B$$

where the second term represents the butanol-associated cell lysis rate. The substrate consumption rate is not at all affected by the concentration of butanol:

$$\frac{1}{X}\frac{dS}{dt} = -k_3 \; S - k_4 \frac{S}{K_s + S}$$

The rates of formation of butyric acid as well as those of acetic acid are

$$\frac{1}{X}\frac{dBA}{dt} = k_5 S \frac{K_i}{K_i + B} - k_6 \frac{BA}{K_{BA} + BA}$$

$$\frac{1}{X}\frac{dAA}{dt} = k_8 \frac{S}{K_s + S} \frac{K_i}{K_i + B} - k_9 \frac{S}{K_s + S} \frac{AA}{K_{AA} + AA}$$

The solvent and gas formation rates are also not influenced by the product butanol:

$$\frac{1}{X}\frac{dB}{dt} = k_7 \; S - 0.841 \frac{1}{X}\frac{dBA}{dt}$$

$$\frac{1}{X}\frac{dA}{dt} = k_{10} \frac{S}{K_s + S} - 0.484 \frac{1}{X}\frac{dAA}{dt}$$

$$\frac{1}{X}\frac{dEt}{dt} = k_{11} \frac{S}{K_s + S}$$

$$\frac{1}{X}\frac{dCO_2}{dt} = k_{12}\frac{S}{K_s + S}$$

$$\frac{1}{X}\frac{dH_2}{dt} = k_{13}\frac{S}{K_s + S} + k_{14} \quad S$$

This model is therefore strikingly dissimilar to that of Leung (1982), yet both have been shown to faithfully represent the respective data of the authors. The model by Votruba et al. (1986) incorporates some of the desirable structural features but violates the observations relating butanol to transport effects. Hence the prospect of considerably more realistic models exists.

C. Models in Anaerobic Digestion

The anaerobic digestion process has been extensively modeled and is covered in greater detail in Chapter 18 (Andrews, 1969; Chen and Hashimoto, 1980; Gaudy and Gaudy, 1980; Hill, 1982; Hill and Barth, 1977; Hobson, 1983; Lawrence, 1971; Murphy et al., 1986). Approaches to the modeling and the extent to which microbial growth is mathematically treated vary greatly and include first-order, Monod, and Contois kinetics for bacterial growth. The simpler treatments are limited to the steady state over a narrow range of operating conditions and do not predict dynamic changes and system failure. Andrews (1969), Hill (1982), Murphy et al. (1986), and others have developed structured, descriptive models with more complex bacterial populations; these models depict the metabolism and interactions found (Bryant, 1969; Iannotti et al., 1986a; Kirsch and Sykes, 1971; McInerney and Gryant, 1981). The more complex models have a greater ability to predict dynamic conditions and to test hypotheses. We have found Hill's model valid for predicting the steady-state and dynamic performance of swine manure digesters, including the point of process failure.

The models must include the inhibition of bacterial growth in order to accurately predict dynamic reactions and process failure under stressed conditions. Chen and Hashimoto (1980) depicted overall inhibition by a general parameter. Andrews (1969) included the inhibitory effect of un-ionized volatile fatty acids. Hill modified the kinetics further to include the inhibitory effect of free ammonia (Hill, 1982; Hill et al., 1983). The growth rate for the methanogens that metabolize acetate is

$$\mu = \mu_{max}\frac{1}{K_A/AA + 1 + Ac/K_{iAc} + Am/K_{iAm}}$$

VI. CONCLUSIONS

Product inhibition in biological systems is more a rule than an exception. Its purpose is to prevent the build-up of toxic chemicals, and it is implemented through metabolic control, when necessary; it is manifested at all the possible levels ranging from genetic expression to transport of molecules across membranes. In this chapter, some of the key metabolic effects have been summarized along with the mathematical models that have been employed to represent product inhibition. As a result of fundamental understanding of the mode and locale of inhibition phenomenon, it is apparent that simple mathematical models cannot succeed in representing product inhibition in its *entire range*. Hence, a trend toward structural modeling is emerging, as suggested by some of the models presented here. It is obvious that considerable work needs to be done to develop models to account for the observed experimental factors associated with inhibition.

VII. NOMENCLATURE

a, a_0	constants
A	acetone concentration
Ac	total acid concentration
Am	concentration of ammonia
AA	acetic acid concentration
B	butanol concentration
BA	butyric acid concentration
CO_2	amount of carbon dioxide formed
D	dilution rate
E	amount of intracellular marker in the system
E_{inv}	amount of invertase per unit weight of cells
Et	ethanol concentration
F	fructose concentration
G	glucose concentration
H_2	amount of hydrogen formed
K_1-K_{14}	constants
K_a	product-related inhibition constant for invertase activity
K_A	saturation constant for growth on acetic acid
K_{AA}	saturation constant for reconsumption of butyric acid by the cells
K_F	saturation constant for growth on fructose
K_F'	saturation constant for product formation on fructose
K_G	saturation constant for growth on glucose

K'_G	saturation constant for product formation on glucose
K_i	butanol-associated inhibition constant
K_{i1}, K_{i2}	monosaccharide-related inhibition parameters for invertase activity
K_{iAc}	acid inhibition constant
K_{iAm}	ammonia inhibition constant
K_{ip}	inhibition constant for product formation
K_{is}	substrate inhibition constant for cell growth
K_{ix}, K_{ix1}, K_{ix2}	inhibition constants for cell growth
K_p	saturation constant for product formation
K_s	saturation constant for cell growth
K_{Sc}	saturation constant for hydrolysis of sucrose
m, n	constant exponents
m_p	maintenance consumption of substrate
p	maximum growth rate per unit amount of intracellular marker y
P	product concentration
P_A	molar concentration of acetic acid
P_s	sum of molar concentrations of butanol and butyric acid
P_{max}	maximum product concentration that reduces cell growth rate to zero
P'_{max}	maximum product concentration that reduces product formation rate to zero
q_s	specific substrate uptake rate
r_{inv}	specific rate of hydrolysis of molasses
r_m	maximum specific rate of hydrolysis per unit amount of invertase
S	substrate concentration
Sc	concentration of sucrose
S_{in}	inlet substrate concentration
S_0	initial substrate concentration
X	cell dry weight concentration
t	time
Y_{px}	yield constant relating product formation to biomass production
Y_{sx}	yield constant relating substrate consumption to biomass production
$Y_{x/s}$	yield of cell mass from substrate

Greek Letters

μ	specific growth rate
ν	specific product formation rate
λ	constant

Superscripts

F, G	property corresponding to fructose and glucose

Subscripts

m, max	maximum value
min	minimum value

REFERENCES

Aiba, S., and Shoda, M. (1969). Reassessment of product inhibition in alcohol fermentation. *J. Ferment. Technol.* *47*:790-794.

Aiba, S., Shoda, M., and Nagatani, M. (1968). Kinetics of product inhibition in alcohol fermentation. *Biotechnol. Bioeng.* *10*:845-864.

Albertson, O. E. (1961). Ammonia nitrogen and the anaerobic environment. *J. Water Pollut. Fed.* *33*:987-995.

Allison, M. J. (1978). Production of branched-chain volatile fatty acids by certain anaerobic bacteria. *Appl. Environ. Microbiol.* *35*:872-877.

Andrews, J. F. (1969). Dynamic model of the anaerobic digestion process. *J. Sanit. Eng.* *95*:95-116.

Balba, M. T., Clarke, N. A., and Evans, W. C. (1979). The methanogenic fermentation of plant phenolics. *Biochem. Soc. Trans.* *5*:302-304.

Baresi, L., Mah, R. A., Ward, D. M., and Kaplan, I. R. (1978). Methanogenesis from acetate: Enrichment studies. *Appl. Environ. Microbiol.* *36*:186-197.

Bazua, C. D., and Wilke, C. R. (1977). Ethanol effects on the kinetics of a continuous fermentation with *Saccharomyces cerevisiae*. *Biotechnol. Bioeng. Symp.* *7*:105-118.

Beaven, M. J., Charpentier, C., and Rose, A. H. (1982). Production and tolerance of ethanol in relation to phospholipid fatty-acyl composition in *Saccharomyces cerevisiae* NCYC 431. *J. Gen. Microbiol.* *128*:1447-1455.

Beesch, S. C. (1952). Acetone-butanol fermentation of sugars. *Ind. Eng. Chem.* *44*:1677-1682.

Benner, R., Maccubbin, A. E., and Hodson, R. E. (1984). Anaerobic biodegradation of the lignin and polysaccharide components of lignocellulose and synthetic lignin by sediment microflora. *Appl. Environ. Microbiol.* *47*:998-1004.

Benshoter, A. S., Dombek, K. M., and Ingram, L. O. (1983). Effects of growth temperature on the membrane composition of *Zymomonas mobilis* strain CP4. Proceedings of the 83rd Annual Meeting of the American Society for Microbiology, p. 240, Abstract O-8.

Biggins, D. R., and Dilworth, M. J. (1968). Control of pyruvate phosphoroclastic activity in extracts of *Clostridium pasteurianum* by ADP and acetyl phosphate. *Biochim. Biophys. Acta 156*:285-296.
Blotkamp, P. J., Takagi, M., Pemberton, M. S., and Emert, G. H. (1977). Enzymatic hydrolysis of cellulose and simultaneous fermentation to ethanol. *AIChE Symp. Ser. 74*:85-90.
Blumberg, P. M., and Strominger, J. L. (1974). Interaction of penicillin with the bacterial cell. Penicillin-binding protein and penicillin-binding enzymes. *Bacteriol. Rev. 38*:291-335.
Boone, D. R., and Bryant, M. P. (1980). Propionate-degrading bacterium, *Syntrophobacter wolinii* sp. nov. gen. nov. from methanogenic ecosystems. *Appl. Environ. Microbiol. 40*:626-632.
Borneman, W. S., Akin, D. E., and Vaneseltine, W. P. (1986). Effect of phenolic monomers on ruminal bacteria. *Appl. Environ. Microbiol. 52*:1331-1339.
Bourgeau, G., and Mayrand, D. (1983). Phenylacetic acid production by *Bacteroides gingivalis* from phenylalanine and phenylalanine-containing peptides. *Can. J. Microbiol. 29*:1184-1189.
Bouwer, E. J., and McCarty, P. L. (1983). Transformations of 1- and 2-carbon halogenated aliphatic organic compounds under methanogenic conditions. *Appl. Environ. Microbiol. 45*:1286-1294.
Bowles, L. K., and Ellefson, W. L. (1985). Effects of butanol on *Clostridium acetobutylicum*. *Appl. Environ. Microbiol. 50*:1165-1170.
Boyd, S. A., Shelton, D. R., Berry, D., and Tiedje, J. M. (1983). Anaerobic biodegradation of phenolic compounds in digested sludge. *Appl. Environ. Microbiol. 46*:50-54.
Brown, A. J. (1905). *J. Chem. Soc. 87*:1395. [Referred by Ingram, L. O. and Buttke, T. M. (1984). Effects of alcohols on microorganisms. *Adv. Microb. Physiol. 25*:253-300.]
Brown, S. W., Oliver, S. G., Harrison, D. E. F., and Righelato, R. C. (1981). Ethanol inhibition of yeast growth and fermentation: Differences in the magnitude and complexity of the effect. *Eur. J. Appl. Microbiol. Biotechnol. 11*:151-155.
Bryant, M. P. (1979). Microbial methane production—theoretical aspects. *J. Anim. Sci. 48*:193-201.
Burrill, H., Doelle, H. W., and Greenfield, P. F. (1983). The inhibitory effect of ethanol on ethanol production by *Zymomonas mobilis*. *Biotechnol. Lett. 5*:423-428.
Buswell, A. M., and Morgan, G. B. (1962). Paper chromatographic methods for volatile acids. *Proceedings of the 17th Purdue Industrial Waste Conference*, Purdue University, Lafayette, Indiana, pp. 377-388.
Carey, V. C., and Ingram, L. O. (1983). Lipid composition of *Zymomonas mobilis*: Effects of ethanol and glucose. *J. Bacteriol. 154*:1291-1300.

Carnarius, E. H. (1940). U.S. Patent 2,198,104.
Chen, M., and Wolin, M. J. (1977). Influence of CH_4 production by *Methanobacterium ruminantium* on the fermentation of glucose and lactate by *Selenomonas ruminantium*. *Appl. Environ. Microbiol. 34*:756-759.
Chen, Y. R., and Hashimoto, A. G. (1980). Substrate utilization kinetic model for biological treatment processes. *Biotech. Bioeng. 22*:2081-2095.
Chesson, A., Stewart, C. S., and Wallace, R. J. (1982). Influence of plant phenolic acids on growth and cellulolytic activity of rumen bacteria. *Appl. Environ. Microbiol. 44*:597-603.
Chung, I. S., and Lee, Y. Y. (1985). Ethanol fermentation of crude acid hydrolyzate of cellulose using high-level yeast inocula. *Biotechnol. Bioeng. 27*:308-315.
Chynoweth, D. P., and Mah, R. A. (1971). Volatile acid formation in sludge digestion. *Adv. Chem. Ser. 105*:41-54.
Colberg, P. J., and Young, L. Y. (1982). Biodegradation of lignin-derived molecules under anaerobic conditions. *Can. J. Microbiol. 28*:886-889.
Converse, J. C., Graves, R. E., and Evans, G. W. (1979). Anaerobic degradation of dairy manure under mesophilic and thermophilic temperatures. *Trans. ASAE 20*:336-340.
Converti, A., Perego, P., del Borghi, M., Parisi, F., and Ferraiolo, G. (1986). Kinetic considerations about the study of alcoholic fermentations of starch hydrolysate. *Biotechnol. Bioeng. 28*:711-717.
Conway, E. J., and Downey, M. (1950). An outer metabolic region of the yeast cell. *Biochem. J. 47*:347-355.
Dasari, G., Roddick, F., Connor, M. A., and Pamment, N. B. (1983). Factors affecting the determination of intracellular ethanol concentrations. *Biotechnol. Lett. 5*:715-720.
Dehority, B. A. (1971). Carbon dioxide requirement of various species of rumen bacteria. *J. Bacteriol. 105*:70-76.
Demain, A. L. (1972). Cellular and environmental factors affecting the synthesis and excretion of metabolites. *J. Appl. Chem. Biotechnol. 22*:345-362.
Demain, A. L., Kennel, Y. M., and Aharonowitz, Y. (1979). Carbon catabolite regulation of secondary metabolism. In *Microbial Technology: Current State, Future Prospects*, 29th Symposium Society for General Microbiology, pp. 163-185.
Doelle, H. W., Preusser, H. J., and Rostek, H. (1982). Electron microscopic investigation of *Zymomonas mobilis* cells grown in low and high glucose concentration. *Eur. J. Appl. Microbiol. Biotechnol. 16*:136-141.

Dombek, K. M., and Ingram, L. O. (1983). Effects of ethanol and temperature on the membrane fluidity of *Zymomonas mobilis* strain CP4. *Proceedings of the 83rd Annual Meeting of the American Society for Microbiology*, p. 241, Abstract O-9.

Egamberdiev, N. B., and Ierusalimsky, N. D. (1968). Effect of ethanol concentration on the growth of *Saccharomyces vini* (strain PR-1). *Mikrobiologiia 37*:686-690.

Emerson, R., and Held, A. A. (1969). *Aqualinderella fermentans* Gen. ET. SP. N., A phycomycete adapted to stagnant waters. II. Isolation, cultural characteristics, and gas relationships. *Am. J. Bot. 56*:1103-1120.

Fedorak, P. M., and Hrudey, S. E. (1984). The effects of phenol and some alkyl phenolics on batch anaerobic methanogenesis. *Water Res. 18*:361-367.

Ferguson, T. J., and Mah, R. A. (1983). Effect of hydrogen-carbon dioxide on methanogenesis from acetate or methanol in *Methanosarcina* sp. *Appl. Environ. Microbiol. 46*:348-355.

Ferry, J. G., and Wolfe, R. S. (1976). Anaerobic degradation of benzoate to methane by a microbial consortium. *Arch. Microbiol. 107*:33-40.

Fiebig, K., and Gottschalk, G. (1983). Methanogenesis from choline by a coculture of *Desulfovibrio* sp. and *Methanosarcina barkei*. *Appl. Environ. Microbiol. 45*:161-168.

Fieschko, J., and Humphrey, A. E. (1983). Effect of temperature and ethanol concentration on the maintenance and yield coefficient of *Zymomonas mobilis*. *Biotechnol. Bioeng. 25*:1655-1666.

Finean, J. B., Coleman, R., and Mitchell, R. H. (1978). In *Membranes and Their Cellular Functions*, 2nd Ed. John Wiley and Sons, New York.

Fischer, J. R., Iannotti, E. L., Porter, J. H., and Garcia, A. (1979). Producing methane gas from swine manure in a pilot-size digester. *Trans. ASAE 22*:370-374.

Fond, O., Matta-Ammouri, G. T., Petitdemange, H., and Engasser, E. (1985). The role of acids on the production of acetone and butanol by *Clostridium acetobutylicum*. *Appl. Microbiol. Biotechnol. 22*:195-200.

Foster, E. J. F., and Hofmeyr, J. H. S. (1984). *3rd Congress South African Society for Microbiology*, Cape Town, South Africa.

Foster, M., and McLaughlin, S. (1974). Complexes between uncouplers of oxidative phosphorylation. *J. Membr. Biol. 17*:155-180.

Freese, E. (1974). Mechanism of growth inhibition by lipophilic acids. In *Symposium on the Pharmacological Effects of Lipids*, J. J. Kabara (Ed.). American Oil Chemists Society, Washington, D.C., pp. 123-131.

Gaudy, A. F., and Gaudy, E. T. (1980). *Microbiology for Environmental Scientists and Engineers.* McGraw-Hill, New York.

Georgacakis, D., Sievers, D. M., and Iannotti, E. L. (1982). Buffer stability in manure digesters. *Agric. Wastes 4*:427-441.

Ghose, T. K., and Tyagi, R. D. (1979). Rapid ethanol fermentation of cellulose hydrolyzate. II. Product and substrate inhibition and optimization of fermenter design. *Biotechnol. Bioeng. 21*:1401-1420.

Gottwald, M., and Gottschalk, G. (1985). The internal pH of *Clostridium acetobutylicum* and its effect on the shift from acid to solvent production. *Arch. Microbiol. 143*:42-46.

Gray, W. D. (1948). Further studies on the alcohol tolerance of yeast: Its relationship to cell storage compounds. *J. Bacteriol. 55*: 53-59.

Grbic-Galic, D. (1984). Anaerobic degradation of coniferyl alcohol by methanogenic consortia. *Appl. Environ. Microbiol. 46*:1442-1446.

Guijarro, J. M. G., and Lagunas, R. (1984). *Saccharomyces cerevisiae* does not accumulate ethanol against a concentration gradient. *J. Bacteriol. 160*:874-878.

Hahn-Haegerdal, B. (1985). Comparison between immobilized *Kluyveromyces fragilis* and *Saccharomyces cerevisiae* coimmobilized with β-galactosidase with respect to continuous ethanol production from concentrated whey permeate. *Biotechnol. Bioeng. 27*:914-916.

Hansson, G. (1979). Effects of carbon dioxide and methane on methanogenesis. *Eur. J. Microbiol. Biotechnol. 6*:351-359.

Hansson, G., and Molin, N. (1981). End product inhibitions in methane fermentations: Effects of carbon dioxide and methane on methanogenic bacteria utilizing acetate. *Eur. J. Appl. Microbiol. Biotechnol. 13*:236-241.

Harold, F. M. (1970). Antimicrobial agents and membrane function. *Adv. Microb. Physiol. 4*:45-104.

Hayashida, S., and Ota, K. (1980). Mechanism of formation of high concentration alcohol in Sake brewing. XII. Effect of phosphatidyl choline or ergosteryl oleate on physiological properties of *Saccharomyces Sake. Agric. Biol. Chem. 44*:2561-2567.

Hayashida, S., Feng, D. D., and Hongo, M. (1975). Mechanism of fermentation of high concentration alcohol in Sake brewing. X. Physiological properties of yeast cells grown in the proteolipic-supplemented media. *Agric. Biol. Chem. 39*:1025-1031.

Healy, J. B., Jr., and Young, L. Y. (1979). Anaerobic biodegradation of eleven aromatic compounds to methane. *Appl. Environ. Microbiol. 38*:84-89.

Herrero, A. A. (1981). The physiology of *Clostridium thermocellum.* The toxicity of its fermentation products in relation to the energy metabolism, Ph.D. Thesis, Massachusetts Institute of Technology, Cambridge, Massachusetts.

Herrero, A. A. (1983). End product inhibition in anaerobic fermentation. *Trends Biotechnol.* *1*:49–53.

Hill, D. T. (1982). A comprehensive dynamic model for animal waste methanogenesis. *Trans. ASAE* *25*(5):1374–1380.

Hill, D. T., and Barth, C. L. (1977). A dynamic model for simulation of animal waste digestion. *J. Water Pollut. Control Fed.* *49*(10): 2129–2143.

Hill, D. T., Tollner, E. D., and Holmberg, R. D. (1983). The kinetics of inhibition in methane fermentation of swine manure. *Agric. Wastes* *5*:105–123.

Ho, C. S., and Shanahan, J. F. (1986). Carbon dioxide transfer in bioreactors. *CRC Crit. Rev. Biotechnol.* *4*:185–252.

Ho, C. S., and Smith, M. D. (1986). Morphological alterations of *Penicillium chrysogenum* caused by carbon dioxide. *J. Gen. Microbiol.* *132*:3479–3484.

Hobson, P. N. (1983). The kinetics of anaerobic digestion of farm wastes. *J. Chem. Tech. Biotechnol.* *33*(B):1–20.

Hobson, P. N., and Shaw, B. G. (1976). Inhibition of methane production by *Methanobacterium formicicum*. *Water Res.* *10*:849–852.

Holzberg, I., Finn, R. K., and Steinkraus, K. H. (1967). A kinetic study of the alcoholic fermentation of grape juice. *Biotechnol. Bioeng.* *9*:413–427.

Hoppe, G. K., and Hansford, G. S. (1982). Ethanol inhibition of continuous yeast growth. *Biotechnol. Lett.* *4*:39–44.

Huang, L., Gibbons, L. N., and Forsberg, C. W. (1985). Transmembrane pH gradient and membrane potential in *Clostridium acetobutylicum* during growth under acidogenic and solventogenic conditions. *Appl. Environ. Microbiol.* *50*:1043–1047.

Huesemann, M., and Papautsakis, E. T. (1986). Effect of acetoacetate, butyrate and uncoupling ionophores on growth and product formation of *Clostridium acetobutylicum*. *Biotechnol. Lett.* *8*:37–42.

Hugo, W. B. (1967). The mode of action of antimicrobial agents. *J. Appl. Bacteriol.* *30*:17–50.

Hutkins, R. W., and Kashket, E. R. (1986). Phosphotransferase activity in *Clostridium acetobutylicum* from acidogenic and solventogenic phases of growth. *Appl. Environ. Microbiol.* *51*:1121–1123.

Iannotti, E. L., and Fischer, J. R. (1984). Effects of ammonia, volatile acids, pH and sodium on growth of bacteria isolated from a swine manure digester. *Dev. Ind. Microbiol.* *25*:741–747.

Iannotti, E., Kafkewitz, D., Wolin, M. J., and Bryant, M. P. (1973). Glucose fermentation products of *Ruminococcus albus* grown in continuous culture with *Vibrio succinogenes*: Changes caused by interspecies transfer of hydrogen. *J. Bacteriol.* *113*:1231–1240.

Iannotti, E. L., Fischer, J. R., and Sievers, D. M. (1982). Characterization of bacteria from a swine manure digester. *Appl. Environ. Microbiol.* *43*:136–143.

Iannotti, E. L., Mueller, R. E., Fischer, J. R., and Sievers, D. (1983). Changes in a swine manure anaerobic digester with time after loading. In *Third Annual Solar and Biomass Workshop*, Atlanta, Georgia, pp. 79-81.

Iannotti, E. L., Mueller, R. E., and Fischer, J. R. (1986a). In Microbiology and biochemistry of methane fermentations. In *Biomass Energy Development*, W. Smith (Ed.). Plenum, New York, pp. 415-438.

Iannotti, E. L., Mueller, R. E., Sievers, D. M., Georgacakis, D. G., and Gerhardt, K. O. (1986b). Phenylacetic acid in an anaerobic swine manure digester. *J. Ind. Microbiol. 1*:57-61.

Ingram, L. O. (1982). Regulation of fatty acid composition in *Escherichia coli*: A proposed common mechanism for changes induced by ethanol, chaotropic agents and a reduction of growth temperature. *J. Bacteriol. 149*:166-172.

Ingram, L. O., and Buttke, T. M. (1984). Effects of alcohols on microorganisms. *Adv. Microb. Physiol. 25*:253-300.

Janoff, A. S., and Miller, K. W. (1982). A critical assessment of the lipid theories of general anaesthetic action. In *Biological Membranes*, Vol. 4, D. Chapman (Ed.). Academic Press, New York, p. 419.

Jayanata, Y. (1983). Einsatz reaktionstechnischer Modelle zur Kennzeichnung and Optimierung der kontinuierlichen Aethanolproduktion mit Biomasserueckfuehrung. Ph.D. Thesis submitted to Department of Food and Biotechnology, Technical University, Berlin, Federal Republic of Germany.

Jeffries, T. W., Fady, J. H., and Lightfoot, E. N. (1985). Effects of glucose supplements on the fermentation of xylose by *Pachysolen tannophilus*. *Biotechnol. Bioeng. 27*:171-176.

Jeris, J. S., and McCarty, P. L. (1965). The biochemistry of methane fermentation using 14C tracers. *J. Water Pollut. Control Fed. 37*:178-192.

Jobses, I. M. L., Egberts, G. T. C., Luyben, K. C. A. M., and Roels, J. A. (1986). Fermentation kinetics of *Zymomonas mobilis* at high ethanol concentrations: Oscillations in continuous culture. *Biotechnol. Bioeng. 28*:868-877.

Jones, D. T., and Wood, D. R. (1986). Acetone-butanol fermentations revisited. *Microbiol. Rev. 50*:484-524.

Jones, R. P., and Greenfield, P. F. (1982). Effect of carbon dioxide on yeast growth and fermentation. *Enzyme Microb. Technol. 4*:210-223.

Jung, H. G. (1985). Inhibition of structural carbohydrate fermentation by forage phenolics. *J. Sci. Food Agric. 36*:74-80.

Jung, H. G., and Fahey, G. C., Jr. (1983). Nutritional implications of phenolic monomers and lignin: A review. *J. Anim. Sci. 57*:206-219.

Jung, H. G., and Sahlu, T. (1986). Depression of cellulose digestion by esterified cennamic Acids. *J. Sci. Food Agric.* *37*:659-665.
Jung, H. G., Fahey, G. C., Jr., and Garst, J. E. (1983). Simple phenolic monomers of forages and effects of in vitro fermentation on cell wall phenolics. *J. Anim. Sci.* *57*:1294-1305.
Jungermann, K., Kern, M., Reibeling, V., and Thauer, R. K. (1976). Function and regulation of ferredoxin reduction with NADH in Clostridia. In *Microbial Population and Utilization of Gases*, H. G. Schlegel, G. Gottschalk, and N. Pfennig (Eds.). Goltze, Gottingen, Federal Republic of Germany, Kaczmarek, and Coykendall, pp. 85-96.
Kashket, E., Blanchard, A., and Metzeger, W. (1980). Proton motive force during growth of *Streptococcus lactis* cells. *J. Bacteriol.* *143*:128-134.
Kaspar, H. F., and Wuhrmann, K. (1978). Kinetic parameters and relative turnovers of some important catabolic reactions in digesting sludge. *Appl. Environ. Microbiol.* *36*:1-7.
Katz, E., and Demain, A. L. (1977). The peptide antibiotics of bacillus: Chemistry, biogenesis, and possible function. *Bacterial. Rev.* *41*:339-374.
Kim, B. H., and Zeikus, J. G. (1985). Importance of hydrogen metabolism in regulation of solventogenesis by *Clostridium acetobutylicum*. *Dev. Ind. Microbiol.* *26*:1-14.
Kim, J. (1986). Influence of redox potential on cell growth and metabolism in pH controlled continuous butanol fermentation. M.S. Thesis, submitted to the Department of Chemical Engineering, University of Missouri-Columbia.
Kirsch, E. J., and Sykes, R. M. (1971). Anaerobic digestion in biological waste treatment. *Prog. Ind. Microbiol.* *9*:155-176.
Kollerup, F., and Daugulis, A. J. (1985). Mathematical model for ethanol production by extractive fermentation in a continuous stirred tank fermentor. *Biotechnol. Bioeng.* *27*:1335-1346.
Kroeker, E. J., Schulte, D. D., Sparling, A. B., an d Lapp, H. M. (1979). Anaerobic treatment process stability. *J. Water Pollut. Control Fed.* *51*:718-727.
Kugelman, I. J., and Chin, K. K. (1971). Toxicity, synergism and antagonism in anaerobic waste treatment processes. In *Anaerobic Biological Treatment Process*, F. G. Pholand (Ed.). Advances in Chemistry Series 105. American Chemical Society, Washington, D.C., pp. 55-90.
Lafon-Lafourcade, S. (1983). Wine and brandy. In *Biotechnology*, Vol. 5, H. J. Rehm and G. Reed (Eds.). Verlag Chemie, Weinheim, pp. 81-165.
Lafon-Lafourcade, S., Geneix, G., and Ribereau-Gayon, P. (1984). Inhibition of alcoholic fermentation of grape-must by fatty acids

produced by yeasts and their elimination by yeast ghosts. *Appl. Environ. Microbiol.* *47*:1246-1249.
Latham, M. J., and Wolin, M. J. (1977). Fermentation of cellulose by *Ruminococcus flavefacieus* in the presence and absence of *Methanobacterium ruminantium*. *Appl. Environ. Microbiol.* *34*:297-301.
Laube, V. M., and Martin, S. M. (1981). Conversion of cellulose to methane and carbon dioxide by triculture of *Acetivibriocellulolyticus*, *Desulfovibrio* sp. and *Methanosarcina barkeri*. *Appl. Environ. Microbiol.* *42*:413-420.
Laudrin, I., and Goma, G. (1982). Ethanol production by *Zymomonas mobilis*: Effect of temperature on cell growth, ethanol production and intracellular ethanol concentration. *Biotechnol. Lett.* *4*:537-542.
Lawrence, A. W. (1971). Application of process kinetics to design of anaerobic processes. In *Anaerobic Biological Treatment Process*, F. G. Pholand (Ed.). Advances in Chemistry Series 105. American Chemical Society, Washington, D.C., 163-189.
Leao, C., and van Uden, N. (1982). Effect of ethanol and other alcanols on the glucose transport system of *Saccharomyces cerevisiae*. *Biotechnol. Bioeng.* *24*:2601-2604.
Leao, C., and van Uden, N. (1983). Effect of ethanol and other alcanols on the ammonium transport system of *Saccharomyces cerevisiae*. *Biotechnol. Bioeng.* *25*:2085-2090.
Leao, C., and van Uden, N. (1984). Effect of ethanol and other alcanols on the general amino acid permease of *Saccharomyces cerevisiae*. *Biotechnol. Bioeng.* *26*:403-405.
Lee, K. J., and Rogers, P. L. (1983). Fermentation kinetics of ethanol production by *Zymomonas mobilis*. *Chem. Eng. J.* *27*:B31-B38.
Lee, K. J., Tribe, D. E., and Rogers, P. L. (1979). Ethanol production by *Zymomonas mobilis* in continuous culture at high glucose concentration. *Biotechnol. Lett.* *1*:421-426.
Leung, J. C. Y. (1982). Production of acetone and butanol by *Clostridium acetobutylicum* using free and immobilized cells, Ph.D. Thesis, submitted to the Department of Nutrition and Food Science, Massachusetts Institute of Technology, Cambridge, Massachusetts.
Leung, J. C. Y., and Wang, D. I. C. (1981). Production of acetone and butanol by *Clostridium acetobutylicum* in continuous culture using free cells and immobilized cells. *Proceedings of the II World Congress of Chemical Engineering*, *1*:348-352.
Levenspiel, O. (1980). The Monod equation: A revisit and a generalization to product inhibition situation. *Biotechnol. Bioeng.* *22*:1671-1687.
Linden, J. C., and Moreira, A. (1982). Anaerobic production of chemicals. In *Biological Basis of New Developments in Biotechnology*,

A. Hollaender, A. I. Laskin, and P. Rogers (Eds.). Plenum Press, New York.

Llorente, P., and Sols, A. (1969). Proceedings of the 6th Annual Meeting of the Federation of European Biochemical Society, p. 123.

Loureiro, V., and Ferreira, H. G. (1983). On the intracellular accumulation of ethanol in yeast. *Biotechnol. Bioeng. 25*:2263-2270.

Loureiro, V., and van Uden, N. (1982). Effect of ethanol on the maximum temperature for growth of *Saccharomyces cerevisiae*: A model. *Biotechnol. Bioeng. 24*:1881-1884.

Loureiro-Dias, M. C., and Peinado, J. M. (1982). Effect of ethanol and other alkanols on the maltose transport system of *Saccharomyces cerevisiae. Biotechnol. Lett. 4*:721-724.

Luong, J. H. T. (1985). Kinetics of ethanol inhibition in alcohol fermentation. *Biotechnol. Bioeng. 27*:280-285.

Mackie, R. I., and Bryant, M. P. (1981). Metabolic activity of fatty acid-oxidizing bacteria and the contribution of acetate, propionate, butyrate and carbon dioxide to methanogenesis in cattle waste at 40 and 60°C. *Appl. Environ. Microbiol. 41*:1363-1373.

Maiorella, B., Blanch, H. W., and Wilke, C. R. (1983). By-product inhibition effects on ethanolic fermentation by *Saccharomyces cerevisiae. Biotechnol. Bioeng. 25*:103-121.

Matsuo, T., Oyama, Y., Tanimoto, H., Hashida, W., and Terramota, S. (1966). Fermentative production of glutamic acid and its application. X. Effects of antibiotics on glutamic acid fermentation. 3. Effects of penicillin and biotin. *Hakko Kogaku Zasshi 44*:915-924.

Mayrand, D., and Bourgeau, G. (1982). Production of phenylacetic acid by anaerobes. *J. Clin. Microbiol. 16*:747-750.

McCarty, P. L. (1964). Anaerobic waste treatment fundamentals. I. Chemistry and microbiology. *Public Works 95*:107-112.

McCarty, P. L., and Brosseau, M. H. (1963). Effects of high concentration of individual VFA on anaerobic treatment. Proceedings of the 18th Industrial Waste Conference, Purdue University, Lafayette, Indiana, pp. 283-296.

McCarty, P. L., and McKinney, R. E. (1961). Salt toxicity in anaerobic digestion. *J. Water Pollut. Control Fed. 33*:399-418.

McInerney, M. J., and Bryant, M. P. (1981). *Review of Methane Fermentation Fundamentals.* CRC Press, Boca Raton, Florida, pp. 19-46.

McInerney, M. J., Bryant, M. P., Hespell, R. B., and Costerton, J. W. (1981). *Syntrophomonas wolfei* gen. nov. sp. nov., anaerobic syntrophic fatty acid oxidizing bacterium. *Appl. Environ. Microbiol. 41*:1029-1039.

Meyer, C. L., McLaughlin, J. K., and Papautsakis, E. T. (1985). The effect of CO on growth and product formation in batch cultures of *Clostridium acetobutylicum. Biotechnol. Lett. 7*:37-42.

Millar, D. G., Griffiths-Smith, K., Algar, E., and Scopes, R. K. (1982). Activity and stability of glycolytic enzymes in the presence of ethanol. *Biotechnol. Lett.* *4*:601-606.
Monot, F., Engasser, J. M., and Petitdemange, H. (1983). Regulation of acetone-butanol production in batch and continuous cultures of *Clostridium acetobutylicum*. *Biotechnol. Bioeng. Symp.* *13*:207-216.
Monot, F., Engasser, J. M., and Petitdemange, H. (1984). Influence of pH and undissociated butyric acid on the production of acetone and butanol in batch cultures of *Clostridium acetobutylicum*. *Appl. Microbiol. Biotechnol.* *19*:422-426.
Moreira, A. R., Ulmer, D. C., and Linden, J. C. (1981). Butanol toxicity in the butylic fermentation. *Biotechnol. Bioeng. Symp.* *11*:567-579.
Moss, C. W., Hatheway, C. L., Lambert, M. A., and McCrossley, L. M. (1980). Production of phenylacetic and hydroxyphenyl acetic acids by *Clostridium botulinum* type G. *J. Clin. Microbiol.* *11*:743-745.
Moulin, G., Boze, H., and Galzy, P. (1980). Inhibition of alcoholic fermentation by substrate and ethanol. *Biotechnol. Bioeng.* *22*:2375-2381.
Mountfort, D. O., and Asher, R. A. (1978). Changes in proportions of acetate and carbon dioxide used as methane precursors during the anaerobic digestion of bovine waste. *Appl. Environ. Microbiol.* *35*:648-654.
Mountfort, D. O., Brulla, W. J., Krumholz, L. R., and Bryant, M. P. (1984). *Syntrophus buswellii* gen. nov., sp. nov.: A benzoate catabolizer from methanogenic ecosystems. *Int. J. Syst. Bacteriol.* *34*:216-217.
Murphy, M. P., Balwin, R. L., and Ulyatt, M. J. (1986). An update of dynamic model of ruminant digestion. *J. Anim. Sci.* *62*:1412-1422.
Nagase, M., and Matsuo, T. (1982). Interactions between amino acid degrading bacteria and methanogenic bacteria in anaerobic digestion. *Biotechnol. Bioeng.* *24*:2227-2239.
Nagodawithana, T. W., and Steinkraus, K. H. (1976). Influence of the rate of ethanol production and accumulation on the viability of *Saccharomyces cerevisiae* in "rapid fermentation." *Appl. Environ. Microbiol.* *31*:158-162.
Nagodawithana, T. W., Whitt, J. T., and Cutaia, A. J. (1977). Study of the feed-back effect of ethanol on selected enzymes of the glycolytic pathway. *J. Am. Soc. Brew. Chem.* *35*:179-193.
Naumann, E., Hippe, H., and Gottschalk, G. (1983). Betaube: New oxidant in the Stickland reaction and methanogenesis from betaine and L-alanine by a *Clostridium sporogens-Methanosarcina barkeri* coculture. *Appl. Environ. Microbiol.* *45*:474-483.

Navarro, J. M., and Durand, G. (1978). Alcoholic fermentation: Effect of temperature and ethanol accumulation in yeast cells. *Ann. Microbiol. (Paris)*. *129B*:215-221.
Nottingham, P. M., and Hungate, R. E. (1969). Methanogenic fermentation of benzoate. *J. Bacteriol.* *98*:1170-1172.
Novak, M., Strehaino, P., Moreno, M., and Goma, G. (1981). Alcoholic fermentation: On the inhibitory effect of ethanol. *Biotechnol. Bioeng.* *23*:201-211.
Nyiri, L., and Lengyel, Z. L. (1965). Studies on automatically aerated biosynthetic processes. I. The effect of agitation and CO_2 on penicillin formation in automatically aerated liquid cultures. *Biotechnol. Bioeng.* *7*:343-354.
Ounine, K., Petitdemange, H., Raval, G., and Gay, R. (1983). Acetone-butanol production from pentoses by *Clostridium acetobutylicum*. *Biotechnol. Lett.* *5*:605-610.
Ounine, K., Petitdemange, H., Raval, G., and Gay, R. (1985). Regulation and butanol inhibition of D-xylose and D-glucose uptake in *Clostridium acetobutylicum*. *Appl. Environ. Microbiol.* *49*:874-878.
Peretti, S. W., and Bailey, J. M. (1986). Mechanistically detailed model of cellular metabolism for glucose limited growth of *Escherichia coli* B/r-A. *Biotechnol. Bioeng.* *28*:1672-1689.
Pritchard, R. H., Barth, P. T., and Collins, J. (1969). Control of DNA synthesis in bacteria. *Symp. Soc. Gen. Microbiol.* *19*:263-297.
Rockey, D. A., Turnacliff, W., and Smith, R. J. (1978). A 1900 M^3 digester for laying-hen manure, Iowa. ASAE 77-4569. American Society of Agricultural Engineers, St. Joseph, Michigan.
Roels, J. A. (1983). *Energetics and Kinetics in Biotechnology*. Elsevier, New York.
Roman, G. N., Jansen, N. B., and Tsao, G. T. (1984). Ethanol inhibition of d-xylulose fermentation by *Schizosaccharomyces pombe*. *Biotechnol. Lett.* *6*:7-12.
Sa-Corriea, I., and van Uden, N. (1983). Temperature profiles of ethanol tolerance: Effect of ethanol on the minimum and the maximum temperature for growth of the yeasts *Saccharomyces cerevisiae* and *Kluyveromyces fragilis*. *Biotechnol. Bioeng.* *25*:1665-1667.
Sampson, F. E., Katz, A. M., and Harris, D. L. (1955). Effects of acetate and other short chain fatty acids on yeast metabolism. *Arch. Biochem. Biophys.* *54*:406-423.
Scheifinger, C. C., and Wolin, M. J. (1973). Propionate formation from cellulose and soluble sugars by combined cultures of *Bacteroides succinogenes* and *Selenomonas ruminantium*. *Appl. Microbiol.* *26*:789-795.

Schwab, D. A. (1978). The molecular basis of acetic acid elicited growth inhibition of *Escherichia coli* K-12 and *Shigella flexneri*, Ph.D. Thesis. Submitted to the Department of Biological Sciences, University of California, Irvine.

Scott, R. (1983). Design and evaluation of experiments to provide scale-up information for the ICI single cell protein process. BIOTECH83, Online Publ. Ltd., Northwood, England, p. 235.

Shelton, D. R., and Tiedje, J. M. (1984). Isolation and partial characterization of bacteria in an anaerobic consortium that mineralizes 3-chlorobenzoic acid. *Appl. Environ. Microbiol. 47*:840-848.

Sievers, D. M., and Brune, D. E. (1978). Carbon/nitrogen ratio and anaerobic digestion of swine waste. *Trans. ASAE 21*:537-549.

Smith, P. H., and Mah, R. A. (1966). Kinetics of acetate metabolism during sludge digestion. *Appl. Microbiol. 14*:368-371.

Thomas, D. S., and Rose, A. H. (1979). Inhibitory effect of ethanol on growth and solute accumulation by *Saccharomyces cerevisiae* as affected by plasma lipid composition. *Arch. Microbiol. 122*:49-55.

van Assche, P. F. D. (1978). Differentiation of *Bacteroides fragilis* species by gas chromatographic detection of phenylacetic acid. *J. Clin. Microbiol. 8*:614-615.

van den Berg, L., Patel, G. B., Clark, D. S., and Lentz, C. P. (1976). Factors affecting rate of methane formation from acetate by enrichment cultures. *Can. J. Microbiol. 22*:1312-1319.

van Uden, N. (1985). Ethanol toxicity and ethanol tolerance in yeasts. *Ann. Rep. Ferment. Proc. 8*:11-58.

van Uden, N., and Da cruz Duarte, H. (1981). Effects of alcohol on the temperature profile of *Saccharomyces cerevisiae*. *Z. Allg. Mikrobiol. 21*:743-750.

van Velsen, A. F. M. (1979). Adaptation of methanogenic sludge to high ammonia-nitrogen concentration. *Water Res. 13*:995-999.

Varel, V. H., and Jung, H. G. (1986). Influence of forage phenolics on ruminal fibrolytic bacteria and in vitro fiber degradation. *Appl. Environ. Microbiol. 52*:275-280.

Vollherbst-Schneck, K., Sands, J. A., and Montenecourt, J. A. (1984). Effect of butanol on lipid composition and fluidity of *Clostridium acetobutylicum* ATCC 824. *Appl. Environ. Microbiol. 47*:193-194.

Votruba, J., Volesky, B., and Yerushalmi, L. (1986). Mathematical model of a batch acetone-butanol fermentation. *Biotechnol. Bioeng. 28*:247-255.

Wang, Y., Suidan, M. T., and Pfeffer, J. T. (1984). Anaerobic biodegradation of indole to methane. *Appl. Environ. Microbiol. 48*:1058-1060.

Watson, N. E., Prior, B. A., Lategan, P. M., and Lussi, M. (1984). Factors in acid treated bagasse inhibiting ethanol production from d-xylose by *Pachysolen tannophilus*. *Enzyme Microb. Technol.* *6*:451-456.

Wolin, M. J. (1975). Interactions between the bacterial species of the rumen. In *Digestion and Metabolism in the Ruminant*, I. W. McDonald and A. C. I. Warner (Eds.). University of New England Publishing Unit, Armidale, Australia, pp. 134-148.

Wolin, M. J. (1982). Hydrogen transfer in microbial communities. In *Microbial Interactions and Communities*, Vol. 1, A. T. Bull and J. H. Slater (Eds.). Academic Press, London, pp. 323-356.

Yerushalmi, L., and Volesky, B. (1985). Importance of agitation in acetone-butanol fermentation. *Biotechnol. Bioeng.* *27*:1297-1305.

Yerushalmi, L., Volesky, B., and Szczesny, T. (1985). Effect of increased hydrogen partial pressure on the acetone-butanol fermentation by *Clostridium acetobutylicum*. *Appl. Microbiol. Biotechnol.* *22*:103-107.

Zeikus, J. G. (1980). Fate of lignin and related aromatic substrates in anaerobic environments. In *Lignin Biodegradation: Microbiology, Chemistry and Potential Applications*, T. K. Kirk, T. Higuchi, and H. M. Chang (Eds.). CRC Press, Boca Raton, Florida, pp. 101-109.

Zinder, S. H., and Mah, R. A. (1979). Isolation and characterization of a thermophilic strain of *Methanosarcina* unable to use hydrogen-carbon dioxide for methanogenesis. *Appl. Environ. Microbiol.* *38*:996-1008.

9

Chemically Structured Models for Anaerobic Growth and Product Formation

MOHAMMAD M. ATAAI* / *Polytechnic Institute, Brooklyn, New York*
MICHAEL L. SHULER / *Cornell University, Ithaca, New York*

I. INTRODUCTION

For nearly any system, modeling offers an excellent method of predicting the system's behavior when experiments are either too inaccurate or too expensive. Modeling can also help to understand the behavior of fundamental mechanisms in a system when trying to design and optimize it for product formation. Models are very important to engineers for the development of control schemes and examination of different bioreactor confirguations for enhanced productivity. The distinction should be made between models that are mechanistic and those that rely on correlations. *Mechanistic models* can be used beyond their original data base and are capable of simulating a wide range of conditions. *Correlations* are essentially fits to the experimental data, and their use is restricted to the original data base used to establish the empirical correlation.

Monod's model (Monod, 1949) is one of the first and most widely used mathematical models of microbial growth. It offers a simple expression for relating the growth rate of a cell to the nutrient concentration outside the cell. The Monod model, however, is an *unstructured model*, which means it does not recognize that the cell can alter its internal biosynthetic machinery in response to

---

**Present affiliation*: University of Pittsburgh, Pittsburgh, Pennsylvania.

environmental stimuli. The growth rate is an explicit function only of the concentration of biomass and nutrient and population density. The Monod model cannot predict differences in cell composition, morphology, or product formation, between various growth phases, the sequential uptake of substrates, or change in mean cell size. Unstructured models are applicable only to balanced growth conditions (Campbell, 1957). Balanced growth pertains only when the cell's composition is unchanging, which occurs in steady-state chemostat and possibly during the exponential growth phase in batch culture. For more accurate predictions structured models must be considered. A chemically structured model divides the biomass into two or more subcompartments. Consequently, *structured models* can make predictions under conditions in which the biosynthetic capabilities of the cell are changing. For unbalanced growth (e.g., dynamic response in a chemostat), a structured model should be used.

A model can be nonsegregated or segregated. A *nonsegregated model* does not recognize the existence of individual cells. Under many circumstances (e.g., plasmid-containing cells) some cells in a population are more productive than others owing to the uneven distribution of biosynthetic components. In such situations the explicit recognition of these distributions is essential since population-averaged values cannot accurately indicate the biosynthetic potential of the culture. Thus, *segregated models*, which explicitly recognize that individuals within the population have distinct and variable properties, are important in predicting the productivity of a population.

Models that contain both structure and segregation and are written from the population-balance perspective have proved to be mathematically intractable. A computationally tractable alternative is to build finite representation models in which subpopulations are represented by a "typical cell." An ensemble of these individual cell models can capture the essential quality of segregation while allowing the inclusion of significant biochemical structure. A single cell model allows the possible inclusion of temporal events during the division cycle and spatial effects with relative ease. Cell structures and the interactions of major metabolic pathways can also be readily and accurately accounted for.

Single-cell models have limitations. To use a set of deterministic equations and concentrations, a single-cell model must represent more than a single cell. Typically, a single-cell model must represent 10^4 cells. If a single-cell model represents the entire population, then it behaves as a structured nonsegregated model. If various subfractions are represented by a single-cell model, then the behavior of the whole population can be represented by using a finite number of single-cell models and the whole collection of single-cell models gives a structured-segregated model.

Several single-cell models have been postulated (Heinmets, 1966; Weinberg et al., 1971; Simon, 1973; Davison, 1975). These models contain one or more arbitrary constraints on the computer "growth" of the cell. Examples of such constraints are the mode of growth (exponential), criteria for cell division (doubling of volume), culture conditions (balanced growth only), no explicit account of the extracellular environment (e.g., nutrient uptake), and cell shape.

Shuler et al. (1979) presented a model that simulates cell behavior more accurately and under a wider range of conditions. A major advantage of this model over others is that it is not constrained. All that needs to be specified is the initial nutrient composition in the medium. Cell growth, DNA initiation, cross-wall formation, changes in cell size and shape, and cell division are all "natural" responses of the model. The model contains 14 components, including pools for protein, RNA, DNA, and cell wall components.

A number of improvements to this model have been made (Domach et al., 1984). Greater detail concerning biologic mechanisms and maintenance energy requirements have been included by more explicit accounting of cellular energy expenditures. A more accurate representation of cellular control is achieved by dividing RNA into fractions for m-RNA and immature and mature stable RNA.

The improved version of the single-cell model has 20 components and is a highly structured model for cell growth. Generally, the complexity of a structured model increases with increase in the number of components included in the model. The simplest structured model divides the biomass into two components; the upper component is nearly unlimited, as even a simple bacterium contains several thousand components. The degree of complexity that one would like to incorporate in a structured model depends on the intended application of the model. Models intent on testing cellular mechanisms rather than on design or process control generally require greater detail. It should be noted, however, that the parameters for a model with a few components may have less physical meaning than those for a more structured model. Simple structured models have been recently reviewed by Harder and Roels (1982). They gave examples for two- and three-component models and the methodology used in constructing such models. The application of structured models to bioreactor design and the evolution of structured models in recent years have been reviewed (Shuler, 1985; Barford et al., 1982; Bazin, 1982). In this chapter we focus our discussion on anaerobic *single-cell* models and *population models*. We particularly describe our recent results with recombinant *Escherichia coli* for the prediction of protein production and gene copy numbers under anaerobic conditions. To our knowledge this is the first treatment that mathematically examines recombinant protein production under anaerobic conditions.

E. coli is a facultative organism, and the ability to model its response under both aerobic and anaerobic growth can be important. For example, immobilized cells for protein production have several advantages over those in suspension. A major problem with immobilized cells is that oxygen transfer is limited to a very short distance from the surface of a bead or membrane. Therefore, the ability to predict recombinant protein productivity under anaerobic conditions is an important contribution to the optimization of such processes.

In the rest of this chapter we first describe the formulation of the anaerobic single-cell model developed by the authors (Ataai and Shuler, 1985a, b) and discuss the models for clostridia fermentation (Votruba et al., 1986; Yerushalmi et al., 1986). We then describe a new structured model for the prediction of plasmid copy number and protein productivity for recombinant cells (*E. coli* in this case).

II. ANAEROBIC SINGLE-CELL AND POPULATION MODELS

The single-cell model developed for aerobic growth of *E. coli* is shown in Figure 1. The cell is viewed as a small homogeneous reactor with its cellular components lumped into the categories indicated in the legend of the figure. Two of the model equations and evaluation of the pertinent model parameters are described in Appendix I. The description of these equations is expected to provide a good example of how this model was constructed and to indicate that it may be possible to extend this modeling approach to other bacterial systems that are not studied as extensively as *E. coli*. The full set of equations is described elsewhere (Domach, 1983; Domach et al., 1984; Leung, 1979). Inspection of the materials in the appendix reveals that the knowledge of cell composition and intracellular concentration of cellular components, included in the model along with cell size and shape, can provide much of the information necessary to evaluate a large portion of the parameters for model components. Further, this type of model formulation easily allows the incorporation of more detailed information on any subcellular mechanism, such as plasmid replication.

To extend the model to anaerobic conditions, the aerobic cell model was altered only in the details of calculating the generation of ATP and the reducing power, which reflect the known stoichiometry of glucose catabolism under anaerobic conditions. The anaerobic single-cell model accurately predicted (±10%) the values of cell growth

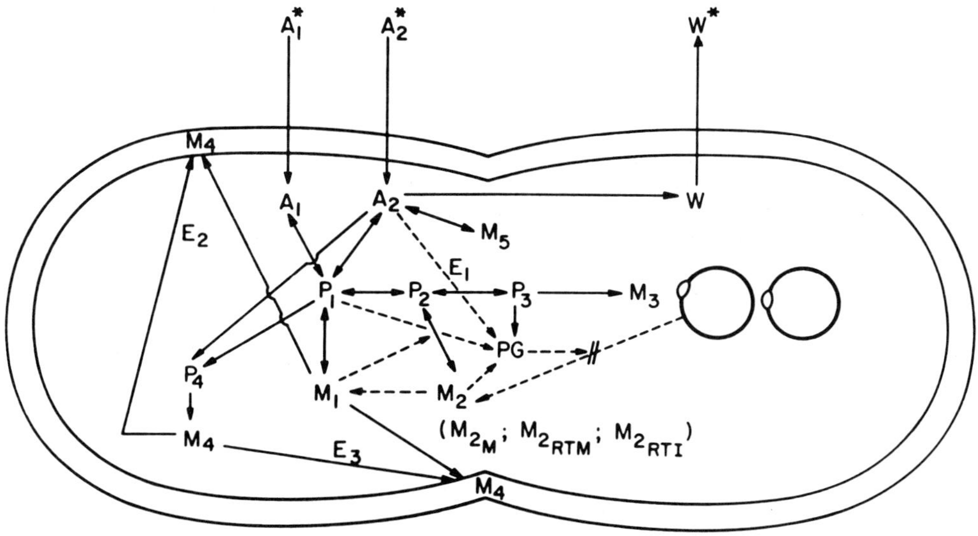

Figure 1 An idealized sketch of the model for *E. coli* B/r A growing in a glucose-ammonium salts medium with glucose or ammonia as the limiting nutrient. At the time shown the cell has just completed a round of DNA replication and initiated cross-wall formation and a new round of DNA replication. Solid lines indicate the flow of material; dashed lines indicate the flow of information. The symbols are as follows: A_1, ammonium ion; A_2, glucose (and associated compounds in the cell); W, waste products (CO_2, H_2O, and acetate) formed from energy metabolism during aerobic growth; P_1, amino acids; P_2, ribonucleotides; P_3, deoxyribonucleotides; P_4, cell envelope precursors; M_1, protein (both cytoplasmic and envelope); M_{2RTI}, immature "stable" RNA; M_{2RTM}, mature "stable" RNA (r-RNA and t-RNA: assume 85% r-RNA throughout); M_{2M}, messenger RNA; M_3, DNA; M_4, nonprotein part of cell envelope (assume 16.7% peptidoglycan, 47.6% lipid, and 35.7% polysaccharide); M_5, glycogen; PG, ppGpp; E_1, enzymes in the conversion of P_2 to P_3; E_2, E_3, molecules involved in directing cross-wall formation and cell envelope synthesis: the approach used in the prototype model was used here, but more recent experimental support is available. An asterisk (*) indicates that the material is present in the external environment. (From Domach et al., 1984.)

rate, cell volume, cell content of DNA, RNA, and protein, substrate yield, ATP yield, timing of the chromosome initiation, and the rate of formation of end products of the fermentation compared to experimental results (Ataai and Shuler, 1985a). All the predictions are the natural model response and only the substrate concentration in the extracellular environment must be specified.

The ability to predict steady-state performance is a far less severe test of a model's capability than predictions of the transient behavior. We performed such tests as building a population model using an ensemble of single-cell models (Ataai and Shuler, 1985b). For example, Figures 2 and 3 show the results of a step change in dilution rate from D = 0.16 h^{-1} to D = 0.55 h^{-1}. As can be seen from these figures, model prediction is in a good agreement with experimental results.

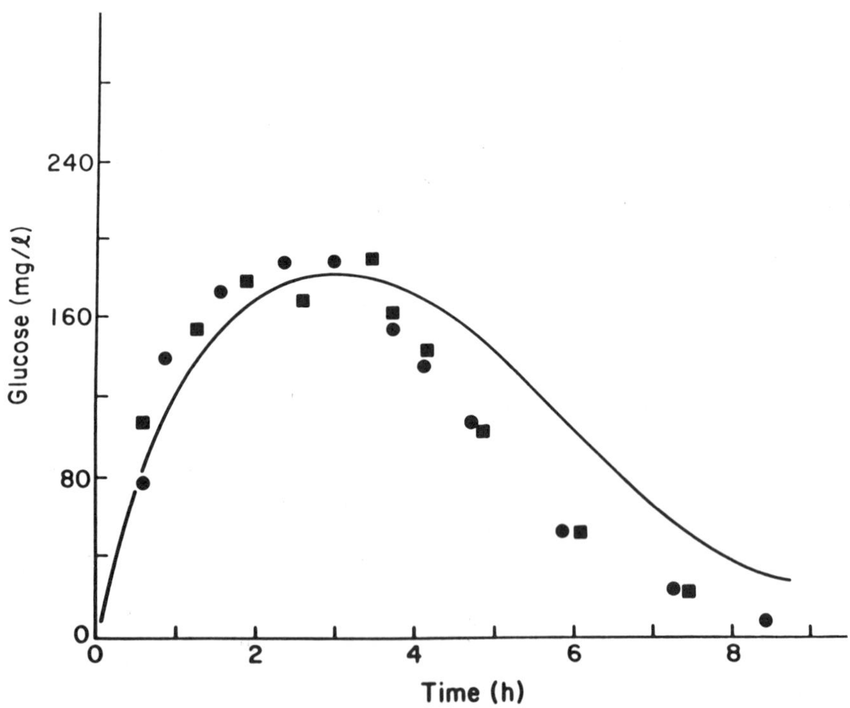

Figure 2 The transient glucose concentration in response to a step change in dilution rate from D = 0.16 to D = 0.55 h^{-1}. The symbols ● and ■ are the results of two duplicate experimental runs. The solid line is the computer prediction. (From Ataai and Shuler, 1985b.)

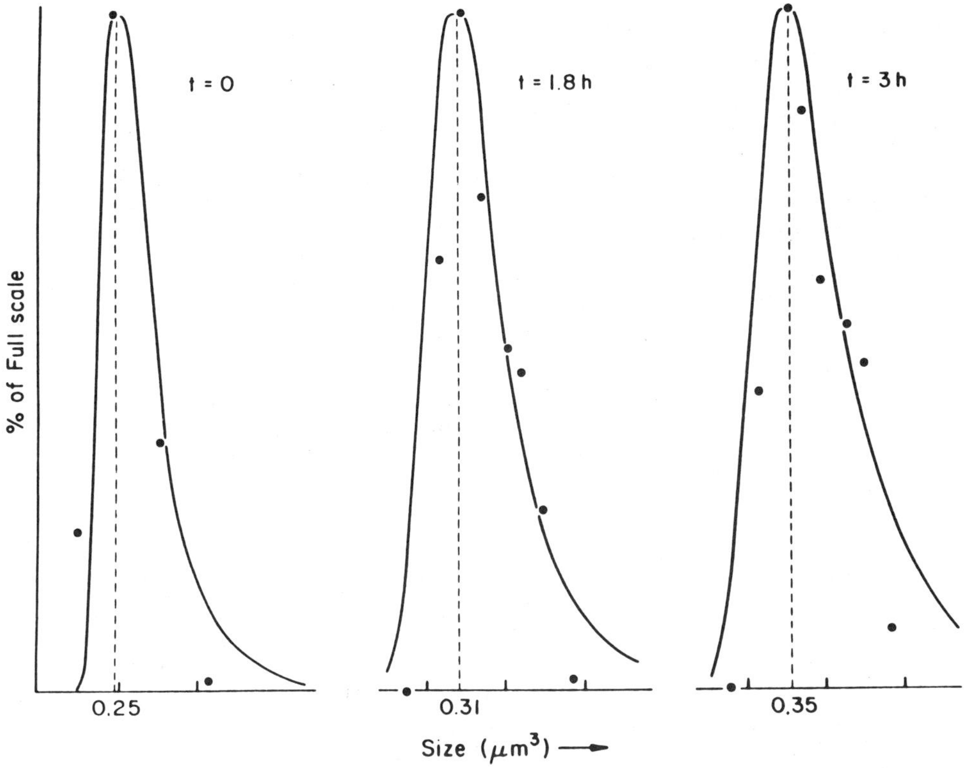

Figure 3 The transient cell size distribution in response to a step change in dilution rate from D = 0.16 to D = 0.55 h^{-1}. The solid line is the Coulter Counter measurement of the size distribution; the computer prediction gives a histogram, and ● represents the fraction of the cells in a given cell class. (From Ataai and Shuler, 1985b.)

All earlier attempts to predict the dynamic behavior of the chemostat by applying unstructured models were unsuccessful (Andrews, 1968; Koga and Humphrey, 1967; Yano and Koga, 1969). Unstructured Monod-type models assume that growth is limited only by the concentration of the limiting substrate and that even while growing slowly, the organism possesses all the required cellular constituents for growth at its maximum growth rate. This means that a cell can accelerate to a high growth rate instantaneously when the substrate concentration in the medium is raised. Several studies have documented that the growth rate of cells in continuous culture cannot instantaneously respond to changes in substrate concentration following a perturbation in the steady state (Aiba et al., 1967a, b; Mateles et al., 1965).

Other researchers have attempted to model the transient behavior of a culture based on the hypothesis that the overall response of the culture is controlled by the response of cellular RNA (Koch, 1971; Maaloe and Kjeldgaard, 1966). Diagger and Grady (1982) suggested that, although the level of RNA significantly affected the nature of transient response, other unidentified components are also important at least for some growth rates. For example, it has been shown that the level of ppGpp (a guanosine phosphate) affects both the protein and RNA synthesis rate and a number of other cellular functions (Cashel, 1975).

We used the anaerobic population model to predict changes in the RNA content of cells in response to a shift in dilution rate in a chemostat (Ataai and Shuler, 1985b). The model predictions shown in Figure 4 indicate that indeed the RNA synthesis rate rather quickly adjusts to that of the postshift steady-state value. This observation implies that employing RNA concentration as a measure of the metabolic activity of the cell may be a reasonable assumption or at least superior to other choices. It should be noted that the real reason for the use of RNA concentration as a growth indicator is not based on the observation that the RNA content of a cell varies linearly with growth rate, as argued by Votruba et al. (1986), but that, as mentioned earlier, RNA synthesis appears to respond rather quickly to changes in intracellular levels of amino acids and energy that correlate with growth rate.

The concept of RNA as a bottleneck component in cell growth and metabolism was used for simulation of the rate of formation of end products of clostridia fermentation (Votruba et al., 1986). The model was based on Powell's idea (1969) that specific growth rate may be proportional to some metabolic function Q and that Q is a quantitative expression of the physiological state of the organism. In this argument, Q is not a unique function of time but is also dependent on the way that substrate concentration has varied in the past. Votruba et al. (1986) used RNA concentration as a marker of the physiological state of the cell to represent the Q function. To predict the rate of end-product formation, differential mass balances for each of the products was written. The equations consisted of two terms: the first term represents the rate of formation of a product from the substrate, and the second term represents its rate of conversion to other products (in the case of intermediates). The model's parameters were obtained using nonlinear regression to minimize the deviation between the model's predictions and actual batch experimental data. The model's predictions of end-product

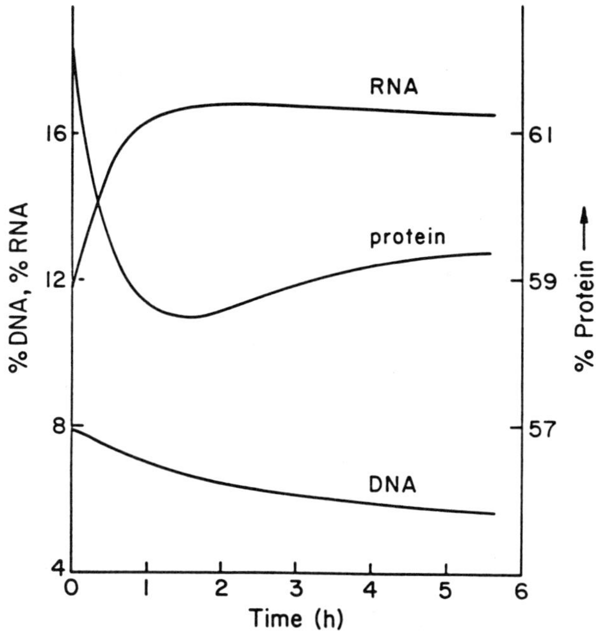

Figure 4 The transient DNA, RNA, and protein content of anaerobic chemostat-grown cells in response to a step change in dilution rate from D = 0.16 to D = 0.55 h^{-1}. The lines are the computer prediction. (From Ataai and Shuler, 1985b.)

concentration in batch culture fermentation are in good agreement with the corresponding experimental data. Correspondence between model predictions and experiment may not indicate much about the plausibility of the structure included in the model owing to extraction of the model's parameters from the batch growth data. In order to test any model for its closeness to the overall regulation that actually occurs at the cellular level, one should apply the model to situations different from the original one used in parameter evaluation. Should, for example, this model prove to be capable of predicting transient behavior in a continuous culture, where the substrate concentration and growth are decoupled, substantial support for the model's structure will be gained.

The approach chosen by Votruba et al. (1986), in our view, may be of considerable interest to fermentation specialists for the purpose of on-line control and/or to test potential designs and operation strategies that will enhance product formation or concentration. This statement is especially true if one considers that estimation of the rate of product formation based on the modeling of the actual enzymatic mechanism involved in the formation of a product is currently difficult.

For example, to extend our current model for *E. coli* to include end-product formation based on enzymatic consideration is not easy. Details of the modeling of glycolysis would involve developing proper kinetic equations for the controlling steps in the glycolysis and related pathways, such as the formation of fructose-6-phosphate and other irreversible reactions catalyzed by glucokinase and pyruvate kinase that play a regulatory role in glycolysis. Then, to predict the rate of end-product formation modeling of the enzyme activity for lactate dehydrogenase and pyruvate formate lyase of *E. coli* is additionally required. For building such a model, an accurate accounting of the internal levels of ATP, NADH, pyruvate, fructose-6-phosphate, and fructose-1,6-diphosphate would be needed. Such an accounting requires rather precise knowledge of the kinetic parameters of each of the intermediates. If these processes were incorporated into a model, the complexity would increase significantly.

Taking into consideration the data base currently available on enzymatic steps leading to the formation of fermentation end products, we believe that the approach adapted by Votruba et al. (1986) may have achieved a balance between the reality of cellular organization and the degree that such cellular processes can be incorporated into a reasonably useful model. The modeling approach described by Votruba et al. (1986) is, however, susceptible to criticism in that it did not make use of stoichiometry balances for carbon, and balances for reducing power and the other possible intermediates which a pseudo-steady-state assumption can be made in regard to their intracellular concentrations. Such balancing has been shown to correlate the concentration of products formed in clostridial or other fermentations (Erickson et al., 1978; Papoutsakis, 1984; Ataai and Shuler, 1985a) and is discussed in Chapters 4 and 5. The use of such correlations reduces the number of equations and consequently the number of model parameters that need to be evaluated. For example, having developed a model that simulates the rate of substrate conversion in the fermentation of *E. coli*, the rate of formation of the two major end products can be predicted simply by applying the carbon and reducing power balances (Ataai and Shuler, 1985a). Similarly for prediction of the rate of ethanol production by yeast, the stoichiometry of fermentation is sufficient to relate the rate of ethanol production to substrate utilization. The challenge is then to develop a model that can account for cell growth, substrate yield, and substrate utilization as a function of environmental conditions (including not only substrate concentration but also the inhibitory effects of ethanol and the role of dissolved oxygen in partitioning of substrate between energy generating and biosynthetic pathways). However, the modeling approach employed for the prediction of the end-product concentration of acetone-butanol fermentation (Votruba et al., 1986) should be viewed as a general approach (but one that may not be easily extended beyond the original data base) that may also be applied to situations in which metabolic pathways are not studied as extensively, for example acetone-butanol, *E. coli*, and yeast fermentations.

So far we have described the products formed in energy-generating pathways during anaerobic growth. Advances in molecular biology and genetics have opened a whole new field for producing valuable protein products using genetically modified cells. The prospects of genetic engineering using anaerobes is the subject of Chapter 2. In the next section we describe mathematical models and their use for processes involving recombinant cells.

III. MODELS FOR RECOMBINANT CELLS

The potential for protein production using recombinant vectors is enormous. The most commonly employed vector for the purpose of introducing new genetic information into bacterial hosts is a plasmid. Plasmids are extrachromosomal pieces of DNA that, because of their small size, high copy (in the case of relaxed control), and the ability to replicate independent of chromosomes, are considered appropriate cloning vectors. The introduction of plasmids carrying a cloned gene into bacterial cells does not necessarily lead to the production of active proteins. The gene promoter and ribosome binding site must be recognized by the host cell RNA polymerase and the ribosome, respectively. These problems, as well as the problem of physical instability of a cloned gene, are generally recognized during construction of a recombinant cell (host-plasmid system). Fusion of a structural gene to promoters different from its own and the possibility of including ribosomal recognition sites has allowed the successful expression of a variety of genes in a number of different cell lines.

Having constructed a host-vector system capable of producing the gene product of interest, segregational instability may arise, which can lead to an unproductive culture. Segregational instability is related to the uneven partitioning of plasmids between daughter cells. Cells with a high plasmid copy number grow more slowly than cells with few or no plasmids owing to the "metabolic burden" imposed on such cells for the biosynthesis of plasmid genes. This growth difference can lead to eventual takeover of a culture by cells free of plasmids.

Mathematical formulations for a description of genetic instability have been developed (Imanaka and Aiba, 1981; Kim and Ryu, 1984; Seo and Bailey, 1985). These models require knowledge of the plasmid copy number, which is related to the rate of reversion of plasmid-containing to plasmid-free cells, and of the ratio of growth rate of plasmid-containing cells to that of plasmid-free cells. The values for such parameters are related to both host cell and plasmid functions. In addition to the problem of genetic instability, the productivity of a cloned gene product is related to plasmid copy number, cell number density, and the level of expression of the gene. Plasmid copy number and cloned gene expression are dependent on the replication control processes regulating the frequency of replication initiation and the strength of cloned gene promoter, respectively. Additionally, these parameters are related to the growth conditions, and there is a limitation on the extent that cellular biosynthetic components can be devoted to the plasmid functions rather than the chromosome. For example, it has been shown that overexpression of β-lactamase due either to high plasmid copy number (Uhlin et al., 1979) or a strong promoter (Georgiou et al., 1985) can lead to cell death.

This discussion clearly indicate the need for mathematical models that can account for host cell and plasmid interactions. Such models may have the potential for the optimization of processes involving protein production using genetically modified cells. Inclusion of a model for plasmid replication and other functions into a highly structured model for host cell functions should allow one to make a priori predictions of the operating conditions needed to control plasmid stability, cloned gene production rate, and the extent that host cell biosynthetic machinery can be devoted to plasmid functions without interfering with normal cell functions.

E. coli is the most widely used host bacterium in genetic engineering. The *E. coli* single-cell model developed by Shuler and coworkers (Shuler et al., 1979; Domach et al., 1984), described earlier, can serve as an effective host cell model in elucidating host-plasmid interactions by incorporating information on plasmid functions into the cell model. The *E. coli* single-cell model has been further extended by Peretti and Bailey (1986). More details of the control of transcription and translation have been included that should account more explicitly for competition between plasmid and chromosome genes and their transcripts for RNA polymerase and ribosomes, respectively. Such an accounting may help to identify the limiting factor leading to cell death when plasmid-harboring cells are forced to produce large quantities of foreign proteins encoded by plasmid genes.

We have recently developed a mathematical model for replication control of ColE1 plasmids and have included the plasmid model in the single-cell model of *E. coli* (Ataai and Shuler, 1986, 1987), which we refer to here as the *recombinant cell model*. The molecular mechanism for the replication of ColE1-type plasmids and a brief description of the model formulation are described in Appendix II. The recombinant cell model accurately simulated the plasmid copy number for various mutations in the regulatory genes of ColE1 origin plasmids (Ataai and Shuler, 1986).

Further, the recombinant cell model predicts values for parameters needed to evaluate plasmid stability and provides needed information for optimization of productivity of a cloned gene product in bioreactors using genetically altered cells. The inclusion of plasmid-host interactions in the recombinant cell model is detailed elsewhere (Ataai and Shuler, 1987). Figure 5 shows our prediction of plasmid

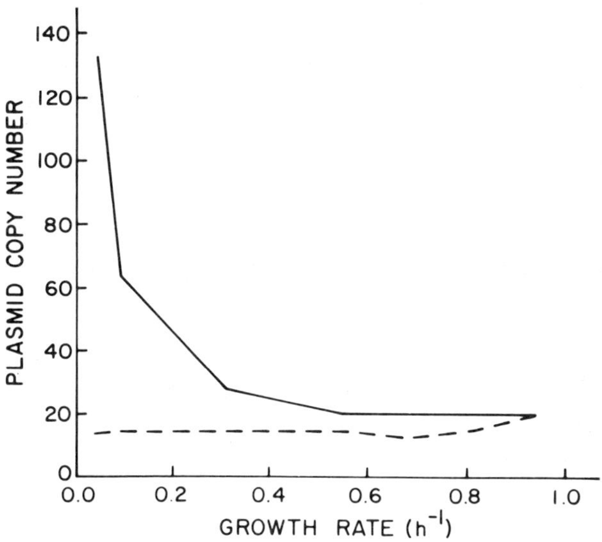

Figure 5 The plasmid copy number is shown as a function of growth rate for a derivative of a ColE1-type plasmid in *E. coli* B/r. The binding constant between RNA I and RNA II for this plasmid was 9.2×10^{-13} in the presence of the ROM protein. The solid line represents the case in which RNA I and RNA II transcription rates vary with growth rate in a manner proportional to the changes in the average transcription rate in the cell. The dashed line represents model predictions if RNA I and RNA II transcription rates are independent of growth rate and set at the observed values obtained under conditions of exponential growth in batch culture. (From Kim et al., 1987.)

copy number at different growth rates. Depending on the assumption made on the dependence of expression of regulatory plasmid genes on growth rate, either plasmid copy number would be almost independent of growth rate (for constant, growth rate independent transcription of regulatory genes) or would decrease with increase in growth rate if the transcription level of regulatory genes varies with growth rate as an average *E coli* gene (Kim et al., 1987). Other factors, such as growth rate-dependent rate of RNA I degradation, could further modulate copy number. Figures 6 and 7 show the model predictions and our experimental results for the plasmid stability of the *E. coli* p17 host-vector system at two different growth rates. p17 is a Co1E1 origin plasmid (Georgiou et al., 1985). The model predictions of stability were made using the assumption that the expression of regulatory plasmid genes is independent of growth rate. Applying the same assumption, the productivity of a plasmid-coded protein was investigated. The results of such simulations indicate that operation of a chemostat at a high dilution rate may

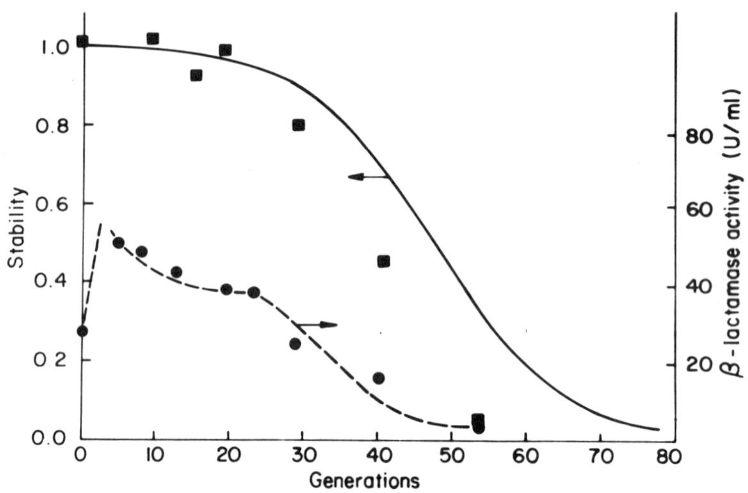

Figure 6 Stability and β-lactamase activity of plasmid p17 in a continuous culture of *E. coli* B/r/p17. Stability is defined as the fraction of plasmid-containing cells in the culture as determined by the ability to form macroscopic colonies on ampicillin-containing plates. The solid line is the model prediction of stability, where ■ denotes our experimental measurement. The ● denotes β-lactamase activity where the dashed line is the best line through the experimental points. Dilution rate was 0.7 h^{-1}. (From Ataai and Shuler, 1987.)

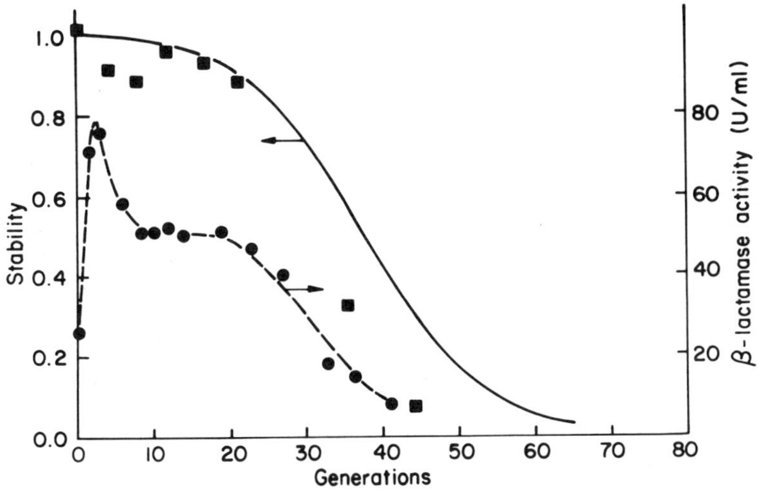

Figure 7 The same as Figure 6 except that the dilution rate is $0.27\ h^{-1}$. (From Ataai and Shuler, 1987.)

lead to higher productivity of plasmid proteins (Ataai and Shuler, 1987). However, other assumptions on the growth rate dependence of transcriptional rates for regulatory genes could lead to different conclusions.

To investigate the effect of anaerobic conditions on plasmid copy number and the productivity of a plasmid-encoded protein, we extended the anaerobic single-cell model to include the information on ColE1 plasmids. Our simulation results indicate that the change in plasmid copy number with growth rate is similar to results obtained under aerobic conditions. The results shown in Figure 5 would be valid for both aerobic and anaerobic conditions. However, it should be noted that maximum growth rate for *E. coli* cells grown anaerobically is less than that of aerobically grown cells ($0.94\ h^{-1}$ for aerobic and $0.68\ h^{-1}$ for anaerobic). Considering that the plasmid copy number would be similar under both aerobic and anaerobic conditions, one would expect overall productivity for plasmid proteins to be much lower under anaerobic conditions than aerobic. The difference in productivity would be mainly due to the low cell density under anaerobic conditions. Over a range of growth rates our simulation results indicate that cell density under anaerobic conditions is about 36% of that under aerobic conditions (calculated using cell yield values in connection with cell size). A comparison of the aerobic and anaerobic conditions described earlier, however, would be meaningful only if the anaerobic conditions did not stimulate or repress the expression of plasmid genes. Although protein productivity may be lower under anaerobic conditions in a suspension culture, such lower productivity may not result in an immobilized cell culture under nongrowing condition in which the maintenance requirements may be the predominant factor in determining the distribution of cell energy content and biosynthetic potential between the cellular and plasmid functions.

IV. CONCLUSION

Single-cell models for anaerobic growth and product formation are described. The current data available on bacterial growth patterns and physiology allow one to construct models that include cellular physiology, growth, and product formation. Structured models for recombinant cells are expected to find a wider application as the processes involving protein production using genetically altered cells are scaled up from bench to large scale.

V. NOMENCLATURE

All the symbols are described in the legend to Figure 1 and the text. The following may serve as a quick reference.

K	= saturation constants
$K_{T_{RNA\ I}}$	= average transcription rate for RNA I (transcripts)
$K_{T_{RNA\ II}}$	= average transcription rate for RNA II; h^{-1} promoter
k_1	= maximum synthesis rate of amino acid
k_2	= the second order binding rate between RNA I and RNA II; ml/molecule h^{-1}.
k_{d_i}	= average degradation rate component i; h^{-1}
v	= cell volume
μ_1	= maximum rate of synthesis of proteins
ε, γ	= stoichiometry coefficients for pseudo-chemical reactions (on mass basis)

APPENDIX I

Model equations for two of the model components (amino acid and protein) are described in detail in this appendix. A full set of equations is given elsewhere (Domach, 1983; Domach et al., 1984). Techniques for the evaluation of model parameters can be found in Dick (1978), Leung (1979), and Domach (1983).

The net rate of amino acid formation is postulated as described by

$$\frac{dP_1}{dt} = k_1 \left(\frac{K_{P_1}}{K_{P_1} + P_1/V}\right)\left(\frac{A_1/V}{K_{P_1A_1} + A_1/V}\right)\left(\frac{A_2/V}{K_{P_1A_2} + A_2/V}\right)V$$

$$- k_{TP_1} \frac{K_{TP_1}}{K_{TP_1} + A_2/V} P_1 - \gamma_1 \left(\frac{dM_1}{dt}\right) - \varepsilon_2\left(\frac{dP_2}{dt} + \gamma_2 \frac{dM_2}{dt}\right.$$

$$\left. + \varepsilon_3 \frac{dP_3}{dt} + \varepsilon_3\gamma_3 \frac{dM_3}{dt}\right) - \varepsilon_4\left(\frac{dP_4}{dt} + \gamma_4 \frac{dM_4}{dt}\right) \qquad (1)$$

The first term on the right-hand side represents the mass rate of formation of P_1 (mass units) from A_1 and A_2. Definitions of the model components are given in the legend to Figure 1. The multiple-substrate rate form has been found to be satisfactory for several enzyme systems (Bader et al., 1975; Fredrickson et al., 1970; Sinclair and Ryder, 1975). In the model, the feedback effects for amino acid have been included by using the term $K_{P_1}/(K_{P_1} + P_1/v)$. Feedback control of amino acid metabolism is well established (Lehninger, 1975). The rate of synthesis is made proportional to the amount of amino acid synthesizing enzymes by assuming the amount of enzyme per unit volume is constant (except for modification due to the feedback inhibition term). A term for the breakdown of amino acids to glucose and nitrogen is given [i.e., $k_{TP_1}(K_{TP_1}/K_{TP_1} + A_2/v)$]; such a term becomes significant only if the cell is starved for glucose. P_1 is largely consumed in protein synthesis (dM_1/dt), but it is also required in the synthesis of ribonucleotides and cell envelope precursors. The last two terms indicate the consumption of P_1 for these other components. The stoichiometry coefficients γ and ε denote the stoichiometry of pseudo-chemical reactions representing the incorporation of amino acids to other cellular components.

The maximum rate of formation of amino acids should equal the steady-state maximum rate of consumption in *E. coli* cell growing exponentially in a glucose-ammonia medium (the model is currently for growth on a minimal medium). After appropriate calculation, $k_1 = 0.390$ g/h cm^3 is obtained. Sakamoto et al. (1975) have reported a saturation constant for ammonia for glutamate dehydrogenase in *E. coli* as 1.1×10^{-3}, which is 2.5×10^{-5} g/cm^3 by our convention. $K_{P_1A_1}$ was set equal to 2.5×10^{-5} g/cm^3. The unmeasured saturation constants for glucose were set as a fraction of normal intracellular concentrations (NIC) for glucose. Generally, they were set equal to 0.04 (1/25) of the intracellular concentration during unrestricted growth. A much higher value would not allow the saturation of enzymatic reactions, and a much lower concentration would result in a cellular control system that would be insensitive to changes in the external environment. For components that are more or less sensitive to energy level values higher or lower than 1/25 of NIC, A_2 were used. The value of $K_{P_1A_2}$ was set equal to 2.5×10^{-4} g/cm^3 (1/7 of NIC A_2). The normal intracellular amino acid concentration for a fast-growing *E. coli* has been given as about 4.5×10^{-3} g/cm^3 (Levy et al., 1973; Watson, 1970). The value of K_{P_1}, the saturation constant for inhibition, was chosen to be equal to 5×10^{-2}, which is 10 times the normal intracellular concentration of P_1. Significant inhibition results only if P_1 levels start to rise much above 4.5×10^{-3} g/cm^3. The values used for saturation constants reflect many literature values for amino acids (Lehninger, 1975) and result in a P_1 pool concentration near that normally measured when amino acids are extracted from bacteria (Tempest et al., 1970). Mandelstam and Halverson (1960) have measured a maximum rate of protein degradation of 5%. The rates of degradation for amino acid vary from nearly zero in actively growing cells to a significant percentage of protein and amino acids. k_{TP_1} was set equal to 0.025 h^{-1}, which is half the value measured for protein. K_{TP_1}, the saturation constant in the degradation term, was assumed to be equal to 0.0025 of the normal intracellular concentration of A_2 (or 1.1×10^{-5} g/cm^3), so that P_1 degradation is not significant until the synthesis of P falls below 5% of its maximal level.

For the purpose of illustration the calculations for obtaining values for γ_1, γ_2, ε_2, and ε_3 are followed. The pseudo-chemical reactions representing the mass balance for protein and RNA formation from amino acids is

$$\gamma_1 P_1 \rightleftarrows M_1 + (\gamma_1 - 1)H_2O$$

$$\gamma_2 P_2 \rightleftarrows M_2 + (\gamma_2 - 1)H_2O$$

A single water molecule is eliminated per peptide bond so γ_1 is equal to 1.167. γ_1 is grams of amino acids consumed for the formation of 1 g of protein. The value 1.167 was calculated using an average molecular weight of amino acid equal to 126 in *E. coli*. Similarly, using the average molecular weight of P_2 equal to 341.4 and noting that one water molecule is lost in the polymerization process, the value of γ_2 is found to be 1.057.

The stoichiometry of pseudo-chemical reactions representing the ribonucleotides and deoxyribonucleotides can be written as

$$\varepsilon_2 P_1 + \beta_2 A_2 \rightarrow P_2$$

$$\varepsilon_3 P_2 \rightarrow P_3 + (\varepsilon_3 - 1)\frac{1}{2} O_2$$

The nitrogen content of P_2 and P_1 is 0.1596 and 0.139 g N_2 per g P_1, respectively (Leung, 1979). These values are used to calculate

$$\varepsilon_2 = \frac{0.1596 \text{ g } N_2/\text{g } P_2}{0.139 \text{ g } N_2/\text{g } P_1} = 1.14 \frac{\text{g } P_1}{\text{g } P_2}$$

ε_3 is calculated in a similar manner as γ_1 and γ_2 using the stoichiometry shown for P_3 formation ($\varepsilon_3 = 1.049$).

The rate of formation of the protein in the model was calculated from the equation

$$\frac{dM_1}{dt} = \mu_1 \left(\frac{A_2/V}{K_{M_1A_2} + A_2/V}\right)\left(\frac{P_1/V}{K_{M_1P_1} + P_1/V}\right) 0.85 M_{2RTM} - k'_{TM_1} M_1 - k_{TM_1}\left(\frac{K_{TM_1}}{K_{TM_1} + A_2/V}\right) M_1 \qquad (2)$$

The hyperbolic term in A_2/v is included because of the energy dependence of the reaction. The rate of formation of protein is made proportional to the amount of r-RNA present (Ezzell and Dobrogosz, 1975; Schaechter et al., 1958).

The maximum rate constant μ_1 can be evaluated knowing that 20 amino acids per second per ribosome can be added to a peptide chain (Lehninger, 1975). With a molecular weight of 1.7×10^6 for r-RNA (Levy et al., 1973), a value of 4.6 g M_1/h g M_2 can be calculated. The maximum rate can also be evaluated from cell composition at maximum growth rates (20 minutes doubling time); a value of 4.7 g M_1/h g M_2 can be obtained. $\mu_1 = 4.6$ was used in the model. Pine (1970) has suggested that protein degradation in *E. coli* results from two different mechanisms: one independent of energy level within the cell and the other dependent. The last two terms in Equation (2) represent the energy-independent and energy-dependent terms for protein degradation. Mandelstam and Halverson (1960) have measured the rate of protein degradation in *E. coli* as 0.05 h^{-1}. The degradation rates k_{TM_1} and k'_{TM_1} were assumed to be 0.025 h^{-1}. Using the same procedure as before, $K_{M_1A_2}$ and $K_{M_1P_1}$, the saturation constants, were assumed to be 1/25 of NIC A_2 and NIC P_1, respectively. The value of K_{TM_1} was set equal to 1.1×10^{-5} g/cm^3 using the same reasoning described for the evaluation of K_{TP_1}.

APPENDIX II

In this appendix, the replication mechanism of ColE1-type plasmids and the mathematical formulation used for simulation of the replication process are briefly reviewed.

A broad group of very useful plasmids utilize the ColE1 origin of replication. Tomizawa and coworkers (Tomizawa, 1984; Tomizawa and Som, 1984; Tomizawa, 1985; Tomizawa and Itah, 1982) have summarized the current understanding of replication of plasmids of ColE1 origin. These plasmids utilize host cell enzymes but control the initiation of replication through the interaction of three chemical species. One is an RNA transcript, RNA II, made 550 bp (base pairs) upstream of the origin and transcribed toward the origin. If RNA II hybirdizes with the DNA template at the origin, then RNase H can cleave the hybrid to yield an RNA primer suitable to initiate a round of plasmid synthesis. A second RNA species, RNA I, can bind with the RNA II to form a complex that does not allow primer formation to occur. The RNA I promoter is located 442 bp upstream from the origin and is transcribed in the opposite direction from that of RNA II. The length of the RNA I transcript is 108 nucleotides. The third chemical species involved in the replication regulation is a plasmid-encoded protein, Rom, which increases the binding rate of RNA I and RNA II. The binding between RNA I and RNA II is a second-order reaction between the two RNA species.

As stated earlier, for initiation of a round of plasmid replication, RNA II transcript must extend to the replication origin without binding with an RNA I molecule. $1/K_{T_{\text{RNA II}}}$, where $K_{T_{\text{RNA II}}}$ is the transcription rate from RNA II promoter, is the average time between two subsequent initiations of transcription from RNA II promoter. One must then evaluate what fraction f of RNA II transcripts initiated from the RNA II promoter would get to the origin prior to binding with an RNA I molecule. The rate equation for deactivation of RNA II can be used to calculate the value of f.

The deactivation rate of RNA II can be represented by the equation

$$\frac{d}{dt}\,\text{RNA II} = -\,k_2\,\frac{\text{RNA I}}{vc}\,\text{RNA II} \tag{3}$$

where k_2 is the second-order binding constant between the two RNA species and vc is the cytoplasmic volume of the cell. Rearranging this equation and assuming that RNA I concentration is almost constant, we obtain

$$f = \frac{\text{RNA II}_{(t=t)}}{\text{RNA II}_{(t=0)}} = e^{-k_2\,\text{RNA I}\,(t-t_0)/vc} \tag{4}$$

Time 0 refers to the time of initiation of an RNA II transcript. As it is shown later, the time required for an RNA II transcript to reach the replication origin is about 7 s. Since the value of t = 7 s is much smaller than the cell cycle time, the assumption of constant RNA I concentration during the transcription of RNA II (i.e., 7 s) is reasonable.

To calculate the time t required for the RNA II transcript to get to the replication origin, values for the average transcription rate of RNA polymerase and the length of the transcript are required. The RNA polymerase transcription rate is 50-80 nucleotides per second (Molin, 1976; Pace, 1973). The length of RNA II transcript is 550 nucleotides. Using an average value of 65 nucleotides per second for the RNA polymerase transcription rate, the value for t can be calculated. However, since RNA polymerase occupies about 70 nucleotides (Glass, 1982), it was assumed that transcription of RNA II has to proceed about 100 nucleotides before it can react with an RNA II molecule to avoid steric effects. Thus, one calculates the time t that RNA II transcription continues up to the replication origin and it is available for reaction with an RNA I molecule to be about 7 s [i.e., (550 - 100)/65]. Therefore the probability that an RNA II transcript would extend to the replication origin prior to binding with an RNA I molecule is

$$\frac{\text{RNA II}_{(t=7)}}{\text{RNA II}_{(t=0)}} = e^{-k_2 \text{ RNA I}/vc\ \Delta t}\ \text{RNA II}_{(t=0)} \tag{5}$$

where Δt is 7 s.

It has been shown that almost half the RNA II transcripts that reach the origin hybridize with the DNA template at the origin and initiate a round of plasmid replication; the other half passes through the origin without hybridizing with the DNA template at the origin (Tomizawa and Itoh, 1982). Thus, the number of plasmids PL after the initiation of transcription of an RNA II is obtained from the equation

$$Pl_t = PL_{t-1}\ 0.5e^{-k_2 \text{ RNA I}/vc\ \ \Delta t} + PL_{t-1} \tag{6}$$

The subscripts t-1 and t indicate the value of PL before and after the latest round of plasmid replication, respectively. Cell growth continues through the cell cycle; when a time interval equal to $1/K_{T_{\text{RNA II}}}$ elapses a new round of transcription from RNA II promoter begins. At that time Equation (6) can be used to calculate the new number of plasmids. Evaluation of the right-hand side of Equation (6) requires the values for RNA I content of the cell and k_2.

Equations that describe the net rate of change in the content of each of the model components involved in the replication process (i.e., RNA I, RNA II, and Rom protein) have been described previously (Ataai and Shuler, 1986). Modification of the host-cell model equations to account for plasmid-cell interactions are detailed elsewhere (Ataai and Shuler, 1987). Only one of the model equations, the equation for RNA I, is described here.

The net rate of change in number of RNA I molecules per cell can be written as

$$\frac{d}{dt}\,\text{RNA I} = K_{T_{\text{RNA I}}}\,\text{PL} - k_2\,\frac{\text{RNA I}}{\text{vc}}\,\text{RNA II} - k_{d_{\text{RNA I}}}\,\text{RNA I}$$

where $K_{T_{\text{RNA I}}}$ and $k_{d_{\text{RNA I}}}$ are the average transcription and degradation rates of RNA I. The first term of the equation is the rate of formation of RNA I. The second term is the rate of deactivation of RNA I by forming a complex with RNA II, and the last term is the rate of degradation for RNA I.

ACKNOWLEDGMENT

This work was supported in part by ONR Grant N00014-85-K-00580.

REFERENCES

Aiba, S., Nagai, S., Nishizaway, Y., and Onodera, M. (1967a). Nucleic approach to some response of chemostatic culture of *Azotobacter vinelandii*. *J. Gen. Appl. Microbiol.* *13*:85-101.

Aiba, S., Nagai, S., Nishizaway, Y., and Onodera, M. (1967b). Energetic and nucleic analyses of a chemostatic culture of *Azotobacter vinelandii*. *J. Gen. Appl. Microbiol.* *13*:73-83.

Andrews, J. F. (1968). A Mathematical model for the continuous culture of microorganisms utilizing inhibitory substrates. *Biotechnol. Bioeng.* *10*:707-723.

Ataai, M. M., and Shuler, M. L. (1985a). Simulation of the growth pattern of a single-cell of *Escherichia coli* under anaerobic conditions. *Biotechnol. Bioeng.* *27*:1027-1035.

Ataai, M. M., and Shuler, M. L. (1985b). Simulation of CFSTR through development of a mathematical model for anaerobic growth of *Escherichia coli* population. *Biotechnol. Bioeng.* *27*:1051-1055.

Ataai, M. M., and Shuler, M. L. (1986). Mathematical model for the control of ColE1 type plasmid replication. *Plasmid* *16*:204-212.

Ataai, M. M., and Shuler, M. L. (1987). A mathematical model for prediction of plasmid copy number and genetic stability in *Escherichia coli*. *Biotechnol. Bioeng.* *30*:389-397.

Bader, F. C., Meyers, J. S., Fredrickson, A. G., and Tsuchiya, H. M. (1975). Comments on microbial growth. *Biotechnol. Bioeng.* *17*:279-283.

Barford, J. P., Pamment, N. P., and Hall, R. J. (1982). Lag phase and transients. In *Microbial Population Dynamics*, M. J. Bazin (Ed.). CRC Press, Boca Raon, Florida, p. 55.

Bazin, M. J. (1982). *Microbial Population Dynamics*. CRC Press, Boca Raton, Florida.

Campbell, A. (1957). Synchronization of cell division. *Bacteriol. Rev.* *21*:263-272.

Cashel, M. (1975). Regulation of bacterial ppGpp and ppGpp. *Annu. Rev. Microbiol.* *29*:301-318.

Davison, E. J. (1975). Simulation of cell behavior: Normal and abnormal growth. *Bull. Math. Biol.* *37*:427-458.

Diagger, G. T., and Grady, C. P. L. (1982). An assessment of the role of physiological adaptation in the transient response of bacterial cultures. *Biotechnol. Bioeng.* *24*:1427-1444.

Dick, C. C. (1978). An individual cell model for *Escherichia coli*. M.S. Thesis, Cornell University, Ithaca, New York.

Domach, M. M. (1983). Refinement and use of a structured model of a single cell of *Escherichia coli* for the description of ammonia limited growth and asynchronous population dynamics. Ph.D. Thesis, Cornell University, Ithaca, New York.

Domach, M. M., Leung, S. K., Cahn, R. E., Cocks, G. G., and Shuler, M. L. (1984). Computer models for glucose limited growth of a single cell of *Escherichia coli* B/r-A. *Biotechnol. Bioeng.* *26*: 203-216.

Erickson, L. E., Minkevich, I. G., and Eroshine, V. K. (1978). Application of mass and energy balances regularities in fermentation. *Biotechnol. Bioeng.* *20*:1595-1621.

Ezzell, J. W., and Dobrogosz (1975). Altered hexose transport and salt sensitivity in cyclic adenosine 3',5'-monophosphate deficient *Escherichia coli*. *J. Bacteriol.* *124*:815-824.

Fredrickson, A. G., Megee, R. D., and Tsuchiya, H. M. (1970). Mathematical models for fermentation processes. *Adv. Appl. Microbiol.* *13*:419-465.

Georgiou, G., Chalmers, J. J., Shuler, M. L., and Wilson, D. B. (1985). Continuous immobilized recombinant protein production from *E. coli* capable of selective protein excretion: A feasibility study. *Biotechnol. Prog.* *1*:75-79.

Glass, R. E. (1982). *Gene Function*. University of California Press, Berkeley.

Harder, A., and Roels, J. A. (1982). Application of simple structured models in bioengineering. *Adv. Biochem. Eng.* *21*:55-107.

Heinmets, F. (1966). *Analysis of Normal and Abnormal Cell Growth*. Plenum Press, New York.

Imanaka, T., and Aiba, S. (1981). A perspective on the application of genetic engineering: Stability of recombinant plasmid. *Ann. N.Y. Acad. Sci. 369*:1-14.

Kim, B. G., Good, T. A., Ataai, M. M., and Shuler, M. L. (1987). Growth behavior and prediction of copy number and retention of ColE1 type plasmids in *E. coli* under slow growth conditions. *Ann. N.Y. Acad. Sci. 506*:384-395.

Kim, S. H., and Ryu, D. D. Y. (1984). Instability kinetics of trp operon plasmid CIEL-trp in recombinant *Escherichia coli* MV12(pVH5) and MV12trpR(pVH5). *Biotechnol. Bioeng. 26*:497-502.

Koch, A. L. (1971). The adaptative response of *E. coli* to a feast and famine existence. *Adv. Microbiol. Physiol. 6*:147-217.

Koga, S., and Humphrey, A. E. (1967). Study of the dynamic behavior of the chemostat system. *Biotechnol. Bioeng. 9*:375-386.

Lehninger, A. L. (1975). *Biochemistry*, 2nd Ed. Worth Publishers, New York.

Leung, S. K. (1979). A mathematical model for the growth of a single cell of *Escherichia coli* and its experimental verification. M.S. Thesis, Cornell University, Ithaca, New York.

Levy, J., Campbell, J. J. R., and Blackburn, T. H. (1973). *Introductory Microbiology*. John Wiley and Sons, New York.

Maaloe, O., and Kjeldgaard, N. O. (1966). *Control of Macromolecular Synthesis*. Benjamin, New York.

Mandelstam, J., and Halverson, H. (1960). Turnover of protein and nucleic acid in soluble and ribosome fractions of nongrowing *Escherichia coli*. *Biochim. Biophys. Acta 40*:43-49.

Mateles, R. I., Ryu, D. D. Y., and Yasuda, T. (1965). Measurement of steady state growth rates of microorganisms. *Nature 208*: 263-265.

Molin, S. (1976). Ribosomal RNA chain elongation rates in *Escherichia coli*. In *Control of Ribosome Synthesis*, O. Maaloe (Ed.). Alfred Benzon Symposium 9, Academic Press, Orlando, Florida, pp. 333-339.

Monod, J. (1949). The growth of bacterial cultures. *Annu. Rev. Microbiol. 3*:371-393.

Pace, N. R. (1973). Structure and synthesis of ribosomal ribonucleic acid of prokaryotes. *Bacteriol. Rev. 37*:562-603.

Papoutsakis, E. T. (1984). Equations and calculations for fermentation of butyric acid bacteria. *Biotechnol. Bioeng. 26*:174-187.

Peretti, S. W., and Bailey, J. E. (1986). Mechanically detailed model of cellular metabolism for glucose-limited growth of *Escherichia coli* B/r-A. *Biotechnol. Bioeng. 28*:1672-1689.

Pine, J. J. (1970). Steady state measurements of the turnover of amino acid in the cellular protein of growing *E. coli*: Existence of two kinetically distinct reactions. *J. Bacteriol. 103*:207-215.

Powell, E. O. (1969). Transient changes in the growth rate of microorganisms. In *Proceedings of the 4th Symposium on Continuous Cultivation of Microorganisms*, I. Malek (Ed.). Publishing House Academia, Prague, pp. 275-284.
Sakamoto, N., Kotre, A. M., and Savageau, M. A. (1975). Glutamata dehydrogenase for *Escherichia coli*: Purification and properties. *J. Bacteriol. 124*:775-783.
Schaechter, M. O., Maaloe, O., and Kjeldgaard, N. O. (1958). Dependency on medium and temperature of cell size and chemical composition during balanced growth of *Salmonella typhimurium. J. Gen. Microbiol. 19*:592-606.
Seo, J. H., and Bailey, J. E. (1985). A plasmid model for plasmid content and product synthesis in unstable binary fission recombinant organisms. *Biotechnol. Bioeng. 27*:156-165.
Shuler, M. L. (1985). On the use of chemically structured models for bioreactors. *Chem. Eng. Commun. 36*:161-189.
Shuler, M. L., Leung, S., and Dick, C. C. (1979). A mathematical model for the growth of a single bacterial cell. *Ann. N.Y. Acad. Sci. 326*:35-55.
Simon, Z. (1973). Bacterial cell model. *J. Theor. Biol. 38*:39-49.
Sinclair, C. G., and Ryder, D. N. (1975). Models for the continuous culture of microorganisms under both oxygen and carbon limiting conditions. *Biotechnol. Bioeng. 17*:375-398.
Tempest, D. W., Meers, J. L., and Brown, C. W. (1970). Influence of environment on the content and composition of microbial free amino acid pools. *J. Gen. Microbiol. 64*:171-185.
Tomizawa, J. (1984). Control of ColE1 plasmid replication: The process of binding of RNA I to the primer transcript. *Cell 38*: 861-870.
Tomizawa, J. (1985). Control of ColE1 plasmid replication: Initial interaction of RNA I and the primer transcript is reversible. *Cell 40*:527-535.
Tomizawa, J., and Itoh, T. (1982). The importance of RNA secondary structure in ColE1 primer formation. *Cell 31*:575-583.
Tomizawa, J., and Som, T. (1984). Control of ColE1 plasmid replication: Enhancement of binding of RNA I to the primer transcript by the Rom protein. *Cell 38*:871-878.
Uhlin, B. E., Molin, S., Gustafsson, P., and Nordstrom, K. (1979). Plasmid with temperature-dependent copy number for amplification of cloned genes and their products. *Gene 6*:91-106.
Votruba, J., Volesky, B., and Yerushalmi, L. (1986). Mathematical model of a batch acetone-butanol fermentation. *Biotechnol. Bioeng. 28*:247-255.
Watson, J. D. (1970). *Molecular Biology of the Gene*, 2nd Ed. W. A. Benjamin, Menlo Park, California.

Weinberg, R., Zeigler, B. P., and Laing, R. A. (1971). Computer simulation of a living cell: Metabolic control system experiments. *J. Cybernetics 1*:34-48.
Yano, T., and Koga, S. (1969). Dynamic behavior of the chemostat subject to substrate inhibition. *Biotechnol. Bioeng. 11*:139-153.
Yerushalmi, L., Volesky, B., and Votruba, J. (1986). Modeling of culture kinetics and physiology for *Clostridium acetobutylicum*. *Can. J. Chem. Eng. 64*:607-616.

10

Control of Carbon and Electron Flows in Acetone-Butanol Fermentations

RATHIN DATTA* / *CPC International, Inc., Summit-Argo, Illinois*

I. INTRODUCTION

Anaerobic fermentation of carbohydrates to the solvents acetone, butanol, and ethanol (ABE) is well known, and these solvents have been commercially produced by fermentation for several decades (Weizmann, 1915; Beesch, 1953). *Clostridium acetobutylicum* has been the organism of choice, and many strains and variants of this type of clostridia have been used. Molasses, corn mash, rice, and wood hydrolyzates were among the many carbohydrate feedstocks utilized. With the advent of petrochemical processes and low-cost petrochemical feedstocks, the fermentation-based processes became economically unattractive and most of the commercial installations were shut down in the 1950s. The increased cost of petrochemical feedstocks and the desire to produce a methanol cosolvent for gasoline (Guibet and Vandercasteele, 1983) have renewed interest in the fermentation process. Several academic and industrial laboratories have been involved in developing an improved process for commercialization. Development of an improved ABE process and an economic comparison with the petrochemical process have been described in a recent paper (Marlatt and Datta, 1986). Further process development directed at the reduction of capital and waste treatment costs and research directed toward increasing yield, production rates, and solvent concentration could make the fermentation process more

---

**Present affiliation*: Michigan Biotechnology Institute, Lansing, Michigan.

attractive (Marlatt and Datta, 1986). Thus, understanding the physiology of *C. acetobutylicum* and the metabolic regulation of solvent production could result in an improved process that in turn would be more economically attractive.

During fermentation, *C. acetobutylicum* produces three major classes of products: (1) solvents—acetone, butanol, and ethanol; (2) organic acids—acetic, butyric, and lactic; and (3) gases—carbon dioxide (CO_2) and hydrogen (H_2). The fermentation is biphasic. During the initial acidogenic phase, organic acids are the primary products, the evolved gas contains a high fraction of H_2 (50-60%), and the cell growth is exponential. Under normal conditions, after accumulation of a certain level of organic acids, the fermentation switches to the solventogenic phase in which the organic acids (especially butyric) and carbohydrates are converted to the solvents, the evolved gas contains a smaller fraction of H_2 (30-40%), and cell growth ceases.

The biphasic nature of this fermentation has been the subject of many studies, some of which are discussed later. Carbon and electron flow is regulated in the complex metabolic pathway of the organism. In this chapter, the concepts of carbon and electron balances from a thermodynamic and kinetic viewpoint are briefly discussed, followed by a discussion of the general metabolic pathway. The regulation of electron and carbon flow and the effects of nutritional factors or metabolic modulators on the fermentation are also discussed. The implications of metabolic modulation on process improvement and potential applications are presented.

II. ELECTRON AND CARBON BALANCES

Concepts of electron and carbon flow in ABE fermentation can be studied at various levels of thermodynamic and kinetic complexity. The electron balance as elaborated by Eroshin, Minkevich, and coworkers (Eroshin and Minkevich, 1982; Erickson et al., 1979) is simply based on the fact that the standard amount of reductivity (available electrons) of any organic compound has practically a constant energy value, about 27-28 kcal (approximately 113-117 kJ) per electron equivalent. This observation in essence allows one to equate any substrate or product on the basis of electron equivalents and conduct an electron balance which also gives an approximation of the overall thermodynamic (combustion energy) efficiency of the fermentation. In anaerobic fermentations, the electron equivalents in the products equals those in the substrate, whereas in aerobic fermentations, the electrons used by oxygen (the terminal acceptor) produce heat. Thus, the simple concept of electron balance provides a powerful insight into the stoichiometry and energetics of a fermentation.

Table 1 ABE Fermentation: Product Yield and Electron Balances[a]

Product (mol)	Weight yield[b]	Electrons (%)[b]
$0.09CH_3COOH$ (acetic)	0.03	3
$0.024C_3H_7COOH$ (butyric)	0.012	2
$1.56H_2$	0.0173 (±0.003)	13 (±2)
$0.3CH_3-COCH_3$ (acetone)	0.097 (±0.01)	20 (±2)
$0.535C_4H_9OH$ (butanol)	0.220 (±0.01)	53.5 (±2.5)
$0.08C_2H_5OH$ (ethanol)	0.0186 (±0.002)	4 (±0.4)
$0.286CH_{1.8}O_{0.5}N_{0.2}$ (cells)	0.039	5
$2.31CO_2$	0.56	0

[a]Fermentable carbohydrate = $C_6H_{12}O_6$.
[b]The numbers reported are average values. Normal variations are shown in parentheses. A positive variation in one product leads to a negative variation in another.
Source: From Datta and Zeikus (1985).

In a complex multiproduct fermentation, such as the ABE fermentation, an electron balance provides a proper understanding of product yield and fermentation efficiency. The average product yield and electron balance for a completed ABE fermentation are shown in Table 1. Thus, about 75-78% of the available electrons in the feed carbohydrate end up in the ABE solvents, and butanol accounts for 50-55% of the electrons. The nonsolvent products, residual organic acids, cells, and H_2 gas account for about 20-25% of the electrons, with H_2 gas accounting for the major fraction (11-15%). This analysis shows that the heterofermentative ABE fermentation is somewhat less efficient than homoproduct anaerobic fermentations, such as ethanol or lactic acid fermentations, in which 90-95% of the electrons are channeled into the product. Electron balances also show the fallacy of lumping all the solvents together, as is done in most published literature, and reporting a solvent yield on this basis, unless the solvent ratio remains invariant.

Thus, the overall solvent yield can be increased by decreasing the production of the most reduced solvent, butanol, and increasing the production of acetone or ethanol, or both. This analysis shows that a reduction in H_2 production could increase solvent yield, if the electrons that would have gone into H_2 could be channeled into metabolic pathways that produce solvents. The work of Datta and Zeikus (1985), who successfully applied this concept, is discussed later.

III. METABOLIC PATHWAY

The electron balance, which is also equivalent to oxidation-reduction (O/R) on the available hydrogen balance, provides only an overall balance of the fermentation. Further insight into the mechanism and interrelationships among various products during the multiproduct fermentation can only be provided by the metabolic pathway of the fermentation. A simplified diagram of the metabolic pathway leading to major catabolic products is shown in Figure 1. The biochemistry of ABE fermentation has been characterized by many researchers. Pyruvate is produced from carbohydrate by the Embden-Meyerhof-Parnas (EMP) pathway. There are three major branch points at acetyl coenzyme A (acetyl-CoA), acetoacetyl-CoA, and butyryl-CoA. Another branch point at pyruvate to lactate was usually ignored in recent literature despite earlier reports of homolactic fermentation by *C. acetobutylicum* (Simon, 1947) under conditions of hydrogenase inhibition. Under normal fermentation conditions, pyruvate is converted to acetyl-CoA by a typical phosphoroclastic reaction with electron transfer to ferredoxin by the enzyme pyruvate ferredoxin oxidoreductase. The reduced ferredoxin can interact with hydrogenase to produce hydrogen or with NADH and NADPH ferredoxin oxidoreductase. The latter reactions have been found to be easily reversed and regulated by the cellular levels of NADH and NADPH (Petitdemange et al., 1976). The physiological function of the clostridial NADH and NADPH oxidoreductases has been elucidated (Jungermann et al., 1973). These enzymes play a crucial role in the regulation of electron flow and overall thermodynamic efficiency of the fermentations.

The pathways for acetate and butyrate formation are well characterized. Phosphotransacylase and acetate kinase are involved in the formation of acetate and ATP from acetyl-CoA. Acetyl-CoA-acetyltransferase catalyzes the formation of acetoacetyl-CoA, which is reduced to butyrate by three enzymes: β-hydroxybutyryl-CoA dehydrogenase, crotonase, and butyryl-CoA dehydrogenase.

The terminal enzymes catalyzing solvent production have been under intensive study, and some have been characterized. Aceto-

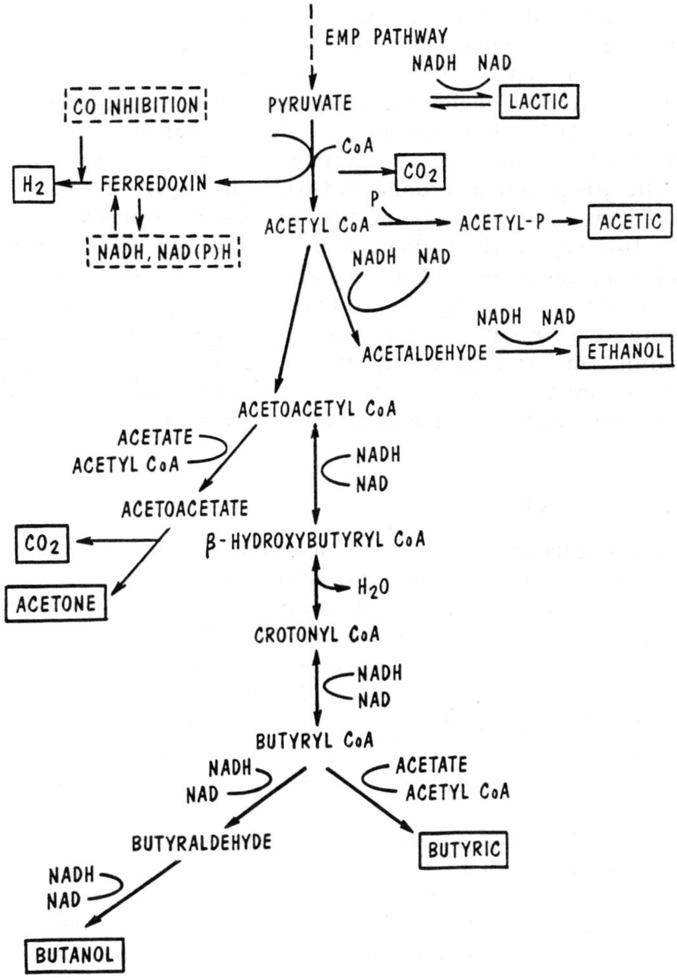

Figure 1 Metabolic pathway and proposed inhibition of hydrogenase of *C. acetobutylicum*. (From Datta and Zeikus, 1985.)

acetate is decarboxylated to acetone by the enzyme acetoacetate decarboxylase, which has been purified and characterized (Davies, 1943; Fridovich, 1972). Recent research (Rogers, 1986; Rogers and Hansen, 1983) on the pathway from butyryl-CoA to butanol has shown that two dehydrogenases are involved. Butyraldehyde dehydrogenase is NADH dependent, whereas the butanol dehydrogenase requires NADPH. Whether the two dehydrogenases producing ethanol are the same or different proteins from those producing butanol has not been determined. In this context, other electron carriers, such as rubredoxin and flavodoxin, have been implicated with oxidoreductases under various growth and culture conditions (Petitdemange et al., 1979; Marczak et al., 1984). For the sake of simplicity, NADH has been used to represent a generalized electron carrier in the proposed metabolic pathway in Figure 1. From the knowledge of metabolic pathway of this organism, researchers have engaged in two types of studies. Studies on metabolic regulation have concentrated on measurement of the activities of key enzymes or intracellular conditions during the course of the fermentation to better understand metabolic control. On the other hand, studies on metabolic modulation have involved the deliberate use of enzyme inhibitors, electron donors and acceptors, or nutritional and environmental factors to alter metabolism and product formation.

IV. METABOLIC REGULATION

During normal batch fermentation by *C. acetobutylicum*, two distinct phases can be distinguished. Davies and Stephenson (1941), in quantitative studies, reported the production of acetate and butyrate and exponential growth in the acidogenic phase followed by the solventogenic phase when growth stopped and acids and carbohydrates were converted to solvents with a concomitant rise in the fermentation pH. A typical fermentation profile of a normal batch fermentation of the CPC-asporogenic strain (ATCC 39236) in a cornstarch-corn steep liquor (CSL) medium is shown in Figure 2. The biphasic nature of the fermentation and metabolic regulation has been the subject of many recent studies. Andersch et al. (1983) studied the levels of enzymes in the acidogenic and solventogenic phases in cells grown in a chemostat at pH 6.0 and 4.3, respectively. They found the specific activities of acetate and butyrate kinase and phosphotransacetylase and phosphotransbutyrylase were reduced 2- to 10-fold in solvent-producing cells. Acetoacetate-CoA transferase and decarboxylase activities were hardly detectable in acidogenic cells but reached high levels in solventogenic cells. The aldehyde and alcohol dehydrogenases were detectable only in the solventogenic cells. Interestingly, hydrogenase activity was approximately the same in both types of cells.

In similar studies, Hartmanis and Gatenbeck (1984) investigated the levels of seven intermediary enzymes involved in acetate and butyrate formation during the acidogenic and solventogenic phases of a batch fermentation. They observed a similar decrease in the activity of acetate kinase and phosphotransacetylase and phosphotransbutyrylase. However, the level of butyrate kinase showed an

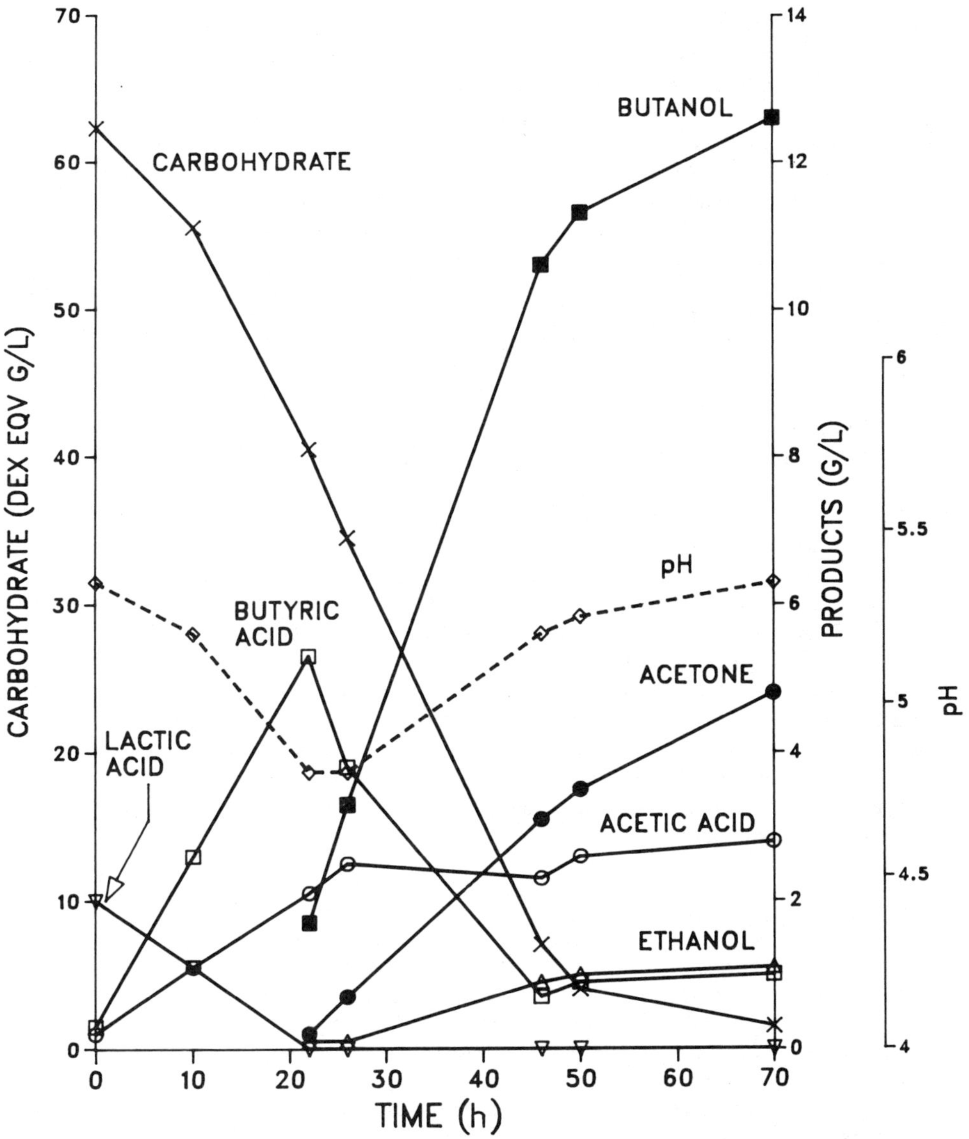

Figure 2 Fermentation profile for normal ABE fermentation. (From Datta and Zeikus, 1985.)

increase in specific activity, suggesting that the uptake of organic acids during the solventogenic phase does not proceed via a complete reversal of phosphate transferase and kinase reactions. The authors calculated the in vivo metabolic fluxes for these enzymes and found them 10 to 1000-fold lower than the specific activities measured in vitro. This implies that the maximal enzyme activities are not rate limiting during the fermentation. Instead, these results suggest that allosteric modification of these enzymes by metabolic intermediates may regulate the fermentation pathway.

The induction of solventogenic fermentation by organic acids as metabolic intermediates has been well known (Gottschal and Morris, 1981; Nakhmanovich and Shcheblykina, 1960; Wood et al., 1945). A recent study by Terracciano and Kashket (1986) showed that total intracellular undissociated acid concentration was at least 40-45 mM before the onset of solventogenesis. Similar results were obtained when cultures were supplemented with 50 mM butyric acid or when a phosphate-buffered medium was used instead of an acetate-buffered medium. The authors conclude that switching from acidogenesis to solventogenesis requires a high concentration of an energy substrate, rapid catabolism, an internal pH greater than 5.5, an internal concentration of total undissociated organic acids >42 mM, and butyric acid of at least 13-14 mM. In a related study, Gottwald and Gottschalk (1985) proposed that the increased butyrate and acetate concentration at the end of the acidogenic phase lead to increased levels of butyryl and acetyl-CoA and butyryl and acetyl phosphate. This results in the decrease in coenzyme A and phosphate pools that triggers the mechanism for activation of solventogenesis. Recent work in the area of metabolic modulation has provided further data to support the coenzyme pool hypothesis.

V. METABOLIC MODULATION

Metabolic modulation involves the deliberate use of inhibitors, electron donors and acceptors, and nutritional or environmental factors to alter metabolism and direct the course of fermentation and product formation. The inhibition of hydrogenase by carbon monoxide (CO) can lead to major changes in fermentation characteristics. Nutritional factors, such as iron limitation or the influence of whey solids or lactate, also lead to metabolic modulation. Various nonphysiological electron donors or acceptors can also modulate the fermentation. Recent studies in these areas are discussed in further detail.

A. Modulation by Carbon Monoxide

The rationale for modulation of the fermentation by CO is based on the regulation of electron flow in the organism. As shown in

Table 1, nonsolvent products account for 20-25% of the electrons, with H_2 gas accounting for the major fraction (11-15%) of these electrons. The specific inhibition of hydrogenase, which is the terminal enzyme for H_2 production, could decrease H_2 production and increase solvent yield. This rationale was espoused by Kim et al. (1984), who showed that CO readily inhibited the hydrogenase activity of cell extracts or of viable cells, and low partial pressures of CO (i.e., up to 0.15 atm) in continuous pH-controlled fermentors did not affect growth. Their results indicated the diversion of carbon and electrons away from H_2, acetate, and butyrate toward the solvents. Based on theoretical calculations, Papoutsakis (1984) derived fermentation equations for *C. acetobutylicum* and rationalized the potential for electron flow away from H_2 to solvent production. However, these studies did not account for the potential of lactate production due to the inhibition of hydrogenase. Simon (1947) had demonstrated that *C. acetobutylicum* fermentation under an atmosphere of CO produced lactic acid instead of the normal fermentation products. Datta and Zeikus (1985) successfully demonstrated metabolic modulation and increased solvent yield under carefully controlled conditions of hydrogenase inhibition and organic acid uptake.

The time course profile of a typical batch fermentation of the CPC-asporogenic strain (ATCC 39236) in cornstarch-CSL medium is shown in Figure 2. The initial production of organic acids (primarily butyric) and a concomitant pH drop is followed by a solventogenic fermentation in which butyric acid is consumed, pH rises, and acetone, butanol, and ethanol are produced. A rapid consumption of lactic acid (from CSL) occurs in the early acidogenic phase. Time course profiles for organic acids and solvents for a batch fermentation sparged with anaerobic gas containing 20% CO, 30% H_2, 48% N_2, and 2% CO_2 are shown in Figures 3 and 4. A dramatic change in fermentation is observed even at a CO partial pressure of 0.2 atm. Lactic acid becomes the predominant organic acid in the acidogenic phase. The pH drops, and alkali is used to maintain it at 5. Butyric acid (from the inoculum) is rapidly consumed in the initial stage of the fermentation. Later, when the pH had risen, the organism began to produce this acid. Thus, instead of a normal butyric acid peak, a lactic acid peak is observed. Solvent production begins early in the fermentation, and only two solvents, butanol and ethanol, are produced. The fermentation rate is slower (about 2.5-fold) than that of the normal fermentation. These results showed that electron flow, and consequently carbon flow and product formation, could be drastically altered by hydrogenase inhibition and by increasing the pool of reduced nucleotides (Fig. 3). It was hypothesized that the addition of extraneous electron acceptors, such as organic acids, could control the electron flow away from lactate during hydrogenase inhibition.

To test this hypothesis, fermentations with butyric acid (approximately 5 g/L) initially added to the broth were conducted. One fermentor was sparged with a gas containing approximately 10% CO, 40% H_2, 40% CO_2, and 10% N_2. Figures 5 and 6 show the time profiles for carbohydrate and organic acids in the two fermentations. The control (Fig. 5) shows a sharp butyric acid peak during the early stages of fermentation. The CO-sparged fermentation (Fig. 6) shows a shallow butyric acid peak on an earlier consumption of this acid. In both cases, lactic acid, initially present, is consumed. The rates

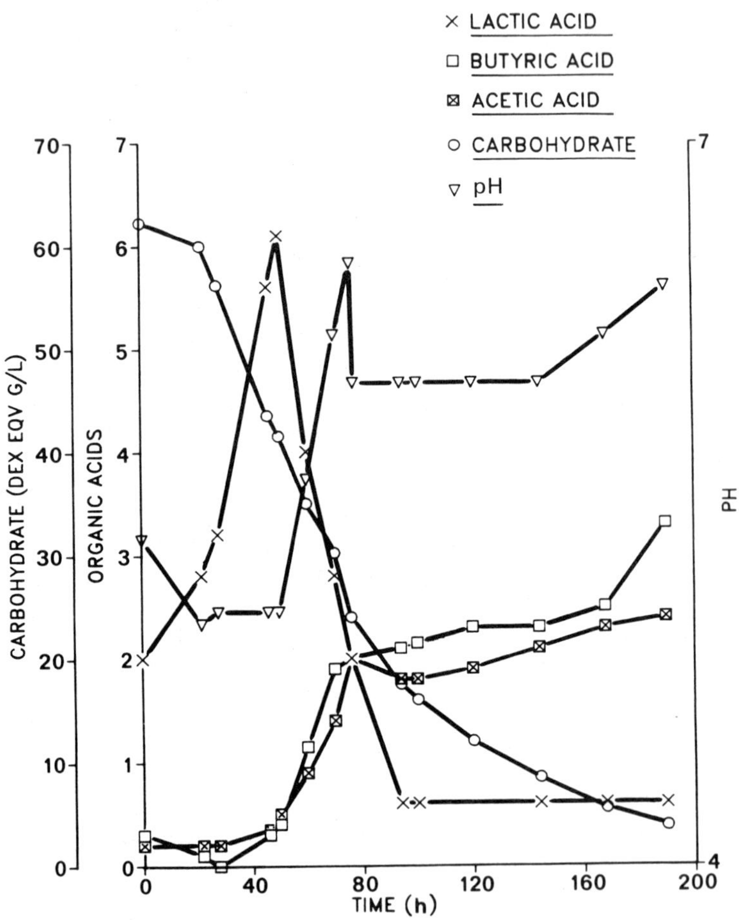

Figure 3 Organic acid and pH profile for CO-sparged ABE fermentation; 20% (v/v) CO. (From Datta and Zeikus, 1985.)

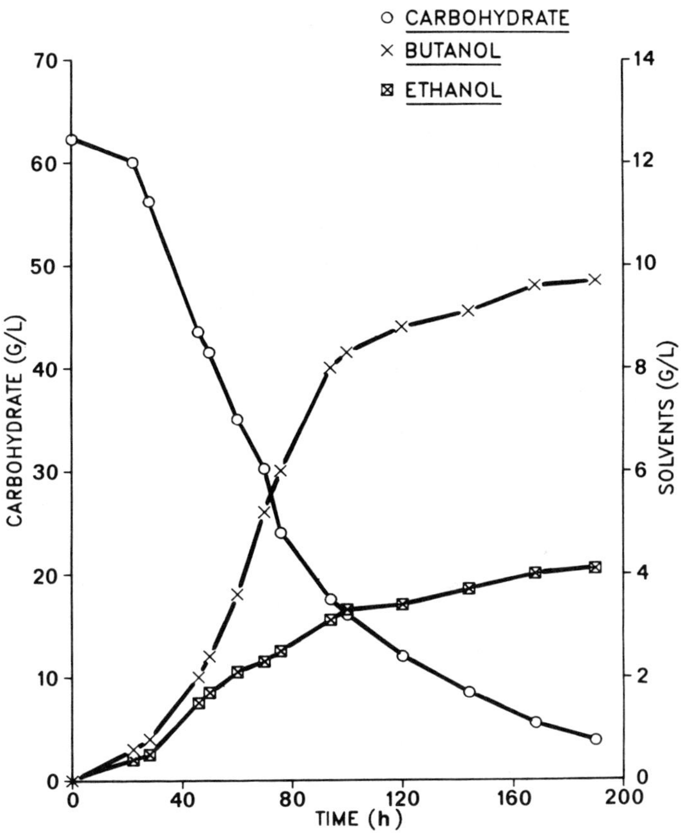

Figure 4 Solvent profile for CO-sparged ABE fermentation; 20% (v/v) CO. (From Datta and Zeikus, 1985.)

of carbohydrate consumption are similar. This is in contrast to the 2.5-fold slower rate for CO-sparged fermentation without butyrate as electron acceptor (Figs. 3 and 4). This shows that proper regulation and modulation of electron flow does not decrease fermentation productivity.

Butyric acid and butanol profiles for a CO- and organic acid-modulated fermentation and the control are shown in Figure 7. The abscissa is normalized to carbohydrate conversion. The data show a faster consumption of butyric acid and an earlier production of butanol for the CO-modulated fermentation. Recent work by Meyer et al. (1985, 1986) has shown similar results in batch and continuous fermentation using another strain (ATCC 824) of *C. aceto-*

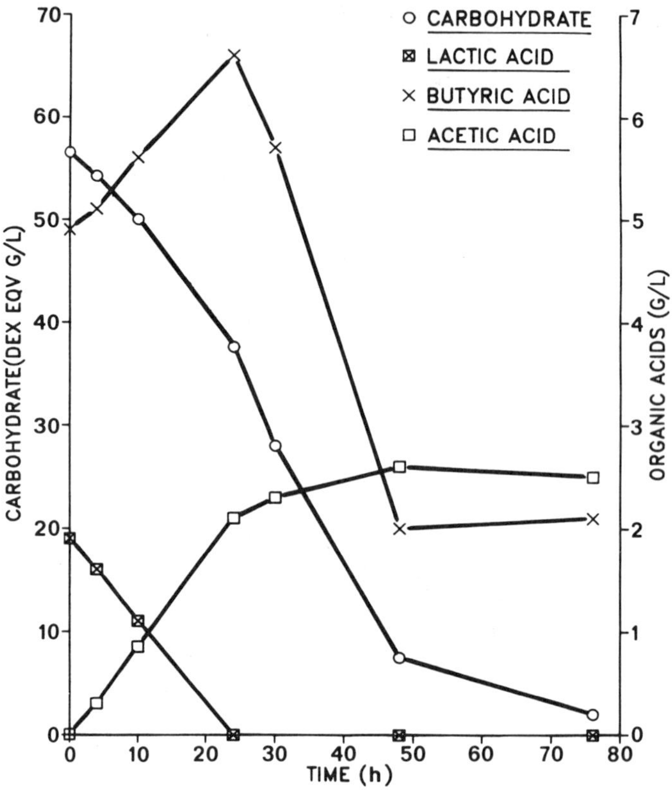

Figure 5 Organic acid profile for ABE fermentation with butyrate addition (control). (From Datta and Zeikus, 1985.)

butylicum. These experiments support the hypothesis of metabolic modulation with hydrogenase inhibition and organic acid electron acceptors.

The conditions for CO modulation and organic acid uptake were further optimized, and the effects on product yield and distribution were obtained. The results are summarized as follows:

1. CO is an effective metabolic modulator for ABE fermentation. The concentration range for dissolved CO should correspond to 0.1-0.2 atm of partial pressure, optimally between 0.1 and 0.15 atm.
2. Metabolic modulation is particularly effective with organic acids, such as butyric acid, as electron sinks.

3. The uptake of organic acids is enhanced by CO modulation, and increases in butyric acid uptake of 50-200% over control can be obtained.
4. Modulation induces solventogenic fermentation at an earlier stage of the fermentation.
5. The production of acetone is decreased, and the production of butanol and ethanol is increased. Acetone production can be reduced to zero.
6. The acetone-butanol solvent ratio can be controlled in the range of 1:4 to 1:2.5 without curtailing fermentation rate.
7. The butanol yield can be increased by 10-15% (compared with control).

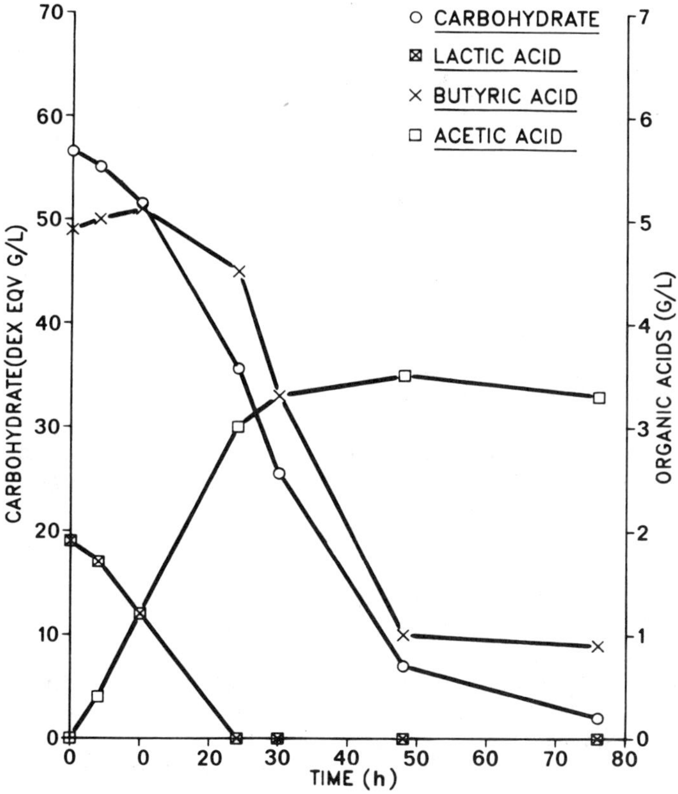

Figure 6 Organic acid profile for CO-sparged ABE fermentation with butyrate addition; 10% (v/v) CO. (From Datta and Zeikus, 1985.)

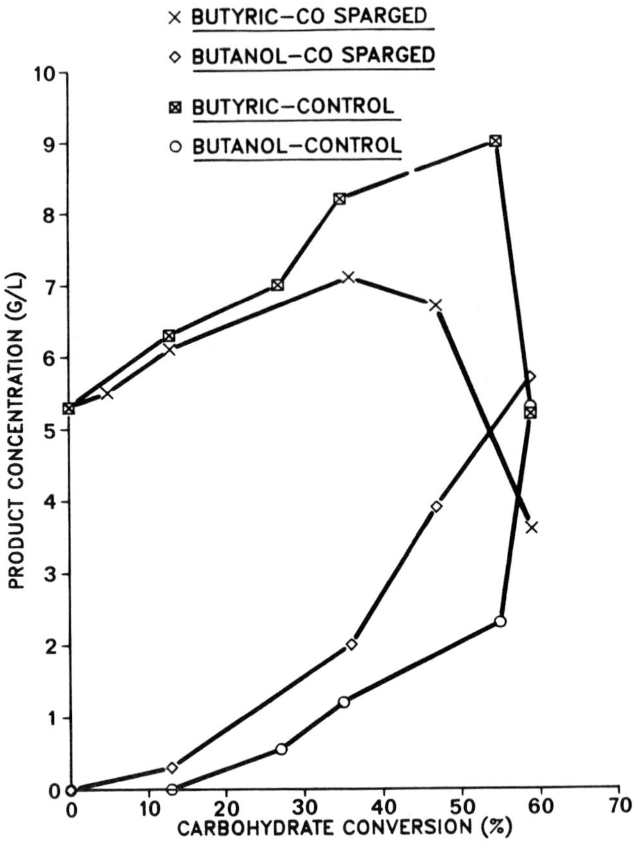

Figure 7 Butyric acid and butanol profiles of control and organic acids plus CO-modulated ABE fermentation; 10% (v/v) CO, 2 g/L acetic, 5 g/L butyric. (From Datta and Zeikus, 1985.)

8. The total solvent yield can be increased by 3-5% and the electron efficiency to solvents increased from 73-78% for control to 80-87% for modulated fermentations. This increase in efficiency is primarily derived from a 40% reduction in H_2 production. Some data on solvent yield and electron balances are summarized in Table 2.
9. When using a medium with excess carbohydrate and nutrients, modulation can lead to a higher final concentration of butanol than the control.

Table 2 CO Modulation with Organic Acids: Yield and Electron Balances

Group	Total electron equivalents consumed from all substrates	Electrons in ABE solvents (%)			Electron efficiency	Yield solvent[a]	Yield butanol[a]
		Acetone	BuOH	EtOH			
Control 1[b]	6.76	21.6	48.8	2.7	73.1	0.33	0.21
Control 2[c]	7.10	22.9	52.7	1.5	77.1	0.34	0.22
Set 4[d]	7.81	12.9	61.1	5.4	79.4	0.37	0.268
Set 7[e]	7.37	19.5	60.2	4.4	84.1	0.39	0.269
Set 8[f]	7.29	19.8	63.1	4.0	86.9	0.40	0.278

[a]Yield = (g product)/(g dextrose equivalent consumed + g butyric acid consumed).
[b]Control 1 (without organic acids, vented).
[c]Control 2 (with 4.8 g/L butyric, vented).
[d]Set 4 (0.1 atm of CO, 4.8 g/L butyric).
[e]Set 7 (0.1 atm of CO, 5 g/L butyric, 2 g/L acetic).
[f]Set 8 (0.15 atm of CO, 5 g/L butyric, 2 g/L acetic).
Source: From Datta and Zeikus (1985).

These results have revealed the dynamic nature of carbon and electron flow in the organism. They are significant from mechanistic, theoretical, and practical viewpoints and are discussed later.

B. Modulation by Nutritional Factors

Changes in key nutritional factors in the fermentation medium can modulate the fermentation and alter electron flow and product formation. Maddox (1980) reported the fermentation of whey filtrate in which the butanol-acetone solvent ratio was 10:1. This was apparently due to some nutritional factors in whey because in a normal fermentation media the same organism produced the normal butanol-acetone ratio of about 2:1. Recent work by Bahl et al. (1986) has elucidated several nutritional factors that may be responsible for the change in solvent ratio. The fermentation of whey permeate produced a butanol-acetone (B/A) ratio of about 100, and the addition of yeast extract did not alter the B/A ratio. In synthetic medium, the fermentation of various sugars (glucose, galactose, or lactose) gave a normal B/A ratio of 2-2.5. The removal of citrate from whey permeate led to the production of more acetone (B/A $\simeq$ 15), and further addition of minerals gave a B/A ratio of 11. They also noted the cofermentation of lactate with reducing sugars and a change in the B/A ratio to 3.8-3.9 owing to lactate conversion. Ferrous ion was found to be another key nutritional factor that had a dramatic effect on product formation. In an iron-limited chemostat under solvent-producing conditions, the B/A ratio increased up to 6 at high dilution rates. Increased pH led to organic acid formation, but under iron limitation, lactate was the predominant product at pH > 5.

This observation of lactate production in iron-limited fermentations is not new. Hanson and Rodgers (1946) reported lactate production in the fermentation of whey at low (0.75×10^{-5} M) iron concentrations. With increased iron concentration, H_2, CO_2, acetone, acetate, and butyrate production increased whereas butanol production was unaffected.

On examination, the results of modulation by nutritional factors appear similar to those observed in CO modulation. The electron transfer cofactors and enzymes—ferredoxin, rubredoxin, and hydrogenase, as well as acetoacetate decarboxylase—contain iron. Under iron-limited conditions the electron flow is altered, which alters the carbon flow. At neutral pH the electrons are readily disposed via reduction of pyruvate to lactate. At lower pH butanol production is enhanced over acetone production. Citrate is an effective chelator of transitional metals. This could explain its effect on the B/A ratio in whey fermentation as observed by Bahl et al.

(1986). Thus, the combined effects of lactate cofermentation, low iron concentration, and citrate chelation could modulate the electron and carbon flow to give the high B/A ratio in whey fermentation.

C. Miscellaneous Metabolic Modulators

Antibiotics such as metronidazole and chloramphenicol can act as electron acceptors from reduced ferredoxin (O'Brien and Morris, 1971, 1972). They can alter electron flow and temporarily halt the formation of hydrogen without actually inhibiting hydrogenase. When the compounds are reduced, hydrogen formation and growth can resume. Other aryl nitro groups in such compounds as nitrophenols, nitrobenzoic acid, and nitrobenzaldehyde can be readily reduced by *C. acetobutylicum* (O'Brien and Morris, 1971).

Hongo (1957) reported that the addition of 6 mM neutral red redox dye to the fermentation media resulted in a shift in the acetone-butanol-ethanol ratio from 26:63:11 to 7:90:3. The physiological basis for this phenomenon could be due to an alteration in intracellular electron metabolism that alters carbon flow in the fermentation (Kim et al., 1984).

VI. CONCLUSIONS

Studies of metabolic modulation and regulation have revealed the dynamic nature of electron and carbon flow in the organism. It appears that regulation of the electron flow can control the carbon flow and product distribution. Thus, channeling electrons away from hydrogen can lead to the production of lactate and the alcohols. However, when a sink for electrons, such as butyric acid, is added to the fermentation, the organism can channel the electrons toward acid reduction to alcohol. This modulation, or deliberate control of flow of electrons from a source to a sink, stabilizes the fermentation and increases productivity and the uptake of organic acids.

It appears that the NAD (or NADP) linked ferredoxin oxidoreductases, which catalyze the easily reversible reactions, play a crucial role in the regulation of electron flow. Lactate production from pyruvate is well known, but the mechanism and regulation of lactate uptake need further study. All the results of metabolic modulation have shown that the cofermentation of lactate with carbohydrates reduces acetone formation.

Although further research needs to be done to determine the factors that trigger the onset of solventogenesis, the data of Datta and Zeikus (1985) and Meyer et al. (1985, 1986) show that solventogenesis can occur in the early stages of a hydrogenase-inhibited fermentation and a sharp acid peak is not necessary. Gottwald and

Gottschalk (1985) have proposed that a decrease in the coenzyme A and phosphate pools trigger the mechanism for activation of solventogenesis. It appears that the concentrations and concentration ratios of the reduced nucleotide pool and coenzyme A or phosphate play a crucial role in metabolic regulation. The metabolic pathway shows that acetone production offers the organism a mechanism for recycling coenzyme A without using reduced nucleotides or generating protons. Thus, when the reduced nucleotide pool is increased in proportion to coenzyme A, the organism switches away from the acetone pathway. Papoutsakis (1984), based on equations and calculations, has also hypothesized that a factor N_fF, defined as the amount of NAD reduced by reduced ferredoxin, can be calculated and used to determine the electron flow and the course of acid or solvent fermentation. Thus, the course of ABE fermentation is not static or invariant but dynamic and variable. The regulation of electron flow controls carbon flow, production rates, and product distribution. The onset of solventogenesis can also be controlled by careful control of electron flow between appropriate sources and sinks.

From a commercial or practical viewpoint, the ability to control electron and carbon flow is economically important. An increased butanol yield reduces the cost of required fermentables. The acetone-butanol ratio can be manipulated, and this allows process flexibility in response to market conditions. The increased uptake of organic acids (especially butyric acid) by metabolic modulation is very significant. The uptake and conversion of organic acids from recycle stillage can reduce the waste treatment load and reduce waste treatment costs, which are a significant factor in the rational price of fermentation butanol (Marlatt and Datta, 1986). Butyric acid can also be produced by the anaerobic fermentation of relatively inexpensive resources, such as methanol and corn stover (Zeikus et al., 1980; Datta, 1981). Under normal fermentation conditions, the uptake and conversion of added organic acids are small. Metabolic modulation can increase uptake substantially and allow the coutilization of less expensive feedstocks. Studies of the control of carbon and electron flow in this and other fermentations have been and will continue to be rewarding for understanding physiology and metabolism and improvement of process economics.

REFERENCES

Andersch, W., Bahl, H., and Gottschalk, G. (1983). Levels of enzymes involved in acetate, butyrate, acetone and butanol formation by *C. acetobutylicum*. *Eur. J. Appl. Microbiol. Biotechnol.* *18*:327-332.

Bahl, H., Gottwald, M., Kuhn, A., Rale, V., Andersch, W., and Gottschalk, G. (1986). Nutritional factors affecting the ratio of solvents produced by *C. acetobutylicum*. *Appl. Environ. Microbiol.* *52*:169-172.

Beesch, S. C. (1953). Microbiological progress report: Acetone-butanol fermentation of starches. *Appl. Microbiol.* *1*:85-95.

Datta, R. (1981). Acidogenic fermentation of corn stover. *Biotechnol. Bioeng.* *25*:991-998.

Datta, R., and Zeikus, J. G. (1985). Modulation of acetone butanol ethanol fermentation by carbon monoxide and organic acids. *Appl. Environ. Microbiol.* *49*:522-529.

Davies, R. (1943). Acetoacetic acid decarboxylase of *C. acetobutylicum*. *Biochem. J.* *37*:230-238.

Davies, R., and Stephenson, M. (1941). Studies on the acetone-butyl alcohol fermentation. 1. Nutritional and other factors involved in the preparation of active suspensions of *C. acetobutylicum*. *Biochem. J.* *35*:1320.

Erickson, L. E., Minkevich, I. G., and Eroshin, V. K. (1979). Utilization of mass-energy balance regularities in the analysis of continuous culture data. *Biotechnol. Bioeng.* *21*:575-591.

Eroshin, V. K., and Minkevich, J. G. (1982). On the upper limit of mass yield of an organic product from an organic substrate. *Biotechnol. Bioeng.* *24*:2263-2265.

Fridovich, I. (1972). Acetoacetate decarboxylase. In *The Enzymes*, Vol. VI, P. D. Boyer (Ed.). Academic Press, New York, pp. 255-270.

Gottschal, J. C., and Morris, J. G. (1981). The induction of acetone and butanol production in cultures of *C. acetobutylicum* by elevated concentrations of acetate and butyrate. *FEMS Microbiol. Lett.* *12*:385-389.

Gottwald, M., and Gottschalk, G. (1985). The internal pH of *C. acetobutylicum* and its effects on the shift from acid to solvent formation. *Arch. Microbiol.* *143*:42-46.

Guibet, J. C., and Vandercasteele, J. P. (1983). Blended fuels containing butyl alcohol acetone and methanol. U.S. Patent 4,398,920.

Hanson, A. M., and Rodgers, N. E. (1946). Influence of iron concentration and attenuation on the metabolism of *C. acetobutylicum*. *J. Bacteriol.* *51*:568-569.

Hartmanis, M. G. N., and Gatenbeck, S. (1984). Intermediary metabolism in *C. acetobutylicum*: Levels of enzymes involved in the formation of acetate and butyrate. *Appl. Environ. Microbiol.* *47*:1277-1283.

Hongo, M. (1957). Butanol fermentation. XI. Change of solvent ratio by addition of redox dye. *Nippon Nogei Kagaku Kaishi* *31*: 735-738.

Jungermann, K., Thauer, R. K., Leimenstoll, G., and Decker, K. (1973). Function of reduced pyridine nucleotide-ferredoxin oxidoreductases in saccharolytic clostridia. *Biochim. Biophys. Acta 305*:268-280.

Kim, B. H., Bellows, P., Datta, R., and Zeikus, J. G. (1984). Control of carbon and electron flow in *C. acetobutylicum* fermentations: Utilization of carbon monoxide to inhibit hydrogen production and to enhance butanol yields. *Appl. Environ. Microbiol. 48*:764-770.

Maddox, I. S. (1980). Production of *n*-butanol from whey filtrate using *C. acetobutylicum* N.C.I.B. 2951. *Biotechnol. Lett. 2*:493-498.

Marczak, R., Ballongue, J., Petitdemange, H., and Gay, R. (1984). Regulation of biosynthesis of NADH-rubredoxin oxidoreductase in *C. acetobutylicum. Curr. Microbiol. 10*:165-168.

Marlatt, J. A., and Datta, R. (1986). Acetone-butanol fermentation process development and economic evaluation. *Biotechnol. Prog. 2*:23-28.

Meyer, C. L., McLaughlin, J. K., and Papoutsakis, E. T. (1985). The effect of CO on growth and product formation in batch cultures of *C. acetobutylicum. Biotechnol. Lett. 7*:37-42.

Meyer, C. L., Roos, J. W., and Papoutsakis, E. T. (1986). Carbon monoxide gassing leads to alcohol production and butyrate uptake without acetone formation in continuous cultures of *C. acetobutylicum*—solventogenesis in glucose limited cultures. *Appl. Microbiol. Biotechnol. 24*:159-167.

O'Brien, R. W., and Morris, J. G. (1971). The ferredoxin-dependent reduction of chloramphenicol by *C. acetobutylicum. J. Gen. Microbiol. 67*:265-271.

O'Brien, R. W., and Morris, J. G. (1972). Effect of metronidazole on hydrogen production by *C. acetobutylicum. Arch. Microbiol. 84*:225-233.

Nakhmanovich, B. M., and Shcheblykina, N. A. (1960). Fermentation of calcium salts of acetic and butyric acids by *C. acetobutylicum. Mikrobiologiia 29*:67-72.

Papoutsakis, E. T. (1984). Equations and calculations for fermentations of butyric acid bacteria. *Biotechnol. Bioeng. 26*:174-187.

Petitdemange, H., Cherrier, C., Raval, G., and Gay, R. (1976). Regulation of the NADH and NADPH—ferredoxin oxidoreductases in clostridia of the butyric group. *Biochim. Biophys. Acta 421*:334-347.

Petitdemange, H., Marczak, R., Blusson, H., and Gay, R. (1979). Isolation and properties of reduced NAD-rubredoxin oxidoreductase of *C. acetobutylicum. Biochem. Biophys. Res. Commun. 91*:1258-1265.

Rogers, P. (1986). Genetics and biochemistry of *Clostridium* relevant to development of fermentation processes. *Adv. Appl. Microbiol.* *31*:1-60.

Rogers, P., and Hansen, W. R. (1983). Regulation of butanol and butyric acid pathway enzymes during fermentation by *C. acetobutylicum*. Presented at the 83rd Annual Meeting, American Society for Microbiology, New Orleans, March 6-11.

Simon, E. (1947). The formation of lactic acid by *Clostridium acetobutylicum* (Weizmann). *Arch. Biochem.* *13*:237-243.

Terracciano, J. S., and Kashket, E. R. (1986). Intracellular conditions required for initiation of solvent production by *C. acetobutylicum*. *Appl. Environ. Microbiol.* *52*:86-91.

Weizman, C. (1915). Acetone and butyl alcohol. British Patent 4845.

Wood, H. G., Brown, R. G., and Werkman, C. H. (1945). Mechanism of the butyl alcohol fermentation with heavy carbon acetic and butyric acids and acetone. *Arch. Biochem.* *6*:243-260.

Zeikus, J. G., Lynd, L. H., Thompson, T. E., Krzycki, J. A., Weimer, P. J., and Hegge, P. W. (1980). Isolation and characterization of a new, methylotropic, acidogenic anaerobe, the Marburg strain. *Curr. Microbiol.* *3*:381-386.

III

BIOENERGETICS OF ANAEROBIC PROCESSES

11

Bioenergetics and Product Yields

MEHMET D. ÖNER / *University of Gaziantep, Gaziantep, Turkey*
LARRY E. ERICKSON and SHIE-SHIEN YANG / *Kansas State University, Manhattan, Kansas*

I. INTRODUCTION

Bioenergetics and biomass and product yields are very important considerations in anaerobic fermentation. Knowledge of bioenergetics and biomass and product yields is very helpful in the design of fermentation conditions and optimum selection of cultures for the fermentation process.

In many anaerobic fermentation processes, high-energy bonds in adenosine triphosphate (ATP) are produced when available electrons are converted from organic substrate into extracellular organic products, such as ethanol, butanol, and lactic acid. The true growth yield and maintenance parameters are directly related to the product formation kinetic parameters for these anaerobic processes.

Available electron concepts, available electron and carbon balances, methods for the estimation of true growth yields, product yields, and maintenance parameters are considered in this chapter. Relationships among the product formation kinetic parameters and the bioenergetic parameters associated with yield and maintenance are presented for those processes in which the formation of ATP from ADP is directly coupled to the anaerobic conversion of the organic substrate to one or more organic products. Considerable prior work provides the background theory and experimental data. A thorough analysis of these experimental data was published previously (Oner et al., 1984, 1986a).

II. MASS BALANCES

When the electron acceptor is the product itself, the balance equation of anaerobic microbial growth and product formation can be written as

$$CH_mO_l + aNH_3 = y_cCH_pO_nN_q + zCH_rO_sN_t + cH_2O + dCO_2 \quad (1)$$

where CH_mO_l denotes the elemental composition of organic substrate, $CH_pO_nN_q$ denotes the elemental composition of biomass, and $CH_rO_sN_t$ represents the average elemental composition of extracellular products. Here, the organic substrate is both the carbon and energy (electron) source for microbial growth.

The carbon balance is

$$y_c + z + d = 1.0 \quad (2)$$

where y_c, z, and d are the fractions of substrate carbon incorporated into biomass, extracellular products, and carbon dioxide, respectively.

Using the valences C = 4, H = 1, 0 = -2, and N = -3, the available electron balance is (Erickson et al., 1979)

$$\eta + \xi_p = 1.0 \quad (3)$$

where η and ξ_p are the fractions of available electrons of organic substrate incorporated into biomass and extracellular products, respectively. η and ξ_p can be calculated directly from carbon yields y_c and z as follows:

$$\eta = y_c \frac{\gamma_b}{\gamma_s} \qquad \xi_p = z \frac{\gamma_p}{\gamma_s} \quad (4)$$

where γ is the reductance degree and the subscripts s, b, and p refer to substrate, biomass, and product, respectively, and can be calculated as

$$\gamma_s = 4 + m - 2l \quad (5)$$

$$\gamma_b = 4 + p - 2n - 3q \quad (6)$$

$$\gamma_p = 4 + r - 2s - 3t \quad (7)$$

These two balance equations (2) and (3) are very useful in checking the consistency of experimental data.

III. YIELD AND MAINTENANCE

In most anaerobic fermentation processes, ATP generation is coupled to extracellular product formation. In these fermentation processes, microorganisms can meet their ATP requirements for growth and maintenance only by producing extracellular products.

If the available electrons that are transferred from organic substrate to products are viewed as being expended to provide energy for growth and maintenance, a true growth yield based on the available electron yield η_{max} can be defined as (Erickson, 1980; Erickson and Oner, 1983)

$$\eta_{max} = \frac{\text{equivalents of available electrons of biomass formed per mol ATP}}{\text{eq of biomass formed per mol ATP} + \text{eq. of available electrons transferred to products to produce 1 g mol ATP from ADP}}$$

$$\eta_{max} = \frac{\sigma_b \gamma_b / 12\ Y_{ATP}^{max}}{\sigma_b \gamma_b / 12\ Y_{ATP}^{max} + \delta} \tag{8}$$

where δ equivalents of available electrons are transferred to products to produce 1 g mol ATP from ADP. The true growth yield η_{max} obtained from Equation (8) gives the fraction of available electrons of substrate that can be incorporated into microbial biomass when maintenance is neglected. Through Equation (8), it is directly related to the growth yield based on ATP, Y_{ATP}^{max}.

The maintenance coefficient m_e in equivalents of available electrons expended per equivalent of available electrons of biomass per hour is related to m_{ATP}, the moles of ATP expanded per gram of cells per hour, according to the relationship (Erickson and Oner, 1983)

$$m_e = \frac{12\,\delta m_{ATP}}{\sigma_b \gamma_b} \tag{9}$$

Considering Pirt's model (Pirt, 1975), if the available electrons that are transferred from substrate to product are viewed as being expended to provide energy for microbial growth and maintenance, the following equations can be written to estimate the maintenance parameter m_e and true growth yield η_{max} (Erickson and Oner, 1983). Pirt's model in available electron units is

$$\frac{\mu}{\eta} = \frac{\mu}{\eta_{max}} + m_e \tag{10}$$

Inserting Equation (3) into the numerator on the left-hand side gives

$$\frac{\mu(\eta + \xi_p)}{\eta} = \frac{\mu}{\eta_{max}} + m_e \tag{11}$$

The carbon balance, Equation (2), can be used to obtain

$$\frac{\mu(y_c + z + d)}{\eta} = \frac{\mu}{\eta_{max}} + m_e \tag{12}$$

where μ is the specific growth rate. Equations (10), (11), and (12) are similar and follow Pirt's model for growth and maintenance. Equation (10) utilizes organic substrate, biomass, and specific growth rate measurements; Equation (11) utilizes biomass, product, and specific growth rate measurements; Equation (12) requires biomass, carbon dioxide, product, and specific growth rate measurements. Equations (11) and (12) would be identical to Equation (10) if the available electron and carbon balances were satisfied at every data point. The derivation of these equations is analogous to that published in the literature for aerobic processes (Erickson et al., 1979; Erickson, 1979; Roels, 1983).

IV. PARAMETER ESTIMATION

Two forms of Equations (10), (11), and (12) were used for parameter estimation; to obtain form I, each term in these equations should be divided by μ; form II is the form of these equations as presented in Equations (10), (11), and (12).

One can use any one of Equations (10), (11), and (12) to estimate m_e and η_{max}. Since different sets of experimental data should be used with each of these equations, depending on the magnitude of experimental error involved in the data, sometimes slightly, sometimes significantly different estimates of m_e and η_{max} were obtained. This problem was solved by using the covariate adjustment method (Solomon et al., 1982, 1983, 1984; Oner et al., 1983, 1984, 1986a) to estimate the true growth yield η_{max} and maintenance parameter m_e when sufficient data are available to use more than one of Equations (10), (11), and (12). This method allows all the data to be used to obtain single and better estimates of m_e and η_{max}.

V. PRODUCT YIELD

Considering Equations (3) and (10), one can derive the equations

$$\xi_p = 1 - \eta \tag{13}$$

$$\eta = \frac{\mu}{\mu/\eta_{max} + m_e} \tag{14}$$

$$\xi_p = 1 - \frac{\mu}{\mu/\eta_{max} + m_e} \tag{15}$$

Equation (15) indicates that product yield ξ_p depends on the parameters m_e and η_{max} as well as on μ. In order to have a large product yield, the maintenance coefficient m_e should be large and true growth yield η_{max} should be small. Satisfaction of these two conditions is possible with a large specific rate of substrate consumption. Figures 1 and 2 illustrate how the product yield ξ_p varies with the specific growth rate for various values of η_{max} and m_e. The values of η_{max} were calculated using Equation (8) and selected values of Y_{ATP}^{max} and δ.

The product formation kinetic parameters α and β in the equation for the specific rate of product formation (Luedeking and Piret, 1959; Luedeking, 1967)

$$\frac{1}{X}\frac{dP}{dt} = \alpha\mu + \beta \tag{16}$$

can be estimated (in terms of available electron units α_e and β_e) using methods similar to those used to estimate m_e and η_{max}.

Rewriting Equation (16) in terms of η and ξ_p gives

$$\frac{\mu}{\eta}\xi_p = \alpha_e\mu + \beta_e \tag{17}$$

Two other equations may also be written by using the available electron and carbon balances, that is,

$$\frac{\mu}{\eta}(1 - \eta) = \alpha_e\mu + \beta_e \tag{18}$$

$$\frac{\mu}{\eta}(y_c + z + d - \eta) = \alpha_e\mu + \beta_e \tag{19}$$

where α_e and β_e are related to the parameters defined above as

$$\alpha_e = \alpha\frac{\sigma_p\gamma_p}{\sigma_b\gamma_b} = \frac{1}{\eta_{max}} - 1 = \frac{12\delta}{\sigma_b\gamma_b\ Y_{ATP}^{max}} \tag{20}$$

and

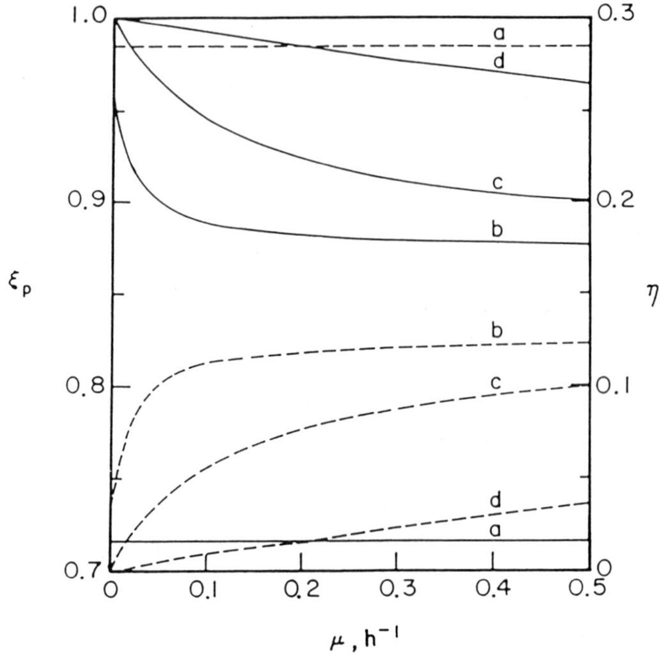

Figure 1 Effect of true growth yield and maintenance coefficient on fraction of available electrons incorporated into biomass (η, ---) and ethanol (ξ_p, ——) when 2 mol ATP are produced per mole of glucose converted to ethanol: (a) $Y_{ATP}^{max} = 28.8$, $\eta_{max} = 0.284$, $m_e = 0$; (b) $Y_{ATP}^{max} = 10.5$, $\eta_{max} = 0.126$, $m_e = 0.1\ h^{-1}$; (c) $Y_{ATP}^{max} = 10.5$, $\eta_{max} = 0.126$, $m_e = 1.0\ h^{-1}$; (d) $Y_{ATP}^{max} = 10.5$, $\eta_{max} = 0.126$, $m_e = 10\ h^{-1}$. (From Oner et al., 1984.)

$$\beta_e = \beta \frac{\sigma_p \gamma_p}{\sigma_b \gamma_b} = m_e = \frac{12\,\delta m_{ATP}}{\sigma_b \gamma_b} \tag{21}$$

Similarly, Equations (17), (18), and (19) can be used together with all the available data by using the covariate adjustment method to obtain better and single estimates of α_e and β_e. Using m_e and η_{max} estimated previously, Equations (20) and (21) also can be used to obtain α_e and β_e. Roels (1980, 1983) has also reviewed the relationships among these parameters.

Equation (20) provides useful information regarding the acceptable range of values of α_e in the Luedeking and Piret model; for example, for anaerobic ethanol production on glucose and mineral salts, the minimum value of δ is 12 equivalents of available electrons per mole ATP

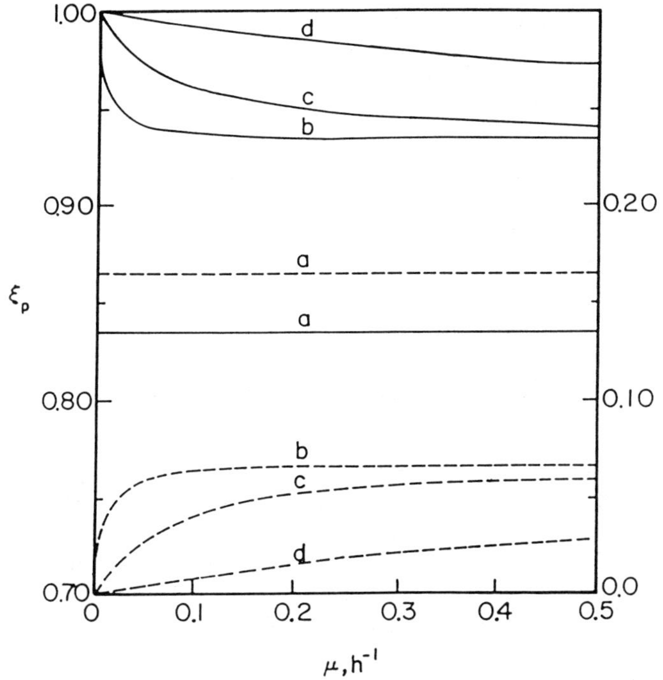

Figure 2 Effect of true growth and maintenance coefficient on fraction of available electrons incorporated into biomass (, ---) and ethanol (ξ_p, ——) when 1 mol ATP is produced per mole of glucose converted to ethanol: (a) Y_{ATP}^{max} = 28.8, η_{max} = 0.165, m_e = 0; (b) Y_{ATP}^{max} = 10.5, η_{max} = 0.0672, m_e = 0.1 h^{-1}; (c) Y_{ATP}^{max} = 10.5, η_{max} = 0.0672, m_e = 1.0 h^{-1}; (d) Y_{ATP}^{max} = 10.5, η_{max} = 0.0672, m_e = 10 h^{-1}. (From Oner et al., 1984.)

generated, and the maximum theoretical value of Y_{ATP}^{max} is 28.8 g cells per g mol ATP. This gives, for σ_b = 0.462 and γ_b = 4.291,

$$\alpha_e \geq 2.52 \tag{22}$$

VI. FREE ENERGY YIELD

Erickson and Oner (1983) have introduced the concepts associated with yield and maintenance parameters based on free energy. The free energy level is taken as zero for NH_3 in aqueous solution, H_2O liquid, and CO_2 gas; only the free energy of the substrate, biomass, and product need be considered.

For anaerobic growth (Erickson and Oner, 1983),

$$\eta_{th} = \frac{g_b \eta}{g_s - g_p \xi_p} \tag{23}$$

$$\eta_{th}^{max} = \frac{g_b \eta_{max}}{g_s - g_p(1 - \eta_{max})} \tag{24}$$

$$m_{th} = \frac{(g_s - g_p) m_e}{g_b} \tag{25}$$

where the subscript th indicates free energy and g is the free energy per equivalent of available electrons relative to aqueous NH_3, CO_2 gas, and liquid H_2O.

The numerical values of the free energy per equivalent of available electrons for the substrate and product may usually be found from tabulated literature data (Roels, 1980, 1983; Weast, 1976). For such substrates as glucose and ethanol that do not contain nitrogen, the free energy of combustion per equivalent of available electrons should be used to find g_s and g_p. For biomass, the value of Roels (1980) of -67.1 kJ/C mol for the free energy of formation of biomass may be used to obtain a value of g_b = 113 kJ per equivalent available electrons for the free energy of the biomass relative to aqueous NH_3, CO_2 gas, and liquid H_2O. This value is in good agreement with the value of 114.7 kJ per equivalent reported by Minkevich (1980, 1982). Values of 117.2 kJ per equivalent available electrons and 110.0 kJ per equivalent available electrons can be obtained for glucose and ethanol, respectively, from these references (Roels, 1980; Weast, 1976).

It should also be noted that these differences in free energies of substrate, product, and biomass are the main driving force for anaerobic fermentation processes.

Equations (23), (24), and (25) allow the free energy equivalent values of η, η_{max}, and m_e to be calculated from the parameter estimation results.

VII. RESULTS AND DISCUSSION

Several sets of experimental data have been analyzed to estimate the true growth yield η_{max}, the maintenance coefficient m_e, and the product formation kinetic parameter α_e and β_e by Oner et al. (1984, 1986a). These results are shown in Tables 1 through 4. Table 5 provides information concerning the consistency of the experimental data for those cases in which the consistency can be checked. Point and 95% confidence interval estimates of m_e and η_{max} for several anaerobic ethanol fermentations are shown in Table 1. The first two results are from the anaerobic growth of *Saccharomyces cerevisiae* on glucose. This organism uses the Embden-Meyerhof-Parnas (EMP) pathway and produces 2 mol of ATP per mole of glucose consumed, which corresponds to δ = 12 equivalents of available electrons transferred to product per mole of ATP produced. Y_{ATP}^{max} values for the anaerobic growth of this organism are given as 10.2 (Bauchop and Elsden, 1960), 11 (Watson, 1970), and 8.6 g cells per mole of ATP (Ghose and Tyagi, 1979). Using Equation (8), values of 0.106, 0.123, and 0.132 are obtained for η_{max} from the values of 8.6, 10.2, and 11 of Y_{ATP}^{max}, respectively. For minimal medium, the estimated maximum value of Y_{ATP}^{max} is 28.8 g cells per mole ATP. This corresponds to η_{max} = 0.284.

As shown in Table 1, the results from the data of Ghose and Tyagi (1979) and Dekkers et al. (1981) differ significantly. As shown in Table 5, both these data sets have reasonably consistent balances. However, the data of Dekkers et al. (1981) cover a relatively narrow range of specific growth rate values (0.012-0.096, h^{-1}). On the other hand, in the work of Ghose and Tyagi (1979), the range of specific growth rate varies from 0.08 to 0.25 h^{-1}, and maximum ethanol productivity is obtained at the specific growth rate of 0.13 h^{-1}. Since smaller values of specific growth rate are needed to obtain good estimates of m_e and large values of specific growth rate are important for η_{max}, the expected values of m_e and η_{max} may lie between the results of these two data sets.

The next seven sets of results in Table 1 are obtained from the data collected during the anaerobic growth of *Zymomonas mobilis* (Lee et al., 1981; Oner, 1982). As is seen from the results, this organism has a higher maintenance coefficient m_e and a lower true growth yield η_{max} compared with *S. cerevisiae*. *Z. mobilis* uses the Entner-Doudoroff (ED) pathway and produces 1 mol of ATP per mole of glucose consumed, which corresponds to a δ value of 24 equivalents of available electrons transferred per mole of ATP produced. Equation (8) yields values of η_{max} = 0.067 and 0.165 for the average (10.5) and maximum (28.8) values of Y_{ATP}^{max} (g cell per mole ATP). Except for the results of run 1 (Oner, 1985; Oner et al., 1984), the values of η_{max} obtained are below the theoretical value of 0.165. As shown in Table 5, the data sets of Lee et al. (1981) and Rogers et al. (1982) have better consistency than the data sets of Oner (Oner, 1982; Oner et al., 1984). By using the covariate adjustment method, Equations (10), (11), and (12) are used with runs 1 and 2 (Oner, 1985; Oner et al., 1984) and the sucrose data of Lee et al. (1981). The rest of the *Z. mobilis* data sets are analyzed by using only Equations (10) and (11) since no CO_2 data are available.

Table 1 Comparison of Point and Confidence Interval Estimates of m_e, η_{max}, α_e, and β_e Obtained from Literature Data (Ethanol Production)[a]

Form of equation	m_e and β_e (h^{-1})	η_{max}	α_e	Micro-organism	Substrate	Source of data
I	0.896	0.164	5.107	*Saccharomyces*		
	(0.663, 1.130)	(0.127, 0.231)	(3.329, 6,874)	*cerevisiae*	Glucose	Ghose and Tyagi
II	1.044	0.195	4.128	NRRL, Y-132		(1979)
	(0.766, 1.322)	(0.147, 0.288)	(2.472, 5.803)			
I	-0.0006	0.088	10.377	*S. cerevisiae*		
	(-0.060, 0.047)	(0.072, 0.1131)	(7.889, 13.870)		Glucose	Dekkers et al.
II	-0.045	0.077	11.987	CRS A26		(1981)
	(-0.168, 0.0079)	(0.064, 0.097)	(9.309, 14.625)			
I	1.970	0.037	26.027			
	(1.410, 2.520)	(0.031, 0.047)	(20.277, 31.258)	*Zymomonas*	Sucrose	(Lee et al. (1981)
II	2.520	0.039	24.641	*mobilis*		
	(0.940, 4.100)	(0.031, 0.054)	(17.519, 31.258)			
I	3.264	0.078	11.879	*Z. mobilis*		Rogers et al.
	(2.459, 4.069)	(0.057, 0.122)	(7.211, 16.548)		Glucose	(1982)
II	3.915	0.098	9.204	ZM4		(60 g/L glucose)
	(2.246, 5.584)	(0.064, 0.203)	(3.926, 14.625)			
I	3.694	0.069	13.558			Rogers et al.
	(3.061, 4.328)	(0.055, 0.092)	(9.862, 17.253)		Glucose	(1982)
II	4.405	0.087	10.567			(100 g/L glucose)
	(3.023, 5.787)	(0.061, 0.147)	(5.809, 15.324)			

I	4.645 (4.033, 5.257)	0.053 (0.038, 0.087)	17.935 (10.487, 25.383)		Glucose	Rogers et al. (1982) (135 g/L glucose)
II	5.452 (4.133, 6.771)	0.078 (0.049, 0.189)	11.812 (4.287, 19.337)			
I	4.112 (3.685, 4.537)	0.045 (0.034, 0.066)	21.357 (14.254, 28.497)		Glucose	Rogers et al. (1982) (170 g/L glucose)
II	5.187 (3.914, 6.459)	0.074 (0.048, 0.162)	12.552 (5.178, 19.926)			
I	3.983 (3.879, 4.087)	0.171 (0.128, 0.259)	4.838 (2.858, 6.813)	*Z. mobilis*	Glucose	Run 1 Oner (1982)
II	4.185 (3.809, 4.561)	0.283 (0.179, 0.664)	2.534 (0.506, 4.587)	ATCC-10988		
I	1.791 (1.242, 2.339)	0.0835 (0.059, 0.143)	10.976 (5.998, 15.949)		Glucose	Run 2 Oner (1982)
II	2.286 (1.602, 2.971)	0.146 (0.092, 0.362)	5.840 (1.762, 9.869)			

[a]The average values $\sigma_b = 0.462$ and $\gamma_b = 4.291$ were used.
Source: From Oner (1985) and Oner et al. (1984).

Table 2 Point and 95% Confidence Interval Estimates of m_e, η_{max}, α_e, and β_e from Data Collected During Anaerobic Growth of *Lactobacillus delbrueckii* NNRL B-445[a]

Form of equation	m_e and β_e (h^{-1})	η_{max}	α_e	Source of data
I	-0.0971 (-0.263, 0.068)	0.203 (0.156, 0.291)	3.928 (2.435, 5.421)	Run 7
II	0.164 (-0.362, 0.689)	0.280 (0.179, 0.638)	2.571 (0.568, 4.574)	(batch culture)
I	0.078 (0.006, 0.220)	0.281 (0.234, 0.351)	2.558 (1.845, 3.271)	Run
II	0.196 (-0.143, 0.536)	0.339 (0.242, 0.566)	1.948 (0.765, 3.129)	(batch culture)
I	0.112 (0.057, 0.166)	0.308 (0.264, 0.371)	2.243 (1.696, 2.791)	Run 10
II	0.178 (-0.035, 0.392)	0.351 (0.274, 0.491)	1.845 (1.038, 2.652)	(batch culture)
I	0.115 (0.072, 0.196)	0.329 (0.295, 0.373)	2.036 (1.684, 2.388)	Run 11
II	0.204 (0.011, 0.397)	0.376 (0.309, 0.478)	1.660 (1.093, 2.228)	(batch culture)
I	0.134 (0.125, 0.144)	0.312 (0.286, 0.342)	2.209 (1.926, 2.493)	Run 12
II	0.162 (0.133, 0.192)	0.351 (0.325, 0.380)	1.853 (1.630, 2.075)	(batch culture, points 1, 6-9 are excluded)
I	0.288 (0.274, 0.302)	0.418 (0.385, 0.457)	1.394 (1.188, 1.600)	Run 14
II	0.287 (0.191, 0.383)	0.424 (0.373, 0.492)	1.357 (1.033, 1.680)	(batch culture)
I	0.013 (-0.075, 0.100)	0.306 (0.251, 0.392)	2.270 (1.552, 2.988)	Run 15
II	0.205 (-0.173, 0.582)	0.394 (0.269, 0.737)	1.537 (0.357, 2.717)	(batch culture)
I	0.057 (0.037, 0.077)	0.329 (0.303, 0.360)	2.042 (1.781, 2.303)	Run 16
II	0.071 (-0.070, 0.211)	0.334 (0.288, 0.398)	1.995 (1.516, 2.473)	(batch culture)
I	0.143 (0.118, 0.168)	0.399 (0.381, 0.419)	1.503 (1.382, 1.624)	Run 17
II	0.156 (0.101, 0.212)	0.408 (0.381, 0.440)	1.450 (1.272, 1.628)	(batch culture)
I	0.172 (0.081, 0.263)	0.402 (0.338, 0.497)	1.487 (1.013, 1.962)	Run 18
II	0.159 (0.053, 0.267)	0.393 (0.333, 0.479)	1.545 (1.089, 2.000)	(continuous culture)

[a]Data of Reilly (1964); σ_b = 0.462 and 4.291 were used in the analysis. The media contained 5% glucose and 3% yeast extract except that 1% yeast extract was used in run 12, 2% yeast extract was used in run 14, and 2.5% glucose was used in run 15.
Source: From Oner (1985) and Oner et al. (1984).

Estimated values of the product formation kinetic parameters α_e and β_e are presented in Table 1. Since the maximum specific growth rate of *Z. mobilis* is approximately 0.2 h^{-1}, a comparison of α_e and β_e values shows that, in most cases, more than 50% of the ethanol production is nongrowth associated or is related to the maintenance requirement. The higher maintenance requirement and lower true growth yield than conventional yeasts are unique to *Z. mobilis*, and they are desirable properties related to the ability of this organism to produce ethanol rapidly.

The estimated yield parameters m_e and η_{max} and the kinetic parameters α_e and β_e from the data of Reilly (1964) for the anaerobic growth of *Lactobacillus delbrueckii* on glucose and yeast extract are shown in Table 2. During lactic acid production, this organism uses the EMP pathway and produces 2 mol of ATP per mole of glucose consumed when the only product is lactic acid (Decker et al., 1979). When yeast extract is used to provide amino acids and nucleic acids for cell growth, Stouthamer (1970) reports that $Y_{ATP}^{max} = 31.9$ is the estimated theoretical maximum value based on biochemical pathway analysis. Using this value and $\delta = 12$ in Equation (8) gives $\eta_{max} = 0.305$. The results in Table 2 and the results of Luedeking and Piret (1959) and Hanson and Tsao (1972) as analyzed in Table 3 show that the estimated values of η_{max} are similar in magnitude to this theoretical value. An average of 0.347 is obtained from the data of Reilly (1964), 0.279 is the average value obtained from the data of Luedeking and Piret (1959), and the Hanson and Tsao data give an average value of 0.339. The average of these values is slightly higher than the theoretical maximum expected value of 0.305. This could be due to a consistent error in the data analysis, such as using an incorrect conversion factor to convert units of optical density (UOD) to biomass dry weight. The conversion factor employed was 0.75 mg/UOD to convert from UOD/mL to mg/mL (Reilly, 1964).

The relatively large true growth yield could also be partly due to small amounts of aerobic growth; that is, the methods used to maintain anaerobic conditions may have allowed small quantities of oxygen to enter the fermenter. Analysis of the batch data of Reilly (1964) shows that the fraction of available electrons of glucose supplied that are converted to lactic acid averages 0.776. Of the remaining fraction, a portion is converted to microbial biomass, a portion may be converted to other products, and some may be transferred to oxygen or other electron acceptors.

The results in Table 5, where the available electron balance is used to check data consistency, are based on the end points of the batch fermentation. Yeast extract is not included in making the balances because no measurement of yeast extract concentration

Table 3 Point and 95% Confidence Interval Estimates of m_e, η_{max}, α_e, and β_e Obtained from Literature Data for Lactic Acid-Producing Organisms

Form of equation	m_e and β_e (h^{-1})	η_{max}	α_e	Micro-organism	Substrate	Product(s)	Source of data
I	0.491 (0.453, 0.530)	0.267 (0.247, 0.290)	2.750 (2.453, 3.049)	*Lacto-bacillus delbrueckii* NRRL - 8445	Glucose	Lactic acid	Luedeking and Piret (1959) (batch culture)
II	0.569 (0.457, 0.681)	0.291 (0.268, 0.318)	2.439 (2.144, 2.734)				
I	0.123 (−0.032, 0.288)	0.336 (0.255, 0.494)	1.976 (1.024, 2.922)		Glucose	Lactic acid	Hanson and Tsao (1972) (continuous culture)
II	0.145 (−0.080, 0.370)	0.342 (0.247, 0.555)	1.924 (0.802, 3.049)				
I	−0.005 (−0.142, 0.132)	0.287 (0.238, 0.362)	2.484 (1.762, 3.202)	*Lacto-bacillus casei* L3	Glucose	Lactate, ethanol, formate, acetate	DeVries et al. (1970)
II	−0.060 (−0.260, 0.139)	0.269 (0.228, 0.328)	2.717 (2.049, 3.386)				
II	0.282	0.071	13.080	*Lacto-bacillus bulgaricus* ATCC 11842	Crude sorghum extract and vetch juice	Lactic acid	Samuel et al. (batch culture) pH = 5.5, run 1
II	0.153	0.064	14.570		Crude sorghum extract and vetch juice	Lactic acid	pH = 6.0, run 2
II	0.226	0.108	8.230		Crude sorghum extract and vetch juice	Lactic acid	pH = 6.5, run 3

II	0.178	0.075	12.400	*Lacto-bacillus plantarum* ATCC 14917	Crude sorghum extract and vetch juice	Lactic acid	pH = 5.5, run 4
II	0.226	0.119	7.370		Crude sorghum extract and vetch juice	Lactic acid	pH = 6.0, run 5
II	0.335	0.074	12.480		Crude sorghum extract and vetch juice	Lactic acid	pH = 6.0, run 6
II	0.331	0.105	8.520		Crude sorghum extract and vetch juice	Lactic acid	pH = 6.5, run 7
II	0.250	0.066	14.260		Crude sorghum extract and vetch juice	Lactic acid	pH = 6.5, run 8
II	0.282	0.050	18.890		Crude sorghum extract and vetch juice	Lactic acid	pH = 7.0, run 9
I	0.396 (0.342, 0.459)	0.206 (0.196, 0.220)	3.861	*L. bul-garicus* (ATCC-11842)	3% NFDM	Lactic acid	Oner et al. (1986a), run 1

(Table 3, continued)

Form of equation	m_e and β_e (h^{-1})	η_{max}	α_e	Microorganism	Substrate	Product(s)	Source of data
I	0.601 (0.373, 1.067)	0.346 (0.196, 0.853)	1.894	*S. thermophilus* (ATCC-19258)	3% NFDM	Lactic acid	Run 2
I	0.331 (0.298, 0.382)	0.199 (0.190, 0.208)	4.028	*L. bulgaricus* (ATCC-11842) + *S. thermophilus* (ATCC-19258)	3% NFDM	Lactic acid	Run 3
I	0.045 (0.146, 0.272)	0.110 (0.083, 0.134)	8.124	*L. bulgaricus* (Microlife)	3% NFDM	Lactic acid acetic acid + butyric acid	Run 4
I	0.106 (0.035, 0.148)	0.253 (0.222, 0.277)	2.948	*S. thermophilus* (Microlife)	3% NFDM	Lactic acid acetic acid	Run 5
I	0.404 (0.339, 0.479)	0.178 (0.150, 0.214)	5.667	*S. thermophilus* (Microlife) *L. bulgaricus* (Microlife)	3% NFDM	Lactic acid	Run 6

Source: From Oner et al. (1984) and Oner et al. (1986a).

Table 4 Point and 95% Confidence Interval Estimates of m_e, η_{max}, α_e, and β_e Obtained from Literature Data

Form of equation	m_e and β_e (h^{-1})	η_{max}	α_e	Microorganism	Substrate	Product(s)	Source of data
I	0.286 (-0.059, 0.630)	0.230 (0.150, 0.494)	3.342 (1.022, 5.661)	*Clostridium aceto-butylicum* ATCC 824	Glucose	Acetone, butanol, ethanol, butyrate	Leung and Wang (1981) (continuous) (from Fig. 3)
II	0.313 (-0.116, 0.741)	0.244 (0.164, 0.473)	3.104 (1.110, 5.097)				
I	0.407 (0.058, 0.755)	0.211 (0.136, 0.464)	3.739 (1.156, 6.321)		Glucose	Acetone, butanol, ethanol, butyrate	from Fig. 3
II	0.540 (0.199, 0.879)	0.258 (0.172, 0.510)	2.875 (0.961, 4.780)				
I	0.317 (0.220, 0.415)	0.173 (0.173, 0.234)	4.780 (3.273, 6.299)		Glucose	Acetone, butanol, ethanol, butyrate	Monot and Engasser (1983) (continuous culture)
II	0.347 (0.199, 0.496)	0.185 (0.148, 0.246)	4.391 (3.065, 5.757)				
II	0.000	0.280	2.571	*Butyri-bacterium methylo-trophicum*	Methanol	Butyrate acetate	Datta and Ogel-tree (1983) (batch culture; Fig. 1)
II	0.052	0.239	3.184		Methanol	Butyrate	Fig. 2
II	0.094	0.348	1.873		Methanol acetate	Butyrate	Fig. 3

Source: From Oner et al. (1984).

Table 5 Available Electron and/or Carbon Balances with 95% Confidence Intervals

Source of data	No. data points	Balance type	Mean and 95%, CI	SD
Ghose and Tyagi (1979)	8	$\eta + \xi_p$	1.0188 (0.9245, 1.1131)	0.0399
Dekkers et al. (1981	14	$\eta + \xi_p$	0.9534 (0.7694, 1.1374)	0.00844
	14	$y_c + z + d$	0.9982 (0.6259, 1.3705)	0.1724
Lee et al. (1981)	11	$\eta + \xi_p$	0.9338 (0.7426, 1.1251)	0.0859
Sucrose	11	$y_c + z + d$	0.9297 (0.7438, 1.1197)	0.0853
Rogers et al. (1982)				
60 g/L glucose	5	$\eta + \xi_p$	1.0204 (0.9211, 1.1197)	0.0539
100 g/L glucose	7	$\eta + \xi_p$	1.0012 (0.9305, 1.0719)	0.0289
135 g/L glucose	6	$\eta + \xi_p$	0.9654 (0.9102, 1.0206)	0.0214
170 g/L glucose	7	$\eta + \xi_p$	0.9684 (0.9086, 1.0282)	0.0244
Luedeking and Piret (1959)	22	$\eta + \xi_p$	1.073 —	—
Hanson and Tsao (1972)	19	$\eta + \xi_p$	1.0019 (-0.1688, 2.1726)	0.5572
DeVries et al. (1970)	6	$\eta + \xi_p$	1.3543 (1.2298, 1.4788)	0.0484
Oner (1982)				
Run 1	20	$\eta + \xi_p$	1.0136 (0.5764, 1.4508)	0.2089
	20	$y_c + z + d$	0.8974 (0.4581, 1.3367)	0.3236
Run 2	25	$\eta + \xi_p$	0.8923 (0.3837, 1.4009)	0.2464
	25	$y_c + z + d$	0.8974 (0.4581, 1.3367)	0.2129
Leung and Wang (1981)				
Fig. 2	8	$\eta + \xi_p$	0.8919 (0.6473, 1.1365)	0.1034
Fig. 3	8	$\eta + \xi_p$	0.8923 (0.3837, 1.4009)	0.0555
Monot and Engasser (1983)	9	$\eta + \xi_p$	0.9814 (0.8079, 1.1546)	0.0752

Reilly (1964)			
Run 7	20	$\eta + \xi_p$	1.045
Run 9	22	$\eta + \xi_p$	1.122
Run 10	20	$\eta + \xi_p$	1.023
Run 11	16	$\eta + \xi_p$	1.048
Run 12	26	$\eta + \xi_p$	0.951
Run 14	17	$\eta + \xi_p$	0.987
Run 15	12	$\eta + \xi_p$	1.198
Run 16	18	$\eta + \xi_p$	1.048
Run 17	15	$\eta + \xi_p$	1.047
Run 18	5	—	—
Oner et al. (1986c)			
Run 1	15	$\eta + \xi_p$	1.351
Run 2	35	$\eta + \xi_p$	0.9629
Run 3	34	$\eta + \xi_p$	1.0225
Run 4	33	$\eta + \xi_p$	1.0474
Run 5	33	$\eta + \xi_p$	1.1547
Run 6	31	$\eta + \xi_p$	1.0046

Source: From Oner (1985) and Oner et al. (1984, 1986c).

is made at the end of the fermentation. That the sum exceeds 1 in several cases is an indication that significant amounts of yeast extract are used in this fermentation.

The parameter estimates in Tables 2 and 3 for the data of Reilly (1964) and Luedeking and Piret (1959) are based on biomass, lactic acid, and specific growth rate measurements; Equation (11) or (18) is used. The sources of available electrons for microbial growth include the glucose and the yeast extract. Both these sources are appropriately considered in the parameter estimation when Equations (11) and (18) are employed. In Equation (11), for example, the sum of the available electrons of biomass and product is used as a measure of the quantity of available electrons of glucose and yeast extract that are consumed.

Equations (10) and (11) are used with the covariate adjustment method to estimate the parameters in Table 3 for the data of Hanson and Tsao (1972). In Equation (11), the quantity of consumed substrate is based on the amount of glucose utilized. The analysis of data consistency in Table 5 indicates that some significant measurement errors may have occurred in this work.

The results of Reilly (1964), Luedeking and Piret (1959), and Hanson and Tsao (1972) show that the true growth yield is relatively large for this organism and that the average maintenance coefficient of 0.145 h^{-1} from the data of Reilly and 0.137 h^{-1} from the data of Hanson and Tsao are comparable to the value of 0.157 h^{-1} reported by Rees and Pirt (1979) and the values for some of the other organisms in Tables 1 through 4. If lactic acid production is the desired goal, the development of organisms with higher maintenance coefficients and lower true growth yields would be desirable.

Results obtained from analysis of the data of de Vries et al. (1970) are shown in Table 3. Relatively large values of η_{max} are also obtained in this work. De Vries et al. (1970) have calculated the production of ATP for each run; these values range from 2.13 to 3.06 mole ATP per mole glucose. They have also estimated that Y_{ATP}^{max} = 24.3 g cells per g mol ATP and that m_{ATP} = 0.00152 mole ATP per g dry cells (h). Both methods of analysis yield very small values for the maintenance coefficient and relatively large values for the true growth yield. Even though small negative values of m_e are estimated in Table 3, the value of the maintenance coefficient must be positive.

The results in Table 3 for the data of de Vries et al. (1970) were obtained by combining the products lactate, acetate, ethanol, and formate and treating them as one product based on the sum of the available electrons of the individual products. Equations (11) and (12) were used with the covariate adjustment method to estimate the parameters. Since the parameters η_{max} and m_e depend on δ, as Equations (8) and (9) indicate, this method of lumping is approximate when δ varies according to the mix of products produced. For the average value of η_{max} = 0.278 from Table 3 and the value of Y_{ATP}^{max} = 24.3 g cells per g mol ATP, Equation (8) may be used to calculate δ = 10.43 equivalents of available electrons per mole of ATP produced from ADP, which corresponds to 2.30 mol of ATP per mole glucose. Using this value of δ and their value of m_{ATP} = 0.00152 mol ATP per g cell (h), one obtains m_e = 0.096 from Equation (9). Thus the results from the two different methods are reasonably consistent, as one would expect.

The results obtained from Samuel et al. (1980) are also given in Table 3. These values have been obtained from the estimates of k_1 and k_2 in Table 4 of Samuel et al. (1980). The estimated growth yields for *Lactobacillus plantarum* and *Lactobacillus bulgaricus* are significantly lower than the other values reported in Tables 2 and 3. There is some evidence in the literature (Doelle, 1969) that *L. plantarum* produces 1 mol of ATP per mole of glucose. The results in Table 3 are appropriate in magnitude in that for δ = 24, one would expect a true growth yield of about 0.067 with a maximum value of about 0.17. For *L. plantarum* the average value is 0.081, and for *L. bulgaricus* it is also 0.081.

The results of Oner et al. (1986a-c; Oner and Erickson, 1986) for *L. bulgaricus* and *Streptococcus thermophilus* are also given in Table 3. The results of *S. thermophilus* (Microlife) in Table 3 indicate that η_{max} is quite large for this organism. Table 10 shows that when 2 mol ATP are produced per mole glucose (δ = 12), the theoretical maximum value of η_{max} is 0.305 for glucose in a rich medium that includes amino acids and nucleic acid bases. For glucose and mineral salts, the theoretical maximum value is η_{max} = 0.284. Thus, the estimated value of 0.253 is fairly large but realistic when compared with the theoretical limiting value for *S. thermophilus* (Microlife). The results clearly indicate that 2 mol ATP must be produced per mole glucose for this culture (δ = 12); the yield is larger than can be achieved with 1 mol ATP per mole glucose (δ = 24). The η_{max} value of 0.206 for *L. bulgaricus* (ATCC-11842) suggests that 2 mol ATP are produced per mole glucose for this strain of *L. bulgaricus*. However, η_{max} was found to be 0.11 for *L. bulgaricus* (Microlife). These results could be produced by an organism that produces only 1 mol ATP per mole glucose. For *S. thermophilus* (ATCC-19258), estimated true growth yields are greater than the theoretical maximum value. Mixed culture growth yields were found to be reasonable and in good agreement with the growth yield from analysis of exponential growth region (Oner et al., 1986b).

Data from the growth of *C. acetobutylicum* from Leung and Wang (1981) and Monot and Engasser (1983) are analyzed, and the results are given in Table 4. Equations (10) and (11) are used with the covariate adjustment method. Maintenance parameters for this organism are higher than most of the values obtained for lactic acid organisms and lower than for *Z. mobilis*. The average value is 0.37 h^{-1}.

The average value of the true growth yield for *C. acetobutylicum* is estimated to be η_{max} = 0.22. This organism produces about 2 mol of ATP per mole of glucose, with the value somewhat higher than this when significant amounts of acetate are produced. The value of η_{max} from Equation (8) is 0.13 for Y_{ATP}^{max} = 10.5 and δ = 12 equivalents of available electrons per mole of ATP.

The consistency of the data is shown in Table 5. The mean value of $\eta + \xi_p$ is closer to 1 for the data of Monot and Engasser (1983) compared with the results for the data of Leung and Wang (1981).

The results of Datta and Ogeltree (1983) are given in Table 4. Data from their Figures 1, 2, and 3 have different methanol-bicarbonate ratios = 2:1, 3.3:1, and 5:1, respectively. Barker (1972) has indicated that methane-producing organisms may produce 0.33 mol of ATP per mole of methanol consumed (δ = 18.18 equivalents of available electrons transferred to products per mole of ATP produced.) Using Equation (8) with Y_{ATP}^{max} = 28.4 g cell per mole ATP (van Dijken and Harder, 1975), a value of 0.205 is obtained for η_{max}. As is seen from the results, all the values obtained are larger than 0.205. This result may be due to experimental error. It appears that further studies with more complete measurements of substrate, biomass, and product concentrations are needed for this system.

The parameter estimation results in Tables 1 through 4 may be used with Equations (24) and (25) and tabulated values for the free energy of the compounds involved to estimate η_{th}^{max} and m_{th}. The free energy value for biomass of Minkevich (1980, 1982) of 114.7 kJ per equivalent of available electrons was used for g_b in calculating the results presented in Tables 6 through 9. In Tables 8 and 9, where a mixture of products was produced, average values of the product free energy were estimated. For the data of de Vries et al. (1970) in Table 8, g_p = 111.95 kJ per equivalent was used. In Table 9, g_p = 108.6 kJ per equivalent was used with the data of Leung and Wang (1981), g_p = 109.5 kJ per equivalent was used with the data of Monot and Engasser (1983), and the values g_p = 110.0 kJ per equivalent in Figure 1 of Datta and Ogeltree (1983) and g_p = 108.7 kJ per equivalent in Figures 2 and 3 were used. The values of the true growth yield η_{th}^{max} should lie between the limits 0 and 1. All the point estimates and all the confidence intervals are within these limits; however, in Tables 7 and 9

Table 6 Point and 95% Confidence Interval Estimates for True Growth Yield η_{th}^{max} and Maintenance Coefficient m_{th} in Free Energy Units for Anaerobic Ethanol Production

Form of equations	m_{th}	η_{th}^{max}	Source of data
I	0.0765 (0.057, 0.096)	0.676 (0.613, 0.754)	Ghose and Tyagi (1979)
II	0.0891 (0.065, 0.113)	0.716 (0.650, 0.797)	
I	-0.0005 (-0.005, 0.004)	0.518 (0.466, 0.583)	Dekkers et al. (1981)
II	-0.0038 (-0.014, 0.007)	0.484 (0.436, 0.544)	
I	0.184 (0.132, 0.236)	0.287 (0.251, 0.339)	Lee et al. (1981)
II	0.215 (0.080, 0.350)	0.318 (0.269, 0.394)	
I	0.2786 (0.210, 0.347)	0.487 (0.407, 0.554)	Rogers et al. (1982)
II	0.3342 (0.192, 0.477)	0.547 (0.436, 0.725)	60 g/L glucose
I	0.3154 (0.261, 0.369)	0.456 (0.398, 0.530)	100 g/L glucose
II	0.3761 (0.258, 0.494)	0.516 (0.424, 0.650)	
I	0.3960 (0.344, 0.449)	0.389 (0.312, 0.516)	135 g/L glucose
II	0.4654 (0.353, 0.578)	0.487 (0.370, 0.709)	
I	0.3510 (0.315, 0.387)	0.350 (0.288, 0.444)	170 g/L glucose
II	0.4428 (0.334, 0.551)	0.474 (0.365, 0.673)	
I	0.3400 (0.331, 0.349)	0.687 (0.615, 0.777)	Oner (1982)
II	0.3573 (0.325, 0.389)	0.794 (0.697, 0.920)	Oner et al., (1984), Run 1
I	0.1530 (0.106, 0.200)	0.505 (0.416, 0.643)	Oner (1982)
II	0.1952 (0.137, 0.254)	0.648 (0.530, 0.838)	Oner et al. (1884), Run 2

Source: From Oner (1985).

Table 7 Point and 95% Confidence Interval Estimates for True Growth Yield η_{th}^{max} and Maintenance Coefficient m_{th} in Free Energy Units for Lactic Acid Production Data of Reilly (1964)

Form of equations	m_{th}	η_{th}^{max}	Source of data
I	-0.004 (-0.012, 0.003)	0.814 (0.770, 0.864)	Run 7
II	0.008 (-0.017, 0.032)	0.859 (0.794, 0.935)	
I	0.004 (0.0003, 0.010)	0.859 (0.835, 0.885)	Run 9
II	0.009 (-0.007, 0.025)	0.881 (0.840, 0.926)	
I	0.005 (0.003, 0.008)	0.870 (0.851, 0.890)	Run 10
II	0.008 (-0.02, 0.018)	0.885 (0.856, 0.916)	
I	0.005 (0.003, 0.009)	0.878 (0.865, 0.891)	Run 11
II	0.010 (0.0005, 0.019)	0.892 (0.871, 0.914)	
I	0.0063 (0.0059, 0.0068)	0.872 (0.861, 0.882)	Run 12
II	0.0076 (0.0063, 0.009)	0.885 (0.876, 0.893)	
I	0.0136 (0.0129, 0.0142)	0.902 (0.894, 0.910)	Run 14
II	0.0135 (0.009, 0.018)	0.903 (0.891, 0.916)	
I	0.0006 (-0.0035, 0.0047)	0.870 (0.845, 0.896)	Run 15
II	0.0096 (-0.008, 0.027)	0.896 (0.854, 0.943)	
I	0.0027 (0.0017, 0.0036)	0.878 (0.868, 0.887)	Run 16
II	0.0033 (-0.0033, 0.0099	0.879 (0.861, 0.897)	
I	0.0067 (0.0056, 0.0079)	0.897 (0.891, 0.902)	Run 17
II	0.0073 (0.0047, 0.010)	0.900 (0.893, 0.906)	
I	0.0081 (0.0038, 0.0124)	0.898 (0.881, 0.917)	Run 18
II	0.0075 (0.0025, 0.0126)	0.896 (0.878, 0.914)	

Source: From Oner (1985) and Oner et al. (1984).

Table 8 Point and 95% Confidence Interval Estimates for True Growth Yield η_{th}^{max} and Maintenance Coefficient m_{th} in Free Energy Units for Several Lactic Acid Fermentations

Form of equations	m_{th}	η_{th}^{max}	Source of data
I	0.023 (0.021, 0.025)	0.861 (0.851, 0.870)	Luedeking and Piret (1959)
II	0.027 (0.022, 0.032)	0.871 (0.861, 0.880)	
I	0.006 (-0.002, 0.014)	0.886 (0.855, 0.919)	Hanson and Tsao (1972)
II	0.007 (-0.004, 0.017)	0.888 (0.851, 0.928)	
I	-0.0003 (-0.010, 0.009)	0.826 (0.794, 0.861)	De Vries et al. (1970)
II	-0.004 (-0.017, 0.009)	0.816 (0.787, 0.847)	
II	0.013	0.622	Samuel et al. (1980), run 1
II	0.007	0.597	run 2
II	0.011	0.714	run 3
II	0.008	0.634	run 4
II	0.011	0.734	run 5
II	0.016	0.631	run 6
II	0.016	0.708	run 7
II	0.012	0.604	run 8
II	0.013	0.536	run 9

Source: From Oner (1985).

Table 9 Point and 95% Confidence Interval Estimates of True Growth Yield η_{th}^{max} and Maintenance Coefficient m_{th} in Free Energy Units for Results in Table 4

Form of equations	m_{th}	η_{th}^{max}	Source of data
I	0.0275 (−0.0057, 0.0606)	0.732 (0.629, 0.876)	Leung and Wang (1981),
II	0.0301 (−0.0112, 0.0713)	0.745 (0.652, 0.869)	Fig. 2
I	0.0392 (0.0056, 0.0726)	0.713 (0.604, 0.866)	Fig. 3
II	0.0520 (0.0191, 0.0846)	0.758 (0.664, 0.880)	
I	0.0282 (0.0195, 0.0369)	0.681 (0.624, 0.750)	Monot and Engasser (1983)
II	0.0308 (0.0177, 0.0441)	0.697 (0.643, 0.760)	
II	0.0000	0.884	Datta and Ogeltree (1983), Fig. 1
II	0.0031	0.835	Fig. 2
II	0.0056	0.893	Fig. 3

Source: From Oner (1985).

Table 10 Comparison of True Growth Yields for Anaerobic Ethanol and Lactic Acid Fermentations for Given Values of δ and Y_{ATP}^{max} (Oner et al., 1984)

Product	δ	Y_{ATP}^{max}	η_{max}	η_{th}^{max}
Ethanol	12	10.5	0.126	0.611
		28.8	0.284	0.795
		31.9	0.305	0.808
Ethanol	24	10.5	0.0674	0.450
		28.8	0.165	0.678
		31.9	0.180	0.698
Lactic acid	12	10.5	0.126	0.744
		28.8	0.284	0.868
		31.9	0.305	0.876

Source: From Oner et al. (1984).

several relatively large values are reported. It is doubtful that the very large values are correct; the reasons for this have been stated in earlier parts of this chapter. Some theoretically estimated values are shown in Table 10. Equations (8) and (24) were used to obtain these values.

The values of the true growth yield η_{th}^{max} for lactic acid fermentations are consistently relatively high. Even in the work of Samuel et al. (1980), where the values of η_{max} are very reasonable, the average value of η_{th}^{max} is 0.64. This appears to be due to the relatively efficient production of ATP from ADP in lactic acid fermentation. In Table 10, the true growth yields on an available electron basis and a free energy basis are compared for given values of δ and Y_{ATP}^{max} for ethanol and lactic acid. It is clear that free energy yields are somewhat larger in the lactic acid fermentation because of the smaller free energy change associated with the conversion of glucose to lactic acid compared with that for converting glucose to ethanol.

VIII. CONCLUSIONS

When product formation is the desired goal, the maintenance coefficient should be as large as possible. Both product yield and specific rate of product formation increase with increasing values of the maintenance coefficient. Small values of true growth yield are desired when product formation is the objective.

IX. NOMENCLATURE

d	Moles of carbon dioxide per quantity of organic substrate containing 1 g atom carbon; g mol/g atom carbon
g_b	Amount of free energy in biomass per g equivalent of available electron in biomass; kJ/g equivalent of available electrons
g_p	Amount of free energy in product per g equivalent of available electron in product; kJ/g equivalent of available electrons
g_s	Amount of free energy in substrate per g equivalent of available electrons in substrate; kJ/g equivalent of available electrons
m_e	Rate of organic substrate consumption for maintenance, g equivalents of available electron per g equivalent of available electrons in biomass (h)
m_{ATP}	Moles of ATP expended per g of cells per hour for maintenance; mol ATP per g hr
m_{th}	Amount of free energy consumed per unit of free energy in biomass per hour; kJ free energy consumed per kJ free energy in biomass (h)
Y_{ATP}	Biomass yield based on ATP; g dry biomass per g mol ATP
Y_{ATP}^{max}	True biomass yield based on ATP; g dry biomass per g mol ATP
y_c	Biomass carbon yield (fraction of organic substrate carbon in biomass); dimensionless
z	Fraction of organic substrate carbon in products; dimensionless
α	Growth-associated kinetic parameter; dimensionless
β	Maintenance-associated kinetic parameter; h^{-1}
γ_b	Reductance degree of biomass; equivalents of available electrons per g atom carbon
γ_p	Reductance degree of products; equivalents of available electrons per g atom carbon
γ_s	Reductance degree of substrate; equivalents of available electrons per g atom carbon
δ	Number of equivalents of available electrons transferred to product to produce 1 g mol of ATP from ADP
η	Fraction of available electrons in organic substrate that is converted to biomass; dimensionless
η_{max}	"True" biomass available electron yield; dimensionless
η_{th}	Biomass free energy yield; see Equation (23); dimensionless
η_{th}^{max}	"True" biomass free energy yield; dimensionless
μ	Specific growth rate; h^{-1}
ξ_p	Fraction of available electrons in organic substrate converted to products; dimensionless
σ_b	Weight fraction carbon in dry biomass; dimensionless

σ_p Weight fraction carbon in products; dimensionless
σ_s Weight fraction carbon in organic substrate; dimensionless

REFERENCES

Barker, H. A. (1972). ATP formation by anaerobic bacteria. In *Horizons of Bioenergetics*, A. S. Pietro and H. Gest (Eds.). Proceedings of a Symposium held at Bloomington, Indiana, October 12-15, 1970. Academic Press, New York, pp. 7-31.

Bauchop, T., and Elsden, S. R. (1960). The growth of microorganisms in relation to their energy supply. *J. Gen. Microbiol. 23*: 457-469.

Datta, R., and Ogeltree, J. (1983). Methanol bioconversion by *Butyribacterium methylotrophicum*—batch fermentation yield and kinetics. *Biotechnol. Bioeng. 25*:991-998.

Decker, K., Jungermann, J., and Thauer, R. K. (1979). Energy production in anaerobic organisms. *Angew. Chem. 9* (2):138-158.

Dekkers, J. G. S., de Kok, H. E. and Roels, J. A. (1981). Energetics of *Saccharomyces cerevisiae* CBS 426: Comparison of anaerobic and aerobic glucose limitation. *Biotechnol. Bioeng. 23*:1023-1035.

De Vries, W., Kapteijn, W. M. C., Von Der Beek, E. G., and Stouthamer, A. H. (1970). Molar growth yields and fermentation balances of *Lactobacillus casei* L3 in batch cultures and in continuous cultures. *J. Gen. Microbiol. 63*:333-345.

Doelle, H. W. (1969). *Bacterial Metabolism.* Academic Press, New York, p. 334.

Erickson, L. E. (1979). Energetic efficiency of biomass and product formation. *Biotechnol. Bioeng. 21*:725-743.

Erickson, L. E. (1980). Growth and product energetic yields of *Rhodopseudomonas sphaeroides* S in dark and aerobic chemostat cultures. *J. Ferment. Technol. 58*:53-59.

Erickson, L. E., and Oner, M. D. (1983). Available electron and energetic yields in fermentation processes. *Ann. N.Y. Acad. Sci. 413*:99-113.

Erickson, L. E., Minkevich, I. G., and Eroshin, V. K. (1979). Utilization of mass-energy balance regularities in the analysis of continuous culture data. *Biotechnol. Bioeng. 21*:575-591.

Ghose, T. K., and Tyagi, R. D. (1979). Rapid ethanol fermentation of cellulose hydrolysate. I. Batch versus continuous systems. *Biotechnol. Bioeng. 21*:1387-1400.

Hanson, T. P., and Tsao, G. T. (1972). Kinetic studies of the lactic acid fermentation in batch and continuous cultures. *Biotechnol. Bioeng. 14*:233-252.

Lee, K. J., Skotnicki, M. L., Tribe, D. E., and Rogers, P. L. (1981). The kinetics of ethanol production by *Zymomonas mobilis* on fructose and sucrose media. *Biotechnol. Lett. 3*:207-212.

Leung, J. C. Y., and Wang, D. I. C. (1981). Production of acetone and butanol by *Clostridium acetobutylicum* in continuous culture using free cells and immobilized cells. Proceedings of the 2nd World Congress of Chemical Engineering and World Chemical Exposition, Montreal, Canada, October 4-9, 1981; Proceeding, ed. C. W. Robinson, Canadian Society for Chemical Engineering, Ottawa, Ontario, Canada, Volume 1, pp. 348-352.

Luedeking, R. (1967). Fermentation process kinetics. In *Biochemical and Biological Engineering Science*, Vol. 1, N. Blakebrough (Ed.). Academic Press, New York, pp. 181-240.

Luedeking, R., and Piret, E. L. (1959). A kinetic study of lactic acid fermentation batch processes of controlled pH. *J. Biochem. Microbiol. Technol. Eng. 1*:393-412.

Minkevich, I. G. (1980). Physical-chemical basis of material-energy balance of cellular metabolism. In *Limitation and Inhibition of Microbial Processes*, V. K. Eroshin (Ed.). Academy of Science Scientific Center for Biological Research, Pushchino, Moscow Region USSR (in Russian), pp. 55-91.

Minkevich, J. G. (1982). Physico-chemical properties of organic compounds and the energetics of metabolism. *J. Theor. Biol. 95*: 569-590.

Monot, F., and Engasser, J. M. (1983). Production of acetone and butanol by batch and continuous culture of *Clostridium acetobutylicum* under nitrogen limitation. *Biotechnol. Lett. 5*:213-218.

Oner, M. D. (1982). Estimation of growth yields in aerobic and anaerobic cultures. M.S. Thesis, Kansas State University, Manhattan, Kansas.

Oner, M. D. (1985). Estimation of yield and kinetic parameters in lactic acid fermentations for yoghurt cultures. Ph.D. Dissertation, Kansas State University, Manhattan, Kansas.

Oner, M. D., and Erickson, L. E. (1986). Anaerobic fermentation of *Lactobacillus bulgarius* and *Streptococcus thermophilus* on 3% nonfat dry milk with pure and mixed culture. *Biotechnol. Bioeng. 28*:883-894.

Oner, M. D., Erickson, L. E., and Yang, S. S. (1983). Estimation of true growth and product yields in aerobic cultures. *Biotechnol. Bioeng. 25*:631-646.

Oner, M. D., Erickson, L. E., and Yang, S. S. (1984). Estimation of yield, maintenance, and product formation kinetic parameters in anaerobic fermentations. *Biotechnol. Bioeng. 26*:1436-1444.

Oner, M. D., Erickson, E. L., and Yang, S. S. (1986a). Estimation of the true growth yield and maintenance coefficient for yoghurt cultures. *Biotechnol. Bioeng. 28*:919-926.

Oner, M. D., Erickson, L. E., and Yang, S. S. (1986b). Analysis of exponential growth data for yoghurt cultures. *Biotechnol. Bioeng. 28*:895-901.

Oner, M. D., Erickson, L. E., and Yang, S. S. (1986c). Utilization of spline functions for smoothing fermentation data and for estimation of specific rates. *Biotechnol. Bioeng.* *28*:902-918.

Pirt, S. J. (1975). *Principles of Microbe and Cell Cultivation*, Wiley, New York.

Rees, J. F., and Pirt, S. J. (1979). The stability of lactic acid in resting suspension of *Lactobacillus delbrueckii*. *J. Chem. Technol. Biotechnol.* *29*:591-602.

Reilly, P. J. (1964). Kinetics of lactic acid and gluconic acid fermentations. Ph.D. Thesis, University of Pennsylvania, Philadelphia.

Roels, J. A. (1980). Application of macroscopic principles to microbial metabolism. *Biotechnol. Bioeng.* *22*:2457-2514.

Roels, J. A. (1983). *Energetics and Kinetics in Biotechnology*, Elsevier, New York.

Rogers, P. L., Lee, K. J., Skotnicki, M. L., and Tribe, T. E. (1982). Ethanol production by *Zymomonas mobilis*. *Adv. Biochem. Eng.* *23*:37-84.

Samuel, W. A., Lee, Y. Y., and Anthony, W. B. (1980). Lactic acid fermentation of crude sorghum extract. *Biotechnol. Bioeng.* *22*:757-777.

Solomon, B. O., Erickson, L. E., Hess, J. L., and Yang, S. S. (1982). Maximum likelihood estimation of growth yields. *Biotechnol. Bioeng.* *24*:633-649.

Solomon, B. O., Erickson, L. E., and Yang, S. S. (1983). Utilization of statistics and experimental design in data collection and analysis. *Biotechnol. Bioeng.* *25*:2683-2705.

Solomon, B. O., Oner, M. D., Erickson, L. E., and Yang, S. S. (1984). Estimation of parameters where dependent observations are related by equality constraints. *AIChE J.* *30*:747-757.

Stouthamer, A. H. (1970). The search for correlation between theoretical and experimental growth yields. *Int. Rev. Biochem.* *21*:1-47.

Van Dijken, J. P., and Harder, W. (1975). Growth yields of microorganisms on methanol and methane. A theoretical study. *Biotechnol. Bioeng.* *17*:15-30.

Watson, T. G. (1970). Effect of sodium chloride on steady-state growth and metabolism of *Saccharomyses cerevisiae*. *J. Gen. Microbiol.* *64*:91-99.

Weast, R. C. (1976). *Handbook of Chemistry and Physics*, 56th Ed. CRC Press, Boca Raton, Florida, pp. D-78-79.

12

Bioenergetics and Yields for Anaerobic Digestion

LARRY E. ERICKSON / *Kansas State University, Manhattan, Kansas*

I. INTRODUCTION

Anaerobic digestion of wastes to produce methane and to biodegrade the waste is widely used; however, the methane produced is not always recovered and utilized. Many reviews consider the various aspects of anaerobic digestion, including the bioenergetics and yields (see for example, Harper and Pohland, 1986; McCarty and Smith, 1986; Kirsop, 1983; McCarty, 1982; McInerney and Bryant, 1981; McCarty, 1972; Thauer and Morris, 1984; Roels, 1983; Thauer et al., 1977; Van den Berg et al., 1985; Stouthamer, 1979; Gottschalk and Andreesen, 1979; Wolfe, 1979; Zeikus, 1983). The research on anaerobic digestion processes can be divided into two types of studies. Studies conducted under carefully defined conditions are most useful in understanding some of the basic bioenergetic aspects of anaerobic digestion. Studies conducted with various types of wastes provide information about the biodegradability of the waste and the methane yield that can be achieved.

II. BASIC BIOENERGETIC CONCEPTS

The free energy change must be negative for the anaerobic digestion process to take place. Thus, relative to carbon dioxide and water, the free energy of the methane produced will be less than that of the substrates from which it is made. Table 1, which is based on the work of Thauer et al. (1977) and Harper and Pohland (1986),

Table 1 Standard-State Free Energy Change for Reactions of Importance in Anaerobic Digestion

Reaction number	Reactants		Products	Free energy change $\Delta G'_0$ (kJ per reaction)
1	$CH_3CH_2COO^- + 3H_2O$	→	$CH_3COO^- + 3H_2 + H^+ + HCO_3^-$	76.1
2	$CH_3CH_2CH_2COO^- + 2H_2O$	→	$2CH_3COO^- + H^+ + 2H_2$	48.1
3	$CH_3CH_2OH + H_2O$	→	$CH_3COO^- + H^+ + 2H_2$	9.6
4	$CH_3CHOHCOO^- + 2H_2O$	→	$CH_3COO^- + H^+ + 2H_2 + HCO_3^-$	-4.2
5	$2HCO_3^- + 4H_2 + H^+$	→	$CH_3COO^- + 4H_2O$	-104.6
6	$HCO_3^- + 4H_2 + H^+$	→	$CH_4 + 3H_2O$	-135.6
7	$SO_4^{2-} + 4H_2 + H^+$	→	$HS^- + 4H_2O$	-151.9
8	$CH_3COO^- + H_2O$	→	$CH_4 + HCO_3^-$	-31.0
9	$CH_3COO^- + SO_4^{2-} + H^+$	→	$2HCO_3^- + H_2S$	-59.9

lists many of the important chemical reactions associated with anaerobic digestion and the free energy changes based on standard-state conditions. The standard state of aqueous solutions is that of a 1 mol/kg solution. The standard-state condition of water is that of the pure liquid. The standard temperature and pressure are 25°C and 1 atm, respectively. The pH is 7, which means that the standard-state condition for hydrogen ion concentration is that of pH 7. The standard state for the gases hydrogen and methane is the gaseous state. The free energy change in Table 1 is the free energy of the products minus the free energy of the reactants. A more complete table is given in Thauer et al. (1977). The actual free energy change $\Delta G'$ is related to the standard-state free energy change $\Delta G'_0$ as follows:

$$\Delta G' = \Delta G'_0 + RT \ln \frac{[C][D]}{[A][B]} \tag{1}$$

for the chemical reaction

$$A + B \rightarrow C + D \qquad (2)$$

where the quantities in brackets are the actual physiological concentrations. In anaerobic digestion, the actual partial pressure of hydrogen is considerably less than 1 atm; this needs to be considered in analyzing the bioenergetics of anaerobic digestion.

Figure 1 shows the effect of hydrogen partial pressure on the free energy change of reaction for several of the reactions in Table 1. Figure 1 is taken from Harper and Pohland (1986), but the basic concepts were presented earlier, as indicated in their work. As shown in Figure 2, hydrogen is an intermediate product in the anaerobic digestion process. Figure 1 shows that the concentration of an intermediate product, such as hydrogen, must take on a value that allows production and consumption to be in reasonable balance. This requires that the concentration of hydrogen be low enough that the free energy change is negative for both its production and consumption.

Some appreciation of the bioenergetics of anaerobic digestion can be obtained by examining the free energy change associated with the oxidation of the compounds of interest to carbon dioxide, nitrogen, and water. Table 2, which contains selected data from that compiled by Roels (1983, p. 40) and Erickson and Patel (1982), gives the free energy of combustion per equivalent of available electrons for several compounds of importance in anaerobic digestion. Table 2 shows that methane has the smallest free energy change per equivalent of available electrons of those in the table. Thus, the free energy change is negative for the conversion of any of the other substances to methane. This fundamental property of methane is the thermodynamic basis for the anaerobic digestion process. The difference in the free energy per equivalent of available electrons provides the driving force for anaerobic digestion. One can easily see that hydrogen has a larger value than some of the reactants that are converted to hydrogen (see Fig. 2 and Table 2). Thus, as shown in Figure 1, the partial pressure of hydrogen must be relatively low for hydrogen to be an intermediate in the pathway, as shown in Figure 2. The actual concentrations of reactants and products for each reaction that occurs must be such that the free energy change is negative. The standard-state free energy changes in Table 1 can be used with Equation (1) to calculate the actual free energy change for a given set of concentrations of reactants and products.

The production of biomass and the conversion of the wastes to methane can be examined using the available electron balance

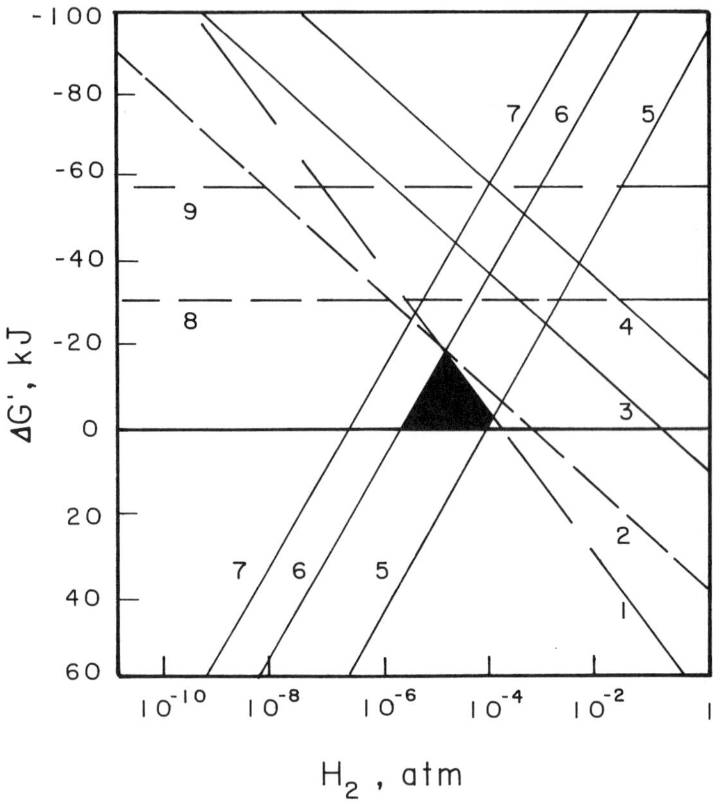

Figure 1 Effect of partial pressure of hydrogen on the Gibb's free energy change per reaction, ΔG' for acetic acid = 25 mM, propionic acid, butyric acid, lactic acid, and ethanol = 10 mM, sulfate and sulfite = 5 mM, bicarbonate = 20 mM, and methane partial pressure of 0.7 atm. The numbers refer to the reactions in Table 1; (1) propionic acid to H_2 and acetic acid; (2) butyric acid to H_2 and acetic acid; (3) ethanol to H_2 and acetic acid; (4) lactic acid to H_2 and acetic acid; (5) H_2 and bicarbonate to acetic acid; (6) H_2 and bicarbonate to methane; (7) H_2 and sulfate to sulfide; (8) acetic acid to CH_4; and (9) acetic acid and sulfate to sulfide and bicarbonate. (From Harper and Pohland, 1986.)

(Erickson et al., 1978). The available electron concept is equivalent to the chemical oxygen demand (COD) and biochemical oxygen demand (BOD) concepts used in environmental engineering. These are all measures of the equivalents of oxygen required to oxidize organic compounds to carbon dioxide and water. The end result of anaerobic digestion is biomass, methane, ammonia, and hydrogen sulfide. The available electrons in the substrates or wastes are converted into biomass and methane under strictly anaerobic conditions; that is, for the chemical balance equation

$$CH_mO_k + aNH_3 \rightarrow y_cCH_pO_nN_q + zCH_4 + dCO_2 \tag{3}$$

the available electron balance is [see Eqs. (3) through (7) of Chap. 11]

$$\gamma_s = y_c\gamma_b + z\gamma_p \tag{4}$$

which may be written in the form

$$\eta + \xi_p = 1.0 \tag{5}$$

where $\eta = y_c\gamma_b/\gamma_s$ is the fraction of available electrons incorporated into biomass and $\xi_p = z\gamma_p/\gamma_s$ is the fraction of available electrons incorporated into methane. It is important to point out that oxygen is readily used as an electron acceptor when it is available, that

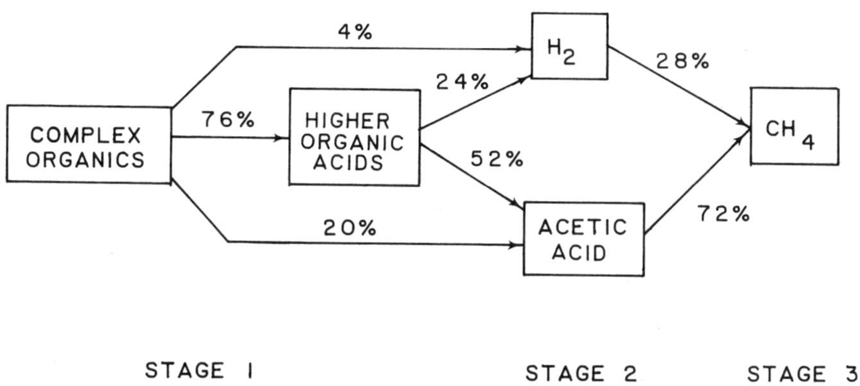

Figure 2 The three stages of methane fermentation and the percentage flow of the energy content of complex organic materials through each stage into methane as represented by the flow of available electrons. Stage 1 involves hydrolysis and fermentation to fatty acids, alcohols, carbon dioxide, ammonia, and hydrogen; stage 2 involved the production of acetic acid, carbon dioxide, and hydrogen by acetogenic bacteria; stage 3 involves methane formation by two groups of methanogens. (From McCarty, 1982.)

Table 2 Free Energy of Combustion per Equivalent of Available Electrons at Standard-State Conditions (Unit Molality for Solutions, 25°C, 1 atm, pH 7)

Compound or group name[a]	$-\Delta G_0'/\gamma$ (kJ per equivalent)
Acetic acid	111.8
Acetone	108.4
n-Butanol	108.0
Butyric acid	108.7
Ethanol	109.9
Glucose	119.7
Hydrogen	119.0
Lactic acid	114.8
Methane	102.3
Methanol	115.5
Microbial biomass	112.8
Palmitic acid	106.5
Propionic acid	109.4
Alcohols	113.2
Aldehydes	115.7
Ketones	108.3
Carbohydrates	120.5
Amino acids	109.9
Aromatics	108.9
Monocarboxylic acids	112.5
Dicarboxylic acids	113.6
Purines	114.7
Normal alkanes	105.9
Normal alkyl cyclopentanes	106.5
Normal alkyl cyclohexanes	106.3
Normal monoolefins	107.6
Normal alkylbenzenes	106.5
Branched alkylbenzenes	105.9
Styrenes	107.6
Cyclopentanes	106.1
Cyclohexanes	105.8
Branched alkanes	105.9
Diolefins	111.1
Branched alkynes	110.7

[a]Values for groups are mean values.

Source: From Roels (1983) and Erickson and Patel (1982).

nitrate can be used as an electron acceptor and is converted to nitrogen or ammonia, and that sulfate can be used as an electron acceptor and is converted to sulfide. When any of these processes occurs, the available electron balance must be altered to include them. The product yield is reduced when other electron acceptors are present because available electrons are transferred to the electron acceptor. For methane production, carbon dioxide is often considered the electron acceptor. The bioenergetics are such that the most energy is liberated when oxygen is the electron acceptor, nitrate is next, and sulfate is slightly greater than carbon dioxide (see Table 3). Experimental data in the literature indicate that microbial cells tend to use the largest free energy driving force available to them. Thus, oxygen is used preferentially, nitrate is selected when oxygen is absent, and sulfate is used when oxygen and nitrate are both absent. Thus, the best methane production conditions occur when oxygen, nitrate, and sulfate are absent as electron acceptors.

The cell yield for a particular organism involved in one of the conversions in Figure 2 may be expressed in terms of a true growth yield and a maintenance parameter; that is,

$$\frac{\mu}{\eta} = \frac{\mu}{\eta_{max}} + m_e \qquad (6)$$

where η is the fraction of available electrons incorporated into biomass, η_{max} is the true growth yield corrected for maintenance, and m_e is the maintenance parameter. The product yield under anaerobic conditions is

$$\xi_p = 1 - \eta \qquad (7)$$

Table 3 Free Energy of Reaction per Equivalent of Available Electrons for Glucose at pH 7 for Several Electron Acceptors

Electron acceptor	Products	$-\Delta G_0'/\gamma$ (kJ per equivalent)
O_2	CO_2, H_2O	119.7
NO_3^-	N_2, CO_2, H_2O	111.9
SO_4^{2-}	HS^-, CO_2, H_2O	18.9
CO_2, H_2O	CH_4, CO_2, H_2O	17.4

Source: Data from Roels (1983), pp. 40 and 62.

Substituting Equation (6) into Equation (7) gives

$$\xi_p = 1 - \frac{\mu}{\mu/\eta_{max} + m_e} \tag{8}$$

Equation (8) shows that the product yield depends on the true growth yield and the maintenance coefficient. The true growth yield is related to the production and utilization of ATP as shown in Equation (8) of Chapter 11; that is,

$$\eta_{max} = \frac{\sigma_b \gamma_b Y_{ATP}^{max}/12}{\sigma_b \gamma_b Y_{ATP}^{max}/12 + \delta} \tag{9}$$

where Y_{ATP}^{max} is the true growth yield based on ATP, δ is the equivalents of available electrons transferred to products to produce 1 mol of ATP from ADP, σ_b is the weight fraction carbon in biomass, and γ_b is the reductance degree of the biomass. Figure 3 shows the variation in the available electron yield corrected for maintenance as a function of δ with the growth yield based on ATP corrected for maintenance as a parameter. Values are presented for Y_{ATP}^{max} = 6.5 and 10 g biomass per mole ATP. The value of 6.5 is for carbon dioxide as carbon source; the value of 10 is for acetate as carbon source (Stouthamer, 1979). The appropriate value of δ for a particular growth process depends on the substrate and products as well as the microorganism. For many of the ATP-producing processes involved in anaerobic digestion, the values fall in the range from 12 to 24 equivalents of available electrons to produce 1 mol of ATP from ADP.

Equation (9) can be substituted into Equation (8) to obtain

$$\xi_p = \frac{1}{[(\mu \sigma_b \gamma_b Y_{ATP}^{max}/12)/(\mu \delta + m_e \sigma_b \gamma_b Y_{ATP}^{max}/12)] + 1} \tag{10}$$

Equation (10) shows that the product yield depends on the specific growth rate μ, the true growth yield Y_{ATP}^{max}, the maintenance coefficient m_e, and the value of δ.

In anaerobic digestion, products produced by one organism are used by other organisms, as shown in Figure 2. When the process is a series arrangement in which the first product is the substrate for the second organism, the observed yield of the second product based on the substrate to the first organism should be the product of the two product yields ξ_{p1} and ξ_{p2}. Thus, the product yields from individual processes, such as methane production from acetate,

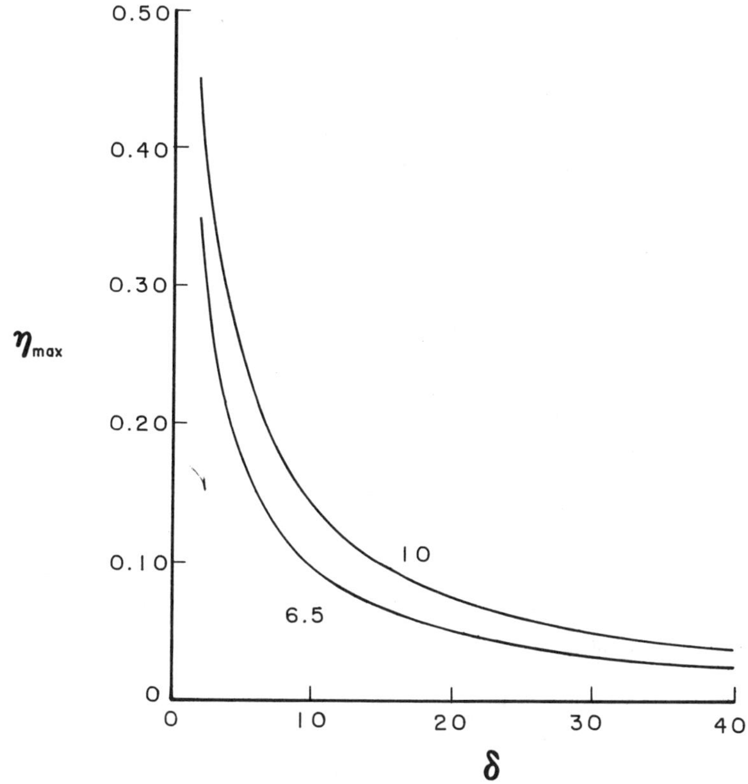

Figure 3 Variation in true available electron yield η_{max} with δ (the equivalents of available electrons transferred to products to produce 1 g mol ATP from ADP). The parameter is Y_{ATP}^{max} in g cells dry weight per mole ATP.

should be larger than the overall observed yield of methane from sewage.

For anaerobic processes, the free energy yield associated with the growth process has been defined as follows (Chap. 11) (Erickson and Oner, 1983):

$$\eta_{th} = \frac{g_b \eta}{g_s - g_p \xi_p} \tag{11}$$

where

$$g = \frac{-\Delta G_0'}{\gamma} \tag{12}$$

is the free energy per equivalent of available electrons relative to the dead state associated with combustion.

III. YIELDS ASSOCIATED WITH ANAEROBIC DIGESTION

In Chapter 11, bioenergetic and yield data associated with the formation of a number of anaerobic products are reviewed. In this section, yield data associated with the anaerobic production of methane from several substrates is presented and reviewed. Table 4 presents yield values for acetate as the organic substrate. The results for several different cultures growing on hydrogen and carbon dioxide are presented in Table 5. Tables 6 and 7 contain yields for growth on methanol and formate, respectively. Mixed substrates may be advantageous since one may be a superior carbon substrate and the other may be a better energy source. Table 8 presents yield values for growth on mixed substrates.

The stoichiometry of Equation (3) is assumed to estimate the biomass available electron yield η from the biomass and methane production data. The available electron balance allows one to write

$$\eta = \frac{\sigma_b \gamma_b Y_{x/p}/12}{8 + \sigma_b \gamma_b Y_{x/p}/12} \tag{13}$$

The average values $\sigma_b = 0.462$ and $\gamma_b = 4.291$ (Erickson and Patel, 1982) are used in Tables 4 through 8.

Stouthamer (1979) has reviewed the theoretical ATP requirements for the formation of biomass from several different substrates. Values of 10 and 6.5 g cells per mole ATP are reported for acetate and carbon dioxide, respectively. Anthony (1978) has considered the assimilation of methanol, formaldehyde, and formate into cell material. Harder and van Dijken (1976) and Harder et al. (1981) have presented several different pathways for the formation of biomass from one-carbon substrates, such as methanol; however, the ATP requirements depend on the pathway, with values ranging from 27.3 to 6.5 g cells per mole ATP.

The experimental yield values can be analyzed to provide additional insight into the bioenergetics of the anaerobic processes under consideration. In Table 4, $\eta = 0.05$ is a typical value and $\eta = 0.087$ is the maximum value reported. Equation (11) may be used to estimate the free energy yield. For the standard-state free energies presented in Table 2, Equation (11) yields $\eta_{th} = 0.39$ for $\eta = 0.05$ and $\eta_{th} = 0.53$ for $\eta = 0.087$. Figure 3 can be used to appreciate the value of δ required to achieve the observed experimental yields. For $\eta = 0.05$ and $Y_{ATP} = 10$ g cells per mole ATP, δ must be less than 30 equivalents per mole ATP.

Table 4 Growth Yields for Methanogenic Bacteria Growing on Acetate

Organism	$Y_{x/s}$ (g cells per g mol acetate)	$Y_{x/p}$ (g cells per g mol CH_4	η	Reference and remarks
Methanosarcina 227	2.0	2.1	0.041	Smith and Mah (1978), complex media
		3.17	0.061	Smith and Mah (1980), minimal media
		2.53	0.050	0.1% Yeast extract, ash in minimal media
		2.21	0.044	0.2% Yeast extract, ash in minimal media
		3.41	0.066	0.1% Yeast extract in minimal media
		2.89	0.056	0.2% Yeast extract in minimal media
		2.81	0.055	0.5% Yeast extract in minimal media
		3.08	0.060	Complex media
		1.72	0.034	Complex media
		2.12	0.042	Complex media
		1.92	0.038	Complex media
		1.96	0.039	Complex media
		2.64	0.052	Weimer and Zeikus ((1978b), complex media
Methanosarcina JM-1		1.8	0.036	Zinder and Mah (1979), complex media
Methanosarcina barkeri	4.2		0.087	Scherer and Sahm (1981), complex media
		1.62	0.032	Weimer and Zeikus (1978b),
		2.20	0.043	complex media
Methanococcus mazei	1.94		0.040	Yang and Okos (1984, 1987), complex media
Methanobacterium soehngenii			0.018	Zehnder et al. (1980), complex media

Table 5 Growth Yields for Methanogenic Bacteria Growing on H_2 and CO_2

Organism	$Y_{x/s}$ (g cells per g mol H_2	$Y_{x/p}$ (g cells per g mol CH_4	η	Reference and remarks
Methanobrevibacter arboriphilus		1.14	0.023	Morii et al. (1987), minimal media
		1.06	0.021	
		1.32	0.026	
		1.42	0.028	
		2.53	0.050	Zehnder and Wuhrmann (1977), complex media
		2.68	0.052	
		2.29	0.045	
Methanobacterium formicicum		3.5	0.067	Schauer and Ferry (1980), complex media
Methanobacterium thermoautotrophicum		3.0	0.058	Schonheit et al. (1980), minimal media
		1.6	0.032	
		0.6-1.6	0.012-0.032	Taylor and Pirt (1977), minimal media
		1.2	0.024	Ronnow and Gunnarson (1981), complex media
Methanosarcina barkeri		6.4	0.117	Weimer and Zeikus (1978a), complex media
Methanobacterium omelianskii M.o. H	0.62		0.051	Robertson and Wolfe (1970), complex media
Methanosarcina 227		8.7	0.152	Smith and Mah (1978), complex media
Methanococcus thermolithotrophicus		0.66-2.5	0.014-0.052	Fardeau and Belaich (1986), minimal media

Table 6 Growth Yields for Methanogenic Bacteria Growing on Methanol

Organism	$Y_{x/s}$ (g cells per g mol CH_3OH)	$Y_{x/p}$ (g cells per g mol CH_4)	η	Reference and remarks
Methanosarcina strain 227		5.10	0.095	Smith and Mah (1980), complex media
Methanosarcina TM-1		4.50	0.085	Zinder and Mah (1979), complex media
Methanosarcina barkeri	3.4-4.5		0.094-0.124	Scherer and Sahm (1981), complex media
		3.3	0.0637	Stadtman (1967)
		7.2	0.129	Weimer and Zeikus (1978a), complex media
		4.36	0.083	Mountfort and Asher (1979), complex media 1.25 mm sulfide
		1.90	0.038	12.5 mm sulfide

Table 7 Growth Yields for Methanogenic Bacteria Growing on Formate

Organism	$Y_{x/s}$ (g cells per g mol formate)	$Y_{x/p}$ (g cells per g mol CH_4)	η	Reference and remarks
Methanobacterium formicicum	1.2	4.8	0.098	Schauer and Ferry (1980), complex media
	1.4		0.116	Chua and Robinson (1983), complex media; true growth yield
Methanobrevibacter arboriphilus	0.81	3.52	0.067	Morii et al. (1983), complex media

Table 8 Growth Yields for *Methanosarcina barkeri* Growing on Mixed Substrates[a]

Substrates	$Y_{x/p}$ (g cells per g mol CH_4)	η
H_2, CO_2, and acetate	12.0	0.198
	12.6	0.206
CH_3OH and acetate	17.6	0.266
	19.2	0.284
	16.3	0.252
CH_3OH, CO_2, and acetate	19.0	0.281
	15.9	0.247
	16.2	0.251
	15.6	0.244
CH_3OH, H_2O, CO_2, and acetate	13.5	0.218
	14.9	0.235
	14.3	0.228
	13.7	0.220

[a]Weimer and Zeikus (1978b), complex media.

Analysis of the results in Table 5 reveals that η_{th} is quite low when the standard-state value for hydrogen is used; however, since the partial pressure of hydrogen is considerably less than 1 atm in the experiments reported in Table 5, the actual free energy yield is somewhat larger. Figure 3 shows that δ needs to be smaller than 20 to account for the results in Table 5 when CO_2 is the sole carbon source and Y_{ATP} = 6.5 g cells per mole ATP is used.

The largest yield reported in Table 6 for methanol as a substrate is η = 0.129 which corresponds to η_{th} = 0.55. The largest available electron yields are in Table 8 for mixed substrates. The largest free energy yields (0.76 for methanol and acetate) are associated with the data in Table 8. The free energy biomass yields show that little additional biomass yield improvement can be expected in Tables 6 and 8.

Sufficient data to estimate the true growth yield and maintenance coefficient in Equation (6) have been obtained by Fardean and Belaich (1986) and Chua and Robinson (1983). Values of the maintenance coefficients are very large in the work of Fardean and Belaich (1986). These authors found values ranging from 0.037 to 0.140 mol CH_4 per g dry weight hr or m_e = 1.79–6.78 h^{-1} for *Methanococcus*

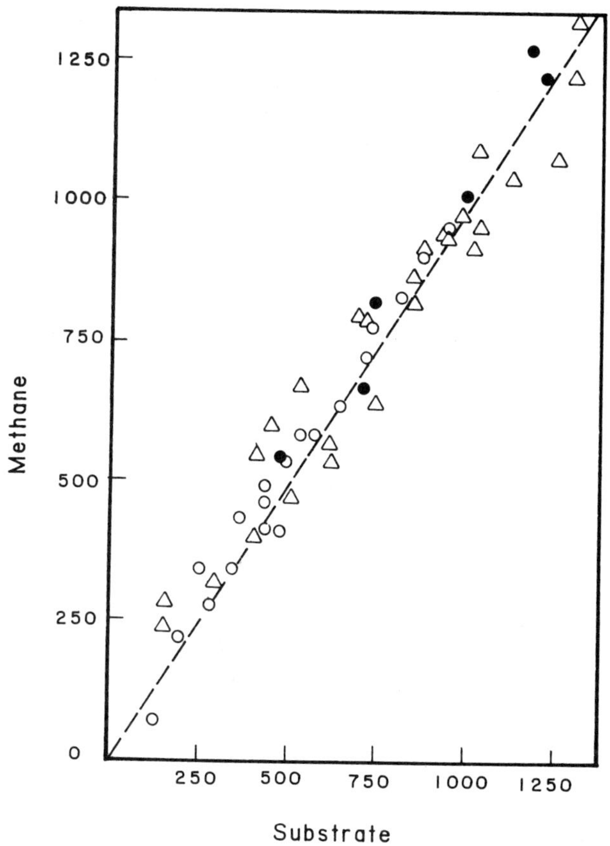

Figure 4 Relationship of methane production to substrate converted for tomato peeling waste, ● ; bean blanching waste, ○ ; and piggery waste, △. Units are equivalents of chemical oxygen demand (or equivalents of available electrons). (From Van den Berg et al., 1985.)

thermolithotrophicus growing on H_2 and CO_2. Chua and Robinson (1983) reported a value for the maintenance coefficient for *Methanobacterium formicicum* growing on formate of 6.8 mmol formate per (g dry weight h, which is equal to $m_e = 0.082\ h^{-1}$. Further research is needed to obtain a good understanding of maintenance requirements in methanogenic cultures.

The yield values and the relationships presented here can help us understand the bioenergetics of anaerobic growth associated with methane production. For more complex substrates, yield data could also be tabulated, but it would not be very helpful to those interested in the bioenergetics of individual growth processes because mixed cultures are required to produce methane from complex substrates.

The methane yield associated with anaerobic digestion can approach 100% or $\xi_p = 1.0$ when maintenance processes predominate and net biomass production is negligible. Figure 4 shows that methane available electron yields can approach 100% in fixed film anaerobic bioreactors. The dashed line corresponds to $\xi_p = 1.0$ or 100% yield of methane. Figure 4 shows that when chemical oxygen demand (or equivalents of available electrons) is used to characterize wastes, the results are similar for several different wastes. When methane production is an important objective, the system should be designed to retain the biomass produced. In this way maintenance processes dominate and product yield is high.

IV. CONCLUSIONS

Yield data for methane production processes is presented using available electron concepts. Most biomass yields associated with methanogens are quite low. When these yields are reported using free energy concepts, the free energy yield is much larger. The free energy yields are used to show that the reported biomass yields are reasonably close to the maximum theoretical yields in several cases.

V. NOMENCLATURE

d	Moles of carbon dioxide per quantity of organic substrate containing 1 g atom carbon; g mol/g atom carbon
$\Delta G'$	Free energy of reaction
$\Delta G_0'$	Standard-state free energy of reaction
g_b	Amount of free energy in biomass per g equivalent of available electrons in biomass; kJ/g equivalent of available electrons
g_p	Amount of free energy in product per g equivalent of available electrons in product; kJ/g equivalent of available electrons
g_s	Amount of free energy in substrate per g equivalent of available electrons in substrate; kJ/g equivalent of available electrons
m_e	Rate of organic substrate consumption for maintenance; g equivalent of available electrons per g equivalent of available electrons in biomass h^{-1}

m_{ATP}	Moles of ATP expended per g of cells per hour for maintenance; mol ATP per g h
m_{th}	Amount of free energy consumed per unit of free energy in biomass per hour; kJ free energy consumed per kJ free energy in biomass (h)
Y_{ATP}	Biomass yield based on ATP; g dry biomass per g mol ATP
Y_{ATP}^{max}	True biomass yield based on ATP; g dry biomass per g mol ATP
y_c	Biomass carbon yield (fraction of organic substrate carbon in biomass); dimensionless
z	Fraction of organic substrate carbon in products; dimensionless
γ_b	Reductance degree of biomass; equivalents of available electrons per g atom carbon
γ_p	Reductance degree of products; equivalents of available electrons per g atom carbon
γ_s	Reductance degree of substrate; equivalents of available electrons per g atom carbon
δ	Number of equivalents of available electrons transferred to product to produce 1 g mol of ATP from ADP
η	Fraction of available electrons in organic substrate converted to biomass; dimensionless
η_{max}	"True" biomass available electron yield; dimensionless
η_{th}	Biomass free energy yield; see Equation (11); dimensionless
μ	Specific growth rate; h^{-1}
ξ_p	Fraction of available electrons in organic substrate converted to products; dimensionless
σ_b	Weight fraction carbon in dry biomass; dimensionless
σ_p	Weight fraction carbon in products; dimensionless
σ_s	Weight fraction carbon in organic substrate; dimensionless

REFERENCES

Anthony, C. (1978). The prediction of growth yields in methylotrophs. *J. Gen. Microbiol. 104*:91-104.

Chua, H. B., and Robinson, J. P. (1983). Formate limited growth of *Methanobacterium formicicum* in steady state cultures. *Arch. Microbial. 135*:158-160.

Erickson, L. E., and Oner, M. D. (1983). Available electron and energetic yields in fermentation processes. *Ann. N.Y. Acad. Sci. 413*:99-113.

Erickson, L. E., and Patel, S. A. (1982). Analysis of regularities and process efficiencies in living systems. *CRC Crit. Rev. Biomed. Eng. 8*:311-333.

Erickson, L. E., Minkevich, I. G., and Eroshin, V. K. (1978). Application of mass and energy balance regularities in fermentation. *Biotechnol. Bioeng. 20*:1595-1621.

Fardean, M. L., and Belaich, J. P. (1986). Energetics of the growth of *Methanococcus thermolithotrophicus. Arch. Microbiol. 144*:381-385.

Gottschalk, G., and Andreeson, J. (1979). Energy metabolism in anaerobes. *Int. Rev. Biochem. 21*:85-115.

Harder, W., and van Dijken, J. P. (1976). Theoretical considerations on the relation between energy production and growth of methane utilizing bacteria. In *Microbial Production and Utilization of Gases*, H. P. Schlegel, N. Pfennig, and G. Gottschalk (Eds.). Erich Goltze Verlag, Gottingen, pp. 403-418.

Harder, W., van Dijken, J. P., and Roels, J. A. (1981). Utilization of energy in methylotrophs. In *Microbial Growth on* C_1 *Compounds*, H. Dalton (Ed.). Heyden, London, pp. 258-269.

Harper, S. R., and Pohland, F. G. (1986). Recent developments in hydrogen management during anaerobic biological wastewater treatment. *Biotechnol. Bioeng. 28*:585-602.

Kirsop, B. H. (1983). Methanogenesis. *CRC Crit. Rev. Biotechnol. 1*:109-159.

McCarty, P. L. (1972). Energetics of organic matter degradation. In *Water Pollution Microbiology*, R. Mitchell (Ed.). Wiley, New York, pp. 91-118.

McCarty, P. L. (1982). One hundred years of anaerobic treatment. In *Anaerobic Digestion 1981*, D. E. Hughes et al. (Eds.). Elsevier, New York, pp. 3-22.

McCarty, P. L., and Smith, D. P. (1986). Anaerobic waste treatment. *Environ. Sci. Technol. 20*:1200-1206.

McInerney, M. J., and Bryant, M. P. (1981). Review of methane fermentation fundamentals. In *Fuel Gas Production from Biomass*, D. L. Wise (Ed.). CRC Press, Boca Raton, Florida, pp. 19-46.

Morii, H., Nishihara, M., and Koga, Y. (1983). Isolation, characterization, and physiology of a new formate-assimilable methanogenic strain (A2) of *Methanobrevibacter arboriphilus. Agric. Biol. Chem. 47*:2781-2789.

Morii, H., Koga, Y., and Nagai, S. (1987). Energetic analysis of the growth of *Methanobrevibacter arboriphilus* A2 in hydrogen limited continuous culture. *Biotechnol. Bioeng. 29*:310-315.

Mountfort, D. O., and Asher, R. A. (1979). Effect of inorganic sulfide on the growth and metabolism of *Methanosarcina barkeri* strain DM. *Appl. Env. Microb. 37*:670-675.

Roberton, A. M., and Wolfe, R. S. (1970). Adenosine triphosphate pools in *Methanobacterium. J. Bacteriol. 102*:43-51.

Roels, J. A. (1983). *Energetics and Kinetics in Biotechnology*. Elsevier, New York, pp. 1-330.

Ronnow, P. H., and Gunnarsson, L. A. H. (1981). Sulfide dependent methane production and growth of thermophilic methanogenic bacterium. *Appl. Env. Microbiol. 42*:580-584.

Schauer, N. L., and Ferry, J. G. (1980). Metabolism of formate in *Methanobacterium formicicum. J. Bacteriol. 142*:800-807.

Scherer, P., and Sahm, H. (1981). Influence of sulfur-containing compounds on the growth of *Methanosarcina barkeri* in a defined medium. *Appl. Microbiol. Biotechnol. 12*:28-35.

Schonheit, P., Moll, J., and Thauer, R. K. (1980). Growth parameters (K_S, μ_{max}, Y_S) of *Methanobacterium thermoautotrophicum. Arch. Microbiol. 127*: (1980).

Smith, M. R., and Mah, R. A. (1978). Growth and methanogenesis by *Methanosarcina* strain 227 on acetate and methanol. *Appl. Environ. Microbiol. 36*:870-879.

Smith, M. R., and Mah, R. A. (1980). Acetate as sole carbon and energy source for growth of *Methanosarcina* strain 227. *Appl. Env. Microbiol. 39*:993-999.

Stadtman, T. C. (1967). Methane fermentation. *Annu. Rev. Microbiol. 21*:121-142.

Stouthamer, A. H. (1979). The search for correlation between theoretical and experimental growth yields. *Int. Rev. Biochem. 21*:1-47.

Taylor, G. T., and Pirt, S. J. (1977). Nutrition and factors limiting the growth of a methanogenic bacterium (*Methanobacterium thermoautotrophicum*). *Arch. Microbiol. 113*:17-22.

Thauer, R. K., and Morris, J. G. (1984). Metabolism of chemotrophic anaerobes: Old view and new aspects. In *The Microbe 1984, Part II, Prokaryotes and Eukaryotes*, D. P. Kelly and N. G. Carr (Eds.). Cambridge University Press, Cambridge, pp. 123-168.

Thauer, R. K., Jungermann, K., and Decker, K. (1977). Energy conservation in chemotrophic anaerobic bacteria. *Bacteriol. Rev. 41*:100-180.

Van den Berg, L., Duff, S. J., and Kennedy, K. J. (1985). Methane production from anaerobic digestion. In *Comprehensive Biotechnology*, Vol. 4, M. Moo Young (Ed.). Pergamon Press, New York, pp. 1051-1058.

Weimer, P. J., and Zeikus, J. G. (1978a). One carbon metabolism in methanogenic bacteria: Cellular characterization and growth of *Methanosarcina barkeri. 119*:49-57.

Weimer, P. J., and Zeikus, J. G. (1978b). Acetate metabolism in *Methanosarcina barkeri. Arch. Microbiol. 119*:175-182.

Wolfe, R. S. (1979). Microbial biochemistry of methane—a study in contrasts. I. Methanogenesis. *Int. Rev. Biochem. 21*:267-300.

Yang, S. T., and Okos, M. R. (1984). Kinetics of anaerobic methanogenesis from acetate. Paper 39d, AIChE Annual Meeting, San Francisco, pp. 1-12.

Yang, S. T., and Okos, M. R. (1987). Kinetic study and mathematical modeling of methanogenesis of acetate using pure cultures of methanogens. *Biotechnol. Bioeng. 30*:661-667.

Zehnder, A. J. B., and Wuhrmann, K. (1977). Physiology of a *Methanobacterium* strain A E. *Arch. Microbiol. 111*:199-205.

Zehnder, A. J. B., Huser, B. A., Brock, T. D., and Wuhrmann, K. (1980). Characterization of an acetate decarboxylating non hydrogen oxidizing methane bacterium. *Arch. Microbiol. 124*:1-11.

Zeikus, J. G. (1983). Metabolism of one-carbon compounds by chemotrophic anaerobes. *Adv. Microb. Physiol. 24*:215-299.

Zinder, S. H., and Mah, R. A. (1979). Isolation and characterization of a thermophilic strain of *Methanosarcina* unable to use H_2-CO_2 for methanogenesis. *Appl. Env. Microbiol. 38*:996-1008.

13

Bioenergetics and Yields with Electron Acceptors Other Than Oxygen

ADRIAAN H. STOUTHAMER / *Vrije Universiteit, Amsterdam, The Netherlands*

I. INTRODUCTION

Many bacteria are able to utilize external hydrogen acceptors during anaerobic growth. The presence and the utilization of a hydrogen acceptor is even necessary for the growth of a large number of bacteria. On the other hand, some bacteria are able to grow in the absence of a hydrogen acceptor. These bacteria obtain energy from fermentation reactions. Free energy transduction in the living cell proceeds via the ATP system. ATP may be synthesized by substrate-level phosphorylation and by electron transport-linked phosphorylation. Heterotrophic bacteria, which can grow anaerobically in the absence of a hydrogen acceptor, obtain energy nearly in all cases exclusively from substrate-level phosphorylation. In some fermentations, additional ATP is gained from electron transport-linked phosphorylation. In the latter case an electron acceptor (fumarate or CO_2) is formed during fermentation of the growth substrate. During growth in the presence of hydrogen acceptors, energy generation by electron transport-linked phosphorylation always occurs. In a number of cases, dependent on the bacterium and on the substrate, substrate-level phosphorylation also takes place.

A. Utilization of Hydrogen Acceptors by Various Bacteria

A large variety of external hydrogen acceptors can be reduced by microorganisms. The reduction of various hydrogen acceptors is listed in Table 1.

Table 1 Standard Free Energy Changes ($\Delta G_0'$) for the Reaction of Hydrogen with a Number of Electron Acceptors[a]

Reaction	Substrates	Products	kJ per reaction	$\Delta G_0'$ (kJ per H_2)
a	$H_2 + N_2O$	$N_2 + H_2O$	-341.4	-341.4
b	$2H_2 + O_2$	$2H_2O$	-474.5	-237.2
c	$5H_2 + 2H^+ + 2NO_3^-$	$N_2 + 6H_2O$	-1120.5	-224.1
d	$H_2 + NO_3^-$	$NO_2^- + H_2O$	-163.2	-163.2
e	$4H_2 + 2H^+ + NO_3^-$	$NH_4^+ + 3H_2O$	-599.6	-149.9
f	H_2 + TMAO	TMA + H_2O	-160.1	-160.7
g	H_2 + fumarate^{2-}	Succinate^{2-}	-86.2	-86.2
h	$4H_2 + H^+ + SO_4^{2-}$	$HS^- + 4H_2O$	-151.9	-38.0
i	$4H_2 + H^+ + HCO_3^-$	$CH_4 + 3H_2O$	-135.6	-34.9
j	$H_2 + S$	$HS^- + H^+$	-28.0	-28.0
k	$4H_2 + H^+ + 2HCO_3^-$	Acetate$^-$ + $4H_2O$	-104.6	-26.1

[a]TMAO and TMA are abbreviations for trimethylamine *N*-oxide and trimethylamine, respectively.
Source: Data derived from Thauer et al. (1977).

The standard free energy changes ($\Delta G_0'$) associated with these reactions are included. For comparative purposes in all cases H_2 is used as electron donor. It is evident that the $\Delta G_0'$ per mole H_2 associated with the reduction of nitrous oxide has the highest value, whereas the $\Delta G_0'$ per mole H_2 for the reduction of CO_2 to acetate has the smallest value. Furthermore, the $\Delta G_0'$ per mole H_2 for respiration (reaction b) and denitrification (reaction c) are about the same. Similarly, the $\Delta G_0'$ per mole H_2 for the reduction of nitrate to nitrite (reaction d), nitrate to ammonia (reaction e), and of trimethylamine *N*-oxide to trimethylamine (reaction f) do not differ very much. However we must realize that these differences in $\Delta G_0'$ per mole H_2 give no indication about the amount of ATP formed by a certain reaction, since this is determined by the nature of the biochemical pathway for the transfer of electrons from donor to acceptor. Since there is great biochemical diversity in such pathways, the amount of ATP synthesized by the same reaction in different organisms may show large differences. However from the data in Table 1 it seems clear that the amount of ATP formed in association with reactions h through k is very low. This becomes even more clear when the $\Delta G_0'$ for the oxidation of acetate with a number of electron acceptors is considered (Table 2). The reactions mentioned are used by various microorganisms to provide the ATP needed for growth. The $\Delta G_0'$ varies from -844.4 kJ per reaction for the utilization of oxygen as oxidant (reaction a) to only -7.4 kJ per reaction for the utilization of sulfur as oxidant (reaction e). Consequently there is also a large difference in ATP production during acetate oxidation in these cases.

Table 2 Standard Free Energy Changes for Oxidation of Acetate with Various Hydrogen Acceptors

Reaction			$\Delta G_0'$ (kJ per reaction)
a	$Acetate^- + 2O_2$	$\rightarrow 2HCO_3^- + 2H_2O + H^+$	-844.4
b1	$Acetate^- + 0.6H^+ + 0.8NO_3^-$	$\rightarrow 2HCO_3^- + 0.4N_2 + 0.8H_2O$	-791.8
b2	$Acetate^- + 4\ fumarate^{2-}$	$\rightarrow 2HCO_3^- + 4\ succinate^{2-} + H^+$	-240.2
c	$Acetate^- + SO_4^{2-}$	$\rightarrow 2HCO_3^- + HS^-$	-48
d	$Acetate^- + H_2O$	$\rightarrow 1HCO_3^- + CH_4$	-31
e	$Acetate^- + 4S + 2H_2O + H^+$	$\rightarrow 2HCO_3^- + 4HS^-$	-7.4

Source: Calculated from the data given by Thauer et al. (1977).

A number of bacteria can utilize a variety of hydrogen acceptors. A number of examples are given in Table 3. Under anaerobic conditions *Escherichia coli* can utilize oxygen, nitrate, nitrite, trimethylamine *N*-oxide, and fumarate as hydrogen acceptors and is also able to grow anaerobically with sugars without hydrogen acceptor

Table 3 Possibility of Utilizing a Number of Hydrogen-Accepting Reactions of Energy Metabolism in a Variety of Microorganisms[a]

Organism	Hydrogen-accepting reactions	Growth without hydrogen acceptor	
Escherichia coli	b, d, e, f, g	F (sugars)	
Paracoccus denitrificans	b, c	—	
Rhodopseudomonas capsulata	a, b, d, f	F (sugars)(1)	PhA PhH
Rhodopseudomonas sphaeroides f. *denitrificans*	b, c		PhA PhH
Propionibacterium	b (2), c (3)	F (sugars, lactate)(1)	
Veillonella alkalescens	b (2), e, g	F (lactate)(1)	
Wolinella succinogenes	a, b, e, g, j	—	
Campylobacter sp.	b (2), e, g, h	—	
Desulfovibrio vulgaris	e, g, h	—	
Desulfuromonas acetoxidans	g, h, j	—	
Desulfotomaculum orientis	h, k	—	
Desulfobulbus propionicus	d, g, h	F (ethanol, lactate)	
Thermoproteus neutrophilus	j	—	
Clostridium formicoaceticum	g, k	F (sugars)	

[a]The letters refer to the formally hydrogen-consuming reactions mentioned in Table 1. The possibility of growing anaerobically without hydrogen acceptor and thus obtaining ATP by fermentation reactions is indicated by F. An example of a possible substrate for growth without hydrogen acceptor is given in parentheses. The possibility of growing under photoautotrophic or photoheterotrophic conditions is indicated by PhA or PhH, respectively. (1) Reduction of fumarate takes place in these fermentations. (2) High concentrations of oxygen are toxic to these organisms. (3) Nitrate is reduced to nitrous oxide by this organism.

(Ingledew and Poole, 1984; Cole et al., 1985, Barrett and Kwan, 1985; Kröger, 1980; Stouthamer, 1980b, 1988; Stouthamer et al., 1980). Since *E. coli* is also able to utilize a large variety of carbon substrates, this organism is very versatile. The systems for the utilization of alternate hydrogen acceptors in *E. coli* are induced by their substrates under anaerobic conditions. The induction of nitrate reductase is repressed by oxygen; the induction of fumarate reductase is repressed by oxygen and nitrate. The general principle that appears to govern the organism's reaction to changes in availability of potential electron acceptors has been expressed by Stouthamer (1976) as follows: "the amount of energy released by catabolism is as large as possible since in all cases the formation of a reductase for a certain substrate is repressed under conditions at which a hydrogen acceptor with a higher energy-yielding potential is also present." The most versatile organism in this table is *Rhodopseudomonas capsulata* (Madigan and Gest, 1979; Vignais et al., 1985). It can grow under the following conditions: (1) under anaerobic conditions in light with H_2 as electron acceptor and CO_2 as carbon source (photoautotrophic growth); (2) under anaerobic conditions in light with organic acids as carbon source (photoheterotrophic growth); (3) under aerobic conditions in the dark with H_2 as electron donor and CO_2 as carbon source (chemolithotrophic growth); (4) under aerobic conditions in the dark with organic compounds as carbon source (heterotrophic growth); (5) under anaerobic conditions in the dark with sugars (e.g., fructose) as carbon source (Schultz and Weaver, 1982); and (6) under anaerobic conditions with N_2O (McEwan et al., 1985a), nitrate (McEwan et al., 1982), trimethylamine *N*-oxide, or dimethylsulfoxide (Yen and Mars, 1977; Madigan et al., 1980; Schultz and Weaver, 1982; McEwan et al., 1983, 1985b) as hydrogen acceptors.

A number of bacteria that utilize reactions with a low $\Delta G_0'$ per mole H_2 (Table 1) are very specialized. They can utilize only a very limited number of electron acceptors and a limited number of carbon substrates.

B. Determination of Growth Parameters

Under growth conditions in which both substrate-level and electron transport-linked phosphorylation occur, it is difficult to establish the relation between ATP production and the amount of growth. This problem is illustrated by two equations derived by Stouthamer (1977b):

$$q_{ATP} = \frac{\mu}{Y_{ATP}} = \frac{\mu}{Y_{ATP}^{max}} + m_e \qquad (1)$$

$$q_{ATP} = \alpha(1 - \beta)\, q_{sub} + q_{el.ac.}\, P/2e^{-}.\gamma. \qquad (2)$$

In these equations q_{ATP} = specific rate of ATP production (mmol/g dry weight h^{-1}), μ = the specific growth rate (h^{-1}), Y_{ATP} = the growth yield per mole/ATP (g dry weight/mole ATP), Y_{ATP}^{max} = the molar growth yield per mole ATP after correction for maintenance energy, m_e = the maintenance coefficient (mmol ATP per g dry weight h^{-1}), α = the number of ATP molecules formed by substrate-level phosphorylation during the conversion of the substrate, β = the part of the substrate that is assimilated, q_{sub} and $q_{el.acc}$ are the specific rates of substrate or electron acceptor consumption, respectively, γ = the number electron pairs accepted by the electron acceptor, and $P/2e^-$ = the number of ATP molecules formed by electron transport-linked phosphorylation during the transfer of two electrons to the electron acceptor. It is evident from chemostat experiments that insufficient information is obtained to allow the determination of Y_{ATP}^{max}, m_e, and $P/2e^-$ with Equations (1) and (2). Consequently, either Y_{ATP}^{max} must be known to determine the $P/2e^-$ ratio or, conversely, the $P/2e^-$ ratio must be known to determine Y_{ATP}^{max}. A good knowledge of the bioenergetics of the organism studied is therefore very essential. In a large number of cases, however, there are large gaps in our knowledge that prevent definite conclusions about the magnitude of Y_{ATP}^{max}, m_e, and $P/2e^-$. This applies to most of the organisms listed in Table 3. Methods to overcome this difficulty are discussed in Section III. Because of the great importance of the bioenergetics of the utilization of electron acceptors, a general discussion of this subject is first given. Then the reduction of various hydrogen acceptors and its importance for the growth of bacteria are treated.

II. MECHANISM OF ATP GENERATION AND UTILIZATION

A. General Discussion

Great progress has been made in our understanding of the mechanism of energy generation under both aerobic (Stouthamer and Van Verseveld, 1985) and anaerobic conditions (Thauer and Morris, 1984). A scheme for the generation and utilization of ATP is given in Figure 1. It is generally accepted that transmembrane gradients play a central role in energy production and utilization. Electron transport to oxygen or alternative electron acceptors yields a protonmotive force ($\Delta\mu_{H^+}$). This is achieved by the vectorial translocation

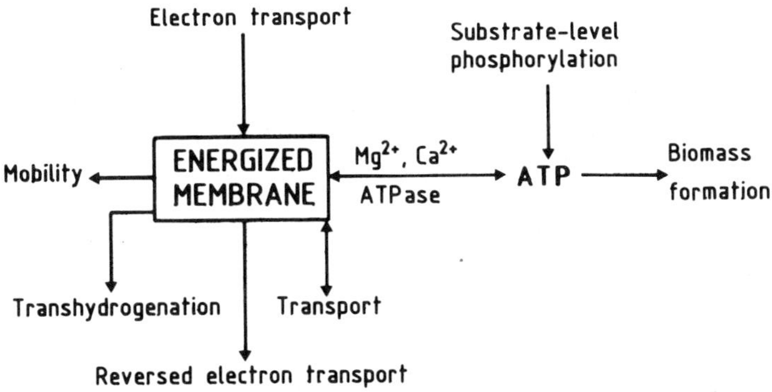

Figure 1 Scheme for the generation and utilization of ATP.

of protons coupled to electron transfer from oxidant to reductant along the respiratory chain. It has been concluded that proton translocation does not occur by the arrangement of respiratory carriers but is due to the presence of conformational proton pumps (Wikstrom and Krab, 1980; Kell et al., 1981). The complexity of the respiratory chain and the number of proton pumps is very different in different microorganisms. The proton-motive force is the driving force for ATP synthesis, transport processes, transhydrogenation, and reverse electron transfer.

The minimum amount of free energy to drive the synthesis of 1 mol ATP in vivo is approximately 44 kJ/mol (Thauer and Morris, 1984). Values reported for the phosphate potential in intact cells are *Paracoccus denitrificans*, -50.4 kJ/mol (Erecinska et al., 1978); *Streptococcus lactis*, -46.8 kJ/mol (Maloney, 1983), *Streptococcus cremoris*, -43.5 kJ/mol (Otto et al., 1985), and *E. coli*, -50.2 kJ/mol (Ugurbil et al., 1982). The $\Delta G_0'$ values of various reactions in Table 1 (reactions h through k) and 2 (reactions d and e) are lower than that required for the synthesis of 1 mol ATP. However, by conservation of free energy in the $\Delta\mu_{H^+}$, a mechanism is provided whose operation could result in a fractional yield of ATP. For the synthesis of ATP, three vectorial protons are required to drive the synthesis of one ATP molecule via the membrane-associated proton-translocating ATPase (Kashket, 1982, 1983; Maloney, 1983; Ferguson and Sorgato, 1982). Processes associated with the translocation of one or two charges yield 0.33 and 0.67 mol ATP, respectively. The idea that fractional stoichiometries of ATP formation are possible is of great importance to our understanding of energy generation associated with certain processes.

Another important step forward in our knowledge of the efficiency of oxidative phosphorylation was the observation that the number of protons translocated by the electron transport chain is different for the different coupling segments of the respiratory chain (Wikström and Krab, 1980; Van Verseveld et al. 1981; Boogerd et al., 1981; Stouthamer et al., 1982; Stouthamer and Van Verseveld, 1985). This implies that the classic sites of electron transport-linked phosphorylation have a different $P/2e^-$ ratio and that the traditional "one ATP per site" concept has been challenged (Stouthamer, 1985).

B. Influence of Enzyme Localization on Energy Generation

It has recently been realized that in an oxidative process the side of the membrane where the reactions take place is of great importance to the amount of energy generated. In its most general way this is illustrated in Figure 2 (Stouthamer, 1985). It is possible that the hydrogen donor donates its electrons at the periplasmic (AH_2) or at the cytoplasmic side (BH_2) of the membrane. Similarly, the hydrogen acceptor may accept electrons from the hydrogen donor at the periplasmic side (XO) or at the cytoplasmic side of the membrane (YO). AH_2 could be hydrogen or formate in *Campylobacter* (de Vries et al., 1982, 1984) or *Desulfovibrio* (Kröger et al., 1980; Kröger and Winkler, 1981). XO can be nitrite in *Campylobacter*

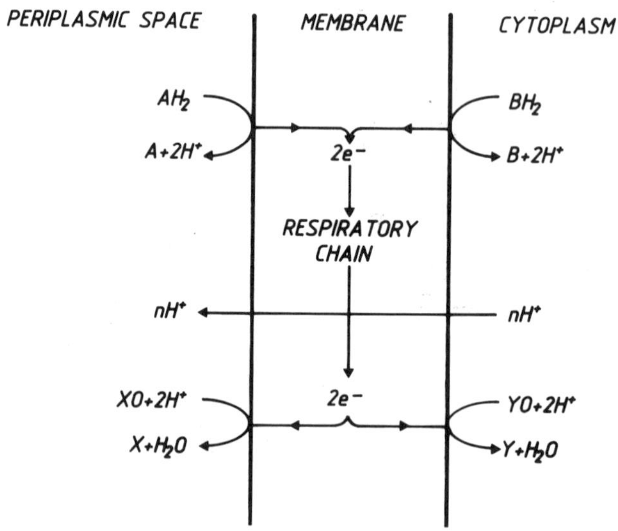

Figure 2 Influence of enzyme localization on the stoichiometry of ATP generation. (From Stouthamer, 1985.)

(de Vries et al., 1982), nitrite or nitrous oxide in *P. denitrificans* (Meijer et al., 1979; Boogerd et al., 1981; Alefounder and Ferguson, 1980; Alefounder et al., 1983); nitrate, nitrite, or nitrous oxide in *Rhodopseudomonas sphaeroides* f. *denitrificans* (Sawada and Satoh, 1980; Urata et al., 1982; Kundu and Nicholas, 1985); or trimethylamine *N*-oxide in *R. capsulata* (McEwan et al., 1985b) and *Alteromonas* sp. (Easter et al., 1983). YO can be fumarate in *Wolinella succinogenes* (Kröger et al., 1980) or oxygen or nitrate in *P. denitrificans* (John, 1977; Meijer et al., 1979; Boogerd et al., 1983a; Alefounder and Ferguson, 1980; Boogerd et al., 1981). It is remarkable that whereas nitrate in most bacteria is reduced at the cytoplasmic side of the membrane as in *E. coli* (Jones et al., 1980) and *P. denitrificans*, it is reduced at the periplasmic side of the membrane in *R. sphaeroides* f. *denitrificans*. Another remarkable feature is that a large number of electron acceptors are reduced at the periplasmic side of the membrane. In the situation of Figure 2 during electron transport, n protons are translocated. Thus the number of charges translocated over the membrane for electron transport from AH_2 to XO and YO and from BH_2 to XO and YO will be n, n + 2, n - 2, and n, respectively. The efficiency of oxidative phosphorylation is directly related to the charge separation. At this moment I must point out that in the literature there are a number of misunderstandings about the effects of enzyme localization on energy generation. In a number of cases it is wrongly considered that the two protons liberated at the periplasmic side of the membrane during the oxidation of AH_2 would be available for ATP synthesis. On the other hand in a number of cases it is wrongly assumed that the two protons consumed at the periplasmic side of the membrane during the reduction of YO would be withdrawn from the amount available for ATP synthesis. In both cases we deal with scalar protons, and only vectorial protons are able to drive ATP synthesis. These data indicate that the effect of enzyme localization on the amount of ATP generated is very large. The reduction of an electron acceptor at the periplasmic side of the membrane has a reducing effect on the amount of ATP generated. The location of the electron-donating site of hydrogenase and formate dehydrogenase at the periplasmic side of the membrane increases the amount of ATP from hydrogen and formate (which are generally available extracellularly). It may be expected that when the $\Delta G_0'$ per mole H_2 is low (Table 1, reactions h through k) no proton translocation occurs, but that due to enzyme localization a charge separation of 2 is reached.

C. Sodium Gradients as Driving Force for ATP Synthesis

A recent development is the realization that in addition to proton gradients in some cases sodium gradients are also of importance.

A number of membrane-associated, biotin-containing enzymes that catalyze the decarboxylation of a number of substrates are sodium dependent. Schink and Pfennig (1982) have isolated a strictly anaerobic sodium-requiring bacterium *Propionigenium modestum* that can grow on succinate as sole energy source by converting it to propionate and carbon dioxide.

$$\text{Succinate}^{2-} + H^+ \rightarrow \text{propionate}^- + HCO_3^-$$

The $\Delta G_0'$ of this reaction is -20.6 kJ/mol, which is not sufficient to allow substrate-level phosphorylation. In this reaction no electron transport takes place. The decarboxylation of methylmalonyl-CoA decarboxylase is doupled with Na^+ ion transport across the membrane. It was concluded that in *P. modestum* ATP synthesis is driven by a "sodium ion motive force" (Hilpert et al., 1984). Other anaerobes that require sodium ions (e.g., methanogens) could also benefit from the operation of similar sodium ion pumps.

D. Reversed Electron Flow

In Figure 1 is indicated that the $\Delta\mu^-{}_{H^+}$ can be utilized for reversed electron flow. Originally this was found to be highly important for the formation of NAD(P)H in many aerobic chemolithotrophic bacteria and anaerobic phototrophs (Aleem, 1977; Knaff, 1978). Recently it has been observed that it also occurs in a number of bacteria, which oxidize acetate with either sulfate or sulfur (Table 2, reactions c and d) as hydrogen acceptor. It has been demonstrated that in *Desulfuromonas acetoxidans* and *Desulfobacter postgatei* the oxidation of acetate proceeds by a modified citric acid cycle (Gebhardt et al., 1985; Brandis-Heep et al., 1983). In both cases the activation of acetate occurs by the reaction acetate + succinyl-CoA → acetyl-CoA + succinate (succinyl-CoA: acetate CoA transferase). Consequently no ATP seems to be obtained from substrate-level phosphorylation in these organisms. In both cases there is a problematic step in the citric acid cycle.

The redox potential under standard conditions (E_0') of the succinate-fumarate couple is +30 mV, whereas the E_0' of the S/H_2S couple is -270 mV and of the HSO_3^-/HS^- couple, -116 mV. This is further illustrated by the $\Delta G_0'$ values of the reactions

$$\text{Succinate}^{2-} + 1/4H^+ + 1/4SO_4^{2-} \rightarrow \text{fumarate}^{2-} + 1/4HS^- + H_2O$$

$$\Delta G_0' = 48.2 \text{ kJ per reaction}$$

$$\text{Succinate}^{2-} + S \rightarrow \text{fumarate}^{2-} + H_2S \qquad \Delta G_0' = 58.2 \text{ kJ per reaction}$$

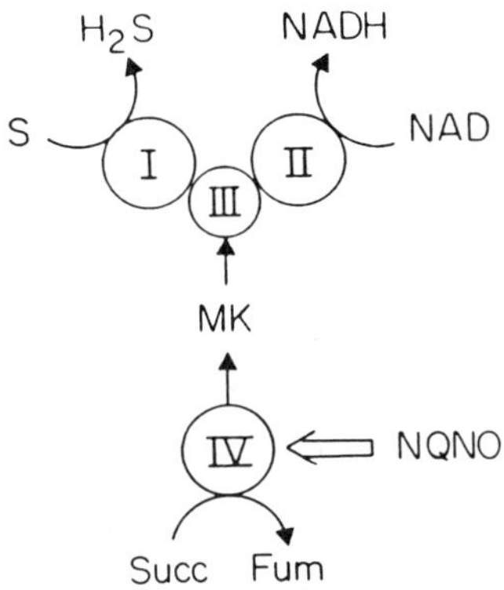

Figure 3 Hypothetical scheme for succinate oxidation in *Desulfuromonas acetoxidans*. (I) Sulfur reductase; (II) NADH dehydrogenase; (III) electron carrier connecting I and II; (IV) succinate dehydrogenase. MK = menaquinone NQNO = 2-(*n*-nonyl)-4-hydroxyquinoline-*N*-oxide. (Reproduced with permission from Paulsen et al., 1986.)

This problem could be solved in two ways: In the first place the $\Delta G'$ could be lower than the $\Delta G_0'$ when the cells would be able to maintain a low fumarate-succinate ratio. In *D. postgatei* malate is oxidized by a membrane-bound enzyme and the malate dehydrogenase reaction is not readily reversible (Brandis-Heep et al., 1983). So possibly this mechanism could be operative in *D. postgatei*. A second possibility would be energy-driven reversed electron transport to make the oxidation of succinate to fumarate possible with S or SO_4^{2-} as hydrogen acceptor. This reversed electron transport consumes part of the free energy conserved in dissimilation, and this amount is already very low as outlined before. For *D. acetoxidans* it has indeed been demonstrated that the oxidation of succinate is ATP driven and dependent on the presence of sulfur and NAD^+ (Paulsen et al., 1986). Menaquinone was shown to be involved in the electron transport chain from succinate to elemental sulfur or NAD. The scheme for succinate oxidation in the membrane of *D. acetoxidans* is shown in Figure 3. In *Desulfobacter hydrogenophilus* the oxidation of acetate most likely also occurs by the citric acid cycle (Schauder et al., 1986). In a number of other sulfate-reducing bacteria, however, the oxidation of acetate takes place by the acetyl-CoA pathway (Schauder et al., 1986). This is treated in Section VI.B.

E. Transport-Coupled Membrane Energization

Under normal conditions the $\Delta\mu_{H^+}$ is the driving force for the uptake of solutes (Fig. 1). Under anaerobic conditions a likely mechanism to generate an electrochemical potential could be the excretion of protons in symport with metabolic end products (Michels et al., 1979). These metabolic end products accumulate in the cell until the outwardly directed driving force supplied by the chemical product gradient exceeds the inwardly directed driving force supplied by the electrochemical gradient of protons. It has been calculated that the efflux of lactate produced by homolactic fermentation of glucose can increase the theoretical ATP yield by about 30% (Michels et al., 1979) to 2.6 mol/mol glucose fermented. When lactate accumulates in the medium, the energy yield obtained by lactate-proton efflux decreases rapidly. The energy gain from lactate efflux is also dependent on the extracellular pH. At low pH no energy can be gained by efflux of metabolites. From measurements of molar growth yields for lactose-limited chemostat cultures of *S. cremoris* in the presence of low and high external concentrations of lactate, it was concluded that the energy gain from lactate efflux amounted to 12% (Otto et al., 1980b). In mixed cultures the molar growth yield of *S. cremoris* on lactose was increased by lactate consumption by *Pseudomonas stutzeri* (Otto et al., 1980a). The lactate consumption of *P. stutzeri* under anaerobic conditions in the presence of nitrate kept the ATP gain of lactate efflux by *S. cremoris* at the maximal level. In *S. cremoris*, lactate efflux could potentiate the uptake of various solutes (Otto et al., 1980b, 1982). Therefore it seems that the generation of a $\Delta\mu_{H^+}$ by lactate efflux in *S. cremoris* is well documented. It is not yet clear, however, how large this contribution is to the establishment and maintenance of the $\Delta\mu_{H^+}$ in growing cells. Recently it has been suggested that energy could be gained by succinate efflux in *Selenomonas ruminantium* (Michel et al., 1987). It was demonstrated that the molar growth yield for glucose was proportional to the amount of succinate formed. It is not yet known whether the efflux of metabolites other than lactate and succinate can generate an electrical potential. Some of these products (formate, acetate, propionate, and butyrate) are weak acids, which can cross the membrane as undissociated acid.

III. METHODS FOR THE DEMONSTRATION OF ENERGY GENERATION DURING ELECTRON TRANSPORT TO HYDROGEN ACCEPTORS

A large number of methods are suitable for the demonstration of energy generation during electron transport to hydrogen acceptors. A number of these methods are based on the various possibilities for ATP generation and utilization outlined in Figure 1. In addition, the effect of the utilization of electron acceptors can be demonstrated by measuring molar growth yields. The following methods can be used.

1. Measurement of the proton-motive force. The $\Delta\mu_{H^+}$ has two components: a ΔpH (the difference between the intracellular and extracellular pH) and $\Delta\Psi$, the transmembrane difference in electrical potential. The ΔpH can be measured by studying the distribution of weak acids and bases; the $\Delta\Psi$ can be measured by studying the distribution of permeant ions. In addition, both components can be measured by the utilization of optical probes. The methods for these measurements and their results have recently been reviewed (Kashket, 1985).
2. Measurement of active transport. From Figure 1 it is clear that the $\Delta\mu_{H^+}$ is the driving force for the uptake of solutes in bacteria. This transport occurs against a concentration gradient. The formation of $\Delta\mu_{H^+}$ by electron transport to a hydrogen acceptor can therefore be indirectly demonstrated by measuring the uptake of a suitable solute.
3. Measurement of $P/2e^-$ ratios. The measurement of the $P/2e^-$ ratio can be performed with whole cells or with membrane vesicles. This method is the most direct for the demonstration of ATP formation by electron transport-linked phosphorylation coupled to the reduction of an electron acceptor. However in most cases a large underestimate of the true $P/2e^-$ ratio is obtained. This is understandable because electron transport generates the $\Delta\mu_{H^+}$ that drives ATP synthesis, and both the $\Delta\mu_{H^+}$ and ATP may be utilized for other purposes (Fig. 1). This method has been used in a large number of cases.
4. Measurement of H^+-electron acceptor ratios. Electron transfer to an electron acceptor leads to the extrusion of protons, which can easily be measured. In most cases a pulse of oxidant is added to an anaerobic suspension of microbial cells and the amount of protons is determined. The H^+-electron acceptor ratio measured must be corrected for the production or utilization of protons at the periplasmic side of the membrane (Fig. 2). In this way the $H^+/2e^-$ ratio can be determined, and this is a good measure for the stoichiometry of energy generation by electron transport-linked phosphorylation.
5. Measurement of molar growth yields. The amount of bacterial dry weight is directly proportional to the amount of ATP that can be obtained from the catabolism of the energy-yielding substrate (Bauchop and Elsden, 1960). The literature on the relation between growth and energy production has been reviewed several times (Stouthamer, 1979; Tempest and Neijssel, 1984).

In Section I.B it was mentioned that it is difficult to measure growth parameters for conditions in which both substrate-level and electron transport-linked phosphorylation occur. In a number of cases it is possible to exploit the versatility of the organism to determine these growth parameters. For that purpose we can compare various growth conditions; for example, for facultative organisms we can compare growth in the presence and absence of a hydrogen acceptor (Stouthamer, 1977b, 1980a, b). In another case we can compare growth with various hydrogen acceptors. The latter method has been applied to aerobic and anaerobic growth of *P. denitrificans* with nitrogenous oxides as hydrogen acceptor (Stouthamer et al., 1982; Van Verseveld et al., 1977) and various other bacteria (Stouthamer, 1988). It is essential that these measurements be made under conditions in which growth is limited by the energy supply. This can be performed most easily by studying growth in the chemostat with the carbon source as the growth-limiting nutrient. The carbon source used should not be too reduced, since growth may then be carbon limited (Stouthamer and Van Verseveld, 1985). Without detailed knowledge of the bioenergetics of the organism studied, only the magnitude of the stoichiometry of ATP generation can be obtained from the measurement of molar growth yields (Thauer and Morris, 1984).

In many cases the correlation between the data obtained by the various methods already outlined is not very good. However, it must be stressed that owing to the limitations of the various methods, good correlations can hardly be expected. Furthermore, methods 1-3 have only a qualitative value.

IV. UTILIZATION OF NITROGENOUS OXIDES

A. General Remarks

Many bacteria can utilize nitrate as the terminal electron acceptor. Nearly all these bacteria can also use oxygen. The process of nitrate reduction is generally called nitrate respiration or dissimilatory nitrate reduction. It generally occurs only under anaerobic conditions. The first step in nitrate reduction is a reduction to nitrite. In a number of organisms nitrite can also be used as the terminal hydrogen acceptor under anaerobic conditions. It has become clear during the last years that two different ways of dissimilatory nitrite reduction can be distinguished. In the first, nitrite can be converted to gaseous products, such as nitrogen or nitrous oxide, a process called denitrification. In addition, nitrite can be converted into ammonia in a dissimilatory process. The $\Delta G_0'$ values for the reduction of nitrate to N_2 and to NH_4^+ are -224.1 and -149.9 kJ/mol H_2 (reactions c and e, Table 1), respectively. However, the amount of ATP produced during denitrification is much larger than during the dissimilatory reduction of nitrate to ammonia than would be expected on basis of the $\Delta G_0'$ per mole H_2 values. The distribution of the capacity of denitrification has recently been reviewed (Betlach, 1982; Knowles, 1982; Payne, 1981). The properties of nitrate reductase, the composition of the electron transport chain toward nitrogenous oxides, and energy production during nitrate respiration have also been treated in a number of recent reviews (Stouthamer et al., 1980; Stouthamer, 1988, Ingledew and Poole, 1984).

B. Enzymology of Nitrate Respiration

1. Nitrate Reductase

Nitrate reductase is a membrane-bound enzyme. The different procedures used for the solubilization of nitrate reductase have led to the isolation of different nitrate reductase complexes. It is remarkable that nitrate reductases purified from a large number of organisms are very similar in subunit structure, molecular weight of the subunits, and other properties. Purified enzyme from *E. coli*, which reduces nitrate to nitrite with such artificial donors as reduced viologens, consists of only two polypeptides. These α and β subunits have molecular weights of about 150,000 and 60,000, and they are present in equimolar amounts. Per 200,000 daltons, *E. coli* nitrate reductase contains about 12 Fe-S groups and one atom of molybdenum (Forget, 1974; MacGregor et al., 1974; Lund and De Moss, 1976). Electron paramagnetic studies clearly show the participation of the Fe-S centers and the molybdenum in the nitrate-reducing activity of the enzyme (reviewed by Stouthamer et al., 1980). It has been demonstrated that the iron-sulfur clusters and the molybdenum are associated with the α subunit, which is therefore the catalytic subunit (Chaudhry and MacGregor, 1983a). The β subunit has been implicated in membrane binding (Lund and de Moss, 1976; MacGregor, 1975a). By various purification procedures a purified nitrate reductase is obtained, which contains three subunits: α, β, and γ. The γ subunit is a cytochrome b with a molecular weight of about 20,000 (Clegg, 1976; Enoch and Lester, 1975; MacGregor, 1975b). This cytochrome is essential to the functional association of the enzyme with physiological electron donor systems, such as formate dehydrogenase (Enoch and Lester, 1974). It is also required for the association of nitrate reductase with the cell membrane (MacGregor, 1975b). A new purification procedure for the isolation of a nitrate reductase complex as $\alpha_2\beta_2\gamma_4$ has recently been described, and this is supposed to be the native form of the enzyme (Chaudhry and MacGregor, 1983b). The molybdenum in nitrate reductase is associated with a pterin cofactor. A structure for this molybdopterin, which is a cofactor of many molybdoproteins, has been recently proposed (Johnson and Rajagopolan, 1982).

2. Nitrite Reductases Involved in Denitrification

Nitrite reductase has been purified from several bacteria and appears to be of two main types: a multiheme enzyme and a copper-containing enzyme. The first type contains two types of prosthetic groups: heme c, which is covalently linked to the protein, and a noncovalently bound chlorin-type heme d1 (Poole, 1983). Therefore this type of nitrite reductase is often referred to as cytochrome cd. The molecular weight of cytochrome cd is about 120,000, consisting of two identical subunits, each of which contains one c-type and one d1-type heme. The end product of nitrite reductase by purified enzymes is a mixture of NO and N_2O (for review see Stouthamer, 1988). This raises the question of whether NO is an obligatory intermediate in denitrification and whether a separate NO-reducing activity plays a role in the main line of the denitrification pathway. Schemes for the reduction of nitrite to N_2O in which NO is not an intermediate have been presented by Averill and Tiedje (1982) and by Stouthamer (1988). A copper-containing nitrite reductase is present in various organisms. The molecular weight of the enzyme, estimated by gel filtration, was 80,000. The enzyme contained two identical subunits and two copper atoms per molecule. The product of this type of nitrite reductase was in all cases NO. Both types of nitrite reductase also have cytochrome oxidase activity, although the K_m for O_2 is much larger than that for nitrite.

3. Nitrous Oxide Reductase

Nitrous oxide reductase is a very labile enzyme, and this has made purification very difficult. It has recently been shown that cells of an *Alcaligenes* sp. and of *Pseudomonas perfectomarina* were devoid of nitrous oxide reductase after growth in copper-deficient medium (Matsubara et al., 1982). This has led to the identification of nitrous oxide reductase as a copper protein (Matsubara and Zumft, 1982). The enzyme has been purified and characterized (Coyle et al., 1985). Earlier conflicting reports on the properties of the enzyme were shown to be due to effects of the presence of oxygen during purification (Zumft et al., 1985). The molecular weight of the enzyme is 120,000, and it contains about eight copper atoms (Coyle et al., 1985). The subunit structure was dimeric with two peptides of equal size. The nitrous oxide reductase of *R. sphaeroides* f. *denitrificans* also consists of subunits with a molecular weight of 60,000 (Michalski and Nicholas, 1984). Although the enzyme of *R. capsulata* had an apparent molecular weight of 76,000, it resembled that of *P. perfectomarina* in a number of properties (McEwan et al., 1985a).

4. *Nitrite Reductase Involved in Dissimilatory Nitrate Reduction to Ammonia*

Dissimilatory nitrate reduction to ammonia takes place in many more bacteria than thought previously. This reaction has been proposed as a short circuit in the nitrogen cycle (Cole and Brown, 1980). In *E. coli* there are two systems for the reduction of nitrite to ammonia. One utilizes NADH as electron donor (Coleman et al., 1978; R. H. Jackson et al., 1981); the other, formate (Abou-Jaoudé et al., 1977). The NADH-dependent enzyme is a soluble enzyme with a molecular weight of 190,000 consisting of two identical subunits. The enzyme contains one noncovalently bound FAD molecule and about five Fe and four acid-labile S atoms per subunit. The enzyme contains in addition siroheme. Consequently one Fe is thought to be associated with the siroheme and the others with a four Fe-S cluster (R. H. Jackson et al., 1981). A nitrite reductase containing 6c-type heme groups per molecule has been purified from *Desulfovibrio desulfuricans*. This is a membrane-bound enzyme. It has a minimal molecular weight of 66,000. NADH and NADPH do not function as direct electron donors for this enzyme. Reduced FAD can act as electron donor. The enzyme from *Achromobacter fischeri* contains only two c-type heme groups and can also utilize reduced flavins (Prakash and Sadana, 1972). The molecular weight is about 80,000 (Husain and Sadana, 1974).

C. Bioenergetics of Nitrate Reduction

1. *Respiratory Chain to Nitrogenous Oxides*

The composition of the respiratory chain to nitrogenous oxides has been reviewed recently (Stouthamer et al., 1980; Stouthamer, 1988). The discussion here is restricted to only a few organisms and to some general phenomena. The composition of the respiratory chain to nitrogenous oxides in a number of organisms is shown in Figure 4.

a. E. coli. An extensive review of the respiratory chain to various hydrogen acceptors in *E. coli* has appeared recently (Ingledew and Poole, 1984). The composition of the electron transport chain is highly dependent on the cultivation medium. c-Type cytochromes generally do not play an important role in electron transport in *E. coli*. It has been concluded that in whole cells NADH is the preferential electron donor for nitrate reduction (Stouthamer et al., 1980). The respiratory chain to oxygen and nitrate is shown in Figure 4a. Nitrate-grown cells mainly contain the cytochrome d oxidase pathway (Ingledew and Poole, 1984). With newer methods

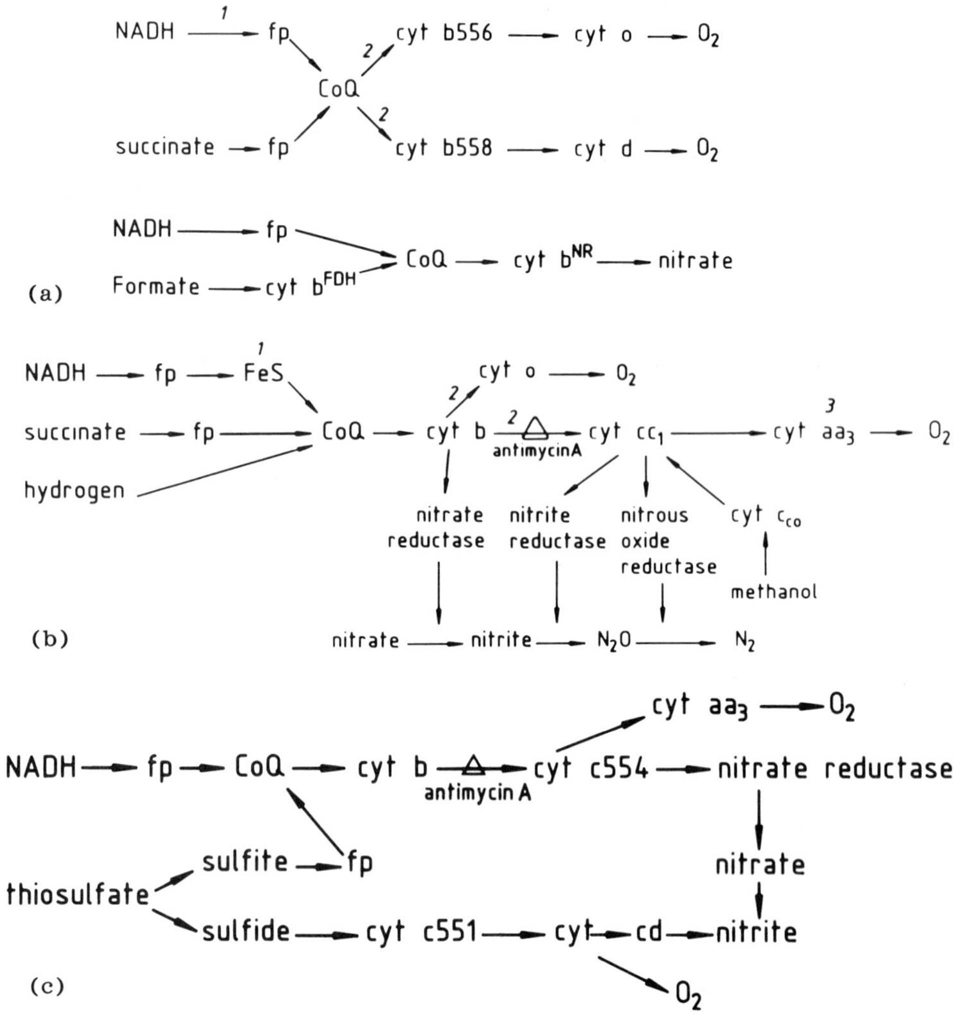

Figure 4 The respiratory chain of *Escherichia coli* (a), *Paracoccus denitrificans* (b), and *Thiobacillus denitrificans* (c) to oxygen, nitrate, nitrite, and nitrous oxide. Proton-translocating segments are indicated by 1, 2, and 3, respectively. For *P. denitrificans* and *T. denitrificans* the site of inhibition by antimycin A is indicated.

for the analysis of cytochromes, it has been demonstrated that the picture in Figure 4a is a simplification (Ingledew and Poole, 1984; Stouthamer, 1988). At least six b-type cytochromes were shown to be present in nitrate-grown cells (Hackett and Bragg, 1983). Several special cytochrome b components are associated with nitrate reductase and formate dehydrogenase. The cytochrome b^{NR} has midpoint potentials of +120 and +20 mV; the cytochrome b^{FDH}, -110mV (Hackett and Bragg, 1983). In contrast to the soluble NADH-nitrite reductase system, membrane-bound cytochromes are involved in the reduction of formate by nitrite (Abou-Jaoudé et al., 1979).

b. *P. denitrificans.* The respiratory chain of *P. denitrificans* to oxygen and nitrogenous oxides is shown in Figure 4b. The chain has a branched structure. There are two cytochrome oxidases: cytochrome aa_3 and most probably cytochrome o (Stouthamer, 1980c; Stouthamer et al., 1982; Van Verseveld and Stouthamer, 1978). The phosphorylation sites are indicated 1, 2, and 3 in Figure 4b. The functioning of these sites is strongly dependent on the growth conditions, which can be varied by continuous cultivation in various media. In cells grown with succinate or mannitol as the carbon and energy source, respiration occurs mainly through cytochrome o, and consequently during the oxidation of NADH two phosphorylation sites are passed. Electrons from methanol enter the cycle at the level of cytochrome c (Stouthamer, 1980c). In cells grown with methanol respiration occurs mainly through cytochrome aa_3, and consequently during the oxidation of NADH three phosphorylation sites are passed. Electron transport to nitrate involves cytochrome b (Lam and Nicholas, 1969). However, electron transport to nitrite (John and Whatley, 1975) and nitrous oxide (Matsubara, 1975; Boogerd et al., 1980) involves cytochrome c. In Figure 4b the respiratory chain is presented as a linear array of electron carriers. This picture suggests that energy production with oxygen, nitrite, and nitrous oxide is the same since the same number of phosphorylation sites is passed. However Figure 4b is misleading in this aspect since it does not take into account the side of the membrane where the reductions take place. This was discussed in Section II.B (Fig. 2).

c. *Thiobacillus denitrificans.* The respiratory chain of this organism to oxygen, nitrate, and nitrite has been studied intensively (Aminuddin and Nicholas, 1973, 1974; Sawnhey and Nicholas, 1977, 1978). This organism is an obligate chemolithotroph that can derive energy from the reaction $5S_2O_3^{2-} + 8NO_3^- + H_2O \rightarrow 10SO_4^{2-} + 4N_2 + 2H^+$. An unusual point in the respiratory chain of this organism (Fig. 4c) is that cytochrome c is involved in nitrate reduction, whereas nitrate reductase normally accepts electrons from cytochrome b (Stouthamer, 1976). There are separate chains for sulfide and sulfite, which both arise from thiosulfate (Fig. 4c). Electrons from sulfite are transferred to nitrate, whereas electrons from sulfide are transferred to nitrite (Sawnhey and Nicholas, 1978). Different c-type cytochromes are involved in these two branches of the respiratory chain.

d. Organisms with dissimilatory reduction of nitrate to ammonia. No information is available about the respiratory chain to nitrate in organisms that show dissimilatory nitrate reduction to ammonia. Cytochrome c is involved in nitrite reduction in *Campylobacter sputorum* (de Vries et al., 1982), *A. fischeri* (Prakash and Sadana, 1972), and *D. desulfuricans* (Steenkamp and Peck, 1981). The respiratory chains in these organisms are very primitive, however. In *Clostridium perfringens* (Hasan and Hall, 1975) and *Clostridium tertium* (Hasan and Hall, 1977), cytochromes are not involved in the reduction of nitrate and nitrite. In these organisms reduced ferredoxin is the electron donor for the reduction of nitrogenous oxides.

2. Stoichiometry of Proton Translocation

a. Enteric bacteria. For the oxidation of endogenous substrates (yielding mostly NADH) by *E. coli*, H^+/NO_3^- ratios of 4.0 and 4.2 have been measured (Brice et al., 1974). For the related *Klebsiella pneumoniae* the H^+/O and H^+/NO_3^- ratios were 3.92 and 3.85, respectively. These values are about the same, indicating that the efficiency of oxidative phosphorylation with oxygen and nitrate is about the same, which was concluded earlier for *Klebsiella aerogenes* from the measurement of molar growth yields (Hadjipetrou and Stouthamer, 1965; Stouthamer, 1967). It has been demonstrated that the oxidation of flavin-linked substrates, such as succinate, D-lactate, and glycerol-3-phosphate, yielded a H^+/O ratio of about 2, whereas the oxidation of a NAD^+-linked substrate, such as L-malate, produced a H^+/O close to 4 (Garland et al., 1975; Lawford and Haddock, 1973). On the basis of these results it can be concluded that in *E. coli* two sites of energy generation are present. The data available on *E. coli* can be represented as in Figure 5. In this scheme the electron transport chain is organized in redox loops. A redox loop is an alternating sequence of hydrogen and electron carriers. As an alternative for the redox loop system of Figure 4, a modification of a proton-motive Q cycle can be proposed (Mitchell, 1976). This is discussed and illustrated in a recent review (Stouthamer, 1988).

The Q cycle is a complex mechanism for the reduction of Q (ubiquinone) to QH_2 and its reoxidation. In all cases Q^- (ubisemiquinone anion) is an intermediate. During the reoxidation of QH_2 one electron is transferred to the cytochrome oxidase and one electron

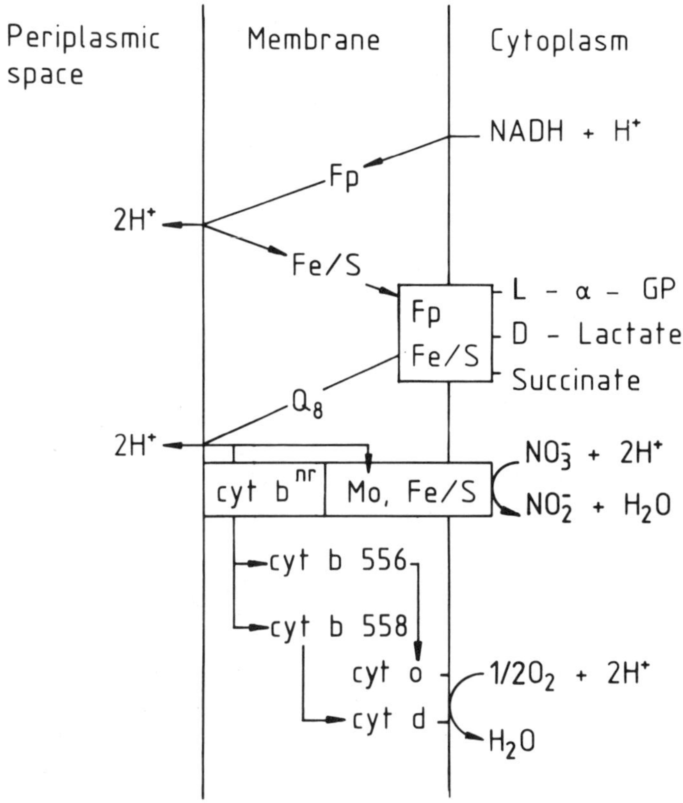

Figure 5 Hypothetical scheme for the functional organization of the redox carriers with oxygen and nitrate as terminal hydrogen acceptors in *Escherichia coli*. Fp, flavoprotein; Fe/S, iron-sulfur center; Q, ubiquinone, L-α-GP, L-α-glycerolphosphate. (The scheme is revised from Haddock and Jones, 1977; Stouthamer et al., 1980.)

is transferred to a b-type cytochrome. The latter electron is transferred to another b-type cytochrome and is used again in the reduction of Q^- to QH_2. There are various modifications of the Q cycle, and the exact mechanism is not yet known. In this chapter the b-cycle modification of Wikstrom and Krab (1980) is used. In this modification two protons are translocated across the membrane by the formation and reoxidation of QH_2. The b-type cytochromes involved in the cycle may function as a proton pump and translocate two protons per two electrons. However, this does not occur in *E. coli*. In *P. denitrificans*, however, the b-type cytochromes are supposed to act as a proton pump (see Sec. IV.C.2.b and Fig. 6).

The NADH-nitrite reaction is not associated with energy production (Cole and Brown, 1980). However, a membrane potential can be generated by the formate-nitrite reaction, since it led to the intracellular accumulation of the lipophilic cation butyltriphenylphosphonium (Motteram et al., 1981; Pope and Cole, 1982).

b. P. denitrificans. It has been demonstrated that two to three protons are extruded at site I and four protons at site II (Van Verseveld et al., 1981). In methanol-grown cells, which contain cytochrome aa_3, it was shown that this cytochrome oxidase acts as a proton pump (Van Verseveld et al., 1981). Electron transport between cytochrome c and oxygen was found to be associated with the extrusion of two protons. In earlier studies it had been demonstrated that iron-sulfur center 2 was essential for proton translocation at site I (Meijer et al., 1977b); possibly this compound acts as a proton pump. Consequently, in *P. denitrificans* there is a difference in the stoichiometry of proton translocation by the three sites of energy conservation (Van Verseveld et al., 1981).

Normally, thiocyanate is included in reaction mixtures for the study of proton translocation. However, thiocyanate inhibits nitrous oxide reductase and has an effect on nitrate transport (Boogerd et al., 1983a). Therefore thiocyanate had to be replaced by another agent able to collapse the membrane potential. Triphenylmethylphosphonium was used for that purpose (Boogerd et al., 1981, 1983b). The stoichiometries of proton translocation associated with electron transfer from various donors to oxygen and nitrogenous oxides are shown in Table 4. A scheme for proton translocation and electron transport is given in Figure 6. In this scheme protons for the reduction of nitrite and nitrous oxide are utilized from the periplasmic side of the membrane (Sec. II.B). In this scheme, two protons arise from ubiquinol (site IIa) and two protons are pumped by cytochrome b (site IIb). Electrons for nitrate reductase do not pass site IIb, since electron transfer to nitrate is not sensitive to inhibition by antimycin A (Boogerd et al., 1983a; Henry and Vignais, 1983). From this scheme it is possible to calculate the theoretical H^+/oxidant ratios, which are included in Table 4. In general there is good correspondence between the experimental values and those calculated from the scheme in Figure 6. The experimental H^+/oxidant ratios can be converted into $H^+/2e^-$ ratios by correcting for scalar proton consumption in the case of reduction of nitrite and nitrous oxide. The $H^+/2e^-$ values for the reduction of nitrite and nitrous oxide with endogenous substrate or H_2 are between 6 and 7 or about 4, respectively. It must be mentioned that H_2 donates its electrons at the cytoplasmic side of the membrane in *P. denitrificans*, in contrast to the situation in a number of other organisms mentioned in Section II.B. The $H^+/2e^-$ ratio for the reduction of nitrate is

Table 4 Respiration-Driven Proton Translocation with Endogenous Substrate (ES) or H_2 as Electron Donor and O_2 or Nitrogenous Oxides as Electron Acceptor in *P. denitrificans*[a]

Acceptor couple	Experimental H^+/oxidant ratio			Theoretical H^+/oxidant ratio (Fig. 6)	
	ES	ES	H_2	ES	H_2
O_2/H_2O	6.32	7.5	3.55, 3.92	6.0-7.0	4.0
NO_3^-/N_2		10.75	4.03	9.0-11.5	4.0
NO_2^-/N_2	5.79	5.55	2.87	5-6.5	2.0
NO_2^-/N_2O	3.37	3.3	1.68	3-4	1.0
N_2O/N_2	4.02	4.5	2.06	4-5	2.0

[a]The theoretical H^+/oxidant ratios are calculated according to Figure 6. For endogenous substrate (NADH), the calculation is made for a translocation of two or three protons at site I. For the H^+/O ratio with H_2, two values are given: the first was obtained with triphenylmethylphosphonium cation as permeant ion, the second with thiocyanate. *Source*: Data from Boogerd et al. (1981, 1983b) and Kristjansson et al. (1978).

lower than for the reduction of nitrite and nitrous oxide. The $H^+/2e^-$ ratios calculated from the experimental H^+/oxidant ratios correspond very well with those calculated from the scheme in Figure 6. The efficiency of oxidative phosphorylation is related to $q^+/2e^-$ (number of charges translocated across the membrane during flow of $2e^-$ to an acceptor), the charge separation. For aerobic electron transport the $q^+/2e^-$ for sites I, II, and III of oxidative phosphorylation are 2-3, 2, and 4, respectively (Van Verseveld et al., 1981). Based on Figure 6, the $q^+/2e^-$ values for electron transport from NADH, succinate, and H_2 to oxygen and nitrogenous oxides can be calculated. The results are given in Table 5. It is remarkable that although the electron transport chain to nitrate is different from that to nitrite and nitrous oxide, the charge separation is the same, which is due to the difference in location of the reductases. The efficiency of oxidative phosphorylation during electron transport from NADH to nitrogenous oxides is 67-71% of that during electron transport to oxygen.

Measurements of oxidative phosphorylation coupled to reduction of nitrate with NADH and succinate yielded $P/2e^-$ ratios of 0.9 and 0.06, respectively (John and Whatley, 1970). $P/2e^-$ ratios as high as 1.8 and 0.6 were measured for electron transport from NADH and succinate, respectively, to oxygen (John and Whatley,

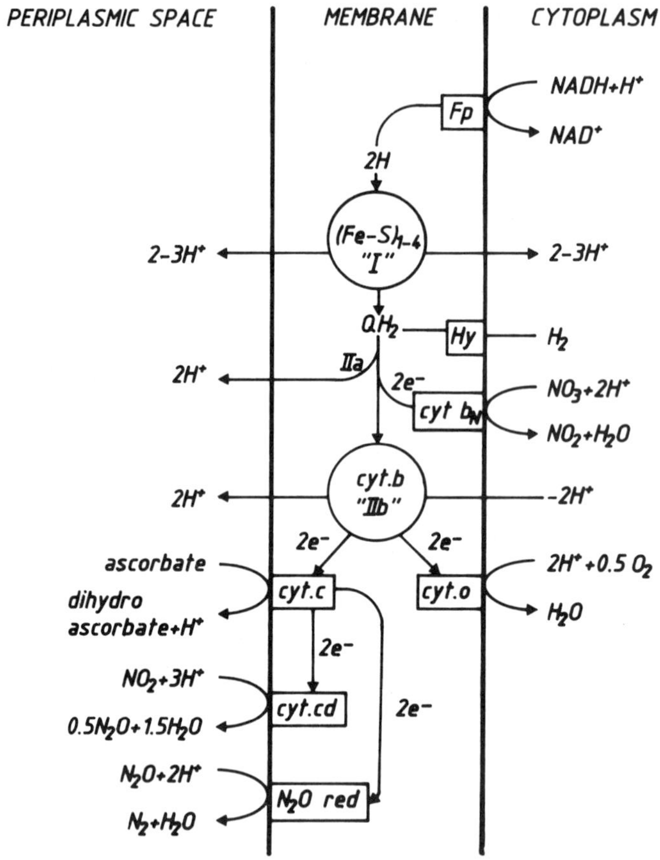

Figure 6 Anaerobic respiratory chain of *Paracoccus denitrificans* and its proton-translocating properties. Fp, flavoprotein; Fe-S, iron sulfur center; QH_2, ubiquinol; Hy, hydrogenase; cyt b_N, cytochrome b associated with nitrate reductase; cyt, cytochrome; N_2O red, nitrous oxide reductase. I, IIa, and IIb are the traditional sites of energy conservation. (Adapted from Boogerd et al., 1983b.)

1970). The observation that almost no ATP synthesis could be detected to accompany electron flow from succinate to nitrate is surprising. However, in recent experiments it was demonstrated that, in accordance with Figure 6 and Table 5, electron flow from succinate to nitrate generates a proton electrochemical gradient (Parsonage and Ferguson, 1983). In addition, a $P/2e^-$ ratio of 0.16 was measured. The scheme in Figure 6 and $q^+/2e^-$ ratios in Table 5 can be considered of general validity for denitrifying organisms.

Table 5 Number of Charges Translocated Across the Cytoplasmic Membrane During Flow of $2e^-$ ($q^+/2e^-$ ratio) from NADH, Succinate, or H_2 to an Electron Acceptor in *Paracoccus denitrificans*[a]

Electron acceptor reaction	$q^+/2e^-$	
	NADH	H_2 or succinate
$O_2 \rightarrow H_2$ (cyt c)	6-7	4
$O_2 \rightarrow H_2O$ (cyt aa_3)	8-9	6
$NO_3^- \rightarrow NO_2^-$	4-5	2
$NO_3^- \rightarrow 0.5N_2O$	4-5	2
$NO_3^- \rightarrow 0.5N_2$	4-5	2
$NO_2^- \rightarrow 0.5N_2O$	4-5	2
$NO_2^- \rightarrow 0.5N_2$	4-5	2
$N_2O \rightarrow N_2$	4-5	2

[a]Cyt aa_3 is not present in cells grown anaerobically.
Source: Data from Boogerd et al. (1981, 1983b) and Stouthamer et al. (1982).

c. C. sputorum subsp. *bubulus*. *C. sputorum* is a microaerophilic organism that grows optimally at reduced oxygen tension. However, it can also grow anaerobically with various hydrogen acceptors (Stouthamer et al., 1979). The same applies to other *Campylobacter* sp. (Table 3) (Laanbroek et al., 1978). The stoichiometries of proton translocation associated with electron transfer from various donors to oxygen and nitrogenous oxides are shown in Table 6. A scheme for proton translocation and electron transport is given in Figure 7. It is evident that the H^+/O ratios for this

Table 6 Respiration-Driven Proton Translocation with Various Substrates and O_2 or Nitrogenous Oxides as Electron Acceptor in *Campylobacter sputorum*[a]

Electron donor	Experimental H^+/oxidant ratio		Theoretical H^+/oxidant ratio		
	H^+/O	H^+/NO_2^-	H^+/O	H^+/NO_3^-	H^+/NO_2^-
Formate	3.51	2.79	3.8	7.2	3.4
Hydrogen	3.74	2.61	4.0	8.0	4.0
Lactate	2.15	-1.64	2.0	2.0	-2.0

[a]The theoretical values are calculated from Figure 7.
Source: Data from de Vries et al. (1980, 1982, 1984).

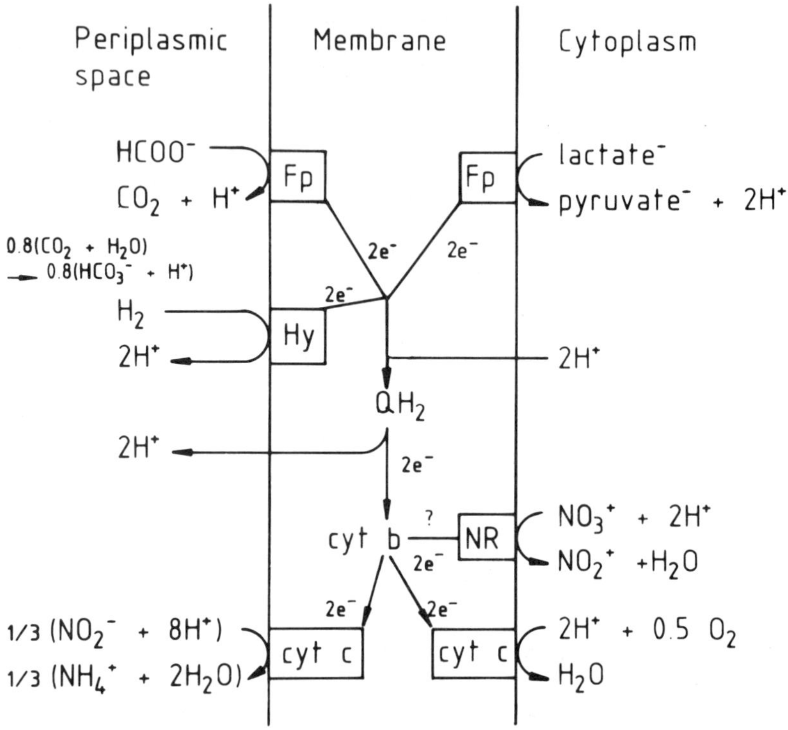

Figure 7 Scheme for proton translocation and electron transport to oxygen and nitrogenous oxides in *Campylobacter sputorum* subsp. *bubulus*. Fp, flavoprotein; QH_2, quinol; Hy, hydrogenase; NR, nitrate reductase. (Revised from de Vries et al., 1982.)

organism are much lower than those for *P. denitrificans*. Consequently it must be concluded that the respiratory chain in this organism is relatively simple, since the $H^+/2e^-$ ratio is only about 2. It has been suggested that a quinone is an intermediate in electron transfer. This quinone is supposed to be the only place in the respiratory chain of *C. sputorum* where two protons are translocated. Consequently proton pumps are supposed not to be present in *C. sputorum* (de Vries et al., 1982). There is a good correspondence between the theoretical H^+/O and H^+/NO_2^- ratios calculated from the scheme in Figure 7 and the experimental values, which can be regarded as strong evidence for the validity of the scheme. However, the theoretical H^+/NO_3^- ratios are much higher than the experimental values. The charge separation ($q^+/2e^-$ values) for electron transport

Table 7 Number of Charges Translocated Across the Cytoplasmic Membrane During the Flow of $2e^-$ ($q^+/2e^-$ ratio) from Various Substrates to an Electron Acceptor in *Campylobacter sputorum* subsp. *bubulus*

	$q^+/2e^-$ Electron acceptor reaction			
Substrate	O_2/H_2O	NO_3^-/NO_2^-	NO_3^-/NH_4^+	NO_2^-/NH_4^+
Formate	4	4	2.5	2
Hydrogen	4	4	2.5	2
Lactate	2	2	0.5	0

Source: Data from de Vries et al. (1980, 1982, 1984).

from various substrates to oxygen and nitrogenous oxides are shown in Table 7. The $q^+/2e^-$ value for the oxidation of formate and hydrogen by oxygen is 4, and notwithstanding the simple respiratory chain of this organism the $q^+/2e^-$ value is the same as that for hydrogen oxidation in *P. denitrificans* (Table 4). This is due to the donation of electrons by hydrogen at the periplasmic side of the membrane in *C. sputorum* (Sec. II.B). For electron transport from lactate to nitrate and nitrite, the $q^+/2e^-$ values are 2 and 0, respectively. Thus although the electron transport chains from formate and lactate to nitrite are similar, only the first is energy yielding and this is merely because formate and lactate donate their electrons to different sides of the membrane.

An important conclusion is that the $q^+/2e^-$ values for electron transport to nitrogenous oxides in *C. sputorum* are much lower than those in *P. denitrificans*. This is due to a difference in the complexity of the respiratory chain. Thus although the standard free energy changes of reduction of nitrate to nitrogen and reduction of nitrate to ammonia are about the same, the ATP yield for the first process is much higher. This is a general phenomenon, which is due to the fact that the respiratory systems associated with the reduction of nitrate to ammonia are less complicated and have no or fewer proton pumps than those in denitrifying organisms.

In recent experiments it was demonstrated that in cell-free extracts of *W. succinogenes* (this bacterium has many properties in common with *C. sputorum*) the reduction of nitrite by H_2 was associated with ATP formation (Bokranz et al., 1983). The observed $P/2e^-$ ratio was 0.11. Nitrite reduction by formate was not associated with ATP formation, however. From growth yields it was concluded that the ATP gain with nitrate or nitrite is about 0.5 mol ATP per mole formate or H_2.

d. Desulfovibrio sp. *Desulfovibrio* sp. are anaerobic, sulfate-reducing bacteria. It is very remarkable that some strains can reduce nitrate. This was published long ago (Senez and Pichinoty, 1955). Recently, similar strains were isolated and it was demonstrated that nitrate and sulfate could be used simultaneously as hydrogen acceptor (Keith and Herbert, 1983; Mitchell et al., 1986). However in some strains of *D. desulfuricans* the reduction of sulfate was suppressed in media containing nitrate plus sulfate or nitrite plus sulfate (Seitz and Cypionka, 1986). It must be remarked that after more extensive studies, it has been concluded that the ability to use nitrate as terminal electron acceptor is restricted to rather few species of sulfate-reducing bacteria (Mitchell et al., 1986; Seitz and Cypionka, 1986). Based on results described in Section II.A, a scheme for proton translocation during the reduction of nitrite by lactate or H_2 in *D. desulfuricans* is presented in Figure 8. In the electron transport chain from lactate and hydrogen two and one proton-translocating sites are assumed to be present, respectively (Steenkamp and Peck, 1981). Both for *D. desulfuricans* (Steenkamp and Peck, 1981) as for *Desulfovibrio gigas* (Barton et al., 1983), $H^+/2e^-$ ratios of about 2 were reported for electron

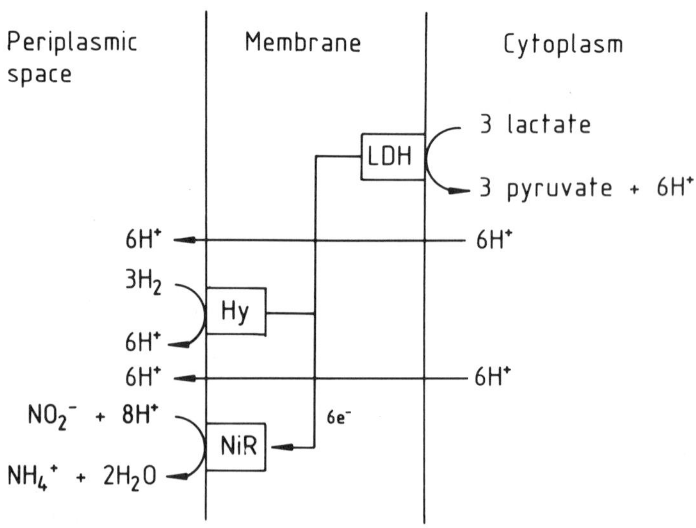

Figure 8 Scheme for proton translocation during the reduction of nitrite by lactate or H_2 in *Desulfovibrio desulfuricans*. LDH, lactate dehydrogenase; Hy, hydrogenase; NiR, nitrite reductase. (Adapted from Steenkamp and Peck, 1981.)

transport from hydrogen to nitrite. Membrane fractions of *D. gigas* formed ATP coupled to the reduction of nitrite by H_2 with a $P/2e^-$ ratio of 0.32 (Barton et al., 1983).

3. Nitrate Transport

In Section II.B it was mentioned that nitrate is reduced at the cytoplasmic side of the membrane. Therefore nitrate has to be imported inside the cell. The ability of the cell to discriminate between nitrate and chlorate (John, 1977) suggests the presence of a specific carrier system in the cytoplasmic membrane. For *P. denitrificans* it has been demonstrated that two systems are present for the entry of nitrate in the cell: an active uptake process driven by the proton-motive force and a nitrate-nitrite antiport system (Boogerd et al., 1983b; Stouthamer et al., 1982). These systems are illustrated in Figure 9. The first system is needed to start nitrate reduction. The uptake by this system is electrogenic, which implies that nitrate is taken up together with two or more protons. The existence of the second system has the advantage that after the reduction of nitrate has started its further uptake has no requirement for energy. The reduction of nitrite at the periplasmic side of the membrane (Sec. II.B) makes this system possible. During nitrate assimilation nitrite is further reduced to ammonia inside the cell. In that case the nitrate-nitrite antiport system cannot function. It has been demonstrated that in *Pseudomonas fluorescens* (Betlach et al., 1981) and *R. capsulata* (M. A. Jackson et al., 1981) the uptake of nitrate is driven by the proton-motive force. The consequence is that nitrate uptake for assimilatory nitrate reduction is always an energy-requiring process.

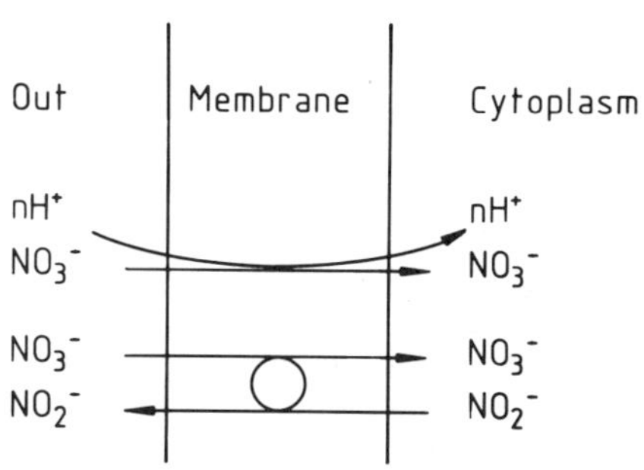

Figure 9 Transport systems for nitrate in *Paracoccus denitrificans*. (From Stouthamer et al., 1982.)

4. *Molar Growth Yields*

a. Enteric bacteria. Enteric bacteria can grow with various hydrogen acceptors (Sec. I.A, Table 3). Therefore it is possible to compare molar growth yields for growth with various hydrogen acceptors. A number of data are given in Table 8. All these data are for batch cultures. Information about growth yields in chemostat culture is very scarce. The growth of *Citrobacter freundii* has been studied in the chemostat with oxygen, nitrate, and nitrite as hydrogen acceptor and without hydrogen acceptor (Keevil et al., 1979). However, during growth with hydrogen acceptors pyruvate and succinate were produced in quantities that were almost equal to those for growth without hydrogen acceptor. In all cases the molar growth yield for a certain substrate and organism is higher when nitrate is present than when no hydrogen acceptor is available. On the other hand, the molar growth yields with oxygen as hydrogen acceptor are always higher than those with nitrate. With the enteric bacteria the increase in the molar growth yields in the presence of nitrate is due to two effects: (1) the occurrence of electron transport-linked phosphorylation during electron transfer to nitrate; and (2) the formation of more acetate and consequently a larger amount of ATP will be produced by substrate-level phosphorylation. The molar growth yield for aerobic growth is higher than for anaerobic growth in the presence of nitrate, since in the latter case the citric acid cycle does not function (Forget and Pichinoty, 1967; Wimpenny and Cole, 1967). It must be remarked, however, that this applies mostly to cultures growing with glucose. During growth of *E. coli* with lactate as carbon source and nitrate as hydrogen acceptor, part of the lactate is converted by the citric acid cycle (Yamamoto and Ishimoto, 1981). From the molar growth yield with glucose and mannitol for *K. aerogenes* for growth without hydrogen acceptor, Y_{ATP} values of 10.2 and 10.0, respectively, can be calculated (Hadjipetrou and Stouthamer, 1965). It is interesting that the molar growth yield for mannitol, which is more reduced than glucose, is higher than that for glucose for growth in the presence of a hydrogen acceptor.

For growth with glycerol and glucose the amount of substrate-level phosphorylation and electron transport-linked phosphorylation has been calculated for growth with various hydrogen acceptors (Table 9). For that purpose the dissimilation equations used are given in the heading of the table. From the H^+/O and H^+/NO_3^- ratios (about 4; Sec. IV.C.2) and the H^+/ATP ratio (3; Sec. II.A) a $P/2e^-$ ratio of 1.33 was calculated for electron transport from NADH to oxygen and nitrate and of 0.67 for electron transport from $FADH_2$. Molar growth yields were calculated using $Y_{ATP} = 10$ with the methods outlined later for *P. denitrificans*. The experimental

Table 8 Influence of Growth with Various Hydrogen Acceptors on Molar Growth Yields in Batch Cultures of Some Enteric Bacteria

Organism	Energy source	Hydrogen acceptor	Molar growth yield (g/mol)	Reference
Klebsiella aerogenes	Glucose	None	26.1	Hadjipetrou et al. (1963)
		Nitrate	45.5	
		Oxygen	72.7	
	Mannitol	None	21.8	Hadjipetrou and Stouthamer (1965)
		Nitrate	50.6	
		Oxygen	95.5	
Proteus mirabilis	Glucose	None	14.0	Stouthamer and Bettenhaussen (1972)
		Nitrate	30.1	
		Oxygen	58.1	
Escherichia coli	Formate	Nitrate	7.8	Ishimoto and Yamamoto (1977)
	Glucose	None	24	
		Nitrate	46	
		Oxygen	160	
Citrobacter freundii	Glucose	None	19.2	Keevil et al. (1979)
		Nitrate	33.0	
		Oxygen	35.5	

growth yield of *K. aerogenes* for aerobic growth with glucose is higher than the calculated value (Table 9). However, we must realize that during the aerobic growth of *K. aerogenes* with glucose about 0.4 mol acetate per mole glucose was formed, and this lowers the molar growth yield. The experimental growth yield of *K. aerogenes* for growth with nitrate as hydrogen acceptor and glucose was 45.5. This is between the values for conversion of nitrate to nitrite and that for conversion of nitrate to ammonia (Table 9). It was indeed concluded from the experiments that nitrate was partly converted to ammonia and that the reduction of nitrite to ammonia did not yield energy (Hadjipetrou and Stouthamer, 1965). If these points are taken into consideration, there is a good correspondence between experimental and calculated growth yields. From aerobic yield studies in the chemostat, $P/2e^-$ ratios have been determined of the same order of magnitude as those used in the calculations in Table 9.

It is necessary to point out that it is many times wrongly assumed that the molar growth yield is directly proportional to the amount of

ATP produced in the dissimilation reaction. This is incorrect for growth in minimal media in which the carbon source is used for both dissimilation and assimilation. The data in Table 9 clearly indicate that the molar growth yield is not a linear function from the ATP yield in the dissimilation reaction.

In Table 9 growth with glycerol and fumarate as hydrogen acceptor is included. This is discussed in Chapter V.

Table 9 ATP Production and Molar Growth Yields of Enteric Bacteria for Anaerobic Growth with Various Substrates and Hydrogen Acceptors[a]

Substrate	Hydrogen acceptor	Substrate-level phosphorylation (mol ATP per mole)	Electron transport-linked phosphorylation (mol ATP per mole)	Y (g/mol)
Glycerol	Fumarate	2	2.0	28.5
	Nitrate	2	3.32	33.7
	O_2	2	8.0	45.4
	None	3	—	25.0
Glucose	Nitrate (a)	4	5.32	57.5
	Nitrate (b)	4	1.33	40.0
	O_2	4	14.62	83.1

[a]The $P/2e^-$ ratio for electron transport from NADH to oxygen and nitrate was supposed to be 1.33; that from $FADH_2$ to oxygen and nitrate, 0.67. The $P/2e^-$ ratio for electron transport to fumarate was supposed to be 0.67 for both NADH and $FADH_2$. For growth with glycerol and glucose the dissimilation reactions were

$$1 \text{ glycerol} + 3 \text{ fumarate} \rightarrow 1 \text{ acetate} + 1CO_2 + 3 \text{ succinate}$$

$$1 \text{ glycerol} + 3NO_3^- \rightarrow 1 \text{ acetate} + 1CO_2 + 3NO_2^-$$

$$1 \text{ glucose} + 4NO_3^- \rightarrow 2 \text{ acetate} + 2CO_2 + 4NO_2^- \quad \text{(a)}$$

$$1 \text{ glucose} + NO_3^- \rightarrow 2 \text{ acetate} + 2CO_2 + NH_4^+ \quad \text{(b)}$$

Under aerobic conditions the substrates were completely converted to CO_2 In all reactions the formation of H_2O is neglected. Under anaerobic conditions the dissimilation reaction for glucose is 1 glucose → 1 ethanol + 1 acetate + 2 formate. Y_{ATP} was assumed to be 10. The molar growth yields were calculated using assimilation and dissimilation reactions as outlined in this section for *P. denitrificans*.

b. Denitrifying bacteria. From Figure 6 and Table 5 it was concluded that oxidative phosphorylation coupled to reduction of nitrate, nitrite, and nitrous oxide have similar efficiencies that are all lower than the efficiency of oxidative phosphorylation with oxygen. The molar growth yield of *Thiobacillus* A2 on glucose was 66.4 and 68.6 g/mol for growth with nitrate or nitrous oxide as hydrogen acceptor, respectively (Wood and Kelly, 1983). This indicates that the efficiency of oxidative phosphorylation is the same for electron transport to nitrate and nitrous oxide. In *Pseudomonas denitrificans* the Y_{max}^{el} values for growth with glutamate (molar growth yield per electron transferred to a hydrogen acceptor after correction for maintenance energy) were 7.7 and 4.5 for electron transport to oxygen and nitrate, respectively (Koike and Hattori, 1975). This indicates that the efficiency of oxidative phosphorylation of electron transport to nitrate is about 60% of that to oxygen. From the chemostat studies with *T. denitrificans* it has been concluded that 4-5 mol ATP and 6-7 mol ATP are produced per mole thiosulfate with nitrate and oxygen as hydrogen acceptor, respectively (Justin and Kelly, 1978). From the experiments with *T. denitrificans* (Justin and Kelly, 1978) and *Thiobacillus* A2 (Wood and Kelly, 1983), it was concluded that energy conservation anaerobically is about two-thirds the aerobic value. These conclusions are in accordance with the data on the $q^+/2e^-$ values in Table 5. The most extensive studies on molar growth yields with various hydrogen acceptors have been performed with *P. denitrificans* (Boogerd et al., 1984; Meijer et al., 1977a; Stouthamer et al., 1982; van Verseveld et al., 1977; van Verseveld et al., 1979). For these data it was even possible to give an exact calculation of the ATP production during growth (Boogerd et al., 1984). In this case the molar growth yields were related to the $q^+/2e^-$ values (Table 6). Hence the oxidation of NADH and succinate by oxygen will yield 2.33 and 1.33 mol ATP, respectively, whereas oxidation of these compounds by nitrogenous oxides yields 1.67 and 0.67 mol ATP, respectively. For the calculation of the ATP production from the molar growth yields shown in Table 10, use is made of assimilation and dissimilation reactions (Stouthamer, 1977a; van Verseveld et al., 1979). For example, the calculations will be given for growth with gluconate ($C_6H_{12}O_7$). The reactions for assimilation [Eq. (3)] and dissimilation [Eq. (4)] are

$$C_6H_{12}O_7 + 1.5NH_3 + 1.25(NADH + H^+) \rightarrow C_6H_{10.8}N_{1.5}O_{2.9} \text{ (cell material)} + 4.1H_2O + 1.25NAD^+ \quad (3)$$

$$C_6H_{12}O_7 + 9NAD^+ + 2FAD + 5H_2O \rightarrow 6CO_2 + 9(NADH + H^+) + 2FADH_2 \quad (4)$$

Table 10 Growth Parameters of *Paracoccus denitrificans* for Different Growth Conditions with Various Substrates as Carbon and Energy Source[a]

Growth condition	H-acc	Y_{sub}	q_{sub}	Y_{2e^-}	q_{2e^-}	v_{H2}	v_{ATP}	$P/2e^-$ theor.	$P/2e^-$ exp.
Gluconate	O_2	73.6	2.04	18.18	8.25	5.00	10.62	2.12	—
Gluconate, *sulfate*	O_2	57.0	2.63	12.70	11.81	6.35	8.45	1.33	1.33
Gluconate	NO_3^-	69.4	2.16	13.00	11.54	5.34	7.84	1.47	1.48
Mannitol	O_2	91.2	1.64	16.26	9.23	5.56	12.17	2.19	—
Mannitol	NO_3^-	74.8	2.00	13.46	11.14	6.90	10.52	1.53	1.72
Mannitol	NO_2^-	76.1	1.97	12.70	11.81	6.79	10.36	1.53	1.63
Succinate	O_2	35.1	4.27	12.70	11.81	4.14	8.17	1.97	—
Succinate, *sulfate*	O_2	29.8	5.03	8.54	17.56	4.57	6.08	1.33	1.28
Succinate	NO_2^-	37.9	3.96	9.30	16.13	3.91	5.10	1.30	1.46
Succinate	O_2	47.7	3.14	15.28	9.81	3.11	5.96	1.92	—
Succinate	NO_3^-	38.1	3.94	10.30	13.27	3.89	5.07	1.30	1.29
Succinate	NO_2^-	38.8	3.87	10.70	14.02	3.84	4.99	1.30	1.36

[a]The specific growth rate is in all cases 0.15 h^{-1}. The growth-limiting factor during growth in the chemostat is in italics. Y_{sub} = g dry weight per mol substrate; q_{sub} = specific rate of substrate consumption (mmol substrate per g per h); Y_{2e^-} = growth yield per electron pair (g dry weight/mol electron pair; q_{2e^-} = specific rate of electron transport to the hydrogen acceptor (mmol electron pair per g per h); v_{H2} = number of reduction equivalents/mole substrate; v_{ATP} = number of moles ATP formed by oxidative phosphorylation per mole substrate; Theoretical $P/2e^-$ = overall $P/2e^-$ ratio calculated with the equation v_{ATP}/v_{H2}; Experimental $P/2e^-$ ratio = overall $P/2e^-$ ratio calculated from the q_{ATP} for the preceding conditions for which no experimental $P/2e^-$ ratio is given.
Source: Data from Boogerd et al. (1984).

The molecular weight of cellular material is thus 150.2. The parts of gluconate that are assimilated and dissimilated are $Y_{gluc}/150.2$ and $(1 - Y_{gluc}/150.2)$, respectively. The amounts of reduction equivalents and ATP produced by oxidative phosphorylation are given by Equations (5) and (6):

$$v_{H2} = \left(1 - \frac{Y_{gluc}}{150.2}\right) 11 - \frac{1.25Y_{gluc}}{150.2} \tag{5}$$

$$v_{ATP} = 2\left(1 - \frac{Y_{gluc}}{150.2}\right) \frac{P}{2e^-} (FADH_2) + \left[9\left(1 - \frac{Y_{gluc}}{150.2}\right) - \frac{1.25Y_{gluc}}{150.2}\right] \frac{P}{2e^-} (NADH) \tag{6}$$

The overall $P/2e^-$ ratio can easily be calculated with Equation (7):

$$\frac{P}{2e^-} = \frac{v_{ATP}}{v_{H2}} \tag{7}$$

It must be emphasized that in these calculations ATP is used in the meaning "ATP equivalent," for the proton-motive force is not only indirectly employed in energy-consuming reactions as ATP, but is also used directly, for example, for transport processes. In these equations the $P/2e^-$ ratios calculated from the charge separations must be substituted. During sulfate-limited growth under aerobic conditions, phosphorylation site I is lost (Meijer et al., 1977b, 1979) and therefore in this case $P/2e^-$ (NADH) = $P/2e^-$ ($FADH_2$). In Table 10 the experimental Y_{sub} and Y_{2e^-} values are given. Since all data are for a specific growth rate of 0.15 h^{-1}, the corresponding q values can be calculated with the equations $q_{sub} = 150/Y_{sub}$ and $q_{2e^-} = 150/Y_{2e^-}$, respectively. The v_{H2} and v_{ATP} and the theoretical $P/2e^-$ ratio are calculated from the Y_{sub} values with Equations 5, 6, and 7 or similar equations for the other substrates. Now the specific rate of ATP production q_{ATP} can be calculated from q_{sub}, q_{2e^-}, and the theoretical $P/2e^-$ ratio with the aid of Equation (8):

$$q_{ATP} = q_{gluc}\left(1 - \frac{Y_{gluc}}{150.2}\right) 3 + q_{2e^-} \frac{P}{2e^-} \tag{8}$$

For aerobic growth with gluconate, a q_{ATP} of 20.61 mmol ATP g^{-1} h^{-1} can be calculated in this way. Consequently, Y_{ATP} for growth with gluconate at μ = 0.15 h^{-1} is 7.3. Now it is possible to postulate that the q_{ATP} for the other growth conditions with gluconate is equal to the q_{ATP} for aerobic gluconate-limited growth, since under these conditions biomass is formed from the same carbon source, gluconate, and at the same rate. From the q_{ATP} for aerobic gluconate-limited growth and the q_{sub} and q_{2e^-} for the other growth conditions it is possible to calculate the P/2e⁻ (called experimental P/2e⁻ ratio) for those conditions. Then the experimental and the theoretical P/2e⁻ ratios can be compared. It is evident from the results in Table 10 that in all cases there is a very good correspondence between the experimental and the theoretical P/2e⁻ ratios. This gives strong support to the suppositions made and to the validity of the scheme in Figure 6. It also indicates that the uptake of nitrate under physiological conditions fully takes place by the nitrate-nitrite antiport system and does not require energy (Sec. IV.C.3). In a number of cases the material and redox balances do not fit very well, which is due to the observation that in some cases 10-15% of the carbon source was used for the production of siderochromes (van Verseveld et al., 1979). This is also the reason that in Table 10 two values are given for aerobic succinate-limited growth. Accumulation of siderochromes does not inferfere with the calculations given above and the results collected in Table 10.

c. C. sputorum. The molar growth yields with various substrates and hydrogen acceptors are given in Table 11. During anaerobic growth with lactate and fumarate, no oxidative phosphorylation occurs and energy generation occurs completely by substrate-level phosphorylation associated with acetate formation (de Vries et al., 1980). Consequently Y_{ATP} is about 16.5. From the growth yields a P/O ratio of 0.63 was calculated and a P/2e⁻ ratio for the reaction nitrate → nitrite of 0.30. In accordance with the scheme in Figure 7 and the results in Table 7, the reduction of nitrite by lactate was concluded not to be associated with oxidative phosphorylation. The observation that with formate the growth yields with nitrate are higher than with nitrite is also in accordance with the scheme in Figure 7 and the results in Table 7.

Table 11 Molar Growth Yields of *Campylobacter sputorum* subsp. *bubulus* Growing with Lactate and Formate in Continuous Culture with Growth-Limiting Amounts of the Hydrogen Acceptor

Hydrogen acceptor	Molar growth yield (g/mol) Lactate	Formate
O_2	42.2	5.9
Fumarate	16.2	3.8
Nitrate	20.2	3.9-4.3
Nitrite	18.0	3.7

Source: Data from de Vries et al. (1980, 1982).

d. Anaerobic bacteria forming propionate or succinate. Anaerobic bacteria forming propionate or succinate contain b-type cytochromes. These cytochromes are involved in fumarate reduction, and this aspect is discussed in more detail in Section V. Many of these bacteria have the capacity to reduce nitrate. This has been demonstrated in *Veillonella alkalescens* (Inderlied and Delwiche, 1973; de Vries et al., 1974), *S. ruminantium* (de Vries et al., 1974), anaerobic actinomycetes (Slack, 1974), and *Propionibacterium pentosaceum* (= *P. acidi propionici*) (van Gent-Ruyters et al., 1975; Kaneko and Ishimoto, 1978). *V. alkalescens* is able to reduce nitrate to ammonia (Inderlied and Delwiche, 1973). It is interesting that *P. pentosaceum* is able to denitrify (van Gent-Ruyters et al., 1975). The end product of denitrification is N_2O, however (Kaspar, 1982), and this has been described as a detoxification mechanism to avoid the accumulation of nitrite. The influence of nitrate on molar growth yields is shown in Table 12. In all cases studied the molar growth yield was strongly increased by the presence of nitrate. In the absence of nitrate electron transport-linked phosphorylation occurs during electron transfer to fumarate (see Sec. V). However, in the presence of nitrate the number of electrons that can be transferred to the ultimate hydrogen acceptor is much larger than when fumarate is the hydrogen acceptor. For *P. pentosaceum* the formation of propionate is strongly repressed in the presence of nitrate (van Gent-Ruyters et al., 1975). The growth curve for anaerobic cultures growing in the presence of nitrate could be differentiated in two phases. In the first period a large part of the lactate was converted into pyruvate. In the second part pyruvate and acetate were utilized. At the end of the experiment in which growth still occurred the following fermentation balances were found: 1 lactate → 0.33 acetate + $2.3CO_2$ and 1 glycerol → 0.4 acetate + $2.2CO_2$. Under these conditions the molar growth yields gradually increase upon longer incubation (indicated by ^{c} in Table 12). It is evident that under these circumstances the citric acid cycle, which was detected before in *P. pentosaceum* (Delwiche and Carbon, 1953), functions. The molar growth yields for the growth of *P. pentosaceum* with glycerol and nitrate are even higher than those calculated for the aerobic growth of enteric bacteria with glycerol (Table 9). It is highly unlikely that the $P/2e^-$ for electron transport to nitrate in *P. pentosaceum* is higher than in enteric bacteria. On the contrary, these anaerobic organisms have very simple respiratory chains. That nonetheless the molar growth yields for these organisms are very high is due to two factors. (1) These anaerobic bacteria are grown in complex medium. For many of those organisms high Y_{ATP} values have been found (Stouthamer, 1979). If we accept a $P/2e^-$ ratio of 0.67 for electron transport-linked phosphorylation for electron transfer to

Table 12 Influence of Nitrate on Molar Growth Yields of Various Anaerobic Bacteria Producing Propionate or Succinate as Fermentation Product[a]

Organism	Substrate	present	Molar growth yield (g/mol)	Reference
Propionibacterium pentosaceum	Lactate	−	12.9	van Gent-Ruyters et al. (1975)
		+	22.0-52[a]	
	Glycerol	−	26.3	
		+	62-81[a]	
Veillonella alkalescens	Lactate	−	8.6; 12.5	de Vries et al. (1977);
		+	19.8-22.9; 22.0	Ruoff and Delwiche (1977)
	Pyruvate	−	Poor growth	de Vries et al. (1977)
		+	25.5	
	Citrate	−	19.3	de Vries et al. (1977)
		+	25.3	de Vries et al. (1977)
Propionibacterium acidi-propionici	Lactate	−	5.3	Kaneko and Ishimoto (1978)
		+	10.8	
	Glucose	−	54.2	
		+	53.9	

[a]With *P. pentosaceum* in the presence of nitrate the molar growth yields were dependent on the incubation time (see text).

fumarate (see Table 9 and Chap. 5), Y_{ATP} values of 14.6 and 15.8 can be calculated for the anaerobic growth of *P. pentosaceum* without hydrogen acceptor and with lactate and glycerol as energy source. For the growth of *V. alkalescens* with citrate without hydrogen acceptor, an Y_{ATP} value of 15.1 can be calculated. Under these circumstances the whole ATP production is by substrate-level phosphorylation (de Vries et al., 1977). (2) Owing to growth in a complex medium, the carbon source can be utilized completely for energy generation and no part or only a very small part of the carbon source is used for assimilation. The growth yields for *P. acidi propionici* for lactate (Kaneko and Ishimoto, 1978) are much smaller than those for *P. pentosaceum*. In this organism the citric acid cycle probably does not function under anaerobic conditions with nitrate.

e. Sulfate-reducing bacteria. Recently, growth yields were reported for the growth of *D. desulfuricans* with H_2 as electron donor, acetate as carbon source, and nitrate, nitrite, or sulfate as hydrogen acceptor (Seitz and Cypionka, 1986). The growth yields per mole electron acceptor at the maximum specific growth rate (between 0.12 and 0.19 h^{-1} for the various growth conditions) were 13.1, 8.8, and 9.7 g cell dry weight per mole nitrate, nitrite, and sulfate reduced, respectively. The ratio between the yields with nitrate and nitrite is approximately 4:3, which is similar to the ratio of H_2 molecules consumed per mole nitrate and nitrite, respectively. This indicates that the ATP production by electron transport-linked phosphorylation during electron transfer to nitrate and nitrite is the same. The yield for growth with sulfate (that just as nitrate is reduced with the consumption of 4 mol H_2; Table 1) is less than that with nitrate but larger than that with nitrite. However we must consider that the activation of sulfate in *Desulfovibrio* sp. requires 2 mol ATP (see Sec. VI.B.1). If this point is taken into account it may be concluded that the stoichiometry of electron transport-linked phosphorylation with nitrate, nitrite, and sulfate is about the same. In Table 1 the $\Delta G_0'$ per mole H_2 values for the reduction of nitrate and sulfate (reactions e and h, respectively) are given. There is a large difference between these values. The conclusion of Seitz and Cypionka (1986) from their results was that the thermodynamic efficiency of growth of *D. desulfuricans* coupled to ammonification of nitrate is far lower than of growth coupled to sulfate reduction. In Section I.A it was mentioned that although the $\Delta G_0'$ per mole H_2 for the utilization of various hydrogen acceptors greatly varies (Table 1), this does not mean that the ATP production in these reactions is also different, since this is determined by the electron transport chain of the organism. The results with *D. desulfuricans* give a nice illustration of this point.

V. UTILIZATION OF FUMARATE

A. General Remarks

Many bacteria can utilize fumarate as terminal electron acceptor. The properties of fumarate reductase, the composition of the respiratory chain to fumarate, and energy production during fumarate respiration have been reviewed recently (Kröger, 1980; Stouthamer, 1980b; Cole et al., 1985; Ingledew and Poole, 1984). All bacteria that reduce fumarate contain menaquinone in the respiratory chain (Ingledew and Poole, 1984; Cole et al., 1985). Organisms that contain only ubiquinone can never reduce fumarate. Among the bacteria that reduce fumarate, two groups can be distinguished: (1) bacteria able to form fumarate from other carbon sources are anaerobic or facultative and form succinate or propionate as fermentation product; in these bacteria the reduction of fumarate is a step that leads to the final fermentation products; and (2) bacteria that cannot form fumarate themselves but are dependent on its presence in the medium.

Most bacteria form propionate from succinate. However some exceptions are known: *Clostridium propionicum*, *Megasphaera elsdenii*, and *Bacteroides ruminicola*, which ferment lactate to propionate with acrylate as intermediate.

B. Enzymology of Fumarate Respiration

Fumarate is reduced by fumarate reductase. This enzyme has been studied in great detail from *E. coli* (Cole et al., 1985) and from *W. succinogenes* (Unden and Kröger, 1986). The enzyme of *E. coli* contains four subunits with molecular weights of 66,052 (602 amino acid residues), 27,082 (244 amino acid residues), 15,000, and 13,000 (Ingledew and Poole, 1984; Cole et al., 1985). The latter two subunits were detected only after purification from strains with amplified levels of the enzyme. The fumarate reductase genes are arranged as an operon, which consists of a promoter-operator region, four cistrons (*frd* A, -B, -C, and -D), and a transcriptional terminator. The whole *frd* operon has been cloned into multicopy plasmids. Expression of the cloned *frd* operon results in amplified levels of enzyme activity. The complete nucleotide sequence of this operon has been determined, and the DNA sequence has been used to predict the primary structure of the subunits. The largest subunit contains the FAD-binding site. The subunit with the molecular weight of 27,082 contains two iron-sulfur centers, which are involved in electron transfer to the FAD in the large subunit. The two smallest subunits are membrane anchors: these two polypeptides are highly hydrophobic. When those subunits are not made, large amounts of soluble enzyme are produced. In *E. coli*, the succinate-fumarate couple may be utilized as either oxidant or reductant. These two reactions are catalyzed by two different enzymes, succinate dehydrogenase and fumarate reductase. The enzyme of *E. coli* does not contain cytochrome b, in contrast to the enzyme of *W. succinogenes*. The latter enzyme consists of three different subunits with molecular weights of 79,000, 31,000, and 25,000 (Unden and Kröger, 1986). These subunits are present in the ratio 1:1:2. The enzyme contains five prosthetic groups: FAD (bound to the largest subunit), a (4Fe-4S) cluster, a (2Fe-2S) cluster, a cytochrome b with a midpoint potential of -20 mV, and a cytochrome b with a midpoint potential of -200 mV. The (4Fe-4S) cluster is located on the largest subunit; the (2Fe-2S) cluster subunit with the 31,000 molecular weight subunit. The smallest subunit is the cytochrome b. For fumarate reduction by reduced viologens only the largest subunit is required, which thus contains the active site of the enzyme.

C. Bioenergetics of Fumarate Reduction

1. *Respiratory Chain to Fumarate*

It has been mentioned before that menaquinone is always involved in fumarate reduction (Thauer et al., 1977; Stouthamer, 1980b). In addition cytochrome b is involved. In the anaerobic bacteria that form succinate or propionate as a fermentation product, cytochrome b is always present (Stouthamer, 1980b). In sulfate-reducing bacteria several c-type cytochromes are present (Odom and Peck, 1984). By cultivation of *D. gigas* with fumarate the proportion of cytochrome b to that of cytochrome c was increased (Hatchikian and LeGall, 1972). In *D. acetoxidans* b-type cytochromes were formed only after growth with ethanol and malate (Pfennig and Biebl, 1976).

The involvement of cytochromes in fumarate reduction has been demonstrated nicely in *Bacteroides fragilis*. This organism is dependent on the presence of hemin in the growth medium for the formation of cytochromes. The influence of the presence of hemin in minimal medium on fermentation balance, generation time, and molar growth yield for glucose is given in Table 13 (Macy et al., 1975). During the slow growth in the absence of hemin no succinate or propionate is formed; instead a large amount of fumarate accumulates. Evidently in the absence of hemin in the growth medium, when no cytochromes can be formed, the *B. fragilis* cells are unable to reduce fumarate to succinate. Growth in the presence of hemin is much faster, and in this case appreciable amounts of succinate and propionate are formed. At the same time the molar growth yield is much higher, indicating that fumarate reduction is indeed associated with additional ATP formation. In *W. succinogenes* the following respiratory chain for electron transfer from formate to fumarate has been proposed

Table 13 Influence of the Presence of Hemin in Minimal Medium on the Fermentation Balance, Generation Time, and Molar Growth Yield for *Bacteroides fragilis* Growing with Glucose

Presence of hemin (+ present, - absent)	Fermentation product (mol/mol glucose)	Generation time (h)	Molar growth yield (g/mol)
-	0.7 lactate + 0.3 acetate + 0.3 formate + 1 fumarate	8	18.1
+	0.4 acetate + 0.4 succinate + 0.2 propionate + 0.4 formate	2	47.0

Source: Data from Macy et al. (1975).

(Kröger, 1980): formate → formate dehydrogenase → cytochrome b (midpoint potential, -200 mV) → menaquinone → cytochrome b (midpoint potential, -20 mV) → fumarate reductase → fumarate. This whole system consists of six proteins, and the purified components have been reconstituted into proteoliposomes (Unden and Kröger, 1986).

It has been demonstrated by using heme-deficient mutants that in *E. coli* fumarate reduction is possible when no cytochromes are present (Singh and Bragg, 1975). However, the rates of fumarate reduction were much lower and the growth rate was strongly decreased. It has been concluded that in the absence of cytochrome fumarate reduction functions solely as an electron sink to allow the reoxidation of reduced pyridine nucleotides and the synthesis of a number of nutrients. It has been demonstrated that fumarate reduction in *E. coli* has an important nutritive function; it acts as a hydrogen acceptor in oxidative steps in the biosynthesis of pyrimidines (Newton et al., 1971) and heme (Jacobs and Jacobs, 1975). Also, for this nutritive function of fumarate reduction cytochromes are not necessary. However, energy production by fumarate reduction is possible only when cytochromes are present.

Bacteroides amylophilus is an exception, since this organism does not contain cytochromes and still seems to obtain energy from fumarate reduction (Wetzstein and Gottschalk, 1985; Wetzstein et al., 1987). For this organism the growth yields are among the highest recorded for bacteria growing fermentatively under anaerobic conditions (Hobson and Summers, 1967; Jenkinson and Woodbine, 1979). It was demonstrated that the NADH-fumarate reductase reaction is associated with proton translocation, $1H^+$ per NADH oxidized

(Wetzstein et al., 1987). The efflux of fermentation products did not contribute to the $\Delta\mu_{H^+}$. The generation of a $\Delta\psi$ was dependent on Na^+. However, Na^+ is not directly involved in the generation of a $\Delta\psi$. It is suggested that a large Na^+ gradient is maintained by an antiporter-catalyzed exchange of $2H^+$ per Na^+. The Na^+ concentration gradient is suggested to be involved in the uptake of the carbon source (Wetzstein et al., 1987). It seems that the biochemical aspects of energy generation in this organism are completely different from those in other bacteria forming propionate or succinate.

2. *Stoichiometry of Proton Translocation*

For *E. coli*, the H^+/fumarate ratio has been measured by a number of authors using various methods and various substrates. The results indicate a H^+/fumarate ratio of about 2 (Ingledew and Poole, 1984). The *W. succinogenes* electron transfer from formate to fumarate is associated with the liberation of two protons on the outside of the membrane (Kröger, 1977). This is a very low value since formate donates its electrons at the periplasmic side of the membrane (see Sec. II.B). This indicates that most likely no additional proton translocation by menaquinone occurs in this electron transport system. We may conclude that the charge separation in the formate-fumarate reaction of *W. succinogenes* is 2.

3. *Molar Growth Yields*

a. Enteric bacteria. Growth of *E. coli* (Bernhard and Gottschalk, 1978) and *C. freundii* (Lütgens and Gottschalk, 1982) has been studied with hydrogen or formate as electron donor and fumarate as hydrogen acceptor under anaerobic conditions. These experiments were performed in complex medium. A molar growth yield of 4.8 g cells per mole fumarate was found for *E. coli* growing with H_2 and fumarate. From the data a $P/2e^-$ ratio of 0.43 was calculated. For growth of *C. freundii* in the chemostat a Y_{fum}^{max} value of 10.5 was found and an ATP yield of 0.6 per mole fumarate reduced was calculated. These values are in reasonable agreement with the expected value of 0.67 based on the charge separation and a H^+/ATP ratio of 3. It is for these reasons that this $P/2e^-$ ratio has been used for the calculation of theoretical molar growth yield for growth with glycerol and fumarate (Table 9). There is no experimental material available for comparison with this theoretical growth yield. Molar growth yields for growth with various substrates have been determined for *Proteus rettgeri* (Kröger, 1974). Molar growth yields and fermentation balances are presented in Table 14. From growth with pyruvate a Y_{ATP} value of 6.3 could be calculated. Originally Y_{ATP} values were calculated using a $P/2e^-$ ratio of 1. These data

Table 14 Molar Growth Yields and Fermentation Balances for Growth of *Proteus rettgeri* with Various Substrates[a]

Fermentation reaction	Y (g/mol)	Y_{ATP} ($P/2e^- = 1$)	Y_{ATP} ($P/2e^- = 0.67$)
1 pyruvate → 1 acetate + 1 formate	6.3	6.3	6.3
3 glucose + $2HCO_3^-$ → 2 succinate + 3 acetate + ethanol + 4 formate	24	8	8.6
2 citrate → 2 succinate + acetate + $2HCO_3^-$	7.8	7.8	9.3
7 fumarate → 6 succinate + $4HCO_3^-$	4.7	5.5	7.6

[a]Y_{ATP} values have been calculated with $P/2e^-$ ratios of 1 and 0.67, respectively for electron transfer to fumarate.
Source: Data from Kröger (1974).

are included in Table 14. On basis of the newer findings (Secs. II.A and V.C.2), recalculation has been performed with a $P/2e^-$ ratio of 0.67. It is evident that the results with a $P/2e^-$ ratio of 0.67 give similar Y_{ATP} values for growth with the various substrates. Therefore we may conclude that the results can also be explained with a $P/2e^-$ ratio of 0.67. As an illustration of the complexity of these fermentation reactions, fumarate fermentation in *P. rettgeri* is outlined in Figure 10.

b. W. succinogenes. Growth of *W. succinogenes* with formate and fumarate has been studied in chemostat culture (Mell et al., 1982). An Y^{max} of 14 g cells per mol formate was found. From theoretical calculations it was concluded that about 18 g cells can be formed per mole ATP (Bronder et al., 1982). Generally the experimental growth yields are about 50% of the theoretical yields (Stouthamer, 1979). From these data the authors concluded that the $P/2e^-$ ratio for electron transfer from formate to fumarate was 0.8 (Mell et al., 1982). Formation of ATP during electron transport from formate to fumarate in whole cells exhibited an ATP-fumarate ratio of about 0.9 (Kröger and Winkler, 1981). A similar value was found with membrane vesicles. In later work these values were considered overestimates. Therefore all these values may be considered in accordance with a $P/2e^-$ ratio of 0.67.

Recently it was demonstrated that *W. succinogenes* can grow at the expense of the reaction H_2S + fumarate → succinate + S (Macy et al., 1986). The $\Delta G_0'$ of this reaction (calculated with H_2S in the gaseous state) is −53 kJ/mol. The growth yield was 6 g dry weight

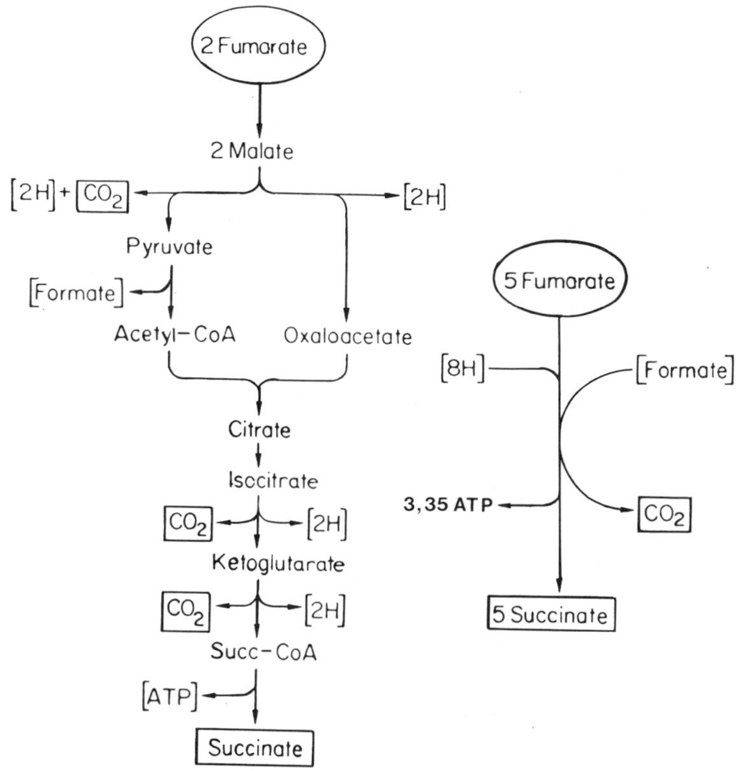

Figure 10 Scheme for the fermentation of fumarate by *Proteus rettgeri*. (Adapted from Kröger, 1974.)

per mol fumarate, which is about the same as the growth yield for cells grown with formate plus fumarate (Y = 6.5 g/mol for batch cultures) (Bronder et al., 1982). Therefore the $P/2e^-$ ratio is very likely the same in both cases. It is very surprising that *W. succinogenes* can also grow at the expense of the reaction formate$^-$ + H^+ + S → CO_2 + H_2S ($\Delta G_0'$ = -37 kJ/mol). A growth yield of 3.5 g/mol formate was calculated. This bacterium may therefore be involved in both the formation and reduction of sulfur.

c. Anaerobic bacteria forming propionate or succinate. Some remarks about the growth yields of these bacteria were made in Section IV.C.4.d. Molar growth yields and fermentation balances for a number of organisms forming propionate or succinate are given in Table 15. In previous publications, Y_{ATP} values were calculated with $P/2e^-$ ratios of 1 or even 2 for electron transport to fumarate

Table 15 Molar Growth Yields and Fermentation Balances for Growth of a Number of Organisms Forming Succinate or Propionate as an End Product of Fermentation with Various Substrates[a]

Organism	Normalized fermentation reaction	Y (g/mol)	Y_{ATP} ($P/2e^- = 0.67$)	Reference
Propionibacterium freudenreichii	1.5 glucose → 1 acetate + 2 propionate + $1CO_2$	82	23.1	de Vries et al. (1973)
	1 glycerol → 1 propionate	27	16.2	
	1 lactate → 0.33 acetate + 0.67 propionate + $0.33CO_2$	10.2	13.2	
Propionibacterium pentosaceum	1.5 glucose → 1 acetate + 2 propionate + $1CO_2$	76.3	21.5	
	1 glycerol → 1 propionate	26.3	15.7	
	1 lactate → 0.33 acetate + 0.67 propionate + 0.33 CO_2	12.9	16.7	
Actinomyces naeslandii	1 glucose → 2 lactate	24.7	12.3	Buchanan and Pine (1967)
	3.5 glucose + $3CO_2$ → 3 formate + 3 acetate + 3 succinate + lactate	49.4	14.3	
	2 glucose + 2 malate → 2 lactate + 2 acetate + 2 formate + 2 succinate	40.4	11.0	

Selenomonas ruminantium	1 glucose → 2 lactate	17	8.5	Hobson (1965); Hobson and Summers (1972)
	1 glucose → 1 acetate + 1 propionate + $1CO_2$ + $1H_2$	62	16.9	
	1 glucose + 1 pyruvate → 1.5 acetate + 1.5 propionate + $1.5CO_2$ + $0.5H_2$	83	18.4	
Bacteroides fragilis	1 glucose → 0.4 acetate + 0.4 succinate + 0.2 propionate + 0.4 formate	47	16.8	Macy et al. (1975)
Veillonella alkalescens	1 malate → 0.5 acetate + 0.5 propionate + $0.5H_2$ + $1.5CO_2$	14.4	14.4	de Vries et al. (1977)
	1 citrate → 2 acetate + $2CO_2$ + H_2	19.3	15.1	

[a]For electron transport to fumarate, a $P/2e^-$ ratio of 0.67 has been used for the calculation of Y_{ATP}. For *V. alkalescens* an ATP production of 1 mol/mol propionate was used (see text).

(Stouthamer, 1980b; Kröger, 1980). All data have been recalculated with a $P/2e^-$ ratio of 0.67 in this chapter. There are a number of uncertainties in the calculation of $P/2e^-$ ratios and Y_{ATP} values from these data. The most reliable data are those for the propionibacteria and for *Actinomyces naeslandii*. In propionibacteria the formation of propionyl-CoA occurs by a transcarboxylation reaction: pyruvate + methylmalonyl-CoA → oxaloacetate + propionyl-CoA. This reaction has been demonstrated only in propionibacteria but was recently also detected in *Desulfobulbus propionicus* (Stams et al., 1984). It has been concluded that this transcarboxylation reaction is characteristic for gram-positive organisms. In the other organisms the formation of propionyl-CoA occurs by decarboxylation of methylmalonyl-CoA. In Section II.C it was mentioned that energy is generated in this reaction by transport of Na^+ ions. This has been demonstrated for *P. modestum* and *V. alkalescens* (Hilpert et al., 1984). For the calculations in Table 15 it has been assumed that this decarboxylation is associated with the translocation of one charge. This is based on the $\Delta G_0'$ of the reaction and on the normal phosphate potential in microbial cells (see Secs. II.A and C).

The fermentation scheme for lactate by propionibacteria is given in Figure 11. The ATP yield for lactate fermentation is 0.33 (acetate formation) + 0.67 (propionate formation) × 0.67 ($P/2e^-$ ratio) = 0.78. For lactate and glycerol the Y_{ATP} values are between 13.2 and 16.7. These values are quite normal for organisms growing in complex medium (Stouthamer, 1979). The values for glucose are very high. However, it has been found that propionibacteria can perform the reaction fructose-6-phosphate + pyrophosphate → fructose-1,6-diphosphate + phosphate (O'Brien et al., 1975). Pyrophosphate is formed in a number of anabolic reactions: the activation of amino acids to aminoactyl-t-RNA, the formation of RNA and DNA from ribonucleotide triphosphates and deoxyribonucleotide triphosphates, respectively, and the activation of sulfate. The possibility of reutilizing pyrophosphate has a very great influence on the energy cost of biomass formation. This is illustrated by the theoretical calculations given in Table 16. Per mole ATP about 20% more cell material can be formed. This may explain the high Y_{ATP} value for growth of these bacteria with glucose. The effect of pyrophosphate reutilization is extremely important in sulfate-reducing bacteria, and more attention to this point is given in Section VI.C.1.

The Y_{ATP} values calculated for *A. naeslandii* are very similar to that for growth with glucose. In the latter growth condition ATP is produced only by substrate-level phosphorylation since lactate is the fermentation product. No explanation is possible for the data with *S. ruminantium*. In batch culture, glucose is completely converted into lactate and consequently Y_{ATP} = 8.5. In glucose-

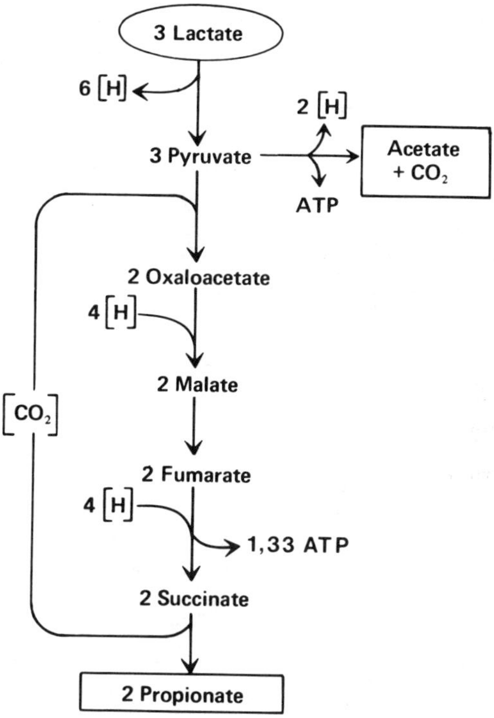

Figure 11 Scheme for the fermentation of lactate by propionibacteria. (Adapted from Kröger, 1980.)

Table 16 Influence of Pyrophosphate Reutilization on the Amount of ATP Required for the Formation of Microbial Cell Material from Glucose in the Presence or Absence of Amino Acids and Nucleic Acid Bases[a]

Medium	Pyrophosphate reutilization –	Pyrophosphate reutilization +
	ATP requirement ($10^4 \times$ mol/g)	
Complex	314.5	258.1
Minimal	347.1	290.7
	Gram cells per mole ATP	
Complex	31.9	38.7
Minimal	28.8	34.4

[a]The theoretical calculations were performed as described by Stouthamer (1973, 1979).

limited chemostat culture no lactate is formed. This is accompanied by a drastic increase in molar growth yield. The occurrence of electron transport-linked phosphorylation during propionate formation cannot explain this large difference. Recently it was suggested that this organism obtained ATP from succinate efflux (Michel et al., 1987), which is formed under some circumstances. The molar growth yield was directly proportional to the amount of succinate formed (see Sec. II.E). The extent to which succinate excretion contributes to the total ATP production is not yet known. In the following calculations this contribution is therefore neglected.

A $P/2e^-$ ratios of 1.0 (assuming the separation of one charge during the conversion of methylmalonyl-CoA to propionyl-CoA) would yield Y_{ATP} values of 15.5 and 16.7 for growth with glucose and growth with glucose and pyruvate, respectively. Evidently there is a large difference in Y_{ATP} between batch cultures and chemostat cultures. A value between 15 and 18 for Y_{ATP} in chemostat-grown *S. ruminantium* seems reasonable. The same applies for the Y_{ATP} value calculated for *B. fragilis*. For *V. alkalescens* a Y_{ATP} value of 15.1 can be calculated for growth with citrate. No electron transport-linked phosphorylation occurs during growth with citrate (see also Sec. IV.C.4.d). A similar value is found for growth with malate, during which 50% of the ATP production is by electron transport-linked phosphorylation. Although the data in Table 13 can be explained in a reasonable way with a $P/2e^-$ ratio of 0.67, it seems necessary that more work be done with these organisms. All the work mentioned in Section V.C.3 was done before the newer insights in the mechanism of ATP generation and utilization (Sec. II) were gained. A fresh look at a number of these organisms may be rewarding. Interesting data have been published for *Clostridium formicoaceticum*. This organism contains menaquinone and cytochromes (Gottwald et al., 1975), which is a very remarkable property for a clostridial species. Molar growth yields of 6.0 and 8.0 g dry weight per mole of fumarate and malate, respectively, were found (Dorn et al., 1978). For fumarate the fermentation balance was 3 fumarate → 2 succinate + acetate + CO_2. For malate, however, 3 malate → 0.16 succinate + 4.3 acetate + $2.9CO_2$ was found. *C. formicoaceticum* is a homoacetogenic bacterium, which means that glucose is fermented to 3 acetate. During malate fermentation acetate must also be formed by the reaction CO_2 + reducing equivalents → acetate (reaction k, Table 3). This means that whereas with fumarate *C. formicoaceticum* performs fumarate respiration, this scarcely occurs with malate. The latter substrate is practically completely converted into acetate. The conversion of fumarate and malate is shown in Figure 12. It seems likely that the observation that the growth yield with malate is higher than with fumarate can

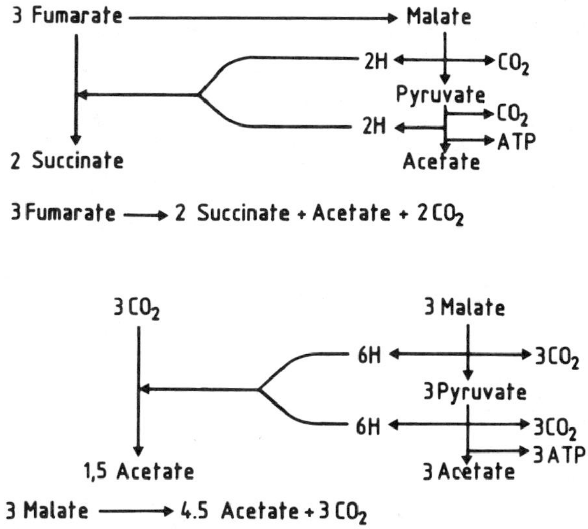

Figure 12 Schemes for the fermentation of fumarate (top) and malate (bottom) in *Clostridium formicoaceticum*.

be explained by the larger amount of substrate-level phosphorylation possible in the scheme shown in the bottom of Figure 12. If we assume a $P/2e^-$ ratio of 0.67 for electron transport to fumarate and that the reaction CO_2 + reducing equivalents → acetate does not yield ATP, we can calculate Y_{ATP} values of 7.7 and 8.0 for growth with fumarate and malate, respectively. In Section VI we encounter a number of other acetogenic bacteria.

d. Sulfate- and sulfur-reducing bacteria. Many sulfate-reducing bacteria can reduce fumarate (see Table 3). A number of them, for example *D. gigas*, can grow by fumarate dismutation (Figs. 10 and 12a) (Miller and Wakerley, 1966). *D. desulfuricans* reduces fumarate even prior to sulfate. In fact, one of the first proofs for electron transport-linked phosphorylation during fumarate reduction was with cell-free extracts of *D. gigas*. A $P/2e^-$ ratio of 0.9 was found for the reduction of fumarate with H_2 (Barton et al., 1970). The molar growth yields for growth of *Desulfovibrio vulgaris* with lactate or pyruvate with fumarate are 2.1 and 2.7 g cells, respectively (Magee et al., 1978). These values are very low and indicate a low efficiency of energy conservation.

D. propionicus is one of the new types of sulfate-reducing bacteria recently isolated and described by Widdel and Pfennig (1982). It is a very versatile microorganism (Table 3). Its most remarkable characteristic is the ability to form as well as to degrade

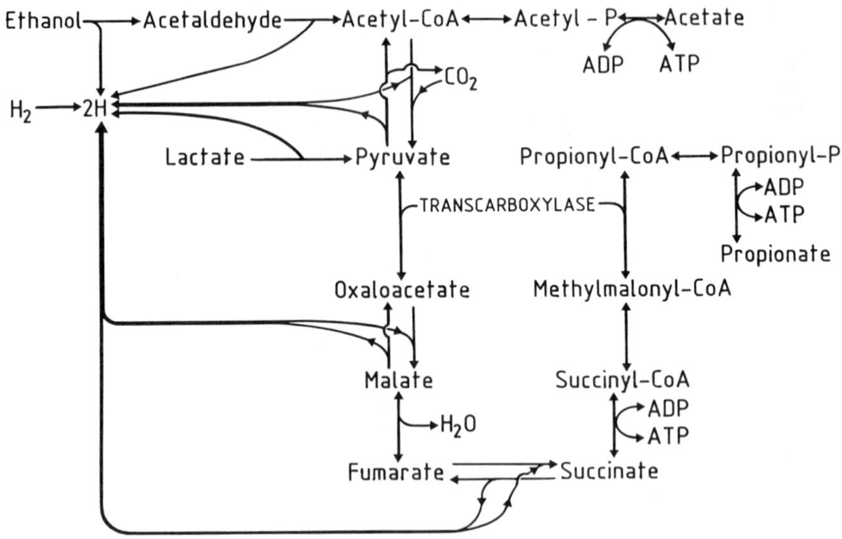

Figure 13 Proposed pathway for the formation and degradation of propionate in *Desulfobulbus propionicus*. During the formation of propionate, reduction equivalents are produced in the dehydrogenation of lactate, pyruvate, ethanol, and H_2. These reduction equivalents are used for the reduction of oxoaloacetate and fumarate. During the degradation of propionate, reduction equivalents are produced in the dehydrogenation of succinate and malate and these reduction equivalents are oxidized by sulfate. (Adapted from Stams et al., 1984.)

propionate. Lactate is fermented as shown in Figure 11 for propionibacteria. The molar growth yield for lactate was 1.45 g/mol. This indicates a very low efficiency of energy conservation, just as was concluded for *D. vulgaris*. However, propionate can also be formed according to the reactions (Widdel and Pfennig, 1982; Laanbroek et al., 1982)

$$\text{Ethanol} + 2/3CO_2 \rightarrow 1/3 \text{ acetate} + 2/3 \text{ propionate} + H_2O$$

$$\text{Ethanol} + CO_2 + 3H_2 \rightarrow \text{propionate}$$

$$\text{Acetate} + CO_2 + 3H_2 \rightarrow \text{propionate} + 3H_2O$$

These reactions are very unusual for organisms forming propionate. The first of these reactions is also performed by the recently isolated *Pelobacterium propionicum* (Schink, 1984), which does not contain cytochromes. Formation and utilization of propionate by *D. propionicus*, however, occurs by the scheme shown in Figure 13 (Stams et al., 1984). In all cases fumarate reductase and methylmalonyl-CoA pyruvate transcarboxylase are involved. Pyruvate is formed from acetate and ethanol by CO_2 fixation. The activation of acetate requires 1 ATP. ATP consumption in the formation of succinyl-CoA is equal to ATP production from propionyl-CoA. According to Stams et al. (1984), this indicates that reduction of fumarate with H_2 yields more than 1 ATP. However it seems more safe to conclude that the reaction acetate + CO_2 + $3H_2$ must yield more than 1 ATP. Propionate degradation can occur by the same scheme. The reaction for this oxidation is: propionate + $3/4SO_4^{2-} \rightarrow$ acetate + CO_2 + $3/4HS^-$.

Growth of *P. propionicum* with ethanol + CO_2 occurs by a fermentation scheme to a large extent similar to that shown in Figure 13 (Schink and Hansen, personal communication). These reaction sequences represent a third way in which CO_2 can serve as electron sink, besides methanogenesis and homoacetogenic fermentation.

Sulfur-reducing bacteria of the genus *Desulfuromonas* can grow in the absence of sulfur if their electron donor is combined with fumarate or malate (Widdel, 1988). An example of this type is the reaction acetate + 4 fumarate + $H_2O \rightarrow 2HCO_3^-$ + 4 succinate$^-$ + H^+ (reaction b2, Table 2).

VI. UTILIZATION OF SULFATE

A. General Remarks

For a long period the sulfate-reducing bacteria were regarded as a small specialized group of strictly anaerobic bacteria that could convert only a limited range of organic substrates; preferred substrates were lactate and pyruvate, which are incompletely oxidized to acetate. The sulfate-reducing bacteria were classified into two genera: *Desulfovibrio* and *Desulfotomaculum*. During the last few years a large number of new types of sulfate-reducing bacteria have been isolated. The microbiology of these organisms has recently been reviewed (Widdel, 1988). At the moment 10 genera comprising 33 species of sulfate-reducing bacteria have been described. Many of these bacteria can perform the complete oxidation of organic substrates as lactate and pyruvate. Consequently a number of these organisms can grow with acetate. A number of these bacteria are even able to consume aromatic compounds, such as benzoic acid, phenol, *p*-cresol, and phenylacetate. Consequently it has become clear that the sulfate-reducing bacteria are a large group of organisms that show a great diversity in biochemical properties. As a consequence the role of sulfate-reducing bacteria in mineralization in anaerobic environments is much larger than was realized until a few years ago.

Studies on the biochemistry and bioenergetics of sulfate-reducing bacteria are nearly all on *Desulfovibrio* and *Desulfotomaculum*. These studies have been reviewed by Thauer and Badziong (1980), Odom and Peck (1984), and LeGall and Fauque (1988). In general it has been assumed that the metabolism would be the same in both genera. However, this seems doubtful on the basis of new results. In *Desulfovibrio* a periplasmic hydrogenase seems of great importance in energy production during sulfate reduction (Odom and Peck, 1981, 1984). However, such an enzyme cannot easily play a role in the Gram-positive *Desulfotomaculum*. In addition, c-type cytochromes are the most important respiratory carriers in *Desulfovibrio*, whereas in *Desulfotomaculum* b-type cytochromes prevail (Widdel, 1988). Therefore the mechanisms in energy transduction seem to be different in *Desulfovibrio* and *Desulfotomaculum*.

Although the energetic aspects of sulfate-reducing bacteria have been intensively studied, there are great gaps in our knowledge and there are a great number of contradictory data in the literature. Growth yield studies are hampered because the growth of these bacteria is inhibited by the end product of sulfate reduction, H_2S. By using different amounts of substrates and sulfate, the amounts of H_2S and therefore the degree of growth inhibition may vary in different studies. In addition, differences in pH may be of great importance. The pH used for cultivation of sulfate-reducing bacteria was generally 7.2-7.6. However, it has been demonstrated that at this pH the bacteria only grow slowly and the growth yields are low (Badziong and Thauer, 1978). The highest growth rates for *D. vulgaris* were observed at pH values between 6.5 and 6.8. Furthermore, the pH must be kept constant owing to the formation of OH^- ions during sulfate reduction: $SO_4^{2-} + 4H_2 \rightarrow H_2S + 2H_2O + 2OH^-$. The picture that thus emerges from growth yield studies is far from conclusive. A continuous culture system has recently been developed that avoids the problem of inhibition by H_2S (Cypionka, 1986). In this system a sulfide electrode is used to control the electron donor supply and the medium pump. The sulfide concentration in the medium is kept at a low level by continuously flushing out H_2S and replacing with CO_2. Cells can therefore be grown at maximal growth rate. This method seems very promising for future studies.

Another factor that complicates the interpretation of growth yield data is that very little is known about the mechanism of uptake of sulfate. Therefore no estimate can be made about the energy cost of sulfate uptake. In this aspect it may be mentioned that Widdel (1988) reports that a number of sulfate-reducing bacteria are unable to grow with sulfite and thiosulfate. As explanation it was suggested that separate transport systems exist for the uptake of sulfate, sulfite, and thiosulfate.

All these factors contribute to the incomplete picture of the bioenergetics of sulfate-reducing bacteria.

Nitrate reduction by sulfate-reducing bacteria was discussed in Sections IV.C.2.d and IV.C.4.e. Fumarate reduction was discussed in Section V.C.3.d.

B. Acetyl-CoA Pathway

Some sulfate-reducing bacteria can grow under autotrophic conditions. It has been demonstrated that in *Desulfovibrio baarsii* the acetyl-CoA pathway is utilized for CO_2 fixation (Jansen et al., 1985). The same occurs in *Desulfobacterium autotroficum*, *Desulfobacterium vacuolatum*, *Desulfococcus multivorans*, *Desulfococcus niacini*, *Desulfosarcina variabilis*, and *Desulfotomaculum orientis* (Fuchs, 1986). We therefore may conclude that autotrophy and CO_2 fixation by the acetyl-CoA pathway is widespread in sulfate-reducing bacteria. It must be remarked, however, that in *D. hydrogenophilus* autotrophic CO_2 fixation occurs by the reductive tricarboxylic acid cycle (Fuchs et al., 1986). This difference illustrates the diversity in sulfate-reducing bacteria. The acetyl-CoA pathway is used for autotrophic CO_2 fixation in methanogenic, acetogenic, and sulfate-reducing bacteria (Wood et al., 1986; Fuchs, 1986). This pathway is utilized as an energy-generating pathway in acetogenic bacteria. These bacteria obtain energy from the reaction $4H_2 + H^+ + 2HCO_3^- \rightarrow$ acetate$^-$ + $4H_2O$ (reaction k, Table 3). The homoacetogenic process has already been described for growth of *C. formicoaceticum* with malate (scheme 12B). It has been demonstrated that *D. orientis* can perform this reaction (Klemps et al., 1985; Cypionka and Pfennig, 1986). Under anaerobic conditions this organism can grow with formate, methanol, and a number of other substrates. Under these conditions acetate is excreted, indicating the function of carbon dioxide as electron acceptor in a homoacetogenic process. In the acetyl-CoA pathway one molecule of CO_2 is first reduced to formate, which is then converted into formyltetrahydrofolate, for which 1 ATP is required. Formyltetrahydrofolate is reduced in two steps to methyltetrahydrofolate. In these reactions ATP may be formed by electron transport-linked phosphorylation. The methyl group of methyltetrahydrofolate is then transferred to a corrinoid (Wood et al., 1986; Fuchs, 1986). This methyl group is converted to the methyl of acetate. The second molecule of CO_2 is reduced to an enzyme-bound carbonyl intermediate. This reaction is catalyzed by carbon monoxide dehydrogenase, which must be regarded as the key enzyme of the acetyl-CoA pathway. The reduction of CO_2 to an enzyme-bound carbonyl intermediate is an energy-requiring step. The methyl group is transferred from the methylcorrinoid enzyme to the carbon monoxide dehydrogenase. The enzyme has binding sites for the methyl, CoA, and CO groups of acetyl-CoA. An acetyl intermediate is formed on the enzyme by transfer of the methyl group to the carbonyl site on CO dehydrogenase. Then the CO dehydrogenase catalyzes the final step of the synthesis of acetyl-CoA synthesis by condensing the acetyl and CoA groups. It is clear that the acetyl-CoA pathway is a linear pathway for CO_2 fixation, which is in contrast to the cyclic pathways for CO_2 fixation, such as the Calvin cycle and the reductive tricarboxylic acid cycle. CO can directly bind to the CO dehydrogenase. *D. orientis* can also grow with CO (Klemps et al., 1985). In homoacetogenic fermentation ATP consumption in the formation of methyltetrahydrofolate is compensated by ATP production from the conversion of acetyl-CoA into acetate. Consequently the net ATP production in this pathway is completely by electron transport-linked phosphorylation. It is estimated that the reaction $4H_2 + H^+ + 2HCO_3^- \rightarrow$ acetate + $4H_2O$ is associated with a net ATP production of 0.67 mol ATP.

The acetyl-CoA pathway is a cheap pathway for autotrophic CO_2 fixation (Thauer and Morris, 1984; Fuchs, 1986). Synthesis of 1 mol triosephosphate from $3CO_2$ via the Calvin cycle requires 9 mol ATP, whereas only 3.33 mol ATP are required via the acetyl-CoA pathway (0.33 mol for the formation of acetyl-CoA, 2 mol for the formation of phosphoenolpyruvate from pyruvate, and 1 mol for the formation of 1,3-diphosphoglycerate from 3-phosphoglycerate). This means that bacteria that obtain energy from reactions with a low $\Delta G_0'$ per mole H_2 (reactions h, i, and k in Table 1) have evolved a cheaper process for autotrophic CO_2 fixation than that occurring in aerobic bacteria (reaction a, Table 1). In bacteria utilizing the reductive tricarboxylic acid cycle for autotrophic CO_2 fixation, 5 mol ATP is required for the synthesis of 1 mol triosephosphate from $3CO_2$. Theoretical calculations of the amount of biomass that can be formed per mole of ATP have been performed by the methods described by Stouthamer (1973, 1979) and by Ingraham et al. (1983). These calculations (Stouthamer, unpublished results) indicate that, per mole of ATP, 6.3 g cell material can be formed from CO_2 via the Calvin cycle, 10.7 g cell material from CO_2 via the reductive tricarboxylic acid cycle, and 14.4 g cell material from CO_2 via the acetyl-CoA pathway. During heterotrophic growth with acetate, 10.1 g cell material can theoretically be formed per mole of ATP. These calculation show that the ATP cost of biomass synthesis from CO_2 via acetyl-CoA is even less than that from acetate in heterotrophs. During heterotrophic growth with malate, 14.9 g cell material can theoretically be formed per mole of ATP, which is about the same amount as that formed from CO_2 via the acetyl-CoA pathway. Part of the relatively high energy demand during heterotrophic growth with acetate and malate is due to the ATP cost for the uptake of the organic substrate (Stouthamer, 1973).

Recently it has been demonstrated that a number of sulfate-reducing bacteria utilize the acetyl-CoA pathway for the degradation of acetate in the presence of sulfate. In this case the acetyl-CoA pathway is involved in energy generation. This has been shown for *Desulfotomaculum acetoxidans*, *D. vacuolatum*, *D. variabilis*, *D. niacini*, *D. multivorans*, *D. baarsii*, and *D. autotrophicum* (Schauder et al., 1986). In these bacteria an incomplete citric acid cycle was found, since no 2-oxoglutarate reductase activity could be detected. After growth with acetate and sulfate, all these species contained high levels of carbon monoxide dehydrogenase and formate dehydrogenase. In Section II.D it was reported that *D. postgatei* and *D. acetoxidans* oxidize actate by the normal citric acid cycle and the problems due to the presence of a step with a relatively high redox potential were outlined. A possible advantage for acetate oxidation by the acetyl-CoA pathway over the citric acid cycle is that a reaction with a relatively high redox potential is not involved. The utilization of two different pathways for the oxidation of acetyl-CoA in sulfate-reducing bacteria again demonstrates the great diversity of this group of bacteria.

It has long been known that a number of sulfate-reducing bacteria produce small quantities of CH_4 (Postgate, 1969). This phenomenon is called minimethane formation. It has been demonstrated that this occurs in all sulfate-reducing bacteria that contain carbon monoxide dehydrogenase (Schauder et al., 1986). It is assumed that the methane is formed from a methyl-X intermediate in the acetyl-CoA pathway.

A number of the reactions of the acetyl-CoA pathway have been suggested as involved in the reaction acetate $\rightarrow CH_4 + CO_2$ (reaction d, Table 1), which occurs in methanogenic bacteria of the genera *Methanothrix* and *Methanosarcina* (Fuchs, 1986). Some small variations in the reactions of the acetyl-CoA pathway are possible, and also differences in the nature of the cofactors involved (e.g., tetrahydrofolate may be replaced by tetrahydromethanopterin) may occur. In this case a methyl-X intermediate is reduced to CH_4.

C. Pathway of Sulfate Reduction

1. *Sulfate Activation and Metabolism of Pyrophosphate*

The initial step of dissimilatory sulfate reduction is the activation of sulfate with ATP to form adenosine-5'-phosphosulfate (APS):

$$SO_4^{2-} + ATP + 2H^+ \rightarrow APS + pyrophosphate$$

The $\Delta G_0'$ of this reaction is + 46 kJ/mol. Thus the equilibrium of the reaction catalyzed by ATP sulfurylase lies in the direction of ATP and SO_4^{2-} and removal of pyrophosphate is of great importance in driving the reaction in the direction of APS synthesis. In *Desulfovibrio* species a strong pyrophosphatase activity is present (Liu and Peck, 1981), and consequently activation of sultate requires 2 ATP. In *D. orientis* pyrophosphatase is also present (Thebrath and Cypionka, unpublished results). However, in some other *Desulfotomaculum* sp. only low levels of pyrophosphatase were detected (Liu and Peck, 1981a). In these organisms the removal of pyrophosphate has been proposed to occur by the reaction pyrophosphate + acetate → acetylphosphate + phosphate. This reaction seems difficult, however, since it is highly endergonic ($\Delta G_0'$ = +23 kJ/mol; Thauer and Morris, 1984). If this reaction occurs only 1 ATP is required for sulfate activation in *Desulfotomaculum*. On this basis, Liu and Peck (1981a) proposed that in *Desulfotomaculum* no electron transport-linked phosphorylation associated with sulfate reduction would occur. This proposal has been shown to be incorrect, however, since growth under conditions that no substrate-level phosphorylation is possible has been demonstrated: *Desulfotomaculum* sp. on H_2 plus sulfate and *D. acetoxidans* on acetate and sulfate (Klemps et al., 1985). Furthermore, addition of pyrophosphate did not give an increase in molar growth yield (Klemps et al., 1985) as reported by Liu et al. (1982). Consequently it seems clear that the way of pyrophosphate removal in *Desulfotomaculum* is still not solved. It must be pointed out, however, that several other possibilities for the utilization of the energy-rich bond of pyrophosphate in bacteria have been described (Reeves, 1976).

Sulfate must be activated since the physiological reducing agents have a too positive redox potential to reduce sulfate itself (SO_4^{2-}/HSO_3^-; E_0' = -516 mV). The E_0' of the analogous reduction of APS is -60 mV.

2. APS Reduction

APS formed in the ATP sulfurase reaction is reduced to sulfite: APS + H_2 → HSO_3^- + AMP + H^+ ($\Delta G_0'$ = -68 kJ/mol). APS reductase is a soluble enzyme with a molecular weight of about 220,000. It contains 1 FAD per molecule and seven to nine irons and acid-labile sulfur atoms. (Bramlett and Peck, 1975). The exact nature and role of the iron-sulfur clusters in this protein are unknown. A similar statement must be made about the nature of the immediate electron donor for this reductase.

3. Sulfite Reduction

For a long period two schemes have been considered for the reduction of sulfite. These schemes are shown in Figure 14. The first mechanism assumes that bisulfite is reduced without the formation of free intermediates (scheme a). The second mechanism assumes that bisulfite is reduced via three two-electron reactions with trithionate and thiosulfate as free intermediates (scheme b). The partial reactions and their $\Delta G_0'$ values are

$$3HSO_3^- + H_2 + H^+ \rightarrow S_3O_6^{2-} + 3H_2O \qquad \Delta G_0' = -46 \text{ kJ/mol}$$

$$S_3O_6^{2-} + H_2 \rightarrow S_2O_3^{2-} + HSO_3^- + H^+ \qquad \Delta G_0' = -123 \text{ kJ/mol}$$

$$S_2O_3^{2-} + H_2 + H^+ \rightarrow H_2S + HSO_3^- \qquad \Delta G_0' = -1.9 \text{ kJ/mol}$$

The last two reactions are catalyzed by trithionate reductase and thiosulfate reductase, respectively. The last reaction cannot be coupled to ATP production (Thauer et al., 1977). Scheme B has been proposed since purified sisulfate reductase irreversibly produced trithionate and thiosulfate in addition to sulfide in proportions depending on the assay conditions. In a recent review, scheme A was favored for the following reasons (LeGall and Fauque, 1988): (1) Labeling experiments with $^{35}SO_4^{2-}$ are not consistent with scheme B (Chambers and Trudinger, 1975). (2) No true trithionate reductase has been characterized so far. Several types of sulfite reductase can be distinguished in sulfate-reducing bacteria. If these enzymes have a pH optimum lower than 7.0, the designation bisulfite reductase is more precise. The reductases differ with regard to their major absorption peaks, EPR spectra, and behavior of their prosthetic group, which is in all cases siroheme. The reductases are called desulfoviridin, desulforubidin, desulfofuscidine, and P582 (LeGall

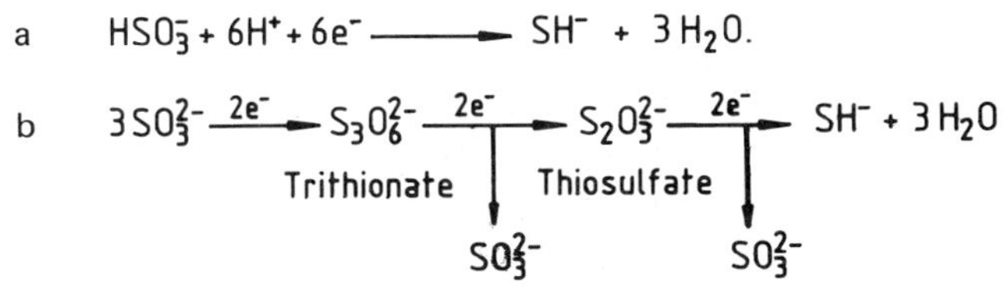

Figure 14 Proposed mechanisms for the reduction of sulfite. (a) No free intermediates involved. (b) Trithionate and thiosulfate occur as free intermediates and are reduced by specific reductases.

and Fauque, 1988). The distribution of these reductases among the sulfate-reducing bacteria is described by Widdel (1988). The first three enzymes have a $\alpha_2\beta_2$ subunit structure and contain three 4Fe- 4S clusters (LeGall and Fauque, 1988). The latter three enzymes react with CO. Desulfovoviridin can be simply identified by a red fluorescence if cell suspensions are treated with alkali under ultraviolet (UV) light (365 nm). Originally this test was used as a diagnostic of *Desulfovibrio* species. However, it has become clear that several *Desulfovibrio* lack desulfoviridin and that it is present in several sulfate-reducing bacteria belonging to other genera (Widdel, 1988). The bisulfite reductases are found in the soluble enzyme fraction after preparation of cell-free extracts. It has been proposed that they are membrane bound, however, to account for a shift in products observed during assay in the presence of membranes (Drake and Akagi, 1977). The enzymes may be only loosely bound to the membrane. Furthermore, it has been speculated that APS reductase and bisulfate reductase, which appear to be present at a ratio close to 1, form *in situ* an enzymatic complex allowing a fast transfer and reduction of the molecule of sulfite (LeGall and Fauque, 1988). Further research is necessary, however, to confirm these suggestions.

D. Bioenergetics of Sulfate Reduction

1. *Respiratory Chain to APS and Sulfite*

A large number of electron transfer proteins have been identified in sulfate-reducing bacteria. Many of them have been purified and have been characterized in detail. Furthermore, their localization (periplasm, membrane, or cytoplasm) has been studied. The results of these studies have been reviewed by Odom and Peck (1984) and LeGall and Fauque (1988). Unfortunately, the function of most of these electron transfer proteins is unknown. An exception is the tetraheme cytochrome c_3, which is a periplasmic protein with a molecular weight of about 13,000. It is an electron donor for periplasmic hydrogenase. In addition it is involved in the reduction of elemental sulfur to H_2S, which can be carried out by several *Desulfovibrio* species (e.g., *D. gigas*) (Biebl and Pfennig, 1977). Various strains of *Desulfovibrio* contain both cytoplasmic and a periplasmic hydrogenase, which have different properties.

2. *Stoichiometry of Proton Translocation*

The stoichiometries of proton translocation of *Desulfovibrio* sp. with nitrite as hydrogen acceptor were treated in Section IV.C.2.d.

These results indicate that proton translocation during electron transfer in these organisms is possible. Proton translocation with sulfite as hydrogen acceptor has been studied with *D. vulgaris* (Kobayashi et al., 1982). With H_2 as electron donor, the H^+/SO_3^{2-} ratio was 12-14. If we assume that $6H^+$ are produced from H_2 at the periplasmic side of the membrane, we can calculate a $H^+/2e^-$ ratio of 2. However, these data must be regarded with great caution since the authors report a H^+/O ratio of 9-14. These ratios are even higher than for aerobic bacteria with a highly effective respiratory chain. This suggests that in this study proton extrusion was overestimated.

3. *Molar Growth Yields and Schemes for Energy Generation in Sulfate-Reducing Bacteria*

The ultimate goal of studies on the physiology and bioenergetics is to come to an integrated picture of the metabolism and energy generation of sulfate-reducing bacteria. For that reason the treatment in this section is different from that in preceding sections. In addition to a discussion of the significance of the molar growth yields, a number of physiological aspects will also be discussed, since they must be integrated in the final hypothetical schemes.

a. Formation of hydrogen from organic substrates and its role in energy generation. It has been demonstrated that *D. vulgaris* evolves hydrogen before sulfate reduction starts (Vosjan, 1975; Tsuji and Yagi, 1980; Traore et al., 1981; Lupton et al., 1984a, b). In a number of cases the H_2 produced initially is later reoxidized. However, for *D. vulgaris* Hildenborough the following balances for growth with lactate and pyruvate were reported (Traore et al., 1981):

$$\text{Lactate} + 0.37SO_4^{2-} + 0.56H^+ \rightarrow CO_2 + 0.98\ \text{acetate} + 0.02\ \text{ethanol} + 0.16H_2S + 0.215HS^- + 0.5H_2O + 0.48H_2$$

and

$$\text{Pyruvate} + 0.2SO_4^{2-} + 0.15H_2O + 0.33H^+ \rightarrow CO_2 + 0.95\ \text{acetate} + 0.05\ \text{ethanol} + 0.087H_2S + 0.113HS + 0.1H_2$$

It is evident that the amount of hydrogen produced is different for growth with lactate and pyruvate. In both cases there is a significant deviation from the theoretical stoichiometries, which require the consumption of 0.5 and 0.25 mol SO_4^{2-} for the degradation of lactate and pyruvate, respectively. In addition it has been shown that lactate can be degraded to acetate by *Desulfovibrio* sp. in the absence of sulfate when cocultured with *Methanosarcina barkeri* (McInerney and Bryant, 1981; Traore et al., 1983). The balance for lactate degradation was 1 lactate → $1.5CO_2$ + $1.5CH_4$. This indicates that the *Desulfovibrio* sp. were able to perform the reaction 1 lactate → 1 acetate + $1CO_2$ + $2H_2$. The hydrogen and acetate are subsequently converted into CH_4 by *M. barkeri*. This is an example of interspecies hydrogen transfer. In a coculture of *Desulfovibrio* sp. with *Methanobrevibacter smithii* (which cannot convert acetate), the balance was 1 lactate → 1 acetate + $1/2CO_2$ + $1/2CH_4$. These data show that hydrogen can be formed by the reaction lactate → pyruvate + H_2, although the $\Delta G_0'$ of this reaction is highly positive (+43.1 kJ per reaction). This is possible since the consumption of hydrogen by methanogenic bacteria keeps the concentration of products low. The $\Delta G = \Delta G_0' + RT \ln$ products/reactants becomes negative, since the second term is very negative. This indicates that in *Desulfovibrio* the electrons from the conversion of lactate to pyruvate can be evolved as H_2. It is unknown which electron acceptors are involved in this process. It has been proposed that energy generation by sulfate-reducing bacteria from organic substrates obligately occurs by evolvement of hydrogen from organic substrates in the cytoplasm, diffusion across the cytoplasmic membrane, and oxidation of hydrogen by the periplasmic hydrogenase coupled to vectorial electron transfer to sulfate (Odom and Peck, 1981, 1984). This proposal is shown in Figure 15. Although these observations above seem to support this model, I propose that it should be rejected for the following reasons.

(1) In this model it is supposed that electron transport-linked phosphorylation associated with the conversion of lactate into pyruvate is the same as that associated with the conversion of pyruvate into acetate. This is unlikely for thermodynamic reasons, since the $\Delta G_0'$ values for these reactions are

$$\text{Lactate}^- + 1/4SO_4^{2-} + 1/4H^+ \rightarrow \text{pyruvate}^- + 1/4HS^- + H_2O \qquad \Delta G_0' = +5.1 \text{ kJ per reaction}$$

$$\text{Pyruvate}^- + 1/4SO_4^{2-} + H_2O \rightarrow \text{acetate}^- + 1/4HS^- + HCO_3^- + 3/4H^+ \qquad \Delta G_0' = -85.3 \text{ kJ per reaction}$$

In accordance with these thermodynamic considerations the molar growth yields for a number of *Desulfovibrio* sp. with lactate are smaller than those with pyruvate (Table 17). An attempt was made to explain this observation by utilizing scheme B of Figure 14 for sulfite

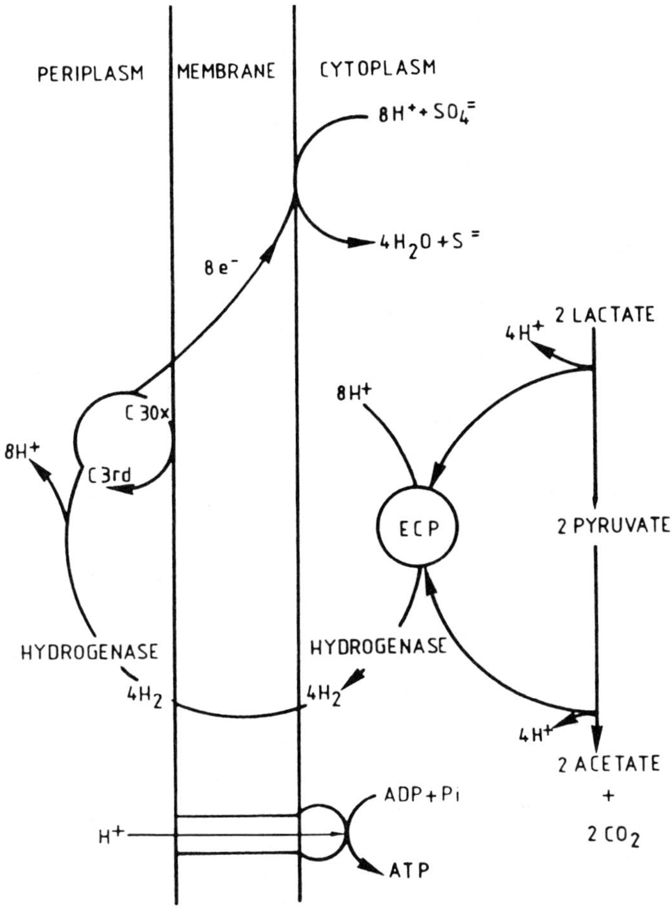

Figure 15 The obligate H_2 cycling model for energy generation in *Desulfovibrio*. Hydrogen is produced by an internal hydrogenase, diffuses freely across the membrane, and is recycled by a periplasmid hydrogenase that assures charge separation, electrons being allowed only to cross the membrane. ECP = electron carrier protein; C_3 = tetraheme cytochrome c_3. (Reproduced with permission from Lupton et al., 1984b.)

reduction and by assuming that different amounts of ATP were produced in the three reduction steps in this scheme (Traore et al., 1981). Since this scheme is now considered less likely, another explanation must be found.

(2) The model requires the presence of a cytoplasmic and a periplasmic hydrogenase. Although these have been demonstrated in a number of *Desulfovibrio* sp., it seems unlikely that they will be present in all sulfate-reducing bacteria. In fact, in *Desulfotomaculum* sp. hydrogenase is reported to be cytoplasmic (Cypionka and Dilling, 1986).

The model outlined in Figure 15 is therefore considered not to give a correct explanation for growth of *Desulfovibrio* sp. with organic substrates. However, it can explain the gain of energy from H_2 evolved from organic substrates by *Desulfovibrio* sp. The formation of H_2 from organic substrates is proposed to have a role in the control of the redox state of electron carriers (Lupton et al., 1984b). This aspect is not treated in detail here. A theoretical calculation of the ATP yield per mole of lactate and pyruvate is presented in Table 18. Two calculations are performed: one with a $P/2e^-$ ratio of 0.67 for the oxidation of both lactate and pyruvate (A) and one with a $P/2e^-$ ratio of 0.67 for the oxidation of pyruvate only. In the model of Figure 15 and all other schemes for energy generation in sulfate-reducing bacteria (Odom and Peck, 1981, 1984; Badziong and Thauer, 1980; Lupton et al., 1984b), the charge separation is 2 per $2e^-$ transferred to APS and sulfite. With a $H^+/ATP = 3$ all these models utilize a $P/2e^-$ ratio = 0.67. It is evident that the ATP yield for lactate in column A is higher than that for pyruvate. Consequently, as outlined earlier, this result can be rejected. In column B the ratio for the ATP yield with lactate and pyruvate is 0.57. The ratio of the growth yields with lactate and pyruvate for *D. vulgaris* is 0.62 (6.0/9.7) or 0.71 (7.8/11.0). These experimental ratios are not very different from the ratio of the theoretical ATP yields shown in Table 18. We can also calculate ATP yields with the Y_{pyr}^{max} value of 10.0 found for growth of *D. vulgaris* with pyruvate in the absence of sulfate (Nethe-Jaenchen and Thauer, 1984). This value indicates a Y_{ATP}^{max} value of 10.0. The growth yields for growth of *D. vulgaris* with lactate and pyruvate

b. Growth with hydrogen and acetate. Growth with hydrogen has been intensively studied in *Desulfovibrio* and *Desulfotomaculum* sp. In addition, growth yields with acetate have been reported for *D. postgatei* (Widdel and Pfennig, 1981). In these cases growth yields have been compared with sulfate, sulfite, and thiosulfate as hydrogen acceptor. For sulfate activation of the hydrogen acceptor at the cost of 2 ATP is necessary, whereas sulfite and thiosulfate

Table 17 Molar Growth Yields of Sulfate-Reducing Bacteria with Lactate and Pyruvate in the Presence of Sulfate[a]

	Substrate			
	Lactate		Pyruvate	
Organism	Y (g/mol)	$Y_{SO_4^{2-}}$ (g/mol)	Y (g/mol)	Reference
Desulfovibrio vulgaris	6.0		9.7	Magee et al. (1978)
Desulfovibrio gigas	1.9		7.6	
D. vulgaris	7.8		11.0	Traore et al. (1981)
D. gigas	3.7		7.7	Traore et al. (1982)
Desulfovibrio desulfuricans	4.1		8.5	
Desulfovibrio africanus	1.7		3.5	
D. desulfuricans	5.1		6.7	Vosjan (1975)
D. vulgaris		11.7		Liu and Peck (1981a)
Desulfotomaculum orientis		3.8		

[a]Values expressed as g protein per mol substrate.

Table 18 Theoretical Calculation of the ATP Yield During Growth of *Desulfovibrio* sp with Lactate and Pyruvate[a]

	ATP yield per mole substrate[a]	
Substrate	A	B
Lactate	1.33	0.67
Pyruvate	1.17	1.17
Lactate-pyruvate ratio	1.14	0.57

[a]In column A a $P/2e^-$ ratio is assumed for both the oxidation of lactate and pyruvate; in column B a $P/2e^-$ ratio of 0.67 is assumed only for the oxidation of pyruvate. For example, the calculation of the ATP yield per mole pyruvate is +1 (acetate formation) - 0.5 (sulfate activation) + 0.67 (oxidative phosphorylation) = 1.17.

can be reduced immediately and therefore do not require ATP for activation. Therefore growth yields are expected to be larger with sulfite and thiosulfate than with sulfate, and from the comparison important conclusions can be drawn. This approach has been applied to growth with lactate and pyruvate by Magee et al. (1978), however, with unclear results. The growth yields of a number of bacteria with H_2 are reported in Table 19. For growth with H_2 acetate was used as the carbon source. With *D. vulgaris* (Marburg strain) 70% of the cell carbon was derived from acetate and 30% from CO_2 (Badziong et al., 1978). The results show that the yields with sulfite and thiosulfate are indeed considerably higher than those with sulfate. Theoretical calculations for the ATP yield per mole electron acceptor with $P/2e^-$ ratios of 0.67 and 1.33 are presented in Table 20. It is evident that for a $P/2e^-$ ratio of 0.67 the theoretical ratio for $Y_{SO_3{}^{2-}}/Y_{SO_4{}^{2-}}$ = 3 and for a $P/2e^-$ ratio of 1.33 the theoretical ratio for $Y_{SO_3{}^{2-}}/Y_{SO_4{}^{2-}}$ would be only 1.2. For *D. vulgaris* an experimental ratio of about 2.8 is found. This is in close agreement with the value for a P/2e ratio of 0.67. On the basis of these results a $P/2e^-$ ratio of 1.33, as suggested by the $H^+/SO_3{}^{2-}$ reported by Kobayashi et al. (1982), seems highly unlikely.

With *D. orientis* the ratio $Y_{SO_3{}^{2-}}/Y_{SO_4{}^{2-}}$ is about 1.7. However is is not possible to draw firm conclusions from the data with *D. orientis* since the values in Table 19 were obtained at very low

Table 19 Molar Growth Yields of Sulfate-Reducing Bacteria Growing with Hydrogen and Various Electron Acceptors[a]

Organism[a]	Hydrogen acceptor	$Y_{el.acc.}$ (g/mol electron acceptor)	Reference
Desulfovibrio vulgaris	$SO_4{}^{2-}$	13	Nethe-Jaenchen and Thauer (1984)
	$SO_3{}^{2-}$	35	
	$S_2O_3{}^{2-}$	36	
Desulfotomaculum orientis	$SO_4{}^{2-}$	9.7	Cypionka and Pfennig (1986)
	$SO_3{}^{2-}$	16.0	
	$S_2O_3{}^{2-}$	17.9	

[a]The values for *D. vulgaris* are Y^{max} determined for chemostat cultures. The values for *D. orientis* are the highest values found in chemostat culture with the electron acceptor as limiting factor.

Table 20 Theoretical Calculations for the ATP Yield per mole Hydrogen Acceptor During Growth with H_2[a]

Hydrogen acceptor	ATP yield per mole electron acceptor	
	$P/2e^- = 0.67$	$P/2e^- = 1.33$
$SO_4{}^{2-}$	0.67	3.32
$SO_3{}^{2-}$	2.0	4.0
$S_2O_3{}^{2-}$	2.0	4.0

[a]The activation of sulfate is supposed to require 2 mol ATP per mole sulfate. The reduction of thiosulfate to sulfite + sulfide does not yield ATP. The calculation is performed with $P/2e^-$ ratios of 0.67 and 1.33.

μ values (about 0.02 h^{-1}) and consequently the influence of possible differences in maintenance coefficient could be of great importance. The conclusions mentioned earlier for *D. vulgaris* are somewhat different from those of the authors. From earlier data it was concluded that dissimilatory reduction of sulfate to sulfide was associated with a net synthesis of 1 mol ATP and that nearly 3 mol ATP were formed during dissimilatory sulfite reduction to sulfide (Badziong and Thauer, 1978). This conclusion was reached from a comparison $Y^{max}_{SO_4{}^{2-}}$ value with the theoretical ATP requirement for biomass synthesis from acetate. However it has been pointed out that there is a discrepancy between the theoretical and experimental ATP requirements for biomass synthesis (Stouthamer, 1979). In later work the growth yields with various hydrogen acceptors were compared with the Y^{max}_{ATP} value determined for anaerobic growth with pyruvate in the absence of sulfate (Nethe-Jaenchen and Thauer, 1984). From this approach it was concluded that dissimilatory sulfate reduction is coupled with the generation of approximately 1.3 mol ATP and sulfite reduction to sulfide with the generation of 3.5 mol ATP. The value for sulfate would correspond with a $P/2e^-$ ratio of 0.82 and that for sulfite with a $P/2e^-$ ratio of 1.17. It is surprising that a relative large difference is found for the $P/2e^-$ ratios calculated for sulfate and sulfite reduction. These values are between the two values used for the theoretical calculations of the ATP yield per mole electron acceptor in Table 20. It is evident that a good explanation for these observations is lacking, which illustrates the great gaps in our knowledge of energy generation during sulfate reduction.

The growth yields of *D. postgatei* with acetate and sulfate, thiosulfate and sulfite are 4.8, 10.4, and 11.2 g/mol acetate, respectively (Widdel and Pfennig, 1981). Again the yields with sulfite and thiosulfate are higher than that with sulfate. As mentioned in Section II.D, during growth of sulfate-reducing bacteria with acetate no ATP is formed by substrate-level phosphorylation. Therefore growth with acetate and sulfate (reaction d, Table 2) is problematic with respect to the generation of ATP, since the yield must be larger than the 2 mol needed for sulfate activation. In this organism a high pyrophosphatase activity is present (Thauer, 1982). During oxidation of acetate, reduced ferredoxin is produced from the oxidation of 2-oxoglutarate and NADPH from the oxidation of isocitrate. Malate and succinate are oxidized by membrane-bound enzymes, and menaquinone may be involved in this oxidation (Brandis-Heep et al., 1983). It is very unlikely that the oxidation of succinate with sulfate (Sec. II.D, $\Delta G_0' = +48.2$ kJ per reaction) will yield ATP; it is even possible that this oxidation requires ATP for reversed electron flow (Sec. II.D). The $\Delta G_0'$ values for the various oxidation steps in the citric acid cycle are very dissimilar: $\text{Malate}^{2-} + 1/4SO_4{}^{2-} + 1/4H^+ \rightarrow \text{fumarate}^{2-} + 1/4HS^- + H_2O$; $\Delta G_0' = +\ 10.1$ kJ per reaction.

$$\text{Isocitrate}^{3-} + H_2O + 1/4SO_4{}^{2-} + 1/4H^+ \rightarrow 2\ \text{oxoglutarate}^{2-} + HCO_{3-} + 1/4HS^- \qquad \Delta G_0' = -23.3 \text{ kJ per reaction}$$

$$\text{2-oxoglutarate}^{2-} + H_2O + 1/4SO_4{}^{2-} + 1/4H^+ \rightarrow \text{succinate}^{2-} + HCO_{3-} + 1/4HS^- \qquad \Delta G_0' = -83.2 \text{ kJ per reaction}$$

The first of these reactions does not seem a likely candidate for the production of ATP. Since the $\Delta G_0'$ values are so highly different it seems unlikely that the $P/2e^-$ electron transfer to sulfate will be same for the various electron donors. However, it seems clear that a higher overall $P/2e^-$ ratio than 0.67 would be necessary to get an ATP production larger than the amount of ATP required for sulfate activation. Recently a solution for this problem was obtained by the finding that in *D. postgatei* citrate is not formed by citrate syntase but by ATP-citrate lyase. This enzyme catalyzes the reaction: acetyl-CoA + oxaloacetate + ADP + P_{au} → citrate + ATP. Consequently, ATP is formed by substrate-level phosphorylation and at least 1 ATP must be formed by electron transport-linked phosphorylation (Möller et al., 1987). Growth yields with acetate and sulfate for organisms utilizing the acetyl-CoA pathway for the oxidation of acetate have not yet been reported. Much more research needs to be done on sulfate-reducing bacteria growing with acetate.

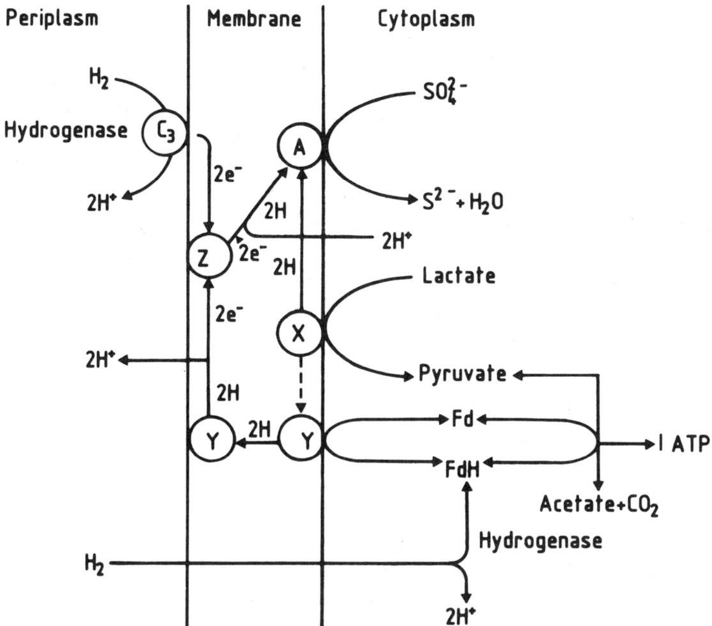

Figure 16 Scheme for energy generation during sulfate reduction in sulfate-reducing bacteria. The compounds A, X, and Y are hydrogen carriers; Z is an electron carrier. Pyruvate donates its electrons via ferredoxin or flavodoxin (Fd) to the hydrogen carrier Y. During transfer from Y to Z, two protons are translocated over the membrane. Electrons donated by hydrogen via the periplasmic hydrogenase are transferred to Z via cytochrome c_3. The hydrogen donor for sulfate reduction is A, which can be reduced by electron transfer from Z. Lactate donates to X, from which the hydrogen can be directly transferred to A. Consequently the P/2e$^-$ ratio for electron transfer from pyruvate and H_2 (both via the periplasmic and the cytoplasmic hydrogenase) is 0.67. Dehydrogenation of lactate to pyruvate with sulfate as hydrogen acceptor does not yield ATP. Formation of hydrogen from organic substrates can be explained with the same scheme. It is reported to be controlled in *Desulfovibrio* sp. by the redox state of the electron carriers. Hydrogen evolution from lactate can be explained by hydrogen transfer from X to Y (dashed line), which is supposed to occur only under very reduced conditions and at low hydrogen pressure.

c. Scheme for energy generation during sulfate reduction. From the description in the preceding sections it has appeared that there are large uncertainties in our knowledge of energy generation in sulfate-reducing bacteria. Furthermore, the sulfate-reducing bacteria are a highly diverse group of organisms and it may be possible that there are differences in the energy-generating mechanisms. Nonetheless a scheme for energy generation is presented in Figure 16. It is a combination of a number of elements from earlier schemes (Odom and Peck, 1981, 1984; Badziong and Thauer, 1980; Lupton et al., 1984b) with a number of new elements. In the scheme a redox loop has been incorporated. This element is not essential, however, and can be replaced by a proton-pumping redox carrier. Explanation of the scheme is given in the legend.

VII. UTILIZATION OF SULFUR

In Section V.C.3.b it was mentioned that *W. succinogenes* is able to reduce elemental sulfur. In Section VI.D.1 it was mentioned that tetraheme cytochrome c is the electron donor for sulfur reduction in *D. gigas* (LeGall and Fauque, 1988). The capacity of sulfate-reducing bacteria to utilize sulfur as hydrogen acceptor has been studied by Biebl and Pfennig (1977). Bacteria have been isolated that reduce sulfur but do not reduce sulfate. Eubacteria with these properties are placed in the genus *Desulfuromonas*, which comprises three species (Widdel, 1988). These bacteria can all accept fumarate in addition to sulfur as hydrogen acceptor (Table 3). The balance for the conversion of acetate with fumarate as hydrogen acceptor for *D. acetoxidans* was treated in Section V.C.3.d. The consequences of growth of *D. acetoxidans* with acetate and sulfur as hydrogen acceptor were treated in Section II.D. The molar growth yields of growth under various conditions are given in Table 21. The results show that the growth yields with fumarate as hydrogen acceptor are much larger than those with sulfur, as expected from the data in Table 1. The yields with ethanol are larger than those with acetate, which is also as expected. With membrane fractions sulfur reduction could be obtained with NADH and NADPH (Paulsen et al., 1986). *Desulfuromonas* sp. are known to contain menaquinone and b- and c-type cytochromes (Widdel, 1988). However, not much is known about the bioenergetics of *Desulfuromonas* and therefore no firm conclusions can be drawn from the data in Table 21.

In addition a number of archaebacterial sulfur reducers have been isolated. Species of six genera have been described: *Thermo-*

Table 21 Molar Growth Yields for Growth of *Desulfuromonas acetoxidans* with Various Carbon Sources and Hydrogen Acceptors

Carbon source	Hydrogen acceptor	Y (g/mol)
Acetate	Sulfur	4
Ethanol	Sulfur	9.8
Acetate	Malate	16.5
Ethanol	Malate	34.2
Pyruvate	Malate	46.2

Source: Data from Pfennig and Biebl (1976).

proteus, *Thermofilum*, *Desulfurococcus*, *Thermococcus*, *Thermodiscus*, and *Pyrodictium*. They are all extreme thermophiles and are nutritionally, morphologically and genetically rather diverse (Widdel, 1987). *Thermoproteus neutrophilus* (Table 3) is obligately autotrophic, growing with hydrogen, sulfur, and CO_2 (Fischer et al., 1983). About the biochemistry and bioenergetics of these organisms virtually nothing is known.

VIII. UTILIZATION OF TRIMETHYLAMINE *N*-OXIDE

A. General Remarks

The utilization of trimethylamine *N*-oxide (TMAO) in bacteria has recently been reviewed by Barrett and Kwan (1985). It has been demonstrated that the ability to reduce TMAO occurs in species of a large number of genera. However, the ability to reduce TMAO is less widespread than that to reduce nitrate. TMAO is a constituent of marine animals. Reduction of TMAO to trimethylamine (TMA) gives rise to the characteristic off-flavor of marine fish in the early stage of decay. Later work has shown that TMA was not the only major indicator of spoilage. Another important reaction in this aspect is demethylation of TMAO to dimethylamine and formaldehyde, a reaction catalyzed by enzymes endogenous in the fish muscle. Dimethylsulfoxide (DMSO) has been regarded as an analog of TMAO. Most of the strains that reduce TMAO also reduce DMSO. However, with nonpurple sulfur bacteria some exceptions were found (Schultz and Weaver, 1982). Rigorous proof that both substrates are reduced by the same enzyme is lacking, however. It has been pointed out that DMSO is common in certain natural environments (Zinder and Brock, 1978).

B. Enzymology of TMAO Reduction

In *E. coli*, several enzymes that can reduce TMAO have been identified. The most important is an inducible membrane-bound enzyme. Its induction is prevented by the presence of oxygen and nitrate. TMAO reductase is a molybdoenzyme since it is not formed in a number of chlorate-resistant mutants that are blocked in the formation of molybdopterin (Barrett and Kwan, 1985). Consequently nitrate reductase and TMAO reductase have the same molybdopterin cofactor. TMAO reductase has been characterized less well than nitrate reductase, fumarate reductase, APS reductase, and sulfite reductase. The enzyme of *Salmonella typhimurium* had a molecular weight of 332,000 and was suggested to consist of four identical subunits (Kwan and Barrett, 1983a). A form of the enzyme containing two subunits was also isolated. Reduced flavin mononucleotide, but not reduced flavin adenine dinucleotide, served as the electron donor. From the analysis of fermentation balances after growth in the presence and absence of TMAO, it can be concluded that in whole cells NADH, formate, and lactate can function as electron donor. Cytochromes were not copurified with the TMAO reductase. For the enzyme of *E. coli* a different molecular weight and differences in subunit composition have been reported (Shimokawa and Ishimoto, 1979). This may be due to differences between *E. coli* and *Salmonella*, differences in experimental methodology, or both. The occurrence of iron-sulfur clusters in TMAO reductase has not yet been studied.

C. Stoichiometry of Proton Translocation

Proton translocation coupled to reduction of TMAO has been studied in *E. coli* (Takagi et al., 1981). With endogenous substrate the addition of a TMAO pulse to an anaerobic cell suspension resulted in a H^+/TMAO ratio between 3 and 4. Thus the H^+/TMAO ratio is about the same as the H^+/O and the H^+/NO_3^- ratio (Sec. IV.C.2.a). In the presence of an uncoupler no proton movement during TMAO reduction is observed. This indicates that TMAO reduction most likely occurs at the cytoplasmic side of the membrane. This conclusion is at variance with that from other experiments (see Sec. VIII.C.1). It was mentioned earlier (Sec. II.B) that TMAO reductase in *R. capsulata* (McEwan et al., 1985b) and an *Alteromonas* sp. (Easter et al., 1983) is located at the periplasmic side of the membrane.

For a *Proteus* sp. it has been demonstrated that TMAO reduction generated a proton-motive force that could drive serine uptake against a concentration gradient (Stenberg et al., 1982). Oxidative phosphorylation associated with TMAO reduction with NADH and formate could be demonstrated with cell-free extracts. The net $P/2e^-$ ratios determined were 0.3 and 0.1, respectively.

D. Bioenergetics of TMAO Reduction: Respiratory Chain to TMAO

E. coli grown anaerobically with TMAO develops a new cytochrome pathway in addition to the aerobic respiratory pathways that are formed (Bragg and Hacket, 1983). Cytochromes with α-absorption peaks at about 548, 552, 554, and 557 nm are rapidly reoxidized by TMAO in a reaction that is not inhibited by 2-*n*-heptyl-4-hydroxyquinoline *N*-oxide. $CuSO_4$ inhibits the oxidation of the first two of these cytochromes. This suggests that the pathway of TMAO reduction is substrates → cytochromes c548, c552 → cytochromes b (?) 554, b557 → TMAO reductase. It is remarkable that c-type cytochromes, which do not play an important role in electron transfer in *E. coli*, are highly important in this electron transfer pathway. Furthermore, it is noticeable that the c-type cytochromes act as electron donors for b-type cytochromes, whereas normally the order is the reverse. It was demonstrated that cytochromes of this pathway were reoxidized by the membrane-impermeant oxidant ammonium persulfate. Cytochromes of the aerobic pathway remained unaffected by this treatment. This indicates that the cytochromes of the TMAO pathway and/or TMAO reductase are located at the periplasmic side of the membrane. This conclusion seems to contradict the data from proton translocation studies (Sec. VII.B). More work is thus necessary to determine the side of the membrane on which TMAO is reduced. Mutants defective in TMAO reduction have been isolated in *S. typhimurium* (Kwan and Barrett, 1983b). One group of mutants was defective in the formation of menaquinone. This quinone thus seems essential for TMAO reduction, just as for fumarate reduction. In accordance with this, these mutants were also deficient in growth with glycerol and fumarate.

E. Molar Growth Yields

1. *Enteric Bacteria*

It has been demonstrated that TMAO supported the growth of *S. typhimurium* on nonfermentable substrates (glycerol) and enhanced anaerobic yields obtained with glucose (Kim and Chang, 1974). Addition of TMAO to an anaerobic glucose medium resulted in increased acetate production; hence the ATP production by substrate-level phosphorylation was increased. However, it was concluded that electron transport-linked phosphorylation associated with electron transport to TMAO also occurred. It was calculated that 1.2 mol ATP and 0.6 mol ATP were produced by oxidative phosphorylation per mole of nitrate and TMAO, respectively. No numbers for molar growth yields were presented, however.

The influence of TMAO on molar growth yields in *E. coli* has been described by Ishimoto and Shimokawa (1978). The molar growth yield for glucose was increased by the presence of TMAO. The data are not discussed in detail here, since in these experiments a number of reduced fermentation products (e.g., ethanol) were formed in large amounts during anaerobic growth with glucose and TMAO. Evidently the rate of electron transfer to TMAO was not sufficient to make the reduction of acetyl-CoA with NADH to ethanol unnecessary. The possibilities of TMAO respiration could therefore not fully be exploited in this organism. The molar growth yield with formate during anaerobic growth in complex medium was about the same with nitrate, fumarate, and TMAO as hydrogen acceptor (Hamamoto and Ishimoto, 1977). The influence of TMAO on molar growth yields and acetate production in a *Proteus* sp. has been studied by Strøm et al. (1979). A number of results are presented in Table 22. Values of 13.6 and 7.8 were determined for Y_{ATP} for growth with glucose and pyruvate, respectively, in the absence of TMAO. These values are similar to those found in other enteric bacteria (Secs. IV.C.4.a and V.C.3.a). With these Y_{ATP} values, $P/2e^-$ ratios of 0.64 and 1.05 can be calculated for electron transport-linked phosphorylation associated with electron transfer to TMAO during growth with glucose and pyruvate, respectively. From the results it was concluded that no ATP is formed during the initial oxidation step of lactate.

2. *Photosynthetic Bacteria*

TMAO was found to act as an accessory oxidant for the dark anaerobic utilization of fructose by *R. capsulata* (Madigan and Gest, 1979; Madigan et al., 1980). These authors were not able to obtain

Table 22 Influence of TMAO on Molar Growth Yields and Acetate Production During Anaerobic Growth of *Proteus* NTHC153

Substrate	Presence of TMAO	Y (g/mol)	Acetate (mol/mol)	Y_{ATP} (g/mol)
Glucose	–	38.5	0.97	13.0
Glucose	+	77.0	1.76	—
Pyruvate	–	5.9	0.76	7.8
Pyruvate	+	14.7	0.75	—
Lactate	+	13.0	0.77	—
Formate	+	10.3	0	—

Source: Data from Strøm et al. (1979).

Table 23 Influence of Various Hydrogen Acceptors on Molar Growth Yields for Dark Growth of *Rhodopseudomonas capsulata* and *Rhodospirillum rubrum*

Substrate	Hydrogen acceptor	Y (g/mol) *R. capsulata*	Y (g/mol) *R. rubrum*
Fructose	—	12.7	15.0
	DMSO	58.5	57.5
	O_2	97.8	102.8
Succinate	DMSO	26.8	21.5
	TMAO	28.4	24.5
	O_2	49.2	42.2
Acetate	DMSO	7.5	—
	O_2	22.4	—

Source: Data from Schultz and Weaver (1982).

growth with nonfermentable substrates, such as malate and succinate, in conjunction with TMAO reduction. However, Schultz and Weaver (1982) were able to demonstrate dark anaerobic growth of *R. capsulata* with fructose and bicarbonate without the addition of an accessory oxidant. Fructose was fermented to a number of products, including succinate and propionate. The influence of TMAO and DMSO on the molar growth yield of *R. capsulata* with a number of substrates is shown in Table 23. The molar growth yields with fructose are strongly increased by the presence of TMAO or DMSO. The effect of TMAO and DMSO is about the same. With O_2 the molar growth yields are even higher. In the presence of TMAO and DMSO, growth with nonfermentable substrates (succinate and acetate) is possible. These findings indicate that the reduction of these compounds is associated with electron transport-linked phosphorylation, in contrast to earlier suggestions that these compounds acted only as an "electron sink" (Madigan et al., 1980). It is impossible, however, to come to a quantitative assessment of the amount of electron transport-linked phosphorylation.

IX. GENERAL CONCLUSIONS

The preceding sections show that there is a great variety in the systems for the utilization of electron acceptors by microorganisms. During the last few years a number of new and highly interesting systems have been discovered, some of which are very ingenious.

In all these cases the occurrence of electron transport-linked phosphorylation has been demonstrated. However, in most cases the amount of ATP produced by electron transport-linked phosphorylation is not precisely known. Our knowledge of the bioenergetics of microbial growth has increased considerably during the last few years. However, there are still a large number of gaps in our knowledge. These gaps cause large uncertainties in the interpretation of the yield data presented in this chapter. A large research effort will be necessary to solve these uncertainties. Study of the bioenergetics of microorganisms must be broadened to a much larger number of organisms and must be directed more to a quantitative understanding of the relation between molar growth yield and ATP yield. Only the combination of carefully planned and performed research on the bioenergetics and on the yields of microorganisms can resolve these uncertainties.

ACKNOWLEDGMENTS

Some parts of Section IV of this chapter (text, tables and figures) are taken from a recent review from the author (Stouthamer, 1988) in *Environmental Microbiology of Anaerobes* (ed. A. J. B. Zehnder) with the permission of John Wiley and Sons, Inc. Publishers.

REFERENCES

Abou-Jaoudé, A., Chippaux, M., Pascal, M.-C., and Casse, F. (1977). Formate: A new electron donor for nitrite reduction in *Escherichia coli* K12. *Biochem. Biophys. Res. Commun. 78*:579-583.

Abou-Jaoudé, A., Pascal M.-C., and Chippaux, M. (1979). Formate-nitrite reduction in *Escherichia coli* K12. 2. Identification of components involved in electron transfer. *Eur. J. Biochem. 95*:315-321.

Aleem, M. I. H. (1977). Coupling of energy with electron transfer reactions in chemolithotrophic bacteria. *Symp. Soc. Gen. Microbiol. 27*:351-381.

Alefounder, P. R., and Ferguson, S. J. (1980). The location of dissimilatory nitrite reductase and the control of dissimilatory nitrate reductase by oxygen. *Biochem. J. 192*:231-240.

Alefounder, P. R., Greenfield, A. J., McCarthy, J. E. G., and Ferguson, S. J. (1983). Selection and organisation of denitrifying electron-transfer pathways in *Paracoccus denitrificans. Biochim. Biophys. Acta 724*:20-39.

Aminuddin, M., and Nicholas, D. J. D. (1973). Sulphide oxidation linked to the reduction of nitrate and nitrite in *Thiobacillus denitrificans. Biochim. Biophys. Acta 325*:81-93.

Aminuddin, M., and Nicholas, D. J. D. (1974). Electron transfer during sulphide and sulphite oxidation in *Thiobacillus denitrificans*. *J. Gen. Microbiol.* *82*:115-123.

Averill, B. A., and Tiedje, J. M. (1982). The chemical mechanism of microbiol denitrification. *FEBS Lett.* *138*:8-12.

Badziong, W., and Thauer, R. K. (1978). Growth yields and growth rates of *Desulfovibrio vulgaris* (Marburg) growing on hydrogen plus sulfate and hydrogen plus thiosulfate as the sole energy sources. *Arch. Microbiol.* *117*:209-214.

Badziong, W., and Thauer, R. K. (1980). Vectoral electron transport in *Desulfovibris vulgaris* (Marburg) growing on hydrogen plus sulphate as sole energy source. *Arch. Microbiol.* *125*:167-174.

Badziong, W., Thauer, R. K., and Zeikus, J. G. (1978). Isolation and characterization of *Desulfovibrio* growing on hydrogen plus sulfate as the sole energy source. *Arch. Microbiol.* *116*:41-49.

Barrett, E. L., and Kwan, H. S. (1985). Bacterial reduction of trimethylamine oxide. *Annu. Rev. Microbiol.* *39*:131-149.

Barton, L. L., LeGall, J., and Peck, H. D., Jr. (1970). Phosphorylation coupled to oxidation of hydrogen with fumarate in extracts of the sulfate reducing bacterium *Desulfovibrio gigas*. *Biochem. Biophys. Res. Commun.* *41*:1036-1042.

Barton, L. L., LeGall, J. Odom, J. M., and Peck, H. D., Jr. (1983). Energy coupling to nitrite respiration in the sulfate-reducing bacterium *Desulfovibrio gigas*. *J. Bacteriol.* *153*:867-871.

Bauchop, T., and Elsden, S. R. (1960). The growth of microorganisms related to their energy supply. *J. Gen. Microbiol.* *23*:457-469.

Bernhard, T., and Gottschalk, G. (1978). Cell yields of *Escherichia coli* during anaerobic growth on fumarate and molecular hydrogen. *Arch. Microbiol.* *116*:235-238.

Betlach, M. R. (1982). Evolution of bacterial denitrifier diversity. *Antonie van Leeuwenhoek* *48*:585-607.

Betlach, M. R., Tiedje, J. M., and Firestone, R. B. (1981). Assimilatory nitrate uptake in *Pseudomonas fluorescens* studied using nitrogen-13. *Arch. Microbiol.* *129*:135-140.

Biebl, H., and Pfennig, N. (1977). Growth of sulfate-reducing bacteria with sulfur as electron acceptor. *Arch. Microbiol.* *112*: 115-117.

Bokranz, M., Katz, J. Robertson, A. M., and Kröger, A. (1983). Energy metabolism and biosynthesis of *Vibrio succinogenes* growing with nitrate or nitrite as terminal electron acceptor. *Arch. Microbiol.* *135*:36-41.

Boogerd, F. C., van Verseveld, H. W., and Stouthamer, A. H. (1980). Electron transport to nitrous oxide in *Paracoccus denitrificans*. *FEBS Lett.* *113*:279-284.

Boogerd, F. C., van Verseveld, H. W., and Stouthamer, A. H. (1981). Respiration-driven proton translocated with nitrite and nitrous oxide in *Paracoccus denitrificans*. *Biochim. Biophys. Acta* *638*:181-191.

Boogerd, F. C., Appeldoorn, K. J., and Stouthamer, A. H. (1983a). Effects of electron transport inhibitors and uncouplers on denitrification in *Paracoccus denitrificans*. *FEMS Microbiol. Lett.* *20*:455-460.

Boogerd, F. C., van Verseveld, H. W., and Stouthamer, A. H. (1983b). Dissimilatory nitrate uptake in *Paracoccus denitrificans* via a $\Delta\mu_{H^+}$ dependent system and a nitrate-nitrite antiport system. *Biochim. Biophys. Acta* *723*:415-427.

Boogerd, F. C., van Verseveld, H. W., Torenvliet, D., Braster, M., and Stouthamer, A. H. (1984). Reconsideration of the efficiency of energy transduction in *Paracoccus denitrificans* during growth under a variety of culture conditions. A new approach for the calculation of $P/2e^-$ ratios. *Arch. Microbiol.* *139*:344-350.

Bragg, P. D., and Hacket, N. R. (1983). Cytochromes of the trimethylamine *N*-oxide anaerobic respiratory pathway of *Escherichia coli*. *Biochim. Biophys. Acta* *725*:168-177.

Bramlett, R. N., and Peck, H. D., Jr. (1975). Some physical and kinetic properties of adenylyl sulfate reductase from *Desulfovibrio vulgaris*. *J. Biol. Chem.* *250*:2979-2986.

Brandis-Heep, A., Gebhardt, N. A., Thauer, R. K., Widdel, F., and Pfennig, N. (1983). Anaerobic acetate oxidations to CO_2 by *Desulfobacter postgatei*. 1. Demonstration of all enzymes required for the operation of the citric acid cycle. *Arch. Microbiol.* *136*: 222-229.

Brice, J. M., Law, J. F., Meyer, D. J., and Jones, C. W. (1974). Energy conservation in *Escherichia coli* and *Klebsiella pneumoniae*. *Biochem. Soc. Trans.* *2*:523-526.

Bronder, M., Mell, H., Stupperich, H., and Kröger, A. (1982). Biosynthetic pathways of *Vibrio succinogenes* growing with fumarate as terminal electron acceptor and sole carbon source. *Arch. Microbiol.* *131*:216-223.

Buchanan, B. B., and Pine, L. (1967). Path of glucose breakdown and cell yields of a facultative anaerobe, *Actinomyces naeslandii*. *J. Gen. Microbiol.* *46*:225-236.

Chambers, L. A., and Trudinger, P. A. (1975). Are thiosulfate and trithionate intermediates in dissimilatory sulfate reduction. *J. Bacteriol.* *123*:36-40.

Chaudhry, G. R., and MacGregor, C. H. (1983a). *Escherichia coli* nitrate reductase subunit A. Its role as the catalytic site and evidence for its modification. *J. Bacteriol.* *154*:387-394.

Chaudhry, G. R., and MacGregor, C. H. (1983b). Cytochrome b from *Escherichia coli* nitrate reductase. Its properties and association with the enzyme complex. *J. Biol. Chem.* *258*:5819-5827.

Clegg, R. A. (1976). Purification and some properties of nitrate reductase (EC.1.7.99.4) from *Escherichia coli* K12. *Biochem. J.* *153*:533-541.

Cole, J. A., and Brown, C. M. (1980). Nitrite reduction to ammonia by fermentative bacteria. A short circuit in the biological nitrogen cycle. *FEMS Microbiol. Lett.* *7*:65-72.

Cole, S. T., Condon, C., Lemire, B. D., and Weiner, J. H. (1985). Molecular biology, biochemistry and bioenergetics of fumarate reductase, a complex membrane-bound iron-sulfur flavoenzyme of *Escherichia coli*. *Biochim. Biophys. Acta* *811*:381-403.

Coleman, K. J., Cornish-Bowden, A., and Cole, J. A. (1978). Purification and properties of nitrite reductase from *Escherichia coli* K12. *Biochem. J.* *175*:483-493.

Coyle, C. L., Zumft, W. G., Kroneck, P. M. H., Körner, H., and Jacob, W. (1985). Nitrous oxide reductase from denitrifying *Pseudomonas perfectomarina*. Purification and properties of a novel multicopper enzyme. *Eur. J. Biochem.* *153*:459-467.

Cypionka, H. (1986). Sulfide-controlled continuous culture of sulfate-reducing bacteria. *J. Microbiol. Methods* *5*:1-9.

Cypionka, H., and Dilling, W. (1986). Intracellular localization of the hydrogenase in *Desulfotomaculum orientis*. *FEMS Microbiol.* *36*:257-260.

Cypionka, H., and Pfennig, N. (1986). Growth yields of *Desulfotomaculum orientis* with hydrogen in chemostat culture. *Arch. Microbiol.* *143*:396-399.

Delwiche, E. A., and Carson, S. F. (1953). A critic acid cycle in *Propionibacterium pentosaceum*. *J. Bacteriol.* *65*:318-321.

De Vries, W., van Wijck-Kapteyn, W. M. C., and Stouthamer, A. H. (1973). Generation of ATP during cytochrome-linked anaerobic electron transport in propionic acid bacteria. *J. Gen. Microbiol.* *76*:31-41.

De Vries, W., van Wijck-Kapteyn, W. M. C., and Oosterhuis, S. K. H. (1974). The presence and function of cytochromes in *Selenomonas ruminantium, Anaerovibrio lipolytica* and *Veillonella alkalescens*. *J. Gen. Microbiol.* *81*:69-78.

De Vries, W., Rietveld-Struyk, T. R. M., and Stouthamer, A. H. (1977). ATP formation associated with fumarate and nitrate reduction in growing cultures of *Veillonella alkalescens*. *Antonie van Leeuwenhoek* *43*:153-167.

De Vries, W., Niekus, H. G. D., Boellaard, M., and Stouthamer, A. H. (1980). Growth yield and energy generation by *Campylobacter sputorum* subspecies *bubulus* during growth in continuous culture with different hydrogen acceptors. *Arch. Microbiol.* *124*: 221-227.

De Vries, W., Niekus, H. G. D., van Berchum, H., and Stouthamer, A. H. (1982). Electron-transport-linked proton translocation at nitrite reduction in *Campylobacter sputorum* subsp. *bubulus*. *Arch. Microbiol.* *131*:132-139.

De Vries, W., van Berchum, H., and Stouthamer, A. H. (1984). Localization of hydrogenase and nitrate reductase in *Campylobacter sputorum* subsp. *bubulus*. *Antonie van Leeuwenhoek 50*:63-73.

Dorn, M., Andreesen, J. R., and Gottschalk, G. (1978). Fermentation of fumarate and L-malate by *Clostridicum formicoaceticum*. *J. Bacteriol. 133*:26-32.

Drake, H. L., and Akagi, J. M. (1974). Bisulfite reductase of *Desulfovibrio vulgaris*: Explanation for product formation. *J. Bacteriol. 132*:139-143.

Easter, M. C., Gibson, D. M., and Ward, F. B. (1983). The introduction and location of trimethylamine-*N*-oxide reductase in *Alteromonas* sp. NCMB 400. *J. Gen. Microbiol. 129*:3689-3696.

Enoch, H. G., and Lester, R. L. (1974). The role of a novel cytochrome b-containing nitrate reductase and quinone in the in vitro reconstruction of formate-nitrate reductase activity of *E. coli*. *Biochem. Biophys. Res. Commun. 61*:1234-1241.

Enoch, H. G., and Lester, R. L. (1975). The purification and properties of formate dehydrogenase and nitrate reductase from *Escherichia coli*. *J. Biol. Chem. 250*:6693-6705.

Erecinska, M., Kula, T., and Wilson, D. F. (1978). Regulation of energy metabolism: Evidence against a primary role of adenine nucleotide translocase. *FEBS Lett. 87*:138-144.

Ferguson, S. J., and Sorgato, M. C. (1982). Proton electrochemical gradients and energy-transduction processes. *Annu. Rev. Biochem. 51*:185-217.

Fischer, F., Zillig, W., Stetter, K. O., and Schreiber, G. (1983). Chemolithotrophic metabolism of anaerobic extremely thermophilic archaebacteria. *Nature 301*:511-513.

Forget, P. (1974). The bacterial nitrate reductases. Solubilization purification and properties of the enzyme A of *Escherichia coli* K12. *Eur. J. Biochem. 42*:325-332.

Forget, P., and Pichinoty, F. (1967). Le cycle trocarboxylique chez *Aerobacter aerogenes*. *Ann. Inst. Pasteur 112*:261-290.

Fuchs, G. (1986). CO_2-fixation in acetogenic bacteria: variations on a theme. *FEMS Microbiol. Rev. 39*:181-213.

Fuchs, G., Länge, S., Rude, E., Schäfer, S., Schauder, R., Scholtz, R., and Stupperich, E. (1986). Autotrophic CO_2-fixation in chemotrophic anaerobic bacteria. In *Microbial Growth with C1 Compounds*, J. A. Duine and H. W. van Verseveld (Eds.). Martinus Nijhoff, Dordrecht, The Netherlands, pp. 39-43.

Garland, P. B., Downie, J. A., and Haddock, B. A. (1975). Proton translocation and the respiratory nitrate reductase of *Escherichia coli*. *Biochem. J. 152*:547-559.

Gebhardt, N. A., Thauer, R. K., Linder, D. P. M., Kaulfers, P. M., and Pfennig, N. (1985). Mechanism of acetate oxidation to CO_2 with elemental sulfur in *Desulfuromonas acetoxidans*. *Arch. Microbiol. 141*:392-398.

Gottwald, M. J., Andreesen, J. R., LeGall, J., and Ljungdahl, L. G. (1975). Presence of cytochrome and menaquinone in *Clostridium formicoaceticum* and *Clostridium thermoaceticum*. *J. Bacteriol. 122*:325-328.

Hackett, N. R., and Bragg, P. D. (1983). Membrane cytochromes of *Escherichia coli* grown aerobically and anaerobically with nitrate. *J. Bacteriol. 154*:708-718.

Haddock, B. A., and Jones, C. W. (1977). Bacterial respiration. *Bacteriol. Rev. 41*:47-99.

Hadjipetrou, L. P., and Stouthamer, A. H. (1965). Energy production during nitrate respiration by *Aerobacter aerogenes*. *J. Gen. Microbiol. 38*:29-34.

Hadjipetrou, L. P., Gerrits, J. P., Teulings, F. A. G., and Stouthamer, A. H. (1963). Relation between energy production and growth of *Aerobacter aerogenes*. *J. Gen. Microbiol. 36*:139-150.

Hasan, S. M., and Hall, J. B. (1975). The physiological function of nitrate reduction in *Clostridium perfringens*. *J. Gen. Microbiol. 87*:120-128.

Hasan, S. M., and Hall, J. B. (1977). Dissimilatory nitrate reduction in *Clostridium tertium*. *Z. Allg. Mikrobiol. 17*:501-506.

Hatchikian, E. C., and LeGall, J. (1972). Evidence for the presence of a b-type cytochrome in the sulfate-reducing bacterium *Desulfovibrio gigas* and its role in the reduction of fumarate by molecular hydrogen. *Biochim. Biophys. Acta 267*:479-484.

Henry, M.-F., and Vignais, P. M. (1983). Electron pathways from H_2 to nitrate in *Paracoccus denitrificans*. Effects of inhibitors of UQ-cytochrome b region. *Arch. Microbiol. 136*:64-68.

Hilpert, W., Schink, B., and Dimroth, P. (1984). Life by a new decarboxylation-dependent energy conservation mechanism with Na^+ as coupling ion. *EMBO J. 3*:1665-1670.

Hobson, P. N. (1965). Continuous culture of some anaerobic and facultatively anaerobic rumen bacteria. *J. Gen. Microbiol. 38*:167-180.

Hobson, P. N., and Summers, P. R. (1967). The continuous culture of anaerobic bacteria. *J. Gen. Microbiol. 47*:53-65.

Hobson, P. N., and Summers, R. (1972). ATP pool and growth yield of *Selenomonas ruminantium*. *J. Gen. Microbiol. 70*:351-360.

Husain, M., and Sadana, J. (1974). Nitrite reductase from *Achromobacter fischeri*: Molecular weight and subunit structure. *Eur. J. Biochem. 42*:283-289.

Inderlied, C. B., and Delwiche, E. A. (1973). Nitrate reduction and anaerobic electron transport in *Veillonella alkalescens*. *J. Bacteriol. 114*:1206-1212.

Ingledew, W. J., and Poole, R. K. (1984). The respiratory chains of *Escherichia coli*. *Microbiol. Rev. 48*:222-271.

Ingraham, J. L., Maaloe, O., and Neidhardt, F. C. (1983). *Growth of the Bacterial Cell*. Sinauer Associates, Sunderland, Massachusetts.

Ishimoto, M., and Shimokawa, O. (1978). Reduction of trimethylamine *N*-oxide by *Escherichia coli* as anaerobic respiration. *Z. Allg. Mikrobiol. 18*:173-181.

Ishimoto, M., and Yamamoto, J. (1977). Cell growth and metabolic products of *Escherichia coli* in nitrate respiration. *Z. Allg. Mikrobiol. 17*:309-320.

Jackson, M. A., Jackson, J. B., and Ferguson, S. J. (1981). Direct observation with an electrode of uncoupler-sensitive assimilatory nitrate uptake by *Rhodopseudomonas capsulata*. *FEBS Lett. 136*:275-278.

Jackson, R. H., Cornish-Bowden, A., and Cole, J. A. (1981). Prosthetic groups of the NADH-dependent nitrite reductase from *Escherichia coli* K12. *Biochem. J. 193*:861-867.

Jacobs, N. J., and Jacobs, J. M. (1975). Evidence for the involvement of the electron transport system at a late step of anaerobic microbiol heme synthesis. *Biochem. Biophys. Res. Commun. 65*:435-441.

Jansen, K., Fuchs, G., and Thauer, R. K. (1985). Autotrophic CO_2 fixation by *Desulfovibrio baarsii*: Demonstration of enzyme activities characteristic for the acetyl-CoA pathway. *FEMS Microbiol. Lett. 20*:311-315.

Jenkinson, H. F., and Woodbine, M. (1979). Growth and energy production in *Bacteroides amylophilus*. *Arch. Microbiol. 120*:275-281.

John, P. (1977). Aerobic and anaerobic bacterial respiration monitored by electrodes. *J. Gen. Microbiol. 98*:231-238.

John, P., and Whatley, F. R. (1970). Oxidative phosphorylation coupled to oxygen uptake and nitrate reduction in *Micrococcus denitrificans*. *Biochim. Biophys. Acta 216*:342-352.

John, P., and Whatley, F. R. (1975). *Paracoccus denitrificans* and the evolutionary origin of the mitochondrion. *Nature 254*:495-498.

Johnson, J. L., and Rajagopolan, K. V. (1982). Structural and metabolic relationship between the molybdenum cofactor and urothione. *Proc. Natl. Acad. Sci. USA 79*:6856-6860.

Jones, R. W., Lamont, A., and Garland, P. B. (1980). The mechanism of proton translocation driven by the respiratory nitrate reductase complex of *Escherichia coli*. *Biochem. J. 190*:79-94.

Justin, P., and Kelly, D. P. (1978). Growth kinetics of *Thiobacillus denitrificans* in aerobic and anaerobic chemostat culture. *J. Gen. Microbiol. 107*:123-130.

Kaneko, M., and Ishimoto, M. (1978). Effects of nitrate reduction on metabolic products and growth of *Propionobacterium acidipropionici*. *Z. Allg. Mikrobiol. 17*:211-220.

Kashket, E. R. (1982). Stoichiometry of the H^+-ATPase of growing and resting aerobic *Escherichia coli*. *Biochemistry 21*:5534-5538.
Kashket, E. R. (1983). Stoichiometry of the H^+/ATPase of *Escherichia coli* cells during anaerobic growth. *FEBS Lett. 154*:343-346.
Kashket, E. R. (1985). The proton motive force in bacteria: A critical assessment of methods. *Annu. Rev. Microbiol. 39*:219-242.
Kaspar, H. F. (1982). Nitrite reduction to nitrous oxide by propionobacteria. Detoxification mechanism. *Arch. Microbiol. 133*:126-130.
Keevil, C. W., Hough, J. S., and Cole, J. A. (1979). Regulation of respiratory and fermentative modes of growth of *Citrobacter freundii* by oxygen, nitrate and glucose. *J. Gen. Microbiol. 113*: 83-95.
Keith, S. M., and Herbert, R. A. (1983). Dissimilatory nitrate reduction by a strain of *Desulfovibrio desulfuricans*. *FEMS Microbiol. Lett. 18*:55-59.
Kell, D. B., Burns, A., Clarke, D. J., and Morris, J. G. (1981). Proteinaceous proton pumps: A minimal model, some properties and their possible universality. *Speculations Sci. Technol. 4*(2): 109-120.
Kim, K. E., and Chang, G. W. (1974). Trimethylamine oxide reduction by *Salmonella*. *Can. J. Microbiol. 20*:1745-1748.
Klemps, R., Cypionka, H., Widdel, F., and Pfennig, N. (1985). Growth with hydrogen and further physiological characteristics of *Desulfotomaculum* species. *Arch. Microbiol. 143*:203-208.
Knaff, D. B. (1978). Reducing potentials and the pathway of NAD^+ reduction. In *The Photosynthetic Bacteria*, R. K. Clayton and W. R. Sistrom (Eds.). Plenum Press, New York, pp. 629-640.
Knowles, R. (1982). Denitrification. *Microbiol. Rev. 46*:43-70.
Kobayashi, K., Hasegawa, H., Takagi, M., and Ishimoto, M. (1982). Proton translocation associated with sulfite reduction in a sulfate-reducing bacterium, *Desulfovibrii vulgaris*. *FEBS Lett. 142*:235-237.
Koike, I., and Hattori, A. (1975). Growth yield of a denitrifying bacterium *Pseudomonas denitrificans* under aerobic and denitrifying conditions. *J. Gen. Microbiol. 88*:1-10.
Kristjansson, J. K., Walter, B., and Hollocher, T. C. (1978). Respiration-dependent proton translocation and the transport of nitrate and nitrite in *Paracoccus denitrificans* and other denitrifying bacteria. *Biochemistry 17*:5014-5019.
Kröger, A. (1974). Electron-transport phosphorylation coupled to fumarate reduction in anaerobically grown *Proteus rettgeri*. *Biochim. Biophys. Acta 347*:273-289.
Kröger, A. (1977). Phosphorylative electron transport with fumarate and nitrate as terminal hydrogen acceptors. *Symp. Soc. Gen. Microbiol. 27*:61-93.

Kröger, A. (1980). Bacterial electron transport to fumarate. In *Diversity of Bacterial Respiratory Systems*, Vol. 2, C. J. Knowles (Ed.). CRC Press, Boca Raton, Florida, pp. 2-17.
Kröger, A., and Winkler, E. (1981). Phosphorylative fumarate reduction in *Vibrio succinogenes*: Stoichiometry of ATP synthesis. *Arch. Microbiol. 129*:100-104.
Kröger, A., Dorrer, E., and Winkler, E. (1980). The orientation of the substrate sites of formate dehydrogenase and fumarate reductase in the membrane of *Vibrio succinogenes*. *Biochim. Biophys. Acta 589*:118-136.
Kundu, B., and Nicolas, D. J. D. (1985). Proton translocation during denitrification in *Rhodopseudomonas sphaeroides f denitrificans*. *Arch. Microbiol. 140*:358-364.
Kwan, H. S., and Barrett, E. L. (1983a). Purification and properties of trimethylamine oxide reductase from *Salmonella typhimurium*. *J. Bacteriol. 155*:1455-1458.
Kwan, H. S., and Barrett, E. L. (1983b). Roles for menaquinone and the two trimethylanine oxide (TMAO) reductases in TMAO respiration in *Salmonella typhimurium*: Mud(Ap'*lac*) insertion mutations in *men* and *tor*. *J. Bacteriol. 155*:1147-1155.
Laanbroek, H. J., Stal, L. J., and Veldkamp, H. (1978). Utilization of hydrogen and formate by *Campylobacter* sp. under aerobic and anaerobic conditions. *Arch. Microbiol. 119*:99-102.
Laanbroek, H. J., Abee, T., and Voogd, I. L. (1982). Alcohol conversions by *Desulfobulbus propionicus* Lindhorst in the presence and absence of sulfate and hydrogen. *Arch. Microbiol. 133*:178-184.
Lam, Y., and Nicholas, D. J. D. (1969). A nitrite reductase from *Micrococcus denitrificans*. *Biochim. Biophys. Acta 178*:225-234.
Lawford, H. G., and Haddock, B. A. (1973). Respiration-driven proton translocation in *Escherichia coli*. *Biochem. J. 136*:217-220.
LeGall, J., and Fauque, G. (1988). Dissimilatory reduction of sulfur compounds. In *Environmental Microbiology of Anaerobes*, A. J. B. Zehnder (Eds.). John Wiley and Sons, New York, pp. 587-639.
Liu, C. L., and Peck, H. D., Jr. (1981a). Comparative bioenergetics of sulfate reduction in *Desulfovibrio* and *Desulfotomaculum* spp. *J. Bacteriol. 145*:966-973.
Liu, M. C., and Peck, H. D., Jr. (1981b). The isolation of a hexaheme cytochrome from *Desulfovibrio desulfuricans* and its identification of a new type of nitrite reductase. *J. Biol. Chem. 256*:13159-13164.
Liu, C.-L., Hart, N., and Peck, H. D., Jr. (1982). Inorganic pyrophosphate: Energy source for sulfate-reducing bacteria of the genus *Desulfotomaculum*. *Science 217*:363-364.
Lund, K., and De Moss, J. A. (1976). Association-dissociation behavior and subunit structure of heat-released nitrate reductase from *E. coli*. *J. Biol. Chem. 251*:2207-2216.

Lupton, F. S., Conrad, R., and Zeikus, J. G. (1984a). CO metabolism of *Desulfovibrio vulgaris* strain Madison: Physiological function in the absence or presence of exogenous substrates. *FEMS Microbiol. Lett.* *23*:263-268.

Lupton, F. S., Conrad, R., and Zeikus, J. G. (1984b). Physiological function of hydrogen metabolism during growth of sulfidogenic bacteria on organic substrates. *J. Bacteriol.* *159*:843-849.

Lütgens, M., and Gottschalk, G. (1982). Cell and ATP yields of *Citrobacter freundii* growing with fumarate and H_2 or formate in continuous culture. *J. Gen. Microbiol.* *128*:1915-1921.

MacGregor, C. H. (1975a). Solubilization of *Escherichia coli* nitrate reductase by a membrane-bound protease. *J. Bacteriol.* *121*:1102-1110.

MacGregor, C. H. (1975b). Anerobic cytochrome b in *Escherichia coli*: Association with and regulation of nitrate reductase. *J. Bacteriol.* *121*:1111-1116.

MacGregor, C. H., Schnaitman, A., Normansell, D. E., and Hodgins, M. G. (1974). Purification and properties of nitrate reductase from *Escherichia coli* K12. *J. Biol. Chem.* *249*:5321-5327.

Macy, J., Probst, I., and Gottschalk, G. (1975). Evidence for cytochrome involvement in fumarate reduction and adenosine-5'-triphosphate synthesis by *Bacteroides fragilis* grown in the presence of hemin. *J. Bacteriol.* *123*:436-442.

Macy, J. M., Schroder, I., Thauer, R. K., and Kröger, A. (1986). Growth of *Wolinella succinogenes* on H_2S plus fumarate and on formate plus sulfur as energy sources. *Arch. Microbiol.* *144*:147-150.

Madigan, M. T., and Gest, G. (1979). Growth of the photosynthetic bacterium *Rhodopseudomonas capsulata* chemoautotrophically in darkness with H_2 as the energy source. *J. Bacteriol.* *137*:524-530.

Madigan, M. T., Cox, J. C., and Gest, H. (1980). Physiology of dark fermentative growth of *Rhodopseudomonas capsulata*. *J. Bacteriol.* *142*:908-915.

Magee, E. L., Ensoley, B. D., and Barton, L. L. (1978). An assessment of growth yields and energy coupling in *Desulfovibrio*. *Arch. Microbiol.* *117*:21-26.

Maloney, P. C. (1983). Relationship between phosphorylation potential and electrochemical H^+ gradient during glycolysis in *Streptococcus lactis*. *J. Bacteriol.* *153*:1461-1470.

Matsubara, T. (1975). The participation of cytochrome in the reduction of N_2O to N_2 by a denitrifying bacterium. *J. Biochem. (Tokyo)* *77*:627-632.

Matsubara, T., and Zumft, W. G. (1982). Identification of a copper protein as part of the nitrous oxide-reducing system in nitrite-respiring (denitrifying) pseudomonads. *Arch. Microbiol.* *132*:322-328.

Matsubara, T., Frunzke, K., and Zumft, W. G. (1982). Modulation by copper of the products of nitrite respiration in *Pseudomonas perfectomarinus*. *J. Bacteriol.* *149*:816-823.

McEwan, A. G., George, C. L., Ferguson, S. J., and Jackson, J. B. (1982). A nitrate reductase activity in *Rhodopseudomonas capsulata* linked to electron transfer and generation of a membrane potential. *FEBS Lett.* *150*:277-280.

McEwan, A. G., Ferguson, S. J., and Jackson, J. B. (1983). Electron flow to dimethylsulfoxide or trimethylamine *N*-oxide generates a membrane potential in *Rhodopseudomonas capsulata*. *Arch. Microbiol.* *136*:300-305.

McEwan, A. G., Greenfield, A. J., Wetzstein, H. G., Jackson, J. B., and Ferguson, S. J. (1985a). Nitrous oxide reduction by members of the family *Rhodospirillaceae* and the nitrous oxide reductase of *Rhodopseudomonas capsulata*. *J. Bacteriol.* *164*:823-830.

McEwan, A. G., Wetzstein, H. G., Jackson, J. B., and Ferguson, S. J. (1985b). Periplasmic location of the terminal reductase in trimethylamine *N*-oxide and dimethylsulfoxide respiration in the photosynthetic bacterium *Rhodopseudomonas capsulata*. *Biochem. Biophys. Acta* *806*:410-417.

McInerney, M. J., and Bryant, M. P. (1981). Anaerobic degradation of lactate by syntrophic associations of *Methanosarcina barkeri* and *Desulfovibrio* sp. and effect of H_2 on acetate degradation. *Appl. Environ. Microbiol.* *41*:346-354.

Meijer, E. M., van Verseveld, H. W., van der Beek, E. G., and Stouthamer, A. H. (1977a). Energy conservation during aerobic growth in *Paracoccus denitrificans*. *Arch. Microbiol.* *112*:25-34.

Meijer, E. M., Wever, R., and Stouthamer, A. H. (1977b). The role of iron-sulfur center 2 in electron transport and energy conservation in the NADH-ubiquinone segment of the respiratory chain of *Paracoccus denitrificans*. *Eur. J. Biochem.* *81*:267-275.

Meijer, E. M., van der Zwaan, J. W., and Stouthamer, A. H. (1979). Location of the proton consuming site in nitrate reduction and stoichiometries for proton pumping in anaerobically grown *Paracoccus denitrificans*. *FEMS Microbiol. Lett.* *5*:369-372.

Mell, H., Bronder, M., and Kröger, A. (1982). Cell yields of *Vibrio succinogenes* growing with formate and fumarate as sole carbon source and energy sources in chemostat culture. *Arch. Microbiol.* *131*:224-228.

Michalski, W. P., and Nicholas, D. J. D. (1984). The adaptation of *Rhodopseudomonas sphaeroides f.* sp. *denitrificans* for growth under denitrifying conditions. *J. Gen. Microbiol.* *130*:155-165.

Michel, T. A., Melville, S. B., and Macy, J. M. (1987). Physiological evidence for the generation of metabolic energy by succinate efflux in *Selenomonas ruminantium*. Abstracts of the Annual Meeting of the American Society for Microbiology, K100, p. 219.

Michels, P. A. M., Michels, J. P. J., Boonstra, J., and Konings, W. N. (1979). Generation of an electrochemical proton gradient in bacteria by the excretion of metabolic end products. *FEMS Microbiol. Lett.* *5*:357-364.

Miller, J. D. A., and Wakerley, D. S. (1966). Growth of sulphate-reducing bacteria by fumarate dismutation. *J. Gen. Microbiol.* *43*:101-107.

Mitchell, G. J., Jones, J. G., and Cole, J. A. (1986). Distribution of nitrate and nitrite reduction by *Desulfovibrio* and *Desulfotomaculum* species. *Arch. Microbiol.* *144*:35-40.

Mitchell, P. (1976). Possible mechanisms of the protonmotive function of cytochrome systems. *J. Theor. Biol.* *62*:327-367.

Möller, D., Schauder, R., Fuchs, G., and Thauer, R. K. (1987). Acetate oxidation to CO_2 via a citric acid cycle involving an ATP-citrate lyase: A mechanism for the synthesis of ATP via substrate-level phosphorylation in *Desulfobacter postgatei* growing on acetate and sulfate. *Arch. Microbiol.* *140*:202-207.

Motteram, P. A. S., McCarthny, J. E. G., Ferguson, S. J., Jackson, J. B., and Cole, J. A. (1981). Energy generation during the formate-dependent reduction of nitrite by *Escherichia coli*. *FEMS Microbiol. Lett.* *12*:317-320.

Nethe-Jaenchen, R., and Thauer, R. K. (1984). Growth yields and saturation constants of *Desulfovibrio vulgaris* in chemostat culture. *Arch. Microbiol.* *137*:236-240.

Newton, N. A., Cox, G. B., and Gibson, F. (1971). The function of menaquinone (vitamin K2) in *Escherichia coli* K12. *Biochim. Biophys. Acta* *244*:155-166.

O'Brien, W. E., Bowien, S., and Wood, H. G. (1975). Isolation and characterization of a pyrophosphate-dependent phosphofructokinase from *Propionibacterium shermanii*. *J. Biol. Chem.* *250*:8690-8695.

Odom, J. M., and Peck, H. D., Jr. (1981). Hydrogen cycling as a general mechanism for energy coupling in the sulfate-reducing bacteria. *Desulfovibrio* sp. *FEMS Microbiol. Lett.* *12*:47-50.

Odom, J. M., and Peck, H. D., Jr. (1984). Hydrogenase, electron-transfer proteins and energy coupling in the sulfate-reducing bacteria *Desulfovibrio*. *Annu. Rev. Microbiol.* *38*:551-592.

Otto, R., Hugenholz, J., Konings, W. N., and Veldkamp, H. (1980a). Increase of molar growth yield of *Streptococcus cremoris* for lactose as a consequence of lactate consumption by *Pseudomonas stutzeri*. *FEMS Microbiol. Lett.* *9*:85-88.

Otto, R., Sonnenberg, A. S. M., Veldkamp, H., and Konings, W. N. (1980b). Generation of an electrochemical proton gradient in *Streptococcus cremoris* by lactate efflux. *Proc. Natl. Acad. Sci.* *77*:5502-5506.

Otto, R., Lageveen, R. G., Veldkamp, H., and Konings, W. N. (1982). Lactate efflux-induced electrical potential in membrane vesicles of *Streptococcus cremoris*. *J. Bacteriol.* *149*:733-738.
Otto, R., Klont, B., and Konings, W. N. (1985). The relation between phosphate potential and growth rate of *Streptococcus lactis*. *Arch. Microbiol.* *142*:97-100.
Parsonage, D., and Ferguson, S. J. (1983). Reassessment of pathways of electron flow to nitrate reductase that are coupled to energy conservation in *Paracoccus denitrificans*. *FEBS Lett.* *153*: 108-112.
Paulsen, J., Kröger, A., and Thauer, R. K. (1986). ATP-driven succinate oxidation in the catabolism of *Desulfuromonas acetoxidans*. *Arch. Microbiol.* *144*:78-83.
Payne, W. J. (1981). *Denitrification*. John Wiley and Sons, Baltimore.
Pfennig, N., and Biebl, H. (1976). *Desulfuromonas acetoxidans* gen. nov. and sp. nov., a new anaerobic sulfurreducing acetate-oxidizing bacterium. *Arch. Microbiol.* *110*:3-12.
Poole, R. K. (1983). Bacterial cytochrome oxidase. A structurally and functionally diverse group of electron transfer proteins. *Biochim. Biophys. Acta* *726*:205-243.
Pope, N. R., and Cole, J. A. (1982). Generation of a membrane potential by one of two independent pathways for nitrite reduction by *Escherichia coli*. *J. Gen. Microbiol.* *128*:219-222.
Postgate, J. R. (1969). Methane as a minor product of pyruvate metabolism by sulphate-reducing and other bacteria. *J. Gen. Microbiol.* *57*:293-302.
Prakash, D., and Sadana, J. C. (1972). Purification, characterization and properties of nitrite reductase of *Achromobacter fischeri*. *Arch. Biochem. Biophys.* *148*:614-332.
Reeves, R. E. (1976). How useful is the energy in inorganic pyrophosphate? *Trends Biochem. Sci.* *1*:53-55.
Ruoff, K. L., and Delwiche, E. A. (1977). Nitrate reductase electron transport cofactors in *Veillonella alkalescens*. *Can. J. Microbiol.* *23*:1562-1567.
Sawada, E., and Satoh, T. (1980). Periplasmic location of dissimilatory nitrate and nitrite reductase in a denitrifying phototrophic bacterium *Rhodopseudomonas sphaeroides* forma sp *denitrificans*. *Plant Cell Physiol.* *21*:205-210.
Sawnhey, V., and Nicholas, D. J. D. (1977). Sulphite- and NADH-dependent nitrate reductase from *Thiobacillus denitrificans*. *J. Gen. Microbiol.* *100*:49-58.
Sawnhey, V., and Nicholas, D. J. D. (1978). Sulphide-linked nitrite reductase from *Thiobacillus denitrificans* with cytochrome oxidase activity: Purification and properties. *J. Gen. Microbiol.* *106*:119-128.
Schauder, R., Eikmann, B., Thauer, R. K., Widdel, F., and Fuchs, G. (1986). Acetate oxidation to CO_2 in anaerobic bacteria via a

novel pathway not involving reactions of the citric acid cycle. *Arch. Microbiol. 145*:162-172.

Schink, B. (1984). Fermentation of 2,3-butanediol by *Pelobacterium carbinolicus* sp. nov. and *Pelobacterium propionicus* sp. nov., and evidence for propionate formation from C2 compounds. *Arch. Microbiol. 137*:33-41.

Schink, B., and Pfennig, N. (1982). *Propiionigenium modestum*, gen. nov. sp. nov. a new strictly anaerobic nonsporing bacterium growing on succinate. *Arch. Microbiol. 133*:209-216.

Schultz, J. E., and Weaver, P. F. (1982). Fermentation and anaerobic respiration by *Rhodospirrillum* and *Rhodopseudomonas capsulata*. *J. Bacteriol. 149*:181-190.

Seitz, H.-J., and Cypionka, H. (1986). Chemolithotrophic growth of *Desulfovibrio desulfuricans* with hydrogen coupled to ammonification of nitrate and nitrite. *Arch. Microbiol. 146*:63-67.

Senez, J. C., and Pichinoty, F. (1955). Sur la réduction du nitrite aux dépens de l'hydrogène moléculaire par *Desulfovibrio desulfuricans*. *Bull. Soc. Chim. Biol. 40*:2099-2117.

Shimokawa, O., and Ishimoto, M. (1979). Purification and some properties of inducible tertiary N-oxide reductase from *Escherichia coli*. *J. Biochem. (Tokyo) 86*:1709-1717.

Singh, A. P., and Bragg, P. D. (1975). Reduced nicotinamide adenine dinucleotide-dependent reduction of fumarate coupled to membrane energization in a cytochrome-deficient mutant of *Escherichia coli*. *Biochim. Biophys. Acta 423*:450-461.

Slack, J. M. (1974). Genus I Actinomyces (Harz, 1877). In *Bergey's Manual of Determinative Bacteriology*, 8th Ed., R. E. Buchanan and N. E. Gibsons (Eds.). Williams & Wilkins, pp. 661-667.

Stams, A. J. M., Kremer, D. R., Nicolay, K., Weenk, G. H., and Hansen, T. A. (1984). Pathway of propionate fermation in *Desulfobulbus propionicus*. *Arch. Microbiol. 139*:167-173.

Stenberg, E., Styrvold, O. B., and Strøm, A. R. (1982). Trimethylamine oxide respiration in *Proteus* sp. strain NTCC 153: Electron transfer-dependent phosphorylation and L-serine transport. *J. Bacteriol. 149*:22-28.

Steenkamp, D. J., and Peck, H. D., Jr. (1981). Proton translocation associated with nitrite respiration in *Desulfovibrio desulfuricans*. *J. Biol. Chem. 256*:5450-5458.

Stouthamer, A. H. (1967). Mutant strains of *Aerobacter aerogenes* which require both methionine and lysine for aerobic growth. *J. Gen. Microbiol. 46*:389-398.

Stouthamer, A. H. (1973). A theoretical study on the amount of ATP required for synthesis of microbial cell material. *Antonie van Leeuwenhoek 39*:545-565.

Stouthamer, A. H. (1976). Biochemistry and genetics of nitrate reduction in bacteria. *Adv. Microb. Physiol. 14*:315-375.

Stouthamer, A. H. (1977a). Theoretical calculation on the influence of the inorganic nitrogen dource on parameters for aerobic growth of microorganisms. *Antonie van Leeuwenhoek 43*:351-367.
Stouthamer, A. H. (1977b). Energetic aspects of the growth of micro-organisms. *Symp. Soc. Gen. Microbiol. 27*:285-315.
Stouthamer, A. H. (1979). The search for correlation between theoretical and experimental growth yields. In *Microbial Biochemistry*, J. R. Quayle (Ed.). University Park Press, Baltimore, pp. 1-47.
Stouthamer, A. H. (1980a). Energetic regulation of microbial growth. *Vierteljahrsch. Natur. Ges. Zurich 125*:43-60.
Stouthamer, A. H. (1980b). Electron-transport linked phosphorylation in anaerobes. In *Anaerobes and Anaerobic Infections*, G. Gottschalk, N. Pfennig, and H. Werner (Eds.). Gustav Fischer Verlag, Stuttgart, pp. 17-29.
Stouthamer, A. H. (1980c). Bioenergetic studies on *Paracoccus denitrificans*. *Trends Biochem. Sci. 5*:164-166.
Stouthamer, A. H. (1985). Towards an integration of various aspects of microbial metabolism: Energy generation, protein synthesis and regulation. In *Proceeding Third European Congress on Biotechnology*, Vol. 4. VCH-Dechema, Weinheim, pp. 223-239.
Stouthamer, A. H. (1988). Dissimilatory reduction of oxidized nitrogen compounds. In *Environmental Microbiology of Anaerobes*, A. J. B. Zehnder (Ed.). Wiley and Sons, New York, pp. 245-303.
Stouthamer, A. H., and Bettenhausen, C. W. (1972). Influence of hydrogen acceptors on growth and energy production of *Proteus mirobilis*. *Antonie van Leeuwenhoek 38*:81-90.
Stouthamer, A. H., and van Verseveld, H. W. (1985). Stoichiometry of microbial growth. In *Comprehensive Biotechnology*, Vol. 1, M. Moo-Young (Ed.). Pergamon Press, Oxford, pp. 215-238.
Stouthamer, A. H., de Vries, W., and Niekus, H. G. D. (1979). Microaerophily. *Antonie van Leeuwenhoek 45*:5-12.
Stouthamer, A. H., Oltmann, L. F., and vant Riet, J. A. (1980). Respiration with nitrate as acceptor. In *Diversity of Bacterial Respiratory Systems*, Vol. II, C. J. Knowles (Ed.). CRC Press, Boca Raton, Florida, pp. 19-48.
Stouthamer, A. H., Boogerd, F. C., and van Verseveld, H. W. (1982). The bioenergetics of denitrification. *Antonie van Leeuwenhoek 48*:545-553.
Strøm, A. R., Olafsen, J. A., and Larsen, H. (1979). Trimethylamine oxide: A terminal electron acceptor in anaerobic respiration of bacteria. *J. Gen. Microbiol. 112*:315-320.
Takagi, M., Tsuchiya, T., and Ishimoto, M. (1981). Proton translocation coupled to trimethylamine *N*-oxide reduction in anaerobically grown *Escherichia coli*. *J. Bacteriol. 148*:762-768.

Tempest, D. W., and Neijssel, O. M. (1984). The status of Y_{ATP} and maintenance energy as biologically interpretable phenomena. *Annu. Rev. Microbiol. 38*:459-486.

Thauer, R. K. (1982). Dissimilatory sulphate reduction with acetate as electron donor. *Philos. Trans. R. Soc. Lond. [8] 298*:467-471.

Thauer, R. K., and Badziong, W. (1980). Respiration with sulfate as electron acceptor. In *Diversity of Bacterial Respiratory Systems*, Vol. II, C. J. Knowles (Ed.). CRC Press, Boca Raton, Florida, pp. 65-85.

Thauer, R. K., and Morris, J. G. (1984). Metabolism of chemotrophic anaerobes: Old views and new aspects. *Symp. Soc. Gen. Microbiol. 36*:123-168.

Thauer, R. K., Jungermann, K., and Decker, K. (1977). Energy conservation in chemotrophic anaerobic bacteria. *Bacteriol. Rev. 41*:100-180.

Traore, A. S., Hatchikian, C. E., Belaich, J. P., and LeGall, J. (1981). Microcalorimetric studies of the growth of sulfate-reducing bacteria: Energetics of *Desulfovibrio vulgaris* growth. *J. Bacteriol. 145*:191-199.

Traore, A. S., Hatchikian, C. E., LeGall, J., and Belaich, J. P. (1982). Microcalorimetric studies of the growth of sulfate-reducing bacteria: Comparison of the growth parameters of some *Desulfovibrio* species. *J. Bacteriol. 149*:606-611.

Traore, A. S., Fardeau, M. L., Hatchikian, C. E., LeGall, J., and Belaich, J. P. (1983). Energetics of growth of a defined mixed culture of *Desulfovibrio vulgaris* and *Methanosarcina barkeri*: Interspecies hydrogen transfer in batch and continuous cultures. *Appl. Environ. Microbiol. 46*:1152-1156.

Tsuji, K., and Yagi, T. (1980). Significance of hydrogen burst from growing cultures of *Desulfovibrio vulgaris* Miyazaki, and the role of hydrogenase and cytochrome c_3 in energy production system. *Arch. Microbiol. 125*:35-42.

Ugurbil, K., Rottenberg, H., Glynn, P., and Shulman, R. G. (1982). Phosphorus-31 nuclear magnetic resonance studies of bioenergetics in wild type and adenosinetriphosphatase (1-) *Escherichia coli* cells. *Biochemistry 21*:1068-1075.

Unden, G., and Kröger, A. (1986). Reconstitution of a functional electron-transfer chain from purified formate dehydrogenase and fumarate reductase complexes. *Methods Enzymol. 126*:387-399.

Urata, K., Shimata, K., and Satoh, T. (1982). Periplasmic location of nitrous oxide reductase in a photodenitrifier, *Rhodopseudomonas sphaeroides* forma sp. *denitrificans*. *Plant Cell Physiol. 23*:1121-1124.

Van Gent-Ruyters, M. L., de Vries, W., and Stouthamer, A. H. (1975). Influence of nitrate on fermentation pattern, molar growth yields and synthesis of cytochrome b in *Propionobacterium pentosaceum*. *J. Gen. Microbiol. 88*:36-48.

Van Verseveld, H. W., and Stouthamer, A. H. (1978). Electron transport chain and coupled oxidative phosphorylation in methanol-grown *Paracoccus denitrificans*. *Arch. Microbiol. 118*:13-20.

Van Verseveld, H. W., Meijer, E. M., and Stouthamer, A. H. (1977). Energy conservation during nitrate respiration in *Paracoccus denitrificans*. *Arch. Microbiol. 112*:17-23.

Van Verseveld, H. W., Boon, J. P., and Stouthamer, A. H. (1979). Growth yields and efficiency of oxidative phosphorylation of *Paracoccus denitrificans* during two (carbon) substrate-limited growth. *Arch. Microbiol. 121*:213-223.

Van Verseveld, H. W., Krab, K., and Stouthamer, A. H. (1981). Proton pump coupled to cytochrome c oxidase in *Paracoccus denitrificans*. *Biochem. Biophys. Acta 635*:525-534.

Vignais, P. M., Colbeau, A., Willison, J. C., and Jouanneau, Y. (1985). Hydrogen, nitrogenase, and hydrogen metabolism in the photosynthetic bacteria. *Adv. Microbiol. Physiol. 26*:156-234.

Vosjan, J. H. (1975). Respiration and fermentation of the sulphate-reducing bacterium *Desulfovibrio desulfuricans* in a continuous culture. *Plant Soil 43*:141-152.

Wetzstein, H. G., and Gottschalk, G. (1985). A sodium-stimulated membrane-bound fumarate reductase system of *Bacteroides amylophilus*. *Arch. Microbiol. 143*:157-162.

Wetzstein, H. G., McCarthy, J. E. G., and Gottschalk, G. (1987). The membrane potential in a cytochrome-deficient species of *Bacteroides*: Its magnitude and mode of generation. *J. Gen. Microbiol. 133*:73-83.

Widdel, F. (1988). Microbiology and ecology of sulfate- and sulfur-reducing bacteria. In *Environmental Microbiology of Anaerobes*, A. J. B. Zehnder (Ed.). John Wiley and Sons, New York, pp. 469-585.

Widdel, F., and Pfennig, N. (1981). Studies on dissimilatory sulfate-reducing bacteria that decompose fatty acids. I. Isolation of new sulfate-reducing bacteria enriched with acetate from saline environments. Description of *Desulfobacter postgatei* gen. nov. sp. nov. *Arch. Microbiol. 129*:395-400.

Widdel, F., and Pfennig, N. (1982). Studies on the dissimilatory sulfate-reducing bacteria that decompose fatty acids. II. Incomplete oxidation of propionate by *Desulfobulbus propionicus* gen. nov. sp. nov. *Arch. Microbiol. 131*:360-365.

Wikstrom, M., and Krab, K. (1980). Respiration-linked H^+ translocation in mitochondria. Stoichiometry and mechanisms. *Curr. Top. Bioenerg. 10*:51-101.

Wimpenny, J. W. T., and Cole, J. A. (1967). The regulation of metabolisme in facultative bacteria. III. The effect of nitrate. *Biochim. Biophys. Acta 148*:233-242.

Wood, A. P., and Kelly, D. P. (1983). Autotrophic, mixotrophic and heterotrophic growth with denitrification by *Thiobacillus* A2 under anaerobic conditions. *FEMS Microbiol. Lett. 16*:363-370.

Wood, H. G., Ragsdale, S. W., and Pezacka, E. (1986). The acetyl-CoA pathway: A newly discovered pathway of autotrophic growth. *Trends Biochem. Sci. 11*:14-18.

Yamamoto, I., and Ishimoto, M. (1977). Anaerobic growth of *Escherichia coli* on formate by reduction of nitrate, fumarate and trimethylamine *N*-oxide. *Z. Allg. Mikrobiol. 17*:235-242.

Yamamoto, I., and Ishimoto, M. (1981). Oxidation of lactate in comparison with that of glucose in nitrate respiration of *Escherichia coli*. *J. Gen. Appl. Microbiol. 27*:11-20.

Yen, H. C., and Marrs, B. (1977). Growth of *Rhodopseudomonas capsulata* under anaerobic dark conditions with dimethylsulfoxide. *Arch. Biochem. Biophys. 181*:411-418.

Zinder, S. H., and Brock, T. D. (1978). Dimethylsulfoxide as an electron acceptor for anaerobic growth. *Arch. Microbiol. 116*:35-40.

Zumft, W. G., Coyle, C. L., and Frunzke, K. (1985). The effect of oxygen on chromatographic behavior and properties of nitrous oxide reductase. *FEBS Lett. 183*:240-244.

IV

DATA COLLECTION AND ANALYSIS

14

Measurements, Data Analysis, and Control

KA-YIU SAN and E. TERRY PAPOUTSAKIS* / *Rice University, Houston, Texas*
GREGORY STEPHANOPOULOS / *Massachusetts Institute of Technology, Cambridge, Massachusetts*

I. INTRODUCTION

Close interactions between the cell population and the growth environment exist in most fermentation processes. In order to develop, optimize, and operate biologic processes in an efficient and reliable manner, it is crucial that the states of the growth environment be monitored and controlled and that the response of the cell culture to its environment be determined and understood. Measurements, data analysis, and control are involved in achieving this goal. A review of the various aspects of these functions is presented in this chapter.

II. MEASUREMENTS

The most desirable parameters to be monitored in anaerobic fermentations are biomass, carbon and nitrogen substrates, phosphate concentration, gas (e.g., CO_2, H_2, and CH_4) evolution, nongaseous product concentrations, pH, and temperature. Measurements of pH and temperature are not considered here. Additional desirable

**Present affiliation*: Northwestern University, Evanston, Illinois.

measurements may include various intracellular parameters (enzyme activities, ATP, NADH, and NAD concentrations, and concentrations of various intracellular intermediates). With the exception of ATP, NADH, and NAD, measurements of the latter set of parameters are not considered here since they are specific to each fermentation under consideration. The various measurements can be carried out off-line and/or on-line. The latter mode of measurement is most desirable, but in many cases it is either very expensive or impossible.

A. Biomass Concentration

The monitoring of biomass concentration has been reviewed recently by Clarke et al. (1986). In the following brief discussion, that review paper should be consulted if additional references are not provided. Off-line biomass measurements are not considered as either routine (turbidometry and dry weight determination) or overly specialized. On-line measurements can be carried out using turbidometry, the collection of a biomass cake on a filter (Thomas et al., 1985), or by an indirect method. The difficulties of nonlinearity and bubble interference in the turbidometric measurement can be minimized by autosampling into a collection cell, where dilutions can be carried out. Among the indirect methods, the fluorescence method based on constant intracellular NADH concentration (when and if applicable to the particular fermentation) appears most promising (Scheper et al., 1986). For simple fermentations on defined media, on-line biomass estimation can be carried out using carbon, nitrogen, and available electron balances (see Chap. 5) by measuring the rates of substrate consumption and product formation (see, e.g., San and Stephanopoulos, 1984). For complex multiproduct fermentations and/or complex media, this estimation procedure is probably unreliable (McLaughlin et al., 1985; Meyer et al., 1986).

B. Gas Analysis

Although somewhat expensive, analysis of gas composition and flow rate is very accurate and reliable. The analysis can be carried out by either mass spectrometry (MS) (see, e.g., Heinzle et al., 1983) or gas chromatography (GC) (McLaughlin et al., 1985; Meyer et al., 1986). Continuous gas analysis by MS is preferably carried out using a capillary gas inlet, although a membrane interface can also be used. MS provides a very fast and robust method for gas analysis that is stable over long time periods. GC is somewhat slower (2-3 minutes required for the analysis after each injection) but is as reliable and less expensive than MS. In order to measure accurately the production rates of the various gases, the flow rate of a tie (inert) gas (such as N_2) must be accurately controlled and measured using an electronic mass-flow controller (e.g., Tylan Corp., Carson, California). Details can be found in the papers by McLaughlin et al. (1985) and Meyer et al. (1986). More conventional metering devices (needle valves) can be also used for the flow rate of the tie gas. Off-line GC gas analysis can be carried out using a gas-tight syringe (Doremus et al., 1985).

C. Liquid Product Analysis

Liquid products can be a variety of volatile and nonvolatile alcohols, ketones, and organic acids, among others. Off-line analysis can be carried out by a combination of GC, high-performance liquid chromatography (HPLC), and various other specialized instrumental and wet chemistry methods. The on-line analysis of volatile products (e.g., ethanol, acetone, and butanol) can be carried out by MS (Doerner et al., 1982; Heinzle et al., 1983; Pungor et al., 1983) or GC (Comberbach and Bu'lock, 1983; Comberbach et al., 1984) analysis of the fermenter headspace gas. These measurements are based on the assumption that the volatile compounds equilibrate between the gas and liquid phases in the fermenter. Thus, linearity and proper calibration require that the fermenter temperature and ionic strength of the broth remain constant during each experiment. The MS method still has a number of problems (especially with multicomponent volatile product mixtures), which derive from the membrane inlet (nonlinear calibration, membrane memory effects, and slow response times, see references above). Gas sensors can be combined with synthetic membranes (e.g., silicone membranes) and a liquid dilution system to provide a reliable sensor for a single volatile component (see, e.g., Mandenius and Mattiasscn, 1983). Interference from other volatile components, severe nonlinearities, and commercial unavailability are serious drawbacks, however. Concepts and development efforts for a variety of sensors for volatile and nonvolatile compounds have been widely reported in the literature. However, for a variety of reasons (see, e.g., Schügerl et al., 1986), with few exceptions, the technology has not reached the point of commercial exploitation. It is nevertheless likely that many of the persistent difficulties will be resolved in the future, and, thus, such sensors will become available for on-line uses.

In the meantime, liquid-phase analysis by GC and HPLC remain two of the most reliable, robust, and general analytic tools for the on- and off-line analysis of volatile and nonvolatile compounds (both substrates and products). The application of these chromatographic techniques requires the sterile production of clear (cell-free) solutions from the fermentation broth containing the compounds of interest at the same concentrations as in the fermentation suspension. An automatic injection valve is used to sample the clear solution for injection in the chromatograph. The clear solution can be typically produced by membrane ultrafiltration or microfiltration by choosing a membrane of desirable porosity or molecular weight cutoff. A continuous and reliable method for a prolonged operation filtration system is desirable. Tangential flow ultrafiltration or microfiltration at various geometric configurations (hollow fibers, flat disks, or flat sheets) or filtration cells can be used. Tangential flow systems typically have better performance characteristics. When polymeric membranes are used, chemical instead of heat sterilization is usually employed. A ceramic hollow filter system produced by Norton Co. (Worchester, Massachusetts, but marketed by Amicon, Danvers, Massachusetts) can be repeatedly heat sterilized. The choice of a particular configuration depends on the fermenter size, other uses of the filtration system (e.g., biomass recycle), cost, and operational robustness. Although not specifically developed for chromatographic analysis, various filtration system designs have

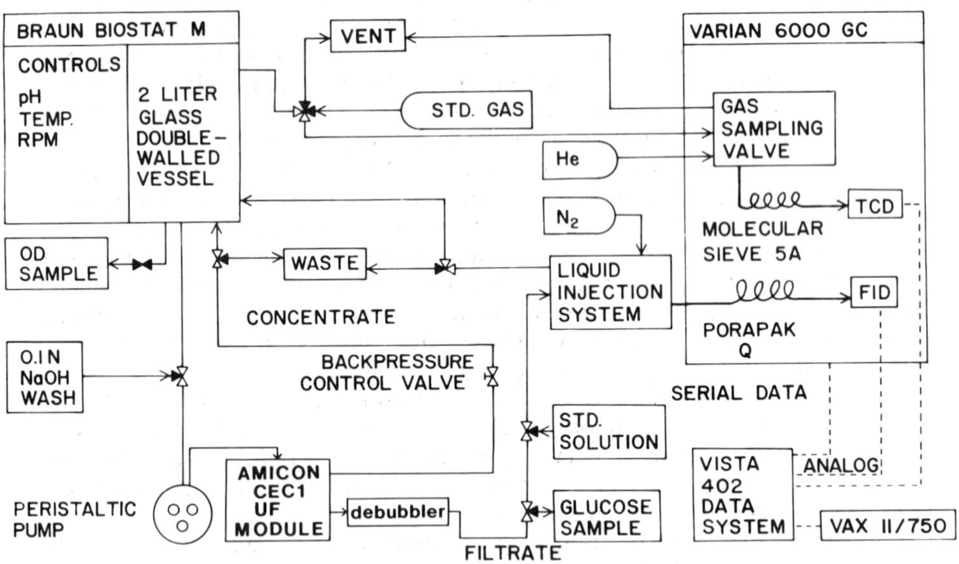

Figure 1 Fermentation monitoring system using an ultrafiltration module (UF) and on-line gas chromatographic (GC) analysis of both the gas and liquid phases of the bioreactor. (From McLaughlin et al., 1985, and Meyer et al., 1986.)

been reported (see, e.g., Chotani and Constantinides, 1982; McLaughlin et al., 1985; Afschar et al., 1985; Meyer et al., 1986; Schlote and Gottschalk, 1986; Pierrot et al., 1986; Ferras et al., 1986). A specific design for the on-line GC analysis of gaseous (CO_2 and H_2) and liquid products (butyrate, acetate, butanol, acetone, ethanol, acetoin, propionate, and propanol) of butyric acid fermentations used in the authors' laboratory (McLaughlin et al., 1985; Meyer et al., 1986) is shown in Figure 1. This system employs an Amicon CEC1 ultrafiltration module (thin-channel, spiral, tangential flow utilizing 90 mm diameter disk membranes of various molecular weight cutoff values) and liquid and gas injection valves (controlled by a dedicated microcomputer) for injection in the GC unit. Depending on the compounds to be analyzed, the liquid injection system must be carefully designed, as described, for example, by McLaughlin et al. (1985). Similar systems can be used for on-line HPLC analysis of nonvolatile substrates and products. The on-line HPLC analysis of antibiotic products has been reported in the literature (Schügerl et al., 1986), and the authors are aware of a number of such systems operating in industrial laboratories and processes for monitoring of antibiotics and various polypeptides in biologic reactions. There is little doubt that with the fast hardware, software, and method developments, GC and HPLC will become widely useful in the analysis and on-line monitoring of many fermentation processes.

D. Analysis of Organic Carbon and Nitrogen Substrates

Simple carbon substrates (such as glucose) can be analyzed on-line by automatic glucose analyzers (Chotani and Constandinides, 1982; Ghoul et al., 1986). More complex sugar mixtures can be analyzed (both on-line and off-line) either by HPLC or by advanced automatic analyzers (Schügerl et al., 1986). Similarly, various organic nitrogen sources (urea and amino acids) and dissolved organic compounds as a lumped parameter can be measured by the aforementioned automatic analyzers (Schügerl et al., 1986) and/or HPLC.

The indirect estimation of simple substrates (e.g., glucose) using elemental balances or advanced stoichiometric equations is also possible [see Chap. 5 (McLaughlin et al., 1985; Meyer et al., 1986)]. In the future, the development of robust biosensors for sugar substrates, amino acids, and other organic compounds is expected to provide a reliable and accessible means for monitoring these parameters.

E. Ammonia and Phosphate Analysis

Ammonia and phosphate concentrations can be measured on-line or off-line by an automatic analyzer (Afschar et al., 1985; Schügerl et al., 1986). NH_3/NH_4 concentrations can be measured on-line or off-line by selective electrodes (for example, Ingold, Willington, Massachusetts, or Orion Research, Boston) that measure either NH_3 or NH_4 directly. the choice of electrode depends on the pH of the fermentation fluid (where the measurement is carried out) and interference from other ions. For example, if direct NH_3 measurement is desirable or necessary (owing to interfering ions in the medium), a clear fluid can be continuously produced by filtration and the pH adjusted automatically to the necessary high value.

F. Other Measurements

Intracellular NAD and NADH concentrations can be determined using acid (for NAD) and alkaline (for NADH) extracts of cells (Wimpenny and Firth, 1972). After extraction and neutralization, NADH can be converted to NAD with glutamate dehydrogenase and α-ketoglutarate. NAD can be assayed by the cycling assay of Bernofsky and Swan (1973). Intracellular NADH concentrations can be correlated with the cultural fluorescence signal (obtained by a microfluorometer probe) for the continuous on-line measurement of intracellular NADH concentrations (Scheper and Schügerl, 1986; Scheper et al., 1986). Cultural fluorescence using microfluorometer probes is a useful measurement for characterizing cell metabolism and can be also used, under proper conditions, for biomass monitoring (Scheper and Schügerl, 1986; Scheper et al., 1986).

Intracellular ATP, ADP, and AMP levels can be measured by a variety of protocols reported in the literature. Two similar protocols have been successfully used in the authors' laboratory. The first is described in detail by Chu and Papoutsakis (1987). In the second protocol, levels of ATP and ADP are measured using the same acid extracts of cells as for NAD. ADP is converted to ATP using phosphocreatine kinase and creatine phosphate. ATP (before and after conversion of ADP) of each sample is assayed with a luminescence method similar to that described by Wulff and Doppen (1985). Calibration and calculation procedures for this assay were modified based on the results of Higashi et al. (1985).

III. DATA ANALYSIS

The measurements obtained from process instruments may not be readily employed for monitoring or control purposes. These raw data are generally subject to certain processing procedures during which meaningful information is extracted. These procedures may include simple noise reduction or more involved and complicated state and parameter estimation schemes. In this section, three different commonly employed procedures are discussed in detail: they are noise reduction, data rectification and reconciliation, and state and parameter estimation.

A. Noise Reduction

The measurement signals generated from the process instruments are often corrupted with noise. These fluctuations in the measurements render most of the instrument output unsuitable for instantaneous use. It is therefore general practice that these signals are subject to noise reduction. When the frequency spectra of the signal and noise lie in different ranges, the signal can be extracted out by suitable bandpass filters (Fig. 2).

These filters can either be analog or digital in nature. In the analog case, the most common and simplest passive filter is in the form of an RC network. Other more sophisticated, special-purpose bandpass filters can also be constructed with active elements. Detailed discussion of this topic is outside the scope of this chapter and can be found in the books by Lam (1979) and Stephenson (1985).

The discrete form has received wide acceptance in recent years because (1) it has a distinct advantage in that it can easily be implemented by a digital computer, and (2) the parameters in the filter can be adjusted with minimal effort. A discrete first-order filter (a counterpart of the RC filter), for example, can take the form of

$$y_k = a_0 u_k + a_1 u_{k-1} - b_1 y_{k-1}$$

where u is the filter input (noisy signal) and y is the filtered output; the subscripts k and k - 1 denote the current and previous values; and a_0, a_1, and b_1 are the parameters that determine the filter characteristics.

In certain cases only the current average value of the signals is of interest, that is, very low frequency component. Various

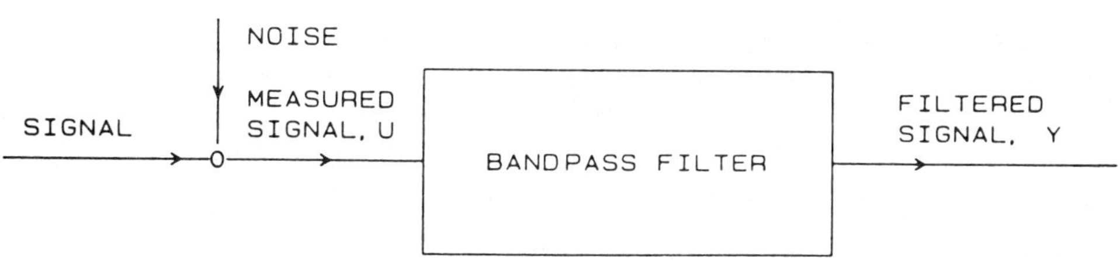

Figure 2 Schematic of a bandpass filter.

Table 1 Examples of Digital Recursive Averaging Algorithms

1. Averaging with infinite memory:

$$y(k) = y(k - 1) + \frac{u(k) - y(k - 1)}{k}$$

2. Averaging with limited memory:

$$y(k) = y(k - 1) + \frac{u(k) - u(k - N)}{N}$$

where N = number of past measurements, averaged with equal weight.

3. Averaging with fading memory:

$$y(k) = y(k - 1) + (1 - \lambda)\, u(k)$$

algorithms have been developed for such a purpose (see, for example, Table 1). Averaging with infinite memory is suitable for the estimation of a constant signal. This is because with increasing numbers of measurements, the new measurements are weighted increasingly less. However, if the signal is slowly time variant and the current average is to be estimated, other algorithms should be used. Of interest is averaging with fading memory. In this algorithm, the parameter λ dictates the amount of weight to be assessed to the incoming measurement. When λ equals 1, the incoming measurement is ignored and has no effect on the estimate. At the other extreme, when λ equals 0, the estimate is equal to the incoming measurement. Thus, the algorithm is capable of estimating time-varying parameters by selecting λ appropriately to provide good tracking.

As a word of caution, however, the digital filters described above cannot effectively filter out noise that is larger than the Shannon frequency because of their sampling nature. In this case, a combination of analog and digital filters should be employed (Isermann, 1981).

B. State and Parameter Estimation

When the spectra of the signal and the noise have overlapping frequency ranges, estimation methods must be used to determine the signal. Among the existing state and parameter estimation schemes, least squares and the Kalman filter or modifications thereof are the most commonly employed approaches.

The essence of the least-squares approach is given below, and without loss of generality, only a single-input/single-output process is presented. Consider a linear process described by a difference equation relating the present output y_n to the previous outputs y_i and the inputs m_i by

$$y_n + a_1 y_{n-1} + a_2 y_{n-2} + \cdots + a_k y_{n-k} = b_1 m_{n-1} + b_2 m_{n-2} + \cdots + b_k m_{n-k}$$

The coefficients a_i and b_i in this model are to be estimated from the measurement of the outputs with known input histories. Then the predicted output at time n is given by

$$\hat{y}_n(n/n-1) = -\hat{a}_1 y_{n-1} - \hat{a}_2 y_{n-2} \cdots - \hat{a}_k y_{n-k} + \hat{b}_1 m_{n-1} + \hat{b}_2 m_{n-2} + \cdots + \hat{b}_k m_{n-k} = \psi^T(n)\hat{\theta}(n-1)$$

where the data vector is

$$\psi^T(n) = [-y_{n-1} \cdots -y_{n-k}\ m_{n-1} \cdots m_{n-k}]$$

and the parameter vector is

$$\hat{\theta}(k-1) = [\hat{a}_1 \cdots \hat{a}_k \hat{b}_1 \cdots \hat{b}_k]$$

When the parameters are changing, it is expected that the measured output will deviate from that of the prediction, thus generating an error signal, which is given by

$$e_{\hat{n}} = y_{\hat{n}} - \hat{y}_{\hat{n}}(n/n-1)$$

A number of least-squares algorithms (recursive or nonrecursive) have been developed to provide the best estimates for these coefficients by minimizing the error signal yielding the general form

$$\hat{\theta}(n+1) = \hat{\theta}(n) + \gamma(n)\,[y_{n+1} - \psi^T(n+1)\hat{\theta}(n)]$$

where $\gamma(n)$ is the correcting vector (Anderson and Moore, 1979; Isermann, 1981). Notice that the new updated estimate $\hat{\theta}(n+1)$ is given by the sum of the old estimate $\hat{\theta}(n)$ and a correction term proportional to the error between the measurement and the prediction,

$y_{n+1} - \psi^T(n+1)\hat{\theta}(n)$. Depending on the type of minimization procedure, different forms of γ are obtained.

The other common state and parameter estimation technique is the Kalman filter or its modification. The theory of filtering is too involved to be described here in detail but can be found in the book by Jazwinski (1970). Only a brief summary is presented here.

The essence of this approach can be stated as follows: consider the state x and its measurements y of a dynamic system satisfying the equations

$$\dot{x} = f(x, u) + \zeta(t)$$

$$y = h(x) + \xi(t)$$

where u is the nonrandom input vector and $\zeta(t)$ and $\xi(t)$ are the random disturbances. The linearized Kalman estimate $\hat{x}$ of the true state x is described by the equations

$$\dot{\hat{x}} = f(\hat{x}, u) + PH^TS^{-1}(y - h(\hat{x}))$$

$$\dot{P} = FP + PF^T + Q - PH^TS^{-1}HP$$

with

$$F = f_x(\hat{x}, u) \qquad H = h_x(\hat{x})$$

where P is the symmetrical covariance matrix of the estimation error; Q and S are the positive definite matrices, which are the measures of the intensities of the noise processes $\xi(t)$ and $\zeta(t)$; and the subscript x refers to the partial derivative with respect to the vector x.

Based on this idea, an adaptive extended Kalman filter supplied with information readily obtained from the on-line material balance was proposed by Stephanopoulos and San (1984). In their approach, time-varying unknown parameters, such as the specific growth rate and the yield factors, are estimated simultaneously by augmenting them into the state vector. The dynamics of these parameters are modeled as random noise processes with adaptive intensity. For example, the governing equation for the specific growth rate μ can be modeled as

$$\dot{\mu} = C + \eta_1(t)$$

$$\dot{C} = \frac{-C}{\tau} + \eta_1'(t)$$

where τ is the characteristic time constant of the system and $\eta_1(t)$ and $\eta_1'(t)$ are white noises whose intensities are determined on-line adaptively based on the incoming data. For more detail, the reader is referred to the original papers (Stephanopoulos and San, 1984; San and Stephanopoulos, 1984). This approach was tested in a variety of cases that are characteristic of the operation of biologic reactors. The results, both numerical and experimental, gave excellent agreement between the on-line estimates and off-line measurements of the corresponding parameters.

It should be pointed out that these estimation algorithms can be used either as stand-alone tools for the study of system dynamics or as an integral part of on-line system identification in adaptive control schemes. A number of these algorithms have actually been used, with great success in the system identification portion of an adaptive control algorithm (see Sec. IV.B).

C. Data Rectification and Reconciliation

When the system is overdetermined, that is, when there are more independent measurements available than absolutely required for the full identification of the system, several techniques have been developed to exploit such a redundancy to enhance data reliability. These techniques can be applied to check for data consistency, to detect the presence of gross errors, to locate the source of gross errors, and so on. Because there are only a limited number of available sensors and very often their reliability is risky, it is highly desirable that these redundancies be fully exploited.

In some cases, there may actually exist a redundancy in instrumentation when well-founded relationships, such as material and energy balances, can be applied to a subset of these measurements. Erickson and coworkers have applied macroscopic elemental balances combined with the regularity found in the oxygen-heat correlation and the degree of reductance of biomass to check for data consistency. Such a consistency check has been applied to the literature data of vast diversity, ranging from single-cell production (Erickson, 1979a) to extracellular product formation (Erickson, 1979b; Erickson et al., 1978). Undoubtedly, this type of analysis is invaluable in the assessment of the reliability of the literature data.

However, these analyses did not consider the variation of errors contained in the measurements, and no criteria were given for assessing the reliability of the measurements. Recently, Wang and Stephanopoulos (1983) have used an on-line method proposed by Romagnoli and Stephenopoulos (1981) that employs a multidimensional chi-square (χ^2) test to indicate the level of inconsistency. Their algorithm is able to detect the presence of gross measurement errors, to identify the source of these errors if they are indeed found to exist, to adjust the inconsistent set of fermentation data, and to rectify the erroneous data in an optimal manner. In their study, they demonstrated that the scheme is capable of identifying the same source of gross errors as suspected by the original investigators, except that, instead of relying on past experience or comparison with other sets of data, the scheme systematically and consistently yielded the correct judgment. The capability of such a scheme to detect and to rectify gross measurement errors is especially valuable in the integration of automated data collection, data analysis, and computerized control of bioreactors.

IV. CONTROL

Control in general serves two major purposes. They are, first, to ensure process reliability (to operate a process in a reproducible and predictable manner) and, second, to maintain process optimality. The performance of the controller depends on a wide variety of factors. These may include the complexity of the processes involved and our knowledge of them (especially their dynamic behavior), the types of control objectives, the number and the quality of the available measurements, and, finally, controller design approaches.

A. Regulatory Control

In most cases, it is desirable to maintain certain process variables in the bioreactor at some specific values (set points) despite fluctuations in process dynamics or parameters and/or external disturbances. These variables may include temperature, pH, and agitation speed. Since most of these variables can be measured directly on-line, their regulation can be accomplished by using a conventional feedback controller (see Fig. 3). These controllers in most cases employ negative feedback in which a measured process output (controlled variable) is subtracted from its desired set point; the difference is known as the error signal e(t). The controller makes use of this error signal to generate an output, which will then be used as an input to the final control element to reduce the error.

1. *On-Off Control*

The simplest type of control is the on-off control. A pure on-off control function produces an output of either zero (off) or 100% (on), according to the sign of the error. However, a perfect on-off device cannot be used for actual control purposes because any infinitesimal level of noise leads to constant chattering. In practice,

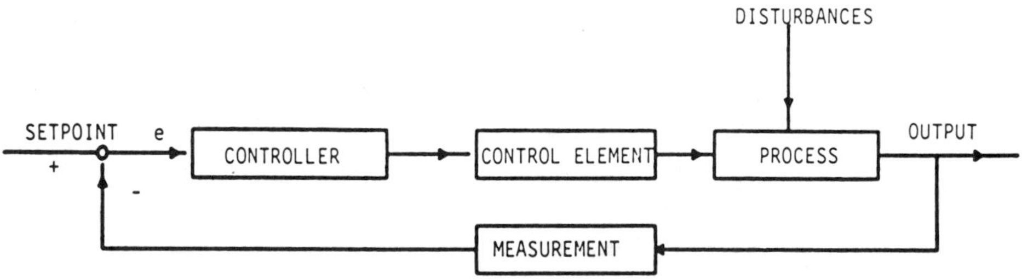

Figure 3 Schematic of a feedback control system.

the controller with hysteresis or a differential dead band is most common. This type of control system is suitable when the loads are relatively constant. However, when there are significant changes in the load during the course of fermentation, sustained oscillation or constant offset from the desired set point may occur. This is best exemplified by considering the regulation of the reactor temperature of a batch fermentation. The heat of fermentation is removed by a stream of cold water through a heat exchanger, and its flow rate is controlled by an on-off controller. It is known that the amount of heat released increases during the fermentation process. Therefore, if the flow rate of the cooling water is adjusted to provide good regulation when the heat load is low, then the fermenter tends to have an offset when the heat load is high. On the other hand, if it is adjusted for good regulation at high heat load, the temperature oscillates at a low heat load.

2. Proportional-Integral-Derivative Control

This problem can be partially alleviated if the control system is modulated in such a way that the control action responds to the magnitude as well as the sign of the error signal. The most common control system of this kind is the proportional-integral-derivative (PID)-type of controller, which utilizes the proportion, integral, and derivative action or a combination thereof. Implementation of the controller can either be analog or digital (direct digital control) in nature. In the analog case, the controller output u(t) is given by

$$u(t) = K_c \left(e(t) + \frac{1}{t_i} \int_0^t e(\tau)\, d\tau + t_d \frac{de}{dt} \right)$$

where K_c, t_i, and t_d are the proportional, integral, and derivative constants, respectively.

The digital counterpart of the PID controller can be approximated by the equation

$$u(k) = K_c \left[e(k) + \frac{T_0}{t_i} \sum_{i=0}^{k} e(i-1) + \frac{t_d}{T_0} (e(k) - e(k-1)) \right]$$

where T_0 is the sampling time, u(k) and e(k) are the current controller output and error, respectively, and e(k − 1) are previous errors. Notice that a rectangular integration has been employed in this equation. A slightly different form is obtained if a different type of integration algorithm is employed, such as the trapezoidal rule. The control algorithm given above is nonrecursive and is commonly known as the position algorithm since it calculates the overall change in u(k). The velocity algorithm, which calculates the current output u(k) based on the previous value u(k − 1) and a correction term, is given by the equation

$$u(k) = u(k-1) + K_c \left(1 + \frac{t_d}{T_0}\right) e(k) - K_c \left(1 + 2\frac{t_d}{T_0} - \frac{T_0}{t_i}\right) e(k-1)$$
$$+ K_c \frac{t_d}{T_0} e(k-2)$$

The velocity algorithm has advantages over the position form in that it does not need initialization and that it is protected against integral "windup."

A properly adjusted or "tuned" controller of this type often provides adequate regulation of the measured variables. A poorly tuned controller can destabilize the system, causing undesirable fluctuations. Detailed discussion of the design of these types of control systems can be found in standard references (Coughanowr and Koppel, 1965; Shinskey, 1979; Stephanopoulos, 1984).

3. *Feed-Forward Control*

Feedback control systems can never achieve perfect control because a feedback controller reacts only after it has detected a deviation in the output value from its desired set point. Another controller design approach is called the feed-forward control, in which the disturbances are measured directly and control actions are taken to eliminate its impact on the process output (see Fig. 4). This type of controller is particularly suitable for processes with slow dynamics. Despite that feed-forward has the potential for perfect control, it also suffers from several inherent weaknesses. These may include (1) the requirement for identification and direct measure-

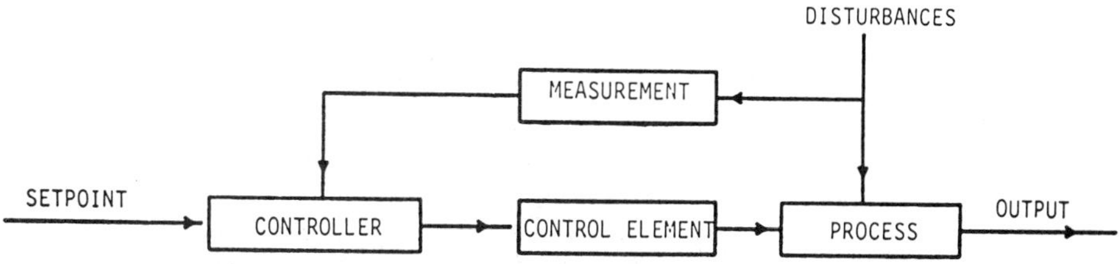

Figure 4 Schematic of a feed-forward control system.

ment of all possible disturbances, (2) the inability to compensate for inherent changes in process dynamics, and (3) the requirement of a good model for the process. Therefore, in practice, a combination of feed-forward and feedback control systems is generally employed. In this scheme, the feed-forward controller is designed with the best available model in an attempt to suppress any impact of external disturbances on the process and the feedback controller takes care of the model mismatch. Indeed, such an approach has been successfully applied to various fermentation processes, yielding a far better performance than the sole feedback controller (Wang et al., 1979; Takamatsu et al., 1985).

B. Adaptive Control

The linear control systems discussed in the preceding section serve their purposes well for the majority of processes. However, there are several fundamental weaknesses often associated with the design of effective controllers of this type. Of particular importance is its heavy reliance on our knowledge of the dynamics of the controlled processes. Therefore, when the process dynamics are poorly understood or the process parameters are poorly known, as in fermentation processes, more sophisticated control algorithms are often required to achieve satisfactory performance. These types of controllers in general require a significant amount of computation, and their implementation to real processes were made possible only very recently by the general availability of computers at reasonable cost.

One of the major problems generally encountered in fermentation processes is the possible occurrence of time-varying system characteristics. As such, if this variation is not compensated for during the course of the fermentation, it eventually leads to deterioration in controller performance or even reactor failure. A controller design approach that specifically tries to address this problem comes under the topic of adaptive control. The main idea behind such an approach is to monitor the changes incurred in the process on-line and then to make necessary adjustments in the controller. That is, two major tasks must be performed on-line: system identification as well as controller design. There are three major controller design approaches: gain scheduling, self-tuning, and model reference. More in-depth discussions on this topic can be found in the review papers by Astrom (1983), Astrom and Wittenmark (1973), and Wittenmark (1975).

The simplest kind of adaptive controller is known as gain scheduling. In this approach, an auxiliary measurable variable that correlates well with changes in process dynamics must first be identified. The controller parameters are then designed as a function of this auxiliary variable in such a way as to reduce the effects of process parameter variations. Such an approach has recently been applied to the control of batch bioreactors, yielding a much improved controller performance (Cardello and San, 1987).

The other two approaches, self-tuning and model reference, have also been applied to control bioprocesses (Williams et al., 1984; Bastin et al., 1983; Takamatsu et al., 1985; Shioya et al., 1986). For example, Dochain and Bastin (1984, 1985) use a self-tuning approach to control an anaerobic digestion process. A recursive least-squares algorithm with a fixed forgetting factor is employed to track the specific growth rate of the culture. The controller design is based on minimum variance. On the other hand, Shioya et al. (1986) employ with great success the model reference approach to regulate the specific growth rate in a fed batch culture. In their scheme, the specific growth rate of the culture was constantly tracked on-line by the adaptive filtering algorithm proposed by San and Stephanopoulos (1984). The estimated growth rate was then fed into a control scheme called the programmed-controller/feedback-compensator (PF) system.

In summary, the development of adaptive controllers for bioreactors is severely hindered by the lack of well-organized estimation studies in the past. However, as a result of recent research efforts, effective system identification schemes are beginning to emerge. This will subsequently enable the development of meaningful adaptive controllers. Recent results, although limited in scope, are very encouraging. More work undoubtedly needs to be done, but nevertheless these results certainly indicate the potential of adaptive control for the effective control of complex systems, such as bioprocesses.

C. Optimization

A major objective of a control system is to achieve process optimality. Within the framework of bioreactor optimization, there are two main optimization groups: static and dynamic optimization.

Static optimization often refers to steady-state continuous operations. The objective is to find a set of operation conditions, that is, the set points of the controlled variables, to maximize a given performance function. These process variables may include the reactor temperature, pH value, and medium feed rate. When reliable models are available, the optimization can be performed off-line using principles of calculus or other search techniques, such as the simplex or complex (constrained simplex) techniques. When the process model is crude or the parameters are unknown or may change with the environmental conditions, the optimization must then be carried out on-line using multivariable optimum seeking techniques (Unden and Heden, 1973; Nyeste et al., 1981). In this approach, upon the establishment of each steady state, a small excursion from the present operating conditions is purposely imposed by a small but predetermined perturbation in the control variables. The resulting performance index is then calculated, and the next move is planned. The same procedure is repeated until no further improvement in the performance index occurs. This is undoubtedly a time-consuming procedure, as a new steady state has to be established before each round of optimization. It is generally believed that an improvement is likely to fade after the first few rounds unless the initial operating conditions are grossly nonoptimal or the system characteristics have undergone drastic changes during the course of operation.

Batch and fed batch bioreactors are operated dynamically and thus require dynamic optimization. The objective here is to find the time profile of the control variables to maximize the performance index. When a process model is available and the parameters are known and constant, the optimization problem can be solved off-line under the framework of optimal control theory (Lee and Markus, 1967; Bryson and Ho, 1975). However, when the dynamics as well as the process parameters are subject to change during the course of fermentation, an adaptive on-line optimization must be performed (Rolf and Lim, 1985). The basic concept of this approach is to combine adaptive control with optimization. That is, the scheme must perform on-line (1) system identification, (2) process optimization, and finally (3) controller design. An extensive review of this topic can be found in the paper by Wang and Stephanopoulos (1984).

V. CONCLUSION

The effective control of biologic processes requires good information about the state of process variables as well as their dynamic behavior. This in turn depends on our capability to monitor and to interpret process data. Recent advances in sensor technology and molecular genetics as well as cell physiology will provide invaluable information that may subsequently enable the development of more accurate process models. In addition, the availability of computers at reasonable cost will further facilitate the development and implementation of new system identification and process control schemes for the full exploitation of this information.

ACKNOWLEDGMENTS

Papoutsakis acknowledges the support of the National Science Foundation through a series of grants related to the subject of this chapter, and in particular support through grants ECE-8613077 and CBT-8518959 for support during the preparation of this chapter.

REFERENCES

Afschar, A. S., Biebl, H., Schaller, K., and Schügerl, K. (1985). Production of acetone and butanol by *Clostridium acetobutylicum* in continuous culture with cell recycle. *Appl. Microbiol. Biotechnol.* *22*:394-398.

Anderson, B. D. O., and Moore, J. (1979). *Optimal Filtering.* Prentice-Hall, Englewood Cliffs, New Jersey.

Astrom, K. J. (1983). Theory and applications of adaptive control—a survey. *Automatica 19*:471-486.

Astrom, K. J., and Wittenmark, B. (1973). On self tuning regulators. *Automatica 9*:185-199.

Bastin, G., Dochain, D., Haist, M., Installe, M., and Opdenacker, P. (1983). Modelling and adaptive control of a continuous anaerobic fermentation process. In *Proceedings of the First IFAC Symposium on Biotechnical Processes*, pp. 299-306. Helsinki, Finland, 1982 (Halme, A., Ed.), Pergamon Press, Oxford.

Bernofsky, C., and Swan, M. (1973). An improved cycling assay for nicotinamide adenine dinucleotide. *Anal. Biochem. 53*:452-458.

Bryson, A. G., and Ho, Y. C. (1975). *Applied Optimal Control.* John Wiley and Sons, New York.

Cardello, R., and San, K.-Y. (1987). Applications of gain scheduling to the control of batch bioreactors. In *Proceedings of the American Control Conference*, Minneapolis, Minnesota, pp. 682-686 (1987).

Chotani, G., and Constantinides, A. (1982). On-line glucose analyzer for fermentation applications. *Biotechnol. Bioeng. 24*:2743-2745.

Chu, I., and Papoutsakis, E. T. (1987). Growth dynamics of a methylotroph (*Methylomonas* L3) in continuous culture. II. Growth inhibition and comparison against an unstructured model. *Biotechnol. Bioeng. 29*:65-71.

Clarke, D. J., Blake-Coleman, B. C., Carr, R. J. G., Calder, M. R., and Atkinson, T. (1986). Monitoring reactor biomass. *Trends Biotechnol.* *4*:173-178.

Comberbach, D. M., and Bu'lock, J. D. (1983). Automatic on-line fermentation headspace gas analysis using a computer-controlled gas chromatograph. *Biotechnol. Bioeng.* *25*:2503-2518.

Comberbach, D. M., Scharer, J. M., and Moo-Young, M. (1984). An improved chromatographic procedure for the rapid determination of volatile solvents via the headspace gas. *Biotechnol. Lett.* *6*:91-96.

Coughanowr, D. R., and Koppel, L. B. (1965). *Process Systems Analysis and Control*. McGraw-Hill, New York.

Dochain, D., and Bastin, G. (1984). Adaptive identification and control algorithms for nonlinear bacterial growth systems. *Automatica* *20*:621-634.

Dochain, D., and Bastin, G. (1985). Stable adaptive controllers for waste treatment by anaerobic digestion. *Environ. Technol. Lett.* *6*:584-593.

Doerner, P., Lehmann, J., Piehl, H., and Megnet, R. (1982). Process analysis of the acetone-butanol fermentation by quadrapole mass spectrometry. *Biotechnol. Lett.* *4*:557-562.

Doremus, M. G., Linden, J. C., and Moreira, A. R. (1985). Agitation and pressure effects on acetone-butanol fermentation. *Biotechnol. Bioeng.* *27*:852-860.

Erickson, L. E. (1979a). Application of mass-energy balances in on-line data analysis. *Biotechnol. Bioeng. Symp.* *9*:49-59.

Erickson, L. E. (1979b). Energetic efficiency of biomass and product formation. *Biotechnol. Bioeng.* *21*:725-743.

Erickson, L. E., Selga, S. E., and Viesturs, U. E. (1978). Application of mass-energy balances regularities to product formation. *Biotechnol. Bioeng.* *20*:1623-1638.

Ferras, E., Minier, M., and Goma, G. (1986). Acetonobutylic fermentation: Improvement of performances by coupling continuous fermentation and ultrafiltration. *Biotechnol. Bioeng.* *28*:523-533.

Ghoul, M., Ronat, E., and Engasser, J. M. (1986). An automatic and sterilizable sampler for laboratory fermentors: Application to the on-line control of glucose concentration. *Biotechnol. Bioeng.* *28*:119-121.

Heinzle, E., Furukawa, K., Dunn, I. J., and Bourne, J. R. (1983). Experimental methods for on-line mass spectrometry in fermentation technology. *Bio/Technol.* *1*:181-188.

Higashi, T., Isomoto, A., Tyuma, I., Kakishita, M., Uomoto, M., and Nagai, K. (1985). Quantitative and continuous analysis of ATP release from blood platelets with firefly luciferase luminescence. *Thromb. Haemos.* *53*:65-69.

Isermann, R. (1981). *Digital Control Systems*. Springer-Verlag, New York.

Jazwinski, A. H. (1970). *Stochastic Processes and Filtering Theory*. Academic Press, New York.

Lam, H. Y. F. (1979). *Analog and Digital Filters—Design and Realization*. Prentice-Hall, Englewood Cliffs, New Jersey.

Lee, E. B., and Markus, L. (1967). *Foundations of Optimal Control Theory*. John Wiley and Sons, New York.

Mandenius, C. F., and Mattiasson, B. (1983). Improved membrane gas sensor systems for on-line analysis of ethanol and other volatile organic compounds in fermentation media. *Eur. J. Appl. Microbiol. Biotechnol. 18*:197-200.

McLaughlin, J. K., Meyer, C. L., and Papoutsakis, E. T. (1985). Gas chromatography and gateway sensors for on-line state estimation of complex fermentations (butanol-acetone fermentation). *Biotechnol. Bioeng. 27*:1246-1257.

Meyer, C. L., McLaughlin, J. K., and Papoutsakis, E. T. (1986). On-line chromatographic analysis and fermenter state characterization of butanol/acetone fermentations. *Biochem. Eng. IV, Ann. N.Y. Acad. Sci. 469*:350-363.

Nyeste, L., Szigeti, L, Pungor, A., Kurucz, I., and Hollo, J. (1981). Automated fermentation equipment. II. Computer-fermentor system. *Biotechnol. Bioeng. 23*:405-417.

Pierrot, P., Fick, M., and Engasser, J. M. (1986). Continuous acetone-butanol fermentation with high productivity by cell ultrafiltration and recycling. *Biotechnol. Lett. 8*:253-256.

Pungor, E., Schaefer, E. J., Cooney, C. L., and Weaver, J. C. (1983). Direct monitoring of the liquid and gas phases during a fermentation in a computer-mass-spectrometer-fermentor system. *Eur. J. Appl. Microbiol. Biotechnol. 18*:135-140.

Rolf, M. J., and Lim, H. C. (1985). Experimental adaptive on-line optimization of cellular productivity of a continuous baker's yeast culture. *Biotechnol. Bioeng. 27*:1236-1245.

Romagnoli, J. A., and Stephanopoulos, G. (1981). Rectification of process measurement data in the presence of gross errors. *Chem. Eng. Sci. 36*:1849-1865.

San, K.-Y., and Stephanopoulos, G. (1984). Studies on on-line bioreactor identification. II. Numerical and experimental results. *Biotechnol. Bioeng. 26*:1189-1197.

Scheper, T., and Schügerl, K. (1986). Culture fluorescence studies on aerobic continuous cultures of *Saccharomyces cerevisiae*. *Appl. Microbiol. Biotechnol. 23*:440-444.

Scheper, T., Lorenz, T., Schmidt, W., and Schügerl, K. (1986). Measurement of culture fluorescence during the cultivation of *Penicillium chrysogenum* and *Zymomonas mobilis*. *J. Biotechnol. 3*:231-238.

Schlote, D., and Gottschalk, G. (1986). Effect of cell recycle on continuous butanol-acetone fermentation with *Clostridium acetobutylicum* under phosphate limitation. *Appl. Microbiol. Biotechnol. 24*:1-5.

Schügerl, K., Lorenz, T., Lubbert, A., Niehoff, J., Scheper, T., and Schmidt, W. (1986). Pros and cons: On-line versus off-line analysis of fermentations. *Trends Biotechnol. 4*:11-15.

Shinskey, F. G. (1979). *Process-Control Systems*. McGraw-Hill, New York.

Shioya, S., Shimizu, F., and Takamatsu, T. (1986). Estimation and control of the specific growth rate in a fed-batch culture. *World Prog. III Chem. Eng. Tokyo* 1986:793-796.

Stephanoupolos, G. (1984). *Chemical Process Control, An Introduction to Theory and Practice*. Prentice-Hall, Englewood Cliffs, New Jersey.

Stephanopoulos, G., and San, K.-Y. (1984). Studies on on-line bioreactor identification. I. Theory. *Biotechnol. Bioeng. 26*:1176-1188.

Stephenson, F. W. (1985). *RC Active Filter Design Handbook*. John Wiley and Sons, New York.

Takamatsu, T., Shioya, S. and Okada, Y. (1985). Profile control scheme in a bakers' yeast fed-batch culture. *Biotechnol. Bioeng. 27*:1675-1686.

Thomas, D. C., Chittur, V. K., Cagney, J. W., and Lim, H. C. (1985). On-line estimation of mycelial cell mass concentrations with a computer-interfaced filtration probe. *Biotechnol. Bioeng. 27*:729-742.

Unden, A., and Heden, G. C. (1973). An approach to on-line continuous culture optimization. *Proceedings of the First European Conference on Computer Process Control in Fermentation*, Dijon, France.

Wang, H. Y., Cooney, C. L., and Wang, D. I. C. (1979). Computer control of bakers' yeast fermentation. *Biotechnol. Bioeng. 21*: 975-995.

Wang, N. S., and Stephanopoulos, G. (1983). Application of macroscopic balances to the identification of gross measurement errors. *Biotechnol. Bioeng. 25*:2177-2208.

Wang, N. S., and Stephanopoulos, G. (1984). Computer applications in fermentation processes. *CRC Crit. Rev. Biotechnol. 2*:1.

Williams, D., Yousefpour, P., and Swanick, B. H. (1984). Online adaptive control of a fermentation process. *IEE Proc. [D] 131*:117.

Wimpenny, J. W. T., and Firth, A. (1972). Levels of nicotinamide adenine dinucleotide and reduced nicotinamide adenine dinucleotide in facultative bacteria and the effect of oxygen. *J. Bacteriol. 111*: 24-32.

Wittenmark, B. (1975). Stochastic adaptive control methods: A survey. *Int. J. Control 21*:705-730.

Wulff, K., and Doppen, W. (1985). Luminometric methods. In *Methods of Enzymatic Analysis*, Vol. VII, 3d Ed., H. U. Bergmeyer (Ed.). VCH Publishers, Weinheim, pp. 357-364.

15

Statistical Estimation of Kinetic and Yield Parameters

LARRY E. ERICKSON and SHIE-SHIEN YANG / *Kansas State University, Manhattan, Kansas*

MEHMET D. ÖNER / *University of Gaziantep, Gaziantep, Turkey*

I. INTRODUCTION

Experimental research with anaerobic cultures is relatively expensive. There is a need to use the best available experimental designs in planning the work and appropriate statistical methods in the analysis of the experimental data. This chapter provides a very brief review of available statistical resources. The application of statistical methods in anaerobic fermentations is illustrated by considering some examples from the published literature.

Considerable progress has been made in on-line data acquisition and analysis (see Chap. 14). Statistical methods can be used in the on-line analysis of data and for parameter estimation. Although the applications reviewed here are off-line examples, many of the methods can be extended and used in on-line data analysis.

Software for many of the statistical methods is readily available and widely used. Statistical analysis system software (SAS, 1979) has been extensively tested and has been applied to many different problems.

II. STATISTICAL ESTIMATION

In this section, we give a review of the statistical problem of estimating common parameters for multiresponse models of which the estimation of kinetic and yield parameters is a special case. Observations obtained from chemical and biological experiments often relate several responses to a common set of unknown parameters. Suppose that in an experimental program, N sets of reaction conditions are run, and at each set of conditions, p responses $(y_1, \ldots, y_p)$ are recorded. Suppose furthermore that we can write a mathematical model for the ith response at the uth set of reaction conditions:

$$y_{iu} = f_{iu}(X_u, \theta) + \varepsilon_{iu} \qquad i = 1, \ldots, p \qquad u = 1, \ldots, N \tag{1}$$

where ε_{iu} is the error in the ith response for the uth run, $\theta = (\theta_1, \ldots, \theta_q)'$ are the q unknown parameters, and $X_u = (x_{1u}, \ldots, x_{ru})'$ are the values of the r input variables defining the reaction conditions for the uth run. The error vectors $\varepsilon_u = (\varepsilon_{1u}, \ldots, \varepsilon_{pu})'$ $(u = 1, \ldots, N)$ are assumed to be independently identically distributed according to a multivariate normal distribution with zero mean vector and unknown covariance matrix Σ. The objective is to find a good estimate of θ (Hunter, 1967).

If there is only one single response, that is $p = 1$, θ can be estimated by the method of least squares. When $f_{iu}(x_u, \theta)$ is a linear function of the components of θ, the least-squares estimate $\hat{\theta}$ of θ can be expressed explicitly as a function of the y_{iu} and the x_{ju}. However, when $f_{iu}(x_u, \theta)$ is a nonlinear function of θ, explicit expression of $\hat{\theta}$ is not available in general and can only be determined numerically. The book on applied regression analysis by Draper and Smith (1966) gives a readable account of the method of fitting single-response models. Also, the book by Bard (1974) gives a detailed account of the method of nonlinear parameter estimation. If there are several (more than one) responses, that is $p > 1$, one standard estimation procedure is the bayesian estimate proposed by Box and Draper (1965). This bayesian estimate of θ is the value of θ that minimizes the determinant of the $p \times p$ matrix with element

$$\sum_{u=1}^{N} \left(y_{iu} - f_{iu}(x_u, \theta)\right) \left(y_{ju} - f_{ju}(x_u, \theta)\right) \qquad i, j = 1, \ldots, p$$

This method yields reasonably good estimates. However, the computation can be cumbersome when there are many responses.

If the covariance matrix Σ of ε_u is known, then a weighted least-squares estimate of θ can be easily determined, which is the value of θ minimizing

$$\sum_{u=1}^{N} \left(Y_u - f_u(x_u, \theta)\right)' \Sigma^{-1} \left(Y_u - f_u(x_u, \theta)\right)$$

where

$$Y_u = (y_{iu}, \ldots, y_{pu})'$$

and

$$f_u(X_u, \theta) = \left(f_{1u}(X_u, \theta), \ldots, f_{pu}(X_u, \theta) \right)'$$

It is the best unbiased estimate of θ. Unfortunately, Σ is usually unknown and is replaced by an estimate. A natural estimate of Σ is given by

$$\hat{\Sigma} = \sum_{u=1}^{N} \hat{\varepsilon}_u \hat{\varepsilon}_u' / N$$

where

$$\hat{\varepsilon}_u = (\hat{\varepsilon}_{1u}, \ldots, \hat{\varepsilon}_{pu})'$$

$$\hat{\varepsilon}_{iu} = y_{iu} - f_{iu}(X_u, \hat{\theta}^i)$$

and θ^i is the least-squares estimate of θ based on the ith response $y_{iu}(u = 1, \ldots, N)$. This is called the two-stage least-squares estimation method in the econometric literature (see, for example, Amemiya, 1985, Chaps. 7 and 8). Solomon et al. (1984) used this technique to estimate parameters for Pirt's model for microbial growth.

Sometimes model (1) can be reparameterized and put into the type of linear model considered by Putthoff and Roy (1964), which is frequently called the general growth curve model; it was originally used to study growth curves. Examples can be found in which transient data are encountered, as in microbial growth and chemical kinetics. Erickson and coworkers (Solomon et al., 1982, 1984; Yang et al., 1984; Oner et al., 1984, 1986a–c) applied this approach to estimate parameters for fermentation data. This approach has the advantage of reducing the multiresponse model to a linear model with one response variable so that standard multiple regression programs can be used to estimate parameters and to test hypotheses concerning the parameters. These authors also derived a covariance adjustment technique to deal with the problem of the responses being related by equality constraints.

In many complex chemical systems, the measured responses are expected to satisfy mass and energy balance constraints (Knepper and Gorman, 1980). However, the given constraints are generally not satisfied because of the presence of random errors. Furthermore, the fact that not all variables are measured because of cost considerations or technical infeasibility causes additional difficulty. Therefore, it is common practice to adjust the measured variables and sometimes estimate the unmeasured variable so that they satisfy the balance constraints. This is known as the data reconciliation problem. A related problem is utilization of the equality constraints to detect gross errors due to malfunctioning of measuring instruments, unexpected leaks, and so forth, so that suitable corrective actions can be taken. This is known as the gross error detection problem. The paper by Tamhane and Mah (1985) gives a detailed review of these two problems and contains many important references. Extreme care should be taken when reconciliated data are used to estimate parameters (de Kwaadsteniet et al., 1976; deKok and Roels, 1980; Geurts et al., 1980) because statistical inference associated with the estimated parameters may be invalidated. Box et al. (1973) considered this problem and recommended that experimenters should try to determine each response independently.

Often, owing to cost or other constraints, only a small number of measurements can be taken and we still want an accurate estimate of the unknown parameter θ. One way to increase the accuracy of the estimate of θ is to choose an optimal set of input variables $X_u = (x_{1u}, \ldots, x_{ru})'$ $(u = 1, \ldots, N)$ defining the reaction conditions from a prespecified region so that the "best" estimate of θ can be estimated. This is essentially the problem of experimental design. The accuracy of the estimate $\hat{\theta}$ of θ, which depends on the number of runs and choice of input variables, can be measured by the sum of the diagonal elements of the covariance matrix A of $\hat{\theta}$ or the determinant of A. The latter measure is preferred because the former ignores the dependence among the $\hat{\theta}_j$ $(j = 1, \ldots, q)$. The (i, j) element of A is

$$N^{-1} \sum_{u=1}^{N} f_u^{(i)'} \Sigma^{-1} f_u^{(j)}$$

where

$$f_u^{(i)} = \left(\frac{\partial f_{1u}(X_u, \theta)}{\partial \theta_i}, \ldots, \frac{\partial f_{pu}(X_u, \theta)}{\partial \theta_i} \right)' \qquad i, j = 1, \ldots, q \quad u = 1, \ldots, N$$

Hence, A is a function of θ, Σ, and X_u. The optimal design is the set of values of X_u $(u = 1, \ldots, N)$ that minimizes the determinant of A. Unfortunately, since θ and Σ are unknown, some initial estimates for θ and Σ must be available. The initial estimates may be determined from data obtained from similar experiments or a small pilot experiment. The ith diagonal element of A is the so-called information index defined by Yoo et al. (1986) for the estimation of parameters from batch culture. If $f_u(X_u, \theta)$ is a linear function of the θ_i, then A does not depend on the θ_i in which case initial estimates for θ are not needed. The theory of optimal design of experiments can be found in the book by Federov (1972). The paper by Box and Draper (1972) considers some design problems for multiresponse nonlinear models.

Many models used for multiresponse data are dynamic models defined by a system of differential equations (Bard, 1974, Chap. 8). The papers by Bates and Watts (1985) and Biegler et al. (1986) treat this type of problem. Their approach involves solving a system of differential equations. In the estimation of parameters for fermentation data, Oner et al. (1986b) and LeDuy and Zajic (1973) adopted a different approach that involves the estimation of derivatives. This is the subject of Section VI.

III. KINETIC MODELS

The most common kinetic model encountered in anaerobic fermentations is the Monod kinetic model, which is identical in form to the Michaelis-Menten enzyme kinetic model. This model may be written in the form

$$\mu = \frac{\mu_{max} S}{K_s + S} \tag{2}$$

The estimation of the kinetic parameters μ_{max} and K_s can be accomplished using a Lineweaver-Burk plot. Equation (2) is rearranged to obtain

$$\frac{1}{\mu} = \frac{K_s}{\mu_{max}} \frac{1}{S} + \frac{1}{\mu_{max}} \tag{3}$$

A plot of $1/\mu$ versus $1/S$ gives a straight line with slope K_s/μ_{max} and intercept $1/\mu_{max}$. This method of analysis separates the two variables, but the slope is dependent upon the data collected at low specific growth rates at which measurement errors are usually relatively large. When batch or fed batch culture is used to collect the data, the specific growth rate must be obtained by differentiating the actual experimental data.

Another linear form of the Monod equation is obtained by writing Equation (2) in the form

$$\frac{S}{\mu} = \frac{S}{\mu_{max}} + \frac{K_s}{\mu_{max}}$$

In this case, the slope provides an estimate of $1/\mu_{max}$ and the intercept is an estimate of K_s/μ_{max}. Equation (2) may also be written in the form

$$\mu = \mu_{max} - K_s \frac{\mu}{S} \tag{4}$$

however, the specific growth rate, which is difficult to measure, appears in both variables when this form is used.

Wilkinson (1961) suggests using a linear form, such as the Lineweaver-Burk plot based on Equation (3), to obtain an initial estimate of the parameters. He then suggests using the original nonlinear form to obtain revised estimates of the parameters. Wilkinson (1961) linearizes the nonlinear form around the estimates of the parameters as follows:

$$\mu = b_1 f(S) + b_2 \frac{df}{dK_s} \tag{5}$$

where

$$f(S) = \frac{\mu_0 S}{K_0 + S} \tag{6}$$

$$\frac{df}{dK_s} = -\frac{\mu_0 S}{(K_0 + S)^2} \tag{7}$$

$$b_1 = \frac{\mu_{max}}{\mu_0} \tag{8}$$

$$b_2 = \frac{(K_s - K_0)\mu_{max}}{\mu_0} \tag{9}$$

The parameters K_0 and μ_0 are the best initial estimates of K_s and μ_{max}. The initial values may be obtained from the Lineweaver-Burk plot, for example.

The parameters b_1 and b_2 are estimated to obtain better estimates of μ_{max} and K_s. This process can be repeated if necessary. The values of b_1 and b_2 tend toward 1.0 and 0, respectively. The standard errors of the estimates of the parameters can be estimated as described by Wilkinson (1961).

Another approach is to treat Equation (2) as a nonlinear model—nonlinear in the parameters μ_{max} and K_s. Nonlinear statistical parameter estimation routines, such as that in SAS (1979), can be used to give least-squares estimates of μ_{max} and K_s. Least-squares estimates from fitting the linear model (3) provide good initial estimates of the parameters for use in the nonlinear parameter estimation computation.

Nonlinear statistical parameter estimation methods can be applied to other closely related kinetic models, such as those that include inhibition and/or endogenous metabolism. The experimental design should include appropriate experiments to obtain the necessary data.

IV. EXPERIMENTAL DESIGN

The design of experiments to estimate the parameters in the Monod kinetic model has been considered by a number of investigators. There are advantages to using batch, fed batch, and continuous culture. Batch culture studies are easy to carry out, and the maximum specific growth rate can be estimated with reasonable precision using batch culture data. Yoo et al. (1986) present a systematic approach to the design of experiments for the estimation of parameters from batch culture. In their work they illustrate the use of a sequential experimental design for batch culture in which the new experiment is selected to improve the estimated values of the unknown parameters.

Continuous culture data has the advantage that the specific growth rate can be estimated from the dilution rate. Thus, the data does not have to be differentiated to obtain estimates of the specific growth rate for use in parameter estimation. Furthermore, the substrate concentration remains constant at a given operating condition. Several measured values of the dilution rate and the substrate concentration can be taken over the period of time that the continuous culture is at a particular steady state value. The disadvantage of continuous culture studies is the time and effort required to collect the data (Solomon et al., 1983).

Fed batch culture has been used to obtain experimental data for parameter estimation. Esener et al. (1981) used fed batch culture to estimate both kinetic and bioenergetic parameters. Solomon et al. (1983) show that experimental design methods may be applied to determine the optimum experimental conditions for collecting the data needed for parameter estimation. Fed batch culture provides a method for conducting experiments at relatively low specific growth rates and relatively low substrate concentrations. These experimental conditions are difficult to achieve in batch culture and require more time in continuous culture; however, the conditions are important for estimation of the saturation constant K_s and the maintenance coefficient m. Webster (1983) reviews the fed batch culture literature and some of the advantages of using fed batch culture.

V. COVARIATE ADJUSTMENT METHOD

In anaerobic fermentation research, the measured variables are related to each other through material balance relationships, such as the carbon balance and the available electron balance. The covariate adjustment method is concerned with the application of statistical methods of parameter estimation and data consistency analysis to growth and product formation in which the measured dependent variables are related by equality constraints. When more than the minimum number of variables are measured in an experiment, parameter estimation results depend on which of the measured variables are selected for use in parameter estimation. A multivariate linear model, which is a special case of the general growth curve model of Putthoff and Roy (1964), can be used to analyze data obtained from fermentation processes. The covariate adjustment method of analysis is based on the earlier work of Khatri (1966), Rao (1965, 1966, 1967), and Grizzle and Allen (1969). The generalized growth curve model can be reduced to a unit variate linear model with covariates (Solomon et al., 1984). This allows standard multiple regression programs to be used to estimate parameters and to test hypotheses concerning the parameters. The covariate adjustment method has been applied to anaerobic fermentation processes to estimate the maximum specific growth rate (Oner et al., 1986a) and bioenergetic yield parameters (Oner et al., 1986a-c). The methods and their application to fermentation processes are described in the above references of Oner et al. and in Solomon et al. (1983, 1984) and Yang et al. (1984).

The covariate adjustment method was used in Chapter 11 to estimate the true growth yield η_{max} and maintenance coefficient m_e in the model of Pirt (1975). To illustrate the covariate adjustment method, consider Equations (10), (11), and (12) of Chapter 11 in the form

$$Y_{1i} = \frac{\mu_i}{\eta_i} = \frac{\mu_i}{\eta_{max}} + m_e + \varepsilon_{1i} \tag{10}$$

$$Y_{2i} = \frac{\mu_i(\eta_i + \xi_{pi})}{\eta_i} = \frac{\mu_i}{\eta_{max}} + m_e + \varepsilon_{2i} \tag{11}$$

$$Y_{3i} = \frac{\mu_i(y_{ci} + z_i + d_i)}{\eta_i} = \frac{\mu_i}{\eta_{max}} + m_e + \varepsilon_{3i} \tag{12}$$

where Equation (10) utilizes biomass and substrate measurements, Equation (11) uses biomass and product measurements, and Equation (12) uses biomass, product, and carbon dioxide measurements.

Let

$$\bar{Y}_{\cdot i} = \frac{1}{3}(Y_{1i} + Y_{2i} + Y_{3i}) \tag{13}$$

and let Z_{1i} and Z_{2i} be a set of covariates that are linear combinations of Y_{1i}, Y_{2i}, and Y_{3i} and have zero expected value in the model

$$\bar{Y}_{\cdot i} = \frac{\mu_i}{\eta_{max}} + m_e + \sum_{k=1}^{2} \alpha_k Z_{ki} + \text{error} \tag{14}$$

An example of a set of two linearly independent covariates that have zero expected value is

$$Z_{1i} = -4Y_{1i} + 3Y_{2i} + Y_{3i} \tag{15}$$

$$Z_{2i} = Y_{1i} + Y_{2i} - 2Y_{3i} \tag{16}$$

If there are no measurement errors and all data satisfy the equality constraints, then the Z_{ki} are identically zero. The covariate adjustment method yields maximum likelihood estimates of the desired parameters when a maximum number of linearly independent covariates are included in the model (14). When no covariates are included in Equation (14), the estimates obtained are the average of the three least-squares estimates from Equations (10) through (12). The optimum number of covariates to use in Equation (14) depends on the experimental data. Covariates that are uncorrelated with $\bar{Y}_{\cdot i}$ should not be included since they contain no information about $Y_{\cdot i}$; inclusion of such covariates decreases the degree of freedom for fitting the model without decreasing the residual variance of Equation (14). A measure of the "usefulness" of a particular set of covariates is given by J, the mean square error for fitting the model, Equation (14), divided by the number of degrees of freedom (Solomon et al., 1983). The width of the 95% confidence interval can also be compared in evaluating the parameter estimates. Solomon et al. (1984) present methods for generating the covariates Z_{ki}.

In the next section, the application of the covariate adjustment method in the analysis of exponential growth data is presented using the work of Oner et al. (1986a) to illustrate the application.

A. Analysis of Exponential Growth

Maximum specific growth rate, growth yield, and product yield are important parameters for many processes. The estimation of these parameters during exponential growth in batch culture under anaerobic conditions is presented.

A procedure to select the exponential growth region is presented. In addition to biomass data, substrate consumption data and product formation data may also be used to estimate the maximum specific growth rate μ_{max}. The covariate adjustment method may be used to obtain a combined estimate using all the data simultaneously. Utilization of all the available data is especially important for systems in which accurate direct biomass measurement is difficult or impossible. The biomass dry weight data used in the example that follows were obtained by an indirect method.

The experimental data analyzed in this work are for the anaerobic growth of *Lactobacillus bulgaricus* and *Streptococcus thermophilus* on 3% dry milk in pure and mixed culture. The experimental details are given elsewhere (Oner and Erickson, 1986).

1. Theory

The methods used to estimate the maximum specific growth rate μ_{max}, the growth yield η, and product yield ξ_p during exponential growth are presented (Oner et al., 1986a). The maximum specific growth rate may be determined from the expression

$$\mu = \frac{1}{X}\frac{dX}{dt} \tag{17}$$

under exponential growth conditions. For exponential growth, $\mu = \mu_{max}$ and the growth yield η is constant; let $\eta = \eta_1$ and $\xi_p = \xi_{p1}$ be the yields during exponential growth. The yield expressions

$$\eta_1 = -\frac{\sigma_b\gamma_b}{\sigma_s\gamma_s}\frac{dX}{dS} \tag{18}$$

$$\eta_1 = \frac{\sigma_b\gamma_b}{\sigma_s\gamma_s}\frac{(X - X_0)}{(S_0 - S)} \tag{19}$$

may be used with Equation (17) to obtain

$$\mu = \frac{-(\sigma_s\gamma_s/\sigma_b\gamma_b)\eta_1 \; dS/dt}{X_0 + \eta_1(\sigma_s\gamma_s/\sigma_b\gamma_b)(S_0 - S)} \tag{20}$$

If we let

$$Z = \frac{X_0\sigma_b\gamma_b}{\eta_1\gamma_s} + (S_0 - S)\sigma_s \tag{21}$$

which gives

$$\frac{dZ}{dt} = -\sigma_s \frac{dS}{dt} \tag{22}$$

the expression

$$\mu = \frac{1}{Z}\frac{dZ}{dt} \tag{23}$$

can be obtained from Equation (20).

Similarly, to utilize product data P, substitute

$$\frac{\eta_1}{\xi_{p_1}} = \frac{\sigma_b \gamma_b}{\sigma_p \gamma_p}\frac{dX}{dP} \tag{24}$$

and

$$\frac{\eta_1}{\xi_{p_1}} = \frac{\sigma_b \gamma_b}{\sigma_p \gamma_p}\frac{X - X_0}{P - P_0} \tag{25}$$

into Equation (17) for X and dX to obtain

$$\mu = \frac{1}{Y}\frac{dY}{dt} \tag{26}$$

where

$$Y = \frac{X_0 \sigma_b \gamma_b \xi_{p_1}}{\sigma_p \gamma_p \eta_1} + P - P_0 \tag{27}$$

When carbon dioxide data are available, one can write

$$\mu = \frac{1}{W}\frac{dW}{dt} \tag{28}$$

where

$$W = \frac{12 y_{c1}}{\sigma_b d_1} \int_{t_0}^{t} Q_c X \, dt + X_0 \tag{29}$$

where $Q_c X$ is the rate of CO_2 evolution per unit volume of broth, y_{c1} is the biomass carbon yield, and d_1 is the carbon dioxide yield.

The biomass carbon yield is related to the biomass available electron yield

$$y_{c1} = \frac{\gamma_s}{\gamma_b} \eta_1 \tag{30}$$

where the subscript 1 refers to exponential growth. Since very little CO_2 is produced when lactic acid is the fermentation product, CO_2 was not measured in the experimental work (Oner and Erickson, 1986).

Equations (21), (22), and (23) utilize substrate data. They can be used as they are when only one substrate is utilized and directly measured. In this specific fermentation process, however, lactose is hydrolyzed to glucose and galactose; then glucose is used as substrate. Since glucose is consumed as soon as formed from the hydrolysis of lactose but galactose is only partially utilized, substrate consumption is determined from measured lactose and galactose concentrations. Therefore, Equation (21) requires modification as follows:

$$Z = \frac{X_0 \sigma_b \gamma_b}{\gamma_s \eta_1} + [\sigma_L(S_{L0} - S_L) + \sigma_{GA}(S_{GA0} - S_{GA})] \tag{31}$$

In these equations, X_0, S_0, S_{GA0}, S_{L0}, and P_0 correspond to the concentrations at time t_0, when the exponential growth process starts, and subscript 1 indicates that these yields are for the exponential growth region, where $\mu = \mu_{max}$.

The available electron balance can be written in integrated form as

$$\frac{[(S_{L0} - S_L)\sigma_L + (S_{GA0} - S_{GA})\sigma_{GA}]\gamma_s}{12} = \frac{\sigma_b \gamma_b}{12}(X - X_0) + \frac{\sigma_p \gamma_p}{12}(P - P_0) \tag{32}$$

Substitution of Equations (27) and (31) into Equation (32) yields (note that $\xi_{p1} + \eta_1 = 1.0$),

$$\frac{\gamma_s}{12} Z = \frac{\sigma_b \gamma_b}{12} X + \frac{\sigma_p \gamma_p}{12} Y \tag{33}$$

The carbon balance can also be written as (note that $y_{c1} + z_1 = 1.0$),

$$\frac{1}{12} Z = \frac{\sigma_b}{12} X + \frac{\sigma_p}{12} Y \tag{34}$$

Estimates of η_1 and ξ_{p1} can be obtained from the basic definitions of these parameters for the time period of exponential growth, based on the first and last data points; that is,

$$\eta_1 = \frac{\sigma_b \gamma_b (X - X_0)}{\gamma_s[(S_{L0} - S_L)\sigma_L + (S_{GA_0} - S_{GA})\sigma_{GA}]} \tag{35}$$

and

$$\xi_{p1} = \frac{\sigma_p \gamma_p (P - P_0)}{\gamma_s[(S_{L0} - S_L)\sigma_L + (S_{GA_0} - S_{GA})\sigma_{GA}]} \tag{36}$$

Better estimates of η_1 and ξ_{p1} may be obtained by applying regression analysis using the exponential growth period data and Equations (35) and (36). One may also use the following equations for this purpose,

$$\eta_1 = \frac{\sigma_b \gamma_b}{\gamma_s} \frac{X}{Z} \tag{37}$$

$$\xi_{p1} = \frac{\sigma_p \gamma_p}{\gamma_s} \frac{Y}{Z} \tag{38}$$

Since prior estimates of η_1 and ξ_{p1} are needed to obtain Z and Y, an iterative procedure is required when Equations (37) and (38) are used to estimate η_1 and ξ_{p1}.

The average values of the regularities for microbial biomass of $\sigma_b = 0.462$ for weight fraction carbon and $\gamma_b = 4.291$ for reductance degree are used throughout this work (Erickson, 1980).

2. Estimation of Parameters

The estimation of maximum specific growth rate μ_{max} and the yield parameters η_1 and ξ_{p1} involves several steps. The time interval of exponential growth must be selected, and the values of η_1 and ξ_{p1} must be found before covariate adjustment is used to estimate the maximum specific growth rate.

Graphic analysis of ln X versus t can be used to select an approximate time interval of exponential growth. Values of η_1 and ξ_{p1} can be obtained from Equations (35) and (36), respectively,

Table 1 Variables Calculated for Estimation of the Exponential Growth Region for Growth of *S. thermophilus*, Microlife, on 3% Dry Milk Under Anaerobic Conditions, Run 5

Data points included	Biomass data				Substrate data				Product data			
	μ_{max} (h^{-1})	T	$\sqrt{MSE}$	R	μ_{max} (h^{-1})	T	$\sqrt{MSE}$	R	μ_{max} (h^{-1})	T	$\sqrt{MSE}$	R
0.0-3.0	0.5550	0.8106	0.0141	0.0079	0.3778	0.7068	0.1118	0.1106	0.6351	0.0138	0.0941	0.0740
0.5-3.0	0.5502	0.6928	0.0148	0.0040	0.3249	0.2252	0.0938	0.0914	0.6267	0.0059	0.0906	0.0877
1.0-3.0	0.5462	0.0539	0.0176	0.0129	0.4905	0.9856	0.0618	0.0105	0.6937	0.0993	0.0534	0.0347
1.5-3.0	0.5720	—[b]	0.0082	0.0034	0.4337	—[b]	0.0607	0.0248	0.7100	—[b]	0.0421	0.0172
2.0-3.0[a]	0.5922	—[b]	—[b]	—	0.2615	—[b]	—[b]	—	0.8157	—[b]	—[b]	—
2.0-3.5	0.4828	—[b]	0.0446	0.0182	0.2996	—[b]	0.0155	0.0063	0.5142	—[b]	0.1231	0.0502
2.0-4.0	0.4429	0.2965	0.0407	0.0170	0.4214	0.1795	0.0794	0.0609	0.4874	0.6072	0.0887	0.0134
2.0-4.5	0.4613	0.9557	0.0373	0.0185	0.4353	0.3341	0.0661	0.0139	0.4681	0.4995	0.0746	0.0193
2.0-5.0	0.4426	0.4177	0.0394	0.0312	0.4413	0.3863	0.0577	0.0100	0.4449	0.2892	0.0704	0.0386
2.0-5.5	0.4184	0.0808	0.0510	0.0604	0.4439	0.4430	0.0517	0.0064	0.4168	0.0873	0.0762	0.0702
1.5-5.0[c]	0.4617	0.1065	0.0458	—	0.4670	0.5611	0.0529	—	0.4793	0.0632	0.0828	—

[a]The initial set of points consisted of points 2.0-3.0
[b]There were not enough degrees of freedom to calculate these variables.
[c]Points 1.5-5.0, 8 data points, are selected as the time interval of exponential growth region and used for μ_{max} estimation. Points 0.0-9.5, 20 data points, are used to estimate the following values of η_1 and ξ_{p1}: $\eta_1 = 0.2365$, $\xi_{p1} = 0.8227$, $\eta_1 + \xi_{p1} = 1.0592$.
Source: From Öner (1985).

using regression analysis for the interval of exponential growth selected above. Equations (21) and (27) can be used to obtain values of Z and Y, respectively, using the estimated values of η_1 and ξ_{p1} and the measured experimental values. Estimated values of maximum specific growth rate μ_{max} may be obtained for this time interval using Equations (17), (23), and (26) in integrated form.

The results of selecting the time interval of exponential growth are presented in Tables 1 and 3; Tables 2 and 4 present the results of estimating the maximum specific growth rate for the selected time interval.

Selection of the interval of exponential growth involves adding one set of data points at a time to the time interval and examining the values of T, $\sqrt{MSE}$, and R for the new time interval. The value of T is the observed significance level of the test statistic for testing whether the quadratic term is needed in each of the following equations

$$\ln X = A_1' + B_1' t + C_1' t^2 \tag{39}$$

$$\ln Z = A_2' + B_2' t + C_2' t^2 \tag{40}$$

$$\ln Y = A_3' + B_3' t + C_3' t^2 \tag{41}$$

A small value of T ($T < 0.1$) indicates that the second-order term is significant. A large value of T indicates that the second-order term is negligible compared with the other terms. Values of T are presented in Tables 1 and 3 for X, Y, and Z. Values of the root mean square error, $\sqrt{MSE}$, from fitting the linear regression line are also

Table 2 Point and 95% Confidence Interval Estimates of Maximum Specific Growth Rate μ_{max} for Run 5

Data used	Covariates included	$\mu_{max}(h^{-1})$ Point	$\mu_{max}(h^{-1})$ Interval	J
Biomass	—	0.4617	0.4210, 0.5024	—
Substrate	—	0.4670	0.4199, 0.5141	—
Product	—	0.4793	0.4058, 0.5528	—
All	Average	0.4693	0.4344, 0.5042	0.000258
All[a]	Z_2	0.4672	0.4341, 0.5003	0.000245
All	Z_3	0.4639	0.4225, 0.5053	0.000330
All	Z_2, Z_3	0.4601	0.4230, 0.4972	0.000261

[a]Selected as best set of results.
Source: From Oner (1985).

Table 3 Variables Calculated for the Estimation of the Experimental Growth Region for Growth of *L. bulgaricus*, Microlife, and *S. thermophilus*, Microlife, Mixed Culture, on 3% Dry Milk Under Anaerobic Conditions, Run 6

Data points included	Biomass data μ_{max} (h^{-1})	Biomass data T	Biomass data $\sqrt{MSE}$	Biomass data R	Substrate data μ_{max} (h^{-1})	Substrate data T	Substrate data $\sqrt{MSE}$	Substrate data R	Product data μ_{max} (h^{-1})	Product data T	Product data $\sqrt{MSE}$	Product data R
0.0–3.5	0.5140	0.0360	0.0711	0.0743	0.3925	0.7487	0.0810	0.0824	0.5405	0.0310	0.1575	0.2042
0.5–3.5	0.4843	0.1780	0.0612	0.0142	0.3814	0.3547	0.0739	0.0243	0.4443	0.2525	0.1027	0.0387
1.0–3.5	0.4758	0.0421	0.0696	0.0501	0.4041	0.4258	0.0844	0.0820	0.4581	0.2788	0.1285	0.1279
1.5–3.5[a]	0.4256	0.0307	0.0642	—	0.4920	0.4889	0.0487	—	0.3287	0.8678	0.0657	—
1.5–4.0	0.3790	0.0527	0.0675	0.0467	0.4759	0.3451	0.0422	0.0161	0.3072	0.8122	0.0572	0.0216
1.5–4.5	0.3559	0.0595	0.0648	0.0387	0.4263	0.0467	0.0701	0.0827	0.2730	0.2903	0.0645	0.0570
1.5–5.0	0.3367	0.0370	0.0649	0.0480	0.4008	0.0164	0.0735	0.0638	0.2736	0.4474	0.0576	0.0011
1.5–5.5	0.3118	0.0045	0.0754	0.0873	0.3952	0.0614	0.0682	0.0194	0.2729	0.5338	0.0526	0.0017
0.5–4.0[b]	0.4464	0.0252	0.0796	—	0.3947	0.2485	0.0692	—	0.4041	0.1766	0.1104	—

[a]The initial set of points consisted of points 1.5–3.5.
[b]Points 0.5–4.0, 8 data points, are selected as the time interval or exponential growth region and used for μ_{max} estimation. Points 0.0–5.5, 12 data points, are used to estimate the following values of η_1 and ξ_{p1}: $\eta_1 = 0.1516$, $\xi_{p1} = 0.8544$, $\eta_1 + \xi_{p1} = 1.0060$.
Source: From Öner (1985).

Table 4 Point and 95% Confidence Interval Estimates of Maximum Specific Growth Rate μ_{max} for Run 6

Data used	Covariates included	μ_{max} (h^{-1}) Point	Interval	J
Biomass	—	0.4464	0.3758, 0.5170	—
Substrate	—	0.3947	0.3334, 0.4560	—
Product	—	0.4041	0.3061, 0.5021	—
All	Average	0.4151	0.3733, 0.4569	0.000368
All[a]	Z_2	0.4106	0.3755, 0.4457	0.000274
All	Z_3	0.4075	0.3407, 0.4743	0.000532
All	Z_2, Z_3	0.4014	0.3430, 0.4598	0.000403

[a]Selected as best set of results.
Source: From Oner (1985).

presented in these tables. Small values are desired. The value of R is the absolute value of the deviation from the predicted line for the observed point being added. A small value indicates that the regression line passes close to the added point.

The results of run 5 in Table 1 may be used to illustrate the process of selecting the time interval of exponential growth. The initial time interval is 2.0-3.0 h. When the points at 3.5, 4.0, 4.5, and 5.0 h are added, the results indicate that each of these points should be included in the exponential growth interval. The values of T exceed 0.1 in all cases; the values of $\sqrt{MSE}$ and R are relatively small, and the estimates of μ_{max} are similar for the three time intervals 2.0-4.0, 2.0-4.5, and 2.0-5.0 h. The intervals 0-3.0, 0.5-3.0, 1.0-3.0 and 1.5-3.0 are also examined, but T is small for the product data. Furthermore, the individual estimates of μ_{max} differ considerably. The interval 1.5 to 5.0 is selected.

Closure of the available electron balance ($\eta_1 + \xi_{p1} = 1.0$) may be considered in selecting the best time interval for regression analysis to estimate the yield parameters η_1 and ξ_{p1} using Equations (35) and (36) and the experimental data. In Tables 1 and 2 the time interval for yield estimation differs from that selected for estimation of maximum specific growth rate because the additional criterion of closure of the available electron balance was considered in selecting the time interval for yield estimation. The time intervals for estimation of η_1 and ξ_{p1} are the same or longer compared to the corresponding intervals for μ_{max} estimation.

After the time interval for exponential growth and the estimates of η_1 and ξ_{p1} are determined, estimates of maximum specific growth rate may be obtained using the covariate adjustment method and the model equations in the form

$$\ln X - \overline{\ln X} = \mu_{max}(t - \bar{t}) \tag{42}$$

$$\ln Z - \overline{\ln Z} = \mu_{max}(t - \bar{t}) \tag{43}$$

$$\ln Y - \overline{\ln Y} = \mu_{max}(t - \bar{t}) \tag{44}$$

where the bar denotes the average value for the data points used in estimating the parameters. Arranging the model in this form satisfies the requirement of covariate adjustment that the right-hand side should be identical for the three equations. Tables 2 and 4 report the estimated values of maximum specific growth rate for runs 5 and 6. In Table 2, for example, individual estimates of μ_{max} are presented based on cell concentration, substrate, and product data, respectively. Combined estimates are presented for zero, one, and two covariates using covariate adjustment. The length of the 95% confidence interval is considered, and the shortest interval is selected as the best result.

The process of selecting the interval of exponential growth based on the values of μ_{max}, T, $\sqrt{MSE}$, and R worked relatively well except that the decision-making process is difficult to reduce to a computer algorithm because 12 values can be examined and used to decide the next computational step. When two of the values of T are less than 0.1 for a run, a further increase in the time interval in the same direction is seldom desirable.

The process of using the available electron balance to help select the time interval of exponential growth for estimation of η and ξ_p worked relatively well. Since values of η and ξ_p change only very slowly with changes in μ, a somewhat wider interval appeared to give more consistent estimates for several of the runs. The average value of the available electron balance for the six runs is 0.98, which is close to 1.0.

Tables 2 and 4 give the point and 95% confidence interval estimates of maximum specific growth rate μ_{max} for runs 5 and 6. In these tables, the first three results were obtained using Equations (42), (43), and (44), with biomass, substrate, and product data, respectively. The results using individual estimates may vary slightly (see run 5) or significantly (see run 6). Most of the runs show differences among the estimates of μ_{max} for the individual equations. These different results provide the motivation for using all the experimental data with the covariate adjustment method to estimate μ_{max}. Results with none, one, and two covariates are given in the tables.

Two criteria may be considered for best estimate selection: the width of the confidence interval and the value of the parameter J are both presented in the tables.

Small values of J are desired as J is an estimate of the generalized variance associated with estimating the parameter (Yang et al., 1984). When the number of data points is larger, selection of the best estimate can be based solely on the value of J. However, introducing covariates reduces the degrees of freedom and this also affects the width of the confidence interval when the number of degrees of freedom is relatively small, as it is in this work. Both the value of J and the width of the confidence interval were considered in selecting the best estimates, which are presented in Table 5. The results for run 3 are also graphically presented in Figure 1. This figure shows that the best estimate (solid line) is between the other lines; it is very close to the regression lines based on the biomass data and the product data. Since the errors associated with fitting the biomass data and product data are considerably smaller than those associated with fitting the substrate data, this result is expected.

B. Yield Analysis

Covariate adjustment methods to estimate the true growth yield η_{max} and the maintenance coefficient m_e for anaerobic cultures were presented in Section V. Results are presented in Chapter 11. In several cases in Chapter 11, the true growth yields are very close to the theoretical maximum yields for anaerobic lactic acid fermentations. In the work of Oner et al. (1986c), six sets of experimental data are analyzed using four different pure cultures and two different mixed cultures of *L. bulgaricus* and *S. thermophilus*. The true growth yield η_{max} and maintenance parameter m_e are estimated for each set of data using the methods presented previously and a new method in which the results of analysis of the exponential growth region are used together with the remaining data (Oner et al., 1986c). The experimental methods and results of the six anaerobic batch followed by fed batch fermentations are described in detail by Oner and Erickson (1986). Fed batch fermentation was used in order to collect data at low specific growth rate. Nonfat dry milk was used as the substrate.

The experimentally measured values of biomass, lactose, galactose, and lactic acid concentration were first used to estimate the maximum specific growth rate, growth yield, and product yield for the exponential growth region as described above. The data were then smoothed using spline functions to obtain estimates of the derivatives of the concentration variables and specific rates of growth, lactose

Table 5 Summary of Estimated Parameters of Yogurt Cultures Based on Exponential Growth Interval

Organisms used	Run no.	$\mu_{max}(h^{-1})$		η_1		ξ_{p1}		$\eta_1 + \xi_{p1}$
		Point	Interval	Point	Interval	Point	Interval	
L. bulgaricus (ATCC-11842)	1	0.3942	0.3503, 0.4381	0.1689	0.1274, 0.2104	0.8317	0.6401, 1.0233	1.0006
S. thermophilus (ATCC-19258)	2	0.2805	0.2666, 0.2944	0.1443	0.1323, 0.1563	0.8038	0.7306, 0.8770	0.9481
L. bulgaricus (ATCC-11842) and *S. thermophilus* (ATCC-19258)	3	0.4131	0.3953, 0.4309	0.1730	0.1512, 0.1948	0.7596	0.6569, 0.8668	0.9326
L. bulgaricus (Microlife)	4	0.3191	0.2982, 0.3400	0.1137	0.0713, 0.1561	0.8410	0.6248, 1.0572	0.9547
S. thermophilus (Microlife)	5	0.4672	0.4341, 0.5003	0.2365	0.2250, 0.2480	0.8227	0.7782, 0.8672	1.0592
L. bulgaricus (Microlife) and *S. thermophilus* (Microlife)	6	0.4106	0.3755, 0.4457	0.1516	0.1282, 0.1750	0.8544	0.7158, 0.9929	1.0060

Source: From Oner et al. (1986a).

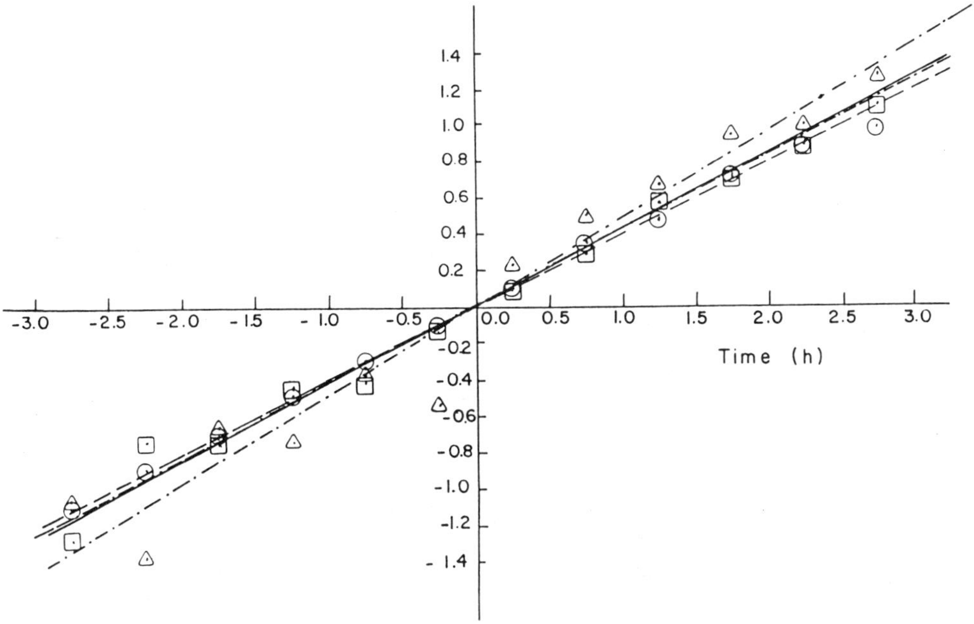

Figure 1 Graphic results of regression analysis for run 3 using Equations (42) (---), (43) (-•-•), and (44) (•---•) and also these equations with covariate adjustment (——). Circles are biomass data, triangles are substrate data, and squares are product data. The ordinate scale is the left-hand side of the appropriate equation; for covariate adjustment it is $(\ln X - \overline{\ln X} + \ln Z - \overline{\ln Z} + \ln Y - \overline{\ln Y})/3$. (From Oner et al., 1986a.)

consumption, galactose production, and lactic acid production (Oner et al., 1986b). The batch and fed batch data are smoothed separately. The covariate adjustment method is used to obtain point and 95% confidence interval estimates of the true growth yield and maintenance coefficient.

1. Theory

The theory associated with the estimation of true growth yield and maintenance parameters for anaerobic fermentations has been developed previously. (See Chap. 11 and the earlier part of this chapter.) For anaerobic fermentation with nonfat dry milk as substrate, only lactose and galactose are considered substrates. When other acid products are formed in addition to lactic acid, the total acid product formation rate is used in the available electron balance:

$$\eta + \xi_p = 1.0 \tag{45}$$

where

$$\eta = \frac{\sigma_b \gamma_b}{Q_L \sigma_L \gamma_L - Q_{GA} \sigma_{GA} \gamma_{GA}} \tag{46}$$

and

$$\xi_p = \frac{\sum_{i=1}^{M} Q_{pi} \sigma_{pi} \gamma_{pi}}{Q_L \sigma_L \gamma_L - Q_{GA} \sigma_{GA} \gamma_{GA}} \tag{47}$$

where M different acid products are produced.

In lactic acid fermentation, small quantities of carbon dioxide are evolved, but CO_2 production was not measured in this work. Since CO_2 was not measured and since the carbon balance in the approximate form

$$y_c + z \simeq 1.0 \tag{48}$$

uses the same measurements as the available electron balance, only the available electron balance was used with covariate adjustment in this work. Pirt's model (1975) may be written as form I,

$$\frac{1}{\eta} = \frac{1}{\eta_{max}} + \frac{m_e}{\mu} \tag{49}$$

$$\frac{\eta + \xi_p}{\eta} = \frac{1}{\eta_{max}} + \frac{m_e}{\mu} \tag{50}$$

and as form II,

$$\frac{\mu}{\eta} = \frac{\mu}{\eta_{max}} + m_e \tag{51}$$

$$\frac{\mu(\eta + \xi_p)}{\eta} = \frac{\mu}{\eta_{max}} + m_e \tag{52}$$

Equations (49) and (51) require lactose, galactose, and biomass measurements, whereas Equations (50) and (52) require acid product and biomass measurements. All the data may be used simultaneously with the covariate adjustment method to obtain combined estimates with either form I or form II.

The estimated values of maximum specific growth rate, biomass yield, and product yield for the exponential growth region may also be used in the process of estimating the true growth yield and maintenance coefficient. When the data for the exponential region are replaced with these estimated values, which may be used as one data point, the exponential growth region data are not weighted as fully as they should be. Greater weighting of the exponential growth data can be obtained by replacing each raw data point in the exponential region with the estimated values of μ_{max}, η, and ξ_p for the exponential region. When only the original data are used, this corresponds to data set A; data set B involves using the exponential region estimates as one data point in place of the data used to estimate μ_{max}; and data set C involves using the exponential region estimates as many times as the number of data points replaced.

2. *Results and Discussion*

Tables 6, 7, and 8 are presented for *L. bulgaricus* and *S. thermophilus* (Microlife), run 6, in order to illustrate the full set of parameter estimates considered for each set of data. Table 6 contains the point and 95% confidence interval estimates based on lactose, galactose, and biomass data using Equation (49) for form I and Equation (51) for form II; the estimates based on biomass and acid product data using Equation (50) for form I and Equation (52) for form II, and the estimates based on all the data using form I and form II with no covariates (average value is obtained) and one covariate. In Table 6 several points have been deleted because of poor available electron balance results or very small specific growth rates. Data points with very low specific growth rates strongly influence the parameter estimates when form I is used. Moreover, because the specific rates are small, these data have a greater chance of being erroneous. With the covariate adjustment method, the best estimate is selected based on the smallest value of

$$J = \frac{\text{mean square error}}{N - p - q - 1} \tag{53}$$

where q is the number of covariates used, p is the number of parameters estimated, and N is the number of data points used in the parameter estimation.

Table 6 gives the results when all the data points except those at 12.5, 15.5, 16.0, and 16.5 are used. In Table 7, points at 0.5-4.0 h are replaced by a single point $\eta_1 = 0.1516$, $\xi_{p1} = 0.8544$, and $\mu_{max} = 0.4106\ h^{-1}$, which was obtained using a separate analysis

Table 6 Point and 95% Confidence Interval Estimates of True Growth Yield and Maintenance Parameter for Run 6: *Lactobacillus bulgaricus* and *Streptococcus thermophilus* (Microlife) and Data Set A[a]

Equations used	Covariates used	m_e (h^{-1})		η_{max}		J
		Point	Interval	Point	Interval	
(49)	—	0.2898	0.2111, 0.3685	0.1397	0.1181, 0.1710	—
(50)	—	0.3751	0.3281, 0.4221	0.1578	0.1404, 0.1800	—
(49), (50)	Average	0.3325	0.2731, 0.3919	0.1482	0.1292, 0.1737	0.133658
(49), (50)	Z	0.3947	0.3387, 0.4507	0.1626	0.1433, 0.1879	0.085240[b]
(51)	—	0.5946	0.4196, 0.7696	0.1995	0.1777, 0.2271	—
(52)	—	0.2979	0.1777, 0.4181	0.1444	0.1361, 0.1537	—
(51), (52)	Average	0.4463	0.3160, 0.5766	0.1675	0.1556, 0.1813	0.002117
(51), (52)	Z	0.3385	0.1819, 0.4951	0.1500	0.1345, 0.1695	0.001902[b]

[a]All data points except 12.5, 15.5, 16.0, and 16.5 are used (N = 27 points).
[b]Selected as the best set of estimates.
Source: From Oner (1985).

Table 7 Point and 95% Confidence Interval Estimates of True Growth Yield and Maintenance Parameter for Run 6, Data Set B[a]

Equations used	Covariates used	m_e (h^{-1})		η_{max}		J
		Point	Interval	Point	Interval	
(49)	—	0.2445	0.1512, 0.3378	0.1194	0.00982, 0.1520	—
(50)	—	0.3774	0.3152, 0.4396	0.1591	0.1336, 0.1966	—
(49), (50)	Average	0.3110	0.2348, 0.3872	0.1364	0.1136, 0.1706	0.236000
(49), (50)	Z	0.4793	0.3811, 0.5775	0.2135	0.1576, 0.3306	0.125470[b]
(51)	—	0.4051	0.1903, 0.6196	0.1435	0.1184, 0.1820	—
(52)	—	0.3931	0.2684, 0.5178	0.1575	0.1388, 0.1820	—
(51), (52)	Average	0.3991	0.2309, 0.5673	0.1502	0.1279, 0.1817	0.002742
(51), (52)	Z	0.3809	0.3064, 0.4554	0.1747	0.1592, 0.1937	0.000563[b]

[a]Points 0.5-4.0 in data set A are replaced by a single exponential point; $\eta_1 = 0.1516$, $\xi_{p1} = 0.8544$, $\eta_1 + \xi_{p1} = 1.0060$, and $\mu_{max} = 0.4106\ h^{-1}$ (N = 20 points)

[b]Selected as the best set of estimates.

Source: From Oner (1985).

Table 8 Point and 95% Confidence Interval Estimates of True Growth Yield and Maintenance Parameter for Run 6, Data Set C[a]

Equations used	Covariates used	m_e (h^{-1})		η_{max}		J
		Point	Interval	Point	Interval	
(49)	—	0.2839	0.2078, 0.3600	0.1369	0.1166, 0.1656	—
(50)	—	0.3869	0.3403, 0.4335	0.1658	0.1469, 0.1904	—
(49), (50)	Average	0.3354	0.2756, 0.3952	0.1500	0.1305, 0.1763	0.135300
(49), (50)	Z	0.4490	0.3900, 0.5080	0.1901	0.1634, 0.2273	0.064994[b]
(51)	—	0.5312	0.3589, 0.7035	0.1798	0.1595, 0.2061	—
(52)	—	0.4554	0.3585, 0.5523	0.1769	0.1652, 0.1902	—
(51, (52)	Average	0.4933	0.3598, 0.6268	0.1783	0.1624, 0.1977	0.001783
(51), (52)	Z	0.3793	0.3239, 0.4347	0.1740	0.1677, 0.1807	0.000278[b]

[a]The exponential point in data set B is used as eight points (N = 27 points).
[b]Selected as the best set of estimates.
Source: From Oner (1985).

of the exponential growth region. The assumption made in estimating this point using the exponential region data is that the culture grows exponentially with constant yield for the time interval selected. In Table 8, the point estimated from the analysis of the exponential region is repeated eight times to replace the eight points in the time interval from 0.5 to 4.0 h.

Tables 9 and 10 give point and 95% confidence interval estimates of the true growth yield parameter η_{max} and the maintenance coefficient m_e for runs 2 and 5, respectively. These parameter estimates utilize the original experimental data, the results of analysis of the exponential growth region, and the results of data smoothing to estimate derivative quantities for specific rate estimation. The estimated values of specific growth rate, specific lactose consumption rate, specific galactose production rate, and specific acid production rate that are used to estimate η_{max} and m_e are tabulated in Oner (1985).

Figure 2 presents in graphic form the results from Table 10 for form II. In Figure 2, the smoothed data for Equation (51) are represented by the circles and the smoothed data for Equation (52) are represented by X. The square and triangle represent the single estimated data point for Equations (51) and (52), respectively, from analysis of the exponential growth conditions. The solid line is the result from data set A, Table 10, form II, with the smallest value of J in which η_{max} = 0.2217 and m_e = 0.0348 h^{-1}. The dashed line is the form II result in Table 10 for data set B in which η_{max} = 0.2544 and m_e = 0.0985 h^{-1}. The intermediate result is for form II, Data set C, Table 10 with no covariates, in which η_{max} = 0.2330 and m_e = 0.0793 h^{-1}. Each of these lines represents a maximum likelihood estimate for the data employed. The points for which the specific growth rate μ is greater than 0.3 h^{-1} are the points that were replaced by the single exponential point.

The results for *S. thermophilus* (ATCC 19258) are shown in Table 9. For this data set significantly different results are obtained using data set A. The true growth yields are greater than the theoretical maximum value for data set A. Much better results are obtained for data sets B and C. The average true growth yield using only data sets B and C is η_{max} = 0.2129; the growth yield from analysis of the exponential growth region is 0.1443. These values are smaller than the values for *S. thermophilus* (Microlife) of η_{max} = 0.2533 and η_1 = 0.2365. The estimated values of the maintenance coefficient are larger for the ATCC culture.

The significant differences in the results from the different treatments of the same original data show that care must be used in smoothing the data and estimating parameters.

The results presented here are the result of careful laboratory experimental work and a rather extensive and complete analysis of

Table 9 Selected Point and 95% Confidence Interval Estimates of True Growth Yield and Maintenance Parameter for Run 2[a]

Data used	Form used	Covariates used	m_e (h^{-1}) Point	m_e (h^{-1}) Interval	η_{max} Point	η_{max} Interval
A	I	Z	0.7199	0.5648, 0.8750	0.3695	0.2874, 0.5171
	II	Average	1.0671	0.8624, 1.2718	0.8525	0.4836, 3.5929
B	I	Z	0.5347	0.4554, 0.6140	0.2338	0.2105, 0.2628
	II	Z	0.4185	0.3590, 0.5311	0.2064	0.1863, 0.2313
C	I	Z	0.4897	0.4258, 0.5536	0.2155	0.2015, 0.2316
	II	Z	0.3728	0.2948, 0.4508	0.1958	0.1850, 0.2079

[a]In data set A, all the data points except 13.0, 13.5, 14.0, 14.5, and 16.0 are used (30 points). In data set B, Points 0.5–5.0 are replaced by single exponential point; $\eta_1 = 0.1443$, $\xi_{p1} = 0.8038$, $\eta_1 + \xi_{p1} = 0.9481$, $\mu_{max} = 0.2805\ h^{-1}$; N = 21. In data set C, above exponential point is used as 10 points; N = 30.
Source: From Oner et al. (1986c).

Table 10 Selected Point and 95% Confidence Interval Estimates of the Growth Yield and Maintenance Parameter for Run 5[a]

Data used	Form used	Covariates used	m_e (h^{-1}) Point	m_e (h^{-1}) Interval	η_{max} Point	η_{max} Interval
A	I	Average	0.1457	0.0973, 0.1941	0.2731	0.2463, 0.3064
	II	Z	0.0348	-0.0684, 0.1380	0.2217	0.1912, 0.2637
B	I	Average	0.1504	0.0837, 0.2135	0.2765	0.2382, 0.3294
	II	Average	0.1061	0.0505, 0.1617	0.2544	0.2360, 0.2758
C	I	Average	0.1270	0.0802, 0.1734	0.2584	0.2390, 0.2812
	II	Average	0.0793	0.0374, 0.1212	0.2416	0.2329, 0.2510

[a]Data set A includes all points except 0.5, 1.0, 1.5, 16.0, and 17.5 h (N = 28). In data set B, points 2.0–5.0 are replaced by a single exponential point: $\eta_1 = 0.2365$, $\xi_{p1} = 0.8227$, $\eta_1 + \xi_{p1} = 1.0592$, and $\mu_{max} = 0.4672\ h^{-1}$ (N = 22). In data set C, the exponential point is used as seven points (N = 28).
Source: From Oner et al. (1986c).

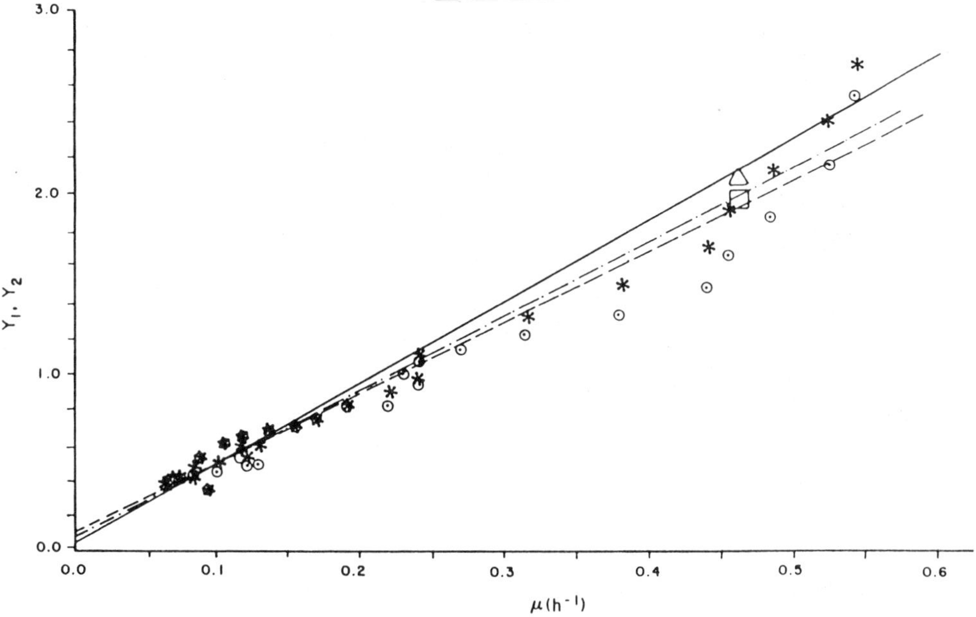

Figure 2 Graphic representation of results from run 5, form II. Y_1 and Y_2 indicate the left-hand side of Equations (51) and (52), respectively. Y_1 is denoted by circle and Y_2 by *. The square and triangle indicate the estimated values from exponential analysis for Y_1 and Y_2, respectively. The solid line (——) indicates the selected result for data set A when the estimated exponential point is not used. The broken line (---) indicates the selected result for data set B when the estimated exponential point is used as one point. The line with dots (-•-•) shows the results obtained for data set C when the estimated exponential point is used as many times as the number of exponential points replaced. (From Oner et al., 1986c.)

the experimental data. Less time is required in the laboratory when batch followed by fed batch culture is utilized instead of continuous culture; however, the results should be analyzed with some care using appropriate consistency checks. Separate analysis of the exponential growth region appears to be desirable as it provides an estimate of specific growth rate and growth yield that does not require data smoothing and derivative estimation. The utilization of spline functions to smooth the data and estimate derivative quantities appears to have contributed positively to the effort to obtain good parameter estimates.

VI. ESTIMATION OF DERIVATIVES

When batch or fed batch data are used, derivatives need to be obtained from the data for use in parameter estimation. LeDuy and Zajic (1973) and Oner et al. (1986b) have presented methods for estimating specific rates. LeDuy and Zajic (1973) presented a simple geometric method that is easy to use and gives useful estimates of the derivative. Three data points are used to estimate each derivative quantity.

Oner et al. (1986b) use spline functions for smoothing fermentation data. Spline functions are piecewise polynomials that can be used for smoothing and differentiation of data (Reinsch, 1967, 1971; Wahba and Wold, 1975; Wold, 1974). Oner et al. (1986b) use a response surface method in which the consistency of the derivative estimates is considered by using the criteria function

$$f = \sum_{i=2}^{N-1} \frac{[1 - (\eta_i + \xi_{pi})]^2}{N-2} \tag{54}$$

in selecting values of the smoothing parameter for each variable. Buono et al. (1986) compare this method with a cross-validation method. Both methods give similar estimates of the derivative quantities. Oner et al. (1986b) report that better results are obtained when batch and fed batch culture are smoothed separately using different values of the smoothing parameters. Since the derivative quantities are not continuous at the point where feeding starts, it is logical to smooth the batch and fed batch data separately.

The computations may be easily carried out using SAS (1979). The response surface method is included in SAS, and the spline function program ICSSCU (1980) may be used as a subroutine in the SAS program.

VII. CONCLUSIONS

Statistical methods can be used effectively in planning experiments and in analysis of data. The availability of SAS and other computer programs and software allows statistical methods to be used with a minimal amount of computer programming. When data are collected and stored using computers, statistical methods can be used in data analysis with little additional effort. The covariate adjustment method is a useful statistical technique to use because maximum likelihood parameter estimates can be obtained and all the data can be used simultaneously even when data are related through equality constraints.

VIII. NOMENCLATURE

A	Covariance matrix of $\hat{\theta}$
A'_1, A'_2, A'_3	Constants; see Equations (39) through (41)
B'_1, B'_2, B'_3	Constants; see Equations (39) through (41)
b_1, b_2	Constants, see Equations (8) and (9)
C'_1, C'_2, C'_3	Constants, see Equations (39) through (41)
d	Moles of carbon dioxide per quantity of organic substrate containing 1 g atom carbon; g mol/g atom carbon
f	Function; see Equations (1), (6), and (54)
J	Mean square error divided by the number of degrees of freedom
K_s	Saturation constant in Monod model
m_e	Rate of organic substrate consumption for maintenance; g equivalent of available electrons per g equivalent of available electrons in biomass (h)
MSE	Mean square error
N	Number of data points
P	Product concentration; g/L
Q	Specific rate
R	Absolute value of residual for the data point that is added
S	Substrate concentration; g/L
T	Level of significance of test statistic for quadratic term in growth curve
t	time; h
W	Defined by Equation (29)
X	Biomass concentration; g/L
X_u	$X_u = (x_{1u}, x_{2u}, \ldots, x_{ru})'$ = input variables
Y	Defined by Equation (27)
Y_{ji}	Dependent variable; see Equation (10)
$\bar{Y}_{\cdot i}$	Mean of responses; see Equation (13)
y	Dependent variable in Equation (1)
y_c	Biomass carbon yield (fraction of organic substrate carbon in biomass); dimensionless.
Z	Defined by Equations (21) and (31)
Z_{ji}	Covariate
z	Fraction of organic substrate carbon in products; dimensionless
α	Parameter in Equation (14)
γ_b	Reductance degree of biomass; equivalents of available electrons per g atom carbon
γ_p	Reductance degree of products; equivalents of available electrons per g atom carbon
ε	Error

η	Fraction of available electrons in organic substrate that is converted to biomass; dimensionless
η_{max}	"True" biomass available electron yield; dimensionless
θ	Unknown parameter
μ_{max}	Maximum specific growth rate; h^{-1}
μ	Specific growth rate; h^{-1}
ξ_p	Fraction of available electrons in organic substrate that is converted to products; dimensionless
Σ	Covariance matrix
σ_b	Weight fraction carbon in dry biomass; dimensionless
σ_p	Weight fraction carbon in products; dimensionless
σ_s	Weight fraction carbon in organic substrate; dimensionless
$\hat{\sigma}^2$	Mean square error of conditional model, Equation (14)

Subscripts

b	Biomass
GA	Galactose
i	Data point; i = 1, . . ., N
L	Lactose
0	Feed condition; initial condition
p	Product; p responses in Equation (1)
q	Number of parameters in Equation (1)
r	Number of input variables in Equation (1)
s	Substrate
u	u = 1, . . ., N sets of conditions

Superscripts

^	Estimated value
'	Transpose

REFERENCES

Amemiya, T. (1985). *Advanced Econometrics*. Harvard University Press, Boston.

Bard, Y. (1974). *Nonlinear Parameter Estimation*. Academic Press, New York.

Bates, D. M., and Watts, D. G. (1985). Multiresponse estimation with special application to linear systems of differential equations. *Technometrics 27*:329-360.

Biegler, L. T., Damiano, J. J., and Blau, G. E. (1986). Nonlinear parameter estimation: A case study comparison. *AIChE J. 32*:29-45.

Box, G. E. P., and Draper, N. R. (1965). The Bayesian estimation of common parameters from several responses. *Biometrika* *52*:355-365.

Box, M. J., and Draper, N. (1972). Estimation and design criteria for multiresponse nonlinear models with non-homogeneous variance. *Appl. Stat.* *21*:13-24.

Box, G. E. P., Hunter, W. G., MacGregor, J. F., and Erjavec, J. (1973). Some problems associated with the analysis of multiresponse data. *Technometrics* *15*:33-51.

Buono, M. A., Yang, S. S., and Erickson, L. E. (1986). Comparison of two methods of selecting smoothing spline functions for estimation of specific rates in fermentations. *Chem. Eng. Commun.* *45*:145-161.

deKok, H. E., and Roels, J. A. (1980). Method for the statistical treatment of elemental and energy balances with application to steady-state continuous-culture growth of *Saccharomyces cerevisiae* CBS 426 in the respiratory region. *Biotechnol. Bioeng.* *22*:1097-1104.

de Kwaadsteniet, J. W., Jager, J. C., and Stouthamer, A. H. (1976). A quantitative description of heterotrophic growth in microorganisms. *J. Theor. Biol.* *57*:103-120.

Draper, N. R., and Smith, H. (1966). *Applied Regression Analysis*. Wiley, New York.

Erickson, L. E. (1980). Biomass elemental composition and energy content. *Biotechnol. Bioeng.* *22*:451-456.

Esener, A. A., Roels, J. A., and Kossen, N. W. F. (1981). Fed-batch culture: Modeling and applications in the study of microbial energetics. *Biotechnol. Bioeng.* *23*:1851-1871.

Fedorov, V. V. (1972). *Theory of Optimal Experiments*. Academic Press, New York.

Geurts, T. G. E., de Kok, H. E., and Roels, J. A. (1980). A quantitative description of the growth of *Saccharomyces cerevisiae* CBS 426 on a mixed substrate of glucose and ethanol. *Biotechnol. Bioeng.* *22*:2031-2043.

Grizzle, J. E., and Allen, D. M. (1969). Analysis of growth and dose response curves. *Biometrics* *25*:357-381.

Hunter, W. G. (1967). Estimation of unknown constants from multiresponse data. *IEC Fund.* *6*:461-463.

ICSSCU (1980). IMSL Library Reference Manual, Vol. 2, IMSL, Inc., Houston.

Khatri, C. G. (1966). A note on a MANOVA model applied to problems in growth curves. *Ann. Inst. Stat. Math.* *18*:75-86.

Knepper, J. C., and Gorman, J. W. (1980). Statistical analysis of constrained data sets. *AIChE J.* *26*:260-264.

LeDuy, A., and Zajic, J. E. (1973). A geometrical approach for differentiation of an experimental function at a point: Applied to growth and product formation. *Biotechnol. Bioeng. 15*:805-810.

Oner, M. D. (1985). Estimation of yield and kinetic parameters in lactic acid fermentations for yoghurt cultures. Ph.D. Dissertation. Kansas State University, Manhattan, Kansas.

Oner, M. D., and Erickson, L. E. (1986). Anaerobic fermentation of *Lactobacillus bulgaricus* and *Streptococcus thermophilus* on 3% nonfat dry milk with pure and mixed culture. *Biotechnol. Bioeng. 28*:883-894.

Oner, M. D., Erickson, L. E., and Yang, S. S. (1984). Estimation of yield maintenance, product formation kinetic parameters in anaerobic fermentations. *Biotechnol. Bioeng. 26*:1436-1444.

Oner, M. D., Erickson, L. E., and Yang, S. S. (1986a). Analysis of exponential growth data for yoghurt cultures. *Biotechnol. Bioeng. 28*:895-901.

Oner, M. D., Erickson, L. E., and Yang, S. S. (1986b). Utilization of spline functions for smoothing fermentation data and for estimation of specific rates. *Biotechnol. Bioeng. 28*:902-918.

Oner, M. D., Erickson, L. E., and Yang, S. S. (1986c). Estimation of the true growth yield and maintenance coefficient for yoghurt cultures. *Biotechnol. Bioeng. 28*:919-926.

Pirt, S. J. (1975). *Principles of Microbe and Cell Cultivation*. Wiley, New York.

Putthoff, R. R., and Roy, S. N. (1964). A generalized multivariate analysis of variance model useful especially for growth curves. *Biometrika 51*:313-326.

Rao, C. R. (1965). The theory of least squares when the parameters are stochastic and its application to the analysis of growth curves. *Biometrika 52*:447-458.

Rao, C. R. (1966). Covariance adjustment and related problems in multivariate analysis. In *Multivariate Analysis*. Academic Press, New York, pp. 87-103.

Rao, C. R. (1967). Least square theory using an estimated dispersion matrix and its application to measurement of signals. *Proc. Fifth Berkeley Symp. Math. Statis. Prob. Vol. 1*:335-372.

Reinsch, C. H. (1967). Smoothing by spline functions. *Numerische Mathematik 10*:177-183.

Reinsch, C. H. (1971). Smoothing by spline functions, II. *Numerische Mathematik 16*:451-454.

SAS (1979). *SAS Users Guide*. SAS Institute, Raleigh, North Carolina.

Solomon, B. O., Erickson, L. E., Hess, J. L., and Yang, S. S. (1982). Maximum likelihood estimation of growth yields. *Biotechnol. Bioeng. 24*:633-649.

Solomon, B. O., Erickson, L. E., and Yang, S. S. (1983). Utilization of statistics and experimental design in data collection and analysis. *Biotechnol. Bioeng.* *25*:2683-2705.

Solomon, B. O., Oner, M. D., Erickson, L. E., and Yang, S. S. (1984). Estimation of parameters where dependent observations are related by equality constraints. *AIChE J.* *30*:747-757.

Tamhane, A. C., and Mah, R. S. H. (1985). Data reconciliation and gross error detection in chemical process network. *Technometrics* *27*:409-422.

Wahba, G., and Wold, S. (1975). A completely automatic french curve: Fitting spline functions by cross validation. *Comm. Stat.* *4*:1-17.

Wang, N. S., and Stephanopoulos, G. N. (1984). Computer applications to fermentation processes. *CRC Crit. Rev. Biotechnol.* *2*: 1-103.

Webster, I. A. (1983). The continuously fed batch reactor for measuring microbial growth rates. *Biotechnol. Bioeng.* *25*:2981-2990.

Wilkinson, G. N. (1961). Statistical estimations in enzyme kinetics. *Biochem. J.* *80*:324-332.

Wold, S. (1974). Spline functions in data analysis. *Technometrics* *16*:1-11.

Yang, S. S., Solomon, B. O., Oner, M. D., and Erickson, L. E. (1984). Estimating and testing common parameters for some multiresponse models associated with microbial growth and bioenergetics. *Technometrics* *26*:355-361.

Yoo, Y. J., Marion-Galarraga, M., Hong, J., and Hatch, R. T. (1986). Experimental design for parameter estimation from batch culture. *Biotechnol. Bioeng.* *28*:836-841.

V

MIXED CULTURE INTERACTIONS

16

Mixed Culture Interactions in Anaerobic Fermentations

DANIEL YEE-CHAK FUNG, MARK ANTHONY BUONO, and LARRY E. ERICKSON / *Kansas State University, Manhattan, Kansas*

MEHMET D. ÖNER / *University of Gaziantep, Gaziantep, Turkey*

I. INTRODUCTION

In the natural environment and in many fermented foods, mixed culture fermentation is the predominant form of fermentation. Pure culture fermentation is the result of understanding the fundamental role of microbes in fermentation and the isolation and cultivation of such cultures for specific reactions. Furthermore, pure culture fermentation can be expanded into mixed pure culture interactions for specific reactions.

This chapter deals with examples of pure mixed culture fermentation and mixed natural culture fermentation in an attempt to show relations between cultures and the resultant product formation.

A. Single-Culture Fermentation

Other chapters of this handbook deal adequately with the subject of single-culture fermentation. The key to success of single-culture fermentation is to provide the culture with a sterile substrate and environment with no contamination during the fermentation. Single-culture fermentation is a man-made situation and is classified as a controlled fermentation since the substrate is prepared and processed in such a way as to minimize contamination. Examples of this type of fermentation are wine making, beer making, bread making, fermented dairy products using single culture, and vinegar production. A detailed discussion of the various forms of food fermentation by pure and mixed cultures can be found in a recent handbook of food fermentation by Steinkraus (1983b).

B. Mixed-Pure Culture Fermentation

For some products it is desirable to have more than one pure culture for proper product development. The mixed pure cultures can be a controlled mixture of bacterium with bacterium or with a combination of yeast and/or mold. The relationships between these pure cultures during growth become very complex. In some processes the cultures are added to the substrate at the beginning of the process. This is the case in yogurt making or mixed culture starters in cheese making. In other cases one pure culture is added later in the fermentation, as in the case of mold-ripened cheeses. A detailed classification of various types of pure and mixed culture fermentations can be found in the books by Pederson (1979) and Steinkraus (1983b).

C. Mixed Natural Culture Fermentation

In many fermented foods in the world, especially in developing countries, the flora in the process are mixed natural cultures. The interactions of those microbes are even more complex than those of mixed pure culture fermentation. The microbial successions in these products follow a special sequence. If the products are treated properly, the desirable cultures provide the characteristic end results.

II. MIXED PURE CULTURE FERMENTATION AND ESTIMATION OF FERMENTATION PERFORMANCE

In this section a detailed analysis of mixed pure culture fermentation in yogurt is presented to illustrate the relationships between cultures in the fermentation process as well as the end products of various combinations of mixed pure cultures. The second portion of this section deals with soymilk fermentation and the various analytical techniques employed in measuring microbial growth during fermentation as well as illustrations of techniques for monitoring various end products. The third portion of this section presents data of interactions of yeast and bacteria in a liquid as monitored by several biomass measurements.

A. Yogurt Fermentation

1. Background

Yogurt is one of the oldest fermented food products. Long ago, milk was produced under primitive conditions. Animals were hand milked, so that contamination by microorganisms was unavoidable. Owing to the lack of refrigeration, there was no way of storing or transporting milk. As a result, people soured milk by natural lactic acid fermentation and thereby prolonged the storage life of milk for probably a couple of days.

Yogurt consumption in the United States increased from 0.04 kg/head-year in 1966 to 1.2 kg/head-year in 1981 (Tamine and Robinson, 1985). Much higher consumption rates are observed in some European countries, such as the Netherlands (16.9 kg/head-year) and Switzerland (14.1 kg/head-year).

In Turkey, where yogurt is used in preparing hot dishes, the yogurt consumption rate is even higher than in the Netherlands. In the Middle East, a variety of yogurt-based foods are consumed, such as ayran, Turkish tarhana, Greek tarhana, and Egyptian kishak (Steinkraus, 1983b).

Nowadays, yogurt is produced by modern fermentation processes. Frozen concentrated yogurt cultures (Gilliland and Speck, 1974) have also improved the product. Although the desired yogurt characteristics change from country to country, the main procedures for yogurt production (see Fig. 1) are the same.

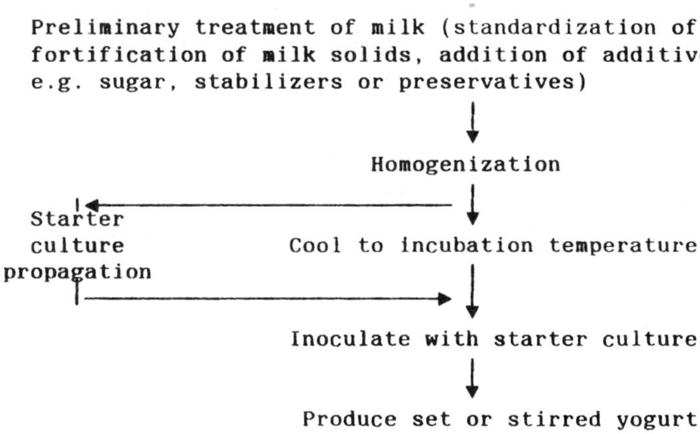

Figure 1 Improved process for yogurt production. (From Tamine and Robinson, 1985.)

The ever-increasing consumption rate of yogurt also encouraged people to work on continuous production processes (Lelieved, 1976, 1984; Driessen et al., 1977a, b; MacBean et al., 1979).

Lactic acid bacteria play an important role in the preservation of food products. Generally they require little or no heat and, therefore, are low-cost fermenters. Lactic acid-fermented foods have an important role in feeding the world today.

Milk serves as a carbon and energy source for lactic acid cultures in yogurt fermentation. Milk from various mammals can be used for yogurt production. Although they contain the same main components, that is, fat, protein, and lactose, the amounts of fat and protein may vary significantly. Table 1 illustrates the chemical composition of milk from several mammals.

The protein content of milk shows a significant variation that affects the quality of yogurt. Various methods to modify the percentage of total solids are applied, that is, the addition of milk powder, buttermilk powder, or casein powder and concentration by evaporation, membrane filtration, and so on. Also, stabilizers and emulsifiers may be added in some countries. Thus, a wide variety of yogurts are produced.

Cow's milk is most commonly used for yogurt production. A typical example of cow milk composition is given in Figure 2. Table 2 gives some information on changes in milk composition that occur during yogurt fermentation (Deeth and Tamine, 1981).

2. Mixed Culture Fermentation

Yogurt fermentation is achieved by a mixed culture of *Streptococcus thermophilus* and *Lactobacillus bulgaricus*. These are homofermentative, lactic acid-producing bacteria. The characteristics of

Table 1 Chemical Composition (%) of Milk from Various Mammals

Species	Water	Fat	Protein	Lactose	Ash
Ass	89.0	2.5	2.0	6.0	0.5
Buffalo	82.1	8.0	4.2	4.9	0.8
Camel	87.1	4.2	3.7	4.1	0.9
Cow	87.6	3.8	3.3	4.7	0.6
Goat	87.0	4.5	3.3	4.6	0.6
Mare	89.0	1.5	2.6	6.2	0.7
Reindeer	63.3	22.5	10.3	2.5	1.4
Sheep	81.6	7.5	5.6	4.4	0.9

Source: Adapted from Tamine and Robinson (1985).

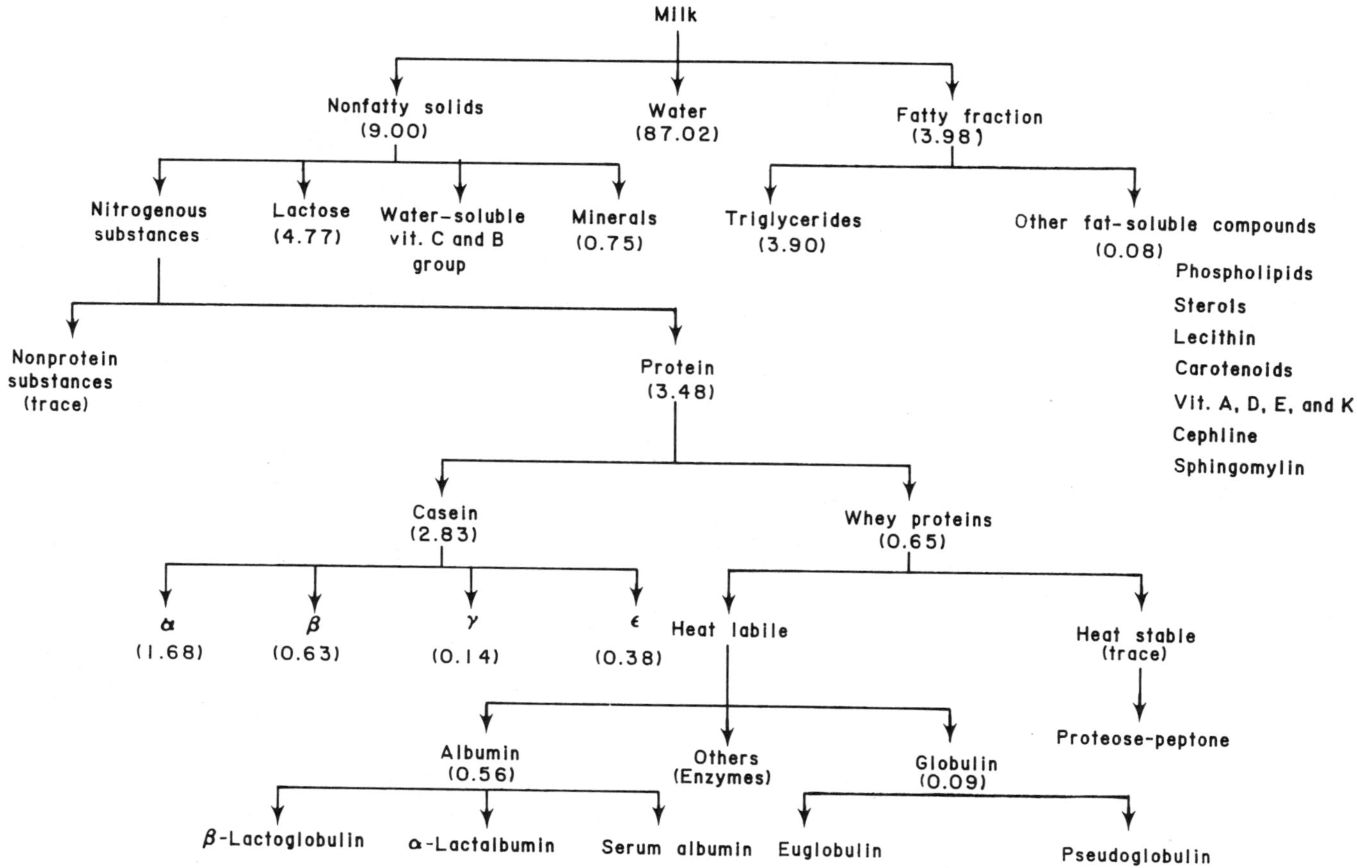

Figure 2 Typical example of the main chemical components of cow's milk, weight percentage. (From Tamine and Robinson, 1985.)

Table 2 Changes in Composition Occurring During Yogurt Fermentation

Decrease	Increase
Lactose	Lactic acid
Protein	Galactose
Urea	Glucose
Fat	Polysaccharides
Some vitamins (e.g., B_{12}, C, biotin, choline)	Peptides
Some organic acids (e.g., hippuric, orotic)	Free amino acids Ammonia Free fatty acids (volatile and longer chain) Some vitamins (e.g., folic acid) Some organic acids (e.g., succinic, fumaric, benzoic) Some nucleotides (e.g., CMP, AMP, UMP, GMP, NAD) Flavor compounds (e.g., acetaldehyde, acetoin, diacetyl) Enzymes (e.g., β-galactosidase, LDH, protease, peptidase) Bacterial mass (including nucleic acids, lipids, carbohydrate, protein)

Source: From Deeth and Tamine (1981).

these two microorganisms are given in Table 3. The possible major metabolic pathways of lactose utilization are given in Figure 3.

Lactose is hydrolyzed to galactose and glucose by β-D-galactosidase or β-D-phosphogalactosidase enzymes and glucose is utilized to produce lactic acid, as shown in Figure 3. Some strains of *S. thermophilus* can utilize or can be adapted to utilize galactose when the medium is deficient in glucose. As shown in Table 4, Somkuti and Steinberg (1979) grouped *S. thermophilus* strains into three groups: a group utilizing lactose, galactose and glucose, a group utilizing lactose and glucose, and a group utilizing lactose. These authors indicated that this phenomenon is related to the transport mechanism of the organisms. The failure by some strains to grow on galactose may also have resulted from the lack of D-tagatose-6-phosphate or the Leloir pathway (Bisset and Anderson, 1974). Similar arguments are made by Thomas and Crow (1984).

Table 3 Main Characteristics Differentiating *S. thermophilus* and *L. bulgaricus* from Other Lactic Acid Bacteria

Characteristic	*S. thermophilus*	*L. bulgaricus*
Morphology	Short chains, rather small cells	Long chains, long straight-sided rods
Colony type in LAB medium	Smooth	Rough
Surface growth (aerobic)	Poor	Nil
Temperature range of growth (°C)	20-50	23-53
Growth in litmus milk	Good: litmus not reduced before clotting	Good: strong acidity and strong reduction
Final acidity in litmus milk (% lactic acid)	~0.6	~1.7 (2-2.7[a])
Growth in bile salt lactose broth	0	0
Fermentation of sugars[b]		
Aesculin	0	0
Arabinose	0	0
Maltose	Varies	0
Mannitol	0	0
Salicin	0	0
Sucrose	Varies	0
Trehalose	0	0

[a]*Lactobacillus jughurt* variant.
[b]The sugar fermentations reflect the "milk" as opposed to the "animal" and "plant" characteristics of these organisms.
Source: Adapted from Davis (1975).

Although each of these microorganisms can produce lactic acid individually, lactic acid is not the only goal in yogurt fermentation. It is also true that the quickest growing or the highest acid-producing strains are not always the best cultures for yogurt production. On the contrary, slow acid production is preferred for good control and some other desired properties. Actually, the relative proportions of the two populations determine the ultimate characteristics of the products, such as acidity, viscosity, and flavor. However, satisfactory yogurt production by single-culture

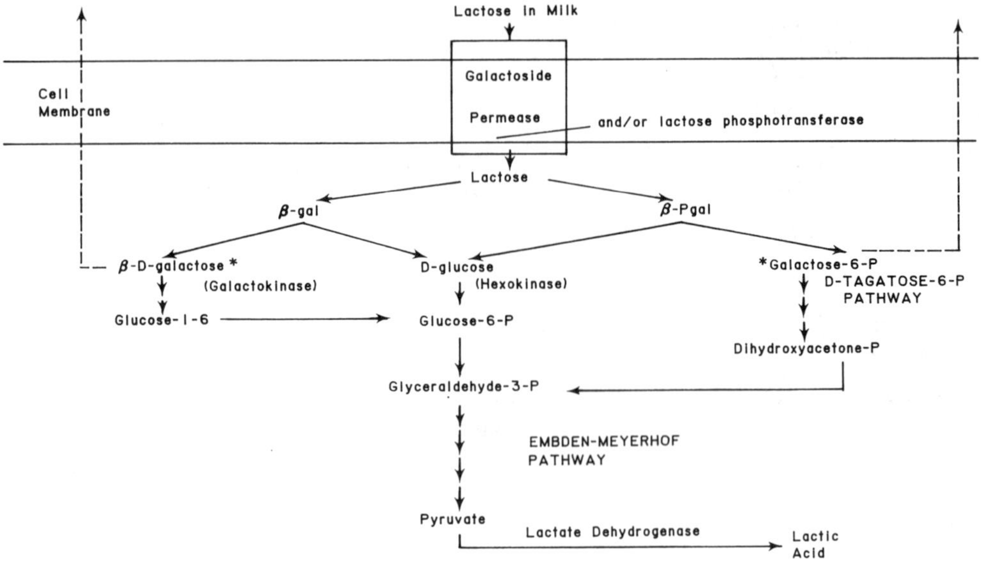

Figure 3 Possible major metabolic pathways of lactose utilization by *S. thermophilus* and *L. bulgaricus*; (*) possible metabolic pathways of strains that can utilize galactose; (---) release of galactose into milk. (From Tamine and Robinson, 1985.)

fermentation of *L. bulgaricus* and *S. thermophilus* is also achieved. Heat- or enzyme-treated milks or milk to which casein hydrolysate or sodium formate was added have been successfully used to produce yogurt with single cultures (Marshal et al., 1982).

The relationship of *L. bulgaricus* and *S. thermophilus* has long been investigated and has been described as symbiotic (Tamine and Robinson, 1985). Fredrickson (1977) has provided good insight into the definitions of interactions of mixed cultures. Whatever it is called, it is known that they grow better as a mixed culture than they do as single cultures. In mixed culture, *S. thermophilus* grows faster at the beginning, reduces the pH by lactic acid production, and also produces formic acid, which is a stimulatory compound for *L. bulgaricus*. *L. bulgaricus* produces amino acids that stimulate the growth of *S. thermophilus*. Higher acid production with mixed culture than pure cultures was indicated by Pette and Lolkema (1950); see Table 5. This same trend was observed in the work of Moon and Reinbold (1976); see Figure 4.

The incubation temperature affects acid and acetaldehyde production. *S. thermophilus* and *L. bulgaricus* produce more acid and

Table 4 Adaptability of *S. thermophilus* to Carbohydrates

Group A (L, G, GA)[a]		Group B (L, G)		Group C (L)	
Strain	Origin[b]	Strain	Origin	Strain	Origin
		4109	Hansen	406	INRA
4074	Hansen	6069	Hansen		
7132	Hansen	7024	Hansen		
9353	Hansen	6097	Hansen		
7	Microlife	9	Microlife		
4	Miles	8	Microlife		
19258	ATCC	YB/ST	Microlife		
371	NIZO	3	Microlife		
L225	NIZO	33	Microlife		
Is	NIZO	ST/AH	Microlife		
Ds	NIZO	19987	ATCC		
MC	UN	3641	NRRL		
		ST	NIZO		
		EB§-8	UN		
		SFi-1	Nestle		
		SFi-3	Nestle		
		391	INRA		

[a]L, lactose; G, glucose; GA, galactose.
[b]Hansen: C. Hansen's Laboratory, Inc., courtesy of R. L. Sellars; Microlife Technics, courtesy of E. R. Vedamuthu; Miles, Miles Laboratories, Inc.; ATCC, American type culture collection; NIZO, Netherlands Institute for Dairy Research, Ede, the Netherlands; UN, University of Nebraska, courtesy of K. M. Shahani; INRA, National Agricultural Research Institute, Jouy-Eu-Josas, France, courtesy of J. Auclair; NRRL, Northern Regional Research Center, U.S. Department of Agriculture; Nestle, Nestle Products Ltd., Lausanne, Switzerland, courtesy of T. Sozzi.
Source: From Somkuti and Steinberg (1979).

acetaldehyde at 37 and 42°C, respectively (Singh et al., 1980). Effects of temperature and pH on mixed cultures of *L. bulgaricus* and *S. thermophilus* were studied by Tayeb et al. (1984). These researchers demonstrated how the automatic acquisition of fermentation parameters of lactic acid bacteria in pure culture can be used to define the parameters for controlling a mixed culture.

Two useful balance equations were given in Chapter 11. The available electron balance is

Table 5 Acid Production and Number of Cells After an Incubation for 3 h at 45°C of the Single and Mixed Cultures of *S. thermophilus* Strain Sts and *L. bulgaricus* Strain Ib

Culture	Acid production (mmol mL^{-1})	Number of cells ($\times 10^6$ organisms mL^{-1})	Kind of cells
4% *S. thermophilus* Sts	24	500	Diplococcus
1% *L. bulgaricus* Ib	20	180	Rod
4% *S. thermophilus* Sts		880	Diplococcus
1% *L. bulgaricus* Ib		170	Rod

Source: Adapted from Driessen (1981).

$$\eta + \xi_p = 1.0 \qquad (1)$$

where η and ξ_p are the fraction of available electrons of organic substrate incorporated into biomass and extracellular products, respectively. The carbon balance is

$$y_c + z + d = 1.0 \qquad (2)$$

where y_c, z, and d are the fractions of substrate carbon incorporated into biomass, extracellular products, and CO_2, respectively. These equations can be used to check the consistency of experimental data.

According to Pirt's model (Pirt, 1975), the available electrons that are transferred from substrate to product are expended to provide energy for microbial growth and maintenance. The following equations were derived in Chapter 11 to estimate the maintenance parameter m_e and the true growth yield η_{max} (Oner et al., 1984):

$$\frac{\mu}{\eta} = \frac{\mu}{\eta_{max}} + m_e \qquad (3)$$

$$\frac{\mu(\eta + \xi_p)}{\eta} = \frac{\mu}{\eta_{max}} + m_e \qquad (4)$$

$$\frac{\mu(y_c + z + d)}{\eta} = \frac{\mu}{\eta_{max}} + m_e \qquad (5)$$

where μ is the specific growth rate per hour.

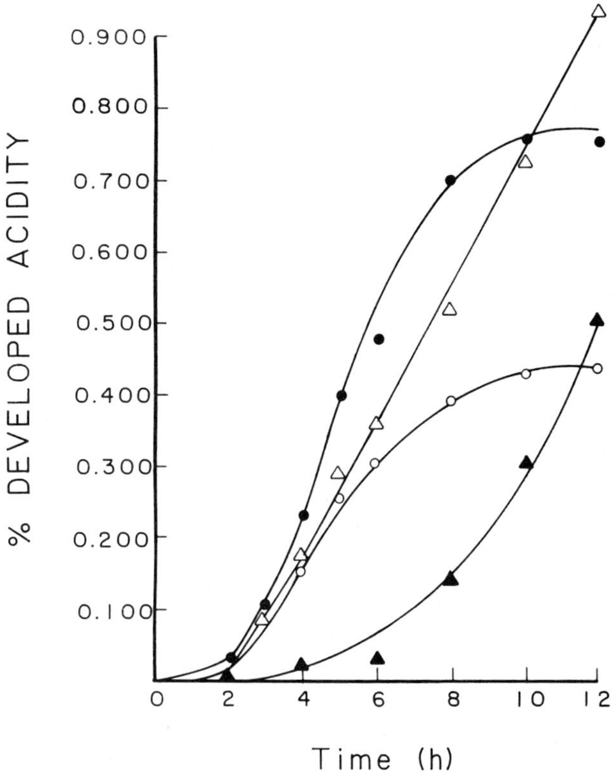

Figure 4 Developed acidity of pure and mixed cultures of *L. bulgaricus* and *S. thermophilus*; (○), *S. thermophilus*; (▲), *L. bulgaricus*, (△), sum of developed acidity of pure cultures; (●), mixed culture. (From Moon and Reinbold, 1976.)

The product formation kinetic parameters α and β in Equation (6) describe the specific rate of product formation (Luedeking and Piret, 1959; Luedeking, 1967):

$$\frac{1}{X}\frac{dP}{dt} = \alpha\mu + \beta \qquad (6)$$

These parameters can be estimated (in terms of available electron units α_e and β_e) from the following equations (Oner et al., 1984):

$$\frac{\mu}{\eta}\,\xi_p = \alpha_e\mu + \beta_e \qquad (7)$$

$$\frac{\mu}{\eta}(1 - \eta) = \alpha_e \mu + \beta_e \tag{8}$$

$$\frac{\mu}{\eta}(y_c + z + d - \eta) = \alpha_e \mu + \beta_e \tag{9}$$

Comparing Equations (3) through (5) with (7) through (9), it was found that

$$\beta_e = m_e \tag{10}$$

and

$$\alpha_e = \frac{1}{\eta_{max}} - 1 \tag{11}$$

These balances and equations are applied to data collected during the fermentation of yogurt cultures. The statistical methods employed in data analysis are described in Chapter 15. *L. bulgaricus* ATCC-11842, *S. thermophilus* ATCC-19258, *L. bulgaricus* (Microlife), and *S. thermophilus* (Microlife) were grown on 3% nonfat dry milk (NFDM) broth in batch followed by fed batch fermentation as pure and mixed cultures (Oner and Erickson, 1986).

Samples were collected every 30 minutes. The method of Kanasaki et al. (1975) was modified and used for cell dry weight measurement (Oner and Erickson, 1986). This consisted of solubilizing milk protein by EDTA and pH adjustment, monitoring absorbance at 480 nm, preparing standard curves for pure cultures, and utilizing these standard curves of absorbance versus dry weight for unknown samples. Lactose, glucose, galactose, and lactic acid concentrations were measured using high-performance liquid chromatography (HPLC) (column, HPLC column for organic acid analysis, Aminex Ion Exclusion HPX-87H 300 × 7.8 mm; detector, RI; oven temperature, 45°C; solvent flow rate, 0.9 mL/min; 2% propionic acid was used as internal standard) (Oner and Erickson, 1986).

Figures 5 through 10 show the experimental raw data and smooth curves obtained through spline function fitting (Oner et al., 1986a). As shown in Figure 9, *S. thermophilus* (Microlife) utilized galactose in addition to glucose.

After smoothing the data and obtaining specific rates, instantaneous carbon and available electron balances can also be calculated (Oner, 1985). Integrated (overall) balances are given in Table 6. The integrated balances indicate that consistent data were obtained for all six runs performed. All the available data are utilized with Equations (3), (4), (5), (10, and (11) by using the covariate adjustment technique described in Chapter 15 (Oner et al., 1986b), and

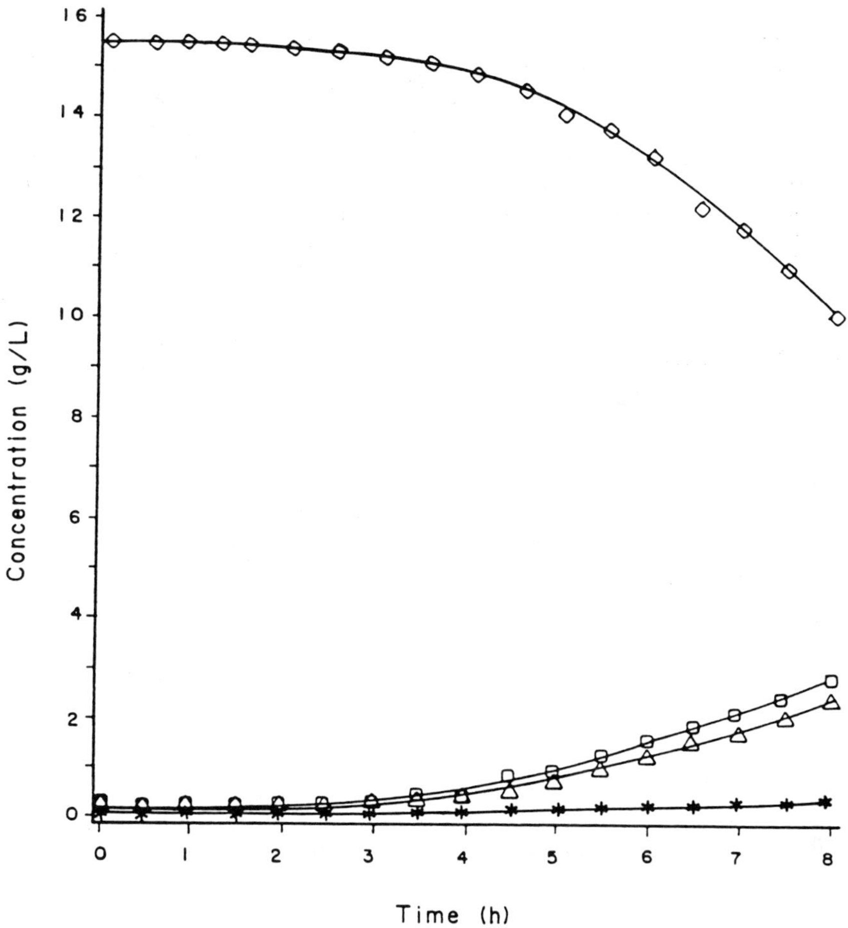

Figure 5 Results of fitting spline functions to batch culture data for *L. bulgaricus* (ATCC-11842) grown anaerobically on 3% nonfat dry milk; biomass concentration (*); lactose concentration (◇); galactose concentration (□); and lactic acid concentration (Δ). Run 1. (Oner et al., 1986a.)

best estimates of m_e, η_{max}, α_e, and β_e are given in Table 7. The results indicate that *S. thermophilus* (ATCC-19258) and *S. thermophilus* (Microlife) have relatively large true growth yields that can be accounted for by a metabolic pathway that produces 2 mol of ATP per mole of glucose consumed. The two strains of *L. bulgaricus* appear to have somewhat different values of true growth yield and maintenance coefficient; the estimated values of both parameters were considerably larger for *L. bulgaricus* (ATCC-11842) compared with the values for *L. bulgaricus* (Microlife). Theoretical values of η_{max} are presented in Table 8. Kinetic and yield parameters were also estimated for yogurt cultures by exponential growth data analysis (Oner et al., 1986c). A procedure to select the exponential growth region was developed. After selection of the exponential growth region, all the available data were utilized to estimate μ_{max}

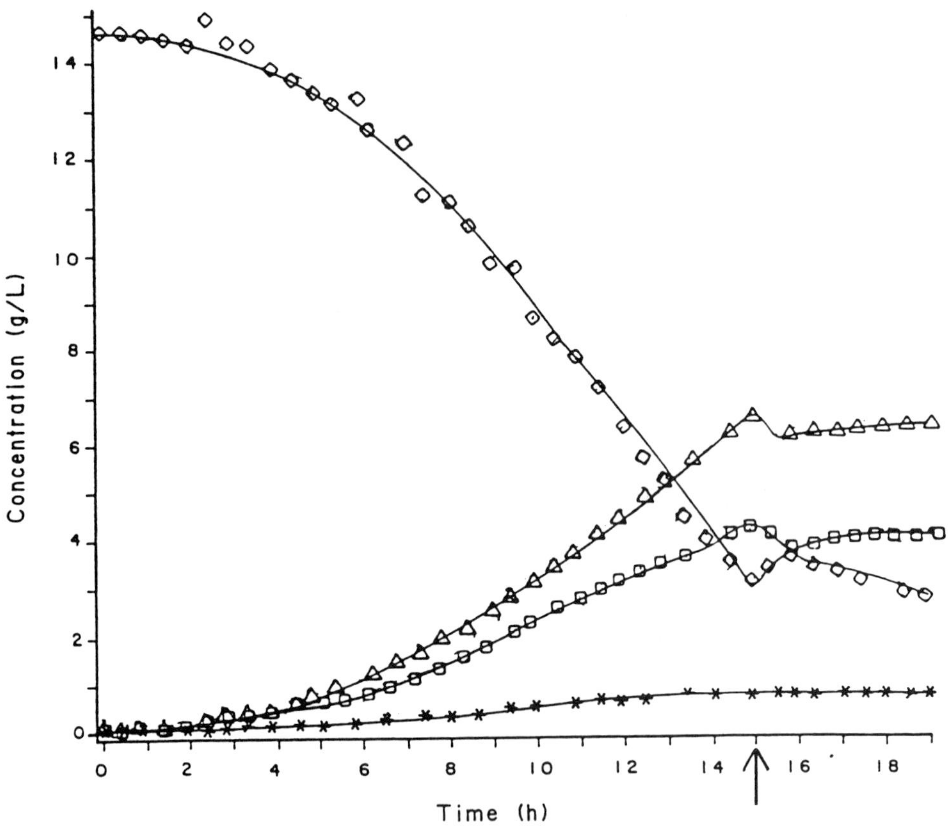

Figure 6 Results of fitting spline functions to batch and fed batch culture data for *S. thermophilus* (ATCC-19258) grown anaerobically on 3% nonfat dry milk; biomass concentration (*); galactose concentration (□); and lactic acid concentration (△); lactose concentration (◇). The arrow indicates when feeding was started. Run 2. (From Oner et al., 1986a.)

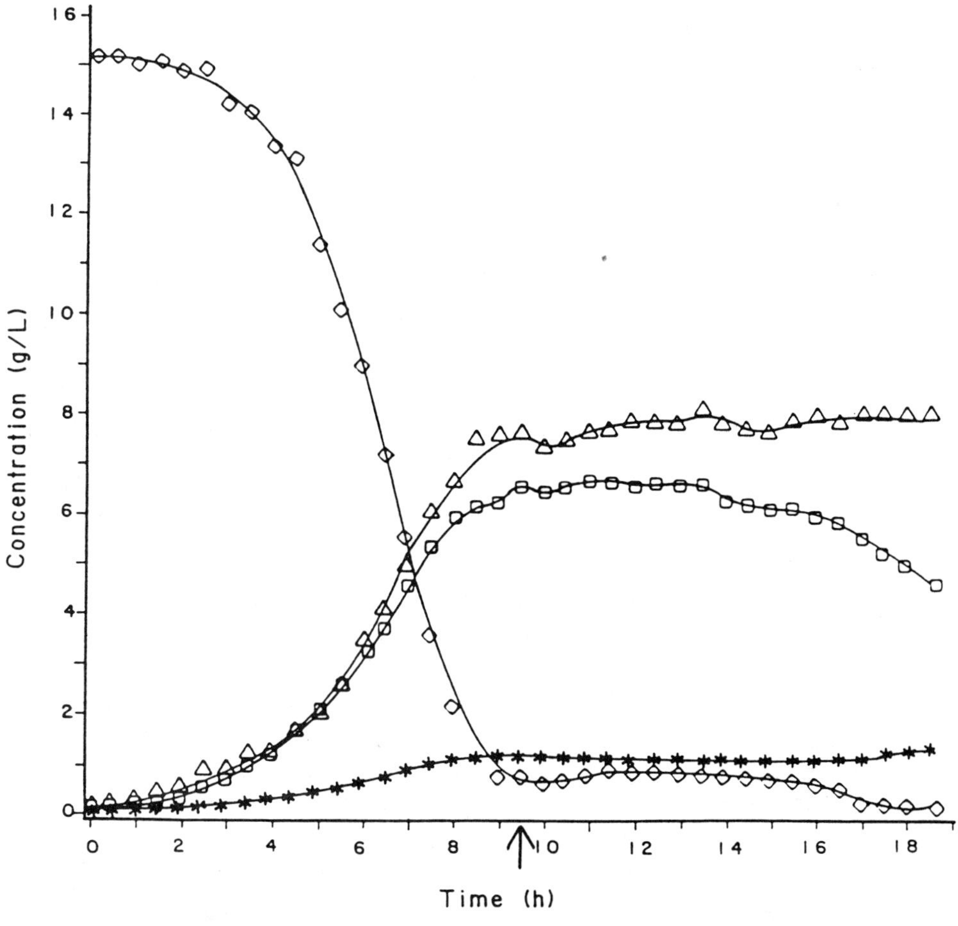

Figure 7 Results of fitting spline functions to batch and fed batch mixed culture data for *L. bulgaricus* (Microlife) grown anaerobically on 3% nonfat dry milk; biomass concentration (*); lactose concentration (◇); galactose concentration (□); and total acid product concentration (△). The arrow indicates when feeding was started. Run 4. (From Oner et al., 1986a.)

by using the covariate adjustment technique described in Chapter 15. Selected best estimates of μ_{max} and exponential region yields (η_1 and ξ_{p1}) for the six runs are given in Table 5 of Chapter 15. Chapter 15 describes the accurate detection of the exponential growth region by a statistical procedure and the utilization of the exponential

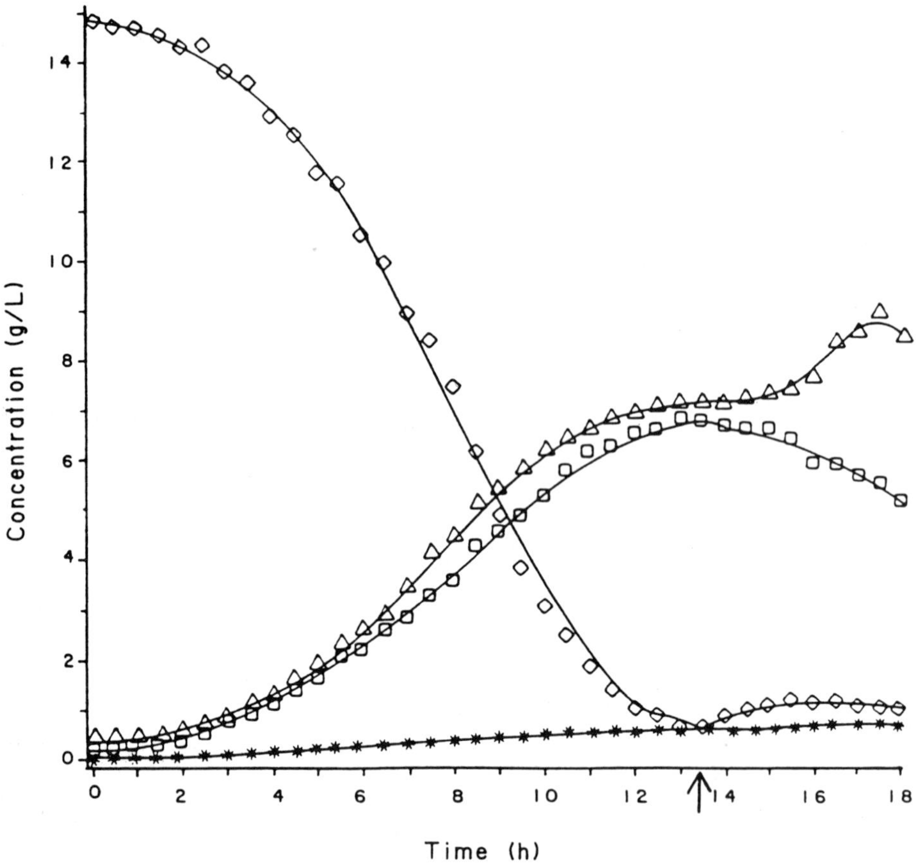

Figure 8 Results of fitting spline functions to batch and fed batch culture data for *L. bulgaricus* (Microlife) grown anaerobically on 3% nonfat dry milk; biomass concentration (*); lactose concentration (◇); galactose concentration (□); and total acid product concentration (△). The arrow indicates when feeding was started. Run 4. (From Oner et al., 1986a.)

growth region substrate and product data in addition to biomass data for estimation of μ_{max}. The estimated values of growth yield and maximum specific growth rate for *S. thermophilus* (Microlife) are larger than those for the other cultures. This can partially be explained by considering the utilization of almost all the galactose, resulting from lactose hydrolysis, by this organism.

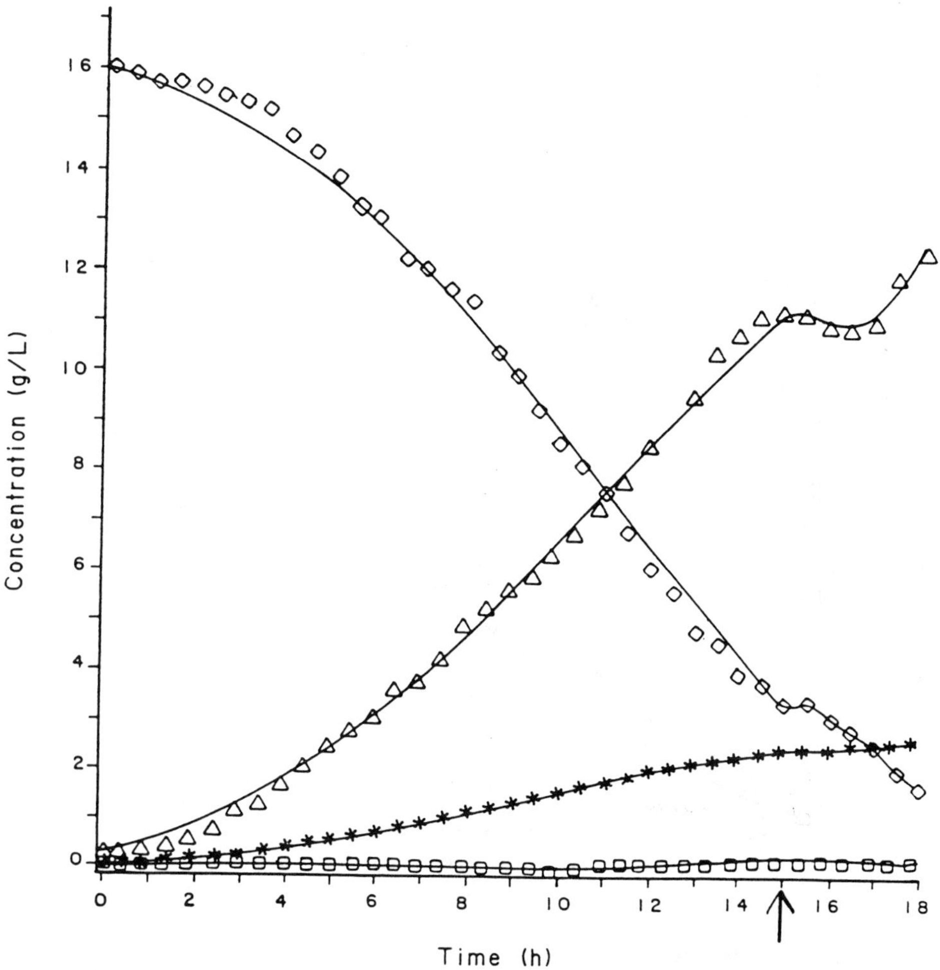

Figure 9 Results of fitting spline functions to batch and fed batch culture data for *S. thermophilus* (Microlife) grown anaerobically on 3% nonfat dry milk; biomass concentration (*); lactose concentration (◇); galactose concentration (□); and total acid product concentration (△). The arrow indicates when feeding was started. Run 5. (From Oner et al., 1986a.)

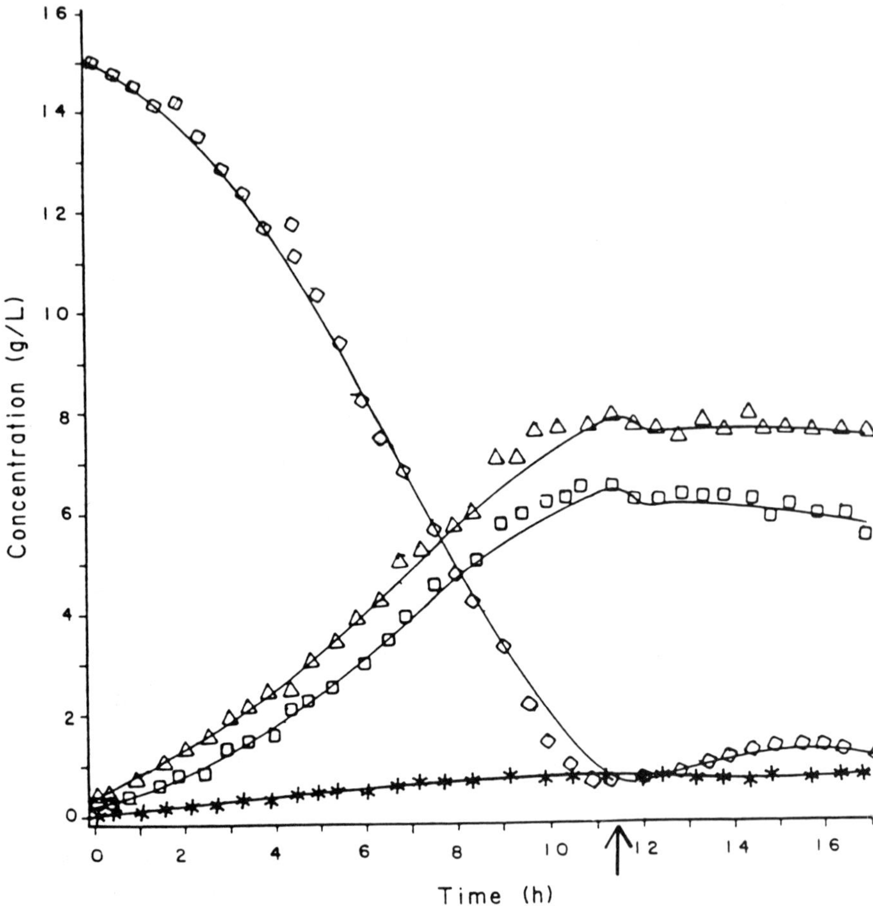

Figure 10 Results of fitting spline functions to batch and fed batch mixed culture data for *L. bulgaricus* (Microlife) and *S. thermophilus* (Microlife) grown anaerobically on 3% nonfat dry milk; biomass concentration (*); lactose concentration (◇); galactose concentration (□); and lactic acid concentration (△). The arrow indicates when feeding was started. Run 6. (From Oner et al., 1986a.)

These cultures were also used to prepare yogurt with whole milk fortified with 5% NFDM. Cultures used for the six different yogurts are indicated in Table 9. The methods are described by Oner (1985). After inoculation, the yogurts were incubated without agitation for 8.5 h at 42°C in jars and then refrigerated.

Table 6 Integrated Available Electron Balances for Runs 1-6 (Galactase Treated as a Substrate)

Run	η	ξ_p	$\eta + \xi_p$
1	0.2232	1.1299	1.3531
2	0.1310	0.8319	0.9629
3	0.1655	0.8570	1.0225
4	0.0951	0.9522	1.0474
5	0.2414	0.9133	1.1547
6	0.1123	0.8923	1.0046

Source: From Oner et al. (1986a).

Table 7 Averages of Selected Point Estimates of True Growth Yields and Maintenance Parameter and Their Minimum and Maximum Values

Run number	m_e (h^{-1})			η_{max}			α_e,	β_e (h^{-1})
	Average	Minimum	Maximum	Average	Minimum	Maximum	Average[a]	Average[a]
1	0.3962	0.3424	0.4589	0.2057	0.1943	0.2198	3.8614	0.3962
2[b]	0.6005	0.3728	1.0671	0.3455	0.1958	0.8525	1.8944	0.6005
3	0.3313	0.2983	0.3821	0.1989	0.1898	0.2077	4.0277	0.3313
4	0.0446	-0.1457	0.2722	0.1096	0.0832	0.1339	8.1241	0.0446
5	0.1062	0.0348	0.1477	0.2533	0.2217	0.2217	2.9479	0.1062
6	0.4036	0.3385	0.4793	0.1775	0.1500	0.2135	5.6667	0.4036

[a]Corresponding average values for the product formation kinetic parameters.
[b]The average values for run 2 without data set A are $m_e = 0.4539\ h^{-1}$, $\eta_{max} = 0.2129$, $\alpha_e = 3.6970$, and $\beta_e = 0.4539\ h^{-1}$.

Table 8 Theoretical Values of True Growth Yield for Anaerobic Lactic Acid Fermentations for Given Values of δ and Y_{ATP}

δ	Y_{ATP} (g biomass per g mol ATP)	η_{max}
12	10.5	0.126
	28.8	0.284
	31.9	0.305
24	10.5	0.067
	28.8	0.165
	31.9	0.180

Table 9 Yogurt Cultures Used

Yogurt 1:	*L. bulgaricus* (ATCC-11842)
Yogurt 2:	*S. thermophilus* (ATCC-19258)
Yogurt 3:	*L. bulgaricus* (ATCC-11842) and *S. thermophilus* (ATCC-19258), mixed culture
Yogurt 4:	*L. bulgaricus* (Microlife)
Yogurt 5:	*S. thermophilus* (Microlife)
Yogurt 6:	*L. bulgaricus* (Microlife) and *S. thermophilus* (Microlife), mixed culture

The analysis of results obtained for yogurts 1-6 are given in Table 10. There were significant differences between cultures in that larger amounts of glucose and galactose were present in ATCC cultures compared with Microlife cultures. The larger glucose concentrations in this work compared with the previous work could be due to much higher initial lactose concentrations in this case (72.4 g/L compared with approximately 15 g/L previously). Galactose was not utilized in the ATCC cultures, which may be due to the availability of excess glucose in the broth. Mixed cultures showed a higher lactic acid production.

Available electron balances are also given in Table 11. Poor balance closures may be due to experimental error during the analysis of samples. The formation of a thick curd made the analysis of the samples difficult. Table 12 compares the ratio of product yield to growth yield for the previous experiments (Tables 6 and 7)

Table 10 Analytic Results Obtained for Yogurts 1-6[a]

Yogurt	Lactose (g/L)	Galactose (g/L)	Glucose (g/L)	Lactic acid (g/L)	pH	Total cell number: *L. bulgaricus* (10^{-8})	Total cell number: *S. thermophilus* (10^{-8})
1	61.500	5.263	1.255	4.065	5.30	3.71	—
2	61.917	5.772	1.339	4.499	5.45	—	4.56
3	58.631	7.158	1.276	6.558	4.90	2.85	3.42
4	67.213	2.309	0.485	2.148	6.50	1.43	—
5	66.284	0.000	0.242	6.452	5.00	—	4.56
6	65.667	0.000	0.220	7.005	4.90	2.57	2.85

[a]Milk used (5% dry milk added to whole milk): lactose = 72.412 g/L, pH = 6.65.

Table 11 Growth and Product Yields and Available Electron Balance for Yogurts 1-6

Yogurt	η	ξ_p	$\eta + \xi_p$
1	0.0505	0.8182	0.8687
2	0.0788	1.1430	1.2218
3	0.0707	1.0800	1.1507
4	0.0282	0.8019	0.8301
5	0.0426	1.0669	1.1095
6	0.0447	1.0182	1.0629

and yogurt experiments. The values of ξ_p/η are much larger for the fermentation with whole milk fortified with 5% NFDM. This could be due to a much larger value for the maintenance coefficient. The higher milk concentration may require the cell to perform more osmotic work. It is also possible that the product continued to be produced after refrigeration with little or no cell growth in order to provide for the maintenance needs of the cells. In agitated experiments with nonfat dry milk, the fermentation was stopped by adding H_2O_2 prior to refrigeration. Since agitation reduces the number of cells in clumps, the absence of mixing in yogurt fermentation may be partially responsible for the low biomass measurement and resulting high ξ_p/η ratio. Much remains to be studied on the kinetics of yogurt fermentation. New studies are needed on the unknown aspects of protocooperation of yogurt cultures and other mixed culture fermentations.

Table 12 Comparison of the Ratio of Product Yield to Growth Yield[a]

Run	3% Dry milk ξ_{p1}/η_1	ξ_p/η from Table 11 (yogurt production)
1	4.92	16.20
2	5.58	14.51
3	4.39	15.28
4	7.37	28.44
5	3.47	25.04
6	5.62	22.78

[a]A pH-controlled experiment using 3% dry milk for yogurt production and an experiment without pH control using whole milk with 5% dry milk added.

B. Soymilk Fermentation

1. Background

The soybean is a most versatile food source. In the Orient, the soybean is used to produce many indigenous foods—tofu, miso, shoyu, soymilk, natto, cheese, and sobean tempeh, to name but a few. The methodology used to produce these food products is quite fascinating, and industrial-scale manufacture will require the combined efforts of the microbiologist, the engineer, the food scientist, and the sensory analyst. Many of the food products made from the sobyean require one or two and in some cases four genera and species of microorganisms growing and producing lactic acid by anaerobic fermentation at one time in a single reaction vessel. The emphasis of this section is on examining mixed culture and multiple culture interactions in soybean products and the analytic methods for data collection during microbial growth and substrate utilization. These data may be applied to the formulation of kinetic models that clearly and concisely represent the complex systems being considered.

Soybeans are a favored plant source of food owing to the higher nutritional value of the soybeans relative to such cereals as rice or wheat. Table 13 compares soybean with rice and wheat for percentage protein, percentage carbohydrate, and percentage oil (Hesseltine, 1985). Soybean has been found to contain a higher concentration of the essential amino acid lysine than such cereals as rice and wheat.

Soybean may be used to produce many indigenous foods, and these processes and products offer limitless possibilities for process and product development studies. Many of the food products made from soybeans are made by fermentation. These fermented food products range in complexity from simple pure culture fermentation, as in tempeh production from the fermentation of soybeans by *Rhizopus oligosporus*, to fermentations that involve several microorganisms acting in symbiosis. In the production of the Chinese

Table 13 Composition of Soybeans, Rice, and Wheat

	Protein (%)	Carbohydrate (%)	Oil (%)
Soybeans	38.7–41.8	35	18–22.5
Rice	6.7–7.5	77–80	1.9
Wheat	9–14	69–75	1.8–2.5

Source: From Hesseltine (1985).

shoyu, for example, four microorganisms are used to ferment defatted soybeans. The organisms are applied to a mixture of soaked and autoclaved beans and roasted wheat. The process requires a mixed culture fermentation, a lactic acid fermentation, and a yeast fermentation. The organisms used in China were two strains of *Aspergillus oryzae*, *Hansenula subpelliculosa*, and *Lactobacillus delbrueckii* (Hesseltine, 1985).

Soymilk is the aqueous extract of the soybean. Soymilk can be heat treated to produce a soymilk drink. In addition, the milk may be fermented by a process not unlike the fermentations performed on cow's milk to produce yogurt or cheese products. Fermentation of the soymilk typically proceeds anaerobically, with lactic acid produced via the glycolytic pathway. The substrates utilized by the microorganisms growing in the soymilk are the sugars sucrose, raffinose, and stachyose, which are naturally present in soymilk. Mital and Steinkraus (1975) stated that the single most important area of concern in the production of a fermented soymilk product for human consumption is the removal of the flatulence-causing sugars stachyose and raffinose. The hydrolysis and metabolism of the tetrasaccharide stachyose and the trisaccharide raffinose by fermenting organisms is the principle criterion used to select organism(s) for use in the fermentation. Of interest to the food scientist, the microbiologist, and the engineer is the formulation of accurate kinetic models that represent the organisms' growth and substrate consumption in soymilk fermentation.

The literature on the topic of soymilk fermentation has been reviewed previously. A paper by Patel et al. (1980) gives an excellent review of the fermentation of soymilk by lactic acid bacteria. The review of lactic fermentation given by Mital and Steinkraus (1974) discusses carbohydrate utilization and focuses on the selection of the microorganisms that consume the greatest amount of stachyose and raffinose sugars in the soymilk.

2. *Analytic Techniques for Carbohydrate Measurements*

The simplest approach used to measure oligosaccharide concentration in soymilk is that discussed by Stern et al. (1977). Simple media containing a single substrate may be used to determine the extent of utilization of that substrate by the organism(s) being considered via titratable acidity. For the single-substrate system, the increase in titratable acidity is proportional to the amount of added sugar that has been consumed by the fermenting organism(s). As lactic acid is produced via the glycolytic pathway, the titratable acidity is measured and the amount of sugar necessary to generate the measured lactic acid concentration is estimated.

A second approach to the measurement of oligosaccharide concentration in soymilk is gas chromatography (GC) (Mital and Steinkraus, 1975). With gas chromatography, the sugars are first converted to their trimethylsilyl derivatives before injection into the GC. Detection is typically with flame ionization. The peaks obtained for raffinose and stachyose showed excellent resolution. Sample chromatograms of these derivitized sugars were presented by Mital and Steinkraus (1975).

A third method used to detect carbohydrates in soymilk is HPLC. Pinthong et al. (1980) discussed the use of an amino-based column packed in their laboratory that gave excellent resolution of the three soymilk sugars stachyose, raffinose, and sucrose. Macrea and Zand-Maghadden (1978) also discussed an HPLC column that was packed in their laboratory and yielded useful information on sugar utilization in soymilk. At present, the packing of HPLC columns in one's laboratory is a less popular art with the availability of more stable, better resolving, and longer lasting commercial HPLC columns. Success with an HPLC column produced by Waters has been reported by Ohtakara et al. (1984) working in Japan. Excellent resolution between the three soymilk sugars was obtained.

A fourth approach that may be considered in the study of soymilk anaerobic fermentation to measure sugar consumption is a total carbohydrate estimation. Dubois et al. (1956) have presented a method that utilizes phenol and sulfuric acid to spectrophotometrically detect the total carbohydrate concentration in cell-free and bacterial fractions of ruminal fluid. The method discussed has become standard operating procedure in many laboratories when information on the total sugar content is needed.

3. *Analytic Techniques for Biomass Measurements*

Estimation of microbial biomass is by far the single most important parameter of any fermentation process. The methods used at present to determine biomass can be classified as chemical methods or physical methods (Harris and Kell, 1985). The chemical methods include convection oven dry weight, microwave oven dry weight, photometric systems, microcalorimetry, filtration methods, impedimetry, electrical counting and sizing, electrochemically based detection systems, and neutron scattering (Harris and Kell, 1985; Buono and Erickson, 1985). Other biomass detection methods, such as radiometry and microscopy, along with the detection methods already listed, are discussed in significant detail in the article by Harris and Kell (1985).

The most widely used biomass detection methods are convection oven dry weights and the transmittance photometric technique. Typically, convection dry weights require 24 h before measurements can be made and the grams of biomass can be determined. Turbidometric measurements, however, are very rapid following sample preparation, with biomass concentrations becoming available within minutes.

As a result of the inability to obtain information rapidly from convection oven dry weight, a calibration curve may be constructed to relate convection oven dry weight with turbidometric measurements. Thus, the state of microbial growth may be estimated with appropriate standard curves (Oner, 1985).

An alternative approach to the rapid measurement of biomass concentration is microwave drying. This procedure has been shown to yield biomass concentrations that approach within 10% of convection oven concentrations (Buono and Erickson, 1985). Microwave results become available in 15 minutes for drying yeast, thus providing rapid information on the state of the fermentation.

A third technique that is most frequently used by the microbiologists to measure microbial growth during fermentation is the viable cell count. The viable cell count method tests only for viable organisms present in the fermenting medium. This method of growth estimation is used quite frequently in soymilk fermentation (Mital and Steinkraus, 1975; Ohtakata et al., 1984; Kanda et al., 1976). The method employs the use of agar for lactic acid bacteria and detects the growth of viable cells by visual examination of the agar plates after appropriate incubation time and temperature.

The calibration of convection oven dry weights with turbidometric measurements has become the standard method used for biomass estimation in research laboratories. The method, however, is sensitive to the content of proteins and suspended solids present in the medium as these solids can interfere with turbidometric analysis by adsorbing a portion of the incident radiation. The problem of interfering solids and proteins is omnipresent in all milk media. Cow's milk, for example, contains the protein casein, which partially blocks the light beam used in the turbidometric technique of spectrophometry. A similar problem exists with soymilks in which oligosaccharides as well as proteins can interfere with spectrophotometric work. A solution to the problem of cow's milk has been presented (Oner, 1985; Oner and Erickson, 1986). These authors modified the Kanasaki method (Kanasaki et al., 1975) and precipitated casein micelles from milk using EDTA and pH adjustment. The procedure was shown to yield reasonable estimates for microbial growth in batch and fed batch yogurt fermentations. A method similar to that used by Oner and Erickson (1986) may be applied to the interfering solids problem in soymilk systems.

For an excellent review of the current methods used for biomass estimation and for a more exhaustive treatment of the techniques discussed here, the reader is referred to Harris and Kell (1985).

C. Interactions of Yeast and Bacteria in Mixed Culture

The interactions of mixed bacterial cultures were discussed in the previous sections. This section presents some data on growth and chemical changes during mixed culture interactions between yeast and bacteria (Shumaker and Fung, 1974).

1. Background

Yeast is one of the most important organisms for industrial use in alcoholic fermentation. Bacterial contamination may occasionally occur during yeast fermentation, which may pose serious problems for the fermentation industry. This section deals with the interaction of *Saccharomyces cerevisiae* with *Acetobacter suboxydans* or *Escherichia coli* in mixed culture growth. The growth characteristics of these mixed cultures as monitored by three biomass evaluation methods (viable cell count, direct microscopic count, and electronic particle counts) during the growth cycle are discussed.

2. Materials and Methods

S. cerevisiae was grown in mixed cultures with either *A. suboxydans* or *E. coli* in a medium containing the necessary growth factors and glucose at 28°C. Growth was monitored by viable cell counts of yeast and bacteria using selective agars, direct count by microscopic reading of cells in a Petroff-Hauser counting chamber, and electronic counts by Coulter Counter. By selecting the appropriate threshold "windows," the electronic instrument can differentiate yeast count and bacterial count in the same liquid.

During the growth period, pH values were monitored by a conventional pH meter and glucose was determined by the Glucostat, a commercial kit designed to detect glucose in solution. A vinometer, which measures the surface tension of the liquid as an estimate of the alcohol content of the liquid, was also used.

3. Results and Discussions

Figure 11 illustrates the interaction between *S. cerevisiae* and *A. suboxydans*. For the yeast, both direct count and electronic count increased with time. The direct count registered about 10 times more cells than the electronic count. Since *A. suboxydans* does not grow well in solid agar medium, viable cell count data were not obtained. For *A. suboxydans*, the direct count increased with time but the electronic count increased to about log 7.8 mL^{-1} and then started to decrease. This decrease in number is due to the adhesion of many of the bacterial cells to yeast cells. The electronic count cannot differentiate these two populations in this condi-

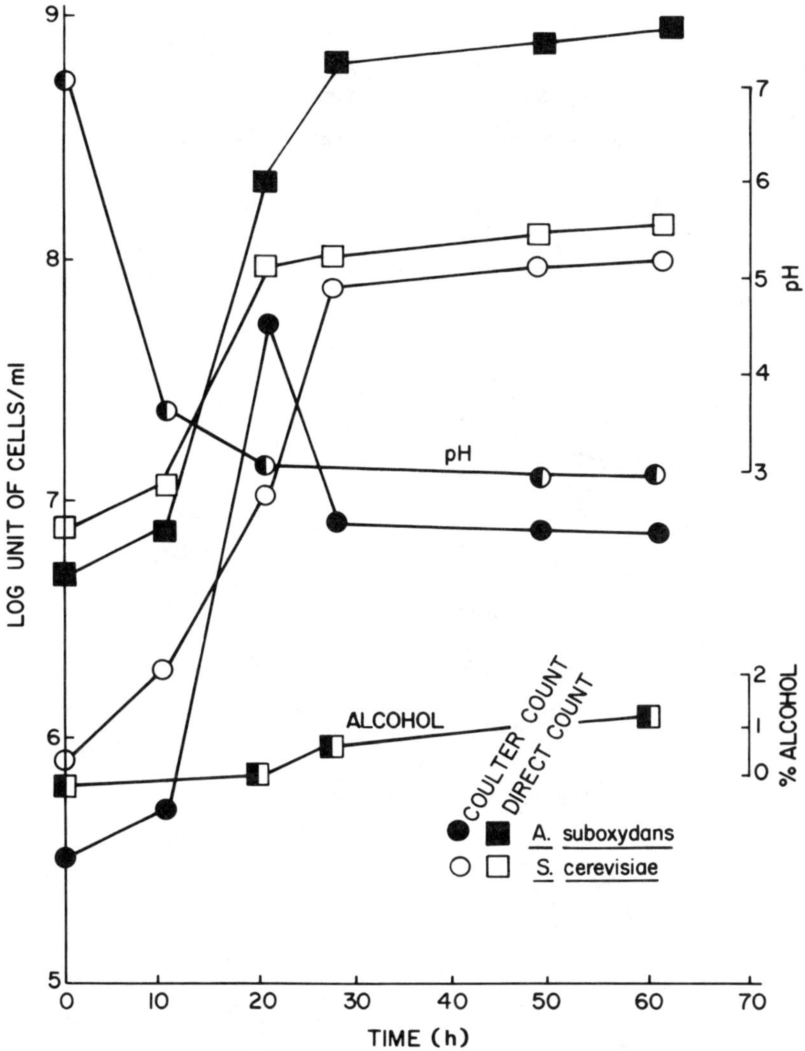

Figure 11 Interaction of mixed cultures of *S. cerevisiae* and *A. suboxydans* in terms of cell numbers monitored by direct count and electronic count as well as product formation (alcohol and acid production).

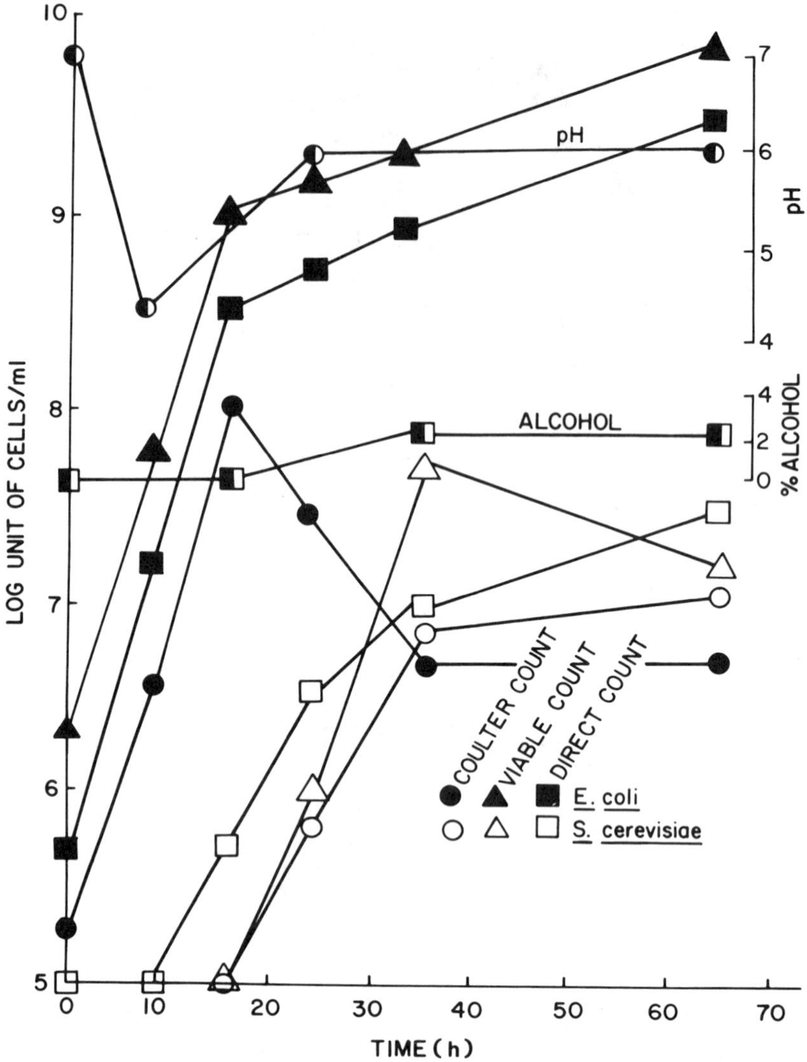

Figure 12 Interaction of mixed culture of *S. cerevisiae* and *E. coli* in terms of cell numbers monitored by direct count, electronic count, and viable cell count as well as product formation (alcohol and acid production).

tion. However, microscopic observations can ascertain the different counts.

The concentration of glucose completely disappeared in the first 10 h of yeast and bacterial interactions. The pH of the medium first decreased and then stabilized at pH 3. This is due to oxidation of alcohol (produced by *S. cerevisiae*) by *A. suboxydans* to acetic acid, making the reaction mixture acidic.

The interactions of yeast and *E. coli* (Fig. 12) showed interesting contrasts compared with the yeast-*Acetobacter* interactions. Growth of yeast as monitored by viable cell count, direct count, and electronic counts all increased with time. After reaching a stationary phase, viable yeast count decreased but direct count and electronic counts registered no reduction in numbers. A likely explanation is that the former two biomass measurements counted both live and dead cells but the viable cell count only registered living cells. After the stationary phase, some of the yeast cells went through the death cycle. The growth curves of *E. coli* showed an increase in the viable cell count and direct cell count. The electronic count reached log 8 mL^{-1} and then declined, exhibiting a trend similar to that observed in the yeast-*Acetobacter* interactions. Alcohol contents increased to about 2%, and the pH first dropped to around 4 and then rose to 6. This pattern differs from that obtained from the interaction of yeast and *Acetobacter*. *E. coli* cannot oxidize alcohol and thus did not create large amounts of acid to counteract the basic metabolites in the reaction vessel. This resulted in a medium that reverted to a more alkaline state.

These two studies indicate that different methods of estimating biomass may provide different results in mixed culture fermentation. The electronic method is useful when a single culture is monitored. In mixed culture interaction that involves clumping, this method is not suitable for differentiating mixed populations.

III. MIXED CULTURE INTERACTIONS IN NATURE AND IN FOOD PRODUCTS

In the natural environment and in man-made fermentation systems, mixed culture interaction is a common occurrence. There is a definite microbial succession process dictated by the indigenous flora of the substrate, the intrinsic parameters of the substrate (i.e., pH, water activity, oxidation-reduction potential, nutrient contents, and naturally occurring inhibitors), and the extrinsic parameters on the substrate (i.e., temperature, environmental gaseous exchange, physical condition of the substrate, and time lapse). Manipulations of the substrate as well as extrinsic parameters dramatically influence the microbial succession of the fermentation process. Several natural processes are described here to illustrate mixed culture interactions.

Table 1 in Chapter 18 details the microorganisms involved in the fermentation and flavor production of cereals, legumes, tubers, vegetables, fruits, milk, meat, and fish products. The *Handbook of Indigenous Fermented Foods* by Steinkraus (1983b) provides detailed descriptions of major indigenous fermented foods throughout the world, such as Indonesian tempeh and related fermentation products, fermented food involving acid production, fermented food involving alcohol as a major product, and the complex fermentation of soy sauce, miso, fish sauce and paste, and other indigenous fermented foods. It is evident that many of these products are the results of mixed culture interactions (Steinkraus, 1983a). In many such products the main microorganisms have been isolated and studied but the precise interactions during fermentation are not well established.

Sauerkraut fermentation is well documented by Pederson and Albury (1969) and Ayres et al. (1980) and can be used as a model to examine mixed culture fermentation and microbial succession.

A good sauerkraut can be made by natural fermentation of shredded cabbage under anaerobic condition with the addition of 2.25% NaCl at 23°C. After mixing the salt and the shredded cabbage, the materials are tightly packed to create an anaerobic condition. Respiration by plant tissue and uptake of oxygen by indigenous bacteria quickly render the system anaerobic, thus killing the strictly aerobic organisms.

Facultative anaerobic enterics, such as *Enterobacter cloacae* and *Erwinia herbicola*, start to grow along with *Leuconostoc mesenteroides*, a heterofermentative bacterium. As *L. mesenteroides* grows, lactic acid is produced. Vegetable juice has a low buffering capacity; thus, the lactic acid formed in the juice by *L. mesenteroides* inhibits the enteric from growing. Other products of the heterofermentor, such as ethanol and esters, also contribute to the desirable flavor of the sauerkraut. Carbon dioxide is generated and appears as bubbles at this stage of fermentation. This gas is undesirable for good sauerkraut and is not produced by the succeeding major bacterium. *Lactobacillus plantarum*, which succeeds *L. mesenteroides*, is a homofermentative organism that produces large amounts of lactic acid. It is able to degrade the bitter mannitol produced by *L. mesenteroides* in the sauerkraut. The final total acidity is about 1.5%, which accounts for the desirable clean crisp quality of good sauerkraut after 21 days of fermentation. In terms of population kinetics during this fermentation, *L. mesenteroides* increases in number to approximately 10^9 mL^{-1} at around the second day of fermentation. Thereafter, the number decreases precipitously. *L. plantarum* then takes over and becomes the predominant organism until the end of the fermentation. Small numbers of *Pediococcus cerevisiae* (a homofermenter) and *(Lactobacillus brevis* (a heterofermenter) also exist during the middle stage of fermentation. The combined biochemical activities of these organisms in proper sequence and proportion produce desirable sauerkraut.

When the salt concentration is varied (higher or lower than 2.25%) while holding the temperature at the desired value of 23°C, the pattern of microbial succession changes. When the sauerkraut contains 1.0% NaCl, the population of *L. mesenteroides* remains high for long periods of time but the population of *L. plantarum* remains low. The resultant sauerkraut has an undesirably high acid (2%). When the salt concentration is increased to 3.0%, bacterial numbers of all species are low and acid development is unsatisfactory.

Changing the temperature (higher or lower than 23°C) while holding the salt at the desirable concentration of 2.25% also dramatically affects microbial successions. When the temperature is dropped to 7.5°C, *L. mesenteroides* is the predominant organism and acid development is slow and unsatisfactory. The number of *L. plantarum* is also low. At 32°C the population of *L. mesenteroides* peaks in 1 day and large numbers of *L. plantarum* take over until the end of the fermentation, resulting in an undesirably high acid content of 2%. Thus, changes in one key component in the substrate (in this case salt) or changes in temperature dramatically influence mixed culture interactions in a naturally fermented food (sauerkraut).

Another example of mixed culture fermentation is ensilage of crops. Dalmacio and Fung (1977) and Dalmacio et al. (1982) studied the influence of ammonia on bacteria and mold during ensilage of high-moisture corn. Bacteria were not affected by ammonia treatment as much as mold at the onset of the treatment. Bacterial populations remained detectable after treatment of 1.0, 1.5, and 2% ammonia. However, no mold was detected for corn treated with 1.0, 1.5, and 2.0% ammonia. Corn treated with 0.5% ammonia had high bacterial and mold counts. After 2 months of storage the bacterial population increased to more than 10^7 g^{-1}. Mold count increased to around 10^4 g^{-1} for the corn treated with 0.5% ammonia. After 4 months of storage the bacterial count reached saturation level (higher than 10^{10} g^{-1}) but the mold population remained around 10^4 or 10^5 g^{-1}. This study indicated that ammonia treatment affected one type of microbe more than another in mixed culture interaction.

In a study of bacterial flora of hot-boned beef (a form of meat processed by removing meat from carcasses before chilling) and conventionally cold chilled beef, Lee et al. (1982, 1985) made a detailed analysis of the population shift between fresh beef and anaerobically cold-stored beef in relationship to the initial chilling rate of the beef cuts. They used computer-assisted numerical taxonomy methods to identify thousands of isolates obtained from fast-chilled, moderately fast chilled, and slow-chilled beef and the corresponding samples after 14 days of anaerobic cold storage. The data indicated that the initial chilling rate had a direct impact on the proliferation of bacteria on hot-boned beef immediately after chilling as well as after cold storage.

Initially, the microbial flora of hot-boned (HB) beef and conventionally processed (CP) beef were heterogeneous. There were, however, population shifts after 14 days of anaerobic cold storage. After storage, the predominant organisms of the fast-chilled HB beef and corresponding CP beef were mainly lactic acid bacteria. The predominant after-storage mesophiles of moderately fast chilled HB beef were lactic acid bacteria. The predominant psychrotrophic organisms were *Streptococcus* and *Psudomonas*, whereas *Pseudomonas* dominated the mesophilic and psychrotrophic flora in corresponding CP beef. In the slow chilling rate experiment, *Lactobacillus* dominated the psychrotrophic and mesophilic populations in HB beef after storage, whereas no particular genus was predominant in the corresponding CP beef. The difference in the number and kinds of phena between HB beef and the corresponding CP beef was greatest in slow-chilled beef, followed by moderately fast chilled beef and fast-chilled beef, which indicated that slow chilling (least desirable procedure) of HB beef resulted in more diversified bacterial types after storage. This experiment demonstrates that initial chilling rates affected the flora of beef before cold storage and that mixed culture interaction produced different flora after anaerobic cold storage.

Other examples of mixed culture interaction in nature and in specialized environments can be found in autoclave-sterilized whey (Bodie et al., 1987), rumen (Hobson and Wallace, 1982), intestinal habitats (Bryant, 1974), anaerobic digestion systems (Zeikus, 1977), and sewage treatment (Green and Kramer, 1979). Useful reference books on this subject include *Microbial Technology: Fermentation Technology* by Peppler and Perlman (1979), *Food and Beverage Mycology* by Beuchat (1978), *Biotechnology* by Crueger and Crueger (1982), and *Bacterial Starter Cultures for Foods* by Gilliland (1985).

IV. CONCLUSIONS

Mixed cultures are widely used in a variety of applications. Although considerable research with mixed cultures has been published, much remains to be done. The examples considered in this chapter are representative of current research with mixed cultures.

ACKNOWLEDGMENT

Contribution no. 81-463-B, Kansas State Agricultural Experiment Station, Manhattan, Kansas.

REFERENCES

Ayres, J. C., Mundt, J. O., and Sandine, W. E. (1980). *Microbiology of Foods*. W. H. Freeman, San Francisco.

Beuchat, L. R. (1978). *Food and Beverage Mycology*. AVI Publishing, Westport, Connecticut.

Bisset, D. L., and Anderson, R. L. (1974). Lactose and D-galactose metabolism in group N-streptococci: Presence of enzymes for both D-galactose-1-phosphate and d-tagatose-6-phosphate pathways. *J. Bacteriol.* *117*:318-320.

Bodie, E. A., Goodman, N., and Schwartz, R. D. (1987). Production of propionic acid by mixed cultures of *Propionibacterium shermanii* and *Lactobacillus casei* in autoclave-sterilized whey. *J. Indust. Microbiol.* *1*:349-353.

Bryant, M. P. (1974). Interaction among intestinal microorganisms. *Am. J. Clin. Nutr.* *27*:1320-1328.

Buono, M. A., and Erickson, L. E. (1985). Rapid measurement of *Candida utilis* dry weight with microwave drying. *J. Food Protect.* *48*:958-960.

Crueger, W., and Crueger, A. (1982). *Biotechnology: A Textbook of Industrial Microbiology*. Science Tech., Madison, Wisconsin.

Dalmacio, I. F., and Fung, D. Y. C. (1977). Microbiology of high moisture corn treated with "cold flow" ammonia. Kalikasan, Philippines. *J. Biol.* *6*:203-216.

Dalmacio, I. F., Fung, D. Y. C., and McNeil, J. H. (1982). Microbiology of high moisture corn treated with a combination of "cold flow" ammonia and propionic acid. *Philippine Phytopathol.* *18*:102-107.

Davis, J. A. (1975). The microbiology of yogurt. In *Proceedings of a Symposium Held at Long Ashton Research Station*, University of Bristol, September 19-21, 1973, J. G. Carr, C. V. Cutting, and G. C. Writing (Eds.). Academic Press, New York.

Deeth, H. C., and Tamine, A. Y. (1981). Yoghurt: Nutritive and therapeutic aspects. *J. Food Protect.* *44(1)*:78-86.

Driessen, F. M. (1981). Protocooperation of yoghurt bacteria in continuous culture. In *Mixed Culture Fermentations*, M. E. Brushel and J. H. Slater (Eds.). Academic Press, New York, pp. 99-120.

Driessen, F. M., Ubbels, J., and Stadhouders, J. (1977a). Continuous manufacture of yoghurt. I. Conditions and kinetics of the prefermentation process. *Biotechnol. Bioeng.* *XIX*:821-839.

Driessen, F. M., Ubbels, J., and Stadhouders, J. (1977b). Continuous manufacture of yoghurt. II. Procedure and apparatus for continuous coagulation. *Biotechnol. Bioeng.* *XIX*:841-851.

Dubois, M., Gilles, K. A., Hamilton, J. K., Rebers, P. H., and Smith, F. (1956). Colorimetric method of determination of sugars and related substances. *Anal. Chem.* *28*:350-356.

Fredrickson, A. G. (1977). Behavior of mixed cultures of microorganisms. *Annu. Rev. Microbiol.* *31*:63-87.

Gilliland, S. E. (1985). *Bacterial Starter Cultures for Foods*. CRC Press, Boca Raton, Florida.
Gilliland, S. E., and Speck, M. L. (1974). Frozen concentrated cultures of lactic starter bacteria. A review. *J. Milk Food Technol. 37*(2):107-111.
Green, J. H., and Kramer, A. (1979). *Food Processing Waste Management*. AVI Publishing, Westport, Connecticut.
Harris, C. M., and Kell, D. B. (1985). The estimation of microbial biomass. *Biosensors 1*:17-84.
Hesseltine, C. W. (1985). Fungi, people and soybeans. *Mycologia 77*(4):505-525.
Hobson, P. N., and Wallace, R. J. (1982). Microbial ecology and activities in the rumen. Part I. *CRC Crit. Rev. Microbiol. 9*(4): 165-320.
Kanasaki, M., Brehesy, S., Hillier, A. J., and Jago, G. R. (1975). Effect of temperature on the growth and acid production of lactic acid bacteria. I. A rapid method for the estimation of bacterial populations in milk. *Aust. J. Dairy Technol. 30*(12):142-144.
Kanda, H., Wang, H. L., Hesseltine, C. W., and Warner, K. (1976). Yoghurt production by lactobacillus fermentation of soybean milk. *Proc. Biochem. 11*(4):23-25.
Lee, C. Y., Fung, D. Y. C., and Kastner, C. L. (1982). Computer-assisted identification of bacteria on hot-boned and conventionally processed beef. *J. Food Sci. 47*:363-367.
Lee, C. Y., Fung, D. Y. C., and Kastner, C. L. (1985). Computer-assisted identification of microflora on hot-boned and conventionally processed beef: Effect of moderate and slow initial chilling rate. *J. Food Sci. 50*:553-567.
Lelieved, H. L. M. (1976). Continuous fermentation in yoghurt manufacture. *Proc. Biochem. 11*(6):39-40.
Lelieved, H. L. M. (1984). Mixed-strain continuous milk fermentation. *Proc. Biochem. 19*(3):112-113.
Luedeking, R. (1967). Fermentation processes kinetics. In *Biochemical and Biological Engineering Science*, Vol. 1. Academic Press, New York, pp. 181-240.
Luedeking, R., and Piret, E. C. (1959). A kinetic study of lactic acid fermentation. Batch processes of controlled pH. *J. Biochem. Microbiol. Technol. Eng. 1*:393.
MacBean, R. D., Hall, R. J., and Linklater, P. M. (1979). Analysis of pH-Stat continuous cultivation and the stability of the mixed fermentation in continuous yoghurt production. *Biotechnol. Bioeng. XXI*:1517-1541.
Macrea, R., and Zand-Moghaddem, A. (1978). The determination of the component oligosaccharides of lupin seeds by high pressure liquid chromatography. *J. Sci. Foods Agric. 29*:1083-1086.

Marshal, V. M. E., Cole, W. M., and Mabbit, L. A. (1982). Yoghurt made from single starter organisms using heat- or enzyme-treated milk or milk to which casein hydrolysate or sodium formate is added. *J. Dairy Res. 49*:147-152.

Mital, B. K., and Steinkraus, K. H. (1974). Growth of lactic acid bacteria in soymilks. *J. Food Sci. 39*:1018-1022.

Mital, B. K., and Stemkraus, K. H. (1975). Utilization of oligosaccharides by lactic acid bacteria during fermentation of soy milk. *J. Food Sci. 40*:114-118.

Moon, N. J., and Reinbold, G. W. (1976). Commensalism and competition in mixed cultures of *L. bulgaricus* and *S. thermophilus*. *J. Milk Food Technol. 39*(5):337-341.

Ohtakara, A., Mitsutomi, M., and Uchida, Y. (1984). Purification and enzymatic properties from *Pycnoporus cinnabarinus*. *Agric. Biol. Chem. 48*:1319-1327.

Oner, M. D. (1985). Estimation of yield and kinetic parameters in lactic acid fermentations. Ph.D. thesis, Kansas State University, Manhattan, Kansas.

Oner, M. D., and Erickson, L. E. (1986). Anaerobic fermentation of *Lactobacillus bulgaricus* and *Streptococcus thermophilus* on 3% nonfat dry milk with pure and mixed culture. *Biotechnol. Bioeng. 28*:883-894.

Oner, M. D., Erickson, L. E., and Yang, S. S. (1984). Estimation of yield, maintenance, and product formation kinetic parameters in anaerobic fermentations. *Biotechnol. Bioeng. 26*:1436-1444.

Oner, M. D., Erickson, L. E., and Yang, S. S. (1986a). Utilization of spline functions for smoothing fermentation data and for estimation of specific rates. *Biotechnol. Bioeng. 28*:902-918.

Oner, M. D., Erickson, L. E., and Yang, S. S. (1986b). Estimation of the true growth yield and maintenance coefficient for yoghurt cultures. *Biotechnol. Bioeng. 28*:919-926.

Oner, M. D., Erickson, L. E., and Yang, S. S. (1986c). Analysis of exponential experimental growth data for yoghurt cultures. *Biotechnol. Bioeng. 28*:895-901.

Patel, A. A., Waghmore, W. M., and Gupta, S. K. (1980). Lactic fermentation of soymilk—a review. *Proc. Biochem. Oct./Nov.*:9-13.

Pederson, C. S. (1979). *Microbiology of Food Fermentations*, 2nd Ed. AVI Publishing, Westport, Connecticut.

Pederson, C. S., and Albury, M. N. (1969). *The Sauerkraut Fermentation*. New York State Agric. Exp. Station Bull. 824.

Peppler, H. J., and Perlman, D. (1979). *Microbial Technology: Fermentation Technology*. Academic Press, New York.

Pette, J. W., and Lolkema, H. (1950). Yoghurt. I. Symbiose en antibiose in meng cultures von *Lb. bulgaricus* en *Sc. thermophilus*. *Netherlands Milk Dairy J. 4*:197-208.

Pinthong, R., Macrea, R., and Rothwell, J. (1980). The development of a soya based yoghurt. III. Analysis of oligosaccharides. *J. Food Technol. 15*:660-665.

Pirt, S. S. (1975). *Principles of Microbe and Cell Cultivation*, Wiley, New York, pp. 1-274.

Shumaker, S. D., and D. Y. C. Fung (1974). Use of Coulter counter in studying kinetics of yeast and bacteria in mixed culture. Proceeding, The Pennsylvania Academy of Science Meeting, Bloomsburg, Pennsylvania, April 18-20, 1974, p. 18.

Singh, J., Khanna, A., and Chander, H. (1980). Effect of incubation temperature and heat treatments of milk from cow and buffalo on acid and flavor production by *S. thermophilus* and *L. bulgaricus*. *J. Food Protect. 43*(5):399-400.

Somkuti, G. A., and Steinberg, D. H. (1979). Adaptability of *Streptococcus thermophilus* to lactose, glucose and galactose. *J. Food Protect. 42*:881-882, 887.

Steinkraus, K. H. (1983a). Lactic acid fermentation in the production of foods from vegetables, cereals, and legumes. *Antonie Van Leeuwenhoek 49*:337-348.

Steinkraus, K. H. (1983b). *Handbook of Indigenous Fermented Foods*. Marcel Dekker, New York.

Stern, N. J., Hesseltine, C. W., Wang, H. L., and Konishi, F. (1977). *Lactobacillus acidophilus* utilization of sugars and production of a fermented soybean product. *J. Inst. Sci. Technol. Aliment. 10*(3):197-200.

Tamine, A. Y., and Robinson, R. K. (1985). *Yoghurt Science and Technology*. Pergamon Press, New York.

Tayeb, J., Bouillane, C., and Desmazeaud, M. J. (1984). Computerized control of growth with temperature in a mixed culture of lactic acid bacteria. *J. Ferment. Technol. 62*(5):461-470.

Thomas, T. D., and Crow, V. L. (1984). Selection of galactose-fermenting *S. thermophilus* in lactose-limited chemostat cultures. *Appl. Environ. Microbiol. 48*:186-191.

Zeikus, J. C. (1977). The biology of methanogenic bacteria. *Bacteriol. Rev. 41*(2):514-541.

17

Interactions Between Hydrogen- and Formate-Producing Bacteria and Methanogens During Anaerobic Digestion

JÜRGEN H. THIELE / *ecoTech—Environmental Biotechnology, Tübingen, Federal Republic of Germany, and Michigan Biotechnology Institute, Lansing, Michigan*

J. GREGORY ZEIKUS / *Michigan Biotechnology Institute, Lansing, Michigan, and Michigan State University, East Lansing, Michigan*

I. INTRODUCTION

The anaerobic digestion process involves the microbially mediated conversion of complex organic matter into methane and CO_2 as final carbon mineralization products. The overall biochemistry of the process involves defined carbon and electron flow pathways for the metabolism of diverse complex organic substrates into simple carbon products. The use of carbon^{14} tracer studies and isolation of prevalent species and bacterial characterization studies have resulted in a very good quantitative picture of carbon flow. However, because of the lack of precise techniques, the electron flow pathway remains speculative and is based on qualitative assessments.

The purpose of this chapter is to provide a critical but non-encyclopedic review of past literature in relation to our new findings that question the quantitative significance of "interspecies hydrogen transfer" as the mechanism for coupling electron transfer during the anaerobic conversion of certain volatile fatty acids, organic alcohols, and aromatic compounds into methane with limited free energy available for the participating syntrophic bacterial populations. We provide quantitative data on measurements of electron transfer reactions in a model anaerobic digester system whose population and metabolism are defined in microbial and biochemical terms. This review compares present and past views of the coupling of electron transfer between H_2- and formate-producing acetogens and methanogens based on thermodynamic and kinetic measurements.

II. ECOPHYSIOLOGICAL CONTEXT: ENVIRONMENT, ORGANISMS, AND METABOLISM

The majority of biochemical transformation reactions that occur in the biosphere involve the cycling of carbon by life processes. Photosynthetic CO_2 reduction to carbohydrates is balanced by oxidative carbon dissimilation to CO_2 via aerobic microorganisms. The microbial ecology of this carbon cycle is very complex because the biochemical reactions involved are performed by a metabolically diverse population. The decomposition of organic carbon in anaerobic ecosystems is characterized by a limited thermodynamic yield of the overall conversion, especially under methanogenic conditions, in which no net carbon oxidation occurs (Table 1). Instead, a disproportionation of organic matter into gaseous CO_2 as the most oxidized carbon compound and gaseous CH_4 as the most reduced end product is the driving force for degradation (Table 2). This anaerobic digestion process requires a unique specialization of the involved bacterial species and results in the formation of a microbial food chain with at least three different trophic groups (Fig. 1):

1. Hydrolytic fermentative microorganisms (HFM)
2. Syntrophic acetogenic bacteria (SAB)
3. Methanogenic bacteria (MB)

Trophic group 1 is composed of bacteria, fungi, or protozoa, depending on the individual ecosystem and its trophic status. In the rumen ecosystem for example, bacteria, fungi, and protozoa function jointly to hydrolyze complex organic matter (carbohydrates, proteins, and lipids) into simple organic molecules (formate, ethanol, acetate, lactate, propionate, butyrate, and succinate) and CO_2 and H_2 gas (Fig. 1). Eubacteria are the prevalent hydrolytic fermentative microorganisms in anaerobic digesters (see also Table 4), whereas protozoa seem to play a more important role in brackish water sediments (van Bruggen et al., 1983, 1985).

The most significant fermentation reactions of hydrolytic fermentative microorganisms are listed in Table 1. The disproportionation of complex organic matter (i.e., glucose) into more oxidized and more reduced carbon compounds is indicated by the ratio of chemical oxygen demand (COD) to carbon of the prevalent fermentation prod-

Table 1 Important Fermentation Reactions in Anaerobic Ecosystems in the Absence of Sulfate or Nitrate[a]

	Reaction	$\Delta G^{0\prime}$ (kJ per reaction)
1	$C_6H_{12}O_6 + 3H_2O \rightarrow 3CH_4 + 3HCO_3^- + 3H^+$	-403.6
	Hydrolytic fermentative bacteria	
2	$C_6H_{12}O_6 + 2H_2O \rightarrow$ 2 ethanol $+ 2HCO_3^- + 2H^+$	-225.4
3	$C_6H_{12}O_6 \rightarrow$ 2 lactate$^-$ $+ 2H^+$	-198.1
4	$C_6H_{12}O_6 + 2H_2O \rightarrow$ butyrate$^-$ $+ 2HCO_3^- + 3H^+ + 2H_2$	-254.4
5	$C_6H_{12}O_6 \rightarrow$ 3 acetate$^-$ $+ 3H^+$	-310.6
6	$C_6H_{12}O_6 + HCO_3^- \rightarrow$ succinate^{2-} + acetate$^-$ + formate$^-$ $+ 3H^+ + H_2O$	-144.0
7	3 lactate$^-$ $\rightarrow$ 2 propionate$^-$ + acetate$^-$ $+ HCO_3^- + H^+$	-164.8
8	2 lactate$^-$ $+ 2H_2O \rightarrow$ butyrate $+ 2HCO_3^- + H^+ + 2H_2$	-56.2
	Syntrophic acetogenic bacteria[b]	
9	Ethanol + 2 $HCO_3^- \rightarrow$ acetate$^-$ + 2 formate$^-$ $+ H_2O + H^+$	+7.0
10	Ethanol $+ H_2O \rightarrow$ acetate$^-$ $+ 2H_2 + H^+$	+9.6
11	Lactate$^-$ $+ 2H_2O \rightarrow$ acetate$^-$ $+ 2H_2 + HCO_3^- + H^+$	-3.96
12	Butyrate$^-$ $+ 2H_2O \rightarrow$ 2 acetate$^-$ $+ 2H_2 + H^+$	+48.1
13	Benzoate$^-$ $+ 6H_2O \rightarrow$ 3 acetate$^-$ $+ 3H_2 + CO_2 + 2H^+$	+53.0
14	Succinate^{2-} $+ 4H_2O \rightarrow$ acetate$^-$ $+ 2HCO_3^- + 3H_2 + H^+$	+56.1
15	Propionate$^-$ $+ 3H_2O \rightarrow$ acetate$^-$ $+ HCO_3^- + 3H_2 + H^+$	+76.1
	Methanogenic bacteria	
16	Acetate$^-$ $+ H_2O \rightarrow$ methane $+ HCO_3^-$	-31.0
17	$4H_2 + HCO_3^- + H^+ \rightarrow$ methane $+ 3H_2O$	-135.6
18	$4HCO_2^- + H^+ + H_2O \rightarrow$ methane $+ 3HCO_3^-$	-130.4

[a]Except for reaction 1, all are partial reactions of anaerobic digestion processes.

[b]Acetogenic CO_2 reduction (reaction 9) yields -1.3 kJ more free energy per 2 mol of reducing equivalents than reaction 10. All syntrophic acetogenic reactions could involve production of either $H_2 + CO_2$ or formate.

Source: Calculated after Thauer et al. (1977).

Table 2 Chemical Composition and Oxygen Demand of Important Anaerobic Fermentation Products[a]

Compound		COD per mole	COD per carbon
Hydrogen	H_2	16	—
Formate	$CHCOO^-$	16	16
Methane	CH_4	64	64
Acetate	CH_3-COO^-	64	32
Pyruvate	$CH_3-CO-COO^-$	80	26.7
Lactate	$CH_3-CHOH-COO^-$	96	32
Ethanol	CH_3-CH_2OH	96	48
Propionate	$CH_3-CH_2-COO^-$	112	37.3
Succinate	$COO^--CH_2-CH_2-COO^-$	112	28
Butyrate	$CH_3-CH_2-CH_2-COO^-$	160	40
Glucose	$C_6-H_{12}-O_6$	192	32

[a]The chemical oxygen demand (COD) is calculated as grams O_2 required for full oxidation to CO_2 and H_2O.

ucts (Table 2). Note that the production of formate or H_2 + CO_2 is redox equivalent in fermentations. Some hydrolytic fermentations yield metabolic energy without the production of CO_2 via an intramolecular rearrangement of H and O atoms (homoacetic and homolactic fermentations, reactions 3 and 5). Other fermentations produce CO_2 and formate or H_2 gas for disposal of electrons generated during the fermentation of organic matter to acetate, succinate, butyrate, or propionate. Certain bacteria ferment C_2 alcohols to propionate (C_3) via CO_2 incorporation (Schink, 1984).

Trophic group 2, the syntrophic (*syn*, Greek: together; *trophein*, Greek: eat) acetogenic bacteria are eubacterial species that metabolize hydrolytic fermentation end products. Individual species are usually very specialized and ferment only a limited number of simple compounds to acetate, formate, bicarbonate, and H_2 (see below). Some SAB also ferment xenobiotic compounds, such as polyethylene glycol (PEG) and phenolic compounds (Schink and Stieb, 1983; Schink, 1985; Szewzyk et al., 1985; Dwyer and Tiedje, 1983; Suflita et al., 1983). The chemical free energy change is positive for most reactions of trophic group 2 (Table 1) under standard conditions. Based on this observation, it was hypothesized that syntrophic acetogenic bacteria suffer from a thermodynamic product inhibition by H_2 gas (Wolin, 1976; Wolin and Miller, 1982) and that their growth rate depends on the simultaneous removal of reduced products and acetate. This would simultaneously provide

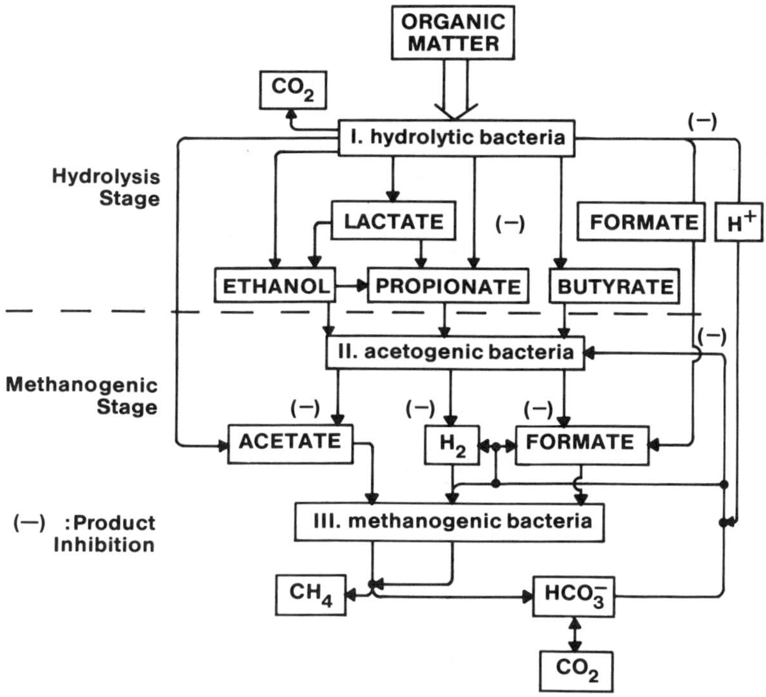

Figure 1 Carbon and electron flow scheme in relation to microbial trophic groups performing anaerobic degradation of organic matter. Note the different product-inhibited steps. Stage separation (---) between hydrolytic and syntrophic methanogenic reactions.

SAB with favorable growth conditions and their metabolic partners with carbon and energy sources (interspecies H_2 transfer). This hypothesis is based on the observation that SAB can usually be maintained in coculture with methanogenic or sulfate-reducing bacteria (Mah, 1982; Wolin and Miller, 1982). Some SAB can be grown in pure culture, however, either with sulfate as terminal electron acceptor (*Desulfovibrio* sp.) or with alternate carbon and energy sources, such as pyruvate (S organism), ethylene glycols (*Pelobacter* sp.), or crotonate (*Syntrophomonas wolfei*) (Bryant et al., 1977; Reddy et al., 1972a; Eichler and Schink, 1985; Schink and Stieb, 1983; Wofford et al., 1986). This observation supports the concept that the exclusive physiological function of the methanogenic partner organism is to remove the inhibitory reduced end products of syntrophic acetogenic metabolism. These data show that SAB can grow in pure culture provided the bioenergetic conditions are favorable, calling into question their previously considered "obligate" syntrophic nature.

Trophic group 3, the methanogenic bacteria, is the most important terminal trophic group in anaerobic digester ecosystems. All methanogenic bacteria belong to the archaebacterial kingdom and differ from eubacteria in their RNA polymerase, ribosome composition, membrane lipids, and cell wall components. Excellent reviews of the structure and physiology of methanogenic bacteria are available (Balch et al., 1979; Jones et al., 1987; Zeikus et al., 1985). Methanogens cannot utilize the fermentation products of trophic group 1 with more than two carbon atoms for growth or methane production. They use simple energy sources, such as acetate, methanol, and methylamines and CO and H_2 or formate (Table 3). It has been shown that some H_2-utilizing methanogens can partially oxidize alcohols, such as ethanol or isopropanol, to acetate or acetone under laboratory conditions. The reducing equivalents are then used for methane formation (Widdel and Wolfe, 1986). Whether these reactions and the oxidation of CO are of any ecophysiological relevance has yet to be shown.

Most methanogens in digesters are very specialized for their growth substrates and can be classified accordingly into acetoclastic methanogens, which disproportionate acetate into methane and CO_2, and CO_2-reducing methanogens, which oxidize H_2 gas, methanol, or formate and methylamines as electron donors (see Table 3 on genera) and reduce CO_2 or activated methyl groups to methane (for review, see Jones et al., 1987).

In anaerobic digesters more than 60% of the biological methane originates from acetate; thus acetoclastic methanogens perform the important ecophysiological function of carbon and electron removal (Jeris and McCarty, 1965; Smith and Mah, 1966; Zeikus, 1979). Methanogenesis from methanol and methylamines is of minor importance, except in marine environments. About 30-40% of the biological methane is produced by CO_2 reduction to methane.

It was not known until recently that strains of the genus *Methanosarcina* can produce H_2 gas upon metabolism of acetate, methanol, methylamines, and carbon monoxide (Table 4) (O'Brien et al., 1984; Lovley and Ferry, 1985; Phelps et al., 1985; Krzycki et al., 1987). Based on this observation, it was hypothesized that *Methanosarcina* would simultaneously produce and reconsume H_2 gas during the metabolism of acetate, methanol, and methylamines (Lovley and Ferry, 1985) to generate a membrane proton-motive force mechanistically similar to that hypothesized for the intracellular cycling of H_2 gas in the genus *Desulfovibrio* (Odom and Peck, 1981). This is theoretically possible, but the available evidence is circumstantial

Table 3 Some Important Methanogenic Bacteria Properties in Pure Culture

Family	Morphology	Substrates	Optimum temperature	pH optimum
Methanobacteriacea				
Methanobacterium formicicum	Rod	H_2, formate	37	7.0
M. bryantii	Rod	H_2,	38	7.0
M. thermoauto-trophicum	Rod	H_2	65–70	7.2–7.6
M. wolfei	Rod	H_2	55–65	7.0–7.5
M. thermoaggre-gans	Rod	H_2	65	7.0–7.5
M. thermoal-caliphilum	Rod	H_2	60	7.5–8.5
M. smithii	Rod	H_2, formate	38	6.9–7.4
Methanobrevibacter ruminantium	Rod	H_2, formate	38	7.2
M. arboriphilus	Rod	H_2	30–37	7.5–8.0
Methanosphera stadtmaniae	Coccus	H_2 + methanol	36–40	6.5–6.9
Methanococcaceae				
M. vannielii	Coccus	H_2, formate	36–40	7.0–9.0
M. voltae	Coccus	H_2, formate	32–40	6.7–7.4
M. maripaludis	Coccus	H_2, formate	38	6.8–7.2
M. deltae	Coccus	H_2, formate	37	ND
M. thermolitho-trophicus	Coccus	H_2, formate	65	6.5–7.5
M. jannaschii	Coccus	H_2	85	6.0
Methanomicrobiaceae				
Methanomicrobium mobile	Rod	H_2, formate	40	6.1–6.9
M. limicola	Planes	H_2, formate	40	7.0
Methanospirillum hungatei	Spirillum	H_2, formate	30–37	6.6–7.4
Methanosarcinaceae				
M. barkeri	Sarcina	H_2, methanol, methylamines, acetate	35	7.0
M. mazei	Coccus	H_2, methanol, methylamines, acetate	40	6.0–7.0
Methanothrix soehngenii	Filamentous rod	Acetate	37	7.4–7.8

Source: After Jones et al. (1987).

Table 4 Selected Examples for H_2- and Formate-Producing and Consuming Microbial Species in Anaerobic Batch Culture and Physiological H_2 or Formate Concentration Ranges[a]

Electron donors or acceptors	Organism	Dissolved intermediates (mM)		Apparent K_M for consumption (μM)		Source[b]
		Formate	H_2	Formate	H_2	
Glucose	*C. pasteurianum*	2-3	0.3-0.4	—	—	Thauer, 1971
Glucose	*E. coli*	20-30	<<	—	—	Thauer, 1971
Lactose	*K. oxytoca*	8	<<	—	—	Chartrain, 1986
Glucose	*R. albus*	10-15	0.5-0.7	—	—	Miller, 1973
Glutamate	*A. hydrogenoformans*	3-7	0.09	—	—	Stams, 1984
Serine	Undefined	3-5	0.013	—	—	Nanninga, 1985
Glucose	*S. ruminantium*	0-1	0.001	—	—	Scheifinger, 1975
Ethanol	*D. vulgaris*	< 0.25	0.002	—	—	Thiele et al., 1987a
Benzoate	*S. buswellii*	Trace-0.8	<<	—	—	Ferry, 1976
Ethanol	(S-organism)	NM	0.019	—	—	Reddy, 1972
Ethanol	(EC1)	NM	0.006	—	—	Bryant, 1977
Sulfate	*D. vulgaris*	NM	NM	NM	1-2	Kristiansson, 1982
Carbonate	*Methanobacterium formicicum*	NM	NM	240-580	4.5-6	Chartrain, 1986 Schauer, 1980, 1982
Carbonate	*Methanospirillum hungatei*	NM	NM	220	3-5	Robinson, 1982 Schauer, 1982
Carbonate	*M. barkeri*	NM	NM	NM	13	Robinson, 1982
Methanol	*M. barkeri*	NM	0.07	NM	NM	Ranalli, 1986
Acetate	*M. barkeri*	NM	0.0003	NM	NM	Krzycki, 1987
CO	*M. barkeri*	NM	> 0.25	NM	NM	O'Brien, 1984
Acetate	*M. acetivorans*	NM	< 0.0004	NM	NM	Lovley, 1985

[a]Values were calculated from mass balances or kinetic experiments: (<<), at the detection limits; NM, not measured.
[b]Only first author of source references listed.

and the observed levels of dissolved H_2 gas in the culture media during methanogenesis from these substrates are extremely low (Table 4). The simultaneous production and consumption of stoichiometric amounts of H_2 during methanogenesis was directly shown for CO utilization and supports a hydrogen cycle mechanism for methanogenesis from this substrate (O'Brien et al., 1984).

Formate is stoichiometrically and energetically equivalent to H_2 for methanogenic CO_2 reduction (Table 1 and 2). Thus, many CO_2-reducing methanogens in nature consume either formate or hydrogen as electron donor (Table 3). The ecophysiological significance of formate or H_2 as an electron donor for methanogenesis has not been well established previous to this review.

Various authors have reported that only a minor fraction of the methane generated in rumen, sediment, and/or digester ecosystems originate from exogenously added formate (Hungate et al., 1970; Phelps and Zeikus, 1985; Chartrain and Zeikus, 1986b). The conclusions were based on the turnover kinetics and the simultaneously measured soluble formate pool size. Likewise, measurement of H_2 turnover and the soluble H_2 pool in digester or sediment systems indicates that this is an insignificant source of reducing equivalents for CO_2 reduction to methane (Conrad et al., 1985). Bacteria in anaerobic ecosystems, such as the rumen (Cheng and Costerton, 1980), contact digesters (Chartrain and Zeikus, 1986b), sludge blanket reactors (Dolfing et al., 1985), and sediments, can be free living, attached to substrates, or self-immobilized in flocs or granules. Thus, we have speculated that local formate or H_2 concentration gradients can exist between juxtaposed bacterial cells within microbial aggregates such that these localized microniches develop high internal turnover constants for these reducing equivalents and they are uncoupled from the external H_2 and/or formate pool of the ecosystem (Conrad et al., 1985; Thiele et al., 1988c). Evidence for the existence of these niches in anaerobic digesters and the quantitative significance of formate versus hydrogen as the electron donor for CO_2 reduction to methane is reviewed here.

An analysis of the dissolved formate and H_2 concentrations found in pure and mixed cultures of trophic groups 1 and 2 (Table 4) indicates the relative importance of formate versus H_2. Fermentative bacteria can simultaneously produce formate and H_2 gas, which are redox-equivalent hydrolytic fermentation intermediates, and 90-99% of the produced and dissolved reducing equivalents (represented by H_2 + formate) can be recovered as formate (Table 4). This is due to the low solubility of H_2 gas and the thermodynamically favorable formate-H_2 equilibrium (see below). In addition, the lower dissolved H_2 concentration is accompanied by much lower apparent K_M values for the methanogenic consumption of H_2 gas versus formate

(Table 4). This means that a simultaneous interspecies electron transfer via H_2 gas or formate ions is theoretically possible. An analysis of the turnover of the dissolved H_2 gas in sediments and digester contents demonstrated that more than 95% of the methanogenic CO_2 reduction was independent of the simultaneously measured in situ H_2 gas consumption (Conrad et al., 1985). This was interpreted as a physical compartmentalization of CO_2 reduction to methane and interspecies H_2 gas transfer within metabolic microniches (i.e., within microbial flocs or granules) based on the assumption that H_2 is the electron donor for CO_2 reduction to methane. Alternatively, one can assume that H_2 is not the electron donor but formate or an unknown (XH_2) was the source of reducing equivalents for methanogenic CO_2 reduction (Thiele et al., 1988c).

Mutually beneficial metabolic interactions between H_2 producers and nonmethanogenic consumers are also observed in anaerobic digestion ecosystems. A classic example is the interaction between *Ruminococcus albus* and *Vibrio succinogenes* in the presence of fumarate (Ianotti et al., 1973). The addition of *V. succinogenes* and fumarate to a glucose fermentation by *R. albus* completely shifted the fermentation balance from the production of ethanol and H_2 to the exclusive formation of acetate and CO_2 by *R. albus*. The mutualistic growth of this coculture provided *V. succinogenes* with the required reducing equivalents for fumarate reduction and a higher cell yield for *R. albus* due to the resulting bioenergetically optimized utilization of glucose. The enhanced production of acetate from glucose provides extra metabolic energy in the form of ATP through the phosphoroclastic cleavage of acetyl-CoA. CO_2-reducing methanogens or sulfate-reducing bacteria have similar effects on metabolism of H_2-producing fermentative bacteria as in the example already described (Weimer and Zeikus, 1977; Chen and Wolin, 1977).

The dynamic interactions between the three anaerobic digestion trophic groups under shock load conditions are an important and characteristic feature of anaerobic ecosystems. Hydrolytic fermentative microorganisms usually grow and metabolize faster than syntrophic acetogenic bacteria or methanogens and tolerate lower pH values. High concentration shock loads of easily hydrolyzable organic matter lead to the rapid growth and metabolism of trophic group 1. This causes an additional accumulation of high concentrations of acidic fermentation end products, because the volumetric conversion rates by trophic group 2 bacteria in the ecosystem are usually much smaller than the fermentation rates by trophic group 1 bacteria. Thus, an acidification of the ecosystem occurs that, if not controlled by base addition, is inhibitory for trophic groups 2 and 3, which have a pH optimum around neutral pH (Table 3). This reduces the relative rates of terminal carbon conversion reactions and, as a consequence, finally leads to a collapse of the ecosystem and to an accumulation of acetate, propionate, butyrate, H_2 gas, and formate. It should be noted that H_2 gas and formate are also inhibitory for the metabolism of the SAB (see below).

Sulfate is a powerful inhibitor of methanogenesis. This has been convincingly demonstrated in many different ecosystems by the addition of sulfate to digesters or sediments (Abrams and Nedwell, 1978; Lovley et al., 1982; Chartrain and Zeikus, 1986b; Winfrey and Zeikus, 1977; Phelps and Zeikus, 1985). The sulfate-reducing bacteria (SRB) can outcompete methanogenic bacteria for H_2 gas in pure culture with nonlimiting sulfate owing to their higher growth yield, lower K_S value for H_2, and higher specific growth rates (Kristiansson et al., 1982; Robinson and Tiedje, 1984; Lovley et al., 1982; Lupton and Zeikus, 1984), and this is assumed to be the physiological reason for sulfate inhibition of methanogenesis.

Syntrophic acetogenesis and sulfate reduction can occur simultaneously in pure culture (McInerney et al., 1981). All SAB can be grown in coculture, with either CO_2-reducing methanogens or SRB functioning as electron acceptors. An apparent lack of competition between methanogenic and sulfidogenic electron flow has been reported in bench-scale reactors operated at steady state with a high COD load in the presence of excess sulfate (Isa et al., 1986). These authors demonstrated that 89% of the carbon COD were channeled to methane production, despite the presence of 5.5 mM excess sulfate. The dissolved H_2 gas concentration was between 0.4 and 1.4 μM, well within the competitive range for SRB (Table 4). Only 50% of the H_2 gas electrons were scavenged by the SRB, however. Bacterial aggregation and the selective attachment of methanogenic bacteria to support materials was used to explain the apparent methanogenic competitiveness, but without direct evidence. An alternative explanation is that SRB were performing syntrophic methanogenesis within the aggregates separated from dissolved sulfate. This view is substantiated later.

Conrad and colleagues (1987) provided evidence to support the hypothesis that in freshwater sediments, sulfate inhibition of methanogenesis is a result of shifting the metabolism of SRB from syntrophic methanogenesis to sulfate reduction. The addition of 10 mM sulfate to sediments from Lake Mendota showed an instantaneous inhibition (>30%) of syntrophic methanogenesis without changes in the turnover constant or pool size of available dissolved H_2 gas. Only long-term incubations with excess sulfate showed the predicted changes in the dissolved H_2 gas pool size and turnover constant, if competition for H_2 occurred between sulfate reducers and methanogens (Conrad et al., 1987). Only 10-14% of the syntrophic reducing equivalents used for CO_2 reduction to CH_4 were available as H_2 gas.

In this ecosystem, the short-term competition between methanogenic CO_2 reduction and sulfate reduction may be explained by a metabolic shift of these SRB from syntrophic life to sulfate reduction (Zeikus, 1983; Phelps and Zeikus, 1985; Conrad et al., 1987). This has been further substantiated (see below) in a defined anaerobic digester ecosystem (Thiele and Zeikus, 1988c). Methanogenic CO_2 reduction has an important control function for the development of dense populations of active syntrophic bacteria and the overall stability of the ecosystem in the absence of sulfate. Facultative syntrophy may be the survival mechanism for sulfate-reducing bacteria in freshwater ecosystems at limiting sulfate conditions.

The H_2-consuming metabolism of the methanogenic partner in a syntrophic consortium is assumed to maintain the H_2 partial pressure far below 10 Pa. However, pure culture studies showed that *Methanobacterium formicicum*, *Methanobacterium bryantii*, and *Methanospirillum hungatei* cultures were unable to maintain H_2 pressures below 6.5 Pa, even in the absence of acetogenic H_2 production (Lovley, 1985). The minimum threshold levels of H_2 in methanogen cultures were about sixfold higher than the H_2 gas levels observed in such environments as freshwater sediments (Table 5). Therefore, it is questionable that methanogens in nature are able to maintain levels of H_2 gas below 6 Pa and maintain their energy metabolism by H_2 consumption.

A bioenergetic analysis of syntrophic methanogenesis in four different anaerobic ecosystems, including a sewage digester, yielded positive free energy changes for the syntrophic propionate and butyrate degradations to acetate and hydrogen under in situ conditions (Conrad et al., 1986). With the reservation that equilibrium thermodynamics are only of limited merit in "open" biological systems, these interpretations were viewed as evidence for the existence of local microniches with lower internal H_2 partial pressures than in the available H_2 gas pool of the ecosystem. However, this is in question owing to the inability of methanogenic bacterial cultures to maintain such low H_2 gas levels (Lovley, 1985; Chartrain et al., 1987). Detailed dynamic reaction diffusion models with measured parameters clearly predicted that syntrophic methanogenic microniches would develop higher H_2 partial pressures inside anaerobic digester flocs (Ozturk et al., 1987).

In essence, these studies document the importance of interspecies electron transfer for the stability and efficiency of the anaerobic degradation of complex organic matter. However, results do not document the quantitative significance of interspecies hydrogen transfer during syntrophic methanogenesis in anaerobic digesters. Interspecies H_2 transfer reactions allow both fermentative hydrolytic bacteria and syntrophic acetogenic bacteria to produce more oxidized

Table 5 Concentrations and Apparent K_s Values for Hydrogen and Formate Utilization in Selected Anaerobic Ecosystems

Ecosystem	Levels (μM)		Apparent K_s (μM)		Reference
	Formate	H_2	Formate	H_2	
Freshwater sediments	10-20	$1\text{-}3 \times 10^{-2}$	NM[a]	3-7	Lovley et al., 1982; Conrad et al., 1985; Phelps et al., 1985; Robinson and Tiedje, 1982
Rumen systems	5-30	0.2-1.5	NM	NM	Hungate, 1967; Hungate et al., 1970
Contact digester	<10-20	0.2-2.5	NM	4-6	Chartrain et al., 1987; Conrad et al., 1985

[a]Not measured.

fermentation products. Thus, they experience a higher potential energy yield but the biochemical nature of the transferred acetogenic product is unknown. The importance of H_2 gas to the transfer reaction has been shown in vitro, but an analysis of the available kinetic and thermocynamic data for this reaction in natural ecosystems and anaerobic digesters results in conflicting evidence with respect to three key parameters:

1. Turnover of the H_2 gas pool in the absence of sulfate accounts for only 5-10% of the simultaneously occurring syntrophic electron transfer reaction rates.
2. The concentration levels of H_2 gas in various ecosystems in the absence of sulfate are below the measured H_2 threshold levels for CO_2-reducing methanogenic bacterial cultures.
3. The calculated thermodynamic free energies for several naturally occurring syntrophic methanogenic reactions (i.e., propionate and butyrate) are positive under in situ conditions.

A simple and consistent explanation for this apparent conflict between ecological and laboratory studies is the existence of compartmentalized syntrophic microniches in natural ecosystems, where interspecies electron transfer occurs independently of the measured available H_2 gas pool. This could be due to a physical compartmentalization through diffusion limitations or a biochemical compartmentalization, if other compounds, such as bicarbonate, were utilized as ecophysiological interspecies electron carriers (see below).

III. MICRONICHE CONCEPTS: THE COMPARTMENTALIZATION OF SYNTROPHIC METHANOGENESIS

Previous studies on freshwater sediments and digesters suggested that interspecies electron transfer between syntrophic acetogens and methanogens is independent of the dissolved H_2 pool concentration in the pore water or the digester fluid (Conrad et al., 1985). This can indicate an electron transfer mechanism independent of H_2 gas or due to the existence of aggregates of juxtaposed H_2 gas-producing and H_2-consuming species in the ecosystems generating an internal pool of dissolved syntrophic reducing equivalents within the aggregate. The diffusion of H_2 in a microbial aggregate can be modeled in analogy to the reaction diffusion problems typical of porous catalysts in industrial engineering applications or immobilized cells, which can be described by the "Thiele modulus" (Satterfield, 1970). This parameter is influenced by the volumetric reaction rate constant and the diffusivity for the transferred compounds.

It is known that substrate depletion in the center of a microbial aggregate can be caused by reduced diffusional fluxes, and this may impair the overall metabolic rates of the involved microorganisms (Atkinson, 1984; Tramper et al., 1984). On the other hand, several examples in nature show that syntrophic relationships are facilitated by the formation of mixed microbial aggregates or consortia in which the involved microorganisms share a common spatial microniche (Cheng and Costerton, 1980; Costerton and Irvin, 1981; Marshall, 1984; Oremland, 1983).

Applying a H_2 diffusion modeling approach (see Fig. 2) to anaerobic digester flocs predicts that if H_2 were the source of reducing equivalents for syntrophic methanogens, then the H_2 concentration in the floc would be higher than in the soluble phase (Ozturk et al., 1987). Only high Thiele moduli in spherical particles caused by large radii, low diffusion rates, or high volumetric reaction rates could cause lower concentrations of H_2 within a microbial aggregate resembling the postulated microniches with a lower internal H_2 concentration than that of the aqueous environment (Conrad et al., 1986). However, since the diffusivity for H_2 gas molecules in aqueous environments is very high (approximately 10^{-5} cm^2/s), these postulated microniches could only exist at large volumetric rates of H_2 gas consumption.

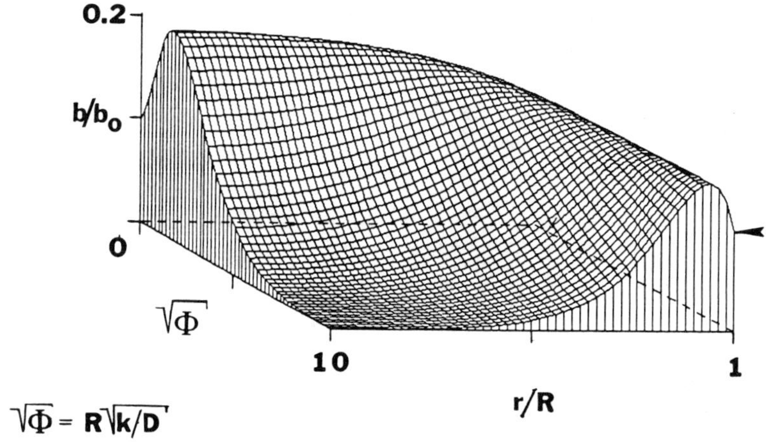

Figure 2 Simulated concentration profiles for the reduced reaction intermediate XH_2 during syntrophic ethanol oxidation in microbial digester flocs. The concentration profiles for the reduced intermediate (XH_2) during the syntrophic ethanol degradation within microbial aggregates were simulated for a first-order reaction mechanism valid at low substrate concentrations (Ozturk et al., 1987). XH_2 was assumed to be H_2 gas. The Thiele modulus ϕ is defined as: $\phi = R^2k/D$, with R the radius of the particle, k the first-order volumetric rate constant, and D the diffusivity of XH_2. b/b_0 indicates dimensionless relative concentrations of XH_2. The arrow indicates the bulk concentration of XH_2 outside the aggregate. Note the different shapes of the profiles at different Thiele moduli. The center of the aggregate corresponds to $r/R = 0$, with r/R the relative radial coordinate of the floc. A high Thiele modulus determines lower concentrations of H_2 inside the flocs.

Microbial flocs from a well-defined whey biomethanation chemostat were used as a model system to examine the niche for syntrophic methanogenesis. The microbial flora in the whey digester was simultaneously present as flocs about 100 μm in diameter and as free-living cells. Chartrain and Zeikus provided detailed studies of the microbial and metabolic properties of lactose biomethanation in this digester (Chartrain and Zeikus, 1986a, b). The trophic group species composition and enumeration, carbon transformation parameters, and intermediate carbon metabolism were reported (see Fig. 3). Here, lactose was fermented by *Leuconostoc mesenteroides*, *Klebsiella oxytoca*, and *Clostridium butyricum* into lactate, ethanol, formate, H_2-CO_2, and acetate as the prevalent intermediary metabolites.

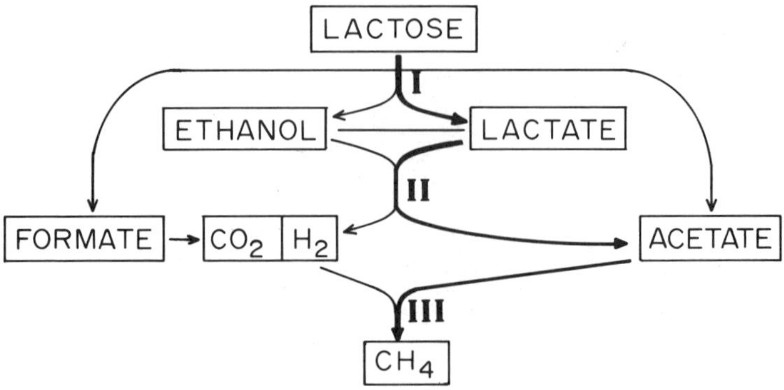

Figure 3 Comparison of the carbon and electron flow pathway in relation to the prevalent species. Trophic group composition for lactose biomethanation in a continuous whey processing digester. (From Chartrain and Zeikus, 1986a.)

Trophic group 2 was predominantly represented by *Desulfovibrio vulgaris*, which in the absence of sulfate metabolized lactate and ethanol to acetate and reducing equivalents in a syntrophic association with *M. formicicum* as the prevalent CO_2-reducing methanogen (group 3). Trophic group 3 acetoclastic methanogenesis included *Methanosarcina barkeri* and *Methanothrix soehngenii*.

The digester contents were separated into pure flocs and free flora, and analysis of the lactose metabolism showed that both fractions contained very similar specific activities for lactose conversion to acetate and methane, with ethanol and lactate as the predominant syntrophic carbon intermediates (Thiele et al., 1988c). Very significant differences were found in the dynamics of gas metabolism displayed by the two cell fractions. The free flora showed a much higher and transient H_2 production during the early stage of lactose conversion (>2000 Pa), whereas the flocs produced lower and more stable H_2 levels (< 920 Pa). The ratio of H_2 to CH_4 formation in the floc preparation during the first 50 h was always < 1.5, whereas the free flora showed ratios >10.

The whey digester flocs represented a preferred compartment for the syntrophic conversion of ethanol and lactate. Figure 4 shows that the onset of syntrophic CH_4 production from ethanol by free flora occurred 10 h later than in the floc preparation and only after the appearance of high H_2 concentrations. In contrast, the floc fraction syntrophically metabolized ethanol at low levels of H_2 gas. Notably, microscopic analysis at the end of the experiments indicated that free flora was not formed in the floc vials and flocs

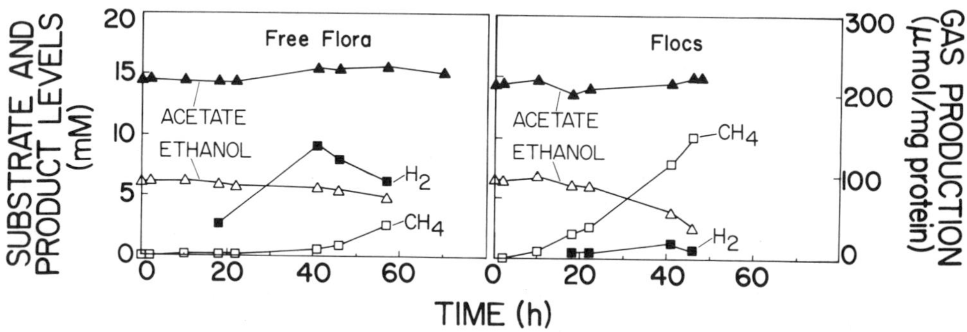

Figure 4 Comparison of metabolic parameters for syntrophic ethanol transformation by free flora versus floc population fractions.

were not formed in free flora vials when ethanol was the sole carbon and electron source. Thus, the organizational structure of the bacteria did not change during the incubation.

Figure 5 represents the cellular compartmentalization predictions for the syntrophic ethanol conversion activity displayed by the digester flocs. Two idealized types of cell arrangements are theoretically possible in the flocs: clusters of microcolonies of H_2-producing syntrophic acetogens and H_2-consuming methanogens (cluster type) embedded in a floc matrix with very low H_2 gas diffusivity or a homogenous dense meshwork of singly juxtaposed syntrophic acetogenic and CO_2-reducing methanogenic cells (lattice type) generating extremely high volumetric reaction rates due to the small interspecies distance and with maximal H_2 transfer. Both mechanisms would generate large Thiele moduli, causing an apparent compartmentalization of the syntrophic metabolism. Thin-section electron and phase-contrast fluorescence microscopy supported the lattice type of cell arrangement (Thiele et al., 1988c).

To precisely identify the ecophysiological mechanisms responsible for the apparent compartmentalization of syntrophic methanogenesis, the functional localization of the three following microbial reactions was determined in flocs versus free flora under substrate saturating conditions (Table 6):

$$2CH_3CH_2OH + 2H_2O \rightarrow 2CH_3COO^- + 2H^+ + 4H_2$$

$$\Delta G_0' = +19.2 \text{ kJ per reaction} \qquad (1)$$

$$4H_2 + HCO_{3-} + H^+ \rightarrow CH_4 + 3H_2O$$

$$\Delta G_0' = -135.6 \text{ kJ per reaction} \qquad (2)$$

$$2CH_3CH_2OH + HCO_3^- \quad 2CH_3COO^- + H^+ + CH_4 + H_2O$$
$$\Delta G_0' = -116.4 \text{ kJ per reaction} \qquad (3)$$

As shown in Table 6, both the total and the specific ethanol oxidation activity and hydrogen consumption activity were much higher in the floc versus the free flora fraction. The floc fraction displayed 87% of the total H_2-consuming methanogenic activity and 76% of the total ethanol-consuming activity in the ecosystem. H_2-consuming acetogenesis and sulfate reduction were insignificant in the flocs owing to the absence of sulfate in the defined media and the lack of acetate formation in the presence of 80% H_2 (data not shown).

To demonstrate whether compartmentalization of syntrophic ethanol metabolism was due to a H_2 gas diffusion limitation between

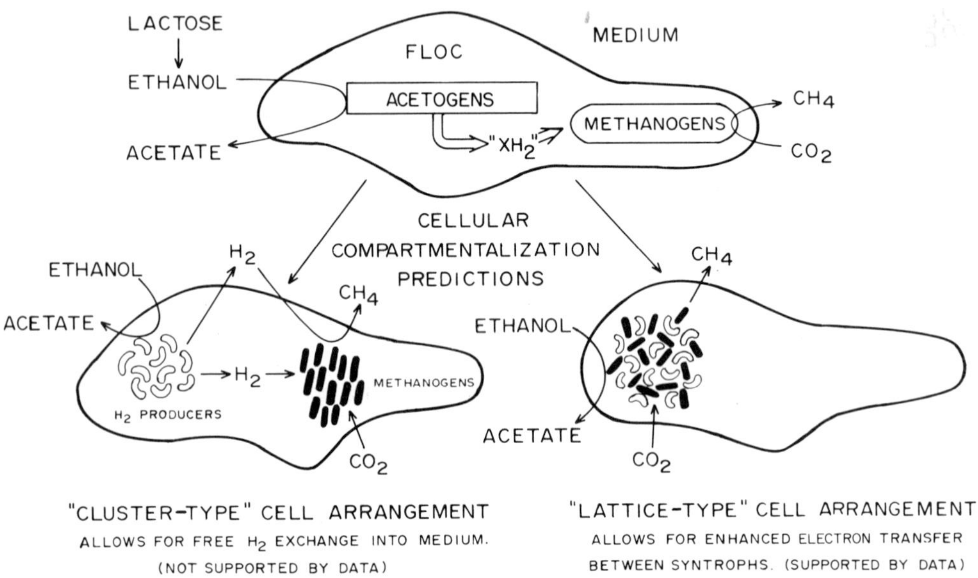

Figure 5 Predictive model for species compartmentalization and electron flow coupling during syntrophic ethanol metabolism in whey digester flocs. The data support a lattice type of cell juxtapositioned between *Desulfovibrio vulgaris*, the principal syntrophic acetogen, and *Methanobacterium formicicum*, the principal CO_2-reducing methanogen. The data also support a mechanism for electron flow coupling from the acetogen to the methanogen via XH_2, which represents reducing equivalents independent of the H_2 pool and interspecies H_2 gas transfer.

Table 6 Comparison of Syntrophic Ethanol Metabolism Substrate Consumption Parameters in Free Flora Versus Floc Population Fractions[a]

Population fraction	Total protein (mg)	Acetogenic ethanol consumption: Specific activity (μmol/h per mg)	Acetogenic ethanol consumption: Total activity (μmol/h)	Methanogenic H_2 consumption: Specific activity (μmol/h per mg)	Methanogenic H_2 consumption: Total activity (μmol/h)
Free flora	49.2	1.2	58.6	0.18	8.7
Flocs	65.3	2.9	188.7	0.88	57.3

[a]Flocs and free flora were freshly prepared from 125 mL of whey-processing digester contents and were incubated at 37°C in 158 mL serum vials that contained 45 mL of PBB medium with a mixture of 6 mM ethanol and either 14 mM acetate or H_2/CO_2 (80:20). The conversion rates for substrate consumption were determined between 10 and 20 h when all rates were linear.

the floc phase and the environment, experiments were designed to test the effect of a cell contact disruption on H_2 gas metabolism, ethanol consumption, and methane production by flocs (Fig. 6). Flocs were incubated in an ethanol-acetate medium under constant shaking in closed pressure bottles with a headspace-liquid ratio of 10:1, ensuring the continuous extraction of 99% of all the H_2 gas in the system into the headspace within 10 minutes. One-half of the preparation was disrupted by a syringe technique into fragments of approximately 1/10 the original size (Thiele and Zeikus, 1988c). It was assumed that if H_2 gas were predominantly exchanged within the flocs, a disruption would destroy the floc integrity and uncouple efficient H_2 gas transfer (i.e., H_2 production and consumption) during syntrophic ethanol conversion.

The diffusional mass transfer of H_2 gas across the surface of the reaction phase into the headspace is proportional to the surface area and the H_2 concentration gradient across the boundary between the headspace and the reaction phase. Thus, in the experimental closed bottle system, a flux of H_2 gas into the headspace during syntrophic ethanol oxidation is proportional to the interfacial area A of the reaction phase and at constant D, d $\mu H_2/dc$ and grad μH_2 is a direct measure of the interfacial area. If syntrophic reducing equivalents were predominantly exchanged between acetogens and methanogens through the environment outside the flocs, then floc disruption should not affect H_2 gas production, ethanol consumption, and methane formation.

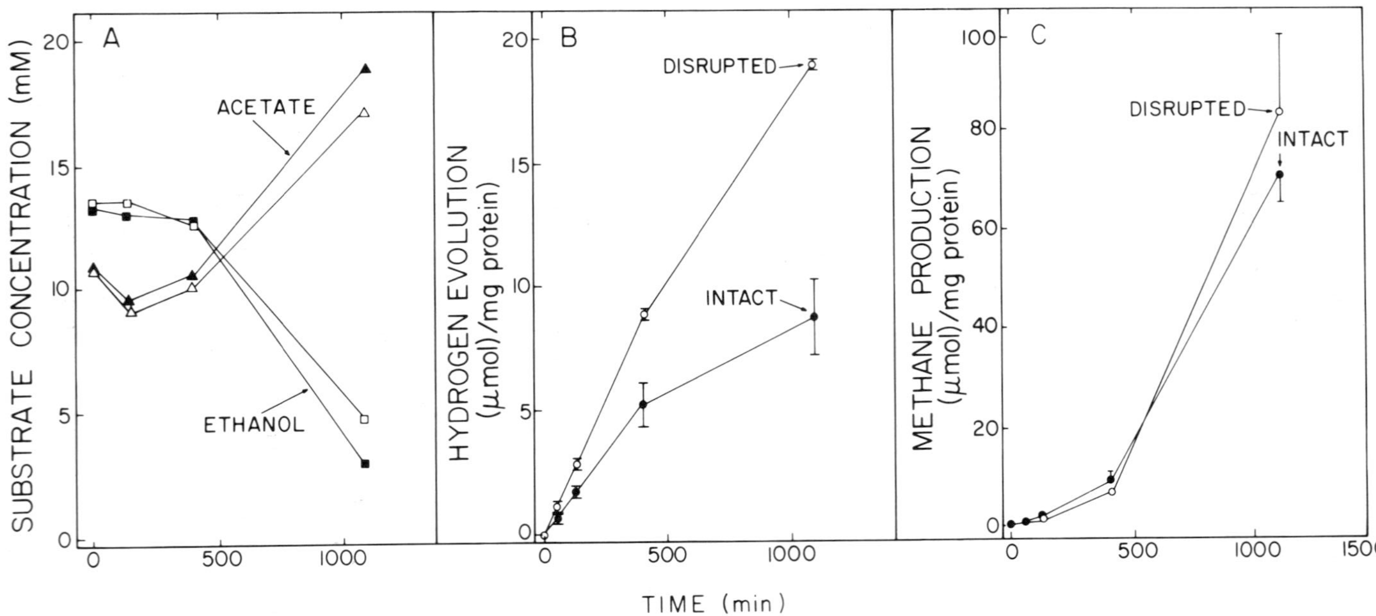

Figure 6 Comparison of ethanol-dependent H_2 and CH_4 production by intact versus disrupted floc preparations.

Figure 6 compares hydrogen and methane production by intact versus disrupted floc preparations incubated in the presence of ethanol and acetate. Identical preparations were incubated with acetate alone as control (data not shown). The measured H_2 partial pressure in the acetate control experiments was 10–30 Pa, whereas in the presence of ethanol the H_2 partial pressure was 300–600 Pa. Ethanol addition enhanced both hydrogen evolution and methane production at least 10-fold over the acetate controls. The total rate of methane production was also 10-fold higher than the rate of H_2 evolution. Notably, floc disruption enhanced H_2 evolution without changing the rate of methane production or ethanol consumption. The rate of increase and the maximum size of the H_2 gas pool in liquid and head phase in disrupted floc preparations at 500 minutes was approximately twice that of intact floc preparations. Thus, disruption reduced interspecies H_2 transfer compartmentalization but did not significantly alter the kinetics of syntrophic ethanol conversion to methane. This indicated a diffusion limitation of the XH_2 exchange between the floc phase and the environment.

The efficiency of interspecies H_2 transfer in the floc compartment can be expressed as the interspecies H_2-transfer ratio (IHT ratio), which is defined as the molar ratio of H_2 used for methane production via CO_2 reduction over the total amount of H_2 produced from the oxidation of ethanol to acetate.

$$\text{IHT ratio} = \frac{H_2 \text{ equivalents consumed for } CO_2 \text{ reduction to } CH_4}{H_2 \text{ equivalents consumed} + H_2 \text{ gas produced}}$$

Disruption of flocs clearly decreased the IHT ratio at all time intervals. This effect was most dramatic during the first 60 minutes, when a 2.8-fold reduction in efficiency was observed upon disruption (24 versus 68% at 0–56 minutes). These results can be explained by the expected increase in the surface area due to floc disruption. Based on the geometry of a sphere, a floc of 50 μm radius, when disrupted into fragments of 10–20 μm radius, yields a 2.5- to 5-fold increase in the total surface area. Notably intact flocs always displayed a very high IHT ratio of >95% during active ethanol conversion.

These results emphasize that syntrophic ethanol conversion to methane occurred predominantly within the floc compartment and was not mediated by the dissolved H_2 gas pool outside the flocs because disruption altered the kinetics of H_2 evolution but not CH_4 production or ethanol consumption. If H_2 gas were exchanged between acetogens and methanogens via the H_2 gas pool outside the flocs, then disruption itself would not have increased the flux of H_2 into the headspace, because the interfacial area and volume of the liquid and gas phase were exactly identical in experimental vials with either disrupted or intact preparations. These data also showed that <5% of syntrophic electron flow could really be accounted for by interspecies H_2 transfer. The balance could be H_2 transfer inside the floc compartment or could depend on some other source of reducing equivalents for interspecies control of electron flow (see below). We thus propose referring to the intermediate of syntrophic methanogenesis as XH_2 in natural and anaerobic digester ecosystems until positive quantitative evidence is shown for its biochemical identity. We showed in a whey biomethanation digester that formate could account for 95% of XH_2 for the syntrophic conversion of ethanol or lactate (see below).

The juxtaposition of bacteria in anaerobic digestors for syntrophic metabolism is somewhat analogous to a coimmobilization of two coupled enzyme reactions to gain high substrate turnover rates. High flux rates by enzyme coimmobilization studies allowed an increase as high as 10-fold in turnover compared with the free enzyme mixture (Mosbach, 1978). Juxtaposition has a clear ecophysiological function, therefore, because the syntrophic growth and energy metabolism of acetogens and methanogens depends on the magnitude of the carbon and electron flux to methane. Thus, syntrophic metabolism is controlled by the interspecies diffusional flux of the inhibitory acetogenic reduced intermediate XH_2, which controls the initial first reaction step of the syntrophic reaction chain. This flux follows Fick's first law of diffusion as $J = -D\ dc/dx$. This means that either a short distance x or a large concentration difference between the acetogen and methanogen cell surface will improve the interspecies transport rates. Pushing the transport by larger values for dc/dx, however, would result in a product inhibition of the acetogenic partial reactions and of the overall carbon and electron flux. Only shorter cell-cell distances allow higher syntrophic fluxes at proportionally lower XH_2 levels.

IV. INTERSPECIES ELECTRON TRANSFER: MECHANISMS AND EVIDENCE

A. Mechanisms

The biochemical mechanism prevalently used to explain syntrophic methanogenesis is interspecies H_2 transfer. This concept is largely based on equilibrium thermodynamic considerations assuming standard conditions, mass balances in closed in vitro systems, and the inhibitory effect of H_2 gas on the acetogenic metabolism in artificial cocultures (Wolin, 1976; Wolin and Miller, 1982). The ecophysiological concept behind interspecies H_2 transfer, that certain anaerobic transformations are only possible in coculture with H_2-consuming bacteria, is supported by numerous excellent studies on the anaerobic degradation of alcohols, fatty acids, and aromatic compounds (Bryant et al., 1967; McInerney et al., 1981; Mountfort an d Bryant, 1982; Boone and Bryant, 1980).

Ecophysiological conclusions about the mechanism of interspecies electron transfer, that H_2 gas functions as an actual reaction intermediate in anaerobic digestion, are based on the following assumptions: (1) ecological phenomena can be explained by equilibrium thermodynamics; (2) standard conditions (unit activity) are close to ecological relevant substrate and product concentrations; (3) all relevant components are homogeneously distributed in the environment, neglecting the poor solubility of such gases as H_2, diffusion gradients, and mass transfer limitations; and (4) H_2 gas, when added to a complex ecosystem or a defined coculture, has only one expected effect.

None of these assumptions alone withstands critical review because of the following contradictions: (1) living cells are thermodynamically undefined and can only be treated by nonequilibrium thermodynamics; (2) the concentrations of substrates and products in natural and applied ecosystems are usually at chemical activities of 10^{-2}-10^{-6}; (3) microorganisms as well as their products are usually spatially organized; and (4) H_2 gas can simultaneously be an inhibitor of syntrophic acetogenesis and a substrate for methanogenesis.

This last point bears closer attention. When H_2 is added to a methanogenic syntrophic coculture, it can inhibit acetogenesis either by direct thermodynamic action or by indirect action via competition with an unknown syntrophic intermediate XH_2 (e.g., formate) at the level of the methanogenic metabolism. This would lead to the accumulation of inhibitory concentrations of XH_2, suggesting an inhibition of syntrophic acetogenesis by H_2 gas. This point clearly shows that it is impossible to use thermodynamics to define the prevalent ecophysiological reaction mechanism of the interspecies electron transfer that is associated with syntrophic methanogenesis.

The thermodynamic features used to support the concept of interspecies H_2 transfer do not allow one to predict or exclude a certain reaction mechanism. At least two alternative routes could theoretically be proposed for syntrophic ethanol degradation—H_2 gas transfer or formate transfer (Table 7)—but additional ecophysiological mechanisms cannot be excluded. It is also theoretically possible that cells in microbial aggregates juxtaposed with biochemical cell contacts use other mechanisms to control intercellular electron transfer (e.g., even direct electron transfer has been speculated). The ecophysiological electron transfer routes for syntrophic methanogenesis are all thermodynamically independent of the identity of XH_2 because the intermediate does not appear in the syntrophic reaction

Table 7 Energetic and Biochemical Comparison of Dehydrogenation and Hydrogenation Reactions for Syntrophic Ethanol Conversion to Methane by Interspecies H_2 Versus Formate Transfer Mechanisms[a]

1. Syntrophic ethanol conversion reaction:	
$2\ \text{ethanol} + HCO_3^- \rightarrow 2\ \text{acetate}^- + H^+ + CH_4 + 2H_2O$	-116
2. Interspecies H_2 transfer reaction:	
$2\ \text{ethanol} + 2H_2O \rightarrow 2\ \text{acetate}^- + 2H^+ + 4H_2$	+19.2
$HCO_3^- + H^+ + 4H_2 \rightarrow CH_4 + 3H_2O$	-135
3. Interspecies formate transfer reaction:	
$2\ \text{ethanol} + 4HCO_3^- \rightarrow 2\ \text{acetate}^- + 2H^+ + 4HCO_2^- + 2H_2O$	+14
$4HCO_2^- + H_2O + H^+ \rightarrow CH_4 + 3HCO_3^-$	-130
4. Biochemical redox reactions	
Formate-H_2 lyase:	
$H_2 + HCO_3^- \rightarrow HCO_2^- + H_2O$	-1.3
Hydrogenase:	
$2Fd^{red} + 2H^+ \rightarrow 2Fd^{ox} + H_2$	+3.08
CO_2 reductase:	
$2Fd_{red} + 2H^+ + HCO_3^- \rightarrow 2Fd^{ox} + HCO_2^- + H_2O$	+1.77
Alcohol-acetaldehyde dehydrogenase:	
$CH_3CH_2OH + 4Fd^{ox} + H_2O \rightarrow CH_3COO^- + 4Fd^{red} + 5H^+$	+3.44

[a]Values given are for $\Delta G_0'$ (kJ per reaction).

(Table 7, equation 1). Since all proposed mechanisms are thermodynamically equivalent under standard conditions, only kinetic studies can decide which of the alternatives is prevalent.

The microbial biochemistry of H_2 gas metabolism in anaerobic ecosystems is very complex. H_2 gas can be produced by hydrolytic fermentative bacteria and syntrophic acetogenic bacteria as a result of proton reduction by hydrogenases that oxidize reduced intracellular cofactors, such as ferredoxins or NADH (Odom and Peck, 1984). Acetoclastic methanogens and acetogens can also produce H_2 gas (Table 4) (Zinder and Koch, 1984). Notably, ferredoxin regeneration can also be achieved by CO_2 reduction to formate via CO_2 reductase (Thauer et al., 1973). In certain members of the genus *Clostridium*, proton reduction to H_2 and CO_2 reduction to formate can coexist as mechanisms of electron disposal (Thauer et al., 1973). It is currently assumed that the main function of this biochemical mechanism is to provide formate as a C_1 compound precursor for anabolic reactions. A catabolic function for CO_2 reductase in the regeneration of oxidized ferredoxin cannot be excluded at present, especially as significant formate levels are excreted by anaerobes (see also Table 4) (Thauer et al., 1971).

Many methanogens utilize formate for CO_2 reduction to methane (Table 3), and methanogens similar to *M. formicicum* or *M. hungatei* are among the prevalent methanogens in anaerobic digesters (Zeikus, 1979; Chartrain and Zeikus, 1986a), eutrophic lake sediments (Lovley et al., 1982), and syntrophic enrichment cultures (Boone and Bryant, 1980; Chartrain and Zeikus, 1986a; Ferry and Wolfe, 1976; Hungate et al., 1970; Mah, 1982; McInerney et al., 1979; Mountfort and Bryant, 1982). Some formate-degrading species can also produce significant quantities of H_2 during methanogenesis (Schauer et al., 1982; Tzeng et al., 1975). The apparent turnover of formate in diverse anaerobic digestion ecosystems is usually very rapid (Hungate et al., 1970; Phelps and Zeikus, 1985), and formate and H_2 gas pools are interconnected by reversible microbial formate-H_2 lyases (Thauer et al., 1973; Woods, 1936). Thus, the conclusion that the biochemical mechanism for the control of interspecies electron flow during syntrophic methanogenesis in nature could be achieved by either an interspecies formate transfer or an interspecies H_2 transfer-coupled process is substantiated by experimental evidence.

Formate-H_2 lyase catalyzes the reversible reaction between formate, H_2 gas, and bicarbonate (Thauer et al., 1973) and establishes a dynamic equilibrium between these reactants. Volumetric reaction rates for this enzyme complex significantly larger than the rates of formate utilization or H_2 production would result in coupled formate and H_2 pools at "quasi-equilibrium." Figure 7 shows the relationship between the bicarbonate and the total dissolved reducing equivalents of H_2 and formate at equilibrium (Thiele and Zeikus, 1988a). The concentration ratio [formate]/[H_2] in the liquid phase is strongly dependent on the bicarbonate concentration. Under equilibrium conditions in anaerobic ecosystems and at bicarbonate levels above 15 mM, more than 95% of the dissolved reducing equivalents would occur as formate owing to its high solubility. With experimental headspaces of 30% CO_2, this ratio would be as high as 99% and consistent with reports (see Tables 4 and 5). The ratio [formate]/[H_2] in the liquid phase is independent of the dissolved H_2 concentration and depends only on the [HCO_3^-]. The molar ratio of formate to hydrogen in the complete system, as used for calculations of mass balances, however, is strongly dependent on the liquid-headspace

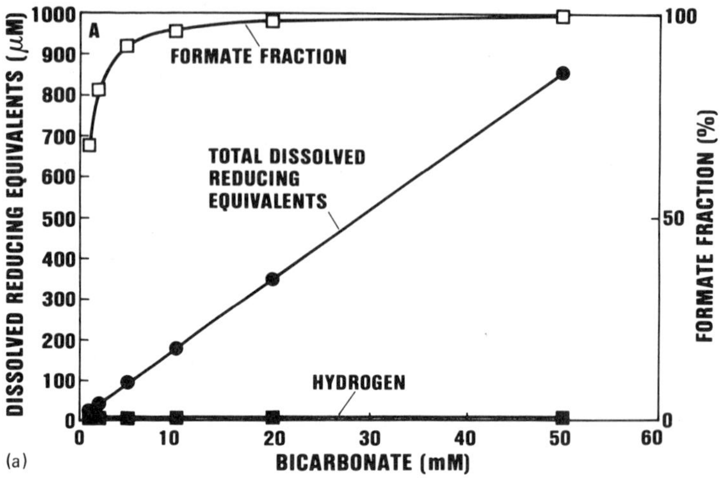

(a)

Figure 7 Biochemical energetic analysis of formate and H_2 metabolism during anaerobic digestion. (a) Dependence of the formate-H_2 equilibrium on the bicarbonate concentration. The data were calculated for a H_2 partial pressure of 1000 Pa. (b) Relative rates of methanogenesis via formate or H_2 as a function of the H_2 partial pressure. It was assumed that H_2, bicarbonate, H_2O, and formate were at chemical equilibrium at 35°C (K_{ass} = 1.66). The concentration of bicarbonate used was 80 mM. A K_M for H_2 of 5 μM and for formate of 220 μM was used. V_{max} was assumed to be identical for both substrates. (c) Dependence of the bioenergetic advantage of CO_2 reduction to formate on the H_2 partial pressure. The free energy difference for acetogenic formate versus H_2 production was calculated according to Thauer et al. (1977) and Table 8. A bicarbonate concentration of 21 mM and equal ethanol and acetate concentrations were used.

ratio. Most studies about interspecies H_2 transfer were based on mass balances at the end of the experiments (see above). In the standard test tubes used for physiological studies, the typical liquid-headspace ratios are as small as 0.3 and the total molar formate-H_2 ratio is close to 1 owing to the poor solubility of H_2 gas. These experimental conditions could artificially introduce H_2 gas-coupled syntrophic interactions because in natural, open systems (e.g., anaerobic digesters or sediments) the total dissolved molar formate-H_2 ratio is between 10 and 200 (Tables 4 and 5).

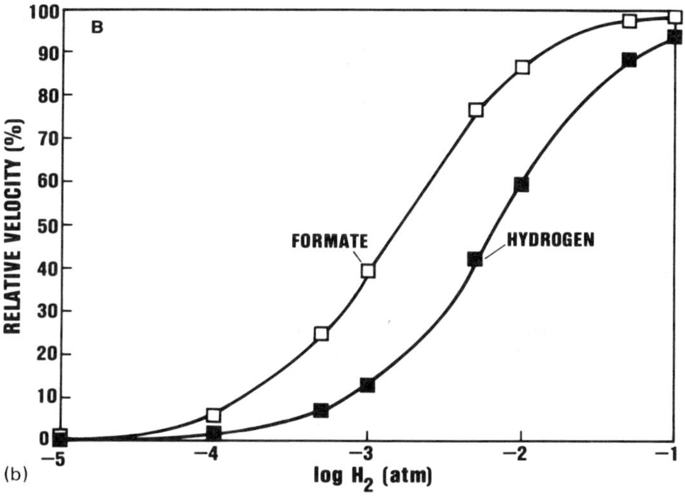
B
RELATIVE VELOCITY (%)
100
90
80
70
60
50
40
30
20
10
0
FORMATE
HYDROGEN
−5
−4
−3
−2
−1
log H2 (atm)

(b)

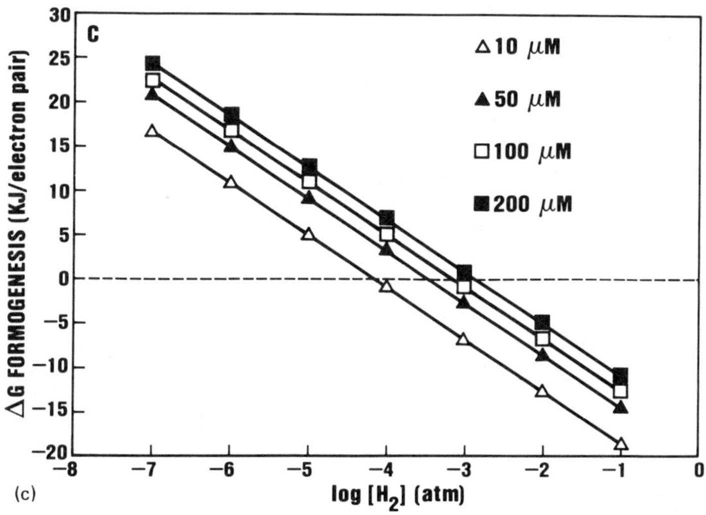
C
ΔG FORMOGENESIS (KJ/electron pair)
△10 μM
▲50 μM
□100 μM
■200 μM
30
25
20
15
10
5
0
−5
−10
−15
−20
−8
−7
−6
−5
−4
−3
−2
−1
0
log [H2] (atm)

(c)

Kinetic support for the hypothesis that formate can be preferentially transferred in syntrophic methanogenic associations in nature is provided by a comparison of the existing kinetic data from pure cultures of H_2 and formate-utilizing methanogens, such as *M. formicicum* or *M. hungatei* (Schauer and Ferry, 1980; Schauer et al., 1982). Figure 7b compares the relative rates of methanogenesis from formate versus H_2, assuming a K_S for dissolved H_2 gas of approximately 5 μM, a K_S for formate of approximately 220 μM, a formate-H_2 + CO_2 equilibrium, and similar maximum methane production rates from formate or H_2/CO_2. Surprisingly, methanogenic utilization of formate instead of H_2 could show three to four times higher rates of methane formation and thus faster interspecies electron flow at physiological H_2 partial pressures below 100 Pa.

B. Evidence for H_2 Transfer

Previously, direct evidence was not available to demonstrate the stoichiometric involvement of H_2 gas as the intermediary source of reducing equivalents that couple syntrophic methanogenesis in certain well-characterized mixed cultures. Interspecies H_2 gas transfer has even been inferred as the mechanism that controls electron flow in syntrophic methanogenic mixed culture systems in which significant activities of formate metabolism have been reported. These include ethanol degradation by the classic "*Methanobacillus omelanskii*" mixed culture (Barker, 1941; Bryant et al., 1967) and the benzoate-degrading consortium (Ferry and Wolfe, 1976). Direct kinetic evidence for syntrophic H_2 gas transfer in these cultures is lacking, and conclusions were based on H_2 detection in the culture headspace.

S. wolfei, a butyrate-degrading acetogen, grew syntrophically with either *M. bryantii* MOH or *Methanobrevibacterium arboriphilum* (McInerney et al., 1981). *M. bryantii* MOH and *M. arboriphilum* use only H_2 gas, and the coculture with *M. bryantii* MOH showed no methane formation from formate. *S. wolfei*, however, grew much better and the syntrophic methane yield was threefold higher, when grown in coculture with *M. hungatei*, a formate-utilizing methanogen (McInerney et al., 1981). This demonstrates that H_2 gas can be utilized as a syntrophic intermediate in the presence of a culture headspace. Formate may also be important during syntrophic butyrate degradation, however.

Most notable to new understanding is a thermophilic syntrophic acetate-degrading coculture comprised of an acetate-oxidizing rod that produced H_2/CO_2 (or formate) and a formate- or H_2/CO_2-utilizing thermophilic methanogen, *Methanobacterium* THF. H_2 gas was presumed as the syntrophic reaction intermediate, but formate could not be excluded in this syntrophic acetoclastic coculture (Zinder and Koch, 1984).

Continuous culture growth and reaction kinetics have also been described for a thermophilic triculture that degrades butyrate and that was comprised of a butyrate-utilizing acetogen, *Methanobacterium thermoautotrophicum*, and an acetate-cleaving rod (Ahring and Westermann, 1987). In this mixed culture, direct evidence was not provided to quantitate the significance of H_2 as the intermediate of syntrophic butyrate degradation to acetate and XH_2. Only the absence of a formate-consuming methanogen implicated H_2 transfer. Furthermore, an additional source of H_2 (i.e., formate) could have been provided by the thermophilic acetate-cleaving rod. Calculations based on the volumetric H_2 consumption (0.48 mMol/L per h) from the dissolved H_2 gas and the K_s of *M. thermoautotrophicum* are 2.4-fold higher than the potential H_2 gas production from butyrate oxidation (0.2 mMol/L per h).

Notwithstanding the lack of direct quantitative evidence for interspecies H_2 transfer in the anaerobic ecosystems cited, one experimental fact is clear: one can easily isolate and construct "artificial" coculture systems comprised of a syntrophic acetogen and an "obligate" H_2-consuming methanogen. In these systems, interspecies H_2 transfer serves as a valid physiological mechanism that controls electron transfer. However, in relation to the organism's ecology (i.e., what it is doing in nature), interspecies H_2 transfer may or may not be valid.

To assess the significance of interspecies H_2 transfer for coupling electron flow during syntrophic methanogenes, we initiated a series of experiments using flocs that were previously shown to represent the preferred niche for ethanol and lactate conversion to methane (Thiele and Zeikus, 1988a, c). Figure 8 summarizes key findings about H_2 metabolism in the whey digester flocs. Syntrophic ethanol conversion to acetate and methane proceeded in the presence of varying hydrogen levels for more than 4 h before a significant inhibition occurred even at 21,000 Pa added H_2. The thermodynamic driving force for an acetogenic H_2 transfer mechanism at 21,000 Pa H_2 was calculated to be +1.61 kJ per reaction, but the initial reaction rate was the same in the presence as in the absence of H_2. The addition of high levels of exogenous H_2 sufficient to thermodynamically inhibit ethanol conversion to methane did not stimulate the total rate of methane formation from H_2 nor did it inhibit ethanol oxidation. CO_2-reducing methanogens in these flocs were metabolically active but apparently quantitatively saturated with XH_2 equivalents during active ethanol oxidation at the low physiological H_2 partial pressures, because H_2 gas addition did not enhance the rate of methane formation. The addition of Na_2SO_4 to floc samples at 51 h blocked methane formation by 80%, but not ethanol consumption, demonstrating that *D. vulgaris* was the prevalent ethanol-oxidizing organism in the flocs

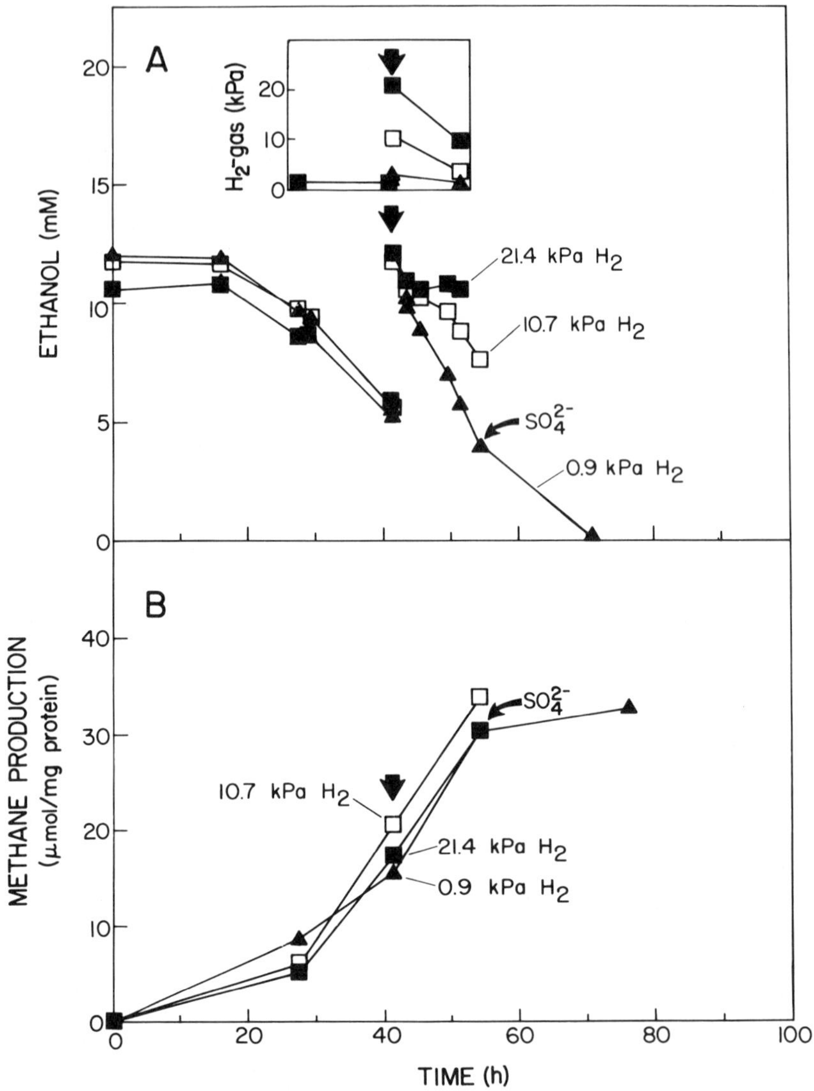

Figure 8 Influence of exogenous H_2 levels and sulfate addition on syntrophic conversion of ethanol to methane.

and that 80% of the methane formed originated from syntrophic metabolism. H_2 diffusion limitations are excluded because we showed it would take less than 120 s for the H_2 gas pools within the floc phase and the environmental liquid phase to reach equilibrium (Ozturk et al., 1987).

Figure 9 compares an electron balance for H_2 and ethanol consumption during the 41-51 h time period of the H_2 pulse experiments shown in Figure 8. These data compare the H_2 equivalents associated with syntrophic ethanol conversion to methane and the consumed H_2 gas as a function of the experimental H_2 concentration with the theoretical H_2 consumption kinetics of *M. formicicum*. *M. formicicum* was apparently fully saturated with reducing equivalents in

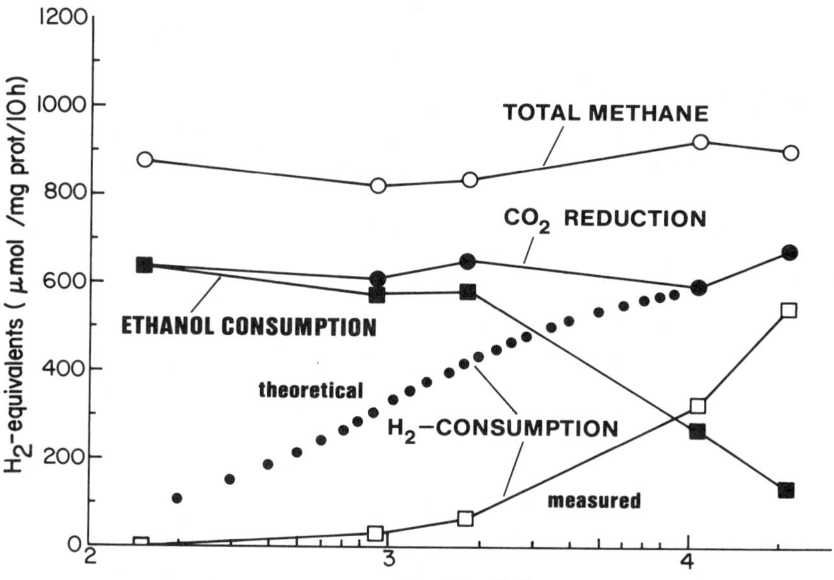

Figure 9 Influence of the H_2 partial pressure on the electron balance of syntrophic ethanol conversion to methane. The unconnected dots represent predicted theoretical values for H_2 consumption by the prevalent H_2-consuming methanogen *M. formicicum* in this digester ecosystem using a K_s of 6.5 µM dissociated H_2 (1000 Pa), and a V_{max} of 70 µmol/mg floc protein per hour. Note the difference between the simulated H_2 gas consumption (•••) and the experimental measured values (□). The total amount of methanogenic CO_2 reduction remained constant during a 100-fold increase in the H_2 partial pressure.

the flocs even at low H_2 concentrations, because the total amount of electrons used for CO_2 reduction originating from either ethanol or H_2 remained constant and independent of H_2 addition. The electron balance for total methane and CO_2 reduction indicated that approximately 80% of the CH_4 was produced from CO_2 reduction, confirming the effect of sulfate addition (see above). Inhibition of ethanol oxidation by H_2 was always accompanied by an identical increase in the consumption of reducing equivalents from H_2 oxidation, indicating a competitive inhibition type of mechanism between H_2 and XH_2, the unknown intermediate. The sum of reducing equivalents originating either from ethanol or H_2 used for CO_2-dependent methane production remained constant between 400 Pa (2.6 μM) and 21,000 Pa (136.5 μM) exogenous H_2.

M. formicicum, the dominant H_2-consuming and CO_2-reducing methanogen in the floc, had a K_s value for H_2 of approximately 5 μM (Table 4). At dissolved H_2 concentrations below one-half of the K_s value, a saturation of methanogenic CO_2 reduction by H_2 is impossible. This demonstrated a H_2-independent syntrophic mechanism for the generation of the reducing equivalents used to apparently saturate the uptake system for methanogenic CO_2 reduction. Thus, H_2 was not a significant source of the reducing equivalents used by *M. formicicum* during syntrophic ethanol oxidation in this floc ecosystem.

C. Evidence for Interspecies Formate Transfer

1. Formate Turnover in Anaerobic Digesters

Formate can be produced by pyruvate-formate lyase during carbohydrate metabolism or CO_2 reduction via the so-called ferredoxin-dependent monocarboxylic acid cycle (Tauer et al., 1973) in the absence of carbohydrate metabolism. CO_2 reductase in *Clostridium pasteurianum* and other clostridia is also able to synthesize formate by direct ferredoxin-dependent CO_2 reduction (Thauer et al., 1973). Reduced ferredoxin is one presumed intermediate of syntrophic ethanol oxidation (Reddy et al., 1972a-c; Thauer et al., 1977). Experiments were designed to examine whether formate was a significant product of $^{14}CO_2$ reduction in association with syntrophic methanogenesis from ethanol or lactate in the whey digester ecosystem (Thiele and Zeikus, 1988a). Figure 10 compares the effect of H_2 and ethanol additions on the $^{14}CO_2$ incorporation into the $^{14}CHOO^-$ pool by whey digester contents. In the presence of either exogenous H_2 or ethanol alone, [^{14}C]formate levels were always below 10 μM formate. Formate was measured as a significant CO_2 incorporation product only in the presence of ethanol or lactate and H_2, but it was not formed as a product of CO_2 reduction by H_2 gas. Since

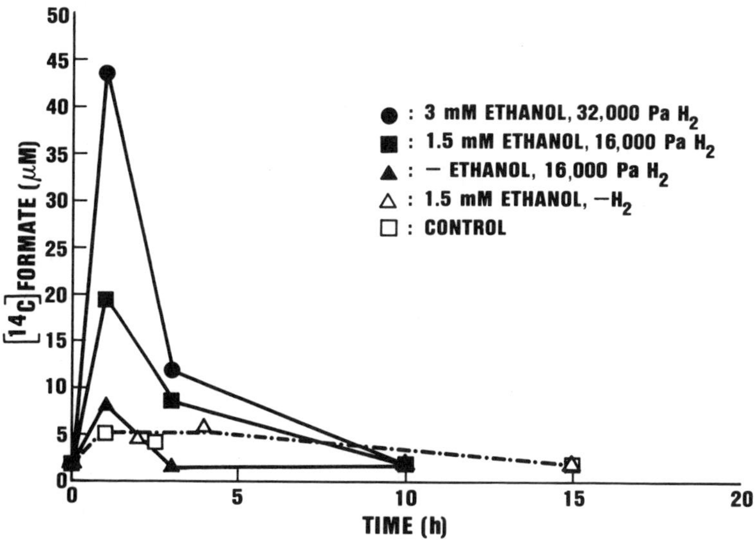

Figure 10 Dependence of syntrophic formate production by anaerobic digester contents on ethanol versus hydrogen.

H_2 was not an electron donor for formate production, its stimulating effect on the [^{14}C]formate levels was most probably due to an inhibition of formate consumption. Further experiments demonstrated that higher levels of formate were dependent on ethanol stimulation of syntrophic CO_2 reduction, not due to an inhibition of formate turnover (data not shown). Because the soluble formate pool sizes in the anaerobic digester were low and methane was not a significant product of formate turnover outside the flocs, it is concluded that formate-H_2 lyase activity is responsible for the turnover of exogenous formate in the environment.

Further experiments were designed to verify this result and to demonstrate the importance of CO_2 reduction to formate during syntrophic ethanol conversion by purified flocs (Fig. 11). In the presence of ethanol, [^{14}C]formate levels reached 30-40 µM (n = 4), whereas in the absence of syntrophic ethanol metabolism [^{14}C]formate levels were below 15 µM. Thus, formate was a direct product of ethanol-dependent syntrophic CO_2 reduction by flocs isolated from the anaerobic digester.

Chloroform and sulfate inhibition experiments were used to demonstrate active methanogenic formate consumption and acetogenic formate production during syntrophic ethanol oxidation in whey digester contents (Thiele and Zeikus, 1988a). Chloroform is a potent

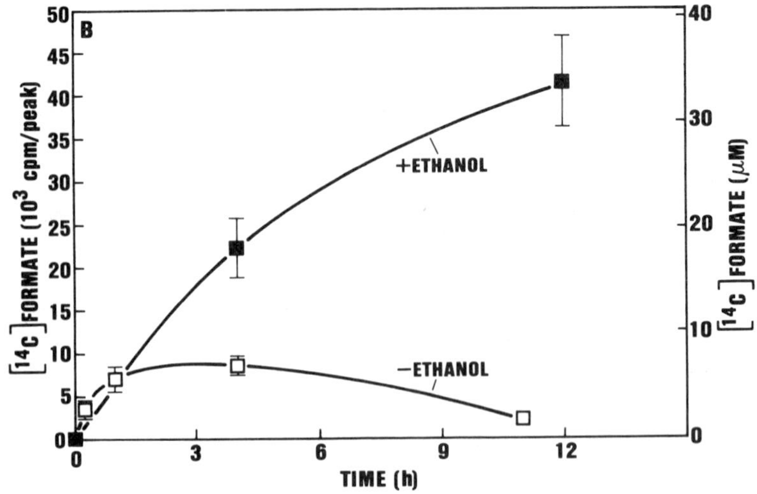

Figure 11 Dependence of syntrophic CO_2 reduction to formate by purified digester flocs on ethanol.

and specific inhibitor of methanogenesis, and sulfate addition alters the physiology or sulfate-reducing bacteria from transferring XH_2 to consuming XH_2 (Conrad et al., 1987; Phelps et al., 1985; Bauchop, 1967). Chloroform addition completely inhibited methanogenesis in the digester flocs. The data shown in Figure 12 demonstrate that chloroform addition resulted in formate accumulation and inhibition of ethanol oxidation. This is suggestive of classic end product inhibition of syntrophic acidogenesis. The further addition of sulfate revived consumption by *D. vulgaris*. These results showed that (1) $CHCl_3$ did not inhibit ethanol-oxidizing acetogens, (2) formate was synthesized from bicarbonate and ethanol, and (3) the lack of electron acceptors inhibited ethanol oxidation under sulfate-limited conditions.

Figure 7c compares the theoretical free energy difference for syntrophic acetogenic bacteria to reversibly excrete reducing equivalents either as H_2 or as formate at various steady-state levels of dissolved H_2 and dissolved formate. Assuming equilibrium thermodynamics, it would be energetically advantageous for syntrophic acetogens to preferentially excrete formate at formate levels between 10 and 200 μM when H_2 partial pressures are $\geq 10^{-3}$ atm and even at $[H_2]$ below 10^{-3} atm when formate levels are ≤ 10 μM.

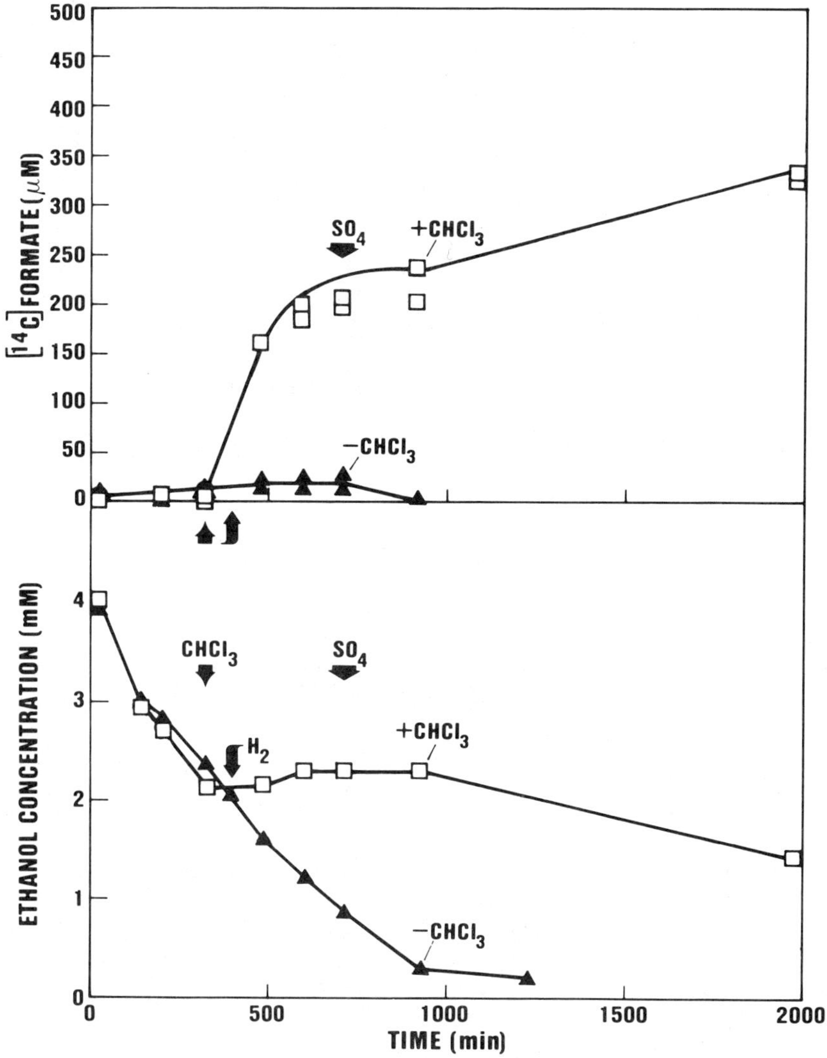

Figure 12 Influence of H_2 and sulfate addition on syntrophic conversion of ethanol to formate in the presence or absence of the methanogenesis inhibitor $CHCl_3$.

2. *Formate-Bicarbonate Electron Shuttle Cycle for Syntrophic Energy Conservation*

Figure 13 illustrates the bicarbonate-formate electron cycle model for control of interspecies electron flow, which is based on our studies. The model accounts for the coupling of electron flow during syntrophic ethanol or lactate conversion to CH_4 and for energy conservation for both species. This model is quite different from that described for interspecies H_2 transfer mechanisms (Wolin, 1976), since bicarbonate is a substrate for the acetogens, not the methanogens. The bicarbonate-formate cycle also couples acetoclastic bicarbonate formation by methanogens with acetogenic formate production by the syntrophic acetogens.

Acetogens, in this case *D. vulgaris*, transport four molecules of bicarbonate into their cytoplasm. CO_2 and OH^- are then produced by putative carbonic anhydrase. CO_2 is the physiological substrate of a putative ferredoxin-CO_2 oxidoreductase (Thauer et al., 1973), which oxidizes reduced ferredoxin molecules from the acetogenic metabolism of two molecules of ethanol. Four molecules of formic acid are produced, which dissociate into formate and protons at the physiological pH of the ecosystem. Methanogenic bacteria, in this case *M. formicicum*, utilize four molecules of formate, one proton, and H_2O to produce CH_4 and three molecules of bicarbonate. One molecule of CO_2 or HCO_3^- is provided by acetoclastic methanogenesis in this ecosystem. If the acetogenic CO_2 reductase is connected to the exterior membrane surface and carbonic anhydrase is in the cytoplasm of the acetogenic bacteria, then catabolic CO_2 reduction would also lead to water cleavage, charge separation, acidification of the periplasmic space, and generation of a proton-motive force (PMF) in the acetogenic cell since formic acid is a much stronger acid than carbonic acid. Thus, acetogenic CO_2 reduction contains the theoretical possibility for acetogenic ATP generation by electron transport-coupled phosphorylation.

Several syntrophic acetogenic conversions are believed to involve substrate-level phosphorylation as the physiological mechanism of energy conservation. This is concluded from metabolic pathways of syntrophic ethanol oxidation by the S organism (Reddy et al., 1972a-c) and characterization of the syntrophic acetogenic butyrate oxidation by *S. wolfei* (Wofford et al., 1986). Substrate-level phosphorylation (SLP) requires in principle the stoichiometric formation of ATP during carbon conversion with a stoichiometry of 1 or greater than 1 (Thauer et al., 1977). In the case of syntrophic acetate formation from alcohols and organic acids, a reasonable mechanism for SLP is the phosphoroclastic cleavage of acetyl-CoA by phosphotransacetylase into acetylphosphate and coenzyme A. Acetate kinase then cleaves acetylphosphate and phosphorylates ADP to ATP (Reddy et al., 1972a-c; Thauer et al., 1977; Wofford et al., 1986).

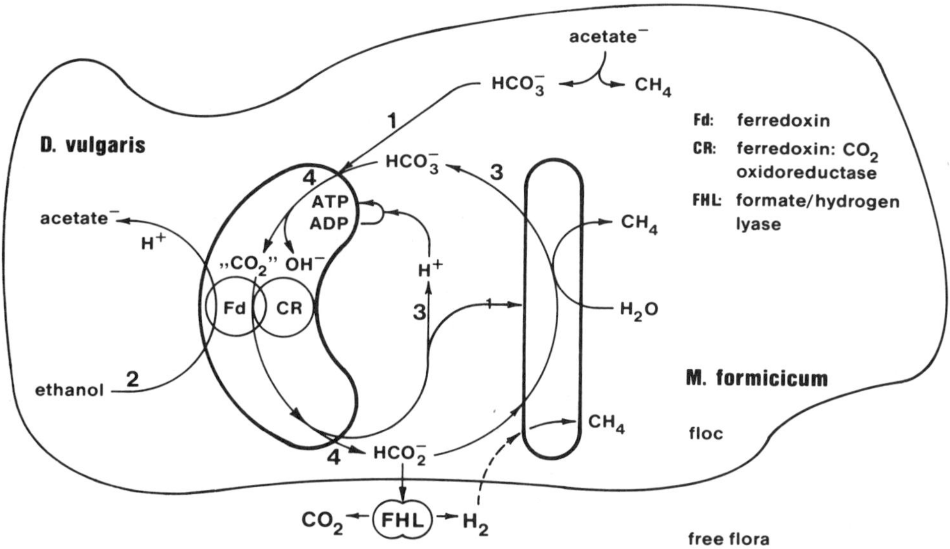

Figure 13 A hypothetical model illustrating the principles of the bicarbonate-formate electron shuttle mechanism for explaining both the control of interspecies electron flow during syntrophic ethanol conversion to methane by anaerobic digester flocs and energy conservation by electron transport-mediated phosphorylation in syntrophic acetogens. The juxtaposition of *D. vulgaris* and *M. formicicum* in flocs enables methanogenic CO_2 regeneration to be coupled with acetogenic CO_2 reduction to formate, which itself serves as the mediator of interspecies electron flow. Energy conservation in the acetogen is postulated to occur as a consequence of putative cytoplasmic carbonic anhydrase and a membrane-linked CO_2 reduction, which leads to generation of an alkaline cell interior, an acidic exterior, and a proton-motive force to drive membrane-coupled ATP synthesis. The model shows the dynamics of formate metabolism in the digester ecosystem. More than 90% of the CO_2-dependent syntrophic ethanol conversion to methane occurs via interspecies formate transfer within flocs, whereas <10% occurs via interspecies H_2 transfer caused by cleavage of floc-execreted formate in the soluble digester phase to H_2 and CO_2.

Phosphotransacetylase and acetate kinase activities have been demonstrated in cell-free extracts from the S organism and from *S. wolfei* (Reddy et al., 1972a-c; Wofford et al., 1986). However acetyl-CoA-dependent ATP formation could not be demonstrated in vitro or in vivo in both syntrophic bacteria; thus positive kinetic evidence for functioning SLP in syntrophic acetogenic bacteria is not available. From a thermodynamic point of view a minimum driving force of more than -44 kJ/mol is required for the phosphorylation of ADP to ATP within an actively metabolizing cell based on the concentrations of ADP, ATP, and phosphate and the ionic conditions in the cytosol (Thauer et al., 1977). This driving force must be provided by the acetyl-CoA-forming metabolic pathway in the acetogenic bacterium under in situ conditions. This assumes that no other net exergonic catabolic reaction occurs within the acetogenic bacterium and that other bacteria have no direct way to contribute energy equivalents (ATP).

Analysis of the thermodynamics of acetogenic H_2 gas production under defined laboratory conditions (Dwyer et al., 1984; Wofford et al., 1986; Thiele et al., 1988c; Ozturk et al., 1987) and under in situ conditions in anaerobic ecosystems (Conrad et al., 1986) showed free energies for the acetogenic reaction that were much less (-15 to -25 kJ per reaction, that is, 1/3ATP to 1/2ATP per reaction). Thus thermodynamic evidence contradicted the functioning of acetogenic energy metabolism via SLP. The possibility that significantly lower acetogenic product levels in our digester flocs (i.e., H_2, acetate, and formate) due to species juxtapositioning would result in lower product concentrations and free energies and thus higher bioenergetic driving forces was excluded by the application of dynamic reaction-diffusion models (Ozturk et al., 1987; Thiele et al., 1988d). H_2 gas, formate, or acetate concentration levels inside these syntrophic microniches were either equal to or larger than the measured concentrations outside the microniches.

A solution to the energy conservation dilemma of syntrophic methanogenesis was suggested by postulating the existence of reversed electron transport in syntrophic acetogenic bacteria (Thauer and Morris, 1984). This hypothesis proposed in essence that ATP generated by SLP via the phosphoroclastic reaction was partly consumed to increase the redox potential of membrane-bound oxidized electron carriers and by this to facilitate endergonic partial redox reactions of the syntrophic acetogenesis. The existence of this mechanism has not yet been demonstrated in syntrophic acetogens, but it is possible.

The proposed formate-bicarbonate electron shuttle cycle is a completely different mechanism that bypasses the requirement for SLP during syntrophic acetogenesis. Here CO_2 is the physiological substrate of ferredoxin-CO_2 oxidoreductase (Thauer et al., 1973), but bicarbonate is the prevalent available form of CO_2. Thus, if carbonic anhydrase activity, which is required for CO_2 production from bicarbonate, is a cytoplasmic enzyme and ferredoxin-CO_2 oxidoreductase is membrane bound, then syntrophic acetogenic CO_2 reduction and formic acid excretion would lead to water cleavage into OH^- (cytoplasm) and a net proton transport out of the cytoplasm via the excreted formic acid (Fig. 13). Thus a proton-motive force would build up that would be sufficient to drive electron transport-coupled phosphorylation (ETP). This would eliminate the bioenergetic dilemma of syntrophic acetogenesis, because fractional values of ATP can be easily utilized by ETP in anaerobes (Thauer and Morris, 1984).

Integration of this model into the entire anaerobic digestion route shown in Figure 14 indicates that three metabolic groups share the free energy of the complete conversion of fatty acids and alcohols to methane and CO_2 in a mutually synergistic way by forming a metabolic microniche. Thus, at least three different metabolic groups may be required for stable syntrophic methanogenesis in anaerobic digestion ecosystems: formogenic acetogens, formate-utilizing methanogens, and acetoclastic methanogens. In return for energy substrates, the methanogens provide conditions for an energetically favorable acetogenesis by formate removal, bicarbonate production, and maintaining the molar bicarbonate-formate ratio at 1000-10,000. This huge bicarbonate-formate ratio functions as a thermodynamic battery for the syntrophic acetogens, and it could maximally provide, under in situ conditions, -5 to -10 kJ per electron pair more free energy than a mechanism dependent on interspecies H_2 transfer. In sum, the acetogens could obtain a larger fraction of the overall free energy change of the syntrophic conversion if they communicated via interspecies formate transfer in lieu of interspecies H_2 transfer.

A precise analytic determination of the formate turnover in a microniche is technically difficult. However, a good estimate can be derived from the measured rate of floc-bound syntrophic ethanol conversion in the digester (330 μmol/L per h) and a stoichiometry of 2 mol formate per mole ethanol. Using a formate concentration in the flocs between 40 and 250 μM, a first-order turnover constant between 0.044 and 0.275 per minute is calculated. However, the flocs constituted only a small volume of the digester contents (1-2% of the total volume). This means that on a volumetric basis the formate flow to methane in the flocs was at least 50 times faster than the potential turnover to H_2 gas in the free flora fraction (Thiele and Zeikus, 1988a).

The whey digester floc system was analyzed with simple diffusion reaction models of the Michaelis-Menten type (Ozturk et al., 1987; Thiele et al., 1988d). Diffusities, apparent K_m values, and volumetric

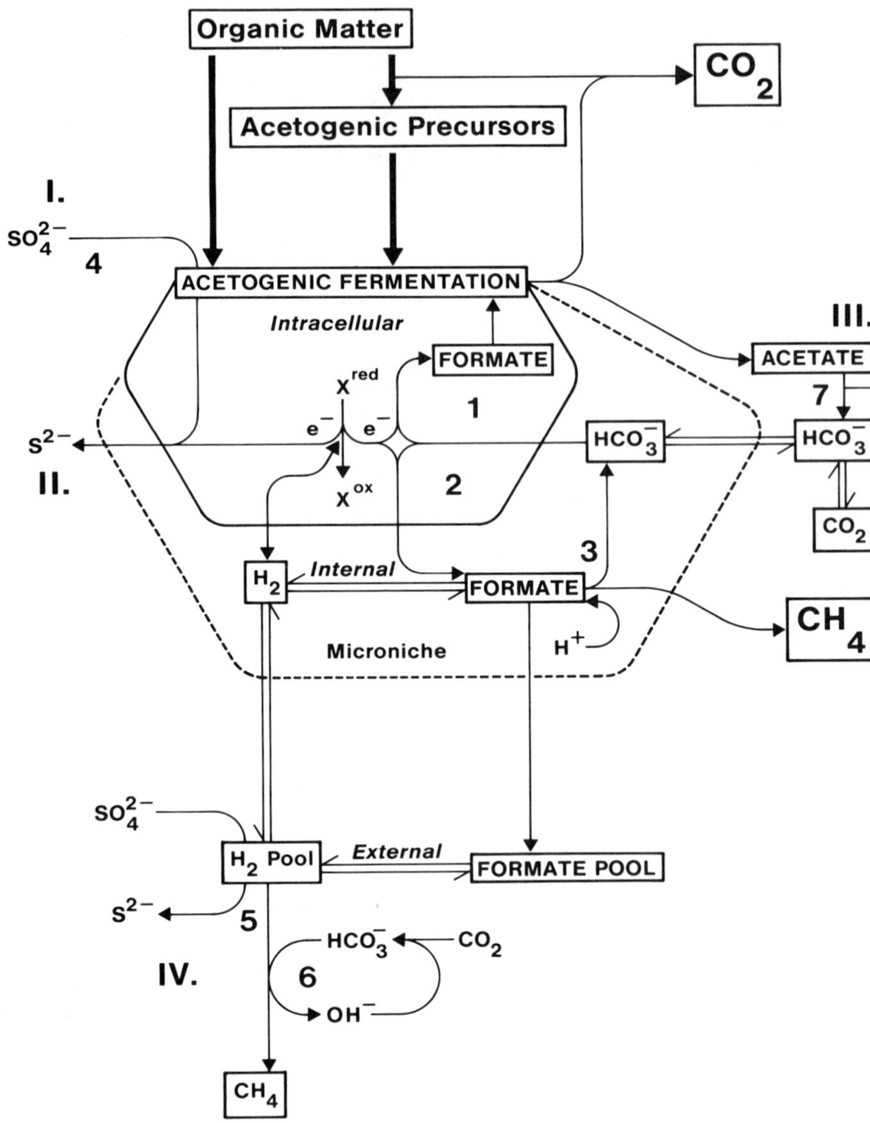
Organic Matter
CO2
Acetogenic Precursors
I.
SO4 2−
4
ACETOGENIC FERMENTATION
Intracellular
III.
FORMATE
ACETATE
X red
7
1
S2−
e−
e−
HCO3−
HCO3−
II.
X ox
2
CO2
H2
Internal
3
FORMATE
CH4
H+
Microniche
SO4 2−
H2 Pool
External
FORMATE POOL
S2−
5
HCO3−
CO2
IV.
6
OH−
CH4

reaction rate constants were determined experimentally (Chartain et al., 1987; Thiele et al., 1988d; Ozturk et al., 1987). It must be stressed that the model simulations were done without a single adjustable parameter. Model predictions showed an excellent agreement with the measured kinetic data for syntrophic reactants in predicting the proposed formate-bicarbonate electron shuttle cycle and the measured H_2 gas production. These modeling studies predicted steady-state formate levels outside the purified flocs of 30-50 μM during syntrophic ethanol degradation. This prediction was in close agreement with the measured data (Fig. 11). The best fit with the experimental data was obtained assuming an in situ ratio of H_2 transfer to formate transfer of 9:91, which was again close

Figure 14 A syntrophic microniche model for anaerobic digestion of organic matter. The hypothetical model integrates present views about the major biochemical routes of anaerobic degradation of organic matter in the presence of high bicarbonate concentrations with available knowledge about the kinetic compartmentalization of syntrophic methanogenesis. Fermentation products of trophic group 1 (I) and undegraded organic matter are metabolized in the microniche (II) as an ecophysiological unit in a synergistic process. CO_2 reductase functions as the key enzyme of the acetogenic metabolism to oxidize reduced coenzymes (X^{red}) and reduce CO_2 (HCO_3^-) to formic acid. Homoacetogenic bacteria (1) utilize the formate as precursor for acetate synthesis, whereas syntrophic acetogens (2) excrete formic acid into the environment. Hydrogenase can function as an alternate "electron valve" of the acetogenic metabolism. Formate-utilizing methanogens (3) are an essential component of the microniche. They are in juxtaposition to the acetogenic bacteria on a very short interspecies distance. This maintains low concentrations of formate in the intercellular space and regenerates the bicarbonate. Acetate-cleaving methanogens (III) provide the required high bicarbonate levels, which equilibrate via diffusion with the internal bicarbonate pool of the microniche. The prevalent effect of sulfate addition to the ecosystem is to alter the acetogenic metabolism from formate production to sulfate reduction (4) independently of the H_2 gas pool of the ecosystem. Sulfate reduction using H_2 gas (5) occurs predominantly outside the microniche, competing with H_2 gas-consuming methanogens (IV). Both trophic groups utilize predominantly H_2 gas produced by formate-H_2 lyase and acetogenic hydrogenases. Thus competition between sulfate reducers and methanogens occurs at two levels, the H_2 gas pool of the ecosystem and the formate pool of the microniche.

to the measured estimates of 10:90 to 5:95 (Thiele and Zeikus, 1988a, c). Thus, experimental biochemical and microbiological evidence together with theoretical modeling based on chemical engineering science clearly demonstrated the ecophysiological existence and the significance of a formate-bicarbonate electron shuttle cycle in this particular ecosystem. In short, >90% of the XH_2 equivalents generated by the syntrophic oxidation of ethanol or lactate were transferred as formate.

3. *Interspecies Formate Transfer in Other Ecosystems*

So far, methanogenic ecosystems have been mostly analyzed with the perspective of interspecies H_2 gas transfer. For example, the long-term inhibitory effect of H_2 gas on syntrophic metabolism in digesters (Kaspar and Wuhrmann, 1978) may not be due to direct thermodynamic or kinetic inhibition of syntrophic acetogens but to effects on methanogens. The importance of formate as an intermediate of syntrophic methanogenic metabolism has been largely underestimated. Formate was suggested as a minor precursor for methane in the rumen (Hungate, 1967). However, the rumen is a heterogeneous environment composed of microbial aggregates (Cheng and Costerton, 1980). Also, a particle-bound formate pool has been reported in the rumen with at least 100-fold higher formate concentrations (Hungate et al., 1970). In analogy to that described here for whey digester flocs, it would not be surprising to learn that microbial aggregates in the rumen utilize formate as a significant electron donor for methanogenic CO_2 reduction.

M. omelanskii, the classic ethanol-oxidizing coculture, utilized formate for methane formation in the presence of ethanol and bicarbonate (Barker, 1941). Ethanol oxidation was bicarbonate dependent, but the rate of ethanol oxidation in the presence of 26 mM formate and 33 mM bicarbonate was also proportional to the rate of formate removal for more than 4 days (Barker, 1941). Formate consumption also stimulated acetate production, and formate appeared inhibitory for ethanol oxidation. The ethanol-oxidizing S organism purified from *M. omelanskii* contained an NAD-coupled formate dehydrogenase (Reddy et al., 1972b). If this enzyme activity is responsible for formate consumption in lieu of H_2 production, then formate consumption should have inhibited the ethanol oxidation. Since the opposite was observed, this could imply that the S organism produced formate by CO_2 reduction. Although *M. bryantii* strain MoH, isolated from *M. omelanskii*, does not utilize formate (Table 3), *M. omelanskii* always contained three colony morphologies: diffuse, hard, and fat colonies (Bryant et al., 1967). It is tempting to speculate that the bacterial species in fat colonies, which were not isolated or characterized, were formate-utilizing methanogens.

Formate-utilizing methanogens have been regularly coenriched in reported enrichments of syntrophic bacteria utilizing either soluble aromatic compounds or volatile fatty acids (Ferry and Wolfe, 1976; Mah, 1982; McInerney et al., 1981; Mountfort and Bryant, 1982).

High formate levels (500-800 μM) were also observed during active benzoate degradation by a syntrophic methanogenic enrichment (Ferry and Wolfe, 1976). Formate was a reaction intermediate of the syntrophic degradation, and it was assumed that it was directly derived from benzoate carbon. The H_2 gas levels were below 2.5 Pa, and formate was converted to methane by *M. formicicum* and *M. hungatei*. It can be calculated from the published maximum rates of H_2 and formate utilization for the purified *M. formicicum* strains, the respective K_m values (Schauer et al., 1982), and the published rates of benzoate degradation by the consortium (Ferry and Wolfe, 1976) that at 2.5 Pa H_2 maximally 0.59 μmol H_2 per h per mg protein would have been utilized, even if the consortium consisted completely of hydrogenotrophic methanogens. With this value it is impossible to explain the observed rate of syntrophic benzoate degradation (3.36 μmol H_2 per h per mg protein) via interspecies H_2 transfer. Using the same published data set for formate (Schauer et al., 1982), 33 μmol formate per h per mg protein could maximally have been utilized by methanogens. Thus, the measured rates would fit with the formate transfer hypothesis, if only 10% of the bacteria in the consortium were *M. formicicum* or *M. hungatei*.

The purified benzoate degrader produced formate in trace amounts in coculture with *M. hungatei* and the H_2 partial pressure was below 5 Pa, but 14% of the added benzoate was also degraded in the presence of 80% H_2 (Mountfort and Bryant, 1982). If H_2 is the major reaction intermediate of syntrophic acetogenic benzoate conversion, this is a thermodynamic impossibility. The proposed formate-bicarbonate electron shuttle cycle is consistent with these results. Formate was not found in a different defined synthetic benzoate-degrading consortium (Dolfing and Tiedje, 1987); thus the possibility of other models for electron transfer coupling, such as interspecies H_2 transfer, cannot be ruled out with the data on hand.

S. wolfei, a butyrate-oxidizing syntrophic acetogen, grew better in defined coculture with a formate-utilizing methanogen than with *M bryantii* MoH, a H_2-utilizing methanogen (McInerney et al., 1979, 1981). Thus, acetogenic CO_2 reduction was a possible electron transfer mechanism during syntrophic butyrate degradation. However, formate was not detected as an intermediate of butyrate degradation by *S. wolfei* (Wofford et al., 1986). It should be noted that the standard assays for formate detection are not very sensitive.

Granular sludge from sugar factory wastewater showed 29% higher maximum rates of methanogenesis from formate than from

H_2 + CO_2 gas (Dolfing and Mulder, 1985). Granular sludge from the same source adapted to propionate or propionate-acetate mixtures for 3-4 months showed 230-240% higher maximum rates of methanogenesis on formate than on H_2 or CO_2 gas. Ethanol-adapted control granules showed basically the same pattern as the inoculum. It has also been shown by ^{14}C-labeled propionate studies that formate is not a biochemical intermediate of acetogenic propionate degradation (Koch et al., 1983). These results therefore imply that propionate degradation is dependent on CO_2 reduction in the granular sludge. It is predictable that the comparatively higher activities of formate-utilizing methanogens in the granules could only have developed if formate was present during the adaptation and a product of syntrophic propionate degradation to methane.

It is thus doubtful whether the term "obligate syntrophic proton-reducing bacteria" is meaningful with regard to the evidence presented. First, there are numerous examples of facultative syntrophic bacteria, such as the sulfate reducer *D. vulgaris* (Bryant et al., 1977). Second, the formate-bicarbonate electron shuttle cycle provides evidence that mechanistically CO_2, not protons, could be reduced by these bacteria. The observed H_2 production in the headspace of pure or mixed cultures may be only a side reaction of very active formate metabolism.

The ability to isolate obligate syntrophic acetogens that grow in coculture with methanogens or sulfidogens that only consume H_2 does not necessarily prove or disprove that this physiological mode is an ecological function practiced in natural ecosystems. It seems more appropriate to ascribe the function of interspecies H_2 transfer or formate transfer to the complete metabolic microniche in the anaerobic digester ecosystem we have characterized. Formogenic acetogens and formate-utilizing and acetoclastic methanogens are the prevalent bacteria that perform syntrophic degradation of organic matter.

The occurrence of saprophytic sediment protozoa containing endosymbiontic bacteria similar to *M. formicicum* also supports the view of formate significance (van Bruggen et al., 1983). These unicellular eukaryotes could function as membrane-covered syntrophic microniches. Methanogens as episymbionts on ruminal protozoa are another example of the microniche concept (Vogels et al., 1980). Finally, we are able to grow hyperactive syntrophic microniches as ecoengineered granules (Thiele and Zeikus, 1988b) consisting of all the above-mentioned trophic groups (see below).

V. APPLICATIONS FOR ADVANCED BIOMETHANATION SYSTEMS

The previous sections showed that methanogenesis can occur in a formate-coupled mode within microbial aggregates. That these aggre-

Table 8 Maximum Specific VFA Conversion Rates and Volumetric Methane Productivities of UASB Granules and Ecoengineered SBR Granules

Reaction	Maximum rates (mmol substrate per g protein per day)		
	UASB granules[a] (VFA adapted)	UASB granules[b] (nonadapted	SBR granules[c] (ecoengineered)
Acetate → methane	14	37	72
Propionate → methane	7-15	9	13
Butyrate → methane	ND[d]	ND	25
VFA → methane[e]	44	52	157

[a]From Dolfing et al. (1985).
[b]From Ten Brummeler et al. (1985).
[c]From Thiele et al. (1988b).
[d]Not determined.
[e]Represents the maximum amount of methane production from total VFA in mmol CH_4 per g protein per day.

gates constituted the major form of syntrophic life in the whey contact digester suggests that syntrophic aggregates had a competitive advantage over associations of free single bacteria in the chemostat environment. More than 75% of the syntrophic metabolism in the whey digester was floc associated, and the flocs were not washed out under steady-state conditions (Thiele et al., 1988c). Thus it is possible to select analogous microniches that degrade volatile fatty acids and to optimize their performance by a general approach we call ecoengineering (Thiele et al., 1988e).

We have ecoengineered high-performance biomethanation catalysts by selecting for granular catalysts that simultaneously degrade acetate, propionate, and butyrate under controlled physical and chemical environmental conditions (unpublished results). After a 6 month selection period, the ecoengineered granular biocatalysts displayed extremely high specific and volumetric rates of syntrophic methanogenesis (see Table 8). The measured maximum conversion rates and steady-state methane productivities during syntrophic volatile fatty acid conversions were two to three times higher than the highest maximum rates reported for comparable upflow anaerobic sludge blanket (UASB) granules (Thiele and Zeikus, 1988b; Dolfing and Bloemen, 1985; Ten Brummeler et al., 1985). In contrast to UASB granules, a very flat pH optimum between pH 6.5 and 7.5 was found for these syntrophic biomethanation granules (Thiele and Zeikus, 1988b). Notably, syntrophic propionate and butyrate degradation to methane was not inhibited when prolonged shock loads were applied for more than one hydraulic retention time. The volumetric methane productivity increased fivefold to values above 200 $L/L_{catalyst\ bed}$ per day, when these ecoengineered biocatalysts were exposed to short shock loads with volatile fatty acid-containing medium (Thiele and Zeikus, 1988b).

Current biomethanation technologies are limited by the environmental requirement of low COD effluents from the systems (20-100 mg COD per L). The important soluble bioreaction intermediates are generally present at levels below the apparent K_S values (<1.0 mM) of the participating methanogenic and acetogenic trophic groups. Therefore, only a small fraction of the microbial trophic group biocatalytic potential is utilized and the volumetric organic conversion rates are underdeveloped. In short, the methane productivities are then rather low and the hydraulic retention times high.

In conventional digesters, slow-growing trophic groups are maintained that catalyze the limiting reactions under nonoptimal conditions. The ecophysiological dynamics of anaerobic ecosystems, however, dictate operational reactor performance. This is the cause for the frequent system instabilities under shock load conditions due to the sequential arrangement of product-inhibited bioconversions. Washout of slow-growing or inhibited trophic groups during the shock loads usually results in long recovery times. This can be partly overcome by immobilization of trophic groups 2 and 3 within granular catalysts with good settling properties in upflow anaerobic sludge blanket reactors (Lettinga et al., 1980, 1985) or within biofilm-anaerobic filter concepts (Young and McCarty, 1969).

Two-stage biomethanation systems were introduced as well to reduce the problem of product inhibition and acidification and to enhance microbial activity by a separate optimization of fatty acid-producing and acid-consuming reaction systems (Pohland and Gosh, 1971). Although these advanced waste treatment concepts have their merits for different types of organic wastes (Sahm, 1984; Pohland and Harper, 1985), many of them do not yet realize the tremendous metabolic potential of anaerobic bacteria.

Practical methane productivities range between 1 and 10 L/L per day and are far below the realized steady-state bed productivities of 80-130 L/L per day that are obtained with the ecoengineered syntrophic biomethanation granules in simple upflow reactors (Thiele and Zeikus, 1988b). These properties and the extremely high conversion rates tempted us to utilize these new granular biocatalysts for advanced biomethanation systems. Figure 15 illustrates a new biomethanation process system that couples a hydrolytic acidogenic reactor (HAR) and a syntrophic biomethanation reactor (SBR) by anion exchange. Both reactors employ ecoengineered catalysts for seeding and start-up (Thiele and Zeikus, 1986).

This process is based on the principles detailed as follows (Fig. 15). Volatile fatty acid anions and protons are the predominant terminal products of the acidogenic stage at pH values above the pK_a of the fatty acids produced. Methanogenic reactors are normally operated at neutral pH. Volatile fatty acids (VFA) are then fully dissociated. Exactly one VFA anion has to be removed from the fermentation liquor for each bicarbonate anion produced in the methanogenic stage (electroneutrality principle). Thus, instead of a bicarbonate release with the effluent and a process base addition to the acidogenic reactor, anion exchange of the VFA anions produced in the acidogenic reactor with the bicarbonate anions produced in the methanogenic reactor is achieved. The protons stayed in the acidogenic reactor and are consecutively removed from the reactor by CO_2 formation from bicarbonate. The shuttle process avoids substrate overloading in the methanogenic reactor, because bicarbonate, a product of the terminal methanogenic reaction, precisely controls the supply of VFA substrate for the next cycle of the biomethanation reactor in an exact 1:1 ratio. As the substrate shuttle occurs between HAR and SBR, it is possible to control and model the carbon and electron flow in both reactors. Thus, process instabilities may be avoided and a complete process automization is feasible.

The substrate shuttle process maximizes the CH_4 content of the gas produced, as illustrated in Table 9. The biogas produced in the methanogenic reactor is partly or completely purified from CO_2 owing to VFA-bicarbonate anion exchange. Based on this property it is theoretically possible to achieve 100% pure methane via the biomethanation of pure acetic acid. The concentration of bicarbonate in the acidogenic stage is lower in the pH range of interest (below 6.0), and the bicarbonate gradient between HAR and SBR functions as the driving force for the shuttle of VFA. The coupling of both process stages without the involvement of an aqueous contact enables very high levels of VFA in the methanogenic stage and simultaneous low levels of VFA in the effluent. A very important feature of this process is the closed-loop design of the syntrophic biomethanation reactor. An aqueous feed stream is not required for the SBR, eliminating dilution and washout of slow-growing biocatalysts.

A major disadvantage of this proposed process is the lack of binding affinity for neutral fermentation products in the hydrolytic

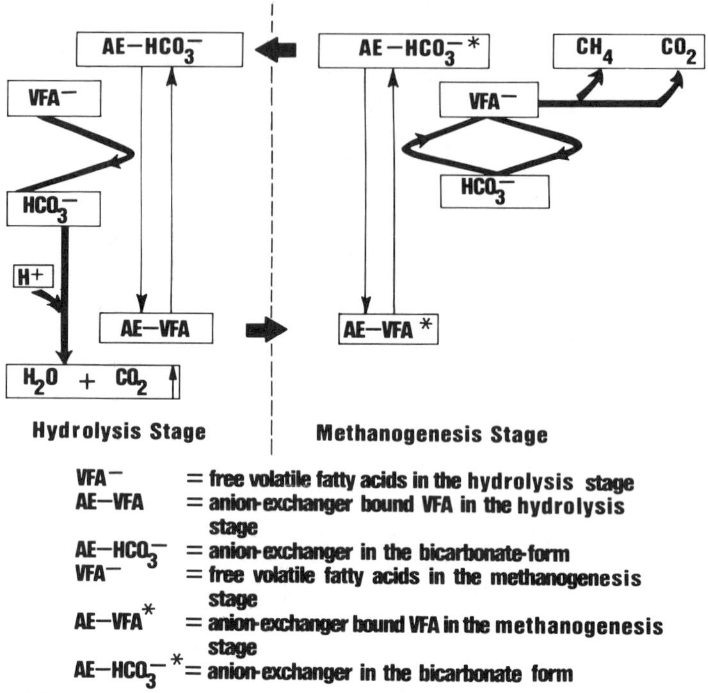
AE—HCO3⁻
AE—HCO3⁻ *
CH4 CO2
VFA⁻
VFA⁻
HCO3⁻
HCO3⁻
H+
AE—VFA
AE—VFA *
H2O + CO2
Hydrolysis Stage
Methanogenesis Stage
VFA⁻ = free volatile fatty acids in the hydrolysis stage
AE—VFA = anion-exchanger bound VFA in the hydrolysis stage
AE—HCO3⁻ = anion-exchanger in the bicarbonate-form
VFA⁻ = free volatile fatty acids in the methanogenesis stage
AE—VFA* = anion-exchanger bound VFA in the methanogenesis stage
AE—HCO3⁻ * = anion-exchanger in the bicarbonate form

(a)

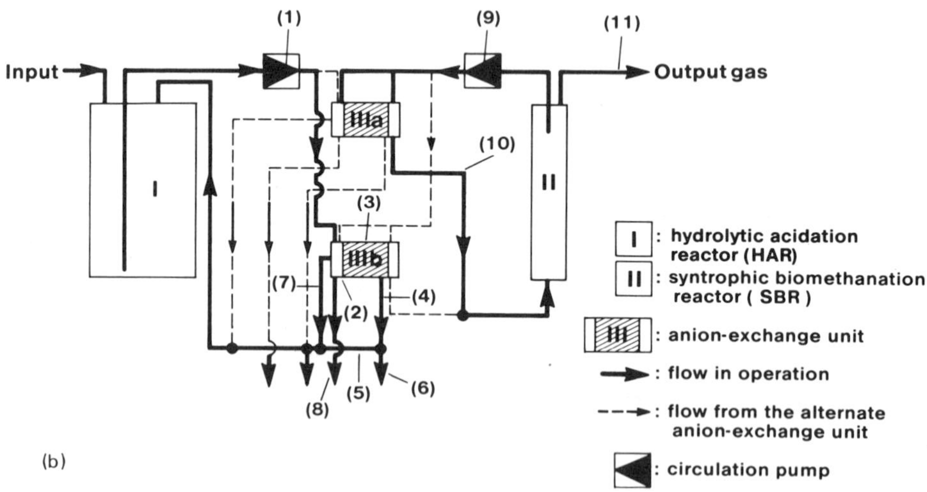
Input
Output gas
(1)
(2)
(3)
(4)
(5)
(6)
(7)
(8)
(9)
(10)
(11)
I
II
IIIa
IIIb
I : hydrolytic acidation reactor (HAR)
II : syntrophic biomethanation reactor (SBR)
III : anion-exchange unit
: flow in operation
: flow from the alternate anion-exchange unit
: circulation pump

(b)

Figure 15 Process principle and conceptual scheme of a preferred embodiment for a continuous substrate shuttle biomethanation process. (a) Process principle. Volatile fatty acid anions (VFA^-) produced during hydrolytic fermentations at pH values of 5-6 are exchanged against bicarbonate, which was bound to an anion-exchange material. The released bicarbonate is protonated by the remaining H^+ and decomposed into H_2O and CO_2. The anion-exchange material with the bound VFA^- is subsequently introduced into a methanogenic reactor at neutral pH values. VFA^- are exchanged against bicarbonate, which originated from the previous methanogenesis of VFA^-. This regenerates the anion-exchange material. Gaseous CH_4, bicarbonate, and/or CO_2 is produced from the syntrophic methanogenesis of the exchanged VFA^-. Reintroduction of the regenerated anion-exchange material into the hydrolysis stage completes the cycle. (b) Preferred embodiment. A continuous two-stage substrate shuttle biomethanation process composed of the hydrolytic acidation reactor (HAR, I) and the syntrophic biomethanation reactor (SBR, II) connected by two substrate shuttle units (IIIa and IIIb) is shown. The thick and thin lines represent the active and nonactive flows, respectively. The HAR is continuously fed with a waste stream (INPUT) and stirred. A circulation pump (1) recycles the reactor contents through two independent anion-exchange substrate shuttle units (IIIa and IIIb) in alternating sequence. IIIb is in bicarbonate form from a previous regeneration cycle in the SBR loop. Bacterial cells are separated from the fermentation broth at (2) and can be released into the environment (bleed stream, 8) or recycled into the HAR (recycle stream, 7). The cell-free broth is passed through the anion-exchange unit (IIIb); volatile fatty acids from the acidogenic fermentation in the HAR are bound to the exchange material, and bicarbonate-rich broth without VFA^- is released from the anion-exchange unit at (4). This stream at neutral pH can be used to titrate the pH in the HAR (process base, 5) or is released at (6) as treated effluent. Simultaneously an identical substrate shuttle unit in the VFA form (IIIa) from a previous load cycle is regenerated in the SBR loop by pumping bicarbonate-rich biomethanation broth by circulation pump (9) through exchanger IIIa. Bicarbonate is bound to the exchange material, and the VFA-rich effluent is biomethanated to bicarbonate, CH_4, and CO_2 in the syntrophic biomethanation reactor (SBR, II). Produced gas is released at (11). When the regeneration is complete a new cycle is started by simultaneously changing the liquid flow from thick to thin lines.

Table 9 Balanced Methanogenic Conversions in the SBR Reactor of Major Selected Organic Acid Anions Produced in the HAR Reactor I Provided by the Anion-Exchange Substrate Shuttle Module[a]

Reaction	Gas composition (% methane)
$Acetate^- + H_2O \rightarrow CH_4 + HCO_3^-$	100
$4\ succinate^{2-} + 6H_2O \rightarrow 7CH_4 + 8HCO_3^- + CO_2$	87
$4\ propionate^- + 6H_2O \rightarrow 7CH_4 + 4HCO_3^- + CO_2$	87
$2\ butyrate^- + 4H_2O \rightarrow 5CH_4 + 2HCO_3^- + CO_2$	83
$8\ lactate^- + 8H_2O \rightarrow 12CH_4 + 8HCO_3^- + 4CO_2$	75

[a]For each molecule of VFA^- converted, one molecule of HCO_3^- is removed from the SBR reactor.

reactor, such as alcohols. These compounds pass through the anion-exchange step and appear in the effluent (Thiele and Zeikus, 1988b). Therefore, ecoengineered catalysts are required that maximize the production of key acids (butyrate and acetate) in the HAR.

The volumetric COD removal rate of a biocatalyst bed in the SBR corresponds to the volumetric methane productivity. The maximum rates of methane production measured at 35°C for ecoengineered SBR granules (370-430 kg COD per m^3 per day) (Thiele and Zeikus, 1988b) are extremely high. Wu et al. (1985) measured maximum values between 40 and 80 kg COD per m^3 per day in UASB granule beds at 35°C. In extremely volatile fatty acid-rich wastewaters, rates up to 133 kg COD per m^3 bed per day could be achieved in a laboratory reactor (Wu et al., unpublished findings). Lettinga et al. (1985) reported maximum bed load rates of 100-115 kg COD per m^3 per day for optimized mesophilic acetate-propionate-grown granules at 37°C.

High methane productivities in syntrophic biomethanation reactors are due to three effects:

1. The uncoupling of the [VFA] in the methanogenic reactor from the effluent of the acidogenic stage
2. The utilization of the complete SBR volume by the methanogenic biocatalysts due to the closed-loop concept
3. The application of ecoengineered syntrophic biomethanation granules

The economic and technological potential of the substrate shuttle biomethanation process is being explored at present at our institute for the biomethanation of food and fermentation wastes. It remains unclear whether this process innovation constitutes an economically viable solution to the intrinsic problems of anaerobic waste treatment. At present, process costs are dominated by the substrate shuttle unit (e.g., microbial fouling of current anion-exchange materials) and in the reactor biocatalyst recycle systems (e.g., membrane fouling). In the long run, the basic bioengineering process concepts employed here to optimize the amount and activity of the biocatalysts will, it is hoped, lead to a practical anaerobic digestion system to treat large-volume industrial-municipal wastewaters and certain toxic waters by a direct high-rate biomethanation process.

VI. CONCLUSIONS

Syntrophic interactions between H_2- and formate-producing acetogenic and CO_2-reducing methanogens are of tremendous importance in controlling the rate-limiting metabolic steps in anaerobic digesters. Upon analysis of the localization and mechanisms for coupling electron transfer between these metabolic groups, we conclude that a syntrophic microniche is dynamically related to the optimized biomethanation performance of the entire ecosystem. This understanding can be used to develop new biomethanation systems for toxic and residual wastes, such as the new substrate shuttle process described here.

Qualitative measurements of electron transfer reactions within the microniche (a floc or a granule) demonstrate the kinetic and thermodynamic preference for interspecies formate transfer in lieu of interspecies hydrogen transfer. Thus, it is concluded that previous considerations of the quantitative significance of interspecies hydrogen transfer in both in vitro mixed cultures and those in anaerobic digester ecosystems, including biomethanation reactors, lake sediments, and the rumen, be re-examined.

We conclude that the speculation provided here on the specific mechanisms for electron flow coupling (interspecies formate transfer) and energy conservation (interspecies formate-bicarbonate cycle) between syntrophic acetogens and CO_2-reducing methanogens are consistent with all data at hand. Nonetheless, further studies of pure and mixed cultures at the physiological and biochemical level are needed to provide this validation.

VII. NOMENCLATURE

D	diffusivity
HAR	hydrolytic acidation reactor

K_m	Michaelis-Menten constant
K_s	substrate constant
μH_2	chemical potential for H_2 gas
PBB	phosphate-buffered basal medium (Kenealy and Zeikus, 1981)
ϕ	Thiele modulus for first-order reaction mechanisms
R	radius of microbial aggregate
SBR	syntrophic biomethanation reactor
UASB	upflow anaerobic sludge blanket
VFA	volatile fatty acids

ACKNOWLEDGMENTS

We acknowledge research support from the State of Michigan and the Kellogg Foundation and the Department of Energy grant DE-FG02 87ER13719. We thank Dr. Susan Lowe for review and comments on this manuscript.

REFERENCES

Abrams, J. W., and Nedwell, D. B. (1978). Hydrogen as a substrate for methanogenesis and sulfate reduction in anaerobic saltmarsh sediments. *Arch. Microbiol. 117*:93-97.

Ahring, B. K., and Westermann, P. (1987). Kinetics of butyrate, acetate, and hydrogen metabolism in a thermophilic, anaerobic, butyrate-degrading tri-culture. *Appl. Environ. Microbiol. 53*:434-439.

Atkinson, B. (1984). Consequences of aggregation. In *Microbial Adhesion and Aggregation*, K. C. Marshall (Ed.). Springer Verlag, New York, pp. 351-373.

Balch, W. E., Fox, G. E., Magrum, L. J., Woese, C. R., and Wolfe, R. S. (1979). Methanogens: Reevaluation of a unique biological group. *Microbiol. Rev. 43*:260-296.

Barker, H. A. (1941). Studies on the methane fermentation. V. Biochemical activities of *Methanobacterium omelanskii*. *J. Biol. Chem. 137*:153-167.

Bauchop, T. (1967). Inhibition of rumen methanogenesis by methane analogues. *J. Bacteriol. 94*:1971.

Boone, D. R., and Bryant, M. P. (1980). Propionate-degrading bacterium, *Syntrophobacter wolinii* sp. nov. gen. nov. from methanogenic ecosystems. *Appl. Environ. Microbiol. 40*:626.

Bryant, M. P., Wolin, E. A., Wolin, M. J., and Wolfe, R. S. (1967). "*Methanobacillus omellanskii*," a symbiotic association of two species of bacteria. *Arch. Microbiol. 59*:20.

Bryant, M. P., Campbell, L. L., Reddy, C. A., and Crabill, M. A. (1977). Growth of *Desulfovibrio* in lactate or ethanol media low in sulfate in association with H_2-utilizing methanogenic bacteria. *Appl. Environ. Microbiol. 33*:1162-1169.

Chartrain, M., and Zeikus, J. G. (1986a). Microbial ecophysiology of whey biomethanation: Characterization of trophic populations and prevalent species composition in continuous culture. *Appl. Environ. Microbiol. 51*:188-196.

Chartrain, M., and Zeikus, J. G. (1986b). Microbial ecophysiology of whey biomethanation. Intermediary metabolism of lactose degradation in continuous culture. *Appl. Environ. Microbiol. 51*:180-187.

Chartrain, M., Bhatnagar, L., and Zeikus, J. G. (1987). Microbial ecophysiology of whey biomethanation: Comparison of carbon transformation parameters, species composition and starter culture performance in continuous culture. *Appl. Environ. Microbiol. 53*: 1147-1156.

Chen, M., and Wolin, M. J. (1977). Influence of CH_4 production by *Methanobacterium ruminantium* on the fermentation of glucose and lactate by *Selenomonas ruminantium. Appl. Environ. Microbiol. 34*:756-759.

Cheng, K.-J., and Costerton, J. W. (1980). The formation of microcolonies by rumen bacteria. *Can. J. Microbiol. 26*:1104-1113.

Conrad, R., Phelps, T. J., and Zeikus, J. G. (1985). Gas metabolism in support of juxtapositioning of hydrogen producing and methanogenic bacteria in sewage sludge and lake sediments. *Appl. Environ. Microbiol. 50*:1285-1291.

Conrad, R., Schink, B., and Phelps, T. J. (1986). Thermodynamics of H_2 consuming and H_2 producing metabolic reactions in diverse methanogenic environments under in situ conditions. *FEMS Microbiol. Ecol. 38*:353-360.

Conrad, R., Lupton, F. S., and Zeikus, J. G. (1987). Hydrogen metabolism and sulfate dependent inhibition of methanogenesis in a eutrophic lake sediment (Lake Mendota). *FEMS Microbiol. Ecol. 45*:107-115.

Costerton, J. W., and Irvin, R. T. (1981). The bacterial glycocalyx in nature and disease. *Annu. Rev. Microbiol. 35*:299-324.

Dolfing, J., and Bloemen, G. B. M. (1985). Activity measurements as a tool to characterize the microbial composition of methanogenic environments. *J. Microb. Meth. 4*:1-12.

Dolfing, J., and Mulder, J. W. (1985). Comparison of methane production rate and co-enzyme F_{420} content of methanogenic consortia in anaerobic granular sludge. *Appl. Environ. Microbiol. 49*:1142-1145.

Dolfing, J., and Tiedje, J. M. (1987). Hydrogen cycling in a three tiered food web growing on the methanogenic conversion of 3-chlorobenzoate. *FEMS Microbiol. Ecol. 38*:293-298.

Dolfing, J., Griffioen, A., van Neerven, A. R. W., and Zeevenhuizenm, L. P. T. M. (1985). Chemical and bacteriological composition of granular methanogenic sludge. *Can. J. Microbiol.* *11*:744-750.

Dwyer, D. F., and Tiedje, J. M. (1983). Degradation of ethylene glycol and polyethylene glycols by methanogenic consortia. *Appl. Environ. Microbiol.* *46*:185-190.

Dwyer, D. F., Shelton, D. R., and Tiedje, J. M. (1984). Abstr. Ann. Meet. Am. Soc. for Microbiol. I27, p. 126.

Eichler, B., and Schink, B. (1985). Fermentation of primary alcohols and diols and pure culture of syntrophic alcohol-oxidizing anaerobes. *Arch. Microbiol.* *143*:60-66.

Ferry, J. G., and Wolfe, R. S. (1976). Anaerobic degradation of benzoate to methane by a syntrophic consortium. *Arch. Microbiol.* *107*:33-40.

Hungate, R. E. (1967). Hydrogen as an intermediate in the rumen fermentation. *Arch. Mikrobiol.* *59*:158-164.

Hungate, R. E., Smith, W., Bandrop, T., Yu, T., and Rabinowitz, J. C. (1970). Formate as intermediate in the bovine rumen fermentation. *J. Bacteriol.* *102*:389-397.

Ianotti, E. L., Kafkewitz, P., Wolin, M. J., and Bryant, M. P. (1973). Glucose fermentation products of *Ruminococcus albus* grown in continuous culture with *Vibrio succinogenes*: Changes caused by interspecies transfer of H_2. *J. Bacteriol.* *114*:1231-1240.

Isa, Z., Grusenmeyer, S., and Verstraete, W. (1986). Sulfate reduction relative to methane production in high rate anaerobic digestion: Microbiological aspects. *Appl. Environ. Microbiol.* *51*:580-587.

Jeris, J. S., and McCarty, P. L. (1965). The biochemistry of methane fermentation using ^{14}C tracers. *J. Water Pollut. Control Fed.* *37*: 178-192.

Jones, W. J., Nagle, D. P., and Whitman, W. P. (1987). Methanogens and the diversity of Archaebacteria. *Microbiol. Rev.* *53*(1):135-177.

Kaspar, H. F., and Wuhrmann, K. (1978). Product inhibition in sludge digestion. *Microb. Ecol.* *4*:241-248.

Kenealy, W., and Zeikus, J. G. (1981). Influence of corrinoid antagonists on methanogen metabolism. *J. Bacteriol.* *146*:133-140.

Koch, M., Dolfing, J., Wuhrmann, K., and Zehnder, A. J. B. (1983). Pathways of propionate degradation by enriched methanogenic cultures. *Appl. Environ. Microbiol.* *45*:1411-1414.

Kristiansson, J. K., Schoenheit, P., and Thauer, R. K. (1982). Different K_S-values for hydrogen of methanogenic and sulfate reducing bacteria: An explanation for the apparent inhibition of methanogenesis by sulfate. *Arch. Mikrobiol.* *131*:278-282.

Krzycki, J. A., Morgan, J. B., Conrad, R., and Zeikus, J. G. (1987). Hydrogen metabolism during methanogenesis from acetate by *Methanosarcina barkeri*. *FEMS Microbiol. Lett.* *40*:193-198.

Lettinga, G., van Vellsen, A. F. M., Hobma, S. W., de Zeeuw, W., and Klapwigk, A. (1980). Use of the upflow sludge blanket reactor concept for biological wastewater treatment, especially anaerobic treatment. *Biotechnol. Bioeng. 22*:1699.

Lettinga, G., de Zeeuw, W., Hulshohh, P. L., Wiegant, W., and Rinzema, A. (1985). Anaerobic digestion 1985. In *Proceedings of the Fourth International Symposium on Anaerobic Digestion*, Guangzhou China, pp. 279-301.

Lovley, D., Dwyer, D. F., and Klug, M. J. (1982). Kinetic analysis of competition between sulfate reducers and methanogens for H_2 in sediments. *Appl. Environ. Microbiol. 43*:1373-1379.

Lovley, D. R. (1985). Minimum threshold for hydrogen metabolism in methanogenic bacteria. *Appl. Environ. Microbiol. 49*:1530-1531.

Lovley, D. R., and Ferry, J. G. (1985). Production and consumption of H_2 during growth of *Methanosarcina* spp. on acetate. *Appl. Environ. Microbiol. 49*:247-249.

Lupton, F. S., and Zeikus, J. G. (1984). Physiological basis for sulfate-dependent hydrogen competition between sulfidogens and methanogens. *Current Microbiol. 11*:7-12.

Mah, R. (1982). Methanogenesis and methanogenic pathways. *Philos. Trans. R. Soc. Lond. [Biol.] 297*:599-616.

Marshall, K. C. (Ed.) (1984). *Microbial Adhesion and Aggregation*. Life Sciences Research Report 31, Dahlem Workshop Reports, Springer, Berlin.

McInerney, M. J., Bryant, M. P., and Pfennig, N. (1979). Anaerobic bacterium that degrades fatty acids in syntrophic association with methanogens. *Arch. Mikrobiol. 132*:129-135.

McInerney, M. J., Bryant, M. P., Hespell, R. B., and Costerton, J. W. (1981). *Syntrophomonas wolfei* gen. nov. spec. nov., an anaerobic syntrophic fatty acid oxidizing bacterium. *Appl. Environ. Microbiol. 41*:1029-1039.

Miller, T. L., and Wolin, M. J. (1973). Formation of hydrogen and formate by *Ruminococcus albus*. *J. Bacteriol. 116*:836-846.

Mosbach, K. (1978). The microenvironment of immobilized multistep enzyme systems. In *Microenvironments and Microcompartmentalization*. Academic Press, New York, p. 381.

Mountfort, D. O., and Bryant, M. P. (1982). Isolation and characterization of an anaerobic syntrophic benzoate-degrading bacterium from sewage sludge. *Arch. Microbiol. 133*:249-256.

Nanninga, H. J., and Gottschal, J. C. (1985). Amino acid fermentation and hydrogen transfer in mixed cultures. *FEMS Microbiol. Ecol. 31*:261-269.

O'Brien, J. M., Wolkin, R. H., Moench, T. T., Morgan, J. B., and Zeikus, J. G. (1984). Association of hydrogen metabolism with unitrophic or mixotrophic growth of *Methanosarcina barkeri* on carbon monoxide. *J. Bacteriol. 158*:373-375.

Odom, J. M., and Peck, H. D., Jr. (1981). Hydrogen cycling as a general mechanism for energy coupling in the sulfate reducing bacteria, *Desulfovibrio* sp. *FEMS Lett. 12*:47-50.

Odom, J. M., and Peck, Jr. (1984). Hydrogenase electron transfer proteins and energy coupling in the sulfate reducing bacterium *Desulfovibrio. Annu. Rev. Microbiol. 38*:551-592.

Oremland, R. S. (1983). Hydrogen metabolism by decomposing cyanobacterial aggregates in Big Soda Lake, Nevada. *Appl. Environ. Microbiol. 45*:1519-1525.

Ozturk, S., Palsson, B., Thiele, J. H., and Zeikus, J. G. (1987). Modeling of interspecies hydrogen transfer in microbial flocs. In *Biotechnology Processes: Scale-up and Mixing*, C. S. Ho and J. Y. Oldshue (Eds.). American Institute of Chemical Engineers, New York, pp. 142-152.

Phelps, T. J., and Zeikus, J. G. (1985). The effect of fall turnover on terminal carbon metabolism in Lake Mendota sediments. *Appl. Environ. Microbiol. 50*:1285-1291.

Phelps, T. J., Conrad, R., and Zeikus, J. G. (1985). Sulfate dependent interspecies H_2 transfer between *Methanosarcina barkeri* and *Desulfovibrio vulgaris* during the coculture metabolism of acetate or methanol. *Appl. Environ. Microbiol. 50*:589-594.

Pohland, F. G., and Gosh, S. (1971). Developments in anaerobic stabilization of organic wastes—the two phase concept. *Environ. Lett. 1*(4):255.

Pohland, F. G., and Harper, S. H. (1985). Biogas developments in North America. In *Anaerobic Digestion 1985*. Proceedings of the Fourth International Symposium on Anaerobic Digestion, Guangzhou, China, pp. 41-81.

Ranalli, G., Whitmore, T. N., and Lloyd, D. (1986). Methanogenesis from methanol in *Methanosarcina barkeri* studies using membrane inlet mass spectrometry. *FEMS Microbiol. Lett. 35*:119-122.

Reddy, C. A., Bryant, M. P., and Wolin, M. J. (1972a). Characteristics of S-Organism isolated from *Methanobacillus omelianskii. J. Bacteriol. 109*:539-545.

Reddy, C. A., Bryant, M. P., and Wolin, M. J. (1972b). Ferredoxin and nicotinamide adenine dinucleotide dependent H_2 production from ethanol and formate in extracts of S-organism isolated from "*Methanobacillus omelianskii.*" *J. Bacteriol. 110*:126-132.

Reddy, C. A., Bryant, M. P., and Wolin, M. J. (1972c). Ferredoxin dependent conversion of acetaldehyde to acetate and H_2 in extracts of S-organism. *J. Bacteriol. 110*:133-138.

Robinson, J. A., and Tiedje, J. M. (1982). Kinetics of hydrogen consumption by rumen fluid, digestor sludge and sediment. *Appl. Environ. Microbiol. 44*:1374-1384.

Robinson, J. A., and Tiedje, J. M. (1984). Competition between sulfate reducing and methanogenic bacteria for H_2 under resting and growing conditions. *Arch. Microbiol. 137*:26-32.
Sahm, H. (1984). Anaerobic wastewater treatment. *Adv. Biochem. Eng. Biotechnol. 29*:84-115.
Satterfield, C. N. (1970). Mass transfer in heterogenous catalysis. MIT, Cambridge, Massachusetts.
Schauer, N. L., and Ferry, J. G. (1980). Metabolism of formate in *Methanobacterium formicicum*. *J. Bacteriol. 42*:800-807.
Schauer, N. L., Brown, D. P., and Ferry, J. G. (1982). Kinetics of formate metabolism in *Methanobacterium formicicum* and *Methanospirillum hungatei*. *Appl. Environ. Microb. 44*:549-554.
Scheifinger, C. C., Linehan, B., and Wolin, M. J. (1975). H_2 production by *Selenomonas ruminantium* in the absence and presence of methanogenic bacteria. *Appl. Microbiol. 29*:480-483.
Schink, B. (1984). Fermentation of 2,3 butanediol by *Pelobacter carbinolicus* sp. nov. and evidence for propionate formation from C_2 compounds. *Arch. Mikrobiol. 137*:33-41.
Schink, B. (1985). Degradation of unsaturated hydrocarbons by methanogenic enrichment cultures. *FEMS Microbiol. Ecol. 31*:69-77.
Schink, B., and Stieb, M. (1983). Fermentative degradation of polyethylene glycol by a strict anaerobic, gram negative, nonsporeforming bacterium. *Appl. Environ. Microbiol. 45*:1905-1913.
Smith, M. R., and Mah, R. A. (1966). Kinetics of acetate metabolism during sludge digestion. *Appl. Microbiol. 14*:368-371.
Stams, A. J. M., and Hansen, T. A. (1984). Fermentation of glutamate and other compounds by *Acidaminobacter hydrogenoformans* gen. nov. sp. nov., an obligate anaerobe isolated from black mud. Studies with pure cultures and mixed cultures with sulfate-reducing and methanogenic bacteria. *Arch. Microbiol. 137*:329-337.
Suflita, J. M., Robinson, J. A., and Tiedje, J. M. (1983). Kinetics of microbial dehalogenation of haloaromatic substrates in methanogenic environments. *Appl. Environ. Microbiol. 45*:1466-1473.
Szewcyk, U., Szewzyk, R., and Schink, B. (1985). Methanogenic degradation of hydroquinone and catechol via reductive dehydroxylation to phenol. *FEMS Microbiol. Ecol. 31*:79-87.
Ten Brummeler, E., Hulshoff, P. L. W., Dolfing, J., Lettings, G., and Zehnder, A. J. B. (1985). Methanogenesis in an upflow anaerobic sludge blanket reactor at pH 6 on an acetate-propionate mixture. *Appl. Environ. Microbiol. 49*:1472-1477.
Thauer, R. K., and Morris, J. G. (1984). Metabolism of chemotrophic anaerobes: Old views and new aspects. *Symp. Soc. Gen. Microbiol. 36*:123-168.
Thauer, R. K., Fuchs, G., and Jungermann, K. (1971). Reduced Ferredoxin:CO_2 oxidoreductase from *Clostridium pasteurianum*: Its role in formate metabolism. *J. Bacteriol. 118*:758-760.

Thauer, R. K., Fuchs, G., and Jungerman, K. (1973). Role of iron-sulfur proteins in formate metabolism. In *Iron Sulfur Proteins*, W. Lowenberg (Ed.). Academic Press, New York, pp. 121-156.

Thauer, R. K., Jungermann, K., and Decker, K. (1977). Energy conservation in chemotrophic anaerobic bacteria. *Bacteriol. Rev. 41*:100-180.

Thiele, J. H., and Zeikus, J. G. (1986). Method and apparatus for biomethanation. U.S. Patent Application Ser. No. 863977.

Thiele, J. H., and Zeikus, J. G. (1988a). Control of interspecies electron flow during anaerobic digestion: The role of formate versus hydrogen transfer during syntrophic methanogenesis in flocs. *Appl. Environ. Microbiol. 54*(1):20-29.

Thiele, J. H., and Zeikus, J. G. (1988b). The anion exchange substrate shuttle process: A new approach to biomethanation of organic and toxic wastes. *Biotechnol. Bioeng*. (In press).

Thiele, J. H., Chartrain, M., and Zeikus, J. G. (1988c). Control of electron flow during anaerobic digestion: The role of floc formation in syntrophic methanogenesis. *Appl. Environ. Microbiol. 54*(1):10-19.

Thiele, J. H., Ozturk, S. S., Palsson, B., and Zeikus, J. G. (1988d). Formate transfer during syntrophic methanogenesis—comparison of experimental data with a dynamic reaction diffusion model. *Biotechnol. Bioeng*. in preparation.

Thiele, J. H., Wu, W., and Zeikus, J. G. (1988e). Ecoengineered biocatalysts for advanced biomethanation systems. I. Composition and performance of granular catalysts for fast syntrophic methanogenesis from volatile fatty acids. *Biotechnol. Bioeng*. in preparation.

Tramper, J., van Groenstign, J. W., Luyben, K. C. A. M., and Hulshof, L. W. (1984). Some physical and kinetic properties of granular anaerobic sludge, pp. 145-155. In *Innovations in Biotechnology*, E. H. Houwink and R. R. van der Meer (Eds.). Elsevier, Amsterdam.

Tzeng, S. F., Bryant, M. P., and Wolfe, R. S. (1975). Factor F_{420} dependent pyridine nucleotide linked formate metabolism of *Methanobacterium ruminantium*. *J. Bacteriol. 121*:192-196.

van Bruggen, J. J. A., Stumm, C. K., and Vogels, G. D. (1983). Symbiosis of methanogenic bacteria and sapropelic protozooa. *Arch. Microbiol. 136*:89-95.

van Bruggen, J. J. A., Stumm, C. K., Zwart, K. B., and Vogels, G. D. (1985). Endosymbiotic methanogenic bacteria of the sapropelic amoeba *Mastigella*. *FEMS Microbiol. Ecol. 31*:187-192.

Vogels, G. D., Hoppe, A. F., and Stumm, C. K. (1980). Association of methanogenic bacteria with rumen ciliates. *Appl. Environ. Microbiol. 40*:608-612.

Weimer, P. J., and Zeikus, J. G. (1977). Fermentation of cellulose and cellubiose by *Clostridium thermocellum* and *Methanobacterium thermoautotrophicum*. *Appl. Environ. Microbiol.* *33*:289-297.

Widdel, F., and Wolfe, R. S. (1986). Growth of methanogenic bacteria in pure culture with 2-propanol as sole hydrogen donor. 86th Annual meeting of the American Society for Microbiology, Abstract I126.

Winfrey, M. R., and Zeikus, J. G. (1977). Effect of sulfate on carbon and electron flow during microbial methanogenesis in fresh water sediments. *Appl. Environ. Microbiol.* *33*:275-281.

Wofford, N. Q., Beaty, P. S., and McInerney, M. J. (1986). Preparation of cell free extracts and the enzymes involved in fatty acid metabolism in *Syntrophomonas wolfei*. *J. Bacteriol.* *167*:179-185.

Wolin, M. J. (1976). Interactions between H_2-producing and methane-producing species, p. 141. In *Microbial Production and Utilization of Gases*, H. G. Schlegel, G. Gottschalk, and N. Pfennig (Eds.). E. Goltz Verlag, Göttingen, Federal Republic of Germany.

Wolin, M. J., and Miller, T. L. (1982). Interspecies H_2 transfer: 15 years later. *ASM News* *48*:561-565.

Woods, D. D. (1936). Hydrogen lyases. The synthesis of formic acid by bacteria. *J. Biol. Chem.* *77*:515-518.

Wu, W., Hu, J., and Xiu-sheng, G. (1985). Properties of granular sludge in upflow anaerobic sludge blanket (UASB) reactor and its performance. In *Proceedings of the Fourth International Symposium on Anaerobic Digestion*, Guangzhou, China, pp. 339-351.

Young, J. C., and McCarty, P. L. (1969). The anaerobic filter for waste treatment. *J. Water Pollut. Control Fed.* *41*:R160-R173.

Zeikus, J. G. (1979). Microbial populations in digestors. In *First International Symposium on Anaerobic Digestion*, D. A. Stafford et al. (Eds.). A. D. Cardiff Scientific Press, Cardiff, U.K., pp. 75-103.

Zeikus, J. G. (1983). Metabolic communication between biodegradative populations in nature. In *Microbes in Their Natural Environment*, J. H. Slater (Ed.). Wiley, New York, pp. 423-462.

Zeikus, J. G., Kerby, R., and Krzycki, J. A. (1985). Single carbon chemistry of acetogenic and methanogenic bacteria. *Science* *227*: 1167-1173.

Zinder, S. H., and Koch, M. (1984). Non-aceticlastic methanogenesis from acetate: Acetate oxidation by a thermophilic syntrophic culture. *Arch. Microbiol.* *138*:263-272.

18

Modeling Mixed Culture Interactions in Anaerobic Digestion

Applications to Digester Start-up and Control

GREGORY STEPHANOPOULOS / *Massachusetts Institute of Technology, Cambridge, Massachusetts*

I. INTRODUCTION

Anaerobic digestion is the biologic stabilization of organic matter under anaerobic conditions. During the process, organic solids are degraded and biomass is formed with a substantial solid mass net reduction. Further, the solids produced are free of pathogenic organisms and odorous compounds, are unattractive to rodents and readily dewaterable, and can sometimes be used as soil conditioners. The chemical oxygen demand (COD) reduction is due to methane production; often the gas produced, a mixture of methane and carbon dioxide, is in excess of the amount required for feed heating.

Low costs, relevant solids reduction, and useful end products make anaerobic digestion a useful process. Civil as well as industrial and agricultural wastes have been successfully treated by anaerobic digestion. Operational difficulties remain, and reactor failures have often been encountered. Poor results refer to low methane production, that is, low organic matter stabilization, as reported by Dague et al. (1970). Failure refers to extreme cases in which the environmental conditions (essentially pH and volatile organic acid concentration) inhibit or even kill microorganisms. In this case the digester becomes "sour" and no methane is produced. Although the immediate cause of these failures is the imbalance in biologic production and consumption rates of volatile acids, it can be caused by different process disturbances, such as poor mixing, large influent solids concentration, large dilution rate, toxic substances, or temperature changes. For this reason an increasing interest in the stability and control strategies of anaerobic digesters has developed recently.

Modeling and simulation have been identified as useful tools for evaluating transient digester performance and control strategies. Models are based on material balances, empirical unstructured growth models, mass transport equations, ionic equilibria equations, and stoichiometric relations.

The first dynamic model, due to Andrews (1969) and Graef and Andrews (1974), included one species (the methanogenic bacteria) and one substrate (acetic acid). In fact, it is generally accepted that methanogenesis is the rate-limiting step and that acetic acid is the acid present in largest concentration. In this model the acid production-consumption imbalance cannot be represented, since the reactor is thought to be fed with acetic acid.

More recently, Kleinstreuer and Poweigha (1982) proposed a two-species (acid and methane formers) model that can reproduce the acid production-consumption imbalance. Organic solids hydrolysis is not modeled and is assumed to be completed before the reactor. The importance of hydrolysis was shown by Eastman and Ferguson (1981), who proved experimentally that this is the rate-limiting step in the overall acidogenesis.

The model presented in this chapter extends the work of the previous researchers in several respects. First, all the basic steps in anaerobic digestion are accounted for: organic solids hydrolysis, acetogenesis, and methanogenesis—one or more of which may be limiting or the cause of process failure. Second, the microorganisms present in the digester are grouped according to the function they perform, and the interactions among these populations, effected through changes that they create in the abiotic medium, analyzed. This analysis allows a better understanding of the causes that change the medium or the biota of a digester and can therefore lead to more efficient control strategies. Finally, realistic rate equations and parameter values from results reported in the literature have been used. Several values are only estimates, so that the model is expected to give semiquantitative predictions. One of the reasons for data insufficiency is that experimentation is troublesome owing to long retention times and extreme sensitivity of methanogenic bacteria to oxygen. On the other hand, this is also a good reason that modeling and simulation studies are so attractive.

Despite these uncertainties, the model is sufficiently accurate to allow anaerobic digester evaluation during transients, which is the rule rather than the exception in the treatment of sewage wastes. In this chapter the model is first developed, then subsequently validated by comparing its predictions to actual digester operation data. Following development, the model is applied to the evaluation of various start-up and control procedures. Improved control strategies are suggested to minimize reactor instabilities and recovery time following perturbations from steady system operation.

II. MODEL DEVELOPMENT

A. Process Description

Solids are fed into digesters in the form of a solid suspension. Only a fraction (60-80% for sewage water) of the total suspended solids (TSS) is organic matter, and this is usually referred to as volatile suspended solids (VSS). Operational definitions of TSS and VSS can be found in wastewater analysis handbooks. The volatile suspended solids consisting of carbohydrates, lipids, and nitrogenous material are hydrolyzed by extracellular enzymes produced by a group of bacteria (B_1) to form lower molecular weight products. These products serve as substrate for the class of B_1 bacteria that produce volatile organic acids C_3-C_6. Volatile acids are the substrate for a second group of bacteria that give methane and carbon dioxide as end products. Small amounts of hydrogen and carbon dioxide are also produced by B_1, and they are probably utilized by methanogenic bacteria for methane production. These bacteria are quite specific, and it is likely that each species grows on a different substrate. Barker (1956) and Lawrence and McCarty (1969) have suggested that the higher molecular weight acids are degraded to methane in successive steps by different species. Thus, there are commensal relationships between acidogenic and methanogenic bacteria and among different methanogenic species.

Methanogenic bacteria are strict anaerobes, grow slowly in comparison with B_1 bacteria, and can be inhibited by large concentrations of their substrate, the volatile acids. The pH of anaerobic digesters is generally 6.8-7.2, which is optimal for methanogens, but B_1 bacteria have lower optimum pH. Since volatile acids and carbon dioxide influence the pH, and this in turn changes the available substrate (undissociated acids), the interaction between cells and abiotic medium is complicated.

B. Model Structure

Owing to the high complexity and inadequate knowledge of the nature of populations that comprise the active culture of a digester, some simplifications must be made in the model description of the latter.

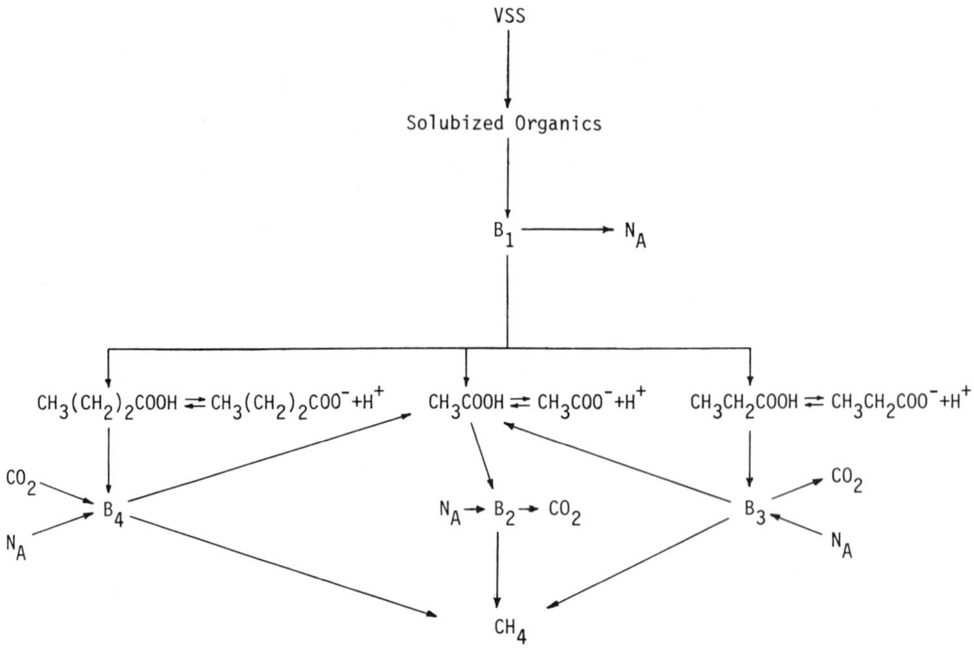

Figure 1 Structure of abiotic microbial interactions in mixed culture anaerobic digestion model. (From Dalla Torre and Stephanopoulos, 1986a.)

Thus, attention is focused on the basic processes taking place in a digester, and the organisms that mediate these processes are grouped together according to the function they perform. Figure 1 is a schematic of these groups and the network of interactions among them. The model recognizes two basic classes of microorganisms, the hydrolyzers and acid formers (B_1) and the acid-consuming methanogens. The latter class comprises three groups of methanogens, B_2, B_3, and B_4, growing, on acetic, propionic, and butyric acid, respectively, that is, the three acids present in the largest concentrations.

The following phenomena are accounted for by the model: (1) enzymatic hydrolysis of organic matter, (2) acidogenesis, (3) methanogenesis, (4) mass transfer between the liquid and gas phases, and (5) pH-dependent dissociation of electrolytes. Common experience has proven that step 3 is the rate-limiting step of anaerobic digestion, and more attention is placed on the modeling of this step. Also, Eastman and Ferguson (1981) have shown experimentally that step 1 is the rate-limiting step in the global acidogenesis reaction.

It is believed that the structure depicted in Figure 1 is a good balance between the present state of knowledge about the complex process of anaerobic digestion and the need for a reliable and flexible model that contains available or observable parameters and is appropriate for dynamic design and control simulations.

For a continuous flow operation the equation for total degradable solids concentration F (see Sec. VI for all definitions) is

$$\frac{dF}{dt} = - r_H + \frac{F_0 - F}{\tau} \tag{1a}$$

Following Eastman and Ferguson (1981), a first-order hydrolysis rate is assumed. Of course, a certain degree of lumping is involved, because during the hydrolysis process, different substrates are hydrolyzed at different rates.

For our purpose a linear relationship is used for rH:

$$r_H = k_H F \tag{1b}$$

as an empirical correlation that describes a very complex process. The value of the rate constant k_H in Table 1 has been obtained for carbohydrates and nitrogenous material hydrolysis and is also used here. Although Eastman and Ferguson (1981) observed no lipid degradation, this may be due to the short retention times used in these experiments. Chynoweth and Mah (1971) observed lipid hydrolysis in their studies and their conclusions are followed here. An attempt to relate the hydrolysis to enzyme synthesis is presented later.

The balance for hydrolyzed organic matter S_1 resulting from hydrolysis is

$$\frac{dS_1}{dt} = r_H - \frac{1}{(YX_1S_1)} \mu_1 X_1 + \frac{S_{10} - S_1}{\tau} \tag{2}$$

indicating that the products of hydrolysis are the limiting substrate for the growth of acidogenic bacteria B_1. Ghosh (1981) considered VSS as the limiting substrate, but such a choice seems less appropriate since hydrolysis is the rate-limiting step in the overall acidogenesis.

The bacterial death rate in acidogenesis was observed and measured experimentally (Eastman and Ferguson, 1981) as proportional to the bacteria cell density X_1. Including this effect, then, the balance for the cell density of the acidogenic bacteria reads

$$\frac{dX_1}{dt} = (\mu_1 - k_d)X_1 + \frac{X_{10} - X_1}{\tau} \tag{3a}$$

Table 1 Parameters and Sources

Symbol	Value	Units	Source
k_H	3.0	day^{-1}	Eastman and Ferguson (1981)
$\tilde{\mu}_1$	3.34	day^{-1}	Ghosh (1981)
K_{S_1}	1.0	g/L	See text
k_d	0.432	day^{-1}	Eastman and Ferguson (1981)
K_{S_2}	8.7×10^{-6}	mol/L	Lawrence and McCarty (1969)
K_{S_3}	2.4×10^{-6}	mol/L	Lawrence and McCarty (1969)
K_{S_4}	0.34×10^{-6}	mol/L	Lawrence and McCarty (1969)
K_{I_2}	0.667×10^{-3}	mol/L	Graef and Andrews (estimate) (1974)
K_{I_3}	0.667×10^{-3}	mol/L	Graef and Andrews (estimate) (1974)
K_{I_4}	0.667×10^{-3}	mol/L	Graef and Andrews (estimate) (1974)
$\tilde{\mu}_2$	0.35	day^{-1}	Lawrence and McCarty (1969)
$\tilde{\mu}_3$	0.26	day^{-1}	Lawrence and McCarty (1969)
$\tilde{\mu}_4$	0.27	day^{-1}	Lawrence and McCarty (1969)
K_{L_a}	100	day^{-1}	Graef and Andrews (estimate) (1974)
YCOD	0.48	$\frac{\text{g of } X_1 \text{ produced as COD}}{\text{g of } S_1 \text{ as COD}}$	Eastman and Ferguson (1981)
YX_1S_1	0.77	$\frac{\text{g of } X_1 \text{ produced}}{\text{g of } S_1 \text{ consumed}}$	Calculated from YCOD[a]
YNX_1	4.6×10^{-3}	$\frac{\text{mol of } C_N \text{ consumed}}{\text{g of } X_1 \text{ formed}}$	Stoichiometry[a]

Symbol	Value	Units	Source
YACET	0.00675	$\frac{\text{mol } S_2}{\text{g}^{\text{COD}} \text{ produced}}$	Eastman and Ferguson (1981)
YPROP	0.00297	$\frac{\text{mol } S_2}{\text{g}^{\text{COD}} \text{ produced}}$	Eastman and Ferguson (1981)
YBUTYR	0.00147	$\frac{\text{mol } S_4}{\text{g}^{\text{COD}} \text{ produced}}$	Eastman and Ferguson (1981)
YX_2S_2	2.56	$\frac{\text{g of } X_2 \text{ produced}}{\text{mol } S_2 \text{ consumed}}$	Stoichiometry
YX_3S_3	2.02	$\frac{\text{g of } X_3 \text{ produced}}{\text{mol } S_3 \text{ consumed}}$	Stoichiometry
YX_4S_4	1.50	$\frac{\text{g of } X_4 \text{ produced}}{\text{mol } S_4 \text{ consumed}}$	Stoichiometry
$YCOX_2$	0.367	$\frac{\text{mol } CO_2 \text{ formed}}{\text{g of } X_2 \text{ formed}}$	Stoichiometry
$YCOX_3$	0.101	$\frac{\text{mol } CO_2 \text{ formed}}{\text{g of } X_3 \text{ formed}}$	Stoichiometry
$YCOX_4$	0.355	$\frac{\text{mol } CO_2 \text{ consumed}}{\text{g of } X_4 \text{ formed}}$	Stoichiometry
$YCHX_2$	0.367	$\frac{\text{mol } CH_4 \text{ formed}}{\text{g of } X_2 \text{ formed}}$	Stoichiometry
$YCHX_3$	0.349	$\frac{\text{mol } CH_4 \text{ formed}}{\text{g of } X_3 \text{ formed}}$	Stoichiometry
$YCHX_4$	0.312	$\frac{\text{mol } CH_4 \text{ formed}}{\text{g of } X_4 \text{ formed}}$	Stoichiometry
WN	0.0083	$\frac{\text{mol of } C_N \text{ consumed}}{\text{g of biomass formed}}$	Stoichiometry
CCOD	1.42	$\frac{\text{g COD}}{\text{g biomass}}$	Stoichiometry
VOLRAT	4.8	$\frac{\text{liquid volume}}{\text{gas volume}}$	

(Table 1, continued)

Symbol	Value	Units	Source
PTOT	1.0	atm	(Total pressure)
RT	25.514	l atm/mol	(Gas constant times temperature)
Parameters for enzyme synthesis model			
k_2	80.3 h^{-1}		Huang (1975)
k_1	1.68 L/g		Huang (1975)
k_3	5.61 L/g		Huang (1975)
X_{1m}	0.305 g/g		Huang (1975)
k_{de}	10.0 h^{-1}		Bailey and Ollis, average (1977)
k	0.440 day^{-1}		Calculation[b]

[a]Using an average S_1 composition from Cynoweth and Mah (1971).
[b]k has been calculated so that at steady state $K_H = 3.0\ day^{-1}$.

with the specific growth rate μ_1 following Monod kinetics as a function of the concentration S_1, as suggested by Eastman and Ferguson (1981):

$$\mu_1(S_1) = \frac{\bar{\mu}_1 S_1}{K_{S_1} + S_1} \tag{3b}$$

The growth of acidogenic bacteria is accomplished by volatile acid production, represented as COD. Assuming that dead cells lyse and lysed cells contribute to the product COD, the following rate equation for product formation can be proposed:

$$r_p = \left[\left(\frac{1}{YCOD} - 1\right)\mu_1 X_1 + k_d X_1\right] CCOD \tag{4a}$$

According to this formulation, the stoichiometry of acidogenesis is based on a COD balance and the distribution of organic acid products is determined by corresponding yield coefficients. Assuming further that (1) maintenance requirements are negligible, (2) acid formation is a growth-associated process, and (3) the consumption of 1 mol of butyric or propionic acid for the growth of B_3 or B_4, respectively, is accompanied by the formation of 1 and 2 mol of acetic acid, respectively, the balances for the main products of acidogenesis, namely, acetic acid S_2, propionic acid S_3, and butyric acid S_4, are written as

$$\frac{dS_2}{dt} = YACETr_p - \frac{\mu_2 X_2}{(YX_2S_2)} + \frac{\mu_3 X_3}{(YX_3S_3)} + \frac{2\mu_4 X_4}{(YX_4S_4)} + \frac{S_{20} - S_2}{\tau} \tag{4b}$$

$$\frac{dS_3}{dt} = YPROPr_p - \frac{\mu_3 X_3}{(YX_3S_3)} + \frac{S_{30} - S_3}{\tau} \tag{4c}$$

$$\frac{dS_4}{dt} = YBUTYRr_p - \frac{\mu_4 X_4}{(YX_4S_4)} + \frac{S_{40} - S_4}{\tau} \tag{4d}$$

These equations also include consumption terms for the growth of the methanogens, as illustrated in Figure 1. Parameters used for these steps have been measured at low pH (5-6) and are summarized in Table 1, except for K_{S1}, which was estimated from simulations discussed later.

For the growth of methanogens, the undissociated acid is assumed to be the limiting substrate. A rate expression similar to that used by Graef and Andrews (1974) is used here, too, exhibiting growth inhibition at high acid concentrations. It should be noted that these functions have not been fully proven. However, they are reasonable and have met with success in digester predictions so they are employed here for the growth of all three types of methanogens, B_2, B_3, and B_4. Hence, for j = 2, 3, 4,

$$\mu_j = \frac{\tilde{\mu}_j}{1 + K_{sj}/f_jS_j + f_jS_j/K_{Ij}} \tag{5a}$$

and the balances of the three types of methanogens are

$$\frac{dX_j}{dt} = \mu_j X_j + \frac{X_{j_0} - X_j}{\tau} \qquad j = 2, 3, 4 \tag{5b}$$

The inhibition constant was estimated by Andrews (1969) and the other kinetic parameters measured by Lawrence and McCarty (1969). They assumed a Monod model and operated at low acid concentration so that inhibition was not significant and allowed a correct parameter estimation.

Balances for total ammonia C_N, total carbon dioxide C_r (dissociated and undissociated), total base C_B, and methane are quite straightforward:

$$\frac{dC_N}{dt} = \mu_1X_1(YNX_1) - (WN)(\mu_2X_2 + \mu_3X_3 + \mu_4X_4 - k_dX_1) + \frac{C_{NO} - C_N}{\tau} \tag{6}$$

$$\frac{dC_T}{dt} = \mu_2X_2(YCOX_2) + \mu_3X_3(YCOX_3) - \mu_4X_4(YCOX_4) - r_{CO_2} + \frac{C_{TO} - C_T}{\tau} \tag{7}$$

$$\frac{dY_{CH_4}}{dt} = \frac{RT(VOLRAT)}{PTOT}\left(r_{CH_4} - \frac{r_{CH_4} + r_{CO_2}}{1 - Y_{H_2O}}\,Y_{CH_4}\right) \tag{8}$$

$$\frac{dC_B}{dt} = A_B - \frac{C_B}{\tau} \tag{9}$$

where the rates for carbon dioxide transfer and methane production are given by

$$r_{CO_2} = -K_La\left([CO_2]_D^* - [CO_2]_D\right) \tag{10}$$

$$r_{CH_4} = \mu_2X_2(YCHX_2) + \mu_3X_3(YCHX_3) + \mu_4X_4(YCHX_4) \tag{11}$$

Dilution caused by base addition was assumed to be unimportant. Also, constant yields were assumed in each balance, either taken from the literature as indicated or calculated from atomic or electron balances (see Christensen and McCarty, 1975, for a fairly general procedure in determining yields from stoichiometry). Half-reactions for methanogens were as proposed by Barker (1956). The formula $C_3H_7O_2N$ was assumed as the average cell composition.

The pH was calculated from a charge balance assuming ionic equilibria for the dissociation of ammonia, volatile acids, carbonic acid, and added electrolytes. Equilibrium constants were calculated at 38°C from data obtained from the *Handbook of Chemistry and Physics* (Weast, 1975) and the *Handbook of Chemistry* (Lange, 1956). Activities were calculated using the semiempirical Davies equation (Butler, 1964) at an estimated ionic strength of 0.1 mol/L.

Well-mixed conditions were assumed for both the solids in liquid suspension and the gas phase. Methane was assumed insoluble, and a constant mass transfer coefficient was used for carbon dioxide.

The material balances of the previous equations describe the operation of a well-mixed anaerobic digester with a microbial culture exhibiting the interactions depicted in Figure 1. Steady states and transients were obtained by numerical integration using a backward differentiation formula. Results are presented in a later section, following a discussion of the validation of the model and the estimation of parameters.

C. Parameter Determination and Model Validation

The majority of the parameters of the model have been determined independently in the references shown in Table 1. The same values are used in the simulation studies reported here. For some parameters, however, either there is no information for their determination or the reported values involve a larger than usual level of uncertainty. The model's sensitivity to such parameters was thus studied in order to assess the expected prediction uncertainties and identify critical research needs.

The only parameter that has not been measured and cannot be reasonably guessed is K_{S_1}. Model predictions were obtained for parameter values covering a range of two orders of magnitude around the value in Table 1. The only significant difference detected in the simulations was the steady-state value of S_1. From an acid former cell balance it can be easily shown that S_1 is proportional to K_{S_1} at steady state for constant feed rate. Eastman and Ferguson (1981) found that hydrolysis products never accumulate and their feed concentration is substantially reduced. This led to the value in Table 1. Any lower value would do, but this would not significantly influence the results.

The hydrolysis rate, which determines the acid formation rate, and the acid consumption rates are of critical importance. The imbalance in these rates is the cause of digester failure. Various values of the rate constant k_H were tried, and the anticipated results were obtained. A decreased k_H slows acidogenesis in transient conditions and decreases VSS reduction at steady state. Since VSS reduction was within the observed range (see below), k_H was not modified.

The adopted growth rate function for the methanogens were proposed by Andrews (1969). It has recently been suggested (Kleinstreuer and Poweigha, 1982) that a function with a plateau rather than a maximum may be more appropriate. The exact shapes of the specific growth rate curves of the methanogens are important for they determine the point of onset of instabilities resulting from an imbalance between acid-forming and acid-consuming processes. Lower values of the inhibition constants K_I result in sharper maxima for the specific growth rates; in this case digester failure is predicted under less severe conditions. One can conclude that a better determination of the growth rate parameters would be of great value in improving the accuracy of model predictions.

Table 2 Steady-State Results

Variable	Steady state	Usual range	Source
Loading g solids per L per day	2.6	1.6-4.8	MOP 16 (1968)
Total alkalinity (mg/L as $CaCO_3$)	3590[a]	1500-5000	Pohland and Gnosh (1971)
Total volatile acids (mg/L as acetic)	78.4	50-300	Pohland and Gnosh (1971)
pH	6.94	6.8-7.2	Pohland and Gnosh (1971)
Volatile solids reduction (%)	73.8	73	MOP 16 (1968)
Methane yield (mol/g VSS degraded)	0.046	0.045-0.050	Eckenfelder and O'Connor (1961)
Methane produced via acetic acid (%)	81	73[b]	MOP 16 (1968)
Methane gas volume percentage (%)	55	65-75	MOP 16 (1968)

[a]Approximate formula from McCarty et al. (1963).
[b]Assume 75% volatile suspended solids, average from Chynoweth and Mah (1971).

The stoichiometry of acid production was taken from Eastman and Ferguson (1981). Constant values were assumed, although yields can vary with pH and feed composition. Such variations, however, have minimal effects on the simulation results that led to the preceding assumptions.

Digester performance was simulated and compared to the results of actual runs under various conditions in an attempt to test the model capability to predict digester behavior. Results are summarized in Table 2, in which steady-state model predictions for typically monitored variables are compared to ranges of values observed in actual digester operation. The simulations were carried out for the feed conditions and initial (start-up) values shown in Table 3. This table also shows the steady-state values of the various model variables from which the quantities of Table 2 were determined.

Good agreement is evident; the only discrepancy is in the gas composition. Note that, both in the model (due to the large assumed

K_La) and in practice, the dissolved carbon dioxide is near equilibrium with the gas phase. Thus, since the pH is in the range observed in practice, the amount of carbon dioxide leaving the reactor in the liquid stream is correct. It is unlikely that the discrepancy is due to relative amounts of produced acids different from those of Table 1, for, as mentioned previously, the results proved to be rather insensitive to yield variations at constant produced COD. Consequently, the low methane fraction observed must be due to the stoichiometry of the methanogenesis step used. The same problem was encountered by Buhr and Andrews (1977), who modified calculated yields to give a better agreement with observed values.

Recently, Bhadra et al. (1984) have proposed a kinetic model for methanogenesis from acetic acid that satisfactorily fits the experimental data. According to this model, a portion of the carbon dioxide produced is transformed into methane via enzymatic reactions. The result is that 28% of the methane is produced in this way. With a biomass yield as in Table 1, a carbon balance gives $YCHX_2 = 0.511$ mol/g and $YCOX_1 = 0.225$ mol/g. Although these yields would give an exceedingly large methane production, no general conclusion can be drawn since the global methane yield also depends on the acidogenesis stoichiometry. Since dissolved carbon dioxide is near equilibrium, the main effect of methanogenesis stoichiometric modifications would be in gas production and composition. Therefore, we decided to leave the calculated yields unchanged.

Table 3 Feed, Steady-State and Initial Start-up Values

Variable	Feed	Steady-state value	Initial conditions for start-up
F (g/L)	26.3	0.848	26.3
S_1 (G/L)	1.86	0.161	1.86
X_1 (g/L)	0	3.930	0.0393
S_2 (mol/L)	0.026	0.892×10^{-3}	0.026
S_3 (mol/L)	0.011	0.291×10^{-3}	0.011
S_4 (mol/L	0.006	0.422×10^{-4}	0.006
X_2 (g/L)	0	1.946	0.01946
X_3 (g/L)	0	0.359	0.00359
X_4 (g/L)	0	0.133	0.00133
C_N (mol/L)	0.041	0.0728	0.041
C_T (mol/L)	0	0.0812	0
Y_{CH_4}	—	0.548	0.55
C_B (mol/L)	0	0	0

Another check of the model is the good qualitative agreement between the calculated sludge composition in a batch digester and the experimental curves reported by Eckenfelder and O'Connor (1961). Also, the model is able to reproduce fairly accurately digester failures following step increases in degradable feed solids concentration or feed rate. A similar satisfactory agreement was obtained for the stability of the reactor and its behavior under the control schemes reported in a later section.

D. A Mechanism for Enzymatic Hydrolysis

Hydrolysis is a key step in the overall anaerobic digestion process. It is catalyzed by enzymes secreted by the acid-forming bacteria B_1 and therefore depends on the cell density of this microbial population. Such dependence must be present in the model or hydrolysis can continue unchecked even in the absence of the species responsible for it. One way to introduce this dependence is through the hydrolysis rate constant k_H and an enzymatic mechanism for solids decomposition described in this section.

The model for enzymatic hydrolysis of amorphous cellulose (Huang, 1975) is adopted:

$$E + A \underset{\substack{\text{equilibrium}\\ \text{constant}\\ K_1}}{\rightleftharpoons} (EA) \xrightarrow{k_2} E + S_1 \underset{\substack{\text{equilibrium}\\ \text{constant}\\ K_3}}{\rightleftharpoons} (ES)$$

According to this model, the enzyme E is adsorbed on the solids A forming a complex EA. The adsorption-complexation process is at equilibrium with an equilibrium constant K_1. The complex irreversibly decomposes with a rate constant k_2 to form the hydrolysate product S_1. Finally, S_1 can combine reversibly with the enzyme to form an inactive complex with an equilibrium formation constant K_3.

The first equilibrium combined with a balance equation for total solids ($c_s = c + a$) yields

$$\frac{c}{c_s} = \frac{K_1 e}{1 + K_1 e}$$

where e is the concentration of free enzyme, c is the concentration of enzyme adsorbed in the form EA, c_s is the value of c at saturation (or total concentration of sites available for enzyme adsorption on the solids), and a is the concentration of unoccupied sites on the solids, all properly expressed per unit mass of solid substrate.

The rate of irreversible complex EA decomposition and, hence, the rate of S_1 formation is proportional to the concentration of EA, or

$$r_H = k_2c = \frac{k_2K_1e}{1 + K_1e} c_s$$

which, when compared to Equation (16), yields for k_H:

$$k_H = \frac{k_2K_1e}{1 + K_1e} \tag{12}$$

The hydrolysis rate constant is thus a function of the free enzyme concentration e. The latter can be related to the total enzyme concentration e_T by accounting for the various forms in which the available enzymes may be present, namely, free, adsorbed, or combined to form the inactive complex (ES_1):

$$e_T = e + K_3S_1e + \frac{K_1e}{1 + K_1e} F \tag{13}$$

It is thus seen that the hydrolysis rate constant depends on e and, through it, on e_T. A balance of total enzyme, in turn, reads

$$\frac{de_T}{dt} = -\frac{e_T}{\tau} + kX_1 - k_{de}e_T \tag{14}$$

where k_{de} is a deactivation constant; we have assumed the enzyme production to be nongrowth associated. To keep complications and parameters at a minimum, we have not considered models for extracellular enzyme synthesis that include genetic control (Van Dedem and Moo-Young, 1973).

To implement this hydrolytic enzyme synthesis model, the state vector is augmented to include the total enzyme concentration e_T, and Equations (12) through (14) are incorporated into the model equations. At each time step, using the current value of e_T provided by the integrator, values for e and then k_H are calculated, using Equations (13) and (12), respectively, for the calculations of the next step. Extensive simulations were performed using the parameters summarized in Table 1. These simulations revealed, at steady state, minor differences from the simpler model presented earlier without the enzyme balance. Such differences become more pronounced during start-up transients and following various perturbations.

III. EVALUATION OF START-UP PROCEDURES

Start-up and operating procedures as applied at present in industry are described in the Water Pollution Control Federation publication (1968) commonly referred to as MOP 16. These procedures have been analyzed on the basis of simulations employing the model of the previous section. Several improvements leading to shorter start-up times and more stable overall operation are presented in this section. Perfect mixing and adequate temperature control were assumed in the simulations.

With a large inoculum, normal operation in terms of flow rate and solids concentration in the feed can resume almost immediately. A straightforward start-up procedure is then to inoculate the digester with a large amount of digesting sludge obtained from other digesters. This procedure was simulated by immediately filling the digester with 25% of raw feed and 75% of digesting sludge (typically, 50-75% of the column is filled with digesting sludge). These conditions are represented in Table 3, columns 2 and 3 for sludge and raw feed composition, respectively. A normal flow rate corresponding to a residence time of 10 days was introduced immediately, and final steady states satisfactorily reproduced observed digester behavior; this agreement can be considered an additional validation of the model.

The procedure cannot be recommended as a standard digester start-up owing to the large amounts of seed sludge required. Two other possibilities can be suggested on the basis of the phase plane schematic of Figure 2 depicting the growth dynamics of a single species (such as B_2, B_3, or B_4 of Fig. 1) with substrate-inhibited kinetics at high substrate (acid) concentrations. Of course, acid production depends on the activities of all types of organisms present so the actual situation is one of a mixed culture commensalistic interaction. However, since hydrolysis and acidogenesis are fast compared with methanogenesis, examination of the latter can give only a representative picture of digester dynamics.

According to Figure 2, there are two stable steady states, one corresponding to culture growth and the other to washout. The stability regions of these steady states are defined by the separatrix that passes through the third possible, unstable (saddle point), steady state. Any initial condition to the right of the separatrix leads to the growth steady state. Any initial condition to the left leads to washout. A large initial biomass concentration produces an initial condition to the right of the separatrix and, hence, a successful start-up, even at high initial acid concentrations. The latter results from a large VSS concentration, indicating that an easy start-up is possible when seed is available. When seed is not available, in order to have an initial point to the right of the

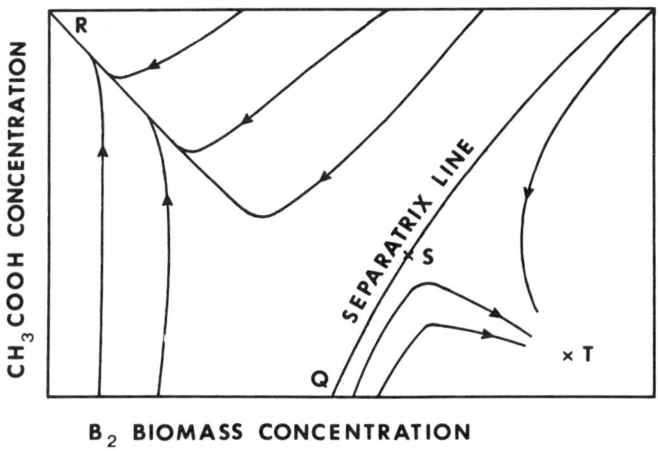

Figure 2 Schematic of phase plane trajectories for substrate inhibited growth kinetics (Dalla Torre and Stephanopoulos, 1986a).

separatrix, the initial substrate must be low. This can be achieved either by controlling the initial VSS level or by manipulating (gradually increasing) the pH in order to control inhibition through the adjustment of the undissociated acid concentration. These two policies are discussed in more detail below.

A. Initial VSS and Flow Rate Control

In order to simulate this procedure, the biomass type and concentration in raw wastewater must be known. This information is essential for start-up evaluation but not as important for continuous operation or start-up with seed, when such biomass can be safely neglected. In our simulation it was assumed that biomass in raw wastewater is of the same composition as the steady-state digesting sludge but is present at much smaller concentrations, namely, 1/100 of the steady-state value (Table 3, column 3). The start-up procedure consisted of filling the digester with raw wastewater and operating at a fraction of the normal feed rate. The hydraulic load was subsequently increased in steps after digestion had been satisfactorily established at each flow rate.

This procedure was not found to be very satisfactory. Unless the initial digester contents were significantly diluted, the reactor could not be started for realistic values of the operating parameters. For example, a start-up with 5% of the normal feed rate could be carried out successfully only when F and S_1 were diluted to 1/100 of the values shown in Table 3 (column 3). For lower initial biomass start-up is impossible. When successful, start-up times are long.

For the operation described in Table 3 start-up fails, causing digester failure. This seems to be in contrast with commonly accepted procedures, although a precise comparison is impossible, because data on actual digesters are lacking. A possible explanation is that the assumed functions for the growth of B_2, B_3, and B_4 populations are not very representative. Simulation results are also rather sensitive to the inhibition constant K_{Ij}. By increasing K_{Ij} enough, growth inhibition by acids is diminished so that a digester could be started from any initial condition. Another possibility is that acid yields could be too large. In this case lower yields could be compensated by larger methane yields for B_2, B_3, and B_4 (as suggested earlier, keeping the overall methane yield constant at steady state. Lower acid yields at constant overall methane yield would also be obtained if gas COD production during acetogenesis is significant at actually used pH and retention times.

B. pH Control

The pH-controlled start-up is accomplished by filling the digester with raw feed and using a fraction of the normal feed rate or even a full load. Full hydraulic load is established in a stepwise operation similar to that discussed in the previous section. A small amount of the steady-state biomass (1/100) and of the same composition is assumed to be present in the raw feed to function as inoculum. The pH is controlled by adding a base until satisfactory conditions are reached. This operation was simulated using NaOH as the base and an ideal on-off controller with band 6.8-7.0. A more sophisticated control action was not needed. The results are shown in Figure 3, where the initial feed rate is 10% of the normal rate (r = 10 days), with normal operation established after 48 days.

Only measurable variables are reported in Figure 3, although simulations provide a number of other variables, such as biomass concentration and composition. Some observations on Figure 3 are in order. The VSS concentration first decreases owing to fast solubilization of feed VSS, then increases owing to B_1 population biomass rise. Then this population tends to a low steady-state value, corresponding to the very high residence time. The slight VSS increase up to 48 days is due to the methane former's biomass. The final fast rise is due to the introduction of the full load and subsequent increase in the B_1 biomass.

The acid accumulation is almost complete before 10 days. Butyric, propionic, and acetic acids then reach a low concentration, in that order. This depends on the different amounts of acids produced and thus on the different inhibition rates. Also, butyric and propionic acid consumption implies the formation of acetic acid; for example, most of the butyric acid is consumed between 20 and 30 days and in this period there is a rise in acetic acid concentration and methane gas percentage and production. The same cannot be observed for propionic acid, since its effects cannot be distinguished from those of the acetic acid fast consumption.

The pH, after a sudden drop, reaches normal values due to the basic substance introduced. The oscillations are typical of the behavior of on-off controller. Between 40 and 48 days the system consumes most of the acids and builds up biocarbonate alkalinity. When the full load is introduced, there is no need for further base addition and the system slowly reaches its steady state. The pH control procedure is generally considered more rapid than the uncontrolled unseeded procedure; full load can usually be introduced within 30-40 days. Also, simulations indicate that with pH control a much wider range of initial conditions can be handled. It was thus established that this procedure is more promising in digester start-up; for this reason the effect of several factors has been studied and some of the results are presented below. Unless otherwise stated, in all reported results the feed and initial values used are the same as those reported in Table 3 (Columns 1 and 3) and full load refers to $\tau = 10$ days. The efficiency of a start-up procedure was evaluated by the number of days t^* after which full load could be introduced without causing any digester upsets.

Lime is commonly used for pH control because of its low cost. The effect of lime is not only to increase the pH but also to bring about a sharp decrease in bicarbonate alkalinity when the carbonate precipitates; redissolution of gaseous carbon dioxide occurs, which may cause dangerous negative pressures in the digester (McCarty, 1964). For lime pH control, t^* was found to be very large (90 days). This may be due to the assumption of ionic equilibria, which can be violated in practice if there is calcium carbonate supersaturation.

Other bases were used. Sodium hydroxide gives $t^* = 48$ days, and sodium bicarbonate gives 40 days. No other significant differences were noted, except an expected larger total carbon dioxide concentration C_T for the latter case. The behavior of NH_4HCO_3 and NH_4OH pH control was remarkably similar to that of $NaHCO_3$ and NaOH, respectively, as could be expected. Maximum NH_4+ concentrations were 0.84 equivalents per L, in the toxic range. To avoid excessively long times, which also mean long computer times and the possible problem of supersaturation, the pH control was actuated with sodium hydroxide.

Although quite effective, the pH control range commonly suggested (6.8-7.0) may not be the most appropriate. From the growth

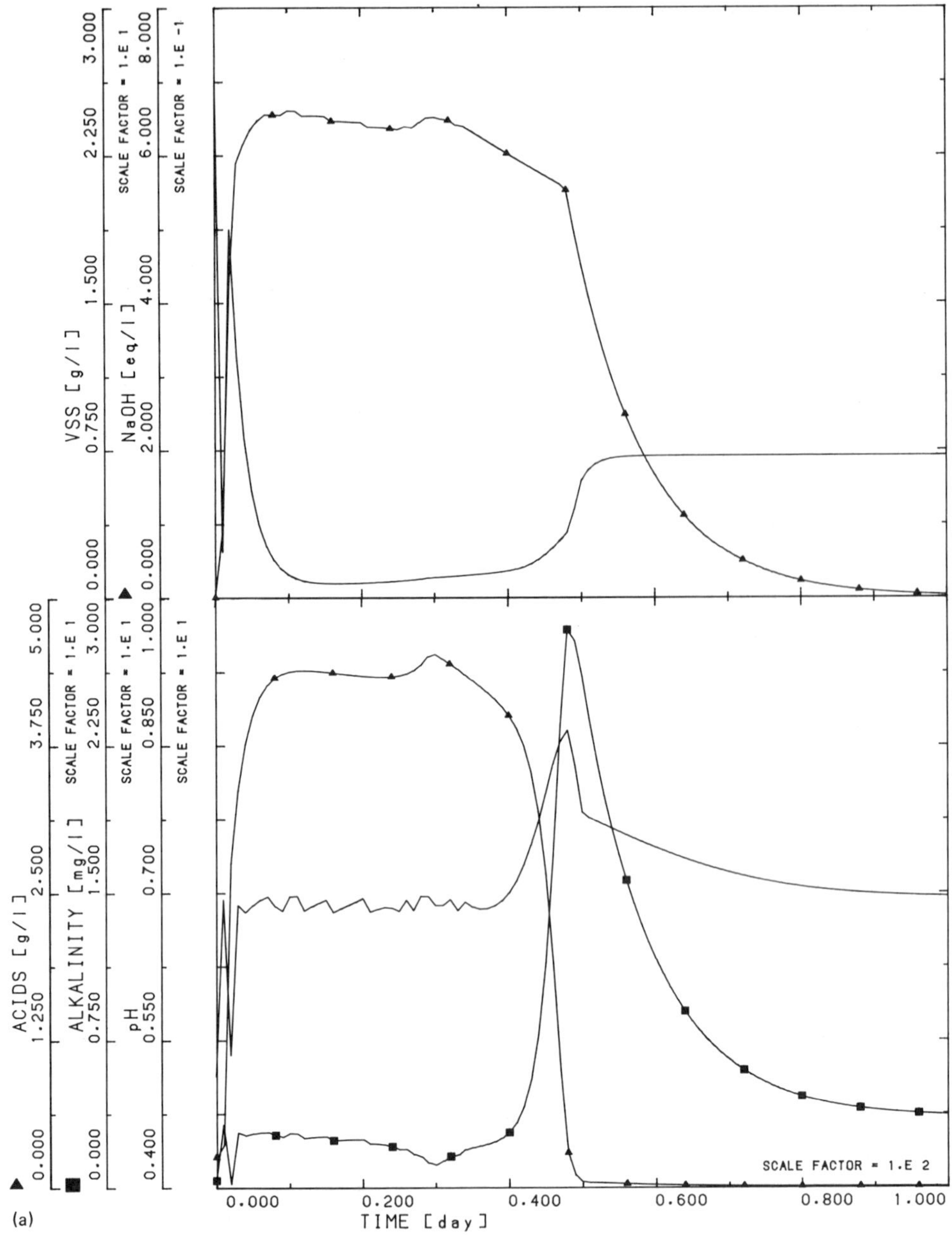

Figure 3 pH-controlled start-up with full load introduced after 48 days: (a) response of system variables; acids refer to volatile organic acids expressed as acetic acid and alkalinity refers to bicarbonate alkalinity expressed as $CaCO_3$; (b) response of additional system variables. (From Dalla Torre and Stephanopoulos, 1986a.)

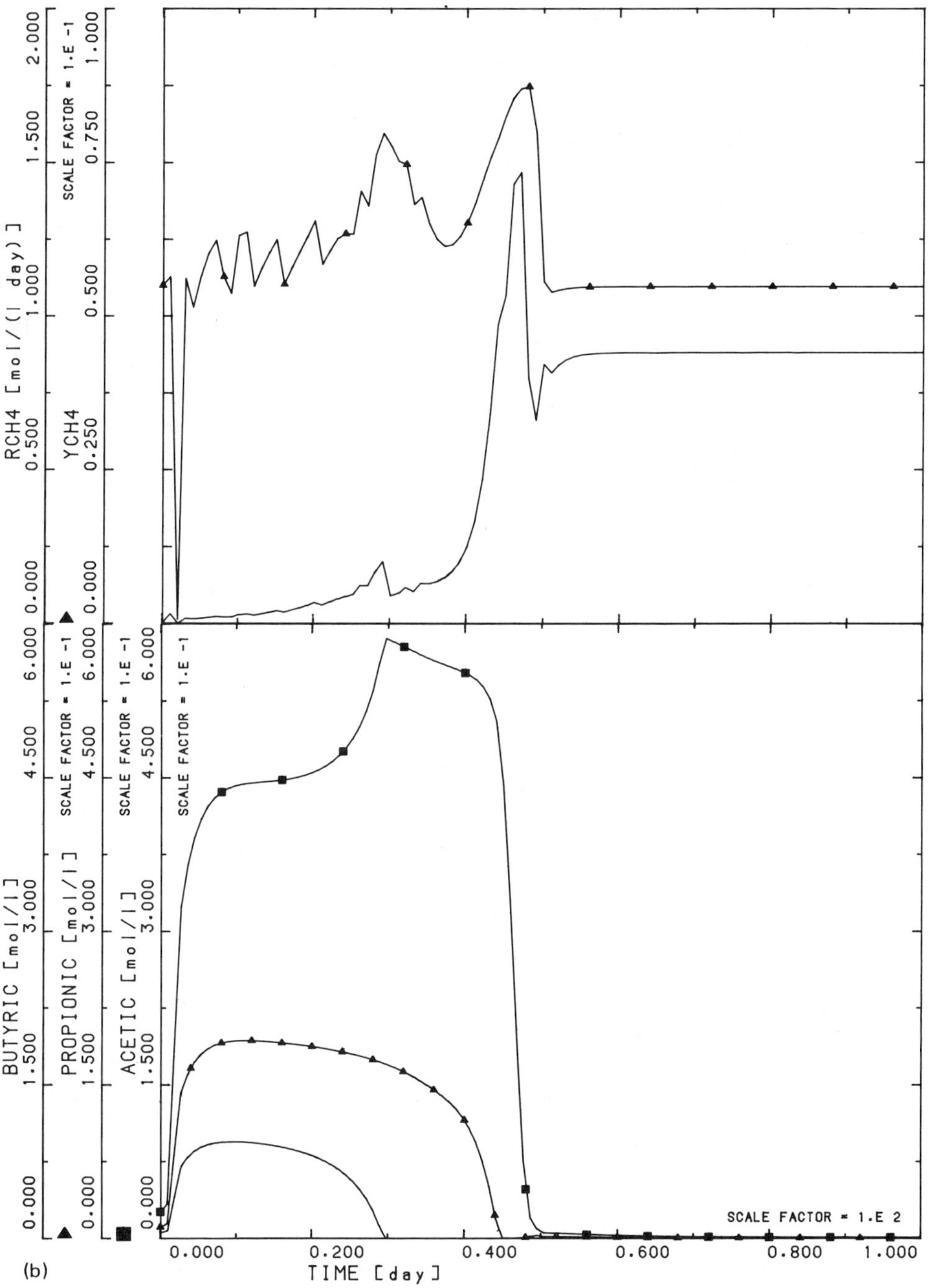
RCH4 [mol/(l day)]
YCH4
SCALE FACTOR = 1.E -1
0.000
0.500
1.000
1.500
2.000
0.000
0.250
0.500
0.750
1.000
BUTYRIC [mol/l]
PROPIONIC [mol/l]
ACETIC [mol/l]
SCALE FACTOR = 1.E -1
SCALE FACTOR = 1.E -1
SCALE FACTOR = 1.E -1
0.000
1.500
3.000
4.500
6.000
SCALE FACTOR = 1.E 2
TIME [day]
0.000
0.200
0.400
0.600
0.800
1.000

(b)

functions used and the maximum acid concentrations of Figure 3, it can be seen that when the concentrations are near the maximum, there is a substantial inhibition of the B_3 population and, to a larger extent, of the B_2 population. A higher pH control range (7.2–7.4) that maximizes the growth rate of B_2 was attempted with an initial load of 10%. A full load could be introduced after only 24 days versus 48 days obtained with the usual pH control range. The optimal pH range thus depends on the initial solids concentration and on the kinetic parameters K_s and K_i. Accurate measures of these parameters could define the best pH control range.

As expected, the lower the initial feed rate the sooner the full load can be introduced. This situation can be readily explained: the main purpose of the initial low feed rate period is to build up the biomass of methane formers. The lower the biomass washout rate, the sooner this biomass buildup occurs. For example, with an initial feed rate equal to 10% of full load, a normal feed rate could be safely applied after 48 days, with 5% after 44 days, or in a batch after 36 days, although failure occurred starting with full load. No other major difference was noticeable among the 5%, 10%, or batch cases.

A change in the feed rate also affects the positions of the steady states and separatrix line in Figure 2. A higher feed rate enlarges the domain of stability of the washout steady state and can make start-up impossible for a broad range of initial conditions. For example, with initial biomass concentrations from Table 3, setting to zero all other initial values and with no pH control, there is success for an initial load equal to 5% of the normal and failure for 10%. Thus, point Q in Figure 2 moves to the right when the feed rate is increased.

Dilution of initial digester content was suggested by the phase plane picture diagram in order to establish digester operation with small amounts of inoculum. Mixing the initial reactor content with water not containing relevant amounts of organics and biomass has two opposing effects. First, the organic concentration and thus the maximum acid concentration is lowered. Second, the reduced biomass concentration delays the formation of substantial biomass amounts. It is obvious that a minimum t^* can be reached by varying the dilution. This is verified by a few simulations that showed that t^* drops from 48 to 24 days by diluting the initial reactor content 4:1 versus no dilution at all. Even larger dilutions did not yield significant advantage.

IV. MODEL APPLICATIONS TO DIGESTER CONTROL

One of the first studies of digester control employed the dynamic model of Andrews (1969) and Graef and Andrews (1974). Several strategies were evaluated in this work, such as pH control through the addition of a base, gas composition control by sludge recycle, and pH control by carbon dioxide-free gas recycle. In another study, Collins and Gilliland (1974) examined pH control by sludge recycle and feed flow rate reduction. These studies investigated both feedback and feed-forward controllers and led to the proposal of a digital adaptive flow rate controller based on the single-species digester model employed. The shortcoming of these controllers was that the models used did not account for the presence of a multi-species consortium in the digester. As a result, delays and related dynamic features were absent and relevant cause-and-effect relationships were inadequately represented.

In this section the model presented earlier is employed for the investigation of existing, proposed, and new control strategies. As before, constant temperature and mixing are assumed. Temperature variations may be included but it was judged that the additional model complexity could not be justified on the basis of available kinetic information and parameter values. Some mixing effects are considered, however. Grit and scum accumulation can be simulated with lower available digester volumes and residence times. Inadequate mixing also limits the maximum solid density that can be fed into the digester.

In the sequel the most frequent causes of digester failure are reviewed and the results of a digester stability study presented. Control strategies are then analyzed, and some useful conclusions about appropriate control policies are presented.

A. Operational Difficulties and Digester Failures

Once successfully started, digesters are quite stable and rather large perturbations are usually required to bring about serious digester upsets. However, once these upsets set in they are very difficult to control and often lead to system failure. Such observations were made in a pilot plant by Torpey (1955) and in a laboratory environment by Andrews and Pearson (1965), Lawrence and McCarty (1969), Ashley and Hurst (1981), Pohland and Ghosh (1971), and Cox (1984).

The most common cause of digester failure is the imbalance between production and consumption of volatile acids. These acids are the substrate for the growth of methanogens and are also inhibitory at high concentrations. In the inhibitory range, a substrate increase causes a decrease in methanogen growth, which in turn decreases the total acid consumption rate, and, finally, a further increase in acid concentration. This unstable situation eventually leads to culture washout (point R, Fig. 2). An additional complication is that usual process indicators, such as pH, gas composition, and acid concentrations, remain within a normal range until failure is incipient and give no early warning of the coming disturbance. This behavior, also accurately reproduced by the model, point to possible shortcomings of feedback controllers and the need to introduce feed-forward action, provided, of course, that either an accurate model or a reliable transfer function (experimentally derived) are available for this purpose.

Several factors can be responsible for the imbalance: poor mixing, low residence time, high solids concentration in the feed, toxic substances, and temperature shocks.

Mixing is accomplished in a number of ways: frequent or continuous feeding, circulation through an external heat exchanger, gas recirculation, mechanical mixing, internal draft tube mixers, and rising bubbles of gas produced during digestion (Eckenfelder and O"Connor, 1961; Rundle and Whyley, 1981). Mixing avoids grit and scum accumulation, which reduce the actual volume of the digester, and also provides uniform substrate and biomass concentrations. It is commonly accepted that good mixing is necessary for satisfactory digester performance and that, with the usual designs, there is a maximum allowable feed solids concentration beyond which adequate mixing is not possible (MOP 21). This limit varies from plant to plant up to 10-12%. Sometimes the maximum solids concentration is determined by the ability of pumps to handle thick sludges. All mathematical models of anaerobic digestion assume perfect mixing. This means that model predictions are not meaningful at high feed density. In any case, such operating conditions should be carefully avoided.

A decrease in the mean residence time can decrease the biomass concentration of methane formers with respect to acid formers, since methanogens have lower specific growth rate maxima. This increases volatile acid concentration. If the inhibitory range is reached, the digester becomes sour because of acid accumulation, and methanogens are washed out.

Owing to very fast solid solubilization and acidogenesis, an increase in volatile suspended solids in the feed causes a fast rise in solubilized organics and acid concentration that, if large enough, can cause inhibition of methanogenic bacteria. The latter are then washed out because of the low growth rates and this makes the digester sour.

Many chemicals have been reported to be toxic to bacteria active in anaerobic digestion: heavy metals, organic bactericides; such oxidants as nitrates and sulfates, and sodium, calcium, and ammonium ions. These ions are important in the pH control of digester operation. Other substances are only occasionally present in significant concentrations in exceptional cases, such as an industrial spill.

Heating can be provided by internal or external heat exchangers or feed preheating and is necessary since digesters are usually operated in the mesophilic temperature range (30–40°C). It is likely that poor temperature control can cause digester failure and that kinetic parameters vary with temperature. Because of insufficient knowledge and data on temperature dependencies, this possibility is not examined here for control purposes. In any case the range for satisfactory operation is quite narrow.

B. Regions of Stable Digester Operation

Since flow rate and total suspended solids are the two operating parameters subject to the broadest variations, digester stability was studied with respect to perturbations introduced by these two parameters. The initial condition before the introduction of a perturbation corresponded to the digester steady state described by columns 1 and 2 of Table 3. As expected, the digester was able to handle small perturbations of these parameters (τ = 5 days; VSS = 75 g/L) but failed when large disturbances were introduced (τ = 2.5 days; VSS = 160 g/L). Representative examples of the responses obtained can be found in Figures 2 through 5 of the original work (Dalla Torre and Stephanopoulos, 1986a, b) on which this chapter is based. These figures show the progress with time of the system variables when subjected to the indicated hydraulic (residence time) and organic matter (VSS) perturbations.

One reason for carrying out these simulations was to identify good indicators of an imminent digester failure. Drops in pH, alkalinity, methane production, rises in volatile acid, and the acid-alkalinity ratio, although evident near failure and in agreement with field observations (Ashley and Hurst, 1981), are not very reliable and can vary significantly even under normal conditions. A parameter that is more generic but also more difficult to measure is the rate of change of acid concentration, also suggested in MOP 16. Simulations confirmed that the methane production rate is not a good process indicator. A more reliable indicator seems to be the distribution of organic acids, with large relative increases toward butyric and higher acids occurring near failure. This behavior is easy to explain on the basis of the functionalities assumed for the growth of the methanogenic bacteria and has also been observed experimentally (Ashley and Hurst, 1981; Cox, 1984; McCarty et al., 1963).

A conclusion that can be drawn from these results is that there is probably no variable that can serve as an accurate real-time indicator of potential digester failure. Some are more reliable than others, but even with these there is little that could be done to avoid a developing acid accumulation with the usual control system.

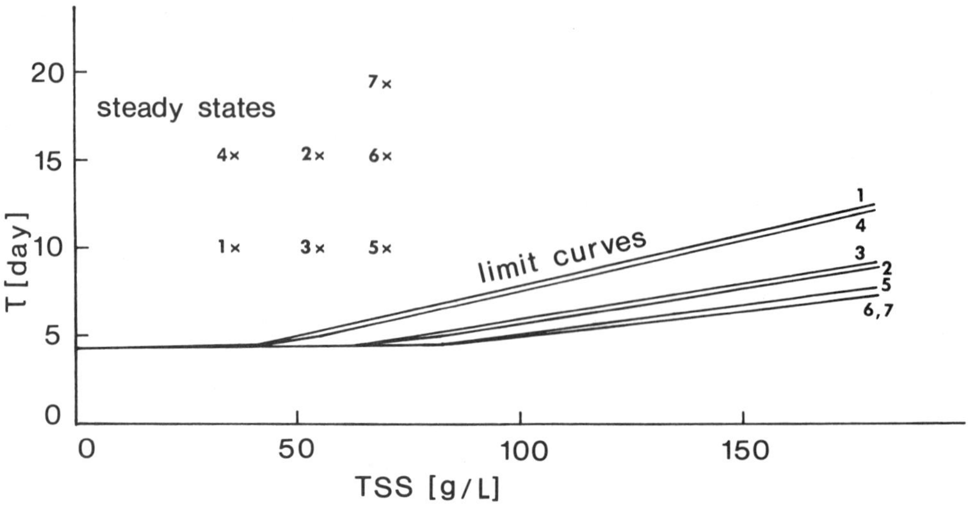

Figure 4 Digester stability to step changes in residence time and feed solids content. Limit curves correspond to indicated steady states. (From Dalla Torre and Stephanopoulos, 1986b.)

The next series of simulations examined the magnitude of step changes in the residence time and feed VSS concentration, alone or in combination, required to bring about digester failure. Step changes in the residence time and VSS feed concentration were used because they provide good tests of system stability. The results are summarized in Figure 4. A digester operating at steady states indicated by points 1-7 in terms of the values of the main operating parameters (residence time and VSS feed concentration) was subjected to various levels of perturbation of these two parameters, and the subsequent dynamics was observed. All other feed values were as in Tables 4 and 5. The limit curves shown in Figure 4 represent the smallest step changes in the two parameters required to make the system unstable when starting from the indicated steady-state point. The abscissa represents total suspended solids rather than VSS, which is better for the interpretation of results. It is assumed that VSS is a constant fraction of TSS; a typical value of 73% was taken from Chynoweth and Mah (1971). If it is further assumed that solids have constant density (a typical value for sewage water is 1.4), then there is a 1:1 correspondence between TSS and sludge density, which, for the usual range, is approximately a proportionality relation. Therefore, any result can be expressed equivalently in terms of density, TSS, or VSS.

Table 4 Primary Settler Model Equations

$$\frac{d}{dt}(VTSS_1) = F_0TSS_0 - (F_{ov} + F_u)TSS_1 - fTSS_1$$

$$\frac{d}{dt}(VTSS_j) = F_{ov}(TSS_{j-1} - TSS_j) - fTSS_j \quad (j = 2, 3, 4, 5)$$

$$\frac{dV}{dt} = \frac{1}{5}(F_0 - F_{ov} - F_u)$$

$$h = \frac{5V}{A} - H$$

$$F_{ov} = 153{,}082\ bh^{3/2} \quad \text{(empirical weir equation)}$$

$$f = F_{ov}[(1 - \beta)^{-1/5} - 1]$$

$$V_h \frac{dTSS_u}{dt} = F_u(TSS_1 - TSS_u) + f\sum_{j=1}^{5} TSS_j$$

Table 5 Sizing Parameters

Parameter	Value
Total volume	1092 m^3
Settling area A	273 m^2
Lengths	34.125 m
Weir length b	48 m
Weir height H	4.0 m
Average load	50 m^3/m^2 per day
Average residence time	1.92 h
Average weir load	283 m^3/m per day

Figure 4 indicates that the effect of a hydraulic load is approximately the same for all steady states of operation. Regardless of the normal operating condition, washout occurs when the residence time drops to the level of 4.3 days or lower. Furthermore, there is nothing that one can do about controlling the system other than reducing the flow rate through the digester to a level that can be accommodated by the culture growth rates.

The effect of the total solids feed concentration is more interesting. The curves in Figure 4 show the magnitude of a disturbance in TSS feed concentration, alone or in combination with a disturbance in residence time, sufficient to destabilize the steady state indicated by the corresponding number. The stabilizing effect of increasing TSS feed concentration is evident: larger perturbations are needed to destabilize a steady state. In comparing the stability results of Figure 4, steady states of equal productivity should be considered. It is easy to show that such steady states lie on a straight line through the origin (e.g., 1, 2, and 7 are all equivalent in terms of digester productivity). Consequently, one should operate at as high a residence time and feed TSS concentration as mixing and other constraints allow in order to enjoy the highest degree of stability with respect to such disturbances.

Concentration of digester feed is generally advised since it reduces tank volumes, heating requirements, and boiler capacity. Here, based on simulation results, it is claimed that the concentration of digester feed also has positive effects on digester stability. The resulting increase in residence time also has expected benefits in VSS reduction and methane production. Therefore, although design efforts should be made to reduce the feed TSS constraints imposed by mixing and pumping, one should choose an operating point close to the maximum allowed density. Since fluctuations in feed characteristics are wide, some density control is necessary.

C. pH Control

The objective of all digester control structures employed to date is the control of pH or other indicators of volatile acid concentration. These controls are basically feedback systems with no anticipatory qualities owing to the lack of quantitative models of digester operation. The most commonly manipulated variables are the feed flow rate, basic substance addition rate, and rate of sludge recycle. The effects of such variables are qualitatively evident. A lower flow rate reduces the amount of solids fed to the digester and thus the amount of acids produced; biomass concentrations also benefit from higher residence time. The addition of a base increases the pH and reduces inhibition, because methanogens are inhibited by undissociated acids. Finally, biomass concentration decreases the likelihood of washout.

Two representative schemes discussed in this section are pH feedback control by base addition and flow rate reduction.

With a pH band of 6.8-7.0 and starting from steady state 1 in Figure 4, the magnitude of the perturbations of residence time and TSS feed concentration were determined that can destabilize the system under pH control. In agreement with Graef and Andrews (1974), no improvement is obtained at low residence times. Some improvement can be obtained at large VSS in the feed. This improvement, however, is quite marginal, and such control is not justified, for these large feed VSS concentrations are at the limit of allowed operation. Any stability enhancement in this region is inconsequential because this region lies outside the operability limits, and in this region model predictions are not very accurate.

Reduction of flow through the digester implies either additional costs for a holding tank before anaerobic digestion or the inconvenience that some feed was discharged without anaerobic treatment, which would be acceptable only if a failure could be avoided this way.

The control needs are as follows: (1) reduce the flow rate only when there is a danger of reactor failure owing to low residence time and (2) bypass the smallest possible amount of feed. This is accomplished by the control scheme shown in Figure 5a, where a flow control is cascaded with a pH control. The outer loop has an on-off band controller such that the flow rate is uncontrolled for normal operation. When the pH is in the low range and the residence time is below a minimum limit τ_{SP}, the controller activates the inner loop with set point V_{AD}/τ_{SP}, where V_{AD} is the digester volume. A simple proportional or PI controller can be used. τ_{SP} = 5 days was chosen on the basis that 0.20 day^{-1} is smaller than the maximum growth rate of all populations present and also leaves a safety margin. After operating for a time with τ_{SP} = 5 days, one could try to reduce this value. The control action avoids an excessive flow reduction ($\tau > \tau_{SP}$) that would not only increase the amount of the bypass, but also would not be significantly beneficial to the process. In fact, the methanogens of the digester are largely affected by the residence time only when the dilution rate $1/\tau$ is close to the maximum specific growth rate.

Effecting pH control by manipulation of the bypass flow rate, represented in Figure 5b, is simpler but does not fully satisfy the control needs just explained. In fact, since the flow rate is not measured, the controller is not able to determine whether the operating difficulties are due to low residence time or to any other process disturbance. The lower limit for the digester feed flow rate (fully open valve) also depends on the total flow rate (before bypass) and may be unnecessarily low. Another advantage of the cascade scheme is that, when the digester already has operating difficulties, any significant hydraulic disturbance is eliminated by the inner loop before it reaches the digester.

The results indicate that neither base addition nor flow rate manipulation can provide a robust and efficient control for anaerobic digesters. Such controls are useless if the maximum allowable density is reached. Finally, they involve extra costs and may cause toxicity problems. For these reasons a new control strategy based on the

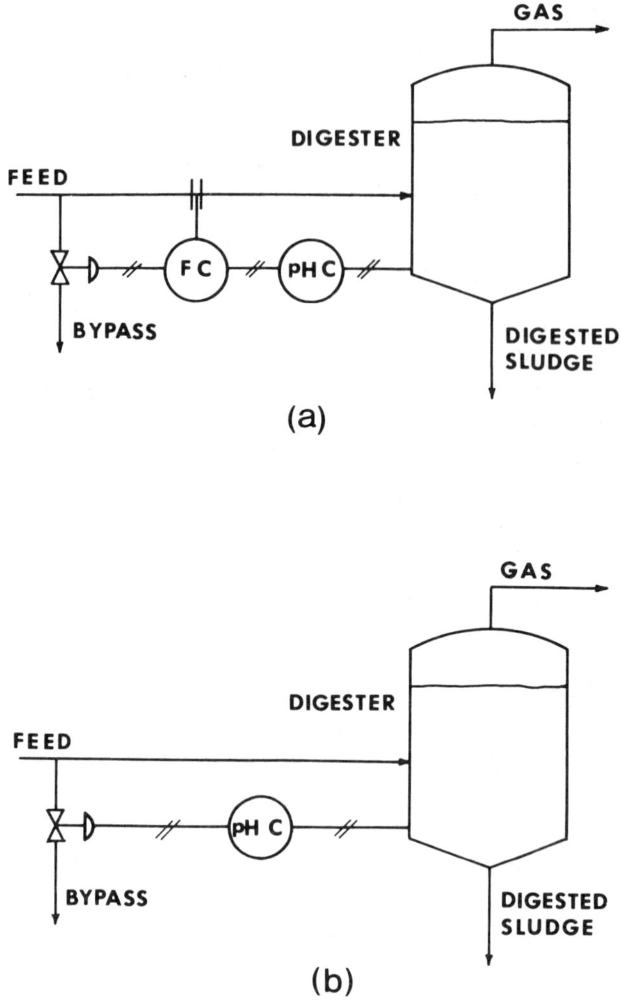

Figure 5 Digester pH control: (a) cascaded with flow control; (b) by flow manipulation. (From Dalla Torre and Stephanopoulos, 1986b.)

density control of primary sedimentation is analyzed here as another possibility for the control of anaerobic digestion systems.

D. Density Control of Primary Sedimentation

The stability results of Figure 4 pointed to several advantages that could be gained by operating at high feed solid densities and the corresponding higher residence times. In order to control these parameters it is necessary to control the operation of the next unit upstream in an anaerobic digestion process. In most cases this is a primary sedimentation unit whose underflow output is the input to the digester. A schematic of the unit arrangement is given in Figure 6, and the operation of these units is briefly described below and used in the sequel for the control of the digester.

Primary sedimentation is accomplished in rectangular or circular basins at retention times of 1.5-2.5 h. The underflow is a concentrated suspension of solids and the overflow a low-solids concentration stream of approximately the same flow rate as the incoming stream due to the very high concentrations achieved.

Let β be the solids removal efficiency:

$$\beta = \frac{F_u TSS_u}{F_0 TSS_0} \tag{15}$$

where F indicates the flow rate and subscripts u and 0 refer to settler underflow and feed. Since the underflow is the feed to the digester, then

$$\frac{\tau}{TSS_u} = \frac{V_{AD}}{\beta F_0 TSS_0} \tag{16}$$

where V_{AD} is the digester volume. Parameter β is at least 0.50 and depends essentially on the overflow velocity, defined as the ratio of the overflow volumetric flow rate to the basin sedimentation area. This dependence is deduced from statistical data by Eckenfelder and O'Conner (1961) and measurements by Thereoux and Betz (1958). Since for constant F_0 the overflow velocity is almost constant, then the efficiency β is constant and Equation (16) represents a straight line through the origin in Figure 7. When F_0 and/or TSS_0 increases, the slope changes according to Equation (16) and the operating line moves from a to b, c, or d in Figure 7. TSS_u and τ are determined by the sedimentation operation along the operating line. However, once they are determined, they define a steady state for the digester indicated by the corresponding point in the τ versus TSS_u diagram. Figure 7 also shows line e of the maximum solids

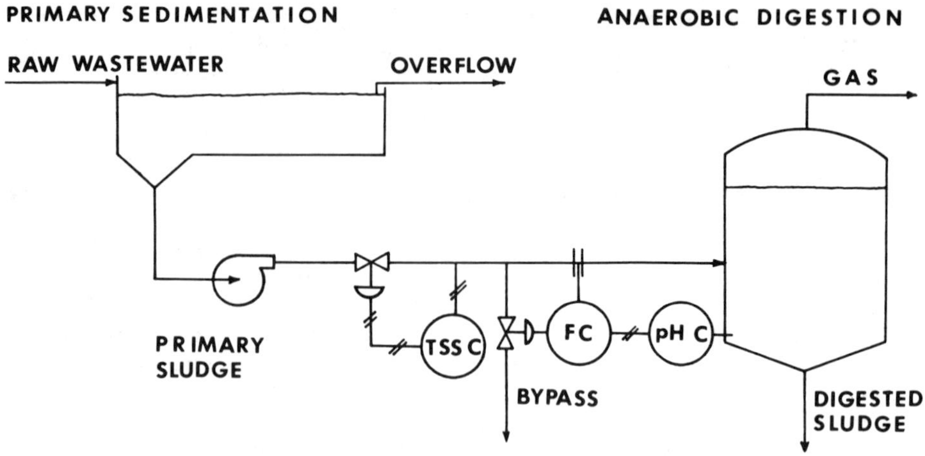

Figure 6 Control scheme for sedimentation and digestion. (From Dalla Torre and Stephanopoulos, 1986b.)

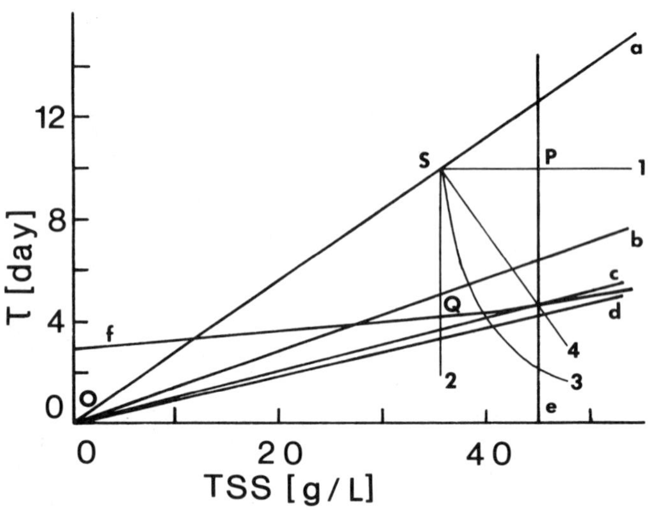

Figure 7 Digester feed characteristics: (a–d) primary sedimentation operating lines; (e) maximum allowable solids concentration; (f) digester limit curve relative to steady state S; (1–4) sedimentation control strategies. (From Dalla Torre and Stephanopoulos, 1986b.)

that can be handled by the digester and the limit curve f for the magnitude of perturbations that can destabilize the system operating at point S.

Various control strategies are possible for the output of the sedimentation unit or the feed to the digester. All these controls are in some respect anticipatory, for they aim at controlling the input to the digester at the desired point instead of responding to indicators of souring when very little is possible to prevent failure. One possibility is to control F_u (using a positive displacement pump), which is equivalent to staying (at steady state) on a horizontal line through S indicated by 1 in Figure 7. Another option is to control TSS_u with a PI controller, which means that the operating variables lie on a vertical line 2 through S. Intermediate control actions are also possible as shown by lines 3 and 4 in Figure 7.

In practice, density control can be more complicated and require an on-off mode to allow solids collection and consolidation in hoppers. The need for time cycles is described by Garrison and Nagel (1959), and most of the conclusions reached here also apply to these types of controllers.

Starting from steady state S, the digester can handle all feed conditions within the limits imposed by lines e and f. If the controller keeps F_u constant (line 1 in Fig. 7), all raw wastewater conditions resulting in lines above OP are acceptable; a TSS_u proportional-integral controller (line 2) can handle all lines above OQ; a TSS_u proportional controller (line 3) would further improve system capabilities. A special-purpose controller (line 4), based on the knowledge of lines e and f, could handle all lines above c.

Raw wastewater conditions corresponding to lines such as those shown in Figure 7d cannot be handled by TSS_u controller, and the addition of a basic substance can do little to correct this problem. On the other hand, flow reduction enlarges the stability domain for values of TSS just around the steady state and seems very well suited to complement density control. The resulting control scheme appears in Figure 6; the bypass can be replaced by a holding tank if undigested sludge discharge cannot be tolerated even in exceptional cases. The position of the holding tank in the wastewater treatment flowchart should be between primary sedimentation and anaerobic digestion, not before the settler, which drastically increases the volumes required and causes more problems with scum and sludge (Tanthapanichakoon and Himmelblau, 1981). The bypass can also be actuated before the settler.

Steady state S corresponds to 1 in Figure 4; similar conclusions can be drawn for steady states 2 or 7. In particular, it should be noted that the system can at least double its normal feed rate without activating the bypass.

Based on these considerations, the control of TSS_u by F_u manipulation is proposed. Proportional control was tested against PI control and steady-state designed feed-forward control. In addition to the mentioned steady-state advantages, proportional control is the fastest and gives no overshoot in response to step input changes in TSS_0 and/or F_0. The proportional controller is defined by

$$F_u = \overline{F}_u + K_p(TSS_u - \overline{TSS}_u) \tag{17}$$

where the overbar indicates set point values. A large value of the proportional gain K_p can be selected with no stability problems; $K_p = 25\ m^3\ L/g$ day gives a suitably low offset; larger values give marginally better performance. The value here was used in the simulations.

$\overline{TSS}_u = 36.03$ g/L (VSS = 26.3 g/L, $\overline{F}_u = 61.685\ m^3$/day, and $V_{AD} = 618.65\ m^3$/day were chosen. For $V_{AD} = 618.65\ m^3$, this is the digester feed corresponding to steady state 1. For a larger digester more settlers could be employed operating in parallel.

The performance of this controller was tested in a variety of dynamic simulations. For the sedimentation unit a rectangular basin primary settler model from Tanthapanickakoon and Himmelblau (1981) was used. According to this model the settler is ideally divided into solids separator with a sludge hopper at the bottom and five completely mixed compartments to account for hydraulic mixing. Only volumetric and suspended solids balances have been included, although it is possible to treat dissolved species in an obvious way. Details of the model can be found in Tanthapanichakoon and Himmelblau (1981). Equations are reported in Table 4. The steady-state solids removal efficiency β was computed after an experimental curve from Thereoux and Betz (1958). It seems that equipment sizing is generous if such a curve is employed. For design purposes F_0 = 13650 m^3/day, TSS_0 = 0.22 g/L, and a peak value of $2F_0$ were assumed.

The performance of the control strategy in Figure 6 was tested with step changes in F_0 and/or TSS_0. It should be noted that these are equivalent. For example, we can get the same underflow straight line with 13,650 m^3/day and 0.4273 g/L, 31,936 m^3/day and 0.22 g/L, or 22,000 m^3/day and 0.2796 g/L. All these conditions give F_uTSS_u = 4330.55 kg/day, which defines line b in Figure 7. Also, the dynamic response of F_u and TSS_u is very similar. Therefore, there is practically no difference for the digester if F_0 or TSS_0 increases. Curves could be drawn in the F_0 versus TSS_0 plane as loci of raw wastewater conditions with constant F_u TSS_u. If the digester can handle a point on this line, it will handle any

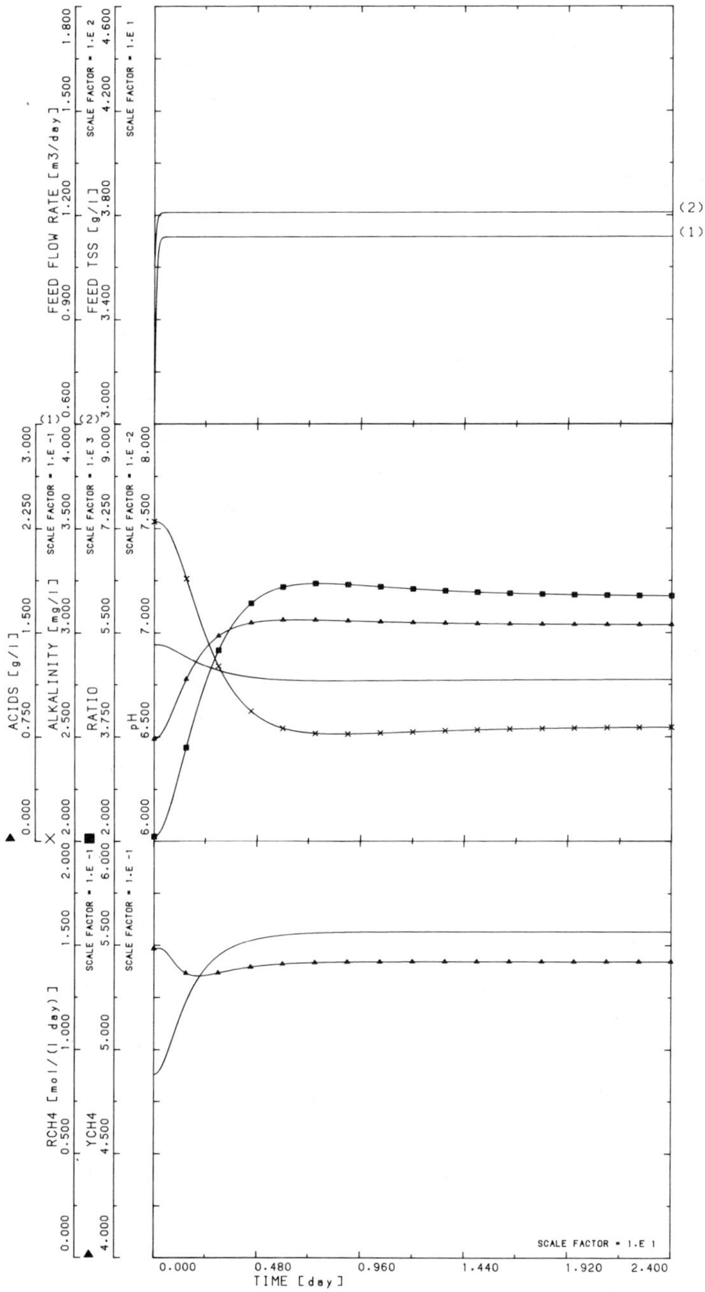

Figure 8 Performance of the control scheme of Figure 6 for a step change in raw wastewater conditions to F_0 = 22,000 m^3/day, TSS_0 = 0.2796 g/L. (From Dalla Torre and Stephanopoulos, 1986b.)

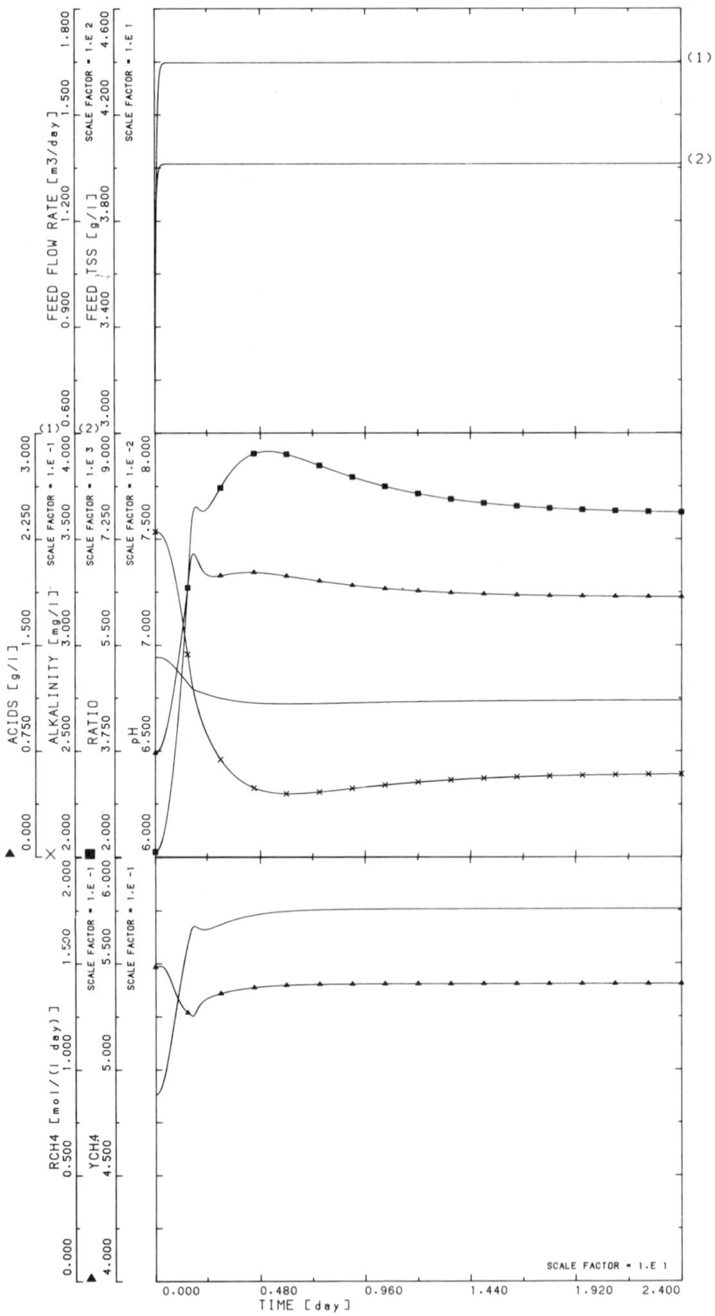

Figure 9 Performance of the control scheme of Figure 6 for a step change in raw wastewater condition to F_0 = 22,000 m^3/day and TSS_0 = 0.350 g/L.

point on and below the line. A curve corresponding to c in Figure 7 would determine conditions that could be handled with suitable control action and no flow reduction.

Results with a step change to 22,000 m^3/day and 0.2795 g/L are reported in Figure 8. Flow reduction is not needed.

Line d in Figure 7 corresponds, for example, to F_0 = 28,000 m^3/day and TSS_0 = 0.350 g/L ($F_u TSS_u$ = 6598.9 kg/day). With such a step increase, Figure 9 was obtained. Flow reduction acts effectively in this case. In both cases the F_u constant control leads to TSS_u beyond the allowable limit of line e.

Low time constants for the settler and speedy action of the proportional controller are such that steady-state conclusions match closely with dynamic performance. Figures 8 and 9 show satisfactory operation of the control scheme in Figure 6.

V. SUMMARY

A dynamic model of anaerobic digestion was presented in this chapter. The model accounts for the main steps of the process, namely, enzymatic hydrolysis of organic matter, biologic acidogenesis, and methanogenesis. Also, the interactions between the main groups of microorganisms and abiotic medium and commensal interactions among different populations were considered. The model satisfactorily predicts observed digester behavior, although a low methane gas percentage was obtained at steady state and large acid concentrations during start-up. These can be corrected by admitting a carbon dioxide conversion to methane even though lower acid yields would be required to maintain a constant overall methane yield. Experimental data on B_1 growth allowing atomic balances (not only a COD balance) would be useful. Also, the B_1 gas production, neglected here, and further utilization by methanogens would lower acid yields.

Other areas needing further investigation have been pointed out. In particular the hydrolysis rate and the methanogenic bacteria growth rates warrant further study. If, as suggested by Kleinstreuer and Poweigha (1982), the growth rate peaks are broader, the maximum acid concentration in a start-up should become lower.

A modified model accounting for hydrolytic enzyme synthesis did not give qualitatively different results. This was somewhat unexpected and may be due to model crudeness and/or to poor parameter estimates.

A qualitative picture in the phase plane explains the dynamic behavior of anaerobic digesters and the reasons behind commonly used start-up procedures. Seeded, unseeded, and pH-controlled start-ups were simulated. Difficulties arise for the case of no seed and no pH control. This may be due to the yields being different from those of Table 1. At any rate, simulations indicate that pH control can handle a much wider range of initial conditions.

Sodium (or ammonium) bicarbonate and hydroxide performed better than lime in digester pH control. In practice, the low toxicity and low cost of lime nullify these advantages. Lime performance may also be handicapped in simulations if in practice there is a significant calcium carbonate supersaturation. The pH control range used can be optimal for one species of methane formers but not for others. The range suitable to B_2 species should be selected, since acetic acid is present in largest concentration. In any case the optimal range depends on acetic acid concentration, and thus a "universal" value of 6.8-7.0 (as commonly suggested) seems inappropriate. A study of B_2 growth rate would be very useful to indicate the best range as a function of acetic acid concentration.

Initial VSS concentration lower than normal feed values can speed up a digester start-up. Avoiding, totally or in part, primary sedimentation or dilution with water could reduce the initial VSS concentration. This would also reduce the required amounts of the base used for pH control, with lower costs and a lower toxicity risk.

Since the purpose of the initial low feed rate period in a start-up is to build up methanogenic bacteria biomass, the lower the initial feed rate, the sooner it is that a full load can be used. A lower initial feed rate also allows a successful start-up for a broader range of initial biomass and solid substrate concentrations.

Digester stability was tested with respect to step input changes in feed flow rate and/or feed concentration of volatile suspended solids. Souring due to acid accumulation following a sufficiently large step input is reproduced by the model. The results can be conveniently expressed as mean residence time versus feed of total suspended solids, where lines can be drawn separating allowed from failure-causing step inputs. The position of these lines depends mainly on the steady state of operation for the corresponding feed VSS; by increasing VSS, the stability domain increases. Based on these results, it is suggested that one should try to operate at a total suspended solids (TSS is assumed to be approximately proportional to VSS) near the maximum allowed by mixing or pumping capabilities. This also saves boiler capacity and the digester column and reduces heat addition.

Some control strategies have been proposed in the past. Essentially, the controlled variable is pH (or another equivalent variable, such as gas composition), and the manipulated variables are feed flow rate or the rate of base addition. The first is effective against low residence times but implies the construction of a holding tank or bypass of a fraction of the feed. The second is ineffective against low residence times and has some effect at high TSS, which is outside the allowed region of operation, however. Although the effect of poor mixing is not included in the model and the performance of this control scheme cannot be fully evaluated, it seems reasonable to act on the cause (high TSS) rather than on the effect (low pH). If lime were used, carbonate precipitation might increase mixing problems. Base addition alone is useful only when a large acid concentration already exists, as in the start-up of a sour digester.

The necessity to guard against high feed TSS and to operate at relatively large VSS suggests the use of a TSS control. It is assumed, as is often the case, that anaerobic digestion follows primary sedimentation, so that TSS control was applied to the settler underflow. Settler steady-state operation can be well approximated by a straight line in the digester residence time versus feed TSS plane, the slope of the straight line depending on raw wastewater flow rate and TSS. Based on the steady-state performance of control strategies, represented in Figure 7, a TSS proportional control was selected. The system can now handle a much wider range of raw wastewater conditions. For exceptional cases a pH control by flow reduction is adequate, since it acts only against large feed rates and at TSS near the set point. The resulting control scheme (Fig. 6) gives good results in a wide range of conditions.

VI. NOMENCLATURE

A	settling area; m^2
A_B	base addition rate; equivalents per L per day
B_1	acid-former population
B_2	methane-former population (acetic acid substrate)
B_3	methane-former population (propionic acid substrate)
B_4	methane-former population (butyric acid substrate)
b	weir length; m
c	enzyme adsorbed per unit mass of solid substrate; g/g
c_S	saturation value of c; g/g
C_B	total added base concentration (dissociated and undissociated); equivalents per L
C_N	total ammonia concentration (dissociated and undissociated); mol/L
C_T	total carbon dioxide concentration (dissociated and undissociated); mol/L
CCOD	biomass COD; g COD per g
$[CO_2]_D$	dissolved CO_2 concentration; mol/L
$[CO_2]_D^*$	dissolved CO_2 concentration in equilibrium with gas (from Henry's law)
e	enzyme (in solution) concentration; mol/L

e_T	total enzyme concentration; mol/L
f	TSS removal rate coefficient; m^3/day
f_j	fraction of undissociated acid; j = 2 (acetic), 3 (propionic), 4 (butyric).
F	degradable solids concentration, g/L
F	flow rate; m^3/day
F_{ov}	settler overflow flow rate; m^3/day
h	liquid level above the settler outlet weir; m
H	weir height; m
K_p	proportional gain; m^3 per L per g per day
k_{Tj}	toxicity constant; L/mol for population j
k	enzymatic synthesis constant; day^{-1}
k_d	B_1 death rate constant; d^{-1}
k_{de}	enzyme deactivation constant; day^{-1}
k_H	hydrolysis rate constant; day^{-1}
K_1	Langmuir adsorption model constant; L/g
k_2	active complex dissociation rate constant; h^{-1}
K_3	inactive complex equilibrium formation constant; L/g
K_{I2}	B_2 growth inhibition constant; mol/L
K_{I3}	B_3 growth inhibition constant; mol/L
K_{I4}	B_4 growth inhibition constant; mol/L
K_La	gas-liquid mass transfer coefficient; day^{-1}
K_{sj}	growth saturation constant of population B_j (j = 1-4); mol/L
N_A	ammonia nitrogen
PTOT	total pressure; atm
RCH4	methane production rate; mol/L per day
r_{CH4}	methane production rate; mol/L per day
r_{CO2}	carbon dioxide production rate; mol/L per day
r_d	B_1 death rate; g/L per day
r_H	hydrolysis rate; g/L per day
r_p	total volatile acid production rate; g COD per L per day
RT	ideal gas constant times temperature; L atm/mol
S_1	hydrolysis product concentration; g/L
S_2	acetic acid concentration; mol/L
S_3	propionic acid concentration; mol/L
S_4	butyric acid concentration; mol/L
T	toxic substance concentration; mol/L
T_f	toxic substance concentration in digester feed; mol/L
t*	time necessary to reach normal feed conditions in start-up procedure; day
TSS	total suspended solids
TSSj	total suspended solids concentration in settler stage j; g/L
V	settler stage volume; m^3

V_{AD}	digester volume; m^3
V_h	hopper volume; m^3
VOLRAT	digester liquid-gas volume ratio
VSS	volatile suspended solids
VSS_{max}	maximum VSS concentration in digester feed (square wave input)
WN	methanogenic biomass nitrogen content; mol/g
X_1	B_1 biomass concentration/ g/L
X_2	B_2 biomass concentration; g/L
X_3	B_3 biomass concentration; g/L
X_4	B_4 biomass concentration; g/L
X_{1m}	adsorbed enzyme-solid substrate ratio at saturation
Y_{CH4}	methane gas molar fraction
Y_{H_2O}	water gas molar fraction
YACET	acetic acid yield; mol/g COD
YBUTYR	butyric acid yield; mol/g COD
$YCHX_2$	B_2 methane yield; mol/g
$YCHX_3$	B_3 methane yield; mol/g
$YCHX_4$	B_4 methane yield; mol/g
YCOD	B_1 biomass yield; g/g COD
$YCOX_2$	B_2 carbon dioxide yield; mol/g
$YCOX_3$	B_3 carbon dioxide yield; mol/g
$YCOX_4$	B_4 carbon dioxide yield; mol/g
YNX_1	B_1 nitrogen yield; mol/g
YPROP	propionic acid yield; mol/g COD
$YX1S_1$	B_1 biomass yield; g/g of S_1
$YX2S_2$	B_2 biomass yield; g/mol of acetic acid
$YX3S_3$	B_3 biomass yield; g/mol of propionic acid
$YX4S_4$	B_4 biomass yield; g/gmol of butyric acid

Greek Letters

β	settler solids removal efficiency
μ_j	specific growth rate of species j; day^{-1}
$\tilde{\mu}_j$	maximum specific growth rate of species j; day^{-1}
τ	mean residence time
τ_{SP}	residence time corresponding to flow controller set point

Subscripts

0	digester feed values
0	settler feed values
u	settler underflow values
-	overbar refers to set point values

ACKNOWLEDGMENTS

The writing of this chapter was supported by the MIT Biotechnology Process Engineering Center and the National Science Foundation, PYI Grant CBT-8514729.

REFERENCES

Andrews, J. F. (1969). Dynamic model of the anaerobic digestion process. *J. Sanit. Eng. Div. ASCE 95*(SA1):95.

Andrews, J. F., and Pearson, E. A. (1965). *Int. J. Air Water Pollut. 9*:439.

Ashley, N. V., and Hurst, T. J. (1981). Acid and alkaline phosphatase activity in anaerobic digested sludge: A biochemical predictor of digestor failure. *Water Res. 15*:633.

Bailey, J. E., and Ollis, D. F. (1977). *Biochemical Engineering Fundamentals*. McGraw-Hill, New York.

Barker, H. A. (1956). *Bacterial Fermentations*. Wiley, New York.

Bhadra, A., Mukhopadhyay, S. N., and Ghose, T. K. (1984). A kinetic model for methanogenesis of acetic acid in a multireactor system. *Biotechnol. Bioeng. 26*:257.

Buhr, H. O., and Andrews, J. F. (1977). The thermophilic anaerobic digestion process. *Water Res. 11*:129.

Butler, J. N. (1964). *Ionic Equilibrium, a Mathematical Approach*. Addison-Wesley, Boston.

Christensen, D. R., and McCarty, P. L. (1975). Multi-process biological treatment model. *J. Water Pollut. Control Fed. 47*:2652.

Chynoweth, D. P., and Mah, R. A. (1971). Volatile acid formation in sludge digestion. *Adv. Chem. Ser. 105*:41.

Collins, A. S., and Gililand, B. E. (1974). Control of anaerobic digestion process. *J. Environ. Eng. Div. ASCE 100*:487.

Cox, D. J. (April 1984). Non-methanogenic bacterial populations in a high rate anaerobic digester. *Proc. Biochem. 19*(2):75.

Dague, G., Hopkins, R. L., and Tonn, R. W. (1970). Digestion fundamentals applied to digester recovery. Two case studies. *J. Water Pollut. Control Fed. 42*:1666.

Dalla Torre, A., and Stephanopoulos, G. (1986a). Simulation study of anaerobic digestion control. *Biotechnol. Bioeng. 28*:1138.

Dalla Torre, A., and Stephanopoulos, G. (1986b). Mixed culture model of anaerobic digestion: Application to the evaluation of start up procedures. *Biotechnol. Bioeng. 28*:1106.

Eastman, J. A., and Ferguson, J. F. (1981). Solubilization of particulate organic carbon during the acid phase of anaerobic digestion. *J. Water Pollut. Control Fed. 53*:352.

Eckenfelder, W. W., Jr., and O'Connor, J. (1961). *Biological Waste Treatment*. Pergamon, New York.

Garrison, W. E., and Nagel, C. A. (1959). A density meter to control sludge pumping. *Sewage Indust. Wastes 31*:1327.
Ghosh, S. (1981). Kinetics of acid-phase fermentation in anaerobic digestion. *Biotechnol. Bioeng. Symp. 11*:301.
Graef, S. P., and Andrews, J. F. (1974). *CEP Symp. Ser. 70*:101.
Huang, A. A. (1975). Kinetic studies on insoluble cellulose-cellulase system. *Biotechnol. Bioeng. 17*:1421.
Kleinstreuer, C., and Poweighha, T. (1982). Dynamic simulator for anaerobic digestion processes. *Biotechnol. Bioeng. 24*:1941.
Lange, N. A. (Ed.). (1956). *Handbook of Chemistry*. Handbook Publishers, Sandusky, Ohio.
Lawrence, A. W., and McCarty, P. L. (1969). Kinetics of methane fermentation in anaerobic treatment. *J. Water Pollut. Control Fed. 41*:R1.
McCarty, P. L. (October 1964). Anaerobic waste treatment fundamentals. II. Environmental requirements and control. *Public Works*, 123-126.
McCarty, P. L., Lewis, J. S., and Murdoch, W. (1963). *J. Water Pollut. Control Fed. 35*:1501.
Pohland, F. G., and Ghosh, S. (1971). Developments in anaerobic treatment processes. *Biotechnol. Bioeng. Symp. 2*:85.
Rundle, H., and Whyley, J. (1981). A comparison of gas recirculation systems for mixing of contents of anaerobic digesters. *Water Pollut. Control 80*:463.
Tanthapanichakoon, W., and Himmelblau, D. M. (1981). Simulation of a time dependent activated sludge wastewater treatment plant. *Water Res. 15*:1185.
Thereoux, R. J., and Betz, J. M. (1958). *Sewage Indust. Wastes 30*:1259.
Torpey, W. N. (1955). *Sewage Indust. Wastes 27*:121.
Van Dedem, G., and Moo-Young, M. (1973). Cell growth and extracellular enzyme synthesis in fermentations. *Biotechnol. Bioeng. 15*:419.
Water Pollution Control Federation (1968). *Anaerobic Sludge Digestion, MOP No. 16*. Water Pollution Control Federation, Washington, D.C.
Weast, R. C. (Ed.) (1975). *Handbook of Chemistry and Physics*. CRC Press, Cleveland, Ohio.

19

Engineering Flavor into Fermented Foods

EBENEZER R. VEDAMUTHU / *Microlife Technics, Sarasota, Florida*

I. INTRODUCTION

A. Flavor Perception

Flavor perception in foods is a complex phenomenon. It encompasses optical, auditory, olfactory, tactile, and psychological senses in an individual. The appearance of the food to a great extent affects its flavor impact. A certain pigmentation or color is associated with a certain food and any deviation from the norm may affect its flavor perception. A burnt piece of meat, a bleached slice of tomato, or a moldy block of cheddar cheese and the lack of carbonation in a glass of beer or aerated soft drink convey subtle signals that affect flavor perception. Auditory responses influencing flavor are exemplified by the crunchiness associated with a fresh stem of celery or a spear of cucumber pickle or the crackling of a potato chip or a cracker. Olfactory responses involve both the sense of smell (aroma) and taste as registered by the nose and the palate. This is exemplified by the stimulation caused by freshly brewed coffee or the delicate bouquet of vintage wine. The tongue is the organ of taste and registers gradations and combinations of the four dimensions of taste, namely, saltiness, sourness, bitterness, and sweetness. Tactile dimension comes into play in flavor perception when mouth-feel is taken into consideration. For example, the smoothness of sour cream or cream cheese as contrasted with the graininess of "grana" cheeses like Parmesan or the slickness of Finnish viilli plays a role in the totality of flavor perception. The psychological dimension of flavor perception, although very important, is often neglected. The psychological factor consists of both internal and external influences. Internal psychological factors are conditioned by the mental blueprint of the flavor profile of a particular food and the cultural mores of the individual. Sometimes, the responses are conditioned by previous pleasant or unpleasant encounters with specific foods. External influences include the temperature of the food or drink at the time of serving, the cleanliness of the surroundings, "the atmosphere" of the serving place, serving etiquette, and the aesthetics comprising the decorative aspects of culinary art. In short, flavor perception is a multidimensional function, and the flavor of a food is the sum of several sensory responses to the food and the environment. These aspects, including the physiology of flavor perception, are discussed in greater detail by Margalith (1981).

B. Flavor Chemistry

Flavor chemistry is an exacting science and involves the use of sophisticated analytic tools that can measure specific chemical compounds present in food in nanogram quantities or less. Even with all the latest sophisticated analytic tools, definition of the flavor of a specified food through chemical analysis is very difficult. Chemical analysis of flavor extracts or concentrates of a specific food in most instances reveals the presence of several compounds, some of which possess very strong odors and flavors; others range from neutral to weak or very weak odors and flavors. In many instances, it is not known if all the identified compounds in a flavor concentrate are necessary for the total flavor impact of the food. Further, there is a lack of true quantitative data on all the compounds, their respective threshold values, synergistic interactions, and other factorial effects on the various identified compounds, such as partitioning factors, lipid solubility, adsorption of flavors by other ingredients, and the degree of dispersion of the aqueous phase in the "native system" as they occur in the food (Kinsella, 1975). Additionally, in the preparation of flavor concentrate or extract for chemical analysis, the distribution of the flavor components in the native system is destroyed or their physico-chemical characteristics may be altered.

The major advantage of chemical analysis is that it affords a means to determine the identity of the dominant notes in a flavor profile, which could through a series of permutations, combinations, and formulations with other less dominant flavor compounds be used to reproduce the "natural flavor" in a bland, neutral food base. Such artificial formulation of flavor is a very exacting science and needs considerable training and experience.

C. Food Flavors

1. *Flavors of Natural Foods*

Flavors of natural foods or food materials can be categorized into fresh (or raw) food flavors and modified food flavors. The flavor of fresh, raw foods in most instances is rather delicate and pleasant, such as the flavor of fresh mushrooms, celery, or raw oysters, the fine blend of acidity and sweetness as in ripe tomato or grapes, the mild sweetness of freshly drawn milk, or the refreshing sourness of lemon.

2. *Flavors Introduced by Processing*

Modifications of the fresh flavor of foods occur as a result of cooking or other means of processing or preservation, such as drying (or concentration of solids), salting, acidification through direct addition of edible acids (e.g., vinegar), biologic acidification through fermentation, and smoking. During cooking, heat denaturation of proteins, caramelization of sugar components, and the protein-sugar interactions resulting in the formation of browning compounds modify the "native, natural flavor" of raw foods.

Processing raw foods not only aids the preservation of valuable nutrients in the food, but also through modification provides variety in foods. Food preparation may involve subjecting food to physicochemical processes, such as heating, drying, soaking in vinegar, dry salting or brining, and smoking. The bioprocessing of certain foods may entail the use of free or immobilized enzymes or fermentation employing intact microorganisms. Flavor enhancement and modification through the use of free enzymes is exemplified by the application of glandular lipase preparations to produce piquant characteristics in Italian cheeses or in the acceleration of flavor development in enzyme-modified cheeses. The use of meat-tenderizing enzymes also falls under this category. Food modifications through the use of microorganisms is an age-old practice and has several advantages over using free enzymes. Shahani et al. (1976) summarized the advantages of using microorganisms in food systems:

> Food flavors are dependent upon odorous compounds derived singly or in combinations from more complex precursors, including proteins, fats, carbohydrates, vitamins and essential oils. Inherent enzymes in food systems or enzymes involved in the metabolism of food constituents by microorganisms in foods play an important role in the development of food flavors. . . .
> Use of microorganisms in the processing of foods has the advantage that microorganisms contain many enzymes which can act in a sequential manner to produce numerous flavor compounds.

Such sequential reactions cannot be achieved by a single enzyme. Additionally, it is more economical to use microbes than a number of purified enzymes.

D. Microbially Induced Flavor Development in Food Systems

The application of microorganisms in food processing and flavor enhancement or modification may involve the direct use of microorganisms in the food or the use of flavorful substances derived through microbial fermentation to enhance flavor(s) or introduce new flavors into the food system. The use of starter cultures in fermented foods illustrates systems in which the direct use of microorganisms is practiced. The flavoring of nondairy spreads (margarine) with "starter distillate" or the fortification of food systems in which fruitlike flavors are required with "fruity" esters derived from microbial fermentation represents the indirect application of microorganisms.

Another facet in microbial enhancement or modification of food flavors relates to prefermentation treatment of the basic food systems to render them "fermentable." Such systems require two-stage fermentations to arrive at the desired end product. The first stage modifies the basic components in the food system to render them accessible to and metabolizable by the microbes used in the major fermentation. Such prebioprocessing is necessary because the desirable microorganisms involved in the second-stage fermentation do not have the enzymes necessary to break the basic components in the food system, such as starch, native proteins, or fats and oils. The use of *Aspergillus* spp. to prepare soybean (koji process) for secondary halophilic, acidifying, and alcoholic fermentation by pediococci and yeasts in the preparation of Oriental soybean sauces and pastes represents such preparative bioprocessing. Table 1 summarizes some of the well-known areas of application of microorganisms in various food and feed commodities in the preservation and improvement of palatability and flavor.

In discussing microbiologically engineered flavoring of foods, it should be mentioned that certain desirable flavors induced by microorganisms in certain foods may be undesirable in another food system. Diacetyl, a diketone formed as a result of citric acid fermentation by "dairy aroma bacteria," is an important flavor component imparting nutmeatlike flavor in cultured creamery butter, sour cream, buttermilk, bread, and coffee. Diacetyl, however, is troublesome off-flavor-causing component in beer and wines. The characteristic flavor of plain yogurt is due to relatively moderate levels of acetaldehyde derived from milk components during fermentation. The presence

Table 1 Use of Microorganisms in the Preservation and Flavor Modification of Food and Feed Materials

Food or feed commodity	Food, beverage, or feed product produced	Predominant microorganisms involved in fermentation and flavor production
Cereals	Leavened breads	Yeasts, primarily *Saccharomyces cerevisiae*
	Sourdough breads	Yeasts, *Lactobacillus* spp.
	Crackers	Yeasts, lactic acid bacteria, primarily *Lactobacillus* spp.
	East Asian leavened steamed cakes (idili)	Lactic acid bacteria, especially *Leuconostoc* spp.; yeasts occasionally involved
	Beer (oats), whiskey (corn), sake and Oriental beers (rice)	*S. cerevisiae* and other *Saccharomyces* spp.; *Lactobacillus delbrueckii* in sour mash whiskey
Legumes	Oriental products made from soybeans, split peas, lentils, and other leguminous pulses	*Aspergillus* spp., lactic acid bacteria comprising *Lactobacillus* spp., pediococci, *Leuconostoc* spp.; various yeasts, *Rhizopus* spp., *Mucor* spp., *Bacillus* spp. (Natto)
Tubers	Vodka (from potatoes); beerlike products of Africa (from cassava)	*Saccharomyces* spp.
Vegetables	Pickles (cucumbers, okra, olives, fresh peas, beets, turnips, radishes, brussels sprouts, mustard leaves, lettuce, vegetable blends)	*Lactobacillus plantarum* and other lactobacilli; *Pediococcus* spp., *Leuconostoc* spp.

(Table 1, continued)

Food or feed commodity	Food, beverage, or feed product produced	Predominant microorganisms involved in fermentation and flavor production
Fruits (pulp and juices)	Wines and ales (grapes and various berries)	*S. cerevisiae* and *Saccharomyces* spp.
	Vinegar	*Acetobacter aceti*
Milk	Cultured milks, yogurt, kefir, koumiss, cultured creamery butter, sour cream, cheeses	Lactic acid bacteria; mainly lactic streptococci, lactobacilli, *Leuconostoc* spp.; *Propionibacterium* spp.; yeasts and certain molds (*Penicillium* spp.)
Meats	Sausages (beef, pork, poultry, game meat)	*Lactobacillus* spp., *Pediococcus* spp., *Micrococcus* spp., certain yeasts and molds
Fish	Oriental fish sauces, pastes, and pickles; fish silage	Lactic acid bacteria, *Bacillus* spp.
Plantation crops	Palm-sap wines	*Zymomonas* spp., *Saccharomyces* spp., lactic acid bacteria
	Tea	Lactic acid bacteria
	Cocoa	Lactic acid bacteria
Food yeasts	Yeast cakes, dried yeasts for leavening, flavoring, and nutrient fortification	*Torula* spp., *Klyveromyces* spp., *Saccharomyces* spp.

of acetaldehyde at similar concentrations in sour cream or cultured buttermilk leads to a defect known as "green flavor." The sharp, peppery ketonic flavor derived through microbial oxidation of milk fatty acids in Roquefort and blue cheeses is something to be avoided in cheddar cheese (Vedamuthu, 1979).

The economic and nutritive benefits of fermented foods in human diet were eloquently stated by Djien (1982):

> Basically, fermented foods are agricultural products which have been converted by enzymic activities of microorganisms into desirable food products whose properties are considered more attractive than those of the original raw materials. Moreover, if the manufacturing procedures are properly followed, the foods are usually safe for consumption. All of these beneficial properties of the final product increase the economic value of the original agricultural commodity.
>
> Generally, traditional methods of manufacturing fermented foods are not complicated, and expensive equipment is not required. Therefore, fermentation of indigenous foods is considered as an inexpensive and effective means of food production that could be utilized in alleviating World food problems.

Fleming (1982) highlighted the importance of fermentation in the preservation of vegetables:

> Brining and fermentation were primary methods for preserving vegetables throughout the World prior to the advent of canning and freezing. Although secondary to modern preservation methods in Western civilization, brining and fermentation remain important methods for preserving certain vegetables in highly developed countries because they: (1) impart certain desired organoleptic qualities in the products; (2) provide a means for extending the processing season for fruits and vegetables; and (3) require comparatively little mechanical energy input, a fact that enhances the potential for these age-old methods of preservation in our modern energy-sensitive world.

From the foregoing introduction, it is evident that flavor development and modification in basic food commodities through the use of microorganisms is a convenient and economical process.

II. ROLES OF MICROORGANISMS IN FLAVOR DEVELOPMENT IN FOODS

The different roles that microorganisms play in the preservation and flavor development in foods are summarized in the following sections (see Table 2).

Table 2 Role of Microorganisms in or Added to Food Systems in the Manufacture and Flavor Development of Fermented Foods

I. Preprocessing treatment using microorganisms to prepare foodstuffs for primary fermentation or to enhance final flavor
   A. Preparative biomechanical processing, e.g., coffee beans
   B. Preparative biochemical processing, e.g., soy products, sour mash

II. Primary fermentation to develop or enhance specific flavors
   A. Direct role
      1. Desired fermentative modification of food components
      2. Pure culture fermentation
      3. Mixed culture fermentation involving associative and/or synergistic activities, or sequential fermentations by different microbial groups
      4. Adventitious microflora and their role
   B. Regulatory role: regulation of microflora
      1. Regulation of desirable microorganisms
      2. Regulation of undesirable microorganisms—spoilage flora, pathogens

III. Secondary fermentation to improve product flavor and quality

IV. Stabilization of food system to deterioration

A. Preprocessing Microbial Treatment

The preprocessing microbial treatment of food materials, referred to earlier, comprise preparative procedures that indirectly or directly affect the final flavor of the food. In some cases, preparative bioprocessing does not materially contribute to flavor per se but is superior to alternative chemical, mechanical, or physical treatments that may be necessary to achieve the same desired preparative treatment. In other cases, preparatory bioprocessing involves primary modification of the food material to render it fermentable by the microorganism(s) used in the main flavor development process.

1. Preparative Biomechanical Processing

The best example of this kind of treatment is coffee fermentation. The fermentation of coffee beans does not directly contribute to the flavor of coffee. The flavor of coffee to a large extent is the result of roasting.

Coffee fruit is classified botanically as a drupe. A drupe is a fruit consisting of a pulpy, leathery, or fibrous epicarp and a hard endocarp enclosing a single seed. The coffee drupe consists of two halves of the bean fused together along the flat sides and surrounded by a thin membrane called the "silver screen." The bean and the silver screen are protected by a hard, horny endocarp called the parchment. The parchment is covered by a pulpy, mucilagenous mesocarp, which in turn is enclosed by fruit skin or pulp (exocarp). The primary objective in coffee processing to the "green bean stage" is dehydration of the bean to a point at which biologic activity is minimized. Dehydration of coffee beans may be achieved by drying directly without pulping (dry processing) or by a three-step process consisting of pulping, fermenting, and drying (wet processing). Pulping refers to the removal of coffee skin. Dry processing of coffee is advantageous only if there is difficulty in disposing of large amounts of wastewater. Wet processing is the most widely used procedure, accounting for 50% of the coffee produced around the world. The popularity of wet processing is due to its simplicity and cheapness. Simple processes are suitable for underdeveloped countries, where much of the world's coffee crop is grown. Additionally, the wet process yields superior quality beans compared with the dry process (Arunga, 1982).

Fermentation of the fruit is the most important step in wet processing. Coffee fermentation is the process by which the mucilaginous mesocarp surrounding the coffee parchment is degraded by enzymes from the fruit tissue itself and from the microorganisms contaminating the fruit. Fermentation achieves the removal of the sticky mucilaginous layer (which makes up 20-25% of the wet weight of the fruit) for quick drying of the parchment. The clean removal of the pulp improves the raw appearance of beans. Both these aspects have a bearing on the flavor of the coffee. From an economic viewpoint, the simplicity and cheapness of fermentation is ideal for a product that is subject to unpredictable price fluctuations.

Mechanical demucilagers are available but require too much motive power and water. Some of the low-power-consuming machines are difficult to adjust or calibrate for differently sized beans. Mechanical demucilage treatments require handling dangerous acids, alkali, and hot water, which are corrosive to equipment and handlers. Also, the disposal of chemicals is a problem.

At the end of natural fermentation, which may last for 24-90 h at pH 5.0-6.0, mucilage detaches from the parchment and is washed off. At this point, the parchment no longer feels slimy. The organisms involved are contaminating yeasts, *Erwinia* spp. and pectinolytic *Enterobacter* spp. Some of the pectinolytic enzymes in the fruit may also be involved. In coffee fermentations, improvements using a trickle-bed fermentation involving recirculation of broth rich in microorganisms and enzymes have increased efficiency. For detailed discussion of these aspects, the reviews by Arunga (1982) and Castelein and Verachtert (1983) should be consulted.

2. *Preparative Biochemical Processing*

Preparative biochemical processing, referred to earlier, involves the actual breakdown of foodstuffs by microorganisms used in the prefermentation stage that make the food components degradable by the microorganism involved in flavor development.

In the production of soy sauce, the starting raw materials are soybeans and grains, such as wheat. Whole soybeans can be used, but the oil must be removed at a later stage to ensure quality. Alternatively, defatted soybean meal or flakes are used. Soybean meal or whole beans are first soaked in water at ambient temperatures for 12-15 h until they swell and increase in weight. During the soaking period, water is circulated or changed often to prevent the growth of undesirable spore-forming *Bacillus* spp. After sufficient soaking, the beans or the meal is boiled or steamed until soft and rapidly cooled to 40°C. Simultaneously, wheat, wheat flour, or wheat bran is roasted to obtain some of the aroma and flavor and the brown color of soy sauce. The roasted wheat kernels are crushed and mixed 1:1 with the mashed soybean portion. The wheat-soybean mixture mainly consists of starch, complex proteins, and oils. Microorganisms involved in the main fermentation leading to the characteristic flavor of soy sauce consist of lactic acid bacteria and yeasts that do not possess the amylases, proteases, and lipases necessary to break down the complexes in the wheat-soybean mash. The mash is thus subjected to a preparative microbial degradation using an inoculum called koji. According to Beuchat (1983), the word *koji* means *bloom of mold* and refers to the enzyme preparation produced on cereals or sometimes pulses and used as a seed or starter for larger batches of plant seed substrates when making a number of traditional fermented foods. In Japan, koji is prepared by culturing a mixed strain starter consisting of *Aspergillus oryzae* or *Aspergillus sojae* on steamed, polished rice. In China, koji is made from a mixture of wheat bran and soy flour. The mold strains used in koji usually have high amylolytic and proteolytic properties.

Another example in which preparative bioprocessing comes into play is in the manufacture of distilled beverages, such as sour mash whiskey. In this system, preparatory fermentation makes the substrate selective and conducive to the primary yeast fermentation. Without the preparatory fermentation, the natural contaminants in the mash, especially the spore-forming and the psychrotrophic bacteria, dominate and impair the desirable yeast fermentation as well as cause off-flavors. The preparative fermentation is carried out with a pure culture inoculum of *Lactobacillus delbrueckii*. This lactic acid bacterium is preferred because it has a high optimum growth temperature (about 50°C), which inhibits mesophilic and psychrotrophic contaminants. Further, *L. delbrueckii* produces high concentration of lactic acid within a relatively short incubation time and, because it is a homofermentative bacterium, does not produce other volatile end products that affect distillate quality. The accumulation of acid and rapid depression of pH caused by the preparatory *L. delbrueckii* fermentation renders the sour mash selective for alcoholic fermentation. For a mash with a high yeast count, a 1:1 mixture of barley malt and rye mash is adjusted to 50°C and inoculated with 0.2-0.4% (v/w) of a 16 to 18 h culture of *L. delbrueckii*. The inoculated mash is incubated at 50°C until the pH drops to 3.8-4.0. At the desired end point, the mash is pasteurized by heating to 86°C and holding at that temperature for 30 minutes. At the end of the holding period, the mash is cooled to 28°C and inoculated with the yeast for the alcoholic fermentation (Bluhm, 1983).

B. Primary Fermentation to Develop or Enhance Specific Flavors

1. Direct Role of Microorganisms in Flavor Development

In most fermented foods, the main fermentative process has a direct role in the development of flavor. Flavor development occurs as a result of modification and/or degradation of basic food components, namely, the carbohydrates, proteins, and fats by microbial metabolic activity.

Fermentation in a physiological sense is anaerobic respiration in which, during the oxidation of an energy-yielding substrate, the final hydrogen acceptor is a short-chain carbon derivative of the original substrate. For example, in lactic acid fermentation, in the oxidation of simple carbohydrates, such as glucose, the final hydrogen acceptor is pyruvic acid, a three-carbon catabolite of glucose. In alcoholic fermentation, similarly, the final hydrogen acceptor is the two-carbon catabolite, acetaldehyde. In fermentative respiration, there is only substrate-level phosphorylation and the net energy yield in terms of high-energy bonds (ATP) is much lower than in aerobic systems. In general, fermentative microorganisms, unless they are facultative, lack catalase and are either anaerobic or microaerophilic.

Apart from fermentative reactions involving carbohydrates, many of the microorganisms involved in flavor development in foods have proteolytic and lipolytic enzymes that also contribute to the pool of flavor compounds.

The major steps involved in microbial flavor development processes in food materials are summarized in Table 3.

Table 3 Major Steps Involved in Microbial Development of Flavor in Food Materials

Selection of raw materials
Processing or preparing raw material
Selection of fermentative microorganisms
Fermentation: optimum conditions; fermentative pathways contributing to food flavors
Curtailing fermentation
Postfermentative processing and preservation

a. Selection of raw materials. The selection of raw materials is extremely important in obtaining a high-quality end product in terms of body, texture, flavor, and shelf life. The raw materials chosen should be chemically and microbiologically of the best quality.

The chemical composition of the raw material used should be normal, and there should be no imbalances in the various chemical components due to adulteration, enzymatic or microbiologic deterioration, or inherent physiological changes associated with the food substances. For example, the composition of cow's milk varies during the lactation period. Early in lactation, for at least 5 days after calving, colostrum is secreted by the mammary glands. Colostrum is quite different in chemical composition from normal milk. It is high in whey proteins, especially immunoglobulins, low in casein, high in lactose, low in chlorides, and easily heat coagulable. Milk tainted with colostrum is unsuitable for fermentation because the antibodies in colostrum may cause clumping of starter bacteria (agglutination) or inhibit starter strains. Sometimes the chemical imbalance found during the postparturition period may persist longer into the lactation period. Late in lactation (closer to the drying of the cow's milk flow), the mineral balance in milk is affected, which influences the rheological properties of curd made from milk.

The handling and storage of a carcass after slaughter affects the pH of the meat, the microbial content and the degree of proteolysis. These aspects affect the quality of fermented sausages made from meats.

Vegetables and forage are prone to autolytic and mechanical damage that makes them vulnerable to microbial deterioration. Such damaged products should be avoided in preparing fermented foods and feeds. The stage of harvesting affects the chemical composition of plant tissue, especially the sugar and fiber contents, which have a direct bearing on the fermentability of the material and the texture and body of the finished product.

In the selection of raw materials for vegetable fermentation, it is important to choose new varieties that have been developed to withstand mechanical damage from current harvesting and handling practices. New hybrid varieties of cabbage with a high solids content have been developed to be more amenable to current harvesting and agronomic practices. Caution should be exercised, however, in the proper selection of hybrid cabbages, because certain varieties appear to contain inhibitors and are deficient in nutrients that impair fermentation. Similarly, newer genetically altered varieties of cucumbers are available that give a uniform end product (Stamer, 1983).

b. Processing or preparation of raw materials. Raw materials for fermentation must usually be processed or treated to make them suitable for vigorous fermentation. The aim of most fermentative processes is to rapidly lower the pH by the accumulation of acid from carbohydrates in the substrate to stabilize the system such that the controlled development of flavor and sequential changes in body and texture occur. Raw agricultural commodities contain the natural ecological microflora and foreign microbial contamination introduced during harvesting, handling, storage, and transportation. If raw materials are not treated before the initiation of desired fermentation, the mixed microflora cause uncontrolled changes that are either undesirable or deteriorative. The preparation of raw materials usually involves treatments that favor predominance of the desired fermentation or to a large measure eliminate the biologic competition for the microflora used in the fermentation.

Another purpose for pretreatment is to destroy naturally occurring inhibitor(s) in the food commodity. In certain cases, preparation involves cutting, shredding, mashing, heating, and so on, to make them suitable for fermentation (for example, releasing the nutrients within the tissues) and to provide the characteristic physical appearance of the finished product.

For the preparation of cultured milks or cheese, raw milk is heated to varying degrees to eliminate natural inhibitors in milk, bacteriophages, and a major portion of the raw milk microflora to provide a suitable environment for the starter culture to grow and produce the *desired* changes. For cultured buttermilk and yogurt, the heat treatment is quite severe (82–87°C for 30–45 minutes) to promote interaction between casein and whey protein to obtain a desirable heavy body, to eliminate natural inhibitors in milk, to stimulate the active growth of starter bacteria due to reduced oxygen tension, formation of stimulatory peptides from milk proteins, and destruction of bacteriophages and competing microflora. For cheeses, milk is subjected to either pasteurization (78°C for 16 s or 63°C for 30 minutes) or subpasteurization (63°C for 17 s) heat treatment to destroy pathogens and considerably reduce the numbers of spoilage, pathogenic, and gas-producing microflora. Milk for certain cheeses may be subjected to hydrogen peroxide-catalase treatment to achieve the same goals.

For sausage manufacture, meat is ground and mixed with spices, fermentable sugar(s), nitrite (curing salt), common salt, and starter culture before stuffing into the casing. The fermentable sugar added to the meat mixture, usually glucose or sucrose, provides a ready source of energy for the starter bacteria, which allows them to predominate in the mixed flora and produce lactic acid. Salt, apart from its role as a flavoring agent, also acts as the selective principle against the contaminant flora of a meat mixture and exerts a preservative effect at the low water activity (a_w) obtainable in the finished sausage. Starter bacteria used in meat fermentations thrive well at the salt-brine concentration of 4-6% in the water phase of the meat mixture.

Vegetables are usually washed to remove soil and dirt and to reduce the microbial load. Further reduction of contaminant bacteria is sometimes achieved with a mild chlorine rinse. Olives contain oleuropein, an extremely bitter phenolic glucoside that is also inhibitory to the lactic acid bacteria involved in olive fermentation. Debittering of olives is achieved by treatment with 1-2% NaOH, which is allowed to penetrate about three-quarters of the distance to the pit. This treatment also inactivates the inhibitory principles. Further treatments may include piercing to allow the gaseous end products of fermentation to escape, which otherwise would cause splitting of the flesh or bloaters. Brining or salting precedes the fermentation step, which provides a selective environment for the desirable lactic acid bacteria.

In sauerkraut production, dry salting of cabbage tissues facilitates the rapid extraction of nutrients from the plant cells necessary for the rapid growth of starter bacteria or desirable indigenous lactic acid bacteria. Additionally, salt provides selection against undesirable microflora, imparts textural integrity to the plant tissue, and also contributes to flavor.

In wine manufacture, it is necessary to wash the grapes well to remove pesticide residues, which affect alcoholic fermentation. The addition of sulfite to grape juice creates a selective and favorable environment for wine yeasts. In some instances, "amelioration" (addition of cane or beet sugar to raise the sugar content of the must) of the grape must is practiced.

c. Selection of fermentative microorganisms. Starter, as defined in the food industry, refers to carefully selected microorganisms that are deliberately added to the primary food commodity to initiate and carry through the desired fermentation process. Functions of starters in fermented foods are fivefold:

Acid or alcohol production
Flavor production
Alteration of body and texture
Creation of the necessary biochemical conditions to favor desirable enzymatic reactions and regulate enzymatic reactions
Suppression of food spoilage, pathogenic microflora, and toxin formation and regulation of the activity of adventitious microflora

A study of the evolution of starter cultures reveals that, for a great number of years, the natural flora of the primary raw material or contaminants acquired during its postharvest handling was relied upon to bring about fermentation. Later, through sustained observation, it was found that when a small portion of the previous lot of fermented food was added to a new batch, the desired changes in fresh batches were faster and relatively consistent. This gave rise to the practice of "back-slopping." With the development of pure culture techniques, the predominant flora of fermented foods were isolated and added singly or in combinations to observe whether the characteristic attributes of fermented foods were duplicated. Based on these observations, culture mixtures or single strains were selected and preserved for repeated use to achieve controlled fermentations. Cultures so isolated were kept active by repeated daily propagations. Later, acid neutralizers, such as $CaCO_3$, were used in conjunction with refrigeration to keep the cultures viable. With the introduction of frozen storage facilities, cultures were held frozen with cryoprotectants. Cultures that were resistant to normal drying processes (e.g., Brewer's yeast) were preserved by drying. Lyophilization and liquid nitrogen storage followed as technology was developed for these processes.

The selection of starter strains for the food industry is based upon several functional and adaptive criteria. These include acid- or alcohol-producing activity (amount and rate), compatibility with other strains in mixed culture operation, flavor-producing potential, activity in the range of temperature used in the process, salt tolerance, acid tolerance, resistance to bacteriophages or other inhibitory principles that may be naturally present in food systems at low concentrations, and the ability to regulate and suppress the growth of pathogens and spoilage flora. It is also necessary to ensure that the strains do not contribute to off-flavors or taints in the food system.

Currently, concentrated frozen and lyophilized cultures for direct addition to product vats are available. Innovations in growth media formulation incorporating internal soluble and insoluble buffer systems and microprocessor-controlled external neutralization systems have boosted starter cell yields and their viability and functional activity. Developments in cell separation technology based on ultrafiltration and centrifugal separation and progress in cryogenic and freeze-drying processes have made starter manufacturing a specialized field. Recently, research on the genetics of food starter bacteria and the potential application of recombinant DNA techniques to these bacterial systems have added a new dimension to starter culture research and development (Gasson and Davies, 1984). For a more complete discussion of the technology of starter culture production, the reviews by Sandine (1979), Porubcan and Sellars (1979), and Vedamuthu and Washam (1983) and a recent book on bacterial starters (Gilliland, 1985) should be consulted.

The various microorganisms used in food fermentations are discussed in a review article by Smith and Palumbo (1981). Detailed discussion of the application of these organisms and their functions in various food systems are found in recent books on food fermentations (Reed, 1983; Rose, 1982).

d. Fermentation. In the production of fermented foods, the basic manufacturing steps, excluding the various technological innovations incorporated into present-day operations, have followed traditional methods. Successful traditional methods empirically incorporated conditions and processes that favored fermentations that yielded products with desirable body, texture, and flavor on a consistent basis.

In food processing involving fermentations, in many instances the fermentation substrate is not sterile and contains a variety of indigenous flora. The major objective is thus to favor the dominance of certain desirable group(s) of microorganisms in the mixed population at the expense of other flora in the system.

In some systems, conditions are manipulated to favor only one group of microflora, in others to promote the simultaneous activity of two associative or synergistic groups of microorganisms, and in yet others, to promote a sequential dominance of different microbial groups.

In the production of short-set cottage cheese, fast-acid-producing lactic streptococci consisting of *Streptococcus lactis* and (or) *Streptococcus cremoris* are used as starters. The starter strains are added to the milk at 30-32°C and held undistrubed at that temperature until an acid coagulum is formed. The setting temperature and the quiescent incubation conditions used are optimal for the lactic streptococci, and they produce the desired lactic acid at a rapid rate to yield a pleasantly acidic, firm curd at the end of cooking. In this process, only one group of organisms is used and the manufacturing conditions adjusted to favor the rapid growth and metabolic activity of these bacteria. Cultured buttermilk production represents a process in which the incubation conditions are adjusted to promote the simultaneous growth of acid-producing lactic streptococci and aroma- and flavor-producing leuconostocs. The final flavor of cultured buttermilk represents the coordinated activity of lactic streptococci and aroma bacteria. Leuconostocs are relatively inert in pure cultures in milk, but they possess the enzymes necessary for the fermentation of milk citrate to the flavor compound, diacetyl. *S. lactis* and *S. cremoris* can actively ferment lactose in milk but cannot ferment citrate. Although leuconostocs contain the enzymes necessary for citrate fermentation, the uptake of citrate via the citrate permease system of leuconostocs is active only below pH 6.0. The normal pH of milk is between 6.4 and 6.6, and to initiate citrate uptake by leuconostocs inoculated into milk, sufficient acid production to depress the pH below 6.0 by lactic streptococci is necessary. These two groups of bacteria are thus used in association in cultured buttermilk production. At temperatures above 28°C, the acid-producing streptococci grow very rapidly. Although the optimum growth temperature for leuconostocs is around 30°C, in mixed cultures in milk they cannot compete with the acid producers above 28°C. As a result the acid-producing streptococci dominate overwhelmingly in milk cultures incubated above 28°C. Under such conditions, very little or no diacetyl is produced. To obtain a fine-flavored buttermilk containing fairly high concentrations of diacetyl, a balanced growth of acid-producing streptococci and leuconostocs is promoted by incubating the vat milk between 21 and 24°C.

An example of a process in which, through manipulation of temperature and promoting the depletion of specific substrate(s), a sequential dominance of different starter flora is favored, is found in Swiss cheese manufacture. In modern rindless Swiss cheese manufacture, four different groups of microorganisms are used as starters. The lactic streptococci comprise one group, and *Streptococcus thermophilus*, *Lactobacillus* spp., and *Propionibacterium* spp. form the remaining three groups. During the ripening of milk and early stages of cooking, acid development in vat milk is almost exclusively contributed by the lactic streptococci, which is followed by *S. thermophilus* after the cooking temperature attains 40°C. The dominance of *S. thermophilus* as the main agent of acid accumulation in vat milk is supplanted by *Lactobacillus* spp. beyond 49°C. The lactobacilli are inhibited above 52-53°C. During pressing of the curd under whey, as the temperature gradually falls to 37-40°C, the dormant *S. thermophilus* cells revive, actively metabolize residual sugar in the curd, and accumulate lactic acid. Most of the acid produced in Swiss cheese manufacture occurs during overnight pressing. When the pressed cheese is salted and cooled in chilled brine, the cheese temperature decreases gradually. As the temperature falls below 20°C, the dormant lactic streptococci revive and scavenge most of the residual sugar in the cheese. The acid-producing activity of lactic streptococci extends through the precooler conditioning of the cheese. The removal of metabolizable sugars to trace levels by sequential dominance of the various groups of lactic acid bacteria prevents undesirable wild fermentations in the cheese when the blocks are moved to the warm-room. During warm-room holding, the propionibacteria added to the milk dominate the fermentation and convert the accumulated lactate to propionic and acetic acids and carbon dioxide. The acetate and propionate contribute to the flavor of Swiss cheese, and the occluded CO_2 in the curd mass gives rise to characteristic eyes of the cheese.

Similar examples from other food fermentations can be cited to illustrate manufacturing steps or ecological setting that promote or curtail different starter microbial groups in the food system (Vedamuthu, 1979; Hobbs, 1986).

2. *Major Fermentative Pathways Contributing to Food Flavors*

a. Fermentation of simple and complex sugars. One of the major end products of fermentative microorganisms that contributes to the flavor of several fermented food groups is lactic acid. Lactic acid not only provides the pleasant tart, acid flavor of many foods but also plays a part in partial dehydration of basic food materials through the promotion of syneresis and also contributes to the body and texture of the finished product. Partial dehydration achieved through lactic acid development in food systems helps to stabilize the food for longer storage by decreasing the a_w of the system.

In fermented dairy products, lactic acid helps in transforming fluid milk into curd and in cheeses aids in the expulsion of whey from the curd. Further, lactic acid contributes to the tart or acid flavor of the end product and by lowering the pH, and as an undissociated molecule lactic acid exerts a regulatory and preservative influence on the microflora of the food system. In sausages (meat system), lactic acid accumulation provides the tart, acid flavor; additionally, the decrease in pH aids in drying of the sausage and developing the chewy texture of the finished product. Another important role of lactic acid is in the preservation of the meat by suppressing spoilage and pathogenic flora and toxin formation. A minor role is involved with the fixation of meat color.

In vegetable fermentation, lactic acid contributes to the acid flavor, crisp texture, and regulation of spoilage flora. In sourdough bread, lactic acid contributes to the sourness. In other food fermentations, lactic acid mainly contributes to acid flavor and regulates the flora in the system.

In most lactic acid fermentations, depending on the organisms involved, other volatile acids, such as acetic, propionic, and formic acids, and traces of ethyl alcohol, aldehydes, ketones, and CO_2 are formed. These trace components have high flavor- and odor-imparting potential. The lactic acid bacteria used in food fermentations are usually divided into homofermentative and heterofermentative types. Homofermentative types produce more than 99% of lactic acid and only traces of other products; heterofermentative types produce about 70% lactic acid and the remaining 30% is composed largely of acetic acid, ethyl alcohol, and CO_2 in almost equal proportions. According to Doelle (1981), lactic acid bacteria can be divided into three groups based upon their enzymatic makeup:

Obligate homofermenters, containing fructose diphosphate aldolase, but no glucose-6-phosphate dehydrogenase or 6-phosphogluconate dehydrogenase

Obligate heterofermenters, containing both dehydrogenases but no fructose diphosphate aldolase

Facultative homofermenters, containing all three enzymes and able to use either of the two pathways

The homofermentative dissimilation of sugars is through the hexose diphosphate pathway (HDP); heterofermentative types metabolize sugars via the phosphoketolase pathway (PK). These pathways are discussed in detail by Doelle (1981).

Propionibacteria, which are associated with propionic acid fermentation in Swiss cheese varieties, utilize sugars up to the pyruvic acid stage through the HDP. Pyruvate and lactate are converted to propionic and acetic acids and CO_2 through a series of complex reactions involving methylmalonyl-COA and a biotin-dependent transcarboxylation reaction. The propionibacteria represent a unique CO_2-fixing system among heterotrophs. Figure 1 shows the propionic acid fermentative pathway among the dairy propionibacteria. For a comprehensive review of propionic acid fermentation, Hettinga and Reinbold (1972) should be consulted.

In dairy products, the primary carbohydrate is lactose. The starter bacteria involved in dairy fermentation possess different enzymatic mechanisms to use the β-galactoside. In the utilization of lactose, the microbial cell needs to possess the specific transport mechanisms to translocate the disaccharide from the exterior into

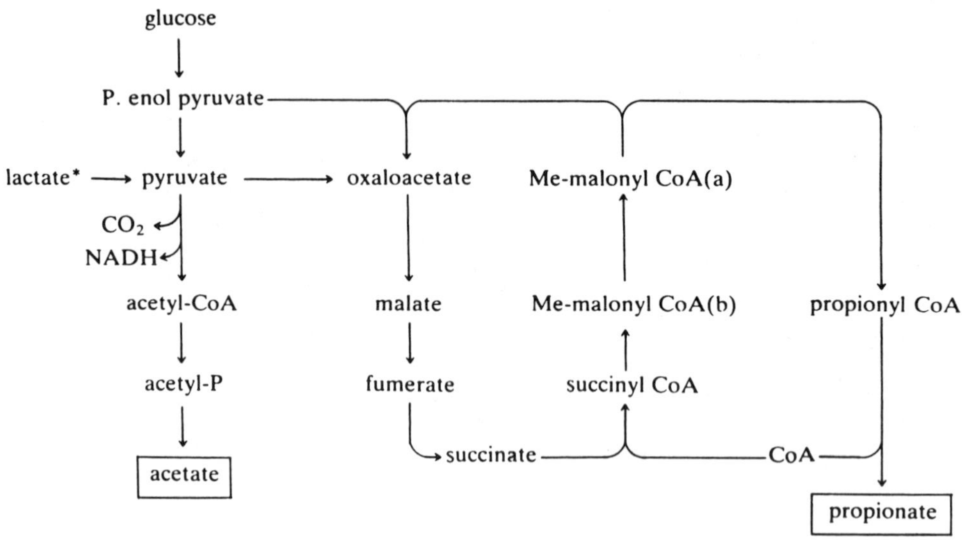

Figure 1 Formation of propionate by *Propionibacterium shermanii*. (From Garvie, 1984.)

the interior of the cell membrane, which is then cleaved to its monosaccharide components and fed into specific dissimilatory pathways. Table 4 summarizes the mechanisms for lactose utilization by various dairy starter bacteria. Figure 2 depicts the pathways for lactose and galactose metabolism in dairy streptococci. For further discussion of lactose metabolism in dairy starter bacteria, the reviews by McKay (1982), Marshall and Law (1984), and Vedamuthu (1978) should be consulted. The mechanism for lactose utilization by *Propionibacterium shermanii*, a species widely used in the production of Swiss cheese varieties, is described by Hartley and Vedamuthu (1975).

The major sugars in plant tissues consist of sucrose, melibiose, and trisaccharide raffinose. Pediococci are commonly found on plant materials, and selected *Pediococcus* spp. are used in the fermentation of vegetables and forage. Recently, the enzymatic mechanisms and genetic elements governing the utilization of sucrose, melibiose, and raffinose by pediococci were described by Gonzalez and Kunka (1986). Among dairy streptococci, certain *S. lactis* strains and *S. thermophilus* ferment sucrose. Leuconostocs, which are also widely distributed in plant materials, have sucrose-utilizing mechanisms. *Leuconostoc dextranicum* and *Leuconostoc mesenteroides*, however, use only the fructose moiety of sucrose and polymerize the dextrose component into dextrans. *Lactobacillus* species (e.g.,

Table 4 Mechanisms for Lactose Utilization by Starter Bacteria

Starter bacteria	Lactose transport system(s)	Lactose cleavage enzyme	Genetic determinant for lactose utilization	Major carbohydrate metabolic pathway	Major end products
Streptococcus cremoris	PEP-phosphotransferase	P-β-gal[a]	Plasmid borne	HDP[f]	Lactic acid
S. lactis	PEP-phosphotransferase	P-β-gal	Plasmid borne	HDP	Lactic acid
S. lactis subsp. *diacetylactis*	PEP-phosphotransferase	P-β-gal	Plasmid borne	PK[g]	Lactic acid, acetic acid, and CO_2
S. thermophilus	β-Galactoside permease	β-gal[b]	—	HDP	Lactic acid
Homofermentative[c] lactobacilli	β-Galactoside permease, PEP-phosphotransferase, or both	β-gal, P-β-gal, or both	Plasmid borne in some species[d]	HDP	Lactic acid
Leuconostoc cremoris[e]	Not known	Not known	Not known	PK	Lactic acid, acetic acid, CO_2, and ethanol

[a]P-β-gal: phospho-β-galactoside galactohydrolase.
[b]β-gal: β-galactosidase.
[c]*L. bulgaricus*, *L. helveticus*, *L. casei*, and *L. lactis*.
[d]Reported for *L. casei* and *L. bulgaricus*.
[e]Very little acid produced in milk.
[f]Hexose diphosphate pathway.
[g]Phosphoketolase pathway.
Source: From Vedamuthu and Washam (1983).

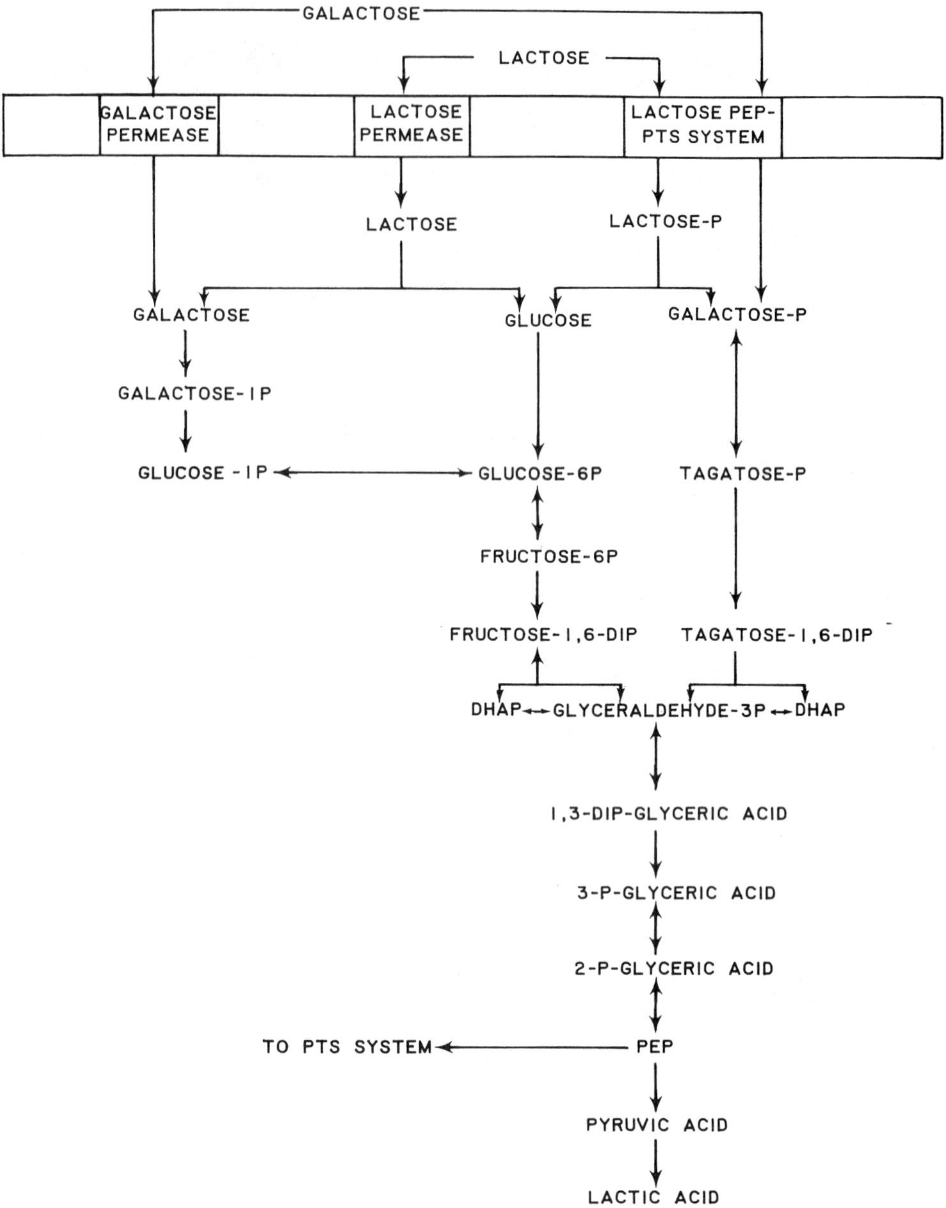

Figure 2 Pathways for lactose and galactose catabolism in dairy streptococci. (From McKay, 1982.)

Lactobacillus plantarum and *Lactobacillus casei* varieties) used in vegetable fermentations also utilize sucrose and the pentose, ribose.

In flour systems, the free sugars consist of glucose, fructose, sucrose, and maltose and glucofructosans (Oura et al., 1982). In sourdough fermentations, the important lactic acid bacteria are *Lactobacillus* spp., *Leuconostoc* spp., *Pediococcus* spp., and streptococci. Both homofermentative and heterofermentative activities are required in the production of high-grade sourdough breads. In San Francisco sourdough bread, a unique group of lactobacilli called *Lactobacillus sanfrancisco* are involved in the fermentation. These bacteria preferentially use maltose over other sugars in the flour system and exist in symbiotic relationship with *Saccharomyces exiguus*, a yeast that does not metabolize maltose. *L. sanfrancisco* is reported to metabolize maltose by the maltose phosphorylase pathway (Wood and Rainbow, 1961). The end products of maltose metabolism by *L. sanfrancisco* comprise 70-80% lactic acid and about 20-30% acetic acid.

In the fermentation of meat emulsions for sausage manufacture, *Pediococcus* spp. and *Lactobacillus* spp. are mainly used for rapid lactic acid accumulation. The pediococci and the lactobacilli used in sausage systems are homofermentative. Heterofermentative lactic acid bacteria are not used as starters in vegetable or meat fermentations because heterofermentative production of CO_2 from sugars would cause gas pockets, splitting of intact vegetable tissue or stuffed sausages, and in extreme cases even explosion of sausages if the gas pressure within the casing is too high. In the production of cheeses, the presence of high numbers of heterofermentative types in milk causes floating curd during cooking and gas fissures, pockets, and splits in molded cheese. The addition of selected heterfermentative lactic acid bacteria is practiced in the manufacture of blue mold cheeses, for which a gas-induced open texture is desirable to promote extensive veining and luxuriant growth and pigmentation of *Penicillium roqueforti*.

Among the minor metabolic end products of lactic acid fermentation, volatile fatty acids, ethyl alcohol, carbonyl compounds, such as acetaldehyde and acetone, and CO_2 are important for their flavor potential. Traces of acetic, formic, and propionic acids are important in the total flavor impact of cultured milks. Sufficient amounts of these acids are produced by both homofermentative and heterofermentative lactic streptococci and lactobacilli to bring out the total flavor profile in cultured milks. Excessive amounts of volatile acids, however, are undesirable in fermented dairy products. In vegetable fermentations, with the exception of sauerkraut and possibly olive fermentation, only traces of volatile acids are produced. In sauerkraut and olive fermentation, the heterolactic dissimilation of carbohydrates results in the accumulation of higher amounts of the volatile acids that contribute to the characteristic flavor of these products. In meat fermentations, excessive amounts of volatile acids cause flavor defects and because homolactic fermentative bacteria are used as starters, very small amounts of volatile acids are produced. The presence of volatile acids above trace levels in wines is an indication of faulty fermentation. In sourdough bread and cracker fermentation in which a combination of homolactic and heterolactic fermentation is desirable, acetic acid imparts sharp tartness and flavor in the finished product. In Oriental cereal, legume, and fish fermentation, a complex flora of mixed lactic acid bacteria contribute to the formation of lactic acid, acetic acid, and other volatile acids, all of which are important to the flavor of the finished products.

Ethanol is produced in small amounts by the aroma bacteria used in dairy fermentation and in heterolactic fermentation. *Leuconostoc* spp. especially are reported to possess high levels of alcohol dehydrogenase. Ethyl alcohol is considered an essential flavor component in cultured buttermilk. Traces of ethanol are found in various cheese varieties and may be involved in the totality of flavor in cheese.

Although acetaldehyde is undesirable in high concentrations in cultured creamery butter, cultured buttermilk, sour cream, cream cheese, cottage cheese, and other dairy products, it is an essential and characteristic flavor component in yogurt. Acetaldehyde is very volatile and has a high flavor potential at very minute concentrations. Certain strains of *S. lactis* and *S. lactis* subsp. *diacetylactis* produce relatively high concentrations of acetaldehyde (Keenan and Lindsay, 1966). Such strains should be avoided in products in which acetaldehyde is considered to impart "green" flavor defect. *Leuconostoc cremoris* scavenges acetaldehyde by converting it into ethanol and hence is the aroma bacteria of choice in cultured buttermilk and sour cream, for example. For excellent yogurt production, selected strains of *S. thermophilus* and *Lactobacillus bulgaricus* capable of good acetaldehyde production during symbiotic growth are necessary (Bottazzi and Dellaglio, 1967).

Acetaldehyde is primarily derived from lactose or other sugars in trace amounts in homolactic and relatively higher amounts in heterolactic fermentations. Acetone is also found in trace amounts in certain lactic fermentations, and the ketone also has important flavor potential in fermented products. Propionibacteria are also reported to produce acetaldehyde and propionaldehyde from carbohydrate dissimilation.

In cucumber fermentation using *Lactobacillus brevis*, *L. plantarum*, *P. cerevisiae*, and *L. mesenteriodes*, Aurand et al. (1965) found increased amounts of formaldehyde, acetaldehyde, acetone, ethyl alcohol, propionaldehyde, and butyraldehyde compared with inoculated control batches. All these compounds are odoriferous and flavorful.

Carbon dioxide plays an essential role in the flavor impact of cultured buttermilk. The gas is entrapped in the thickened milk and provides effervescence to the fermented milk; its flavor impact is akin to that of carbonation in soft drinks. Carbon dioxide is derived from lactose by heterolactic bacteria. The accumulation of CO_2 during warm-room growth of propionibacteria in Swiss cheese contributes to the eyes in the cheese. Carbon dioxide formed during propionic acid fermentation diffuse to weak spots in the kitted curd and from the eyes. The precooler conditioning of the cheese prior to warm-room holding provides elasticity to the cheese curd to allow the formation of rounded eyes.

Minor metabolic products of lactic acid fermentation, although found only in small or even trace concentrations, are important in maintaining a desirable flavor balance in cultured milks and cured cheeses. Any shift in the flavor balance may result in the organoleptic perception of off-flavors. These aspects are covered in reviews by Vedamuthu and Washam (1983) and Sandine et al. (1972).

b. Fermentation of citric acid. The characteristic flavor of fermented dairy foods, which is often described as nutmeatlike, is contributed by diacetyl. Diacetyl is a diketone and is one of the most potent flavor compounds known. At high concentrations, the dicarbonyl has an overpowering coarse odor and flavor, but at low concentrations (about 3-5 μg/g), it imparts a fine, delicate flavor. Diacetyl is derived from the fermentation of native citrate in milk. Fluid milk contains 0.2% citrate on the average but exhibits seasonal fluctuations in citrate concentration. Among dairy starter bacteria, *Leuconostoc* spp. and *S. lactis* subsp. *diacetylactis* possess the enzymes necessary for the fermentation of citrate to diacetyl and other neutral four-carbon compounds. The uptake of citrate in these organisms is mediated by a specific permease that is active below pH 6.0. *Leuconostoc* spp., because of their inability to actively ferment lactose in milk, do not produce diacetyl in pure cultures. In mixed cultures with non-citrate-fermenting *S. lactis* and/or *S. cremoris* strains, because of the active lactose-fermenting ability of the acid-producing lactic streptococci, the leuconostocs are able to utilize citrate actively below pH 6.0 and accumulate diacetyl. *S. lactis* subsp. *diacetylactis*, however, possess mechanisms for both lactose and citrate utilization and hence are capable of diacetyl production in pure cultures.

Table 5 summarizes the genetic and enzymatic systems for citrate utilization among aroma bacteria. Figure 3 depicts the pathway of diacetyl biosynthesis in these bacteria. Other dairy starter organisms

Table 5 Enzymatic Mechanisms for Citrate Utilization by Flavor Bacteria in Starters

	Species	
	Streptococcus lactis subsp. *diacetylactis*	*Leuconostoc cremoris*
Citrate transport enzyme (CT)	Citrate permease active below pH 6.0	Citrate permease active below pH 6.0
Genetic determinant for CT	5.5 Md plasmid	Not known
Mode of CT expression	Inducible	Inducible
Citrate cleaving (CC) enzyme	Citratase	Citratase
Genetic determinant for CC	Chromosome	Chromosome
Mode of CC expression	Constitutive	Inducible

Source: From Vedamuthu and Washam (1983).

that are capable of diacetyl production from citrate include *S. thermophilus*, dairy lactobacilli, and *P. shermanii* (Vedamuthu, 1979). In dairy fermentations, diacetyl does not accumulate and does not remain unchanged indefinitely. As can be seen in Figure 3, diacetyl formed by the condensation of active acetaldehyde and acetyl coenzyme A is reduced to acetoin with concomitant oxidation of reduced nicotinamide adenine dinucleotide (NADH + H^+). Acetoin may be further reduced to 2,3-butanediol. The reduction products of diacetyl are flavorless, and the reductase reaction results in the loss of the desirable nutmeatlike flavor. The reduction of diacetyl to acetoin by the enzyme diacetyl reductase is an irreversible reaction coupled to the regeneration of oxidized nicotinamide cofactor.

Different approaches are used to obtain relatively high levels of diacetyl and circumvent the loss of accumulated dicarbonyl via the reductive reactions. Kinetic studies have shown that diacetyl synthesis and its reduction to acetoin occur in tandem, but in early stages when the citrate level (precursor level) is high, the synthetic phase is more rapid than the reductive phase such that there is an increase in the concentration of the diketone. Once the citrate level falls below a critical point ($<0.07\%$), the reductive phase is more prominent and the concentration of diacetyl falls more precipi-

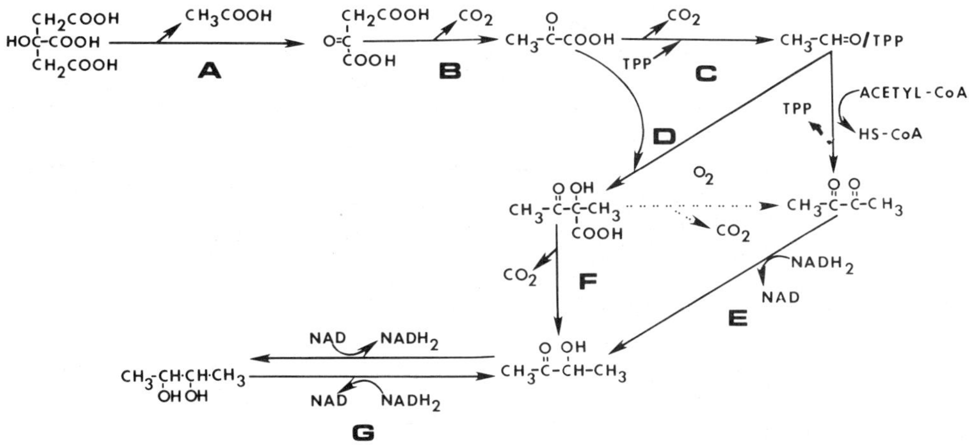

Figure 3 Biosynthetic pathway among dairy lactic acid streptococci for the production of diacetyl and its reduction products from citric acid: (A) citratase; (B) oxaloacetate decarboxylase; (C) pyruvate decarboxylase; (D) α-acetate synthetase; (E) diacetyl reductase; (F) α-acetate decarboxylase; (G) 2,3-butanediol dehydrogenase. Broken line represents the step on which disagreement exists in the literature. It is thought of as either a nonenzymatic reaction or an enzymatic step catalyzed by α-acetolactate oxidase. (From Vedamuthu, 1979.)

tously with time. The abrupt decrease in diacetyl concentration can be arrested by prompt cooling. Fortification of milk with sodium citrate provides additional precursor that helps in obtaining a higher diacetyl concentration and in delaying the onset of the declining phase, which allows a safety margin for cooling large volumes of fermented product to arrest diacetyl reductase activity.

Gentle agitation or aeration after desired acidity and flavor development also helps in retaining high diacetyl flavor. The possible physiological explanations for this phenomenon are as follows:

Flavor bacteria, especially *S. lactis* subsp. *diacetylactis*, possess an oxygen-linked NAD-oxidase system. With gentle aeration, the operation of the NAD-oxidase system spares diacetyl from the role of hydrogen acceptor to regenerate NAD^+ (Collins, 1972).

The oxygen-linked NAD-oxidase system also indirectly influences the formation of diacetyl by cycling the oxidative capacity needed for regeneration of lipoic acid, which is involved in acetyl-COA formation (Collins, 1972).

The increase in the diacetyl concentration in milk pretreated with hydrogen peroxide-catalase and cultured with aroma bacteria after treatment may be attributed to the stimulation of NAD-oxidase activity in these bacteria, which favors accumulation of the diketone (Collins, 1972). Another way to increase the flavor level in fermented milks is to hold the product under refrigeration for 4-5 days. During refrigerated storage there is a gradual buildup of diacetyl content (Pack et al., 1968). The exact mechanism by which this occurs is unknown. Under practical operating conditions, the diacetyl content of fermented dairy products can be maintained by rapid cooling of the material once the titratable acidity reaches 0.75-0.80% (as lactic acid), which roughly coincides with the peak level of diacetyl attained during the synthetic phase. At lower incubation temperatures (20-24°C), although the attainment of desirable titratable acidity is slower, the activity of dicaetyl reductase is muted. Hence, for products in which relatively higher levels of diacetyl are desirable, incubation between 21 and 24°C is better than at 30°C. *S. lactis* subsp. *diacetylactis* is capable of gradual biosynthesis of diacetyl at refrigeration temperature if sufficiently large numbers of viable bacteria are present. Based on this observation, concentrated frozen cultures of *S. lactis* subsp. *diacetylactis* are used for direct addition to chilled cream dressing applied to cottage cheese. During cold storage of the dressed cottage cheese, diacetyl is produced by the aroma bacteria.

Diacetyl is also an important component in the flavor profile of other fermented foods. In the fermentation of tea leaves and plant materials and in the production of soy sauce and related products in which mixed lactic acid bacterial activity occurs, diacetyl is produced. In some of these products, diacetyl is derived from citric acid and in others through carbohydrate fermentation. In wines and in beer, diacetyl causes off-flavors leading to a condition known as "sickness." In these alcoholic beverages, diacetyl may be synthesized by certain yeasts, leuconostocs, and pediococci. In wines, diacetyl may be partly derived from the citric acid present in grape or berry juices as well as from the sugars.

c. Alcoholic fermentation. Yeasts are the primary agents involved in the production of alcoholic beverages. Selected strains of *Saccharomyces cerevisiae* are the most commonly used wine yeasts. Strains of yeasts are selected on the basis of their alcohol tolerance, rapidity of cell lysis during storage of wine, which is a reflection of their capability to produce a higher concentration of esters and higher alcohols, and ability to modify the organoleptic properties of wines made from less aromatic vintages (Lafon-Lafourcade, 1983). In bottom-fermented lager beers *Sacharomyces uvarum* is used, and for top-fermented ales *S. cerevisiae* is used.

In alcoholic fermentation, the classic anaerobic glycolytic pathway, the Embdan-Meyerhof-Parnas cycle, is involved. The final net balance for the entire fermentative system is

$$1 \text{ glucose} \rightarrow 2CO_2 + 2 \text{ ethyl alcohol} + 2ATP$$

Less than 20% of the glucose metabolized by the yeast is diverted through the HMP pathway. In wort fermentation, oxidative metabolism through the tricarboxylic acid cycle (TCA) and the yeast mitochondrial cytochrome system are suppressed once the dissolved oxygen is depleted.

Alcohol production by baker's yeast contributes to the final flavor in bread, and the evolution of CO_2 during fermentation provides the "rise" necessary in the dough. Alcoholic fermentation by yeasts also imparts characteristic flavor attributes to Oriental fermented foods. In addition to the flavor contribution of alcohol per se, chemical and enzymatic reactions involving alcohol and volatile fatty acids present in fermentation systems result in the formation of highly flavorful esters, many of which have pleasant, fruity odors and flavors.

Among fermented fluid milk products, kefir and koumiss derive their characteristic flavors through alcoholic fermentation. The flora of kefir consists of yeasts called *Saccharomyces kefir* and *Torula kefir* and lactobacilli previously known as *Lactobacillus caucasicus*. The yeasts and the lactobacilli exist in a unique symbiotic relationship forming what are known as kefir grains. Kefir grains are largely made up of a polysaccharide called kefiran, with some denatured milk protein associated with the polysaccharide matrix. Kefir grains are used as starters to make the product and after the fermentation can be recovered and reused. Other microbial flora associated with kefir grains include lactic streptococci and *Leuconostoc* spp. Kefir grains are insoluble in water and swell when soaked in water to form a shiny, jellylike product. The grains are structured like the meat of a walnut, with folds and involutions. The microbial flora of kefir grains is found within the folds. The flavor of kefir may be described as mildly alcoholic and yeasty-sour with a tangy effervescence. Kefir is made now from cow's milk and, after fermentation at 21-23°C, consists of a smooth soft curd that when stirred or agitated foams and fizzles like beer. The major end products of fermentation are ethyl alcohol (about 1.0%), lactic acid (about 0.8%), CO_2, and traces of acetaldehyde, diacetyl, and acetone.

Koumiss is another alcoholic milk product made from mare's milk. The flora of koumiss consists of lactose-fermenting yeasts and *L. bulgaricus*. Because mare's milk does not coagulate at the isoelectric point of casein, finished koumiss, which contains about 0.7-1.8% lactic acid and 1.0-2.5% ethanol, is a grayish white liquid product. Because of CO_2 production during fermentation, this product also fizzles and effervesces like kefir. Koumiss is prized as a health drink in Eastern Europe.

d. Metabolism of proteins and other nitrogenous materials. The modification and degradation of proteins by microorganisms is involved in the development of body, texture, and flavor in fermented foods. The desirable degree of microbial proteolysis varies considerably in different fermented foods. In general, excessive proteolysis results in body and testural defects and loss of nutritive value. Control of proteolytic changes by adventitious flora in fermented foods is one of the major functions of starter cultures.

Many of the important functional attributes of starter culture in fermented dairy products are closely tied to their proteolytic ability. The conversion of lactose to lactic acid is the single most important function of dairy starters. Dairy starter bacteria growing in milk need specific proteolytic systems(s) to derive the required amino acids from casein and whey proteins to synthesize the enzymes involved in pathways of lactose dissimilation. Hence, the efficiency of the proteolytic systems of these bacteria is directly related to their acid-producing activity. In lactic streptococci, milk proteolytic function is coded on unstable genetic elements (plasmids) and the loss of these genetic determinants renders the bacteria unusable in dairy fermentation. There are, however, special instances in which high numbers of nonproteolytic or weakly proteolytic variants of lactic streptococci have been successfully used to overcome problems associated with antibiotic-tainted milk, phage inhibition, and bitter flavor production in cheddar cheese.

In hard cheeses that are internally cured by bacteria to obtain the desired body, texture, and flavor, the primary changes in the casein curd are brought about by rennet. This effect is mainly related to the modification of k-casein, which leads to the coagulation of milk. During ripening, however, rennet acts very slowly on α_{sl}-casein and contributes to the pool of high-molecular-weight peptides in the cheese. The major proteolytic changes are brought about by microbial proteases. The transformation of green, rubbery cheese curd to the soft, pliant, waxy body during curing is the result of the proteolytic activity of cheese microflora.

Several different proteolytic systems in starter streptococci are known. These include cell surface-bound protease, intracellular caseinolytic systems, extracellular cell-free and cell-bound peptidases, peptide transport systems, and intracellular peptidases. The proteases and peptidases of lactobacilli have also been described (Marshall and Law, 1984). In cheeses that are internally cured by bacteria, the amino acid pool derived through the gradual degradation of casein is considered to provide the essential "brothy background flavor" over which the other prominent flavor notes are superimposed. Sulfur-bearing aldehydes (e.g., methional) and other sulfhydral compounds derived from sulfur-containing amino acids, amines from various amino acids, and phenolic compounds produced from aromatic amino acids are considered very important in the flavor of cured cheeses. Additionally, the deamination and decarboxylation of amino acids by bacterial systems contribute to aldehydes and ketones. The production of volatile fatty acids from casein by lactic streptococci has also been reported. In Swiss cheese, propionibacteria produce dimethyl sulfide and proline from casein. The sweet nutmeat-like flavor of Swiss cheese is partly attributed to its proline content, and dimethyl sulfide is an essential component of Swiss cheese flavor (Vedamuthu, 1979).

The metabolism of threonine and deoxyribonucleotides in yogurt starter bacteria augments the acetaldehyde content of yogurt. For a comprehensive review of acetaldehyde metabolism in dairy starter bacteria, the reader is referred to the paper by Lees and Jago (1978).

In cheeses that are surface ripened by "smear flora," the extent of proteolysis near the surface layers is considerable. The smear flora consists of *Brevibacterium linens*, micrococci, and yeasts. The mixed flora produce extracellular proteases that seep into the surface layers of the cheese, contributing to the strong flavor of these cheeses. Proteolytic systems of *B. linens* have been described. The strong flavors of these cheeses are due to a combination of mercaptans, amines, ammonia, and the various bitter peptides derived from casein. An extensive review of the literature on the proteolytic systems of bacteria involved in fermented dairy products is provided by Schmidt et al. (1976), Desmazeaud and Gripon (1977), and Law (1982). The proteolytic systems of *Penicillum caseicolum*, the mold used in the surface curing of Camembert, Brie, and related cheeses have also been described (Vedamuthu and Washam, 1983).

In sourdough systems, only 30% of the protein is soluble in unfermented flour-water doughs. After 17 h of fermentation, the solubility increases to 62.1% of the protein; with strong acidification, solubility attains 82.7% of the protein. There is also an increase in the free amino acid content, especially of leucine, phenylalanine, methionine, tyrosine, lysine, and isoleucine. The lactobacilli are especially very active in proteolytic changes in dough systems (Spicher, 1983).

In wines and beer, higher alcohols that are important in flavor, are derived by deamination and decarboxylation of amino acids by yeasts during the exponential growth phase (Lafon-Lafourcade, 1983; Hardwick, 1983).

The role of proteolysis in various Oriental fermented foods is undefined, but by-products of proteolytic changes no doubt contribute to the flavor, texture, and nutritive value of these foods.

e. Metabolism of lipids. Of the various food components, lipids are the most stable to microbial deterioration. Stability of the lipid system of a food depends upon its dispersion in the food, the moisture level in the food, the type of fat, the pH of the system, and the types of microorganisms present in the food. Contamination by metal ions, such as copper, prolonged exposure to light, and undue agitation can induce catalytic chemical reactions that result in the oxidation of fats. Compounds liberated through the microbial degradation of fats, however, have a very significant flavor potential. Fatty acids, carbonyl compounds derived by the oxidation of fatty acids, and fatty acid esters are extremely important components of flavor in various food systems. The entire spectrum of changes brought about by the microbial degradation of lipids involves complex reactions, starting with the hydrolytic cleavage of fat to fatty acids and subsequent oxidation of fatty acids to carbonyls and peroxides by microbial enzymes and metabolically generated hydrogen peroxide in the system. According to Haymon and Acton (1978), it is very difficult to separate the reactions caused by the action of microbes on lipids, because the liberation of free fatty acids by hydrolytic cleavage is accompanied by an increase in the total carbonyl and peroxide values. They depicted the complex chain of events as shown in Figure 4. The primary hydrolytic breakdown of fat by lipases is shown in Figure 5. Different microbial lipases have different specificities, and the specificity is related to the position of attachment of a fatty acid to the triglyceride molecule and to the structure of the fatty acid being hydrolyzed (Haymon and Acton, 1978). Some microbial lipases exhibit end-product inhibition. Some microorganisms degrade fatty acids by the β-oxidation scheme shown in Figure 6. Decarboxylation of fatty acids by microorganisms results in the formation of carbonyl compounds. The breakdown of fats in food systems is usually measured by the increase in thiobarbituric acid value, peroxide value, or total carbonyls. The lipolytic activity in a food system is directly related to the carbohydrate and nitrogen content. Higher carbohydrate and protein concentrations tend to inhibit lipase activity.

In most food systems, extensive lipolysis is not desired, because it leads to rancid, soapy, and in extreme cases fishy off-flavors. In some foods, when a sharp, biting, and soapy flavor is desired, as in Provolone cheese, extensive lipolysis of milk fat is promoted through the use of crude free enzymes derived from animal glands (from calves, kids, or lambs). In some other systems, for example the blue-veined cheeses, moderate to extensive fat breakdown is

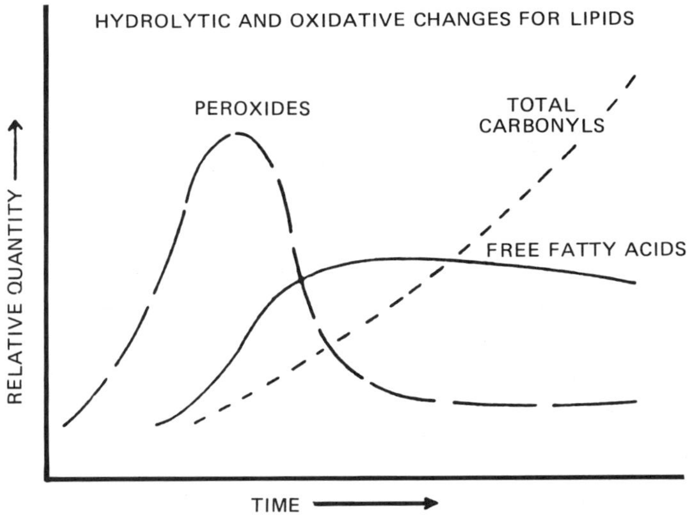

Figure 4 General sequence of lipid fraction changes attributed to hydrolysis and oxidative activities. (From Haymon and Acton, 1978.)

achieved by using selected microorganisms. In blue cheese production, the primary degradation of milk fat to fatty acids is caused by the native lipase system of milk and the lipases of the blue mold *P. roqueforti*. It appears that these initial changes are caused by enzymes liberated by the mold mycelia rather than the mold spores. As the free fatty acid and CO_2 concentrations increase, the mycelial activity is suppressed and the spore enzymes are more predominantly involved in fatty acid metabolism. In the widely used blue cheese manufacturing procedure, whole milk is separated into heavy cream and skim milk. The cream is homogenized and in some cases bleached and pasteurized. The homogenized, pasteurized cream is then co-mingled with raw skim milk in the vat in required propro-tions to obtain the desired fat level in the finished cheese. In milk, butterfat occurs as an emulsion dispersed as innumerable individual fat globules of varying sizes. The individual fat globules are enveloped within a membrane consisting of a phospholipid-protein complex. The presence of the intact fat globule membrane protects the milk fat from native milk lipase. When cream or milk is homogenized by forcing the fluid through small orifices under pressure, the membrane is disrupted and the fat globules are broken down to small, uniformly sized spheres. This achieves several purposes:

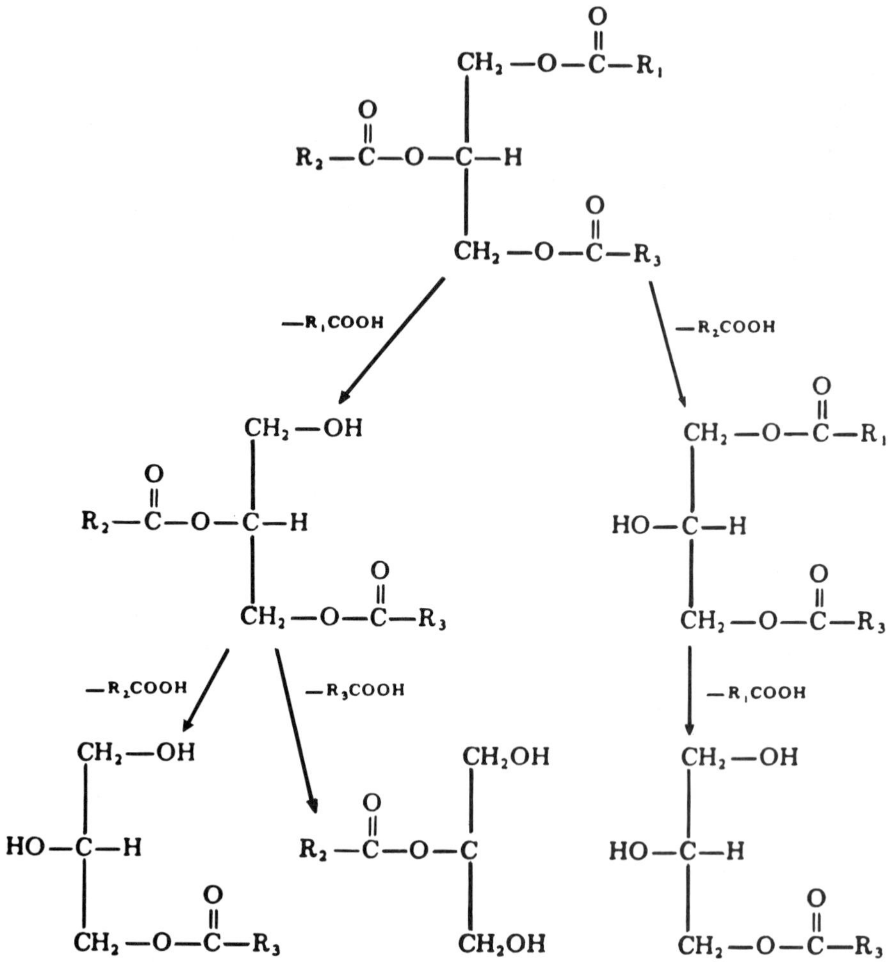

Complete hydrolysis will yield glycerol and free fatty acids.

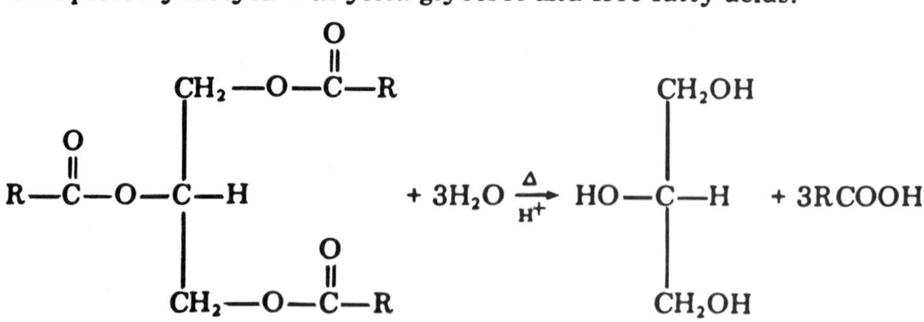

Figure 5 Hydrolytic cleavage of fats by lipase. (From Haymon and Acton, 1978.)

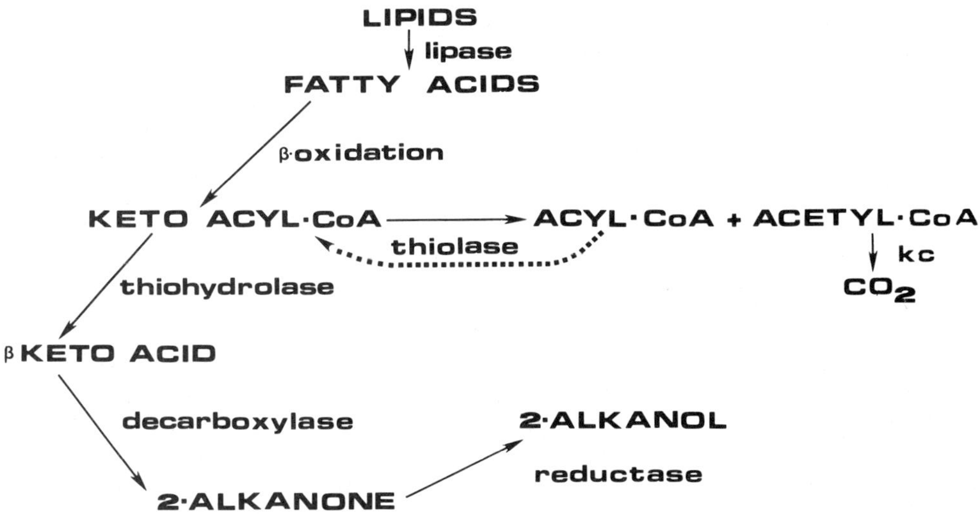

Figure 6 Metabolism of fatty acids by *Penicillium roqueforti* in the blue cheese system. Acyl group may contain from 4 to 16 carbons; kc denotes Kreb's cycle. Broken line represents the possible inhibition of thiolase by acyl-CoA that allows the other branch of the dissimulative pathway to proceed continuously. (According to Kinsella and Hwang, 1976.)

Disruption of fat globules to smaller, uniformly sized particles disperses the fat throughout the fluid and does not allow coalescing and migration of the fat to the top of the fluid to form a "cap" or "cream line."

The breakdown of fat globules to smaller, uniformly sized particles increases the surface area of individual fat globules.

Disruption of the globule membrane exposes the milk fat and allows easy access to the substrate by lipases.

In blue cheese, extensive veining and good growth of *P. roqueforti* is encouraged by punching the loaves of cheese with needles, which allows air to diffuse into the cheese. Also, spaces between curd pieces are not compressed by mechanically pressing the cheese. Free fatty acids liberated by the hydrolytic cleavage of milk fat are further metabolized by the mold enzymes through the β-oxidation scheme. Methyl ketones derived by β-oxidation of milk fatty acids are considered very important in blue cheese flavor. For an extensive review of flavor development in blue cheese, the paper by Kinsella and Hwang (1976) should be consulted.

In most cured cheeses, a slight degree of lipolysis is necessary for full flavor development. The adventitious flora of cheese probably has a greater role in lipolytic changes. The most important among the adventitious flora are the low-temperature lactobacilli, enterococci, and the micrococci. Esterases of lactobacilli (El Soda et al., 1986) and the role of enterococci added as supplemental starters in lipolytic changes in cheddar cheese (Jensen et al., 1975) have been described. Micrococci in cheese have been reported by several investigators to play a part in milk fat degradation in cheeses. Esters of C_2, C_3, C_4, and C_6 fatty acids impart fruity flavors to food systems, and their presence in proper concentrations and proportions is considered necessary for full cheddar cheese flavor. Excess C_6 fatty acid esters, however, have been reported to cause fruity off-flavors in cheeses. Esterases of propionibacteria have also been described, and longer chain fatty acids are found in fully ripened Swiss cheese. Cured cheeses made from raw milk are reported in general to develop flavor at a faster rate than cheeses made from fully pasteurized or heat-treated milk. This observation is attributed to the activity of the native milk enzymes that are retained undenatured in raw milk cheeses and also to the presence of greater numbers and varieties of adventitious flora. It should, however, be emphasized that although excessive lipolysis is undesirable, the right amount of lipolytic changes needed for desirable flavors is to a large measure regulated by starter flora and their metabolic activities. Law and Sharpe (1977) proposed that flavor development in cheeses is the result of "correct conditions" created by starter bacteria in cheese systems:

- A supply of flavor precursors derived from the enzymatic breakdown of lactose, protein, and fat in milk
- A low pH, which prevents most enzyme-catalyzed reactions from proceeding too quickly, and the suppression of spoilage flora
- A low redox potential (-150 to -200 mV) resulting from lactose metabolism

In the Finnish fermented milk cheese viilli, in certain regions a thin layer of growth of the white mold *Geotrichum candidum* is favored. This mold has strong lipolytic enzymes, and there may be a certain degree of fat breakdown in the upper layers of viilli consumed in those areas.

In sausage making, selected strains of *Micrococcus* spp. and/or certain nonpathogenic, nonenterotoxigenic strains of *Staphylococcus* spp. are used as adjunct starters to obtain distinctive flavors. The distinctive flavors are the result of the limited lipolysis of fat in the sausage mixture. The micrococci additionally are good nitrate reducers, which helps in retaining the red color of sausages. Further, the catalase activity of micrococci dissipates any accumulated hydrogen peroxide generated by the catalase-negative, microaerophilic lactic acid bacteria used widely for acid production in sausages. Excessive hydrogen peroxide accumulation results in "greening color" defects.

According to Haymon and Acton (1978), the type of fat used and the properties of different fat sources (e.g., bovine or porcine) influence the flavor characteristics of dry sausages. The distinctive flavor of dry sausages, however, is partly attributable to hydrolysis and oxidative changes that occur in the lipid fraction during ripening and drying. There appear to be two distinct phases in lipid changes in sausage systems. In the early ripening phase, lipase activity releases free fatty acids, followed by oxidation of fatty acids by peroxides to carbonyl compounds during the drying phase. This is supported by the finding that the concentration of free fatty acids declines as the concentration of carbonyls and peroxide values increases in the early stages and remains high before decreasing rapidly with holding beyond 15 days at 15°C. From long-chain fatty acids both saturated and unsaturated carbonyls are formed in the sausage, but the rate of disappearance due to oxidation is faster with unsaturated than with saturated carbonyls. The long-chain carbonyls are probably hydrolyzed by bacterial enzymes to yield short-chain volatile compounds that contribute to flavor. In sausages containing pork fat, unsaturated carbonyls, such as 2-alkenals and 2,4-alkadienals, which are very flavorful, are formed by the oxidation of pork fat; these compounds are found only in limited amounts in oxidized beef fat.

The lactobacilli and pediococci widely used as starters in sausages generate hydrogen peroxide, which is highly reactive, and the hydrogen peroxide causes oxidative changes in fats. Other adventitious flora, such as enterococci, pseudomonads, and yeasts, also possess lipases and esterases and partly contribute to the pool of flavor compounds derived from fat in sausages. For certain European sausage varieties, an external growth of white mold is desired, and in such varieties there may be moderate lipolytic changes in the outer layers contributing to characteristic flavors.

In several of the Oriental fermented fish products, in which the activities of a mixed flora consisting of lactic acid bacteria, yeast, and sometimes molds are involved, lipolytic changes in the systems no doubt add to the flavor of these products. The fungi especially possess very active lipases.

Although the major flavor components in baked goods are derived from the dough ingredients, yeasts, fermentation, the baking process, and the accompanying browning reactions, the lipolysis of shortenings used in breads may be beneficial in obtaining distinctive flavors (Haymon and Acton, 1978).

f. Flavor additives obtained through microbial metabolism. Several flavor additives and fortifiers for foods are obtained through microbial metabolism of suitable precursors or natural substrates. These include extracts of edible mushrooms, fruity flavor compounds, monosodium glutamate, nucleotides, and yeast autolysates and extracts. Although microbial fermentation and metabolism are involved in the production of such flavoring materials, these aspects cannot be considered here as in situ development of flavor in food systems through the use of microorganisms. For details on these aspects, the reader should consult Margalith's book on flavor microbiology (1981).

3. *Role of Adventitious Flora*

The term "adventitious flora" refers to the widely divergent microorganisms that inadvertently gain access to the material being fermented during its handling and processing as opposed the starter flora, which are deliberately added to the system. The exact role of adventitious flora in various fermented food systems is not completely defined, but under normal regulated fermentation conditions they largely appear to play a role in accentuating, intensifying, and accelerating the development of characteristic flavors. Their role is better understood in cheese systems (Law, 1982) than in any other fermented food group. The work pertaining to the contribution of nonstarter flora in cheese has largely been done with aseptic cheese-manufacturing procedures (Law, 1982).

4. *Curtailing Fermentations*

In processes involving biologic systems in which intact microorganisms are used to develop specific changes in a food system, the growth and metabolic activities of microorganisms must be regulated and/or curtailed after the desired changes in the food system have been achieved. Continued activity of microorganisms beyond the desired level of change in body, texture, and flavor in the food leads to defects as well as loss in keeping quality. The most commonly used procedure for arresting microbial action is refrigeration and, in some instances, heat processing or cooking. Cooling retards microbial activity and expands the shelf life of the fermented food. Heat processing when applicable is very effective in curtailing continued metabolism of microorganisms used in food fermentation.

5. *Indirect or Regulatory Role of Starter Cultures*

The indirect role of starter cultures in flavor development primarily relates to the regulation of the enzymatic and microbiologic activity of adventitious flora. The unregulated activity of adventitious flora leads to defects in body, texture, and flavor. For example, the psychrotrophic bacteria possess very active lipolytic, proteolytic, and diacetyl reductase enzyme activities. If psychrotrophs grow unregulated, they cause deterioration in quality. Starter microorganisms are selected to dominate in the fermentation substrate and direct the course of changes occurring within the system. The end products of metabolism and the ecological condition imposed by the starter culture activity are largely responsible for their regulatory role. Some of the factors involved in starter regulatory activity are as follows:

- Depression of pH and accumulation of lactic, acetic, and propionic acids, which are generally inhibitory to pathogenic and spoilage flora—their growth, enzyme activity, and toxin production; propionic acid is especially inhibitory to fungi
- Accumulation of alcohol, for example in wines, making the fluid selective for only certain types of microorganisms
- Production of metabolic H_2O_2, which is inhibitory to spoilage bacteria (Vedamuthu, 1982) especially in combination with natural inhibitory systems in foods, for example, the lactoperoxidase system of milk (Reiter and Harnulv, 1984)
- Production of antibiotic substances, such as nisin, diplococcin, acidophilin, and bulgarican, that are inhibitory to Gram-positive and Gram-negative spoilage and pathogenic bacteria (Vedamuthu, 1982).
- Production of undefined inhibitory substances (Vedamuthu, 1982) and other compounds, including benzoic acid (Chandan et al., 1977), which are bacteriostatic and fungistatic
- Production of bacteriocins active against undesirable flora (Fleming et al., 1975)
- Reduction in a_w brought about by syneresis induced by acid produced by starter bacteria in the food system
- Depression of oxidation-reduction potential E_h, which is detrimental to the growth of aerobic spoilage flora
- Competitive removal of essential cofactors at a neutral pH range that makes them unavailable for certain undesirable bacteria, for example *Staphylococcus aureus* (Haines and Harmon, 1973)
- Competitive depletion of energy and carbon sources from the fermentation media (Vedamuthu, 1982; Lawrence et al., 1976)

C. Secondary Fermentation to Improve Product Flavor and Quality

Unlike primary fermentation processes, which are mainly used to modify basic food commodities to obtain desirable flavors, variety in textural and flavor qualities, and preservative attributes, secondary fermentations comprise processes applied *selectively* to certain batches of fermented products to improve their flavor, marketability, and keeping quality. Secondary fermentations are usually necessary only for products made from food materials that because of compositional deficiencies or differences caused by agronomic or climatic conditions do not yield top-quality end products on completion of the primary fermentations. Secondary fermentations usually help either to tone down harsh flavors or to remove interfering notes that mask the perception of a full bouquet of the flavors desired in the final product. The malolactic fermentation of wines is the best example of such a process.

In northern wine-growing areas, where the climate is cool and moist as in Switzerland, parts of France during certain exceptionally cold seasons, and in upper New York State, the grapes have a higher ratio of malic to tartaric acid. Wine made from such grapes has a higher malic acid content that imparts a harsh, acidic flavor. The biologic conversion of malate to lactate by secondary malolactic fermentation of such wines improves the flavor and bouquet of the product (Benda, 1982). According to Lafon-Lafourcade (1983), in northern viticultural areas malolactic fermentation is definitely beneficial in upgrading the quality of red wines for the following reasons:

The bacterial transformation of malic acid to lactic acid provides a convenient method of deacidifying red wines, which imparts mellowness to the wine.

The conversion of malate to lactate provides a biologic stability to the wine by preventing wild fermentation from occurring in the wine.

Malolactic fermentation yields wines that are more mellow and have a more complex aromatic quality.

Malolactic fermentations, however, must be accomplished by using selected strains and under controlled conditions. Unregulated, wild malolactic fermentations lead to the accumulation of volatile compounds that impart a "lactic odor" to the wine (Lafon-Lafourcade, 1983). Wild malolactic fermentations are usually prevented by controlling fermentation conditions such that all the sugars are completely depleted by the yeasts used in the primary alcohol fermentation. In warm regions, grape musts have only small amounts of malic acid and relatively higher pH. The higher pH values allow the growth of varieties of lactic acid bacteria capable of malolactic fermentation, and excessive deacidification can occur, which is detrimental to the quality and stability of wines.

As a microbiologic growth medium, wine is rather restrictive and selective. The selective property of wines is contributed by the alcohol concentration, the CO_2 content, the low pH, and the level of SO_2. Among the microbial flora found in wines, the lactic acid bacteria possess the enzymatic mechanisms for the conversion of malic acid to lactic acid. Lafon-Lafourcade (1983) and Benda (1982) have summarized the taxonomy of the lactic acid bacteria found in wines. The species most widely encountered in wines belong to the genera *Pediococcus*, *Leuconostoc*, and *Lactobacillus*. Although the pediococci and the lactobacilli possess the enzymes necessary for malolactic fermentation, leuconostocs are considered primary determinants in the malolactic fermentation of wines because of their ability to grow appreciably under the conditions obtainable in wines after the primary alcoholic fermentation. According to Lafon-Lafourcade (1983), the kinetics of malolactic fermentation is directly related to the bacterial biomass formed and the biochemical reactions carried out during the active growth phase of the bacteria. Among the leuconostocs, *Leuconostoc oenos* is considered the primary agent in the transformation of malate to lactate. Three types of *L. oenos* are recognized: *L. oenos* A+, which ferments arabinose, *L. oenus* X+, which ferments xylose, and *L. oenos* P+, which ferments both arabinose and xylose.

Three possible enzymatic mechanisms may be involved in the conversion of malate to lactate:

$$\text{Malic acid} + NAD^+ \xrightleftharpoons[\text{dehydrogenase}]{\text{malate}} \text{oxaloacetic acid} + NADH + H^+$$

$$\text{Oxaloacetic acid} \xrightleftharpoons[\text{decarboxylase}]{\text{oxaloacetate}} \text{pyruvic acid} + CO_2 \qquad (1)$$

$$\text{Pyruvic acid} + NADH + H^+ \quad \overset{\text{lactate}}{\underset{\text{dehydrogenese}}{}} \quad \text{lactic acid} + NAD^+$$

$$\text{Malic acid} + NAD^+ \xrightleftharpoons[\text{enzyme}]{\text{malate}} \text{pyruvic acid} + CO_2 + NADH + H^+$$

$$(2)$$

$$\text{Pyruvic acid} + NADH + H^+ \xrightleftharpoons[\text{dehydrogenase}]{\text{lactate}} \text{lactic acid} + NAD$$

$$\text{Malic acid} \xrightleftharpoons[\text{enzyme}]{\text{malolactate}} \text{lactic acid} + CO_2 \qquad (3)$$

It is still unknown which of the three pathways is actually involved in malic acid conversion to lactic acid in wines. Evidence points, however, to the operation of the third pathway because the resulting lactic acid is always the L(+) isomer. Operation of the other two schemes would involve the intermediate formation of pyruvic acid, which could be reduced to L(+) or D(-) lactic acid depending upon the specificity of the lactic dehydrogenase systems of the bacteria involved (Benda, 1982). It is also unclear whether malolactic enzyme system consists of a single enzyme or an enzyme complex. The role of NAD^+ in this reaction is assumed to be in the exchange of hydrogen atoms. For a more detailed discussion of the malolactic fermentations of wines, the reviews by Lafon-Lafourcade (1983), Benda (1982), and Radler (1975) should be consulted.

Currently, selected strains of *L. oenos* are commercially available for initiating malolactic fermentation in wines. The strains are selected on the basis of functional attributes and the ability to grow under conditions found in wines after primary alcoholic fermentation. Commercial cultures, however, need to be "activated" in a suitable grape juice-yeast extract medium before addition to large vats or barrels of wine to complete the malolactic fermentation relatively quickly. Percolating wine through an immobilized column of *L. oenos* has been suggested but the technology is still wanting (Lafon-Lafourcade, 1983).

D. Stabilization of Fermented Foods Against Deterioration

Although fermentation in itself provides preservative qualities to food materials, the moisture content of most fermented foods is still high enough to support the growth of a wide variety of microflora. Postfermentative contamination also introduces additional spoilage flora. Further, if the fermented food is not consumed fresh or within a reasonable period of refrigerated storage (10-20 days) and is intended for long shelf life, further treatments are necessary to stabilize it. Even if serious spoilage does not readily occur in fermented foods, the loss of flavor or freshness or the hint of staleness reduces market appeal.

Many of the fermented foods that are acidic and contain a high salt concentration are prone to deterioration by fungi, which grow at lower pH and lower a_w values than bacteria. Among the fungi, molds grow much better than yeasts at lowered a_w environments, but because of their aerobic metabolism are a problem mainly on the surface of foods.

Five principal categories of preservation strategies are used in conjunction with fermented foods:

Heat processing is used, for example, for cheese. Processing of natural cheeses or cheese blends assures uniformity in quality with regard to color, body, texture, meltability, and flavor. It further provides good shelf life.

Acidification is used to lengthen the shelf life of relatively high moisture fermented products, such as processed cheese foods, especially cheese spreads and cheese dips.

Packaging helps to protect fermented foods from external contamination, surface dehydration, and other problems. Packaging can be combined with exclusion of air, which prevents surface growth of molds; flushing with CO_2, which helps to control mold growth and in some cases to retain freshness by inhibiting aerobic psychrotrophic bacteria; and heat processing. For example, pickled cucumbers are packaged under brine. In the addition of protective cultures to cottage cheese dressing, selected strains of aroma bacteria are added not only to impart and maintain desirable levels of diacetyl but also to provide inhibition against psychrotrophic bacteria and enteropathogens (Sing, 1976; Vedamuthu et al., 1966)

Approved chemicals, which are bacteriostatic and fungistatic, include sorbates, propionates, benzoates, and such antioxidants as butylated hydroxyanisole (BHA) and butylated hydroxytoluene (BHT) (Furia, 1972; Raccach, 1984; Fung et al., 1985). These products are used in sausages, breads, crackers, and cheeses.

III. GENETIC ASPECTS OF MICROORGANISMS USED IN FERMENTED FOODS

Fermentations represent biologic processes, and variation is the rule in biologic systems. Because of the short generation times of microorganisms and the high populations that can be attained in a short interval of time, mutational events could be manifested relatively rapidly such that they may become detrimental to fermentation processes. However, the short generation times and the high populations that can be obtained in a short interval of time with microbes under controlled conditions within the constrained space of a laboratory have afforded fundamental studies on the genetic mechanisms that govern the metabolic traits important to industrial fermentation. Knowledge of the fundamental genetic mechanisms has allowed the manipulation of genetic material to advantage in fermentative processes. The food industry has been rather slow in exploiting the modern strides in microbial genetics. With the discovery of plasmid-coded traits, advances in restriction enzyme studies and cloning techniques and progress in electronic control equipment leading to improved biologic processes, developing novel products more efficiently and copiously through fermentations seem assured.

Among the microorganisms used in food fermentation, most of the genetic work to date has been done with the lactic acid bacteria. More is known about the genetic mechanisms of importance to food fermentations among mesophilic lactic acid bacteria than among the thermophilic types. In general, the food fermentation industry is poised for earlier improvements in processes involving prokaryotes than those using eukaryotes. However, research activity in unraveling and understanding genetic systems of yeast and molds is progressing steadily.

A. Lactic Acid Bacteria

Mesophilic lactic acid bacteria are widely used in dairy fermentation. Many of the metabolic traits important in milk fermentation have been associated with the plasmids found in lactic streptococci. Plasmid-associated traits are usually unstable because the curing of plasmids in dairy streptococci occurs with repeated transfers or temperature abuse or growing in nonmilk media. The association of important functional attributes with plasmid DNA, however, offers advantages in "genetic engineering." The plasmid-encoded functional properties of lactic streptococci include lactose fermentation (McKay, 1982, 1983); citric acid uptake mediated by citrate permease (McKay, 1983); milk proteolysis (McKay 1982); fermentation of other sugars, such as glucose, galactose, mannose, xylose, and sucrose (McKay, 1983); production of bacteriocins against other lactic streptococci affecting compatibility of strains in mixed-strain starters (McKay, 1983); elaboration of nisin (Gonzalez and Kunka, 1985) and diplococcin (McKay, 1983), which also adversely affect the compatibility of strains in mixed cultures, resistance to phages (Klaenhammer, 1984), assuring uninterrupted fermentations; and production of mucoid coagulum in milk (Vedamuthu and Neville, 1986) which affects textural properties and flavor perception. A majority of these plasmid-encoded phenotypic characteristics have been introduced into suitable recipient strains lacking these traits by specific conjugative mobilization of functional plasmids (Gasson, 1983; Gasson and Davies, 1984; Kondo and McKay, 1985). The conjugative transfer of functional plasmids has been achieved between strains belonging to different species of lactic streptococci. Transductional transfer of lactose-utilizing ability (Lac^{+}) has also been demonstrated (McKay, 1982). Transfer of lactose plasmid by transformation has also been reported (McKay, 1982). Other techniques for gene transfer, such as protoplast fusion, have also been applied to lactic streptococci (Gasson, 1983). Research on cloning strategies and selection and testing of suitable vectors to transfer genes among lactic streptococci has advanced considerably (Kondo and McKay, 1985).

Recently, Tsai and Sandine (1986) reported the conjugal transfer of plasmids encoding for lactose utilization and nisin production from *S. lactis* to dairy *Leuconostoc* spp. and their phenotypic expression in the resultant transconjugants. The *L. dextranicum* transconjugant containing the nisin plasmids from *S. lactis* 7962 produced much greater amounts of nisin than the parent *S. lactis* strain. In these studies, the natural vancomycin resistance of *Leuconostoc* spp. (Orberg and Sandine, 1984) was used as the selection marker. These studies have opened new avenues in starter culture research in developing more efficient aroma-producing *Leuconostoc* strains and in studying the physiology and regulation of nisin biosynthesis.

Among the thermophilic lactic acid bacteria, more studies have been conducted on the genetics of *Lactobacillus* spp. than that of *S. thermophilus*. The association of lactose utilization in *L. casei* with a plasmid and its conjugative transfer to other lactose-negative lactobacilli has been reported (Chassy, 1985). The transformation of lactose-fermenting ability in *Lactobacillus lactis* was described recently (Hofer, 1985). The chromosomal linkage of genes coding for bactriocin production by *Lactobacillus helveticus* was reported by Joerger and Klaenhammer (1986). Genetic work relating to the derivation of virulent phages for *L. casei* S-1 from a temperate phage carried by the same strain was described by Japanese workers. A summary of the Japanese work and other techniques used in gene transfer among lactobacilli is given in the review by Davies and Gasson (1984). Recently, the transfer of broad host-range vector pAMβ 1 to *L. plantarum* was reported by Sharago et al. (1986). Chassy (1985) discussed the avenues available for the genetic improvement of lactobacilli important in food fermentations.

Only recently, the genetic work pertaining to *S. thermophilus* has appeared in the literature. Preliminary characterization of plasmids in *S. thermophilus* and discussion of the possible use of some of the smaller cryptic plasmids for vector construction are found in recent papers (Herman and McKay, 1986; Somkuti and Steinberg, 1986). The cloning of the β-galactosidase gene from *S. thermophilus* into *Escherichia coli* was reported by Herman and McKay (1986).

Conjugative transfer of a broad host-range vector pIP501 between pediococci and other streptococci was demonstrated by Gonzalez and Kunka (1983). Recently, the same authors (Gonzalez and Kunka, 1986) described plasmid-associated sugar metabolism in pediococci, which is of great significance in sausage and vegetable fermentation. Graham and McKay (1985) associated a plasmid with the production of a bacteriocin by *Pediococcus pentosaceus*.

B. Yeasts and Molds

Genetic engineering of eukaryotic microorganisms presents greater challenges than that of prokaryotic systems. Until recently, most of the genetic work with fungi important in food fermentations have employed classic techniques, such as screening of strains from various sources and habitats to isolate cultures with better fermentation performance, nutritional and adaptive studies to boost desired end products or tolerance to end-product accumulation, induced mutations using ultraviolet (UV) rays and other short-wave radiation, and chemical mutagens to isolate hyperproducers, leaky mutants, metabolic blocks, and so on. Techniques for protoplast fusion to introduce new genes have been developed for fungal systems (Peberdy, 1979).

Because molds produce a variety of powerful enzymes that could be advantageously used in flavor enhancement and modification of foods and to improve fermentation efficiency, enzyme cloning techniques are being developed with fungal systems (Wasserman, 1984; White et al., 1984).

With yeasts, most of the work has been done with the strains used in brewing. In the brewing industry, most of the effort is directed toward improving fermentation efficiency and cost. Additional effort is directed toward identifying and eliminating variants that produce ketonic compounds that cause off-flavors in beer. Genetic approaches to the latter problem are also being sought. Currently, different genetic techniques are being tried to introduce genes from other yeast species or genera that would confer on the brewer's yeast strains the ability to utilize a wide variety of complex carbohydrates so that the malting process can be eliminated (Panchal et al., 1984).

IV. SUMMARY AND CONCLUSIONS

Food fermentation represents one of the oldest means of food preservation. Besides the preservative advantage of food fermentation, the microbial processing of foods also provides protection against food-borne pathogens and introduces variation and improvement in the body, texture, flavor, and nutritive value of food materials. From a processing viewpoint, food fermentation offers safe and economic intermediate technology suitable for developing countries.

Over the years, the fermentation of foods has moved more and more from the realm of art to science. More is known now about the proper selection of raw materials, selection of suitable fermentative microbes, and optimization of fermentation conditions to arrive at a consistent high-quality product day after day. Modern studies in basic sciences and interdisciplinary studies leading to biotechnology involving biology, recombinant DNA techniques, engineering design, analytic techniques, computer control of fermentation, and so on, coupled with consumer awareness and demand for safe, nutritive, and "natural" foods, offer a promising future for the fermented food industry.

REFERENCES

Arunga, R. O. (1982). Coffee. In *Fermented Foods*, Vol. 7, *Economic Microbiology*, A. H. Rose (Ed.). Academic Press, London, pp. 259-274.

Aurand, L. W., Singleton, J. A., Bell, T. A., and Etchells, J. L. (1965). Identification of volatile constituents from pure-culture fermentations of brined cucumbers. *J. Food Sci. 30*:288-295.

Benda, I. (1982). Wine and brandy. In Prescott and Dunn's *Industrial Microbiology*, 4th Ed., G. Reed (We.). AVI Publishing, Westport, Connecticut, pp. 393-402.

Beuchat, L. R. (1983). Indigenous fermented foods. In *Biotechnology*, Vol. 5, G. Reed (Ed.). Verlag Chemie, Deerfield Beach, Florida, pp. 477-528.

Bluhm, L. (1983). Distilled Beverages. In *Biotechnology*, Vol. 5, G. Reed (Ed.). Verlag Chemie, Deerfield Beach, Florida, pp. 447-476.

Bottazzi, V., and Dellaglio, F. (1967). Acetaldehyde and diacetyl production by *Streptococcus thermophilus* and other lactic streptococci. *J. Dairy Res. 34*:109-113.

Castelein, J., and Verachtert, H. (1983). Coffee fermentation. In *Biotechnology*, Vol. 5, G. Reed (Ed.). Verlag Chemie, Deerfield Beach, Florida, pp. 587-598.

Chandan, R. C., Gordon, J. F., and Morrison, A. (1977). Natural benzoate content of dairy products. *Milchwissenschaft 32*:534-537.

Chassy, B. M. (1985). Prospects for improving economically significant *Lactobacillus* strains by "genetic technology." *Trends Biotechnol. 3*(11):273-275.

Collins, E. B. (1972). Biosynthesis of flavor compounds by microorganisms. *J. Dairy Sci. 55*:1022-1028.

Davies, F. L., and Gasson, M. J. (1984). Bacteriophages of dairy lactic-acid bacteria. In *Advances in the Microbiology and Biochemistry of Cheese and Fermented Milk*, F. L. Davies and B. A. Law (Eds.). Elsevier Applied Science Publishers, London, pp. 127-151.

Desmazeaud, M. J., and Gripon, J. C. (1977). General mechanism of protein breakdown during cheese ripening. *Milchwissenschaft 32*:731-734.

Djien, K. S. (1982). Indigenous fermented foods. In *Fermented Foods*, Vol. 7, *Economic Microbiology*, A. H. Rose (Ed.). Academic Press, London, pp. 15-38.

Doelle, H. W. (1981). Basic metabolic processes. In *Biotechnology*, Vol. 1, H. J. Rehm and G. Reed (Eds.). Verlag Chemie, Deerfield Beach, Florida, pp. 113-210.

El Soda, M., Korayem, M., and Ezzat, N. (1986). The esterolytic and lipolytic activities of lactobacilli. III. Detection and characterization of the lipase system. *Milchwissenschaft 41*:353-355.

Fleming, H. P. (1982). Fermented vegetables. In *Fermented Foods*, Vol. 7, *Economic Microbiology*, A. H. Rose (Ed.). Academic Press, London, pp. 227-258.

Fleming, H. P., Etchells, J. L., and Castilow, R. N. (1975). Microbial inhibition by an isolate of *Pediococcus* from cucumber brines. *Appl. Microbiol. 30*:1040-1042.

Fung, D. Y. C., Lin, C. C. S., and Gailani, M. B. (1985). Effects of phenolic antioxidants on microorganisms. *CRC Crit. Rev. Microbiol. 12*:153-183.

Furia, T. E. (1972). *CRC Handbook of Food Additives*, 2nd Ed., Vol. 1. CRC Press, Boca Raton, Florida, pp. 190-197.

Garvie, E. I. (1984). Taxonomy and identification of bacteria important in cheese and fermented dairy products. In *Advances in the Microbiology and Biochemistry of Cheese and Fermented Milk*, F. L. Davis and B. A. Law (Eds.). Elsevier Applied Science Publishers, London, p. 59.

Gasson, M. J. (1983). Genetic transfer systems in lactic acid bacteria. *Antonie van Leuwenhoek 49*:275-282.

Gasson, M. J., and Davies, F. L. (1984). The genetics of dairy lactic-acid bacteria. In *Advances in the Microbiology and Biochemistry of Cheese and Fermented Milk*, F. L. Davies and B. A. Law (Eds.). Elsevier Applied Science Publishers, London, pp. 99-126.

Gilliland, S. E. (1985). *Bacterial Starter Cultures for Foods*. CRC Press, Boca Raton, Florida.

Gonzalez, C. F., and Kunka, B. S. (1983). Plasmid transfer in *Pediococcus* spp.: Intergeneric and intrageneric transfer of pIP5ϕ1. *Appl. Environ. Microbiol. 46*:81-89.

Gonzalez, C. F., and Kunka, B. S. (1985). Transfer of sucrose-fermenting ability and nisin production phenotype among lactic streptococci. *Appl. Environ. Microbiol. 49*:627-633.

Gonzalez, C. F., and Kunka, B. S. (1986). Evidence for plasmid linkage of raffinose utilization and associated α-galactosidase and sucrose hydrolase activity in *Pediococcus pentosaceus*. *Appl. Environ. Microbiol. 51*:105-109.

Graham, D. C., and McKay, L. L. (1985). Plasmid DNA in strains of *Pediococcus cerevisiae* and *Pediococcus pentosaceus*. *Appl. Environ. Microbiol. 50*:532-534.

Haines, W. C., and Harmon, L. G. (1973). Effect of selected lactic acid bacteria on growth of *Staphylococcus aureus* and production of enterotoxin. *Appl. Microbiol. 25*:436-441.

Hardwick, W. A. (1983). Beer. In *Biotechnology*, Vol. 5, G. Reed (Ed.). Verlag Chemie, Deerfield Beach, Florida, pp. 165-229.

Hartley, J. C., and Vedamuthu, E. R. (1975). Beta-galactosidase of *Proprionibacterium shermanii. Appl. Microbiol. 29*:74-80.

Haymon, L. W., and Acton, J. C. (1978). Flavors from lipids by microbiological action. In *Lipids as a Source of Flavor*, ACS Symposium Series No. 75, M. K. Supran (Ed.). American Chemical Society, Washington, D.C., pp. 95-115.

Herman, R. E., and McKay, L. L. (1986). Cloning and expression of the beta-D-galactosidase gene from *Streptococcus thermophilus* in *Escherichia coli. Appl. Environ. Microbiol. 52*:45-50.

Hettinga, D. H., and Reinbold, G. W. (1972). The proprionic-acid bacteria—a review. II. Metabolism. *J. Milk Food Technol. 35*:358-372.

Hobbs, G. (1986). Ecology of food microorganisms. *Microb. Ecol. 12*:15-30.

Hofer, F. (1985). Transfer of lactose-fermenting ability in *Lactobacillus lactis. N.Z. J. Dairy Sci. Technol. 20*:179-183.

Jensen, J. P., Reinbold, G. W., Washam, C. J., and Vedamuthu, E. R. (1975). Role of enterococci in cheddar cheese: Free fatty acid appearance and citric acid utilization. *J. Milk Food Technol. 38*:78-83.

Joerger, M. C., and Klaenhammer, T. R. (1986). Characterization and purification of helveticin J and evidence for a chromosomally determined bacteriocin produced by *Lactobacillus helveticus* 481. *J. Bacteriol. 167*:439-446.

Keenan, T. W., and Lindsay, R. C. (1966). Removal of green flavor from ripened butter cultures. *J. Dairy Sci. 49*:1563-1564.

Kinsella, J. E. (1975). Butter flavor. *Food Technol. 29*(5):82-98.

Kinsella, J. E., and Hwang, D. H. (1976). Enzymes of *Pencillium roqueforti* involved in the biosynthesis of cheese flavor. In *CRC Crit. Rev. Food Sci. Nutr.*, 191-228.

Klaenhammer, T. R. (1984). Interactions of bacteriophage with lactic streptococci. *Adv. Appl. Microbiol. 30*:1-29.

Kondo, J. K., and McKay, L. L. (1985). Gene transfer systems and molecular cloning in group N. streptococci: A review. *J. Dairy Sci. 68*:2143-2159.

Lafon-Lafourcade, S. (1983). Wine and brandy. In *Biotechnology*, Vol. 5, G. Reed (Ed.). Verlag Chemie, Deerfield Beach, Florida, pp. 81-163.

Law, B. A. (1982). Cheeses. In *Fermented Foods*, Vol. 7, *Microbiology*, A. H. Rose (Ed.). Academic Press, London, pp. 147-198.

Law, B. A., and Sharpe, M. E. (1977). The influence of the microflora of cheddar cheese on flavor development. *Dairy Indust. Int. 42*(12):10-14.

Lawrence, R. C., Thomas, T. D., and Terazaghi, B. E. (1976). Reviews of the progress of dairy science: Cheese starters. *J. Dairy Res. 43*:141-193.

Lees, G. J., and Jago, G. R. (1978). Role of acetaldehyde in metabolism: A review. 2. The metabolism of acetaldehyde in cultured dairy products. *J. Dairy Sci. 61*:1216-1224.

Margalith, P. Z. (1981). *Flavor Microbiology*. Charles C. Thomas, Springfield, Illinois.

Marshall, V. M., and Law, B. A. (1984). The physiology and growth of dairy lactic-acid bacteria. In *Advances in the Microbiology and Biochemistry of Cheese and Fermented Milk*. F. L. Davis and B. A. Law (Eds.). Elsevier Applied Science Publishers, London, pp. 67-98.

McKay, L. L. (1982). Regulation of lactose metabolism in dairy streptococci. In *Developments in Food Microbiology*, Vol. I, R. Davies (Ed.). Applied Science Publishers, Barking, Essex, England, pp. 153-182.

McKay, L. L. (1983). Functional properties of plasmids in lactic streptococci. *Antoinie Van Leuwenhoek 49*:259-274.

Orberg, P., and Sandine, W. E. (1984). Common occurrence of plasmid DNA and vancomycin resistance in *Leuconostoc* spp. *Appl. Environ. Microbiol. 48*:1129-1133.

Oura, E., Suomalainen, H., and Viskari, R. (1982). Breadmaking. In *Fermented Foods*, Vol. 7, *Economic Microbiology*, A. H. Rose (Ed.). Academic Press, London, pp. 87-146.

Pack, M. Y., Vedamuthu, E. R., Sandine, W. E., and Elliker, P. R. (1968). Effect of temperature on growth and diacetyl production by aroma bacteria in single- and mixed-strain lactic cultures. *J. Dairy Sci. 51*:339-344.

Panchal, C. J., Russell, I., Sills, A. M., and Stewart, G. G. (1984). Genetic manipulation of brewing and related yeast strains. *Food Technol. 38*(2):99-106, 111.

Perberdy, J. F. (1979). *Protoplasts—Applications in Microbial Genetics. A Handbook of Experimental Methods*. Department of Botany, University of Nottingham, England.

Porubcan, R. S., and Sellars, R. L. (1979). Lactic starter culture concentrates. In *Microbial Technology*, 2nd Ed., Vol. 1, H. J. Peppler and D. Perlman (Eds.). Academic Press, New York, pp. 59-92.

Raccach, M. (1984). The antimicrobial activity of phenolic antioxidants in foods. *J. Food Safety 6*:141-170.

Radler, F. (1975). The metabolism of organic acids by lactic acid bacteria. In *Lactic Acid Bacteria in Beverages and Food*, J. G. Carr, C. V. Cutting, and G. C. Whiting (Eds.). Academic Press, London, pp. 17-27.

Reed, G. (Ed.). (1983). *Biotechnology*, Vol. 5. Verlag Chemie, Deerfield Beach, Florida.

Reiter, B., and Harnulv, B. G. (1984). Lactoperoxidase antibacterial system: Natural occurrence, biological functions and practical applications. *J. Food Protect.* *47*:724-732.

Rose, A. H. (1982). *Economic Microbiology*, Vol. 7. Academic Press, London.

Sandine, W. E. (1979). *Lactic Starter Culture Technology*. Pfizer Cheese Monographs, Vol. 6. Pfizer Inc., New York.

Sandine, W. E., Daly, C., Elliker, P. R., and Vedamuthu, E. R. (1972). Causes and control of culture—related flavor defects in cultured dairy products. *J. Dairy Sci.* *55*:1031-1039.

Schmidt, R. H., Morris, H. A., Castberg, H. B., and McKay, L. L. (1976). Hydrolysis of milk proteins by bacteria used in cheesemaking. *J. Agric. Food Chem.* *24*:1106-1113.

Shahani, K. M., Arnold, R. G., Kilara, A., and Dwivedi, B. W. (1976). Role of microbial enzymes in flavor development in foods. *Biotechnol. Bioeng.* *18*:891-907.

Sharago, A. W., Chassy, B. M., and Dabrogosz, W. J. (1986). Conjugal plasmid transfer (pAM 1) in *Lactobacillus plantarum*. *Appl. Environ. Microbiol.* *52*:574-576.

Sing, E. L. (1976). Preparation of cottage cheese. U.S. Patent 3,968,256.

Smith, J. L., and Palumbo, S. A. (1981). Microorganisms as food additives. *J. Food Protect.* *44*:936-955.

Somkuti, G. A., and Steinberg, D. H. (1986). Distribution and analysis of plasmids in *Streptococcus thermophilus*. *J. Indust. Microbiol.* *1*:157-163.

Spicher, G. (1983). Baked goods. In *Biotechnology*, Vol. 5, G. Reed (Ed.). Verlag Chemie, Deerfield Beach, Florida, pp. 1-80.

Stamer, J. R. (1983). Lactic acid fermentation of cabbage and cucumbers. In *Biotechnology*, Vol. 5, G. Reed (Ed.). Verlag Chemie, Deerfield Beach, Florida, pp. 364-397.

Tsai, H. J., and Sandine, W. E. (1986). Conjugal transfer of plasmid genes between lactic streptococci and *Leuconostoc*. Abstracts 1986 Annual Meeting, Society of Industrial Microbiology. *SIM News* *36*(4):A-6.

Vedamuthu, E. R. (1978). Natural (unhydrolyzed) milk versus lactose-hydrolyzed milk for cultured dairy products: Physiological and practical implications to the starter industry. *J. Food Protect.* *41*:654-659.

Vedamuthu, E. R. (1979). Microbiologically induced desirable flavors in the fermented foods of the West. *Dev. Indust. Microbiol. 20*: 187-202.

Vedamuthu, E. R. (1982). Fermented milks. In *Fermented Foods*, Vol. 7, *Economic Microbiology*, A. H. Rose (Ed.). Academic Press, London, pp. 199-226.

Vedamuthu, E. R., and Neville, J. M. (1986). Involvement of a plasmid in production of ropiness (mucoidness) in milk cultures by *Streptococcus cremoris. MS Appl. Environ. Microbiol. 51*:677-682.

Vedamuthu, E. R., and Washam, C. J. (1983). Cheese. In *Biotechnology*, Vol. 5, G. Reed (Ed.). Verlag Chemie, Deerfield Beach, Florida, pp. 231-313.

Vedamuthu, E. R., Hauser, B. A., Henning, D. R., Sandine, W. E., Elliker, P. R., and Day, E. A. (1966). Competitive growth of *Streptococcus diacetilactis* in mixed strain lactic cultures and cheese. *Int. Dairy Cong. Proc. Munich, D*(2):611-618.

Wasserman, B. P. (1984). Thermostable enzyme production. *Food Technol. 38*(2):78-89, 98.

White, T. J., Meade, J. H., Shoemaker, S. P., Koths, K. E., and Innis, M. A. (1984). Enzyme cloning for the food fermentation industry. *Food Technol. 38*(2):90-98.

Wood, B. J. B., and Rainbow, C. (1961). The maltose phosphorylase of beer lactobacilli. *Biochem. J. 78*:204-209.

VI

DESIGN AND APPLICATIONS OF ANAEROBIC SYSTEMS

20

Anaerobic Degradation of Toxic and Hazardous Wastes

LARRY E. ERICKSON and L. T. FAN / *Kansas State University, Manhattan, Kansas*

I. INTRODUCTION

The biodegradation of organic compounds is part of the natural carbon cycle. Microorganisms develop pathways for the catabolism of newly synthesized organic compounds. Both aerobic and anaerobic biodegradation are important industrial and natural processes. However, the emphasis of this chapter is on anaerobic processes.

Considerable biodegradation research under anaerobic conditions has been reported; several substantial reviews have appeared (Evans, 1977; Kobayashi and Rittmann, 1982; Ghisalba, 1983; Speece, 1983; Sleat and Robinson, 1984; Young, 1984; Fannin et al., 1986; Berry et al., 1987; Rochkind-Dubinsky et al., 1987). A number of books and monographs have been published that consider both aerobic and anaerobic biodegradation of pesticides and other chemicals (Maki et al., 1980; Leisinger et al., 1981; Chakrabarty, 1982; Matsumura and Krisha Murti, 1982; Gibson, 1984; Schönborn, 1986; Rochkind-Dubinsky et al., 1987).

Sleat and Robinson (1984) have pointed out that the biodegradation of organic compounds may occur as the result of aerobic respiration, anaerobic respiration, and fermentation. According to them, nitrate and sulfate are common electron acceptors for anaerobic respiration, methane and carbon dioxide are the normal end products under fermentation conditions, and photometabolism is another mechanism resulting in biodegradation by photosynthetic bacteria. These authors have also reviewed the anaerobic biodegradation of aromatic compounds giving separate attention to photometabolism, nitrate-dependent respiration, sulfate-dependent metabolism, and methanogenic fermentation. Nitrate-dependent respiration has also been reviewed by Delwiche and Bryan (1976), Focht and Verstraete (1977), and Payne (1981). All four of these processes contribute to the biodegradation of organic compounds.

Several aspects of microbial biodegradation processes for toxic and hazardous compounds must be considered in the design and control of useful systems. These include the genetics, biochemical pathways, kinetics, bioenergetics, nutrition, stability, and reliability, which are all discussed in this chapter.

II. GENETICS

The genetics of microorganisms involved in the biodegradation of toxic substances have been investigated in a number of studies (see, e.g., Ghosal et al., 1985). A typical microorganism has three classes of genes: essential, often useful, and occasionally useful. Essential and often useful genes are most likely located on the chromosome; however, occasionally useful genes may be found in plasmids. Plasmids are small circular molecules of the genetic material DNA that exist outside the bacterial chromosome. The toluene (TOL) plasmid, coding for enzymes to degrade toluene and xylene, has been widely studied (Keil et al., 1985). Plasmids essential for the biodegradation of halogenated compounds have been reviewed recently (Ghosal et al., 1985). Plasmids involved in the biodegradation of chlorinated phenoxy herbicides, such as 2,4-D(2,4-dichlorophenoxyacetic acid), have been described by Fisher et al. (1978) and Don and Pemberton (1981). Plasmids can be transferred from one organism to another in the process of developing enrichment cultures that can metabolize new substrates (Hopwood, 1981). Recent work by Kim and Maier (1986) indicates that the development of acclimated cultures for the biodegradation of 2,4-D may require a substrate concentration greater than a certain threshold level. In their work, acclimated cultures were developed at 10 and 100 mg/L but not at 1 mg/L.

As cells divide and reproduce, plasmids essential for biodegradation may not be present in some of the new daughter cells. Thus, if a toxic substance is not biodegraded by the cells continuously, but only occasionally, their ability to degrade such a substance may be lost if a special effort is not made to maintain it. The microorganisms developed to degrade a specific toxic substance may be unable to compete well with other organisms in the absence of the toxic substance. For example, 2,4,5-trichlorophenoxyacetic acid

(2,4,5-T) can be biodegraded by *Pseudomonas cepacia* strain AC1100. In recent study (Kilbane et al., 1983; Ghosal et al., 1985), weekly applications of this organism to soil contaminated with 10,000-20,000 ppm of 2,4,5-T resulted in biodegradation of the 2,4,5-T and restoration of the soil to the point that it could be used to grow broad-leaved plants normally sensitive to low concentrations (about 10-15 ppm) of 2,4,5-T. However, in its absence this organism did not compete well with other microorganisms present in the soil. After 6 weeks in soil without 2,4,5-T, it was difficult to detect the presence of the *P. cepacia* in it. When genetic material associated with plasmids is essential for biodegradation, maintenance of the culture must include efforts to maintain these plasmids.

Some knowledge of genetics is highly desirable in the development of strains and in the design and control of processes for the biodegradation of toxic substances. Recent advances in genetic engineering and biotechnology can be applied to solve important biodegradation problems.

III. BIODEGRADATION CHEMISTRY

Knowledge of biochemical pathways is helpful in understanding biodegradation processes. Evans (1977) and Young (1984) have reviewed the literature associated with some important anaerobic biochemical pathways. This section deals with the biochemical pathways of specific substrates or classes of substrates.

A. Chlorinated Aliphatic Hydrocarbons

One of the meaningful applications of biodegradation is the treatment of chlorinated aliphatic compounds. Table 1 lists some of the chlorinated aliphatic compounds of interest. These compounds are frequently found in groundwater. Westrick et al. (1984) report the results of a national effort to evaluate the quality of groundwater through random sampling of both large and small supplies. The three most commonly detected volatile organic compounds were trichloroethylene, tetrachloroethylene, and 1,1,1-trichloroethane. The percentages of water supplies containing at least one volatile organic compound at a concentration above the quantification limit were 16.8% for supplies serving fewer than 10,000 people and 28% for supplies serving more than 10,000 people.

The volatility of chlorinated aliphatic compounds varies considerably with the number of chlorine atoms per molecule and with the number of carbon atoms per molecule. The boiling points are included in Table 1 in order to provide information on the relative volatility of these compounds. Of the most volatile substances—chloromethane,

Table 1 Chlorinated Aliphatic Hydrocarbons and Their Biodegradation Characteristics

Compound	Formula	Normal boiling point (°C)	Biodegradation characteristics	References
Chloromethane (methyl chloride)	CH_3Cl	-24.2		
Dichloromethane (methylene chloride)	CH_2Cl_2	40.1	Aerobic degradation by *Pseudomonas* and *Hydrophomicrobium*	LaPat-Polasko et al. (1984); Stucki et al. (1981); Rittman and McCarty (1980c)
Trichloromethane (chloroform)	$CHCl_3$	61.3	Anaerobic degradation Aerobic degradation	Bouwer and McCarty (1983a); Strand and Shippert (1986)
Tetrachloromethane (carbon tetra-chloride)	CCl_4	76.8	Anaerobic degradation	Bouwer and McCarty (1983a, b); Parsons and Lage (1985)
Chlorethane (ethyl chloride)	CH_3CH_2Cl	12.3		
1,1-Dichlorethane (ethylidene chloride)	CH_3CHCl_2	57.3		
1,2-Dichlorethane (ethylene chloride)	$ClCH_2CH_2Cl$	83.5	Anaerobic degradation Aerobic degradation	Bouwer and McCarty (1983a); Stucki et al. (1983)
1,1,1-Trichloroethane (methyl chloroform)	CH_3CCl_3	74.1	Anaerobic degradation	Bouwer and McCarty (1983a); Parsons and Lage (1985)
1,1,2-Trichloroethane	$ClCH_2CHCl_2$	113.8		

1,1,1,2-Tetrachloroethane	$ClCH_2CCl_3$	130.5		
1,1,2,2-Tetrachloroethane	$Cl_2CHCHCl_2$	146.2	Anaerobic degradation	Bouwer and McCarty (1983a)
Chloroethene (vinyl chloride)	CH_2CHCl	-13.4	Anaerobic degradation	Vogel and McCarty (1985); Fogel et al. (1986)
1,1-Dichloroethane (vinylidene chloride) (1,1-Dichloroethylene)	CH_2CCl_2	37	Anaerobic degradation Aerobic degradation	Vogel and McCarty (1985); Fogel et al. (1986); Barrio-Lage et al. (1986)
1,2-Dichloroethene (1,2-dichloroethylene)	$ClCHCHCl$	60.3 (cis)	Anaerobic degradation Aerobic degradation	Vogel and McCarty (1985); Fogel et al. (1986); Barrio-Lage et al. (1986)
Trichloroethene	$ClCHCCl_2$	87	Anaerobic degradation Aerobic degradation	Bouwer and McCarty (1983a); Parsons and Lage (1985); Nelson et al. (1986); Kloepfer et al. (1985); Wilson and Wilson (1985); Vogel and McCarty (1985); Fogel et al. (1986); Barrio-Lage et al. (1987)
Tetrachlorethene (perchloroethylene, tetrachloroethylene)	Cl_2CCCl_2	121	Anaerobic degradation	Bouwer and McCarty (1983a); Parsons and Lage (1985); Vogel and McCarty (1985)

chloroethene (vinyl chloride), and chloroethane—only chloroethene has been investigated in carefully controlled studies to determine its biodegradability. Fogel et al. (1986) have found that methane-utilizing cultures are able to biodegrade chloroethene under aerobic conditions. Vogel and McCarty (1985) have found evidence to indicate that chloroethene is biodegraded anaerobically; nevertheless, further studies are needed to reveal definitive answers. Because of the volatility of chlorinated aliphatic hydrocarbons with one or two carbon atoms per molecule and with only one chlorine atom per molecule, these substances are easily removed by stripping. Although stripping is commonly used to remove chlorinated aliphatic compounds from water, it results in the compound being found in another phase where further processing may be required.

Anaerobic biodegradation studies with chlorinated aliphatic hydrocarbons have been reported by Bouwer et al. (1981), Bouwer and McCarty (1983a, b), Kloepfer et al. (1985), Parsons and Lage (1985), Vogel and McCarty (1985), and Barrio-Lage et al. (1986, 1987). Bouwer and McCarty (1983a, 1983b) found that methanogenic conditions were generally better than denitrification conditions. Table 2 also presents some of the anaerobic biodegradation results of Bouwer and McCarty (1983a) under methanogenic conditions. They conducted continuous flow experiments in columns containing glass beads with microbial cultures attached to these supports and acetate in the influent to help maintain the biologic activity of the column. Table 2 shows that trichloromethane, tetrachloromethane, 1,1,1-trichloroethane, 1,1,2,2-tetrachloroethane, and tetrachloroethene can be anaerobically biodegraded under methanogenic conditions. Under denitrification conditions, Bouwer and McCarty (1983b) found that tetrachloromethane was biodegraded but trichloromethane and 1,1,1-trichloroethane were not.

Radioactive tracer studies by Bouwer and McCarty (1983a) indicated that trichloromethane and tetrachloromethane were oxidized to carbon dioxide. In these studies a small portion of the trichloromethane (about 9%) was converted to methane. Moreover, 1,2-dichloroethane was found to be oxidized to carbon dioxide at a rate that was much slower than the rates observed for trichloromethane and tetrachloromethane.

The continuous flow column experiments indicated that 1,1,2,2-tetrachloroethane was transformed to 1,1,2-trichloroethane by reductive dechlorination. Similarly, tetrachloroethene was converted to trichloroethene by reductive dechlorination (Bouwer and McCarty, 1983a). Vogel and McCarty (1985) reported that tetrachloroethene and trichloroethene were transformed by reductive dehalogenation to dichloroethene and chloroethene, as shown in Figure 1. Carbon

Table 2 Performance of Glass Bead Column Under Methanogenic Conditions[a]

Compound	Acclimation period (weeks)	Influent concentration (mg/L)	Removal (%)
Trichloromethane	10	0.033	96
Tetrachloromethane	0	0.017	99
1,2-Dichloroethane	16	0.022	0
1,1,1-Trichloroethane	10	0.025	98
1,1,2,2-Tetrachloroethane	0	0.027	97
Tetrachloroethane	10	0.018	86

[a]Acetate was fed at the rate of 600 mg/L for the first 12 months and 100 mg/L after 13 months. The feeding of 1,2-dichloroethane was started after 15 months of operation. The biodegradation product, 1,1,2-trichloroethane, was found in the effluent when 1,1,2,2-tetrachloroethane was fed; trichloroethene was found 10 weeks after the feeding of tetrachloroethene was started.
Source: From Bouwer and McCarty (1983a).

dioxide was also found as one of the transformation products. Vogel and McCarty (1985) suggest that chloroethene may be transformed to carbon dioxide by the addition of water to form 2-chloroethanol, followed by oxidation to form the aldehyde. Further oxidation to carbon dioxide or reduction to methane may follow. The rate of radioactive carbon dioxide production was found to decrease after nonlabeled chloroethene was added to the column; this supports the notion that carbon dioxide is one of the products of anaerobic chloroethene biodegradation.

The reductive dehalogenation of chlorinated aliphatic hydrocarbons has also been studied by Parsons and coworkers (Parsons et al., 1984; Parsons and Lage, 1985; Barrio-Lage et al., 1986). According to them, tetrachloromethane was transformed to trichloromethane, 1,1,1-trichloroethane was transformed to 1,1-dichloroethane, and tetrachloroethene and trichloroethene to dichloroethenes and chloroethene. Kloepfer et al. (1985) also reported that trichloroethene was transformed to 1,2-dichloroethene by reductive dechlorination. Chloroethane was produced from *cis*-1,2-dichloroethene but not from *trans*-1,2-dichloroethene (Barrio-Lage et al., 1986), thereby indicating that reactions other than reductive dechlorination also occurred.

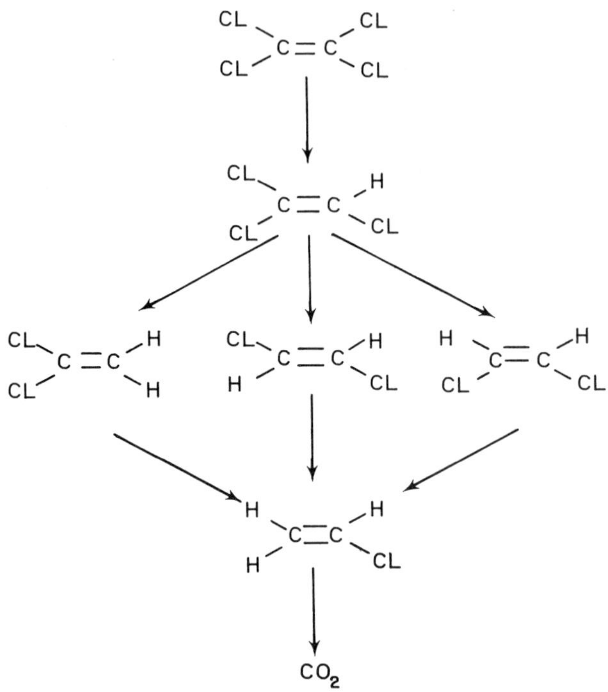

Figure 1 Reductive dehalogenation under methanogenic conditions for tetrachloroethene, trichloroethene, dichloroethene, and chloroethene. (From Vogel and McCarty, 1985.)

B. Aromatic Compounds

The anaerobic biodegradation of aromatic compounds has been reviewed recently by Sleat and Robinson (1984), Young (1984), and Berry et al. (1987). Benzoate and phenol are two of the most widely studied compounds because they are intermediate products for numerous other aromatic compounds biodegraded anaerobically. Young (1984) has reviewed the biochemical pathways and points out that anaerobic biodegradation of benzoate involves reduction of the ring to obtain cyclohexane carboxylic acid followed by reductive cleavage of the ring to aliphatic acids (see Fig. 2). A number of investigators have reported the anaerobic biodegradation of benzoate under denitrification conditions, but Fabig et al. (1980) found that with nitrate biodegradation did not occur under strictly anaerobic conditions. Trace amounts of oxygen appear to be required (Sleat and Robinson, 1984).

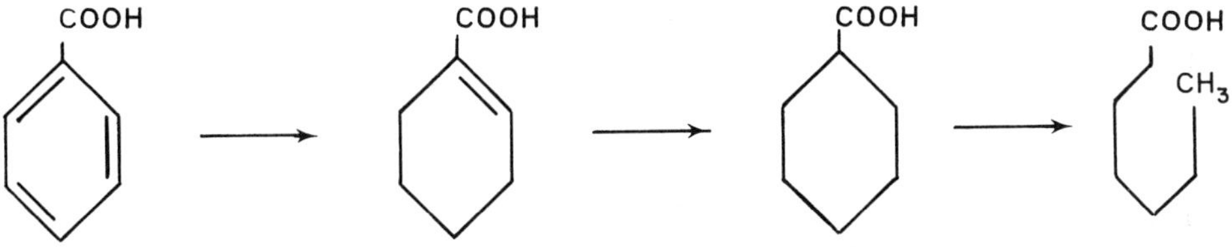

1 2 3 4

Figure 2 Biodegradation of benzoate (1) to obtain cyclohexene-1-carboxylic acid (2), cyclohexane carboxylic acid (3), and heptanoic acid (4). (From Young, 1984.)

Biodegradation of phenol involves reduction of the ring followed by cleavage of the ring, as shown in Figure 3. Table 3 presents information on the anaerobic biodegradation of a number of aromatic compounds by means of photometabolism (P), anaerobic respiration using nitrate (N) or sulfate (S), and methanogenic fermentation (M). An earlier version of Table 3 was presented by Sleat and Robinson (1984). Blum et al. (1986) have reported the biodegradation of coal conversion wastewater under methanogenic conditions, as shown in Table 4. Shelton and Tiedje (1984) have also examined the anaerobic biodegradation of various organic compounds under methanogenic conditions, as shown in Table 5.

Many aromatic compounds are inhibitory at concentrations above about 300 mg/L; however, acclimation may allow anaerobic biodegradation to take place at concentrations six to eight times higher (Young, 1984). Lag times to allow the culture to adapt are observed frequently. Figure 4 illustrates the lag times for phenol (a) and catechol

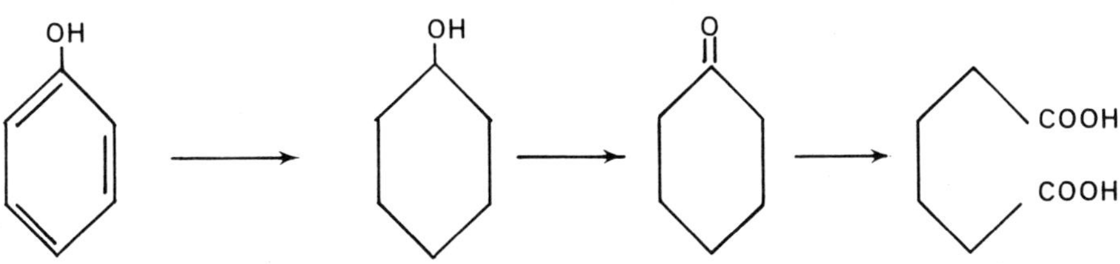

1 2 3 4

Figure 3 Biodegradation of phenol (1) to obtain cyclohexanol (2), cyclohexanone (3), and adipic acid (4). (From Young, 1984.)

Table 3 Aromatic Compounds Degraded Under Anaerobic Conditions[a]

Substrate	Mode of metabolism	References
Benzene	M	Vogel and Grbic-Galic (1986)
Benzoate	P	Scher (1960); Procter and Scher (1960); Guyer and Hegeman (1969); Dutton and Evans (1967, 1969)
	N	Scher (1960); Taylor et al. (1970); Williams and Evans (1973, 1975); Aftring and Taylor (1981)
	S	Balba and Evans (1980); Widdel (1980)
	M	Tarvin and Buswell (1934); Clark and Fina (1952); Fina and Fiskin (1960); Nottingham and Hungate (1969); Ferry and Wolfe (1976); Balba and Evans (1977); Fina et al. (1978); Shlomi et al. (1978); Abram and Nedwell (1978); Healy and Young (1979); Sleat and Robinson (1983)
o-Hydroxybenzoate	N	Taylor et al. (1970)
	S	Widdel (1980)
	M	Balba et al. (1979)
m-Hydroxybenzoate	P	Dutton and Evans (1969)
	N	Taylor et al. (1970); Williams and Evans (1975); Bakker (1977b)
	S	Widdel (1980)
p-Hydroxybenzoate	P	Dutton and Evans (1969)
	N	Oshima (1965); Taylor et al. (1970); Taylor and Heeb (1972); Williams and Evans (1975); Bakker (1977b)

Substrate	Mode of metabolism	References
	S	Widdel (1980)
	M	Balba et al. (1979); Healy and Young (1979)
Protocatechuate	N	Oshima (1965); Taylor et al. (1970); Williams and Evans (1975); Bakker (1977b)
	M	Balba et al. (1979); Healy and Young (1979)
Gallic acid	M	Schink and Pfennig (1982)
2,4,6-Trihydroxybenzoic acid	M	Schink and Pfennig (1982)
Catechol	M	Healy and Young (1978, 1979); Balba et al. (1979)
Benzaldehyde	N	Williams and Evans (1975)
	M	McCarty et al. (1977)
Benzyl alcohol	N	Williams and Evans (1975)
Phenol	N	Bakker (1977a, b)
	M	Healy and Young (1979); Balba et al. (1979); Young and Rivera (1985)
Pyrogallol	M	Schink and Pfennig (1982)
Hydroquinone	M	Young and Rivera (1985)
Phloroglucinol	M	Young and Rivera (1985); Schink and Pfennig (1982)
Phenylacetate	N	Williams and Evans (1975)
	S	Widdel (1980)
	M	Tarvin and Buswell (1934); Balba and Evans (1979)
Hippurate	S	Widdel (1980)
o,*m*,*p*-Cresol	N	Bakker (1977b)

(Table 3, continued)

Substrate	Mode of metabolism	References
p-Cresol	N	Bossert and Young (1986)
	M	Young and Rivera (1985)
Toluene	M	Vogel and Grbic-Galic (1986)
	N	Zeyer et al. (1986)
1,2-Dimethylbenzene (*o*-xylene)	N	Kuhn et al. (1985)
1,3-Dimethylbenzene (*m*-xylene)	N	Zeyer et al. (1986)
1,4-Dimethylbenzene (*p*-xylene)	N	Kuhn et al. (1985)
o,*m*,*p*-Methoxybenzoate	M	Balba et al. (1979)
o,*m*,*p*-Phthalic acid	N	Aftring et al. (1981); Aftring and Taylor (1981)
2-Aminobenzoate (anthrunilic acid)	N	Braun and Gibson (1984)
Cinnamic acid	N	Williams and Evans (1975)
	M	Tarvin and Buswell (1934); Healy and Young (1979)
p-Hydroxycinnamate	N	Williams and Evans (1975)
Caffeic acid	N	Williams and Evans (1975)
Syringic acid	M	Healy and Young (1979); Kaiser and Hanselmann (1982)
Vanillic acid		Healy and Young (1979); Zeikus et al. (1982)
Syringaldehyde	M	Healy and Young (1979)
Vanillin	M	Healy and Young (1979)
Ferulic acid	M	Healy and Young (1979)
Fernitrothion	M	Liu et al. (1981)

[a]P, photometabolism; N, nitrate-dependent; S, sulfate-dependent; M, methanogenic fermentation.
Source: From Sleat and Robinson (1984).

Table 4 Summary of the Biodegradability of Coal Conversion Wastewater Constituents in Serum Bottle Studies of Blum et al. (1986)

	Theoretical gas production achieved in given seed culture (%)			
Compound[a]	Concentration (mg/L)	Northeast plant sludge (%)	Southwest plant sludge (%)	Phenol-acclimated culture (%)
Phenol	500	65[b]	89[b]	
	1000	22		91[b]
	2000[c]	0		0[b]
Resorcinol	500	36		83[b]
	1000[c]	0		4
	2000	0		0[b]
Catechol	500	0	106[b]	69[b]
	1000[c]	0		47[b]
	2000	0		0[b]
Acetic acid	400[c]			94[b]
p-Cresol	180-250[c]	96[b]		91[b]
	500	63[b]	75[b]	82[b]
	1000	0		0[b]
3,4-Xylenol	100	0		0[b]
	250[c]	0		2
	500	0		
Pyridine	50	0	240[b]	
	100	105		58[b]
	200-250	53[b]		9[b]
	400			0
Benzoic acid	200			73[b]
	500	29[b]		76[b]
	1000	37[b]		66[b]
	2000	11[b]		
4-Methylcatechol	150			28[b]
	250	0		36[b]
	500	0		23[b]
	1000	0		23[b]
Acetophenone	50[c], 100	0		
	250	20		

(Table 4, continued)

Compound[a]	Concentration (mg/L)	Theoretical gas production achieved in given seed culture (%): Northeast plant sludge (%)	Southwest plant sludge (%)	Phenol-acclimated culture (%)
Indole	50[c], 100	0		
	250	66		
3-Ethylphenol	100	10		
	250	2[b]		
	500	19[b]		
Aniline	50	0	97	
	100	3		
	250	6		
Hexanoic acid	500	56[b]		
	1000	51[b]		
	2000	45[b]		
1-Naphthol	100	0	4	
	250, 500	0		

[a]The following were not biodegraded: *p*-cresol, 2,3-xylenol, 3,5-xylenol, 4-ethylpyridine, 2-indanol, 5-methylresorcinol, 2-naphthol, 2,3,5-trimethylphenol, and 2-methylquinoline.
[b]Gas significantly more or less than control at p = 0.05.
[c]Concentration in typical coal conversion wastewater.

Table 5 Organic Compounds Mineralized Under Anaerobic Conditions in 10% Sludge[a]

Conditions	Organic compounds
Mineralized in 10% sludge from a secondary digester (greater than 75% of theoretical methane production)	Acetylsalicylic acid
	Acrylic acid
	p-Anisic acid
	o-Anthranilic acid
	Benzoic acid
	Benzyl alcohol
	Catechol
	m-Cresol
	p-Cresol

(Table 5, continued)

Conditions	Organic compounds
	Di-*n*-butylphthalate Dimethylphthalate Ethyl acetate 2-Hexanone *o*-Hydroxybenzoic acid *p*-Hydroxybenzoic acid 3-Hydroxybutanone 3-Methylbutanol 1-Octanol Phenol Phloroglucinol Phthalic acid Polyethylene glycol Protocatechuic acid Pyrogallol
Partially mineralized in 10% sludge from a secondary digester (30-75% of theoretical methane production)	*p*-Aminobenzoic acid Butylbenzylphthalate 4-Chloroacetanilide *m*-Chlorobenzoic acid Diethylphthalate Geraniol 4-Hydroxyacetanilide *p*-Hydroxybenzyl alcohol 2-Octanol Propionanilide
Mineralized in 10% sludge from a primary digester (greater than 75% of theoretical methane production)	Butylbenzylphthalate *m*-Chlorobenzoic acid *o*-Chlorophenol Diethylphthalate *o*-Methoxyphenol *m*-Methoxyphenol *p*-Methoxyphenol *o*-Nitrophenol *p*-Nitrophenol

[a]The following compounds were inhibitory: 5-chlorosalicylic acid, *p*-nitrophenol, *m*-chlorophenol, *p*-chlorophenol, pentachlorophenol, chloroform, hexachloropentadiene, trichloroacetic acid, 4-phenoxybutyric acid, and atrizine.

Source: Shelton and Tiedje (1984).

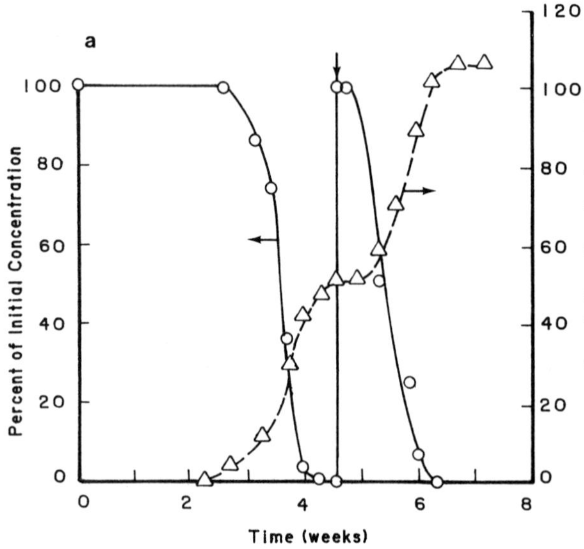

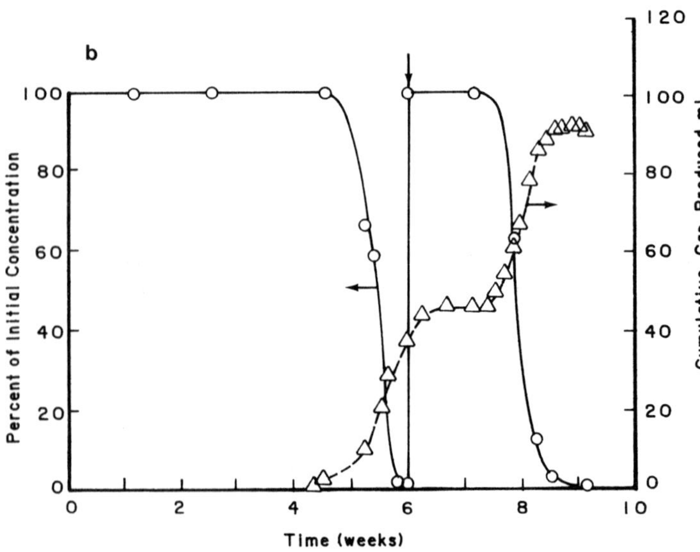

Figure 4 Anaerobic degradation of phenol (a) and catechol (b) under methanogenic conditions. At the point indicated by the arrow, fresh substrate was added to the culture vessels: (○-○-○) substrate; (△-△-△) cumulative gas production. (From Young, 1984.)

(b). For catechol, a lag is also observed after substrate is added to the adapted culture.

C. Halogenated Aromatic Compounds

The biodegradation of halogenated aromatic compounds has been investigated under aerobic and anaerobic conditions in a number of studies. Rochkind-Dubinsky et al. (1987) have reviewed the literature on the microbial decomposition of chlorinated aromatic compounds in a book devoted to this topic. Anaerobic biodegradation under methanogenic conditions has been successfully used to biodegrade 3,4-dichlorobenzoate, 3,5-dichlorobenzoate, pentachlorophenol, 3,4,5-trichlorophenol, 3,5-dichlorophenol, 2-chlorophenol, 3-chlorophenol, 4-chlorophenol, 2,4-D-[4-chloro-*o*-tolyl)oxy]acetic acid (MCPA), and 2,4,5-T, as well as other chlorinated aromatic compounds (Buisson et al., 1986; Gibson and Suflita, 1986; Mikesell and Boyd, 1986; and Rochkind-Dubinsky et al., 1987). Methanogenic conditions appear to be superior to environmental conditions in which nitrate or sulfate are present when reductive dechlorination is to be carried out (Bouwer and McCarty, 1983a, b; Kuhn et al., 1985; Gibson and Suflita, 1986). Figure 5 shows the reductive dechlorination of 3,4-dichlorobenzoate (1) to 3-chlorobenzoate (2), and benzoate (3). Figure 6 displays the experimental results of Mikesell and Boyd (1986) for reductive dechlorination of pentachlorophenol. The intermediate products 3,4,5-trichlorophenol, 3,5-dichlorophenol, and 3-chlorophenol are shown to illustrate the progress of the biodegradation pathway shown in Figure 7.

Various other halogenated aromatic compounds have been metabolized anaerobically. Schennen et al. (1985) and Taylor et al. (1979) investigated the biodegradation of fluorobenzoates under denitrifica-

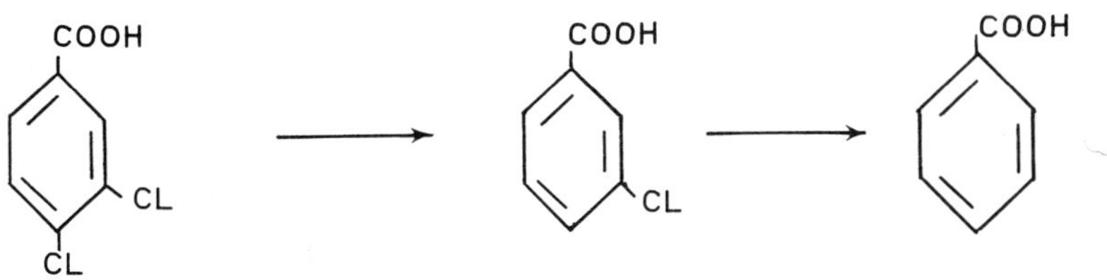

1 2 3

Figure 5 Pathway observed by Gibson and Suflita (1986) for anaerobic biodegradation of 3,4-dichlorobenzoate (1) to 3-chlorobenzoate (2) and benzoate (3). Benzoate is metabolized to CH_4 and CO_2.

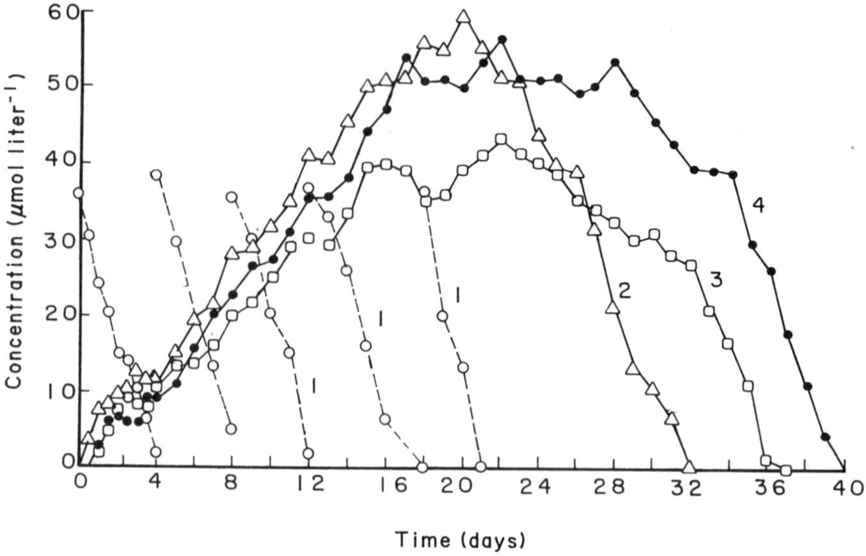

Figure 6 Anaerobic biodegradation of pentachlorophenol (0, 1) under methanogenic conditions in a mixture of 2-, 3-, and 4-chlorophenol-acclimated sludges. The sequential appearance of 3,4,5-trichlorophenol (△, 2), 3,5-dichlorophenol (□, 3), and 3-chlorophenol (●, 4) is shown. (From Mikesell and Boyd, 1986.)

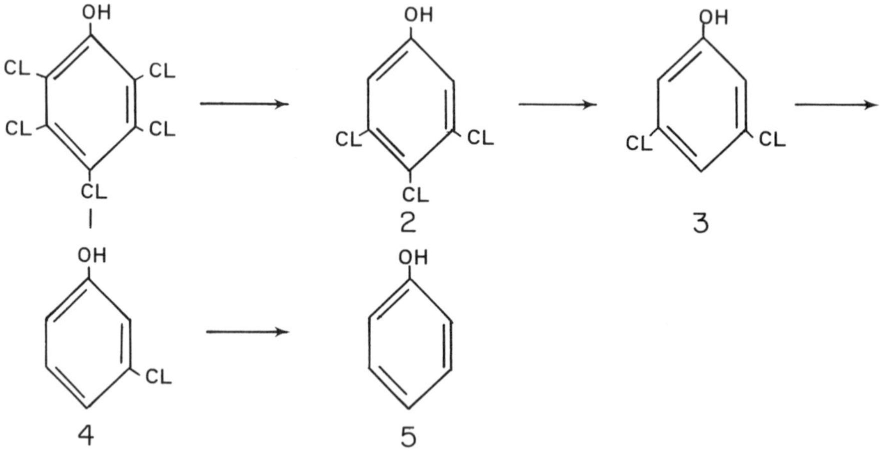

Figure 7 Anaerobic biodegradation of pentachlorophenol (1) under methanogenic conditions to 3,4,5-trichlorophenol (2), 3,5-dichlorophenol (3), 3-chlorophenol (4), and phenol (5). Phenol is metabolized to CH_4 and CO_2. (From Mikesell and Boyd, 1986.)

tion conditions. Suflita et al. (1982) reported that 2-, 3-, and 4-iodobenzoate and 2-, 3-, and 4-bromobenzoate can be metabolized to methane and CO_2.

D. Biodegradation of Pesticides

Much of the anaerobic biodegradation research just discussed is related to the biodegradation of pesticides. In addition, numerous studies have been directed toward the fate of particular pesticides under aerobic and anaerobic conditions (Bromilow et al., 1986; Kiene and Capone, 1986; Lal et al., 1986; Ozaki et al., 1986).

IV. KINETICS

The kinetics of biodegradation must be known for the design of biodegradation processes and for predicting the performance of the resultant operation. Several important rate processes must be considered. The rate of transport of the xenobiotic substance to the enzymes responsible for biodegradation may be crucial in some applications, especially when solubility of the substrate is very low. The biochemical kinetics of anaerobic growth on complex organics is extremely significant; however, this has not been widely considered.

Figure 8 shows some kinetic data of Suflita et al. (1982) for the anaerobic biodegradation of several halobenzoates. The observed rates are zero order with the following rate constants in micromoles per hour per milligram of protein: 0.314 ± 0.012 for 3-chlorobenzoate, 0.114 ± 0.003 for 3,5-dichlorobenzoate, 0.036 ± 0.002 for 4-amino-3-chlorobenzoate, and 0.043 ± 0.001 for 4-amino-3,5-dichlorobenzoate.

The kinetic data of Braun and Gibson (1984) are illustrated in Figure 9 for the anaerobic biodegradation of 2-aminobenzoate in the presence of nitrate. A region of linear growth appears to follow the initial period of exponential growth for these data. These kinetic characteristics are yet to be fully explained.

Figure 10 indicates that the biodegradation kinetics depend on pH and temperature, with a relatively narrow range of growth conditions for the anaerobic biodegradation of 2-aminobenzoate in the presence of nitrate. Finding the proper pH and temperature may be necessary in the search for biodegradability in some applications.

Since many xenobiotic substrates are inhibitory, this effect must be considered in some applications. The kinetics of phenol biodegradation by an immobilized methanogenic consortium was investigated by Dwyer et al. (1986). The maximum rate of biodegradation for the immobilized mixed culture was 68% of that of the native

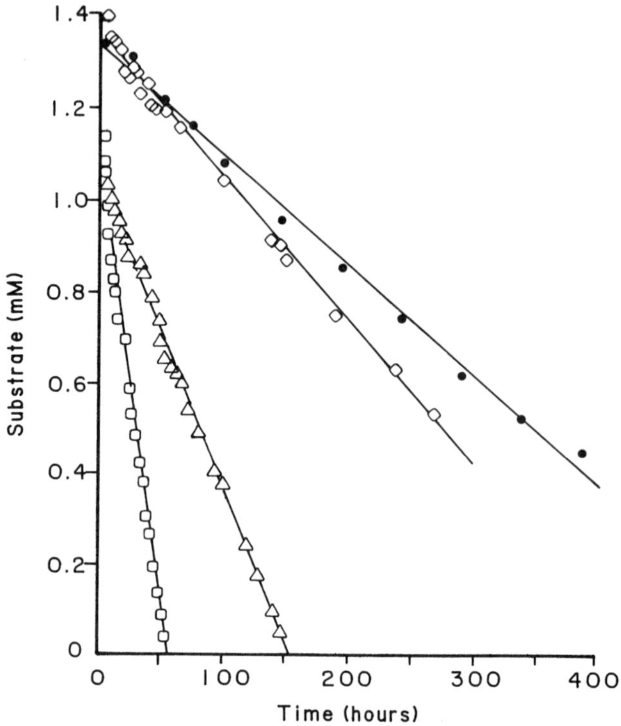

Figure 8 Zero-order biodegradation rates under methanogenic conditions for 4-amino-3-chlorobenzoate (●, 1), 4-amino-3,5-dichlorobenzoate (◇, 2), 3,5-dichlorobenzoate (△, 3), and 3-chlorobenzoate (□, 4). (From Suflita et al., 1982.)

culture; however, the apparent saturation constant was reduced from 90 to 46 mg/L of phenol by immobilization. The reduced value of the saturation constant was attributed to the retention of acetate, H_2, and CO_2 in the proximity of the immobilized mixed culture. The immobilized cells were protected from substrate inhibition caused by high phenol concentration; the apparent value of the inhibition parameter was 900 mg/L of phenol for the native culture and 1725 mg/L for the immobilized mixed culture. Figure 11 compares the experimental data obtained by Dwyer et al. (1986) with the model

$$v = \frac{V_{max}}{1 + K_m/SX + (S/K_i)^n}$$

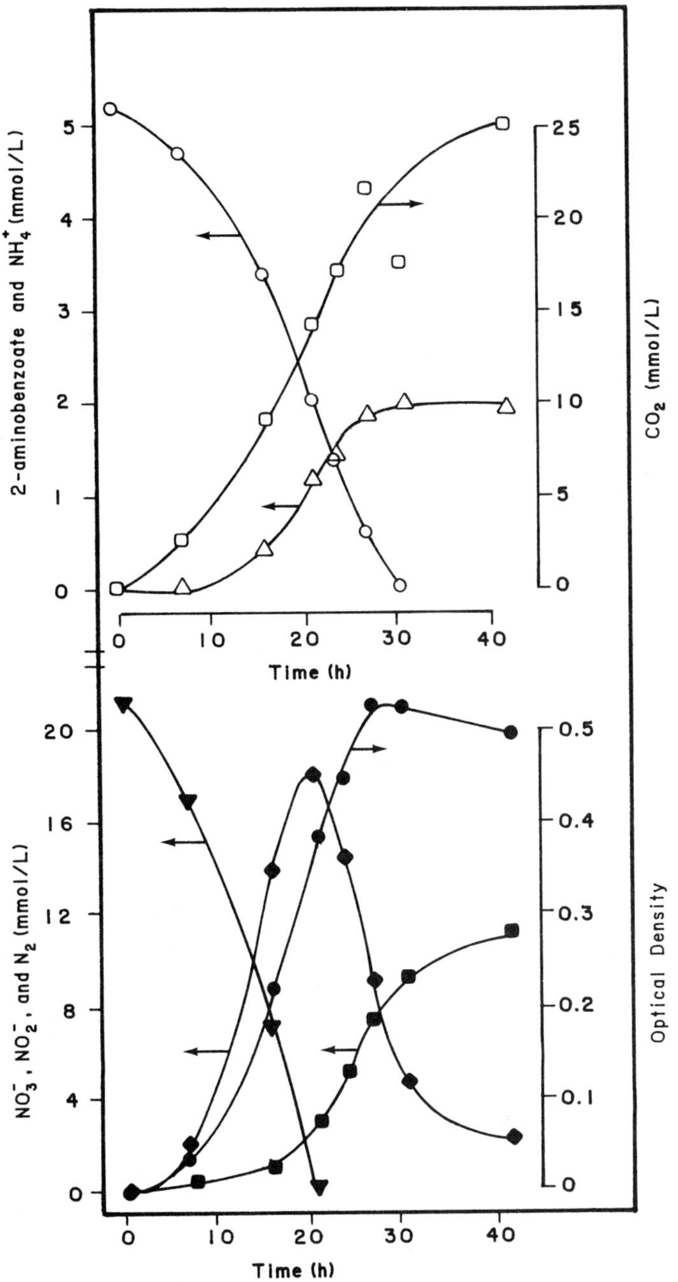

Figure 9 Anaerobic degradation of 2-aminobenzoate in the presence of nitrate by *Pseudomonas* sp. strain KB740: (○) 2-aminobenzoate; (□) CO_2; (△) NH_4^+; (▼) NO_3^-; (◆) NO_2^-; (■) N_2; (●) optical density. (From Braun and Gibson, 1984.)

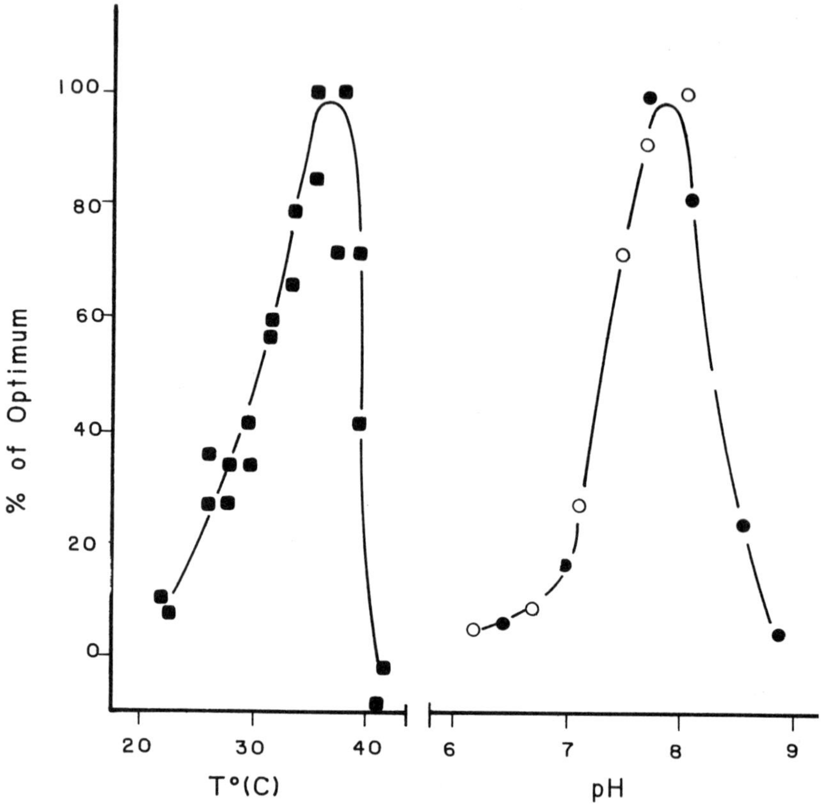

Figure 10 Effect of temperature and pH on growth of *Pseudomonas* sp. strain KB740. Cells were grown anaerobically on 2-aminobenzoate in 10 mL of mineral medium with nitrate in Hungate tubes. Data are mean values of percentage optical density reached and percentage substrate utilized after 20 h of growth. Optimum values are 37°C and pH 7.8: (■) temperature dependence (two experiments); (○) pH dependence (KH_2PO_4-K_2HPO_4 buffer); (●) pH dependence (KH_2PO_4-Na_2CO_3 buffer). (From Braun and Gibson, 1984.)

where v is the specific rate of biodegradation, and V_{max}, K_m, K_i, X, and n are parameters, and S is the phenol concentration.

The kinetics of adaptation of the culture to the xenobiotic substrate has received little attention. Lee and Bailey (1984a-c) have recently published a series of papers on the modeling of genetic events leading to the production of biodegradative enzymes. Never-

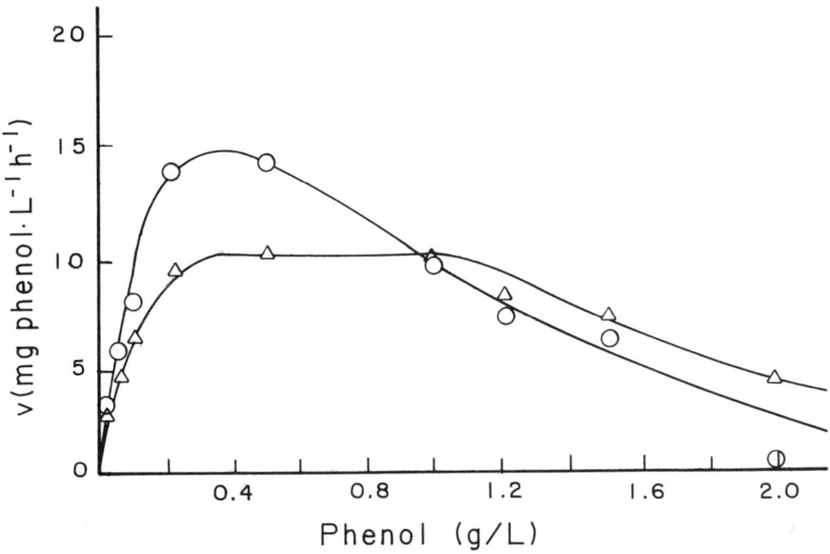

Figure 11 Simulated curves of anaerobic phenol degradation versus phenol concentration for both native and immobilized cells. Experimental data for native (○) and immobilized (△) cells are presented to show their fit with the model-derived curves. Each datum point is a mean of triplicate determinations. (From Dwyer et al., 1986.)

theless, much remains to be done to develop rate expressions useful in modeling the exchange of plasmids by microorganisms and the development of biodegradation capability by a mixed culture.

Another area of kinetic modeling requiring further investigation is that of biodegradation by pure or mixed cultures in which more than one substrate is present. Because toxic and hazardous substances are often present in very minute concentrations, growth in the presence of other substrates is often desirable. Since some organisms exhibit diauxic growth phenomena (Bailey and Ollis, 1986) under these conditions, prediction of performance frequently requires experimental work to determine the kinetics. In several recent studies, xenobiotic substrates have been supplied together with other substrates (Bouwer and McCarty, 1983a, 1985).

The kinetics of biodegradation of complex organic compounds when cometabolites are present has been considered by Schmidt et al. (1985); however, kinetic models useful for a variety of anaerobic biodegradation processes are yet to be developed.

V. BIOENERGETICS AND NUTRITION

A. Nutrition and Evaluating Biodegradability

Microorganisms require an adequate supply of carbon, energy, nitrogen, sulfur, phosphorus, water, and trace quantities of many other substances for growth. For aerobic growth, oxygen is needed. For anaerobic growth, an electron acceptor other than oxygen, such as nitrate or sulfate, may be present. In any experimental effort to investigate biodegradation, it is necessary to ensure that all the essential nutrients for growth are present. Grady (1985) presents an excellent plan for evaluating the biodegradability of chemical compounds. He divides the evaluation into four different sets of experiments to minimize cost and effort and maximize the utility of the information obtained. Screening tests are used first to identify readily biodegradable compounds. Acclimation and enrichment experiments are performed to investigate the biodegradability of those compounds that were not degraded in the screening test. The third phase involves evaluating the extent of biodegradation after a mechanism for biodegradation has been identified. Finally, the kinetics of biodegradation are investigated. Grady (1985) presents many useful suggestions and references for those wishing to undertake an experimental evaluation.

Shelton and Tiedje (1984) have investigated several methods for determining the anaerobic biodegradation potential under methanogenic conditions. They critically evaluate the proposed American Society for Testing and Materials (ASTM) standard method (Gledhill, 1979) and the method of Owen et al. (1979). Based on the results of numerous experiments, Shelton and Tiedje (1984) have recommended the following protocol for evaluating the anaerobic biodegradation potential.

1. Use primary anaerobic sludge with a 15 to 30 day retention time and total organic solids of approximately 1.0-2.0%. They have shown that sludge can be stored for up to 4 weeks at 4°C in tightly capped containers; however, fresh sludge should be used whenever possible. Since sludges vary in their selected populations, it may be advantageous to use more than one sludge when working with more persistent compounds.
2. Autoclave revised anaerobic mineral medium (RAMM) (per liter) for 5-10 minutes: phosphate buffer, 0.27 g of KH_2PO_4 and 0.35 g of K_2HPO_4 (adjusted to pH 7.0); mineral salts, 0.53 g of NH_4Cl, 75 mg of $CaCl_2 \cdot 2H_2O$, 100 mg of $MgCl_2 \cdot 6H_2O$, 20 mg of $FeCl_2 \cdot 4H_2O$, 0.5 mg of $MnCl_2 \cdot 4H_2O$, 0.05 mg of H_3BO_3, 0.05 mg of $ZnCl_2$, 0.03 mg of $CuCl_2$, 0.01 mg of

$NaMo_4 \cdot 2H_2O$, 0.5 mg of $CoCl_2 \cdot 6H_2O$, 0.05 mg of $NiCl_2 \cdot 6H_2O$, and 0.05 mg of Na_2SeO_3. The object of the autoclaving is to drive off O_2. The RAMM is then cooled to approximately 35°C while sparging with a 10% CO_2-90% N_2 gas mixture passed through copper fillings at 300°C to remove traces of O_2. After cooling, 1.2 g of $NaHCO_3$ and 0.5 g of $Na_2S \cdot 9H_2O$ (optional) per liter may be added to the medium. RAMM has been thoroughly evaluated by Shelton and Tiedje (1984) under the test conditions; it minimizes precipitation. If another anaerobic mineral medium is used, a 4 mM phosphate buffer and 1.2 g of $NaHCO_3$ per liter with a 10% CO_2-90% N_2 headspace are recommended. To ensure reducing conditions, the addition of sulfide in a concentration not exceeding 1 mM is recommended but not required.

3. Incubate a 10% homogeneous sludge solution with 50 μg of C per milliliter of test chemical in 160 mL serum bottles with new butyl rubber stoppers and aluminum crimp seals. Each chemical should be tested in triplicate. A standard deviation of $<15\%$ is expected.
4. Incubate serum bottles at 35°C for at least 8 weeks or until biodegradation is complete. If gas production is still in progress at week 8, then incubations should be continued until gas production is complete.
5. Measure gas production with a pressure transducer and multimeter, which is connected to the test bottle with a three-way valve and 20 gauge needle. Measurements should be made on a weekly basis; excess gas pressure should be vented.
6. Incubate a reference chemical(s) with each series of incubations to confirm sludge activity. *p*-Cresol and phthalic acid are suggested as a more rigorous test of biodegradation capacity. Ethanol, which is easily degraded, may be used as a positive control.

This general screening method measures the potential of biodegradation for anaerobic fermentation conditions. It may be important to conduct other experiments because the results of biodegradation may vary depending on whether nitrate or sulfate is present as electron acceptor. The method of Shelton and Tiedje (1984) provides a first-level evaluation of chemical fate and corresponds to the screening phase of Grady (1985). A strong positive correlation has been found to exist between the results obtained using primary anaerobic digester sludge and lake sediments (Shelton and Tiedje, 1984).

Some of the results obtained are summarized in Tables 5 and 6. Table 5 lists organic compounds that were mineralized totally or

Table 6 Theoretical Methane Production (%) from Four Substrates by Sludges from Nine Municipalities[a]

Sludge	*p*-Cresol	*m*-Cresol	2-Octanol	*m*-Chlorobenzoic acid
Adrian	91	82	0	85
Jackson	88	103	22	0
Ann Arbor	101	85	58	0
St. Johns	79	0	0	0
Ionia	80	0	0	0
Holt	88	77	0	85
Mason	94	91	87	0
Chelsea	62	ID[b]	0	0
Portland	77	0	0	0

[a]Fresh sludge (10%) from primary digesters incubated for 8 weeks, after which methane was measured by gas chromatography.
[b]Insufficient data; generally due to leaky bottles or only one bottle showing evidence of degradation.
Source: From Shelton and Tiedje (1984).

partially. Compounds that exhibited some inhibitory properties are identified. The substrate concentration of 50 μg of carbon per milliliter is recommended because 50 μg is sufficiently large to obtain good analytic results and yet small enough to limit inhibition. The results in Table 6 indicate that variations in anaerobic digester sludge may result in significant differences in biodegradation performance.

Gibson and Suflita (1986) have compared the biodegradation data obtained with samples taken from four different anoxic habitats, consisting of sewage sludge, pond sediment, a methanogenic aquifer, and a sulfate-reducing aquifer. Table 7 shows that methanogenic conditions are superior to sulfate-reducing conditions for the biodegradation of chlorinated organic compounds. Based on the results of additional experiments, it has been concluded that the inorganic sulfate inhibits the dehalogenation of chloroaromatic compounds.

As noted in Table 7, Gibson and Suflita (1986) observed intermediate biodegradation products in several of their experiments. Figure 12 illustrates that the intermediates observed depend on the source of the microorganisms. The pond sediment habitat had the ability to transfer 3,4-dichlorobenzoate primarily to 3-chlorobenzoate, which is subsequently mineralized; however, the methanogenic aquifer site habitat produces more 4-chlorobenzoate. In both habitats, there is little or no biodegradation of 4-chlorobenzoate.

Table 7 Substrate Disappearance and Initial Metabolites in Samples from Four Anoxic Habitats

	Substrate disappearance (%)[a] in habitat				
Substrate	Sewage sludge	Pond sediment	Methanogenic aquifer	Sulfate-reducing aquifer	Initial products and comments[b]
Phenoxyacetate	77	100	100	2	Phenol detected in SS, PS, MA
2,4-D	100	100	99	0	2,4-Dichlorophenol, 4-chlorophenol, and phenol detected in SS, PS, MA
2,4,5-T	100	100	100	2	2,4,5-Trichlorophenol, dichlorophenol, and two possible products (unidentified) detected in SS; 2,5-dichlorophenoxyacetate, a dichlorophenol-3-chlorophenol, and phenol in PS and MA; 2,4-D and 4-chlorophenol also detected in MA
Benzoate	100	100	100	100	
3-Chlorobenzoate	2	93	100	3	Benzoate detected in MA
4-Chlorobenzoate	17	6	14	11	No evidence of degradation in any of the incubations
3,4-Dichlorobenzoate	8	97	96	15	3- and 4-chlorobenzoate detected in PS and MA
3,5-Dichlorobenzoate	6	55	100	5	3-Chlorobenzoate detected in PS and MA
Phenol	100	100	100	99	
2-Chlorophenol	57	100	100	20	Loss in SRA not above control
3-Chlorophenol	98	98	100	18	Loss in SRA not different from control; phenol detected in PS and MA but not in SS

(Table 7, continued)

Substrate	Substrate disappearance (%)[a] in habitat				Initial products and comments
	Sewage sludge	Pond sediment	Methanogenic aquifer	Sulfate-reducing aquifer	
4-Chlorophenol	11	100	100	26	Loss in SS no different from control; no products detected in SRA; phenol detected in MA and PS
2,4-Dichlorophenol	93	100	100	39	4-Chlorophenol and phenol found in MA and PS; 4-chlorophenol detected in SS; loss in SRA no different from control
2,5-Dichlorophenol	32	100	83	48	3-Chlorophenol and phenol detected in PS and MA; 3-chlorophenol detected in SS; loss in SRA not above control
3,4-Dichlorophenol	32	100	12	29	Possible product (unidentified) detected in SS; no products observed in either MA or SRA and loss no different from controls; 3- and 4-chlorophenol found in PS
2,4,5-Trichlorophenol	38	100	39	52	No products observed in MA or SRA, and losses no different from controls; two possible intermediates (unidentified) found in SS; dichlorophenol and 3-chlorophenol detected in PS

[a]Substrate disappearance after 3 months of incubation relative to the quantity in the liquid phase at initial sampling.
[b]SS, Sewage sludge; PS, pond sediment; MA, methanogenic aquifer; SRA, sulfate-reducing aquifer.
Source: From Gibson and Suflita (1986).

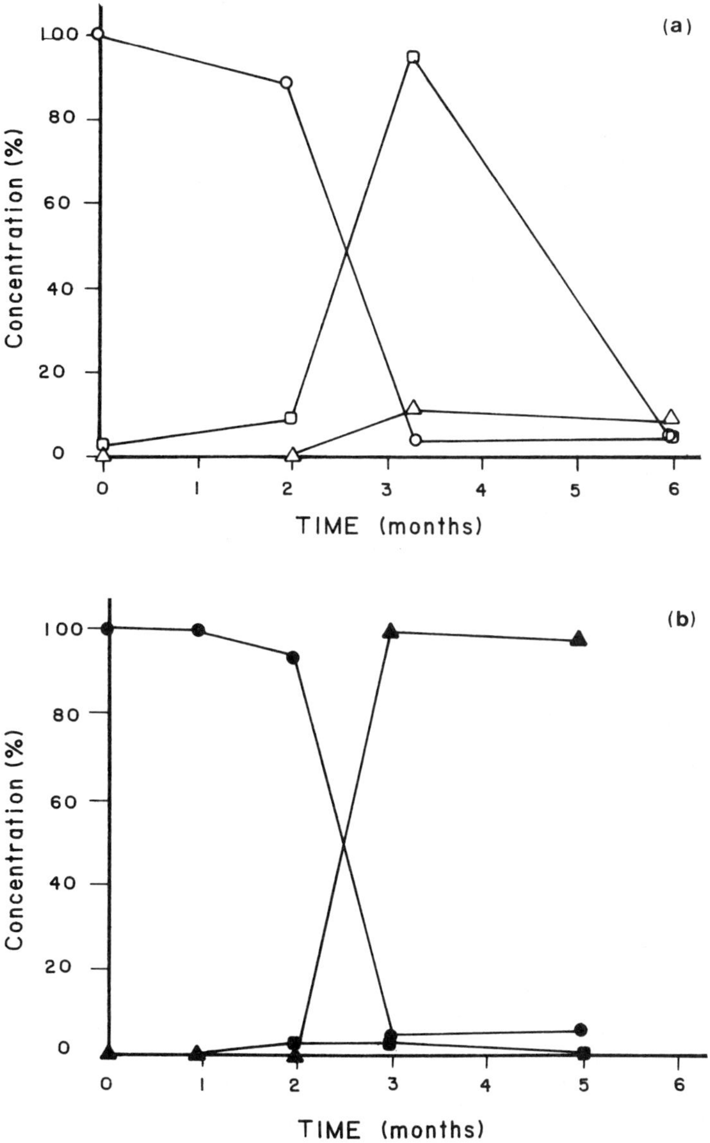

Figure 12 Disappearance of 3,4-dichlorobenzoate (○) and appearance of intermediates (3-chlorobenzoate (□) and 4-chlorobenzoate (△) in samples from (a) pond sediment or (b) a methanogenic-aquifer site. Concentrations are the percentages of initial quantity detected. (From Gibson and Suflita, 1986.)

Some research on predicting biodegradability has been reported (Dearden and Nicholson, 1986), with aerobic biodegradation receiving more attention than anaerobic biodegradation. The need exists for systematic prediction of inhibition as well as biodegradation. The *n*-octanol/water partition coefficient, molecular connectivities, lipophilicity, van der Waals radius of substituents in the aromatic ring, Hammett substituent constant (an electronic factor), Sterimol length parameter, accessible molecular surface area, and atomic charge across important bonds are examples of variables used in attempts to predict biodegradability (Dearden and Nicholson, 1986). Further research is needed to develop this technology.

B. Bioenergetics

The bioenergetics of growth on low concentrations of xenobiotic substances have been investigated by several researchers. The search for the minimum concentration of a xenobiotic substrate required to sustain growth of the microbial culture is a significant concern because of the need to reduce the substrate concentration to a very low level. Rittmann and McCarty (1980a, b) have analyzed aerobic growth on biofilms and found that effluent concentrations of about 1 mg/L can be achieved when the substrate being degraded is the sole source of carbon and energy. Methods to maintain a sufficient concentration of biomass to metabolize the substrate at a reasonable rate depend on the bioenergetics of growth. Biodegradation in the presence of cometabolites to help sustain the organisms has also been examined. In several studies, acetate has been the primary substrate to support growth when numerous secondary compounds are biodegraded. Bouwer and McCarty (1983a, b) have shown that effluent concentrations of 0.1-4 μg/L can be obtained for the secondary substrate when biofilms are maintained using the primary substrate. Some results are shown in Table 2 for several different chlorinated aliphatic compounds that are biodegraded most effectively under anaerobic conditions. These compounds were not biodegraded under aerobic conditions, and biodegradation under denitrification conditions was not comparable to that under methanogenic conditions (Bouwer and McCarty, 1983a, b).

Cometabolism is defined as the transformation of a nongrowth substrate in the obligate presence of a growth-supporting substrate (Grady, 1985). Some hazardous organic compounds may not support microbial growth, and thus they fall into the class of compounds that must be cometabolized for biodegradation. Meager information exists on the bioenergetics of cometabolism.

VI. STABILITY, RELIABILITY, AND CONTROL

In handling and treating toxic and hazardous substances, it is of the utmost importance to prevent their release to the environment owing to process failure, human error, or other causes. An essential aspect of process control is process monitoring. Measurements must be made regularly to verify that the effluent from the biodegradation process meets specifications. In the interim, the effluent may be held in a tank with a recycle pump until chemical analyses show that the product may be released.

The stability and control of the process depend on the genes and enzymes necessary for biodegradation. Catabolite repression may act to suppress the synthesis of the enzymes needed for biodegradation if the concentration of the growth-supporting substrate is excessive. Genes found in plasmids, coding for biodegrading enzymes, may not be produced in a sufficient number under some operational conditions. In mixed cultures, the competition among the different strains may lead to undesirable results. In general, the system gives a superior response when a fairly uniform concentration of waste is fed continuously compared with shock loadings of toxic substances.

Because of the complex nature of the process, fuzzy logic control may be advantageous. Tong et al. (1980) have developed a fuzzy control algorithm for the activated sludge process. Modern systems engineering methods, including knowledge engineering and artificial intelligence, can also be used to examine reliability and failure possibilities; these methods can be employed to design a reliable or robust system.

REFERENCES

Abram, J. W., and Nedwell, D. B. (1978). Inhibition of methanogenesis by sulphate reducing bacteria competing for transferred hydrogen. *Arch. Microbiol. 117*:89-92.

Aftring, P. R., and Taylor, B. F. (1981). Aerobic and anaerobic catabolism of phthalic acid by a nitrate-respiring bacterium. *Arch. Microbiol. 130*:101-104.

Aftring, P. R., Chalker, B. E., and Taylor, B. F. (1981). Degradation of phthalic acids by a denitrifying, mixed culture of bacteria. *Appl. Environ. Microbiol. 41*:1117-1183.

Bailey, J. E., and Ollis, D. F. (1986). *Biochemical Engineering Fundamentals*, 2nd Ed. McGraw-Hill, New York.

Bakker, G. (1977a). Anaerobic degradation of aromatic compounds in the presence of nitrate. *FEMS Lett. 1*:103-108.

Bakker, G. (1977b). Degradation of aromatic compounds by microorganisms in dissimilatory nitrate reduction. Doctoral Thesis, Technical University, Delft.

Balba, M. T., and EVans, W. C. (1977). The methanogenic fermentation of aromatic compounds. *Biochem. Soc. Trans.* 5:302-304.

Balba, M. T., and Evans, W. C. (1979). The methanogenic fermentation of ω-phenylalkanecarboxylic acids. *Biochem. Soc. Trans.* 7:403-405.

Balba, M. T., and Evans, W. C. (1980). The anaerobic dissimilation of benzoate by *Pseudomonas aeruginosa* coupled with *Desulfovibrio vulgaris*, with sulphate as terminal electron acceptor. *Biochem. Soc. Trans.* *8*:624-627.

Balba, M. T., Clarke, N. A., and Evans, W. C. (1979). The methanogenic fermentation of plant phenolics. *Biochem. Soc. Trans.* 7:1115-1116.

Barrio-Lage, G., Parsons, F. Z., Nassar, R. S., and Lorenzo, P. A. (1986). Sequential dehalogenation of chlorinated ethenes. *Environ. Sci. Technol.* *20*:96-99.

Barrio-Lage, G., Parsons, F. Z., and Nassar, R. S. (1987). Kinetics of the depletion of trichloroethene. *Environ. Sci. Technol.* *21*:366-370.

Berry, D. F., Francis, A. J. and Bollag, J. M. (1987). Microbial metabolism of homocyclic and heterocyclic aromatic compounds under anaerobic conditions. *Microbiol. Rev.* *51*:43-59.

Blum, D. J. W., Hergenroeder, R., Parkin, G. F., and Speece, R. E. (1986). Anaerobic treatment of coal conversion wastewater constituents: Biodegradability and toxicity. *J. Water Pollut. Control Fed.* *58*:122-131.

Bossert, I. D., and Young, L.Y. (1986). Anaerobic oxidation of *p*-cresol by a denitrifying bacterium. *Appl. Environ. Microbiol.* *52*:1117-1122.

Bouwer, E. J., and McCarty, P. L. (1983a). Transformations of 1- and 2-carbon halogenated aliphatic organic compounds under methanogenic conditions. *Appl. Environ. Microbiol.* *45*:1286-1294.

Bouwer, E. J., and McCarty, P. L. (1983b). Transformations of halogenated organic compounds under denitrification conditions. *Appl. Environ. Microbiol.* *45*:1295-1299.

Bouwer, E. L., and McCarty, P. L. (1985). Utilization rates of trace halogenated organic compounds in acetate-grown biofilms. *Biotechnol. Bioeng.* *27*:1564-1571.

Bouwer, E. J., Rittmann, B. E., and McCarty, P. L. (1981). Anaerobic degradation of halogenated 1- and 2-carbon organic compounds. *Environ. Sci. Technol.* *15*:596-599.

Braun, K., and Gibson, D. T. (1984). Anaerobic degradation of 2-aminobenzoate (anthrunilic acid) by denitrifying bacteria. *Appl. Environ. Microbiol.* *48*:102-107.

Bromilow, R. H., Briggs, G. G., Williams, M. R., Smelt, J. H., Tuinstra, L. G. M., and Traag, W. A. (1986). The role of ferrous ions in rapid degradation of oxamyl, methomyl, and aldicarb in anaerobic soils. *Pesticide Sci. 17*:535-547.
Buisson, R. S. K., Kirk, P. W. W., Lester, J. N., and Campbell, J. A. (1986). Behavior of selected chlorinated organic micropollutants during batch anaerobic digestion. *Water Pollut. Control 85*: 387-394.
Chakrabarty, A. M. (Ed.) (1982). *Biodegradation and Detoxification of Environmental Pollutants*, CRC Press, Boca Raton, Florida, pp. 1-147.
Clark, F. M., and Fina, L. R. (1952). The anaerobic decomposition of benzoic acid during methane fermentation. *Arch. Biochem. Biophys. 36*:26-32.
Deardon, J. C., and Nicholson, R. M. (1986). The prediction of biodegradability by the use of quantitative structure-activity relationships: Correlation of biological oxygen demand with atomic charge difference. *Pesticide Sci. 17*:305-310.
Delwiche, C. C., and Bryan, A. B. (1976). Denitrification. *Annu. Rev. Microbiol. 30*:241-262.
Don, R. H., and Pemberton, J. M. (1981). Properties of six pesticide degradation plasmids isolated from *Alcaligenes paradoxus* and *Alcaligenes eutrophus*. *J. Bacteriol. 145*:681-686.
Dutton, P. L., and Evans, W. C. (1967). Dissimilation of aromatic substrates by *Rhodopseudomonas palustris*. *Biochem. J. 104*:30-31.
Dutton, P. L., and Evans, W. C. (1969). The metabolism of aromatic compounds by *Rhodopseudomonas palustris*. *Biochem. J. 113*:525-535.
Dwyer, D. F., Krumme, M. L., Boyd, S. A., and Tiedje, J. M. (1986). Kinetics of phenol biodegradation by an immobilized methanogenic consortium. *Appl. Environ. Microbiol. 52*:345-351.
Evans, W. C. (1977). Biochemistry of the bacterial catabolism of aromatic compounds in anaerobic environments. *Nature 270*:17-22.
Fabig, W., Ottow, J. C. G., and Muller, F. (1980). Failure of denitrifying bacteria to utilize benzoic acid under anaerobic conditions with nitrate as the only terminal electron acceptor. *Eur. J. Appl. Microbiol. Biotechnol. 9*:133-135.
Fannin, K. F., Conrad, J. R., Srivastava, V. J., Chynoweth, D. P., and Jerger, D. E. (1986). Anaerobic Processes. *J. Water Pollution Control Fed. 58*:504-510.
Ferry, J. G., and Wolfe, R. S. (1976). Anaerobic degradation of benzoate to methane by a microbial consortium. *Arch. Microbiol. 107*:33-40.
Fina, L. R., and Fiskin, A. M. (1960). Anaerobic decomposition of benzoate during methane fermentation. II. Fates of carbons one and seven. *Arch. Biochem. Biophys. 91*:163-165.

Fina, L. R., Bridges, R. L., Coblentz, T. H., and Roberts, F. F. (1978). The anaerobic decomposition of benzoic acid during methane fermentation. III. The fate of carbon four and the identification of propanoic acid. *Arch. Microbiol.* *118*:169-172.

Fisher, P. R., Appleton, J., and Pemberton, J. M. (1978). Isolation and characterization of the pesticide degrading plasmid PJP1 from *Alcaligens paradoxus*. *J. Bacteriol.* *135*:798-804.

Focht, D. D., and Verstraete, W. (1977). Biochemical ecology of nitrification and denitrification. *Adv. Microbial. Ecol.* *1*:135-172.

Fogel, M. M., Taddeo, A. R., and Fogel, S. (1986). Biodegradation of chlorinated ethenes by a methane-utilizing mixed culture. *Appl. Environ. Microbiol.* *51*:720-724.

Ghisalba, O. (1983). Microbial degradation of chemical waste, an alternative to physical methods of waste disposal. *Experientia* *39*:1247-1257.

Ghosal, D., You, I. S., Chatterjee, D. K., and Chakrabarty, A. M. (1985). Microbial degradation of halogenated compounds. *Science* *238*:135-142.

Gibson, D. T. (Ed.) (1984). *Microbial Degradation of Organic Compounds*. Marcel Dekker, New York, pp. 1-523.

Gibson, S. A., and Suflita, J. M. (1986). Extrapolation of biodegradation results to groundwater aquifers: Reductive dehalogenation of aromatic compounds. *Appl. Environ. Microbiol.* *52*:681-688.

Gledhill, W. E. (1979). Proposed standard practice for determination of the anaerobic biodegradability of organic chemicals. Working document. Draft 2, No. 35.24, American Society for Testing and Materials, Philadelphia.

Grady, C. P. L. (1985). Biodegradation: Its measurement and microbiological basis. *Biotechnol. Bioeng.* *27*:660-674.

Guyer, M., and Hegeman, G. (1969). Evidence for a reductive pathway for the anaerobic metabolism of benzoate. *J. Bacteriol.* *99*:906-907.

Healy, J. B., and Young, L. Y. (1978). Catechol and phenol degradation by a methanogenic population of bacteria. *Appl. Environ. Microbiol.* *35*:216-218.

Healy, J. B., and Young, L. Y. (1979). Anaerobic degradation of eleven aromatic compounds to methane. *Appl. Environ. Microbiol.* *38*:84-89.

Hopwood, D. A. (1981). The genetic programming of industrial microorganisms. *Sci. Am.* *245*(3):91-102.

Kaiser, J. P., and Hanselmann, K. W. (1982). Fermentative metabolism of substituted monoaromatic compounds by a bacterial community from anaerobic sediments. *Arch. Microbiol.* *133*:185-194.

Keil, H., Keil, S., Pickup, R. W., and Williams, P. A. (1985). Evolutionary conservation of genes coding for meta pathway enzymes within LTOL plasmids pWWO and pWW53. *J. Bacteriol.* *164*:887-895.

Kiene, R. P., and Capone, D. G. (1986). Stimulation of methanogenesis by aldicarb and several other n-methyl carbamate pesticides. *Appl. Environ. Microbiol.* *51*:1247-1251.

Kilbane, J. J., Chatterjee, D. K., and Chakrabarry, A. M. (1983). Detoxification of 2,4,5-trichlorophenoxyacetic acid from contaminated soil by *Pseudomonas cepacia*. *Appl. Environ. Microbiol.* *45*:1697-1700.

Kim, C. J., and Maier, W. J. (1986). Acclimation and biodegradation of chlorinated organic compounds in the presence of alternate substances. *J. Water Pollut. Control Fed.* *58*:157-164.

Kloepfer, R. D., Easley, D. M., Haas, B. B., Diehl, T. G., Jackson, D. E., and Wurrey, C. J. (1985). Anaerobic degradation of trichloroethylene in soil. *Environ. Sci. Technol.* *19*:277-280.

Kobayashi, H., and Rittmann, B. E. (1982). Microbial removal of hazardous organic compounds. *Environ. Sci. Technol.* *16*:170A-183A.

Kuhn, E. P., Colberg, P. J., Schnoor, J. L., Wanner, O., Zehnder, A. J. B., and Schwarzenbach, R. P. (1985). Microbial transformations of substituted benzenes during infiltration of river water to groundwater: Laboratory column studies. *Environ. Sci. Technol.* *19*:961-968.

Lal, R., Lal, S., and Shivaji, S. (1986). Use of microbes for detoxification of pesticides. *CRC Crit. Rev. Biotechnol.* *3*:1-16.

LaPat-Polasko, L. T., McCarty, P. L., and Zehnder, A. L. B. (1984). Secondary substrate utilization of methylene chloride by an isolated strain of pseudomonas sp. *Appl. Environ. Microbiol.* *47*:825-830.

Lee, S. B., and Bailey, J. E. (1984a). Genetically structured models for *lac* promoter-operator function in the *Escherichia coli* chromosome and in multicopy plasmids: *Lac* operator function. *Biotechnol. Bioeng.* *26*:1372-1382.

Lee, S. B., and Bailey, J. E. (1984b). Genetically structured models for *lac* promoter-operator function in the chromosome and multicopy plasmids: *Lac* promotor function. *Biotechnol. Bioeng.* *26*:1383-1389.

Lee, S. B., and Bailey, J. E. (1984c). A mathematical model for λdV plasmid replication: Analysis of wild type plasmid. *Plasmid* *11*:151-165.

Leisinger, T., Hutter, R., Cook, A. M., and Nuesch, J. (1981). *Microbial Degradation of Xenobiotics and Recalcitrant Compounds.* Academic Press, New York.

Liu, D., Strachan, W. M. J., Thomson, K., and Kwasniewska, K. (1981). Determination of the biodegradability of organic compounds. *Environ. Sci. Technol.* *15*:788-793.

Maki, A. W., Dickson, K. L., and Cairns, J. (Eds.) (1980). *Biotransformation and Fate of Chemicals in the Aquatic Environment*. American Society for Microbiology, Washington, D.C.

Matsumura, F., and Krisha Murti, C. R. (1982). *Biodegradation of Pesticides*. Plenum, New York.

McCarty, P. L., Young, L. Y., Gossett, J. M., Stuckey, O. C., and Healy, J. B. (1977). Heat treatment for increasing methane yields from organic materials. In *Microbial Energy Conversion*, H. G. Schlegel and J. Barnes (Eds.). Pergamon Press, Oxford, pp. 179-199.

Mikesell, M. D., and Boyd, S. A. (1986). Complete reductive dechlorination and mineralization of pentachlorophenol by anaerobic microorganisms. *Appl. Environ. Microbiol. 52*:861-865.

Nelson, M. J. K., Montegomery, S. O., O'Neill, E. J., and Pritchard, P. H. (1986). Aerobic metabolism of trichloroethylene by a bacterial isolate. *Appl. Environ. Microbiol. 52*:383-384.

Nottingham, P. M., and Hungate, R. E. (1969). Methanogenic fermentation of benzoate. *J. Bacteriol. 98*:1170-1172.

Oshima, T. (1965). On the anaerobic metabolism of aromatic compounds in the presence of nitrate by soil micro-organisms. *Z. Allg. Mikrobiol. 5*:386-394.

Owen, W. F., Stuckey, D. C., Healy, J. B., Young, L. Y., and McCarty, P. L. (1979). Bioassay for monitoring biochemical methane potential and anaerobic toxicity. *Water Res. 13*:485-492.

Ozaki, M., Tanaka, Y., and Kawatsuka, S. (1986). Reductive degradation of isouron in soils under anaerobic conditions. *J. Pesticide Sci. 11*:409-414.

Parsons, F. Z., and Lage, G. B. (1985). Chlorinated organics in simulated groundwater environments. *J. Am. Water Works Assoc. 77*(5):52-59.

Parsons, F. Z., Wood, P. R., and DeMarco, J. (1984). Transformations of tetrachloroethene and trichloroethene in microcosms and groundwater. *J. Am. Water Works Assoc. 76*(2):56-59.

Payne, W. J. (1981). *Denitrification*. Wiley, New York.

Proctor, M. H., and Scher, S. (1960). Decomposition of benzoate by a photosynthetic bacterium. *Biochem. J. 76*:33p.

Rittmann, B. E., and McCarty, P. L. (1980a). Model of steady-state-biofilm kinetics. *Biotechnol. Bioeng. 22*:2343-2357.

Rittmann, B. E., and McCarty, P. L. (1980b). Evaluation of steady-state-biofilm kinetics. *Biotechnol. Bioeng. 22*:2359-2373.

Rittmann, B. E., and McCarty, P. L. (1980c). Utilization of dichloromethane by suspended and fixed-film bacteria. *Appl. Environ. Microbiol. 39*:1225-1226.

Rochkind-Dubinsky, M. L., Sayler, G. S., and Blackburn, J. W. (1987). *Microbiological Decomposition of Chlorinated Aromatic Compounds*. Marcel Dekker, New York.

Schennen, U., Braun, K., and Knackmuss, H. J. (1985). Anaerobic degradation of 2-fluorobenzoate by benzoate degrading, denitrifying bacteria. *Appl. Environ. Microbiol. 49*:321-325.

Scher, S. (1960). Photometabolism of benzoic acid by *Athiorhocaceae*. In *Progress in Photobiology. Proceedings of the Third International Congress on Photobiology*, B. Christensen and B. Buchman (Eds.). Elsevier, London, pp. 583-585.

Schink, B., and Pfennig, N. (1982). Fermentation of trihydroxybenzenes by *Pelobacter acidigallici* gen. nov. sp. nov., a new strictly anaerobic, nonsporeforming bacterium. *Arch. Microbiol. 133*:195-201.

Schmidt, S. K., Simkins, S., and Alexander, M. (1985). Models for the kinetics of biodegradation of organic compounds not supporting growth. *Appl. Environ. Microbiol. 50*:323-331.

Schönborn, W. (1986). *Microbial Degradations*, Vol. 8 of *Biotechnology* (a comprehensive treatise edited by H. Rehm and G. Reed). VCH Publishers, New York.

Shelton, D. R., and Tiedje, J. M. (1984). General method for determining anaerobic biodegradation potential. *Appl. Environ. Microbiol. 47*:850-857.

Shlomi, E. R., Lankhorst, A., and Prins, R. A. (1978). Methanogenic fermentation of benzoate in an enrichment culture. *Microb. Ecol. 4*:249-261.

Sleat, R., and Robinson, J. P.(1983). Methanogenic degradation of sodium benzoate in profundal sediments from a small eutrophic lake. *J. Microbiol. 129*:141-152.

Sleat, R., and Robinson, J. P. (1984). The bacteriology of anaerobic degradation of aromatic compounds. *J. Appl. Bacteriol. 57*:381-394.

Speece, R. E. (1983). Anaerobic biotechnology for industrial wastewater treatment. *Environ. Sci. Technol. 17*:416A-427A.

Strand, S. E., and Shippert, L. (1986). Oxidation of chloroform in an aerobic soil exposed to natural gas. *Appl. Environ. Microbiol. 52*:203-205.

Stucki, G., Galli, R., Ebersold, H. R., and Leisinger, T. (1981). Dehalogenation of dichloromethane by cell extracts of hypomicrobium DM2. *Arch. Microbiol. 39*:1225-1226.

Stucki, G., Krebser, U., and Leisinger, T. (1983). Bacterial growth on 1,2-dichloroethane. *Experientia 39*:1271-1273.

Suflita, J. M., Horowitz, A., Shelten, O. R., and Tiedje, J. M. (1982). Dehalogenation: A novel pathway for the anaerobic biodegradation of haloaromatic compounds. *Science 218*:1115-1117.

Tarvin, D., and Buswell, A. M. (1934). The methane fermentation of organic acids and carbohydrates. *J. Am. Chem. Soc. 56*:1751-1755.

Taylor, B. F., and Heeb, M. J.(1972). The anaerobic degradation of aromatic compounds by a denitrifying bacterium. *Arch. Mikrobiol. 83*:165-171.

Taylor, B. F., Campbell, W. L., and Chinoy, I. (1970). Anaerobic degradation of the benzene nucleus by a faculatively anaerobic microorganism. *J. Bacteriol. 102*:430-437.
Taylor, B. F., Hearn, W. L., and Pincus, S. (1979). Metabolism of monofluorobenzoates and monochlorobenzoates by a denitrifying bacterium. *Arch. Microbiol. 122*:301-306.
Tong, R. M., Beck, M. B., and Latten, A. (1980). Fuzzy control of the activated sludge wastewater treatment process. *Automatica 16*:695-701.
Vogel, T. M., and Grbic-Galic, D. (1986). Incorporation of oxygen from water into toluene and benzene during anaerobic fermentation transformation. *Appl. Environ. Microbiol. 52*:200-202.
Vogel, T. M., and McCarty, P. L. (1985). Biotransformation of tetrachloroethylene to trichloroethylene, dichloroethylene, vinyl chloride, and carbon dioxide under methanogenic conditions. *Appl. Environ. Microbiol. 49*:1080-1083.
Westrick, J. J., Mello, J. W., and Thomas, R. F. (1984). The groundwater supply survey. *J. Am. Water Works Assoc. 76*(5): 52-59.
Widdel, F. (1980). Anaerober Abbau von Fettsäuren und Benzoesäure durch neu isolierte Arten sulfatreduzierender Bakterien. Dissertation Georg-august-Universität, Göttingen, West Germany.
Williams, R. J., and Evans, W. C. (1973). Anaerobic metabolism of aromatic substrates by certain microorganisms. *Biochem. Soc. Trans. 1*:186-187.
Williams, R. J., and Evans, W. C. (1975). The metabolism of benzoate by *Moraxella* sp. through anaerobic nitrate respiration. *Biochem. J. 148*:1-10.
Wilson, J. T., and Wilson, B. H. (1985). Biotransformation of trichloroethylene in soil. *Appl. Environ. Microbiol. 49*:242-243.
Young, L. Y. (1984). Anaerobic degradation of aromatic compounds. In *Microbial Degradation of Organic Compounds*, D. T. Gibson (Ed.). Marcel Dekker, New York, pp. 487-523.
Young, L. Y., and Rivera, M. D. (1985). Methanogenic degradation of four phenolic compounds. *Water Res. 19*:1325-1332.
Zeikus, J. G., Wellstein, A. L., and Kirk, T. K. (1982). Molecular basis for the biodegradative recalcitrance of lignin in anaerobic environments. *FEMS Microbiol. Lett. 15*:193-197.
Zeyer, J., Kuhn, E. P., and Schwarzenbach, R. P. (1986). Rapid microbial mineralization of toluene and 1,3-dimethylbenzene in the absence of molecular oxygen. *Appl. Environ. Microbiol. 52*: 944-947.

21

Performance of Immobilized Cell Reactors

JOSE L. VEGA, EDGAR C. CLAUSEN, and JAMES L. GADDY / *University of Arkansas, Fayetteville, Arkansas*

I. INTRODUCTION

The application of bioreactors to the production of large volumes of chemicals or fuels may find wide commercial application in the near future. A number of factors in the design of these bioreactors will be important in the economic success of this technology. Raw materials constitute a major portion of the product cost; therefore, reactors that achieve high conversions will be necessary. The capital and energy costs of purifying the products are significant, and bioreactors that can operate at high substrate and product concentrations will be most economical. The capital cost of the reactor is a major part of the equipment cost, and reactors that enable high cell concentrations and fast reaction rates are needed.

Among the various bioreactor designs, immobilized cell reactors (ICR) offer distinct advantages in productivity, improved inhibition characteristics, and low cost. The use of cell immobilization has been a common laboratory practice within the last few years as a method to improve the performance and the economics of many fermentation processes. Although immobilization techniques differ widely, the common objective can be summarized as an attempt to confine the biocatalyst within the reactor system, at the same time allowing the media containing the substrate and the products to flow in a continuous manner. The overall effect is to transform a pseudohomogeneous catalytic reaction ("pseudo" referring to the suspension of the cells in the media) into a typical heterogeneously catalyzed system.

The immediate advantages of this technique come from purely chemical reaction concepts: higher concentrations of biocatalyst can be achieved, which leads to a faster reaction rate, and the separation of the desired products is more easily carried out in the absence or at lower concentrations of chemically complex substances, such as cell components. Continuous operation at comparatively high flow rates without washout of the reactor is possible, thus enhancing productivity. There is also some evidence to indicate that bound cell systems generally give higher product yields and allow a better removal of toxic metabolites, as well as freedom from nutrient depletion (Venkatasubramanian and Vieth, 1979). Performance under heterogeneous catalytic conditions can be carried out in the plug flow mode, as opposed to well-mixed systems, which in some cases (such as ethanol fermentation, in which product inhibition may be a factor) leads to more efficient operation(Ghose and Tyagi, 1979).

This chapter provides a summary of the performance and recent applications of immobilized cell reactors. Immobilization methods, reactor start-up, and steady-state performance are described. Modeling of biofilm reactors is discussed, and methods to improve operational stability are presented. In most cases, the data for ethanol production are utilized as exemplary.

II. WHOLE-CELL IMMOBILIZATION TECHNIQUES

Whole-cell immobilization techniques can be classified into two major groups, entrapment and carrier binding. Entrapment includes enclosure of a catalyst both behind a membrane or within a gel structure (Rosevear, 1984). Carrier binding includes all methods in which there is direct binding of cells to water-insoluble carriers by physical adsorption or by ionic and/or covalent bonds (Chibata et al., 1983). Potential mass transfer limitations are always present when dealing with entrapment methods, either across the gel matrix or gel occlusion or across the system membrane in membrane reactors. On the other hand, the carrier-binding method allows direct contact between the fermentation broth and the biocatalyst, and the medium flows in and out of the system without restriction, thus minimizing mass transfer problems. Although the start-up of systems employing carrier-binding methods is influenced by the nature of the support and the specific technique employed, after some time of operation a biocatalyst film is grown and common characteristics to all systems are found, thus allowing grouping as so-called biofilm reactors.

Biofilm reactors present several advantages over entrapped systems. First, as pointed out by Black et al. (1984), it is highly advantageous to be able to inoculate a fermenter in the normal way and to allow cells to become immobilized simply as a consequence of their growth. This compares favorably with the requirement of special techniques to entrap cells into matrices prior to use in the reactors. Second, growth is a requirement for entrapped systems, owing to the loss of viability of the cells in the matrices, which makes reactivation with richer media necessary (Lee et al., 1983). Even so, the stability of such systems is generally not good unless some of the biocatalyst is replaced regularly (Wang and Hettwer, 1982). In biofilm reactors, growth is required and must be promoted during start-up. However, as soon as a high cell density is achieved, it is no longer necessary to promote cell growth, so that a savings in nutrients can be obtained. In gel entrapment systems the most active cells are at the gel surface, so that agitation of the beads in some instances has led to a 50% loss of activity due to the leakage of the outer layer. Other disadvantages of cell entrapment include the requirement in some cases of a continuous supply of chemicals to maintain the hardness of the beads (such as $CaCl_2$ for Ca-alginate matrices) (Cheetham et al., 1979), the lack of strength of some beads to withstand hydrostatic pressures in packed bed columns, and rupture of the matrices due to CO_2 pressure (Shiotani and Yamane, 1981).

In membrane-contained fermenters, entrapment takes place over the system as a whole and their performance is quite different from that of gel occlusion systems. The behavior of the reactors is entirely dependent upon the membrane characteristics, often to the point that the limiting factor is the solute exchange capacity of the membrane. In ethanol production, it has been found that membranes often do not allow the removal of secondary metabolites, which soon accumulate and thus become toxic (Maiorella et al., 1983). Conversion levels have been found to be low, and the achieved productivities have not been too high (Kyung and Gerhardt, 1984). Also, membrane plugging by microbial cells and macromolecules is usually a problem in these type of reactors that cannot be overcome except with complex laboratory equipment.

Another major approach to increasing the cell concentration that has been widely utilized lately is cell flocculation in tower fermenters (Prince and Barford, 1982; Hsiao et al., 1983; Jones et al., 1984; Admassu et al., 1985; Netto et al., 1985). Since no support is required, the cell density that can be achieved in the reactor is higher, yielding better productivity. Although this method presents several advantages in some situations, it cannot be viewed as an immobilization process but as an improvement in cell retention against washout through an increase in particle size. Cell washout is not prevented completely and takes place at a certain dilution rate (Prince and Barford, 1982). Furthermore, the dilution rate at which washout occurs cannot be easily controlled since it is primarily a function of the density and size attained by the cell aggregates. These types of reactors require the utilization of naturally flocculating strains or methods to induce such flocculation that are not applicable to all kinds of microorganisms and thus, in some cases, that may impede the utilization of more efficient organisms (Hsiao et al., 1983). Some other disadvantages are the high mixing effect that takes place at high flow rates or high CO_2 production, converting the system into a less efficient mixed flow reactor (Admassu et al., 1985), as well as the special reactor configurations needed for settling regions or recycle of flocculated aggregates (Netto et al., 1985).

III. BIOFILM REACTORS

As mentioned, the term "biofilm reactor" is used here to represent a group of systems in which growth takes place on the outer layer of a film of microorganisms in direct contact with the surrounding medium. Several common characteristics of such systems, as well as particular differences, are reviewed in an attempt to reach a generalization of their behavior to facilitate understanding, modeling, and control.

Biofilm systems may be classified on the basis of the degree of mixing in which they operate: beds approximating plug flow and beds closer to mixed flow. Packed beds, in general, approximate plug flow, although deviations often occur owing to channeling and plugging. Fluidized beds and tower fermenters can operate in the two regimes depending on such operating conditions as liquid flow rates, rate of gas evolution, and degree of aeration. In ethanol production, for example, some characteristics favor one or the other type of operation. Substrate inhibition favors mixed flow; product inhibition makes plug flow more productive. Since ethanol inhibition is the more predominant of the two, plug flow represents a better operating mode in this case.

By approximating plug flow, the productivity of packed beds is highly dependent on the distance along the axis or, more generally, on the residence time in the column. Figure 1 presents such behavior in a column with *Saccharomyces cerevisiae* at different inlet glucose concentrations (Vega, 1985). As noted, very high productivities can be obtained for very low residence times, and a linear decrease of overall productivity is observed at longer residence times. On the other hand, a bed that operates close to a constant environment

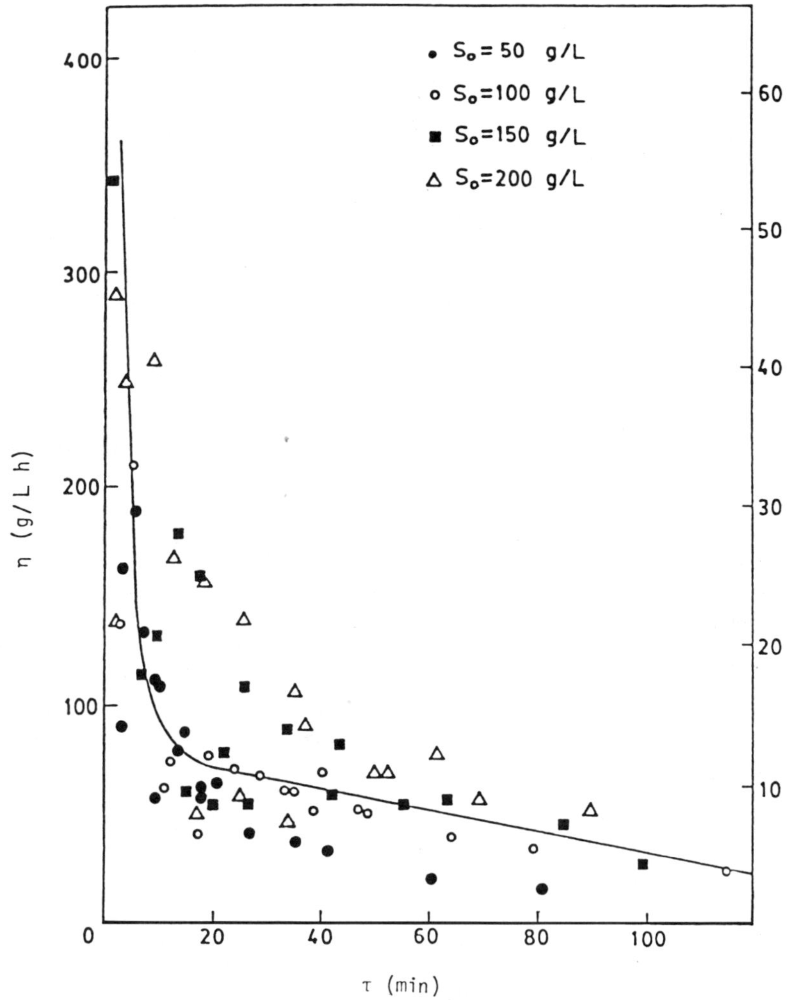

Figure 1 Ethanol productivity in a biofilm reactor as a function of residence time. Residence time is based on liquid holdup on both axes.

will have to have a high ethanol concentration and an almost depleted glucose medium, dictated by economic considerations. The productivity of these reactors should then be close to the values obtained for the final section of a packed bed and constant throughout the system. Although this is a clear disadvantage in terms of specific production rates, these beds usually permit much higher viability of the cells, an easier scaleup, and much better stability, characteristics that in some cases may be more important (Black et al., 1984). Nevertheless, the attention in this review is focused on packed bed bioreactors, in which the intrinsic potential is higher and the application of engineering principles to improve other deficiencies is of most value.

A. Start-up

The start-up of a biofilm reactor for ethanol production is relatively simple. The reactor is packed and sterilized, and a seed culture grown in batch outside the reactor is pumped into the system. When cells are attached by covalent bonding to the support, a short period of time (4-8 h) is allowed for complete contact and bonding before the flow of fresh medium is started (Sitton and Gaddy, 1980). In some cases, the natural physical characteristics of the microorganisms are used to achieve immobilization by passing a highly concentrated seed culture through the bed for an extended period of time (Gencer and Mutharasan, 1980). The result is an attachment due to such effects as diffusion into the pores, direct interception and entrapment, and gravity or surface forces (Herzig et al., 1970). This last method also allows estimation of the initial cell loading obtained during the inoculation by comparing the cell concentration of the seed culture before and after cell deposition in the column (Krug and Dauglis, 1983).

Once the reactor has been inoculated, biofilm growth begins and the concentration of microorganisms increases continuously. The conditions for growth are much better at the entrance of the column (some oxygen is dissolved in the feed, the product concentration is low, the nutrient level is high, and so on) and thus the cell concentration increases more rapidly at the bottom of an upflow vertical packed bed bioreactor. The microorganisms continue to fill the interstitial spaces in the packing by the same mechanisms as mentioned for the seeding of the reactor (filtration effect), as well as by growth onto the outer layer of the biofilm.

Experiments carried out with the yeast *S. cerevisiae* in a packed bed, using glutaraldehyde on gelatin as the covalent bonding agent, show the importance of the packing characteristics in the achievement of a high cell density system (Gollinger et al., 1986). Figure 2

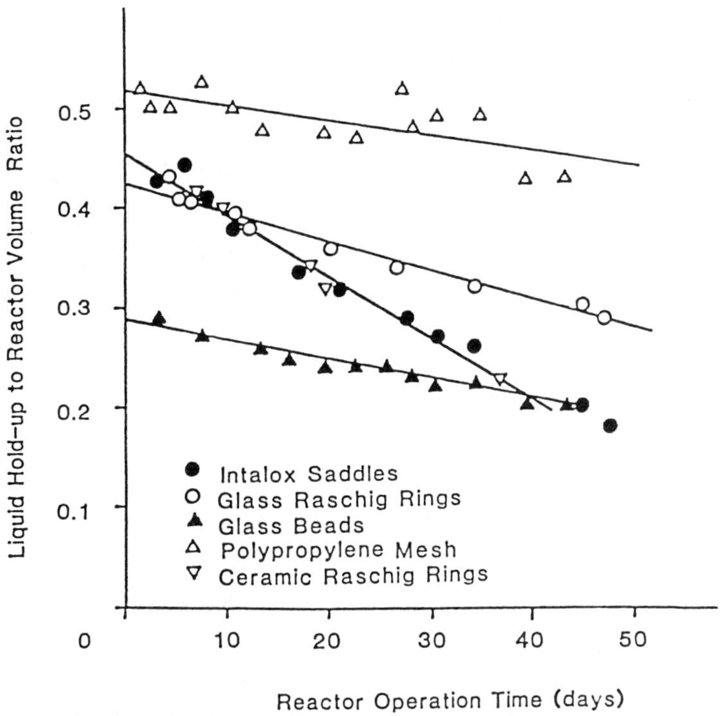

Figure 2 A comparison of the change in liquid porosity with packing material in a biofilm reactor employing covalently attached *S. cerevisiae* for ethanol production.

shows the change in liquid holdup to reactor volume with time for five different packing materials. Since conversion was kept constant during the experimental runs, the gas holdup did not change significantly, and thus the decrease in liquid holdup is due primarily to biomass growth. The biomass loading rates for each packing can then be obtained from the slopes of the lines in Figure 2. Packings that provide adequate interstitial spaces for cell entrapment, such as saddles and Raschig rings, result in faster biomass loading rates. Also, the roughness of the packing is a factor, as noted in comparison of glass and ceramic rings. Although all packing surfaces were coated with gelatin to assure a monolayer of cells, the rough ceramic materials enable a better coating and provide added spaces for subsequent layers of cells to accumulate. Therefore, for quick start-up and high initial productivity in biofilm reactors, two factors are essential: high surface-volume ratio (for large initial monolayer attachment) and adequate interstitial spaces for entrapment of cells (high biomass loading capability). Also, for prolonged periods of operation, packings with high initial porosity are capable of loading the highest cell volumes, provided that reactor plugging is controlled.

Other packing characteristics are important parameters in the start-up of reactors. When adsorption is the method used for establishing a monolayer of cells, the surface properties of carrier and cells are the most important factors. The ζ potential has been useful in some cases in explaining the adhesion of yeast cells to supports (Thonart et al., 1982). For example, the use of aluminum ions to pretreat the yeast cells to decrease the negative character of the yeast surface (Van Haecht et al., 1985) has led to a dense layer of cells strongly attached to a glass support. Many types of materials have been used as supports: glass (Navarro and Durand, 1977), hollow fiber (Mehaia and Cheyran, 1984), wood (Robinson et al., 1980), ion-exchange resin (Krug and Dauglis, 1983), and spongy materials (Del Borghi et al., 1985), among others. The support has sometimes led to a modification of metabolic rates (Navarro and Durand, 1977) that in some cases increased and in other cases decreased the activity of the cells (Durand and Navarro, 1978).

In establishing covalent bonds between the support and the cells, the choice of an appropriate coupling agent depends upon the surface properties of both the support and the cells. Sitton and Gaddy (1980) used glutaraldehyde on gelatin-coated supports to covalently bond the yeast *S. cerevisiae*. Glutaraldehyde reacts with amine groups at the surface of the cells. This immobilization technique has also been successfully applied to bacteria. *Propionibacterium acidi propionici* (Clausen et al., 1982) and *Clostridium acetobutylicum* (Conheady et al., 1981) were attached to coated supports using glutaraldehyde. In these latter cases gelatin alone could not be used as the coating agent owing to the hydrolytic capability of these bacteria. Instead, mixtures of gelatin and agar or agar alone were employed.

Other bifunctional reagents, such as isocyanate and toluene di-isocyanate, may be used in the same manner as glutaraldehyde. Okita et al. (1985) have employed cyanuric chloride to couple yeast and bacteria on a cellulose support by a double bond of the hydroxyl groups of the cells and the cellulose. Hu (1986) has used epoxypolyamine to cross-link *Actinoplanes missouriensis*, a microorganism with glucose isomerase activity. Epoxypolyamine is capable of forming two types of bonds with the cell surfaces: protonated epoxy groups linked to amino groups and imino groups of epoxypolyamine forming hydrogen bonds to surface peptides. Other covalent coupling techniques involve the use of metal-link or chelation processes with several hydrous metal oxides, such as titanium (IV), zirconium (IV), iron (III), vanadium (III), and tin (II) (Kennedy and Cabral, 1985). Immobilization in these instances is based on the formation of partial covalent bonds between hydroxyl groups on the surface of the metal hydroxide and suitable ligands from the cell surface, such as hydroxyls, carboxyls, or amino groups.

B. Steady-State Behavior

The performance of the reactor under different operating conditions is usually followed by monitoring concentration or other profiles along the length of the column. Several parameters may be measured, such as substrate and product concentrations, pH, temperature, cell concentration, gas holdup, and cell viability. These parameters are usually expressed as a function of the axial distance from the feed entrance; however, residence time may also be used. Residence time must be related to a volume basis. The possible liquid bases include empty column volume V_r; initial void volume V_0 ; and liquid holdup volume V_{lh}. A schematic of the distribution of these volumes in a vertical biofilm reactor in continuous operation is given in Figure 3. This system is similar to that shown by Shiotani and Yamane (1981) for gel occlusion systems, except for the necessity to include the cell volume, independent of the support.

Three residence times result from the three volume bases: τ (based on liquid holdup), τ' (based on initial void volume), and τ'' (based on empty column volume). The use of one or another depends on the purpose of the study under consideration. Profiles of variables dependent upon liquid characteristics, such as substrate, product, pH, and temperature along the column, should be based on the actual time of contact between the medium and the biocatalyst, that is, on liquid holdup volume. An advantage in using residence time over axial distance is that the data are independent of location in the column. When bed characteristics are to be considered, residence times based on empty column and initial void volume should be used. The use of initial void volume in calculating the productivities of the reactor allows independence of packing porosity and helps in the comparison of systems that employ different supports. Margaritis et al. (1981) have recommended use of τ' for this purpose. On the other hand, productivity based upon empty column τ'' would allow comparison of all system characteristics, including packing type, and is perhaps the best measure of performance in comparing other immobilization techniques with free cell systems.

The profiles of substrate and product concentrations in a packed bed bioreactor are usually found to be exponential. For example, Figures 4 and 5 present these profiles obtained when employing the

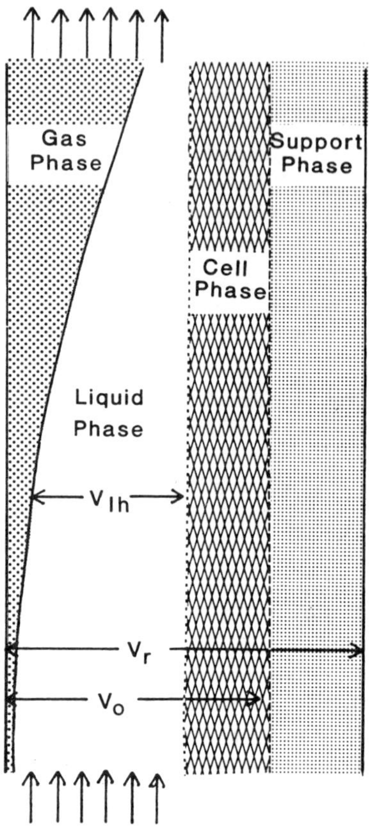

Figure 3 Volumetric distributions in a vertical biofilm reactor: V_r, empty reactor volume; V_0, initial operating volume; V_{lh}, liquid holdup volume.

yeast *S. cerevisiae* in a vertical reactor packed with spherical beads coated with gelatin and glutaraldehyde (Vega et al., 1985). The continuous lines in the figures are the result of fitting the data to equations of the type

$$S = a\ e^{-b\tau} \tag{1}$$

$$P = c(1 - e^{-d\tau}) \tag{2}$$

The data corresponding to each inlet glucose concentration was obtained under four different flow rates, the fastest being at least

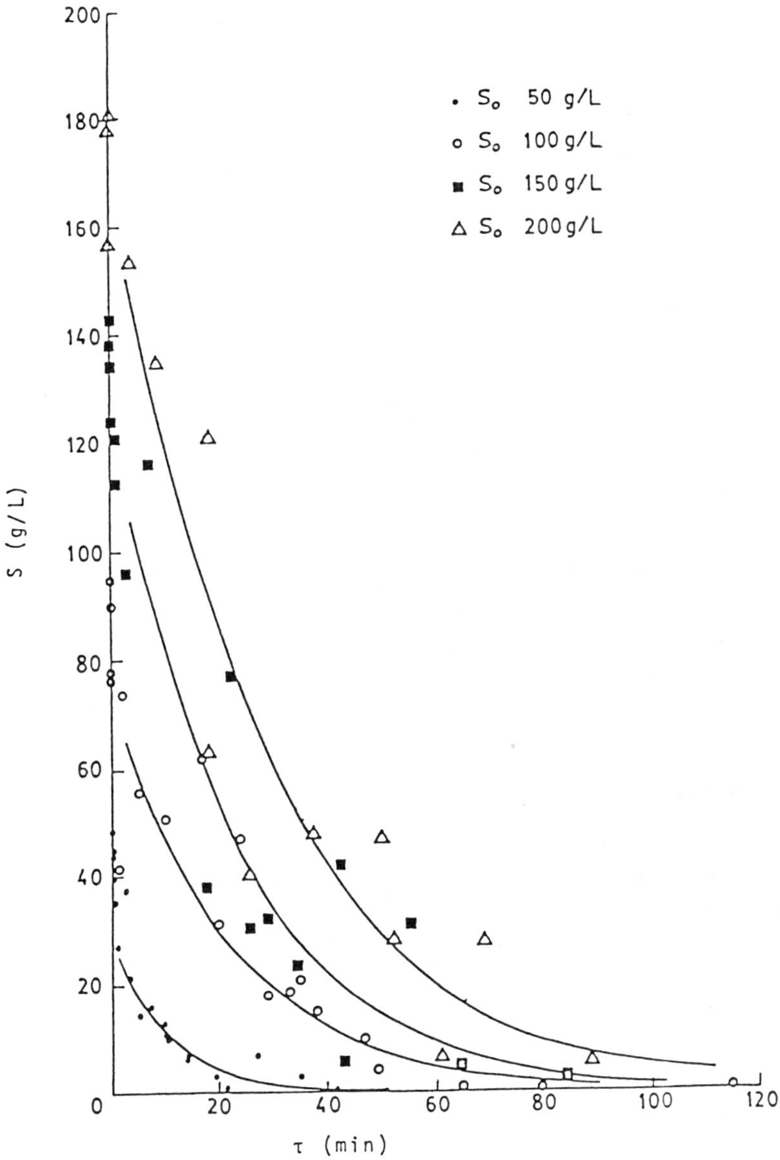

Figure 4 Glucose profiles in a biofilm reactor for ethanol production as a function of residence time based on liquid holdup. The experimental data correspond to various flow rates for each substrate concentration.

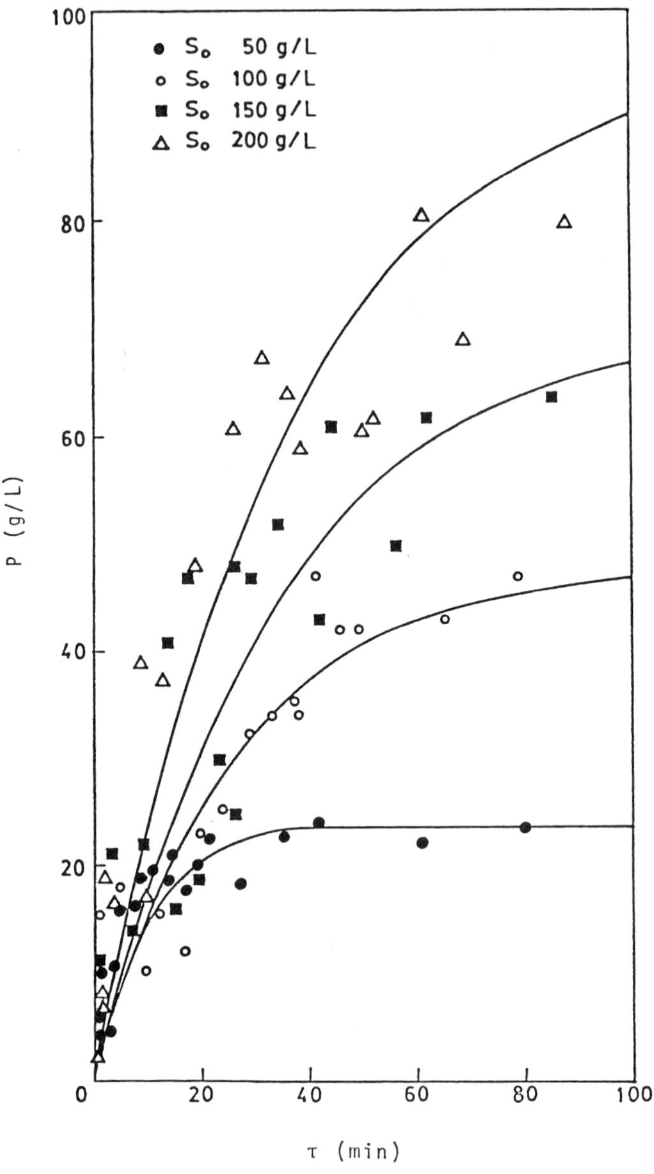

Figure 5 Ethanol concentration in a biofilm reactor as a function of residence time based on liquid holdup. The data correspond to the same experimental runs as in Figure 4.

twice as high as the slowest. The residence time was obtained by draining the column under continuous operation and assuming constant porosity throughout the bed.

Typical overall productivity curves as a function of flow rate employed in the reactor are shown in Figure 6 (Vega et al., 1985). By increasing the dilution rate, the outlet ethanol concentration and, thus, the overall product inhibition are reduced. An increase in productivity is then obtained but at the expense of a decrease in conversion. An increase in the substrate inlet concentration also results in an increase in productivity for a given dilution rate. This behavior can only be explained in terms of mass transfer and cell viability characteristics of the system, as is more carefully examined below. Therefore, productivity in a biofilm reactor is a function of conversion and substrate concentration, and the specification of productivity values should be accompanied by these parameters.

C. Mathematical Modeling

Mathematical analyses of biofilm processes are common in the literature, especially in the area of wastewater treatment (Benefield and Molz, 1985; Shieh et al., 1982; Tanaka and Dunn, 1982; Denac et al., 1983). In ethanol production, several approaches to the modeling of packed beds have been presented (Dale et al., 1985a, b; Tyagi and Ghose, 1982; Gencer and Mutharasan, 1983). Essentially, the basis of mathematical treatment is that of a plug flow heterogeneous reactor at steady state. Assuming no axial dispersion, constant porosity ε and velocity μ, mass balances for biomass, substrate, and product over a differential section of the column yields

$$u \frac{dX}{dz} = r_x \tag{3}$$

$$u \frac{dS}{dz} = -r_s \tag{4}$$

$$u \frac{dP}{dz} = r_p \tag{5}$$

where r_x, r_s, and r_p are the rates of biomass formation, substrate uptake rate, and product formation per unit volume. The associated boundary conditions are typically

$$X(z = 0) = O \tag{6}$$

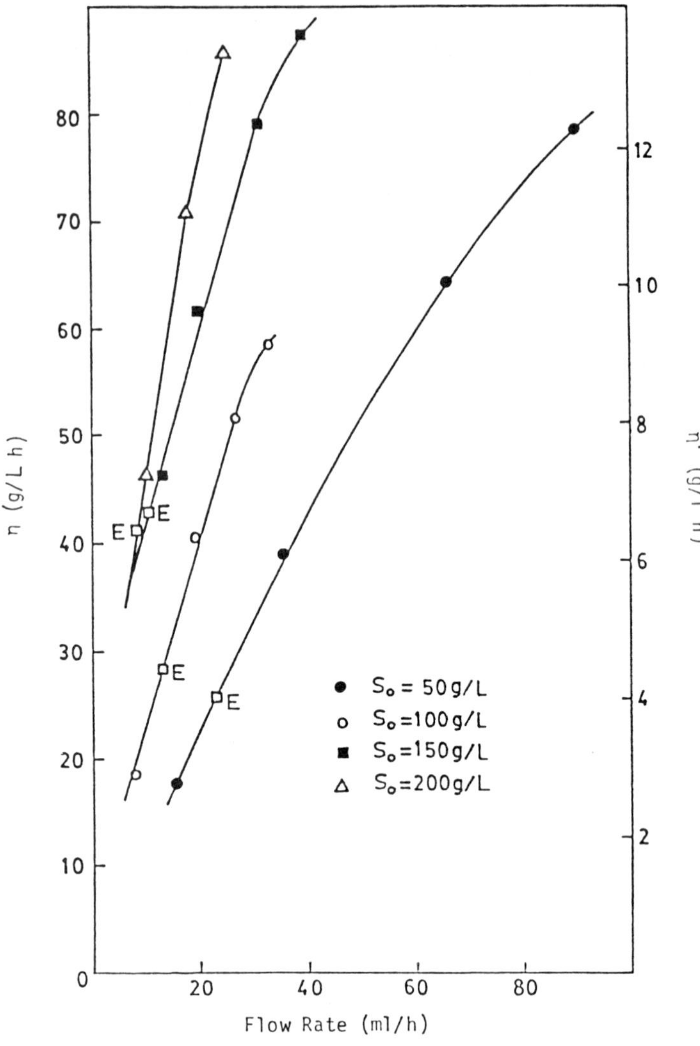

Figure 6 Overall ethanol productivity in a biofilm reactor as a function of the liquid flow rate. Productivities are based on liquid holdup η and on initial operating volume η'.

$$S(z = 0) = S_0 \tag{7}$$

$$P(z = 0) = O \tag{8}$$

In order to develop solutions to these equations, suitable expressions for the cell and product formation rates are needed. The most common approach for the production of primary metabolites is to use the Monod model (Monod, 1949), modified to include product inhibition:

$$r_x = \frac{\mu_m SX}{K_s + S} f(P) \tag{9}$$

$$r_p = \frac{\nu_m SX}{K_s' + S} f'(P) \tag{10}$$

where f(P) and f'(P) are the product inhibition relationships. The substrate uptake rate can be expressed in terms of cell or product formation rates:

$$r_s = - \frac{r_x}{Y_{x/s}} = - \frac{r_p}{Y_{p/s}} \tag{11}$$

where $Y_{x/s}$ and $Y_{p/s}$ are the apparent yields of cell and ethanol on substrate. In most cases, the value of $Y_{p/s}$ has been found to be constant during batch and continuous fermentations under the same medium compositions and is preferred over $Y_{x/s}$, which is dependent upon such factors as product concentration. Moreover, the cell concentration in immobilized cell reactors is not easy to determine, and thus a value of $Y_{x/s}$ is usually not available experimentally. Combining Equations (3), (4), and (5) with Equations (9), (10), and (11) gives

$$u \frac{dX}{dz} = \frac{\mu_m SX}{K_s + S} f(P) \tag{12}$$

$$u \frac{dS}{dz} = - \frac{1}{Y_{p/s}} \frac{\nu_m SX}{K_s' + S} f'(P) \tag{13}$$

$$u \frac{dP}{dz} = \frac{\nu_m SX}{K_s' + S} f'(P) \tag{14}$$

The values of the saturation constants K_s and K_s' are usually small compared with the value of the substrate concentration, throughout almost the entire reactor. Also, residence time can be used to replace column length according to

$$\tau = \varepsilon \frac{z}{u} \tag{15}$$

This substitution yields

$$\frac{dX}{d\tau} = \frac{\mu_m X}{\varepsilon} f(P) \tag{16}$$

$$\frac{dS}{d\tau} = \frac{\nu_m X}{\varepsilon Y_{p/s}} f'(P) \tag{17}$$

$$\frac{dP}{d\tau} = \frac{\nu_m X}{\varepsilon} f'(P) \tag{18}$$

In order to obtain solutions for these equations, suitable inhibition relationships are required. Considerable data have been published on product inhibition during ethanol fermentation. Exponential relationships were proposed by Aiba et al. (1968), where

$$f(P) = e^{-K_1 P} \tag{19}$$

$$f'(P) = e^{-K_2 P} \tag{20}$$

These equations have been used in some cases for modeling ethanol inhibition in biofilm reactors (Dale et al., 1985a; Gencer and Mutharasan, 1983). Aiba and Shoda (1969) proposed a hyperbolic relation based on the model for noncompetitive inhibition of enzyme reactions, yielding

$$f(P) = \frac{K_p}{K_p + P} \tag{21}$$

$$f'(P) = \frac{K'_p}{K'_p + P} \tag{22}$$

These expressions have been used recently by Novak et al. (1981) and Hoppe and Hansford (1982). More frequently used inhibition functions, proposed by Levenspiel (1980) and recently reassessed and slightly modified by Luong (1985), are given by

$$f(P) = \left(1 - \frac{P}{P_m}\right)^{\alpha} \tag{23}$$

$$f'(P) = \left(1 - \frac{P}{P'_m}\right)^{\beta} \tag{24}$$

The numerical values of the constants that quantify ethanol inhibition in these models differ widely in the literature. In fact, the variation is so large that the models overlap, so that no model can be recommended as best a priori. Inhibition in continuous culture seems to be lower than in batch culture (Hoppe and Hansford, 1982). Also, inhibition due to extracellularly added ethanol is found to be lower than due to autogeneously produced ethanol (Novak et al., 1981). Vega et al. (1987b) showed that cell concentration is an important parameter to be considered when selecting the appropriate inhibition function. This effect is shown in Figures 7 and 8, where specific growth and production rates are plotted as functions of the inoculum size employed in batch cultures of *S. cerevisiae* on glucose. As the cell concentration increases, the inhibition relationships move from strongly exponential to more linear behavior in the case of growth rates. For specific production rates, the relationships are clearly linear after a certain value of the initial cell concentration.

At high cell concentrations attained in immobilized cell reactors, Equations (23) and (24) seem to be the most appropriate with $\alpha = \beta = 1$. For these expressions, P_m and P'_m bear the physical meaning of ethanol concentrations at which growth and production rates are stopped (Ghose and Tyagi, 1979). Although their values depend upon the ethanol tolerance of the strain employed, values around $P_m = 96$ g/L and $P'_m = 114$ g/L, which were first reported by Ghose and Tyagi (1979), have agreed with others in the literature (Luong, 1985). It is also worth noting that after combining Equations (24) and (18) and integrating, exponential profiles for ethanol concentration in the immobilized cell reactors are attained, agreeing well with experimental observations.

Substrate inhibition can also be incorporated in the models. The most common approach has been the use of a model proposed by Andrews (1969) as a modification of the Monod model:

$$\mu = \frac{\mu_m S}{K_s + S + S^2/w} \tag{25}$$

$$\nu = \frac{\nu_m S}{K'_s + S + S^2/w'} \tag{26}$$

Substrate inhibition of the yeast *S. cerevisiae* has not been observed in batch culture below approximately 50-100 g/L of glucose (Vega et al., 1985; Bonami et al., 1981). Above this level, the Andrews model has been shown to be satisfactory up to an initial glucose concentration of 200 g/L. The effect of substrate inhibition on packed bed bioreactors should be to decrease productivity for the

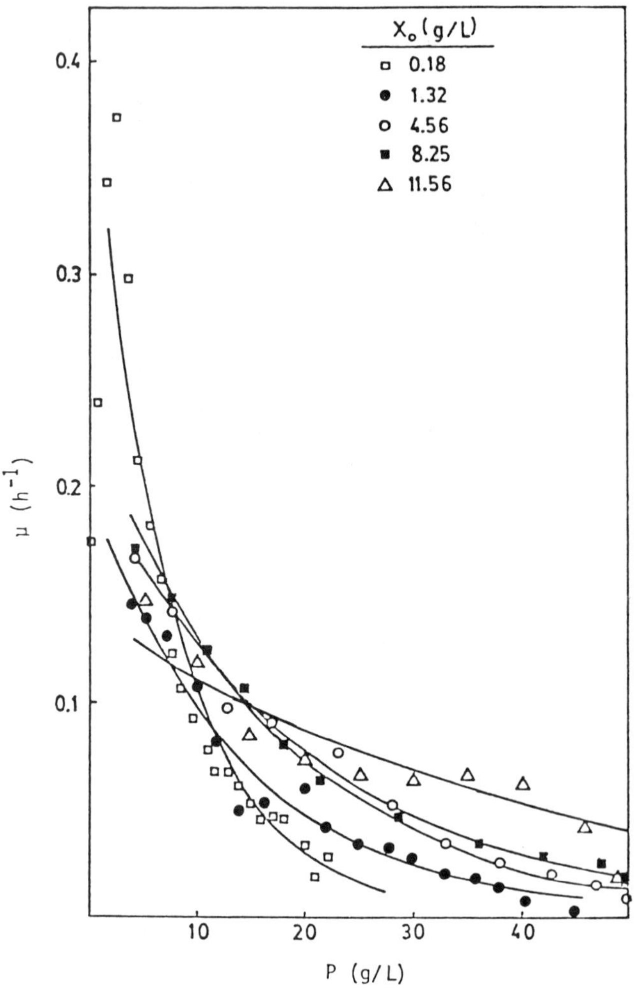

Figure 7 Ethanol inhibition of specific growth rate for different inoculum levels in batch culture.

entrance portion of the bed. Lee et al. (1983) showed that this effect appears to be more significant at lower residence times.

Mass transfer resistances in biofilm reactors have also been considered in some mathematical analyses. Radovich (1985) presented an excellent review of mass transfer limitations on fermentation reactions in immobilized cell reactors. Both external (to the biofilm) and internal (within the biofilm) mass transfer limitations occur in

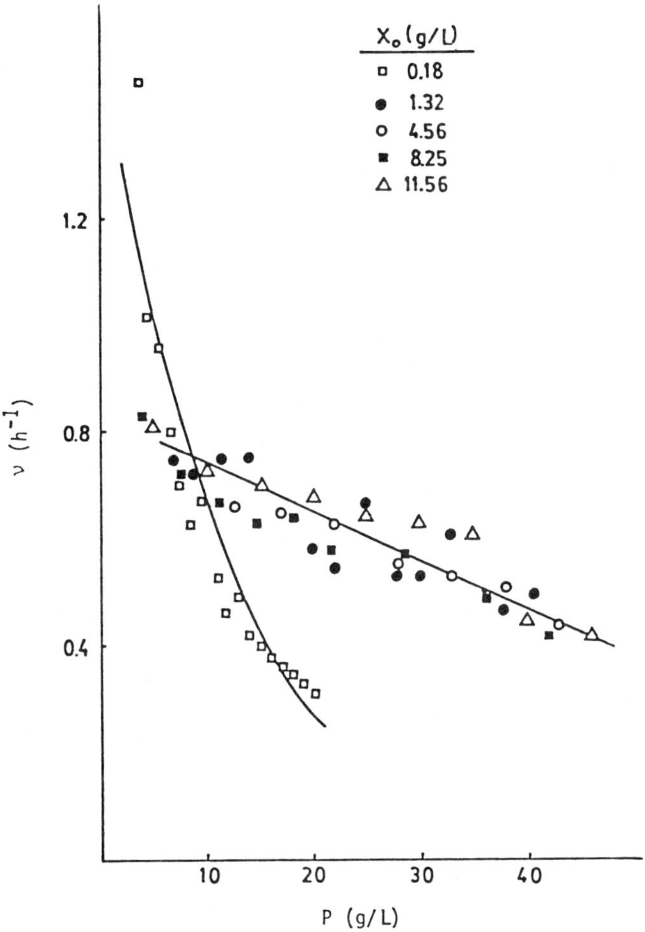

Figure 8 Ethanol inhibition of specific production rate for different inoculum levels in batch culture.

this type of reactor, although the latter should be much less than in gel occlusion systems. Mass transfer in biologic reactors can be modeled through the use of an effectiveness factor, similar to heterogeneous catalysis. For anaerobic ethanol fermentation and the high substrate concentrations employed, it is likely that external mass transfer is not limiting, at least for a large portion of the packed bed.

Internal mass transfer in biofilm reactors is due to substrate diffusion through the cell layers. For thin biofilm layers, such as

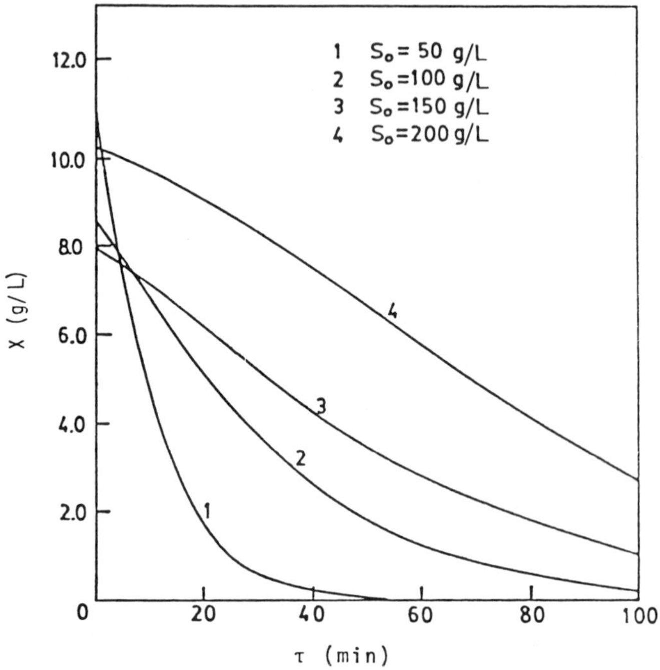

Figure 9 Theoretical cell profile in a biofilm reactor employing *S. cerevisiae* for ethanol production. The cell concentration at each residence time corresponds with that required in the biofilm reactor in order to obtain the same rate of fermentation as in batch culture.

those used in wastewater treatment systems, diffusional resistance is not an important consideration (La Motta, 1976). However, for the thick layers of cells that are achieved in some packed beds for ethanol production, mass transfer may be limiting. Experimentally, Jones et al. (1984), found an increase in the saturation constant K'_S in a tower fermenter operating close to plug flow behavior. The effect of diffusional resistance has the effect of increasing K_S and K'_S but not affecting the maximum growth and production rates (Korus and O'Driscoll, 1974). Vega et al. (1985) obtained the theoretical profiles of cells that should exist in the reactor as a function of the residence time, if the cells were fermenting at the same rates as in batch culture. These profiles are shown in Figure 9 and were obtained by combining Equation (2) with Equations (18) and (24) (with $\beta = 1$) for the data of Figure 5. As noted, cell concentration increases with inlet substrate concentration and decreases with residence times. These results imply that internal mass transfer is less under these conditions and that a deeper layer of cells is active when the substrate concentration and flow rates are high.

Other refinements of the modeling include temperature effects, axial dispersion, and cell viability. Temperature changes in a packed bed reactor have been considered by Ghose and Bandyopadhyay (1982) in gel occlusion systems. Axial dispersion may also be included in such models (Tyagi and Ghose, 1982). Typical values for the dispersion number (D/u) range between 0.04 and 0.25 (Navarro et al., 1985) but are strong functions of the bed porosity, which in turn changes with time. Cell viability is a more difficult variable to include in the models and is usually either assumed or obtained by trial and error for the best fit of the model to experimental data.

In general, the extension of batch kinetic models with modifications to account for deviations from ideal behavior have been the most widely used models. These models usually require numerous parameters that cannot be determined a priori, such as cell concentration. Therefore, although the use of such models has been useful in understanding reactor behavior, they have not allowed predictive estimations and scaleup. Development of more realistic models that may not require measuring the cell mass are probably more likely to succeed. Approaches of this type have been applied to the control of bioreactors for ethanol fermentation (Bovee et al., 1984).

D. Operational Stability

The operational stability of the biocatalyst is an important aspect in evaluating continuous systems. Immobilized cells represent an important advance in attaining long operational half-lives compared with immobilized enzymes (Venkatasubramanian and Vieth, 1979). Nevertheless, stability problems are often encountered when trying to obtain high cell densities. In ethanol production, cell overgrowth has proven to be a major problem to overcome for long-term operation in packed beds (Sitton and Gaddy, 1980; Krug and Dauglis, 1983). In other operational modes, such as fluidized beds (Nagashima et al., 1983), the catalytic action of the cells can be controlled by continuous replacement of a portion of the suspended particles. In tower fermenters (Admassu et al., 1985), the removal of excess bed material resulting from cell growth can be readily accomplished, thus maintaining a constant bed height. In packed bed operation, the removal of cell overgrowth has been successfully achieved with a vigorous gas purge (Sitton and Gaddy, 1980). However, this method interrupts operation, requiring a period for regeneration of the biomass

Cell growth is always necessary to maintain the catalytic activity at a certain level. High cell yields, on the other hand, result in a decrease in product yield, excessive channeling in the reactor (Sitton and Gaddy, 1980), and eventually a pressure buildup that inhibits feed pumping (Krug and Dauglis, 1983). Several methods of maintaining cell productivity while slowing cell growth have been proposed. The most common method has been nutrient limitation, usually nitrogen (Wada et al., 1980; Nilsson and Ohlson, 1982). This practice has not proven satisfactory, and in most cases a decrease in production was observed.

Methods that tend to increase the maintenance energy of the cells, such as a temperature above optimum growth levels or the addition of metabolic inhibitors, have been more successful (Hahn-Hagerdal and Mattiasson, 1982; Amin et al., 1984; Vega et al., 1987a). An example for ethanol fermentation is the use of acetate in the feed (Vega et al., 1987a). Acetate and short-chain fatty acids in general affect phosphate transport across the membrane (Samson, 1955), resulting in a more energetically favorable process. The direct consequences of this effect are increases in the ethanol conversion efficiency and the specific ethanol production rate. Depending upon the concentration of the fatty acid and its chain length, promotion or inhibition of growth occurs (Vega et al., 1986b). It should also be noted that the improvement in immobilized cell-specific productivities over free cell productivities (Doran and Bailey, 1986) may also be explained by interference with the membrane transport process of the cells.

The use of acetate in the feed of a biofilm packed bed bioreactor led to the effects shown in Figure 10. The ethanol concentration profiles obtained by fitting the data to equations of the form of Equation (2) show a conversion that increased with the acetate ion concentration, as well as an increase in the product yield. No significant decrease in cell growth rate was observed during the runs, probably owing to the somewhat low concentration of fatty acid employed. Long-term viability was not affected by the presence of acetate. Important increases of up to 50% in the overall reactor productivities resulted, as illustrated in Figure 11. Longer chain fatty acids have been found to produce similar effects at lower concentrations (Vega et al., 1986b). Figure 12 shows the effect of the addition of nonanoic acid (in a final concentration of 0.20 mM) on the change of the liquid holdup volume V_{1h} and total void volume V_0' in the reactor with time (Vega et al., 1986a). The biomass loading rate, as obtained from the slopes of the change in total void volume with time, is higher for the system without fatty acid. Such a decrease in the cell growth rate is accompanied, however, by an improvement in productivity as can be inferred from the more negative slope observed for liquid holdup with time when nonanoic acid is present (larger gas holdup indicates faster CO_2 production).

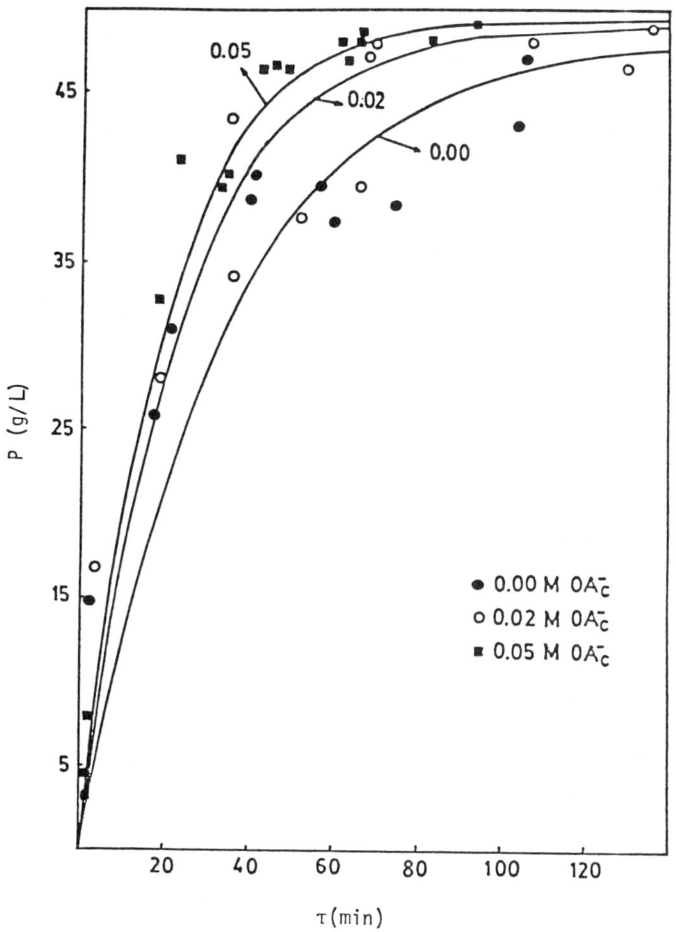

Figure 10 Ethanol concentration profiles in a biofilm reactor using *S. cerevisiae* with various concentrations of acetate ion (OA_c^-) in the feed.

In many processes, such as ethanol fermentation, gas is released and the accumulation of gas in the reactor creates dead spaces, increases the pressure inside the reactor, decreases the liquid-solid contact area, and creates channels, all of which can negatively affect the conversion of substrate to product (Shiotani and Yamane, 1981). The use of horizontal packed bed bioreactors instead of vertical operation has been investigated by Shiotani and Yamane

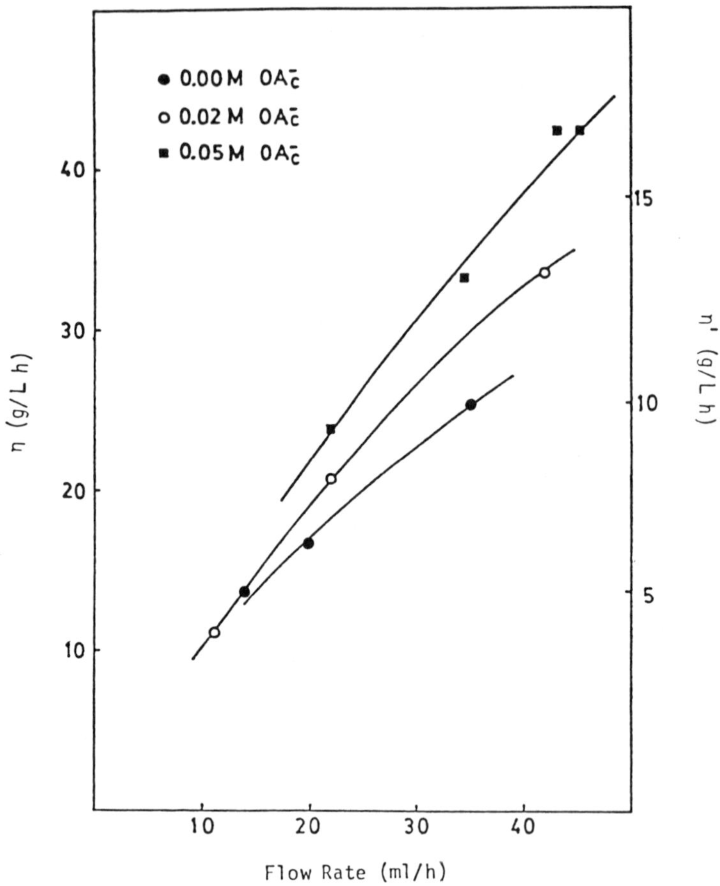

Figure 11 Overall ethanol productivity in a biofilm reactor as a function of the flow rate for various concentrations of acetate ion (OA_c^-) in the feed. Productivities are based on liquid holdup η and on initial operating volume η'.

(1981) for cells entrapped in calcium alginate gel and polyacrylamide gel. This study found that horizontal operation allowed the lateral removal of the CO_2 released during the fermentation, resulting in longer liquid retention time and an improvement in the productivity of the system.

Similar studies in a biofilm reactor (Navarro et al., 1985) showed no increase in the productivity but a loss of stability. The reason for such differences in behavior is probably the different mass transfer effects created by the CO_2 in both systems. Mass transfer within the bead is controlling, and an increase in external mass transfer has little effect. Therefore, an increase in liquid retention time improves the productivity. However, in the biofilm reactor, this advantage is offset by an increase in external mass transfer. Turbulence created by gas evolution is necessary to open channels, reduce plugging, and promote substrate distribution and mixing to increase the mass transfer between cells and substrate.

The use of the cell cross-flow reactor with immobilized yeast entrapped in sodium alginate has been reported as enhancing the

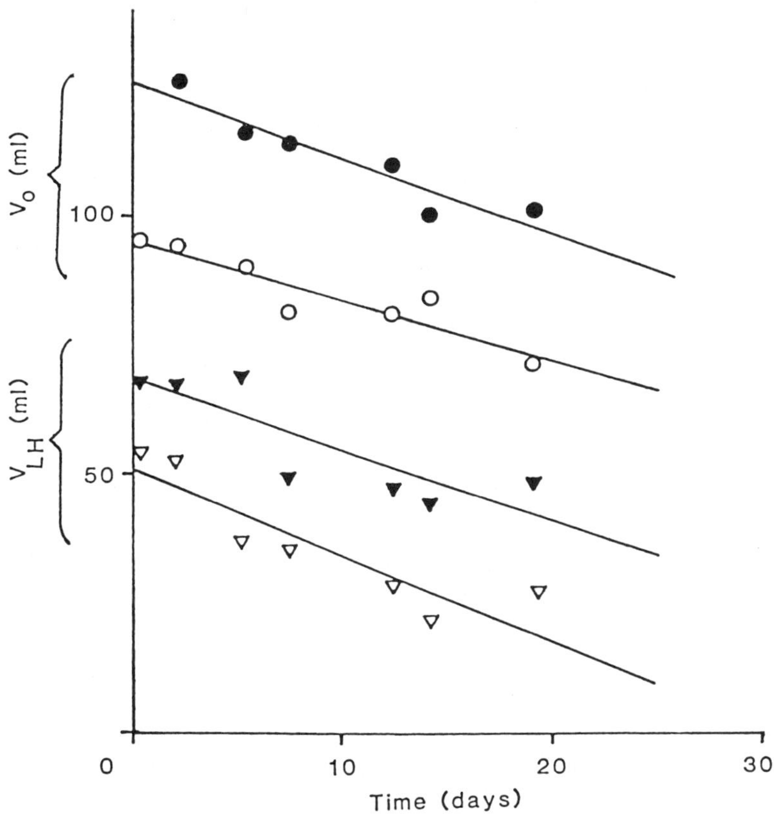

Figure 12 Effect of the addition of nonanoic acid to the feed of a biofilm reactor employing *S. cerevisiae* on the change in void volume V_0 and liquid holdup volume V_{lh} with time. (●, ▼) Feed without acid. (○, ▽) Feed containing 0.20 mM nonanoic acid.

production of ethanol (Chotani and Constantinides, 1984). There have been no studies reported with biofilm cross-flow reactors. Recently, a reactor that uses the internal gelation of alginate has been proposed to overcome the problems observed with gel beads due to released CO_2 (Johansen and Flink, 1986).

IV. CONCLUSIONS

The combination of high conversion, high product concentrations, and high productivity is essential to the economical production of chemicals by fermentation. Immobilized cell reactors, especially biofilm reactors, have been shown to offer the combination of these factors. Biofilm reactors result in high cell densities, high productivities, reduced diffusional resistance, and less inhibition, Problems with these systems include long-term operational stability and development of reliable mathematical models for scaleup.

Desirable characteristics for a biofilm reactor are a short start-up period, high productivity, and long and stable operating life. The choice of the packing material affects each of these areas. The eventual performance of the reactor is dominated by the quantity of cells present. However, to achieve a fast start-up and a rapid biomass loading rate, a packing with a high ratio of surface area to volume with sufficient interstitial spaces for cell entrapment is necessary. These factors must be combined with a high initial porosity for a high cell volume and high productivity. For systems in which cells are attached to surfaces by chemical bonds, Intalox saddles and Raschig rings have been found to have the best characteristics.

The steady-state behavior of packed bed biofilm reactors is never achieved, since growth of the biocatalyst cannot be stopped without the loss of catalytic activity and physical removal is difficult. Batch kinetics, including substrate and product inhibition functions, along with expressions for mass transfer resistances, has usually proven satisfactory in explaining reactor behavior. Still, independent determination of the viable biomass concentration and activity along the reactor has not been accomplished. Mathematical models are therefore constrained to the analysis of results without adequate prediction of the reactor behavior a priori.

Cell overgrowth in the interstitial spaces provided by the packing represents the major problem for long-term operation in packed beds. Removal of cell overgrowth by gas purging has been successful, although operation is interrupted. Among the various methods proposed to maintain cell productivity while slowing cell growth, modifications that tend to increase the maintenance energy of cells have been most successful. Horizontal column operation to reduce gas holdup has shown no improvement in biofilm reactor performance.

V. NOMENCLATURE

a	parameter in Equation (1); g/L
b	parameter in Equation (1); minute^{-1}
c	parameter in Equation (2); g/L
d	parameter in Equation (2); minute^{-1}
D	dispersion coefficient; cm^2/s
f(P)	product inhibition relationship for growth rate
f'(P)	product inhibition relationship for production rate
ICR	immobilized cell reactor
K_1	ethanol inhibition constant for growth rate in exponential model; g/L
K_2	ethanol inhibition constant for production rate in exponential model; g/L
K_p	ethanol inhibition constant for growth rate in Equation (21); g/L
K'_p	ethanol inhibition constant for production rate in Equation (22); g/L
K_s	Monod saturation constant for growth rate; g/L
K'_s	Monod saturation constant for rate or production; g/L
L	total fermenter (bed length; cm
P	extracellular ethanol concentration; g/L
P_m	extracellular ethanol concentration at which growth rate becomes zero; g/L
P'_m	extracellular ethanol concentration at which production rate becomes zero; g/L
r_s	rate of substrate consumption; g/L per h
r_p	rate of product formation; g/L per h
r_x	rate of biomass production; g/L per h
S_0	substrate concentration; g/L
S	initial substrate concentration; g/L
t	time; h
u	superficial velocity, cm/minute
V_{lh}	liquid holdup volume; L^3
V_0	initial void volume; L^3
V'_0	void volume in the reactor; L^3
V_r	reactor volume; L
W	substrate inhibition constant for growth rate; g/L
W'	substrate inhibition constant for rate of production; g/L
X	cell (biomass) concentration; g/L
X_0	initial cell (biomass) concentration; g/L
$Y_{p/s}$	apparent yield factor for product on substrate; g ethanol per g glucose
$Y_{x/s}$	apparent yield factor for product on substrate; g ethanol per g glucose

z	column length; cm
ε	bed porosity
η	ethanol productivity based on liquid holdup; g/L per h
η_{sp}	specific ethanol productivity; g/L per h
η'	ethanol productivity based on initial void volume; g/L per h
μ	specific growth rate; h^{-1}
μ_m	maximum specific growth rate; h^{-1}
ν	specific rate of product formation; h^{-1}
ν_m	maximum specific rate of product formation; h^{-1}
τ	residence time based on V_{lh}; minutes
τ'	residence time based on V ; minutes
τ''	residence time based on V_r; minutes

REFERENCES

Admassu, W., Korus, R. A., and Heimsch, R. C. (1985). Ethanol fermentation with a flocculating yeast. *Chem. Eng. J. 31*:B1-B8.

Aiba, S., and Shoda, M. (1969). Reassessment of the product inhibition in alcohol fermentation. *J. Ferment. Technol. Jpn. 47*: 790-798.

Aiba, S., Shoda, M., and Nagatani, M. (1968). Kinetics of product inhibition in alcohol fermentation. *Biotechnol. Bioeng. 10*:845-864.

Amin, G., Standaert, P., and Verachert, H. (1984). Effects of metabolic inhibitors on the alcoholic fermentation by several yeasts in batch or in immobilized cell systems. *Appl. Microbiol. Biotechnol. 19*:91-98.

Andrews, J. F. (1969). A mathematical model for the continuous culture of microorganisms utilizing inhibitory substrates. *Biotechnol. Bioeng. 10*:707-723.

Benefield, L., and Molz, F. (1985). Mathematical simulation of a biofilm process. *Biotechnol. Bioeng. 27*:921-931.

Black, G. M., Webb, C., Matthews, T. M., and Atkinson, B. (1984). Practical reactor system for yeast cell immobilization using biomass support particles. *Biotechnol. Bioeng. 26*:134-141.

Bonomi, A., Aboutboul, H., and Schmidell, W. (1981). Simulation of the continuous fermentation of manioc hydrolysate. *Biotechnol. Bioeng. Symp. 11*:333-357.

Bovee, J. D., Strehaiano, P., Goma, G., and Sevely, Y. (1984). Alcoholic fermentation: Modelling based on sole substrate and product measurement. *Biotechnol. Bioeng. 26*:328-334.

Cheetham, P. S. J., Blunt, K. W., and Bucke, C. (1979). Physical studies on cell immobilization using calcium alginate gels. *Biotechnol. Bioeng. 21*:2155-2168.

Chibata, I., Tosa, T., and Sato, T. (1983). Immobilized cells in preparation of fine chemicals. *Adv. Biotechnol. Proc. 1*:203-222.

Chotani, G. K., and Constantinides, A. (1984). Immobilized cell cross-flow reactor. *Biotechnol. Bioeng. 26*:217-220.

Clausen, E. C., Shah, R. B., Najafpour, G., and Gaddy, J. L. (1982). Production of organic acids from biomass by acid hydrolysis and fermentation. *Biotechnol. Bioeng. Symp. 12*:67-72.

Conheady, J. A., Clausen, E. C., and Gaddy, J. L. (1981). Performance of immobilized *Clostridium acetobutylicum*. 74th Annual AIChE Meeting, November, New Orleans.

Dale, M. C., Okas, M. R., and Wankat, P. C. (1985a). An immobilized cell reactor with simultaneous product separation. I. Reactor design and analysis. *Biotechnol. Bioeng. 27*:932-942.

Dale, M. C., Okas, M. R., and Wankat, P. C. (1985b). An immobilized cell reactor with simultaneous product separation. II. Experimental reactor performance. *Biotechnol. Bioeng. 27*:943-952.

Del Borghi, M., Converti, A., Parisi, F., and Ferraiolo, G. (1985). Continuous alcohol fermentation in an immobilized cell rotating disk reactor. *Biotechnol. Bioeng. 27*:761-768.

Denac, M., Uzman, S., Tanaka, H., and Dunn, I. J. (1983). Modeling of experiments of biofilm penetration effects in a fluidized bed nitrification reactor. *Biotechnol. Bioeng. 25*:1841-1861.

Doran, P. M., and Bailey, J. E. (1986). Effects of immobilization on growth fermentation properties and macromolecular composition of *Saccharomyces cerevisiae* attached to gelatin. *Biotechnol. Bioeng. 28*:73-87.

Durand, G., and Navarro, J. M. (1978). Immobilized microbial cells. *Proc. Biochem. 13*:14-23.

Gencer, M. A., and Mutharasan, R. (1980). Ethanol fermentation in a yeast immobilized column fermentor. *Adv. Biotechnol. 1*:627-633.

Gencer, M. A., and Mutharasan, R. (1983). Ethanol fermentation in a yeast immobilized tubular fermentor. *Biotechnol. Bioeng. 25*:2243-2262.

Ghose, T. K., and Bandyopadhyay, K. K. (1982). Studies on immobilized *Saccharomyces cerevisiae*. II. Effect of temperature distribution on continuous rapid ethanol formation in molasses fermentation. *Biotechnol. Bioeng. 24*:797-804.

Ghose, T. K., and Tyagi, R. D. (1979). Rapid ethanol fermentation of cellulose hydrolysate. II. Product and substrate inhibition and optimization of fermentor design. *Biotechnol. Bioeng. 21*:1401-1420.

Gollinger, G. B., Vega, J. L., Clausen, E. C., and Gaddy, J. L. (1986). The effects of packing and immobilization characteristics on the performance of a biofilm reactor for ethanol production. AIChE Summer National Meeting, Boston.

Hahn-Hagerdal, B., and Mattiason, B. (1982). Azide sterilization of fermentation media. *Eur. J. Appl. Microbiol. Biotechnol. 14*:140-143.

Herzig, J. P., Leclerc, D. M., and Legoff, P. (1970). *Flow Through Porous Media*. American Chemical Society, Washington, D.C., pp. 131-157.

Hoppe, G. K., and Hansford, G. S. (1982). Ethanol inhibition of continuous anaerobic yeast growth. *Biotechnol. Lett. 4*:39-44.

Hsiao, H. Y., Chiang, L. C., Yang, C. M., Chen, L. F., and Tsao, G. T. (1983). Preparation and performance of immobilized yeast cells in columns containing no inert carrier. *Biotechnol. Bioeng. 25*:363-375.

Hu, J. (1986). Immobilization of cells containing glucose isomerase using a multifunctional crosslinking reagent. *Biotechnol. Lett. 8*:127-130.

Johansen, A., and Flink, J. M. (1986). A new principle for immobilized yeast reactors based on internal gelation of alginate. *Biotechnol. Lett. 8*:121-126.

Jones, S. T., Korus, R. A., Admassu, W., and Heimsch, R. C. (1984). Ethanol fermentation in a continuous tower fermentor. *Biotechnol. Bioeng. 26*:742-747.

Kennedy, J. F., and Cabral, J. M. S. (1985). Immobilization of biocatalysts by metal-link/chelation processes. In *Immobilized Cells and Enzymes, A Practical Approach*. IRL Press, Washington, D.C., pp. 19-37.

Korus, R. A., and O'Driscoll, K. F. (1974). The effects of intraparticle diffusion on the kinetics of gel entrapped enzymes. *Can. J. Chem. Eng. 52*:775-780.

Krug, T. A., and Dauglis, A. J. (1983). Ethanol production using *Zymomonas mobilis* immobilized on an ion exchange resin. *Biotechnol. Lett. 5*:159-164.

Kyung, K. H., and Gerhardt, P. (1984). Continuous production of ethanol by yeast "immobilized" in a membrane-contained fermentor. *Biotechnol. Bioeng. 26*:252-256.

La Motta, E. J. (1976). External mass transfer in a biological film reactor. *Biotechnol. Bioeng. 18*:1359-1370.

Lee, T. H., Ahn, J. C., and Ryu, D. D. Y. (1983). Performance of an immobilized yeast reactor system for ethanol production. *Enzyme Microb. Technol. 5*:41-45.

Levenspiel, O. (1980). The Monod equation. A revisit and a generalization to product inhibition situations. *Biotechnol. Bioeng. 22*: 1671-1687.

Luong, J. H. T. (1985). Cell immobilization in K-carrageenan for ethanol production. *Biotechnol. Bioeng. 27*:1652-1661.

Maiorella, B. L., Blanch, H. W., and Wilke, C. R. (1983). Byproduct inhibition effects on ethanolic fermentation by *Saccharomyces cerevisiae*. *Biotechnol. Bioeng. 25*:103-121.

Margaritis, A., Bajpai, P. K., and Wallace, J. B. (1981). High ethanol productivities using small Ca-alginate beads of immobilized cells of *Zymomonas mobilis* *3*:613-618.

Mehaia, M. A., and Cheryan, M. (1984). Hollow fibre bioreactor for ethanol production: Application to the conversion of lactose by *Kluyveromyces fragilis*. *Enzyme Microb. Technol.* *6*:117-120.

Monod, J. (1949). The growth of bacterial cultures. *Annu. Rev. Microbiol.* *3*:371-394.

Nagashima, M., Azuma, M., and Noguchi, S. (1983). Technology developments in biomass alcohol production in Japan: Continuous alcohol production with immobilized microbial cells. *Ann. N.Y. Acad. Sci.* *413*:457-468.

Navarro, A. R., Vega, J. L., Clausen, E. C., and Gaddy, J. L. (1985). Comparison of the vertical and horizontal ICR for ethanol fermentation. AIChE Meeting, Seattle, Washington, August, Paper 29C.

Navarro, J. M., and Durand, G. (1977). Modification of yeast metabolism by immobilization onto porous glass. *Eur. J. Appl. Microbiol.* *4*:243-254.

Netto, C. B., Destruhaut, A., and Goma, G. (1985). Ethanol fermentation by flocculating yeast: Performance and stability dependence on a critical fermentation rate. *Biotechnol. Lett.* *7*:355-360.

Nilsson, I., and Ohlson, S. (1982). Columnar denitrification of water by immobilized *Pseudomonas denitrificans* cells. *Eur. J. Appl. Microbiol.* *14*:86-90.

Novak, M., Strehaiano, P., Moreno, M., and Goma, G. (1981). Alcoholic fermentation on the inhibitory effect of ethanol. *Biotechnol. Bioeng.* *23*:201-211.

Okita, W. B., Bonham, D. B., and Gainer, J. L. (1985). Covalent coupling of microorganisms to a cellulosic support. *Biotechnol. Bioeng.* *27*:632-637.

Prince, I. G., and Barford, J. P. (1982). Continuous tower fermentation for power ethanol production. *Biotechnol. Lett.* *4*:263-268.

Radovich, J. M. (1985). Mass transfer effects in fermentations using immobilized whole cells. *Enzyme Microb. Technol.* *7*:2-10.

Robinson, C. W., Moo-Young, M., and Lamptey, J. (1980). Ethanol production by immobilized yeast cells in a packed tower. *Adv. Biotechnol.* *2*:105-111.

Rosevear, A. (1984). Immobilized biocatalysts—a critical review. *J. Chem. Technol. Biotechnol.* *34B*:127-150.

Samson, F. E. (1955). Effects of acetate and other short-chain fatty acids on yeast metabolism. *Arch. Biochem. Biophys.* *54*:406-423.

Shieh, W. K., Mulcahy, L. T., and LaMotta, E. J. (1982). Mathematical model for the fluidized bed biofilm reactor. *Enzyme Microb. Technol.* *4*:269-275.

Shiotani, T., and Yamane, Y. (1981). A horizontal packed-bed bioreactor to reduce CO_2 gas holdup in the continuous production of ethanol by immobilized yeast cells. *Eur. J. Appl. Microbiol. Biotechnol.* *13*:96-101.

Sitton, O. C., and Gaddy, J. L. (1980). Ethanol production in an immobilized cell reactor. *Biotechnol. Bioeng.* *22*:1735-1748.

Tanaka, H., and Dunn, I. J. (1982). Kinetics of biofilm nitrification. *Biotechnol. Bioeng.* *24*:669-689.

Thonart, P., Custine, M., and Paquot, M. (1982). Zeta potential of yeast cells: Application in cell immobilization. *Enzyme Microb. Technol.* *4*:191-194.

Tyagi, R. D., and Ghose, T. K. (1982). Studies on immobilized *Saccharomyces cerevisiae*. I. Analysis of continuous rapid ethanol fermentation in an immobilized cell reactor. *Biotechnol. Bioeng.* *24*:781-795.

Van Haecht, J. L., Bolipombo, M., and Rouxhet, P. G. (1985). Immobilization of *Saccharomyces cerevisiae* by adhesion: Treatment of the cells by A1 ions. *Biotechnol. Bioeng.* *27*:217-224.

Vega, J. L. (1985). Ethanol fermentation studies in an immobilized cell reactor. M. S. Thesis. University of Arkansas.

Vega, J. L., O'Malley, J. D., Clausen, E. C., and Gaddy, J. L. (1985). Generalized characteristics of a cross-linked immobilized cell reactor for ethanol production. *Biotechnol. Bioeng. Symp. Ser.* *15*:263-276.

Vega, J. L., Prieto, S., Clausen, E. C., and Gaddy, J. L. (1986a). Optimal kinetic parameters for immobilized *Saccharomyces cerevisiae*. 192nd ACS National Meeting, Anaheim, California, paper 78.

Vega, J. L., Wood, F. D., Clausen, E. C., and Gaddy, J. L. (1986b). Improved ethanol production by *S. cerevisiae* with organic acid addition. *Biotechnol. Bioeng. Symp.* *17*:401-411.

Vega, J. L., Clausen, E. C., and Gaddy, J. L. (1987a). Acetate addition to an immobilized yeast column for ethanol production. *Biotechnol. Bioeng.* *29*:429-435.

Vega, J. L., Navarro, A. R., Clausen, E. C., and Gaddy, J. L. (1987b). Effects of innoculum size on ethanol inhibition modeling and other fermentation parameters. *Biotechnol. Bioeng.* *29*:633-638.

Venkatasubramanian, K., and Vieth, W. R. (1979). Immobilized microbial cells. *Prog. Indust. Microbiol.* *15*:61-75.

Wada, M., Kato, J., and Chibata, I. (1980). Continuous production of ethanol using immobilized growing yeast cells. *Eur. J. Appl. Microbiol. Biotechnol.* *10*:275-287.

Wang, H. Y., and Hettwer, D. J. (1982). Cell immobilization in K-carrageenan with tricalcium phosphate. *Biotechnol. Bioeng.* *24*: 1827-1838.

22

Design of Fixed Film and Fluidized Bed Bioreactors

GRAHAM F. ANDREWS / *Idaho National Engineering Laboratory, Idaho Falls, Idaho*

I. INTRODUCTION

A. Advantages and Applications of Fluidized Bed Bioreactors

The advantages of the fluidized bed* bioreactor can best be illustrated by showing how it can be derived, at least conceptually, via a series of improvements to the simplest type of continuous fermenter, the stirred tank (Fig. 1a). The productivity* of the continuous stirred tank bioreactor is limited because all the cells wash out of the reactor if the hydraulic residence time* (HRT) is less than $1/\mu_0$ where μ_0 is the specific growth rate of the biomass evaluated at the inlet conditions. This can be avoided by separating cells from the effluent and recycling them to the reactor, which makes the mean cell residence time (MCRT) greater than the HRT. Since washout now occurs only if MCRT $< 1/\mu_0$, reactor productivity can be much higher. The easiest technique for separating the cells is gravity settling, either in an external settler as in the activated sludge process (Fig. 1b) or in a lamellar settler arranged so that the cells fall

---

*The fields of fermentation and wastewater treatment often use different terminology. Exact definitions of terms marked with an asterisk, as used in this chapter, are given in the glossary, Sec. V.

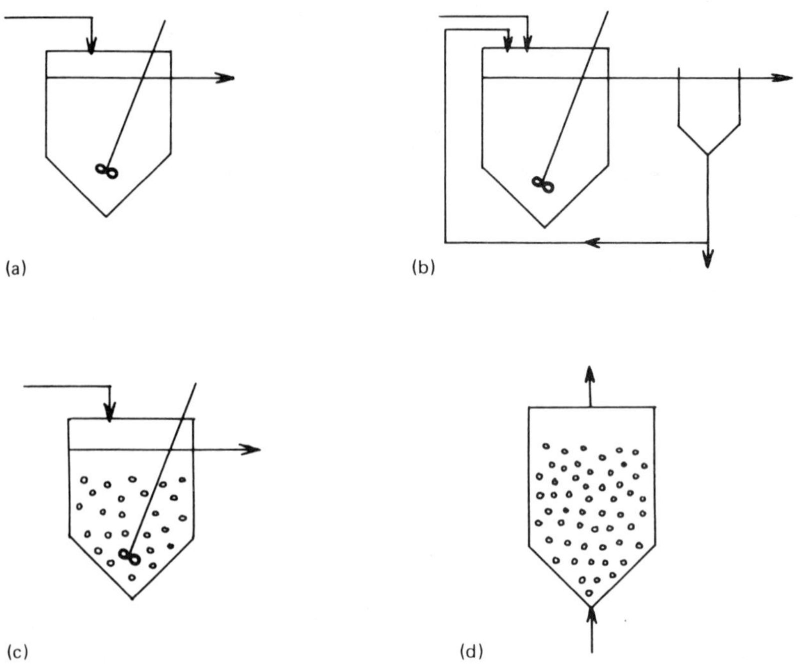

Figure 1 Types of continuous fermentor: (a) the continuous stirred tank fermentor; (b) with cell recycle by settling; (c) combined reactor/settler with immobilized biomass; (d) fluidized bed of immobilized biomass.

straight back into the reactor (Stephanopoulos et al., 1985). Settling is improved if the biomass occurs not as individual cells but as macroscopic "clumps." These may be microbial flocs or pellets, films of biomass attached to solid support particles, or, for microbial strains that are not naturally flocculent,* cells immobilized in beads of synthetic gel or in porous support particles. Particles with a high settling velocity naturally stay near the bottom of the reactor, so the reactor and settler can be combined into a single unit (Fig. 1c). The function of the agitator in this unit is to promote mass transfer by keeping the particles in suspension. It can be eliminated if the liquid is pumped up through the reactor to keep the particles in suspension in a fluidized bed (Fig. 1d). This final step has three advantages. It reduces reactor cost. It changes the liquid-phase mixing condition from completely mixed toward plug flow, which tends to increase reactor productivity (except for substrate inhibition, which is an example of negative-order reaction kinetics). Finally, it reduces the shear stress to which the particles are exposed and thus the rate at which cells are stripped from the particles and washed out of the reactor. This makes the MCRT very large. Cell washout is not a problem even with an HRT of a few minutes because it is prevented by gravity.

The same advantages of plug flow and a high MCRT can be obtained in a packed bed of bioparticles, but there are some difficulties. Most biochemical processes involve a gas phase as either a substrate (O_2) or a product (N_2 in denitrification, CO_2 in ethanol production, and CH_4 in anaerobic wastewater treatment), and gas bubbles pass more easily through a fluidized bed than a packed bed. Furthermore, the growth of biomass tends to clog a packed bed but a fluidized bed can expand to accommodate the extra volume. To prevent clogging and gas binding, the particles in a packed bed must be large, which means a low surface area of biomass per unit reactor volume and long diffusion paths for substrates and products between the liquid and the center of the particle. Fluidized beds can use much smaller particles and therefore usually have higher productivities.

Fixed film and fluidized bed reactors have been proposed for several aerobic fermentations, including penicillin production (Smith and Greenshields, 1974; Behie and Gaucher, 1988), but they have not been adopted commercially mainly because of doubts about carrying out relatively low volume fermentations with highly mutated microbial strains in continuous rather than the traditional batch reactors. Also, laboratory-scale tests have shown the difficulty of transferring sufficient oxygen. Either air must be bubbled directly through the bed or there must be considerable liquid recycle through an external aerator. Both methods have drawbacks, and in either case the concentration of dissolved oxygen in the liquid is so low that it can only diffuse a few hundred micrometers into a bioparticle before it is all consumed. The rest of the particle remains anoxic. This explains why applications have been limited to aerobic wastewater treatment in which substrate concentrations are low and continuous operation is essential (Cooper and Atkinson, 1981; Fan et al., 1987) and to tissue culture in which the low metabolic rate puts less demand on the oxygen transfer system (Dean et al., 1988). It also explains why these reactors are best suited to large-scale anaerobic processes, such as wastewater treatment, biologic denitrification, and the production of bulk chemicals, mainly ethanol, by fermentation. The following sections review the literature on how these processes have been carried out in different types of reactors.

B. Fluidized Beds of Biomass Particles

In the 1960s, several large-scale (3 m diameter × 30 m high) tower fermenters were built by the APV company in England for the continuous brewing of beer (Fig. 2). They contained fluidized beds of flocs of a specially chosen strain of *Saccharomyces cerevisiae* (Greenshields and Smith, 1971). Technically they were successful, being operated continuously for periods of several months with few problems of clogging or contamination. However, some small difference between batch and continuous brewing, perhaps the need to choose a flocculent yeast strain, affected the taste of the beer so the process was discontinued. It may be revived for the production of fuel ethanol from corn or sugarcane, for example. Laboratory-scale reactors containing flocculent *S. cerevisiae* strains have shown reactor productivities of the order of 10 g/L per h with virtually complete conversion of a 15% glucose feed (Jones et al., 1984; Prince and Barford, 1982), much higher than the 2 g/L per h expected from batch reactors. The flocculent strain of the bacterium *Zymomonas mobilis* looks even better. It grows faster than yeast, produces more cohesive and uniform flocs (2-3 mm in diameter), and gives a slightly higher ethanol yield since it uses the less efficient Entner-Duodoroff catabolic pathway. Beds of this organism have shown productivities of the order of 50 g/L per h (note that much higher productivity values have been quoted but they are for glucose conversions that would be uneconomical in practice) (Strandburg et al., 1982; Scott 1983), which translates into reactor volumes at least an order of magnitude smaller than the batch reactors now used in many places.

A feature of all these reactors is their resistance to contamination by foreign organisms; laboratory reactors have been run for several weeks with unsterilized feed with no problem. This is due to some extent to conditions in the ethanol fermentation, which are such that few contaminant organisms can grow. Andrews and Fonta (1988) reported more difficulties when they switched to lactic acid fermentation. However, it also reflects the fact that foreign organisms are washed out of the reactor unless they can attach themselves to the walls or the particles in the bed.

Beds of biomass particles for anaerobic wastewater treatment have been developed (Lettinga et al., 1980) and are known as upflow anaerobic sludge blanket (UASB) reactors. They have been successfully applied on a large scale to the treatment of various food-processing wastes, landfill leachates, and raw domestic sewage (Fernandes et al., 1985; Lettinga et al., 1984). The beds are typically 2 m high with an HRT of 3-12 h. Reactor productivities vary from 1 g COD per L per h for slaughterhouse waste at 30°C

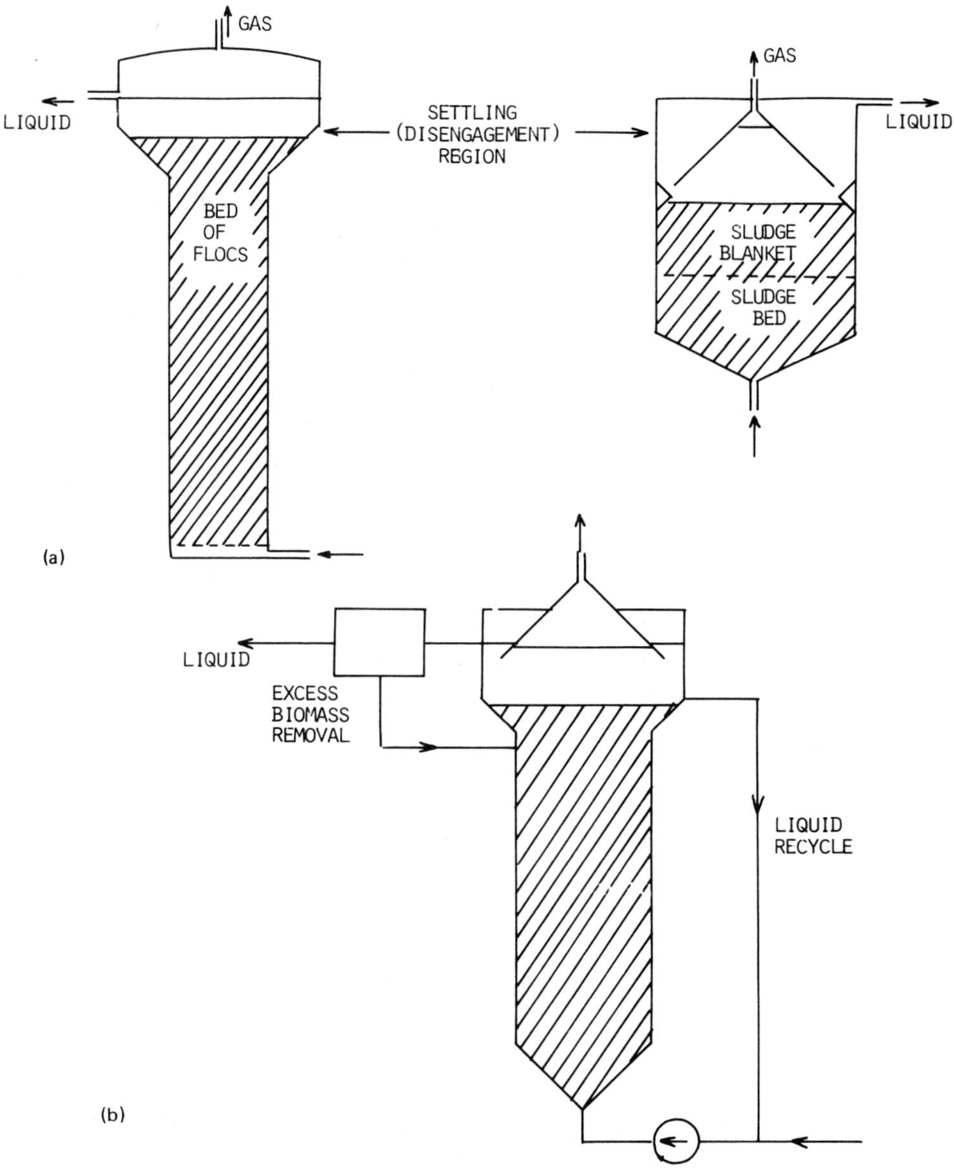

Figure 2 Types of fluidized bed bioreactors: (a) tower fermentor; (b) upflow anaerobic sludge blanket reactor.

to 0.03 g COD per L per h for the more dilute domestic sewage at 20°C (Lettinga et al., 1982). A similar concept has been proposed for denitrification of wastewater (Klapwijk et al., 1981).

The term "flocs" is actually inappropriate in these reactors because the light, variable flocs formed by anaerobic digestion organisms are undesirable; they wash out of the reactor too easily, particularly when gas bubbles formed in the process stick to them. (Lettinga et al., 1982, give some discussion of gas-solid separation devices.) Under the right conditions, the biomass forms into more regular and denser "granules" in the millimeter size range, which give a much more controllable bed. Details of the granulation process are not well understood, but it is known to be affected by the type of methogenic bacteria present, the concentration of bivalent cations, the liquid velocity, and the inlet substrate concentration (Wiegant and de Man, 1986). The presence of inorganic particles, either natural precipitates of sulfides or carbonates or added gravel or brick dust (Fernandes et al., 1985), apparently helps the process by providing nuclei around which granules can form.

A feature of all these reactors is the way the solids in the bed stratify. Large, dense flocs settle to the base of the bed and tend to grow even bigger owing to the large substrate concentrations they encounter. Small, light particles are left farther up the bed, and because they fluidize easier than larger particles, the solids concentration is much lower there. Fernandes et al. (1985) measured an almost uniform decrease in solids concentration from 150 g/L to zero between the base and top of a 2 m high UASB. Greenshields and Smith (1971) found a drop from 300 to 7.5 g/L (approximately) up a laboratory-scale tower fermenter containing yeast flocs. The production of gas in the bed makes the situation worse. With large volumes of gas near the top of the bed, the solids holdup* is even less than would be expected from these measurements of the mass of flocs per unit volume of liquid.

This stratification is undesirable. The low solids holdup near the top of the bed leads to low reactor productivity there. The flocs at the base of the bed can become so large they are no longer fluidized and they agglomerate into a packed mass (Jones et al., 1984). The expanded section on a tower fermenter and the settling section built into UASB reactors (Fig. 2) are ways to maintain a high solids holdup at the top of the bed and to prevent the washout of very small flocs. A bed that tapers uniformly all the way up is the best solution (Scott, 1983) but is difficult to construct on a large scale. The use of support particles is an alternative way for the designer to control the size and density of the particles and thus keep the bed more uniformly fluidized.

C. Packed Bed, Fixed Film Reactors

The simplest way to ensure a more uniform concentration of cells through the reactor is to grow them as a film on solid supports in a packed bed. Several supports have been used successfully for ethanol production reactors. Sutton and Gaddy (1980) immobilized a monolayer of *S. cerevisiae* on 1/4 inch Rasching rings coated with gelatin and glutaraldehyde. However, the subsequent growth of this film clogged the bed and required gas scouring of the bed every 2 weeks. Producitivity of 7.4 g/L per h was observed for 100% conversion of a 3% glucose feed. Tyagi and Ghose (1982) claimed to encounter no clogging problems because their cell growth rate was balanced by washout from the reactor. However, their support material and immobilization procedures were not detailed for proprietory reasons. They claim a productivity of 29 g/L per h with a 20% sugar feed, but this is at a conversion of only 80%, which would be unacceptable commercially since the feedstock is a major cost item in ethanol production. Minier and Goma (1982) used 5 mm brick supports and circulated dodecanol through the reactor to extract ethanol as it was produced, thus minimizing product inhibition. They experienced no clogging problems, perhaps because the CO_2 gas production stripped excess cells from the supports. They productivity of 1 g/L per h is low, but the increased reactor cost implied by the low value would be offset by the lower separation cost resulting from the high ethanol concentrations produced by their 100% conversion of a 26% glucose feed. More details of these reactors are given in Chapter 21.

The use of fixed film, packed bed reactors for anaerobic wastewater treatment is usually traced to the "anaerobic filters" of Young and McCarty (1969), although some earlier work does exist (Coulter et al., 1957). The results from many types of wastewater are given in the review by Henze and Harremoes (1982). Early beds were usually packed with 2-5 cm rock or coke, but high-porosity synthetic plastic media, pall rings, or perforated spheres are now favored because they give a higher surface area (Table 1) with less clogging (Wheatley and Cassel, 1985; Young and Dahab, 1982a; Genung et al., 1982). Both upflow and downflow operation have been tried, but the former is favored because it allows clumps of biomass knocked off the supports to be retained in the filter by gravity. Young and Dahab (1982b) found the best performance with upflow through a packing of modular corrugated blocks and attributed this to the large amount of suspended biomass in the lower part of the reactor. However, this excess biomass may clog the reactor after long-term operation.

Hydraulic residence times in anaerobic filters are typically 10-50 h, or four to five times greater than in USAB reactors. Their

Table 1 Specific Surface Areas of Various Media

Median	Specific surface area (m^{-1})	Bed porosity (%)
25–75 mm rock	62	46
90 mm pall rings	102	95
25 mm ceramic Rasching rings	190	74
20 mesh sand	3800	60 (fluidized)

productivities are smaller by a similar factor. The same comparison between packed beds and floc beds holds for ethanol production. [Compare, for example, the results of Sutton and Gaddy (1980) with those of Jones et al. (1984).]

D. Fluidized Beds with Solid Support Particles

Wastewater treatment units consisting of a fluidized bed of solid support particles coated with a film of anaerobic biomass have been developed from similar aerobic reactors and successfully applied to domestic sewage and many industrial wastes. They have been given a number of acronyms including AFB (anaerobic fluidized bed), AAFEB (anaerobic attached-film expanded bed), and IFCR (immobilized fluidized cell reactor). A typical reactor is shown in Figure 2, and details can be found in Cooper and Atkinson (1981) and Henze and Harremoes (1982). Productivities for the removal of organic matter are reported to be of the order of 1 g COD per L per h, similar to UASB reactors and considerably higher than for the anaerobic filter, a result of the lower surface area of biomass in the packed bed reactor (Table 1).

Many different particles have been used as supports, including sand, gravel, coal, ion-exchange resins, polyvinyl chloride (PVC), ground glass, and garnet, with particle diameters ranging from 50 to 1000 μm. This represents a thousandfold variation in particle settling velocity, always an important variable in fluidization, and there have been few attempts to find an optimum particle size and density (Andrews, 1986). The usual procedure has been to select a particle arbitrarily and to set the liquid velocity in the bed to fluidize this particle. The problem is that fixing the reactor height at a convenient value (typically 5 m) now fixes the HRT for one pass through the bed, so the required total hydraulic residence time (typically 1-10 h) must be made up by liquid recycle. Choosing a support particle with a large settling velocity can produce recycle ratios of 100 or more, which is inefficient, because although some recycle is necessary to damp out changes in wastewater concentration or flow rate, very high values only increase pumping costs and reduce the substrate concentrations to which the particles are exposed. High recycle is not needed, as it is in some aerobic reactors, to provide aeration.

Very similar reactors containing denitrifying bacteria have been built for the removal of nitrate from wastewaters (Francis and Hancher, 1981). Bland et al. (1982) have demonstrated a reactor containing vermicullite particle coated with *Z. mobilis* for ethanol production.

The particles in these fluidized beds have been observed to stratify. If there is a variation in support particle size, the larger particles move to the base of the bed and the smaller to the top. If the supports are monosized, stratification is based on the amount of biofilm on the particle, but in contrast to the floc beds, the largest particles now move up the bed. The reason for this involves the way the biofilm affects the particle's settling velocity.

Based on a drag coefficient correlation of the form $C_D = a/Re_t^e$,

$$\frac{U_t}{U_{ts}} = \frac{(1 + Bx)^{1/(2-e)}}{(1 + x)^{1/3}} \qquad (1)$$

This is plotted in Figure 3 for e = 0.85, a suitable value for the range of Re_t for these particles in water. For low-density supports or floc particles, B = 1 (where B is the ratio of the buoyant density of the biomass to that of the support) so adding biomass (increasing x) increases the settling velocity because the particle diameter is increasing. For very dense supports, $B \rightarrow 0$, the dominant effect of the biofilm is on the particle's density rather than its diameter, so the settling velocity is decreased and heavily coated particles move up the bed. Note that this stratification, unlike that in a floc bed, is unstable; thinly coated particles at the bottom of the bed grow faster because they are exposed to higher substrate concentrations and will then tend to move up the bed.

During the start-up period while biofilm is developing on the particles, the fluidized bed expands to accommodate the volume of the biomass and the gas produced and also, to a lesser extent, because the biomass changes the particle settling velocity (Andrews and Tien, 1979). If there is enough room in the reactor the bed reaches a steady-state height when the growth of new biomass is balanced by the washoff of biofilm from the particles and endogenous metabolism in the biofilm. Some beds are operated at this steady state, particularly in anaerobic wastewater treatment where the biofilm is slow-growing and relatively uncohesive. An expanded section is provided on top of the reactor to ensure that the bed

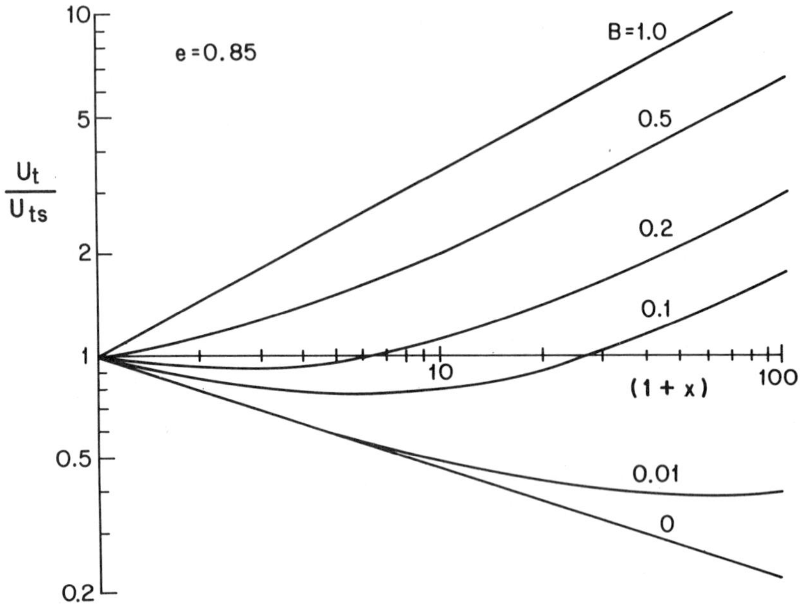

Figure 3 The effect of biofilm on settling velocity.

does not reach the effluent port, although some particles are always carried into this region by gas bubbles.

Other types of bacteria produce large volumes of more cohesive biomass, which quickly expands the bed beyond the effluent port. Andrews and Fonta (1988) found that a bed containing lactic acid bacteria would expand from 7 to 15 feet overnight, partly because this fermentation produces none of the gas bubbles that tend to strip biofilm from the particles. In these cases, there must be some external provision for stripping excess biomass from the particles and returning them to the bed. This can be done by putting a mechanical agitator in the top of the reactor or by pumping the effluent over vibrating screens or through a high-shear zone. When liquid recycle is used, this high-shear zone can be the recycle pump itself as long as it is designed to handle solids. An alternative is to periodically remove the top part of the bed from the reactor and strip off the biomass, but this is only possible when sterility is not required.

Note that excess biomass is removed from the top of the bed because this is where stratification puts the most heavily coated particles. It would be inappropriate for a floc bed or for low-density support particles because the overlarge particles would then all be at the base of the bed (they could be drained off periodically, although this control strategy has not been reported). Furthermore, the external removal of excess biomass only works as a way of controlling the thickness of the biofilm in the bed if the support particles are monosized. If they are not then the largest support particles stay near the base of the bed, accumulating an uncontrolled thickness of biofilm, and never go through the biofilm stripping device.

E. Other Types of Particle

There are two types of porous support particles. In adsorbent supports, such as activated carbon, the pores are too small for the bacteria to enter. Their advantage is the ability to adsorb organic molecules that may inhibit the biofilm. Cross et al. (1982) showed that a fluidized bed of activated carbon coated with an anaerobic biofilm could successfully treat a phenolic wastewater. A particular advantage of this system is that shock loads of phenolics that would normally be toxic can be adsorbed by the carbon and subsequently released slowly (so-called bioregeneration) for degradation by the biomass. Andrews and Fonta (1988) have described a reactor in which a fermentation product (ethanol or lactic acid) is adsorbed as it is produced, thus reducing its inhibitory effect on the biofilm. The adsorbent must be cycled through the reactor and regenerated externally.

In macroporous supports, the cells are trapped inside the particle. There are two types. In some the objective is to entrap flocculent biomass to control the particle size and density in a way that is not possible with a floc particle. These particles need the highest possible internal porosity. Balls of stainless steel or plastic mesh have been used (Atkinson et al., 1981). In the other type the object is to provide a surface for the immobilization of a monolayer of nonflocculent biomass. This requires the largest possible internal surface area obtainable with pores much larger than the cells. Porous glass and other ceramics and particles of collagen sponge are suitable.

Cells immobilized in synthetic gel beads are not considered at length in this chapter. They are usually put in upflow packed beds rather than fluidized beds. There are some exceptions, especially when considerable gas is generated or when the cells tend to leak out of the gel and grow as a film on the outer surface, thus clogging a packed bed (Margaritis and Wallace, 1982).

II. THEORY

A. The Limiting* and Rate-Controlling* Components*

Consider a solid spherical support particle of radius R coated with a thickness L of biofilm (Fig. 4). If it is assumed that the properties

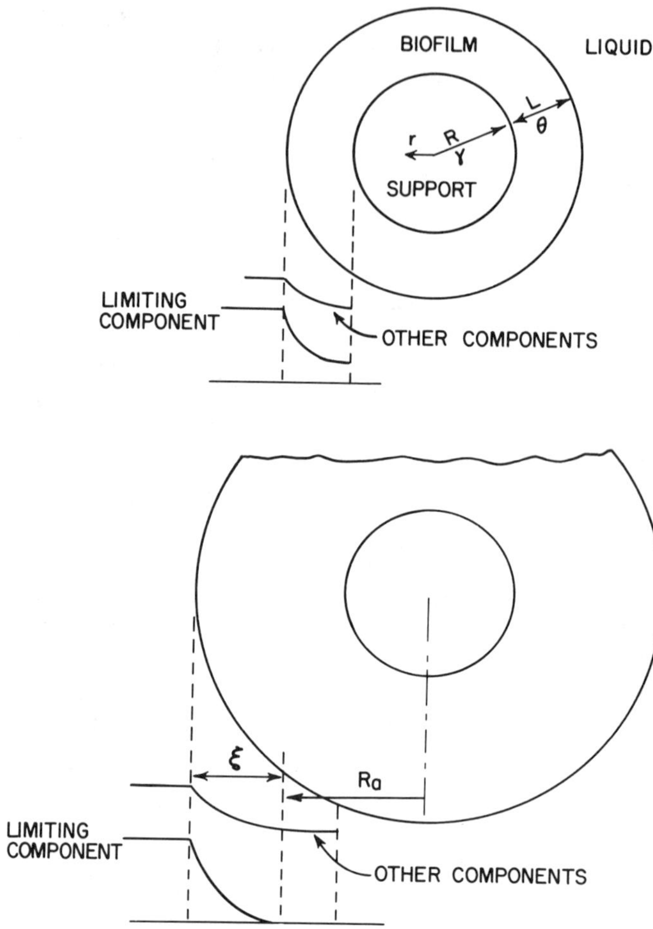

Figure 4 Concentration profiles in a biofilm.

of the biomass (diffusivity D, cell concentration, and so on) do not vary with depth, then an equation can be written to describe the diffusion and consumption of a component in the biofilm:

$$\frac{D}{r^2}\frac{d}{dr}\left(r^2\frac{dC}{dr}\right) = \frac{\mu(C_j)}{Y} \tag{2}$$

where

$$\frac{dC}{dr} = 0 \qquad \text{at } r = R$$

and

$$C = S \qquad \text{at } r = R + L$$

Several points must be made about this equation: $\mu(C_j)$ is the function that describes how the biomass growth rate depends on the concentrations of all the components (substrates and inhibitory products) in the biofilm. An equation (2) exists for each component. For products, the right-hand side should be negative, but this is avoided by the variable transformation $C = \bar{p} - C_p$, where C_p = actual product concentration and $\bar{p}$ = completely inhibitory concentration (Note that C = 0 now stops growth just as it would for a substrate). Y is the cell yield expressed as cubic centimeters of biofilm produced per gram of component consumed (or produced). Uptake of a substrate for maintenance (or non-growth-associated product formation) is ignored in Equation (2). Andrews and Przezdziecki (1986) have shown how they can be included.

Equation (2) with slight modifications applies to all types of spherical immobilized cell particles. Flocs or granules of biomass and porous supports are covered by setting R = 0. Adsorbent supports are included by modification of the first boundary condition for any component that adsorbs (Andrews and Fonta, 1986). For gel-immobilized cells, R = 0 and Y is replaced by Y_x/X, where Y_x is the yield as gram cells per gram and X is the local cell concentration in gram cells per cubic centimeter. This case is complicated since X becomes a function of both r and time because cell growth in the gel is faster near the surface where component concentrations are higher.

Strictly speaking there should always be a time differential term in Equation (2) because L increases with time. However, this growth is slow so the quaesi-steady state approximation is justified.

The form of the second boundary condition implies that mass transfer resistance in the liquid surrounding the particle is negligible. Greenshields and Smith (1971) and Andrews and Przèzdziecki (1986) concluded that this was valid in the high-concentration conditions found in anaerobic fluidized bed fermenters. It would be less valid in the low-concentration, high-rate conditions found in aerobic wastewater treatment systems.

There being an Equation (2) for each component, we can eliminate $\mu(C_j)$ between any two of them, integrate, and apply the first boundary condition:

$$DY \frac{dC}{dr} = D_i Y_i \frac{dC_i}{dr} \tag{3}$$

Integrate again and apply the second condition:

$$DY(S - C) = D_i Y_i (S_i - C_i) \tag{4}$$

This equation shows how the concentrations of all the nutrients and products in the biofilm are related to each other. The mathematical procedure from this point is to take the function $\mu(C_j)$ that predicts how all these components affect the metabolic rate, eliminate all but one of the concentrations using Equation (4), and then solve Equation (2) for this component. Computer solutions make this possible for very complex forms of the function $\mu(C_j)$ or even for structured kinetic models of the biomass. However, in view of the uncertainty about many parameter values and the approximate approach used in the design and control of real bioreactors, this type of mathematical complexity is not justified, except perhaps for anaerobic wastewater treatment in which several groups of bacteria are involved and a structured model has obvious advantages (Droste and Kennedy, 1986). An alternative procedure, adopted here, is to identify the two components that have the greatest effect on the overall metabolic rate of the biomass and then to solve Equation (2) analytically for the resulting simplified forms of the function $\mu(C_j)$.

The most important component is the limiting component, defined as the one whose concentration first reaches zero inside a thick film (Fig. 4) while all other C_i are positive. It follows from Equation (4) that this component is the one with the smallest value of DYS.

To say that a component is limiting in this sense does not necessarily mean that its concentration is the main factor controlling the local metabolic rate in the biofilm. For example, oxygen is almost always the limiting component in aerobic films owing to its low S_i value, but it is very efficiently scavenged by bacteria. Even at the point deep inside the film where its concentration is only 0.2 mg/L, many bacteria would be capable of growing at close to their maximum rate. However, at the same point, the concentration of organic matter, although much higher, may be restricting their metabolic rate. In terms of the Monod model a component starts to restrict the growth rate when its concentration becomes of the same order as its half-velocity constant K. We can therefore define the rate-controlling component as that one whose concentration first reaches 4K in the film. Equation (4) shows that this is the component with the smallest value of DY(S - 4K). Choosing 4K, not 2K or 10K, is arbitrary but reasonable; it is the concentration at which growth is limited to 80% of its maximum value.

The distinction between limiting and rate-controlling components is particularly important in the production of inhibitory catabolic products. Consider, for example, the homolactic fermentation of glucose (no subscript) to lactic acid (subscript p). The product yield from substrate is close to 100%, so $Y/Y_p = 1$; the diffusivities are proportional to their values in water, so $D/D_p = 0.73$. Suppose the organism follows Monod kinetics for glucose with K = 10 mg/L and linear inhibition kinetics for lactic acid with $\bar{p} = 1.5$ g/L. At a point where the glucose concentration S = 1 g/L and the lactic acid concentration is zero ($S_p = \bar{p} = 1.5$ g/L), the ratio DYS/(DYS) = 0.49 so glucose is the limiting component. However, $DY(S - 4K)/[DY(S - 0.8\bar{p})]_p = 2.34$, so lactic acid is the rate-controlling component. Note that $0.8\bar{p}$ replaces 4K because, with the variable transformation used for inhibitory products, linear inhibition kinetics can be written $\mu = \bar{\mu}S_p/\bar{p}$, so $0.8\bar{p}$ is the value of S_p at which metabolism is restricted to 80% of its maximum value. To understand why the product has a greater effect on the rate, consider the point inside the biomass where the glucose concentration C = 100 mg/L. This level of glucose restricts the growth rate to $\mu = \bar{\mu}C/(K + C) = 0.91\bar{\mu}$. At the same point, however, $C_p = 0.84$ g/L [Eq. (4)] so product inhibition restricts the rate to $\mu = \bar{\mu}C_p/\bar{p} = 0.56\bar{\mu}$.

The general effect of the rate-controlling component (no subscript) and the limiting component (subscript l) on the kinetics can be described by a modified Monod function:

$$\mu = \frac{\bar{\mu}C}{K + C} H(C_l) \tag{5}$$

Here $H(C_l)$ is the Heavyside function, which equals 1 when C_l is positive and zero when $C_l \leq 0$. When the rate-controlling component is the same as the limiting component, the $H(C_l)$ term is superfluous. Linear inhibition kinetics appear as the low-C asymptote of Equation (5) with K replaced by $\bar{p}$.

B. Effectiveness Factors

Solving Equation (2) with the kinetics of Equation (5) presents three main difficulties. The first is to correctly define the dimensionless distances (Thiele moduli). Bischoff (1965) showed that the best definition for Monod-type kinetics is to multiply the distances by $(\bar{\mu}/DYKf(S/K))^{1/2}$, where

$$f\left(\frac{S}{K}\right) = 2\left[\frac{S}{K} - \ln\left(1 + \frac{S}{K}\right)\right]\left(1 + \frac{K}{S}\right)^2 \tag{6}$$

The dimensionless variable $\bar{r}$ is measured from the biofilm surface, not the particle center (Fig. 4).

The second difficulty arises from the discontinuous nature of the function $H(C_l)$. Thin films (those for which $C_l \geq 0$ at $\bar{r} = \theta$; Fig. 4a) present no problems; $C_l \geq 0$ everywhere so $H(C_l) = 1$ everywhere. However, in thick films C_l may reach zero, and it is assumed that there is no metabolic activity below this point (this may not be true; if the nitrogen source is limiting maintenance activity could continue, producing concentration gradients for the carbon source and catabolic products not shown in the idealized picture of Fig. 4b). This difficulty is overcome by defining a dimensionless "active depth" ξ as the point where all $dC/dr = 0$. For thin films $\xi = \theta$ (the dimensionless film thickness); for thick films $\xi < \theta$. Equation (2) can now be rewritten:

$$\frac{d}{d\bar{r}}\left[(\gamma + \theta - \bar{r})^2 \frac{dC}{d\bar{r}}\right] = Kf\left(\frac{S}{K}\right)(\theta + \gamma - \bar{r})^2 \frac{C}{K + C} H(C_l) \tag{7}$$

where

$$\frac{dC}{d\bar{r}} = 0 \qquad \text{at } \bar{r} = \xi$$

and

$$C = S \qquad \text{at } \bar{r} = 0$$

The third difficulty is that this equation is nonlinear and cannot be solved analytically. Various asymptotic solutions are therefore derived and the results are expressed as effectiveness factors defined by

$$\eta = \frac{\text{actual consumption (or production) of component in particle}}{\text{consumption (or production) if whole particle were biomass exposed to liquid-phase conditions}}$$

$$= \frac{3DY}{(R + L)\mu(S_j)} \left.\frac{dC}{dr}\right|_{(R + L)} = - \frac{3(K + S)}{KSf(S/K)(\theta + \gamma)} \left.\frac{dC}{d\bar{r}}\right|_0 \tag{8}$$

Note that this definition is based on total particle volume, not biomass volume as used by Shieh et al. (1981) and others. This definition is preferred because it is directly proportional to the local productivity of the reactor and it facilitates comparison between flocs and supported film particles.

1. Zero-Order Kinetics: $\mu = \bar{\mu}H(C_1)$; $S >> K$

For large S/K, $\ln(1 + S/K) << S/K$, so the function f approaches 2S/K. Equation (7) is solved for the limiting component, which, since $K \to 0$, is necessarily the same as the rate-controlling component:

$$1 - \frac{C}{S} = \frac{2\bar{r}}{3}\left[\theta + \gamma - \frac{(\theta + \gamma - \xi)^3}{(\theta + \gamma)(\theta + \gamma - \bar{r})} - \frac{\bar{r}}{2}\right] \tag{9}$$

The effectiveness factor can now be derived by substituting this in Equation (8). It is simpler to remember that the top layer of biofilm ($\bar{r} < \xi$) has substrate available and is metabolizing it at its maximum rate. The effectiveness factor must therefore equal the fraction of the particle volume in this active layer:

$$\eta = 1 - \left(1 - \frac{\xi}{\theta + \gamma}\right)^3 \tag{10}$$

ξ is found from the requirement that $C = 0$ at $\bar{r} = \xi$. Equation (9) gives this as

$$\xi^2 \left[1 - \frac{2\xi}{3(\theta + \gamma)}\right] = 1 \tag{11}$$

If this equation has no solutions in the range $0 < \xi < \theta$, then the film is thin and $\xi = \theta$. The exact criterion for the thin film is

$$\theta^2 \left(\frac{\gamma + \theta/3}{\gamma + \theta}\right) < 1 \tag{12}$$

A plot of η versus the dimensionless particle radius $\theta + \gamma$ for different support particle sizes γ is shown in Figure 5.

2. First-Order Kinetics: $\mu = \bar{\mu}CH(C_1)/K$: $S << K$

For small S/K, $\ln(1 + S/K) = (S/K) - (S/K)^2/2 + \cdots$, so the function f approaches 1.

Equation (7) is solved for the rate-controlling component using the variable transformation $y = C(\theta + \gamma - \bar{r})$.

$$\frac{C(\theta + \gamma - \bar{r})}{S(\theta + \gamma)} = \frac{(\theta + \gamma - \xi)\cosh(\xi - \bar{r}) + \sinh(\xi - \bar{r})}{(\theta + \gamma - \xi)\cosh\xi + \sinh\xi} \tag{13}$$

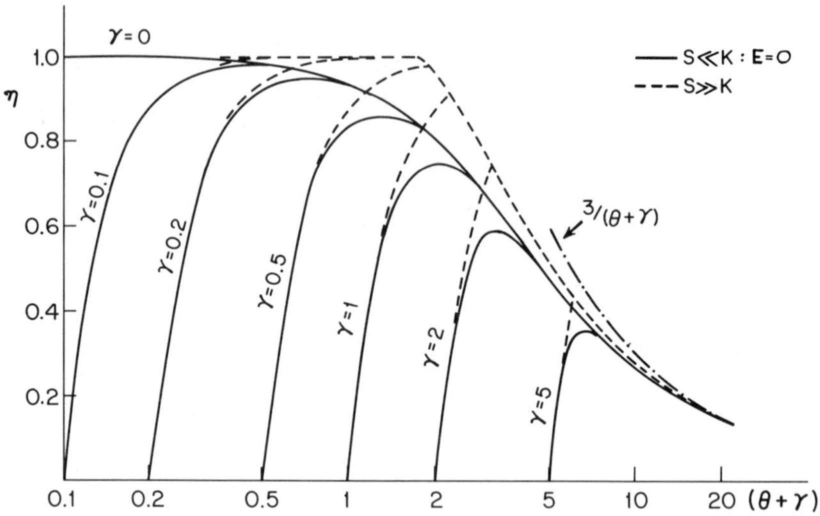

Figure 5 Particle effectiveness.

Substituting in Equation (8),

$$\eta = \frac{3}{\theta + \gamma} \left[\frac{(\theta + \gamma - \xi) \tanh \xi + 1}{(\theta + \gamma - \xi) + \tanh \xi} - \frac{1}{\theta + \gamma} \right] \tag{14}$$

For thick films the limiting component concentration is zero at $\bar{r} = \xi$. Equation (4) shows that $C_1 = 0$ corresponds to $C = ES$ where $E = 1 - (DYS)_1/DYS$. Putting this in Equation (13) gives

$$E[(\theta + \gamma + \xi) \cosh \xi + \sinh \xi] = \theta + \gamma \tag{15}$$

Once again, if this has no solutions in the range $0 < \xi < \theta$, then the film is thin and $\xi = \theta$.

$$E < \frac{\theta + \gamma}{\sinh \theta + \gamma \cosh \theta} \tag{16}$$

The effectiveness factor is plotted in Figure 6 for various values of the support particle radius γ and the parameter E. The case $E = 0$, when the limiting component and rate-controlling component are the same, is plotted in Figure 5 for comparison with the zero-order kinetics case. (Note that Equation (16) shows that the film is always "thin" in this case.) The important feature of Figure 5 is that, with the Thiele moduli (θ and γ) as defined here, the

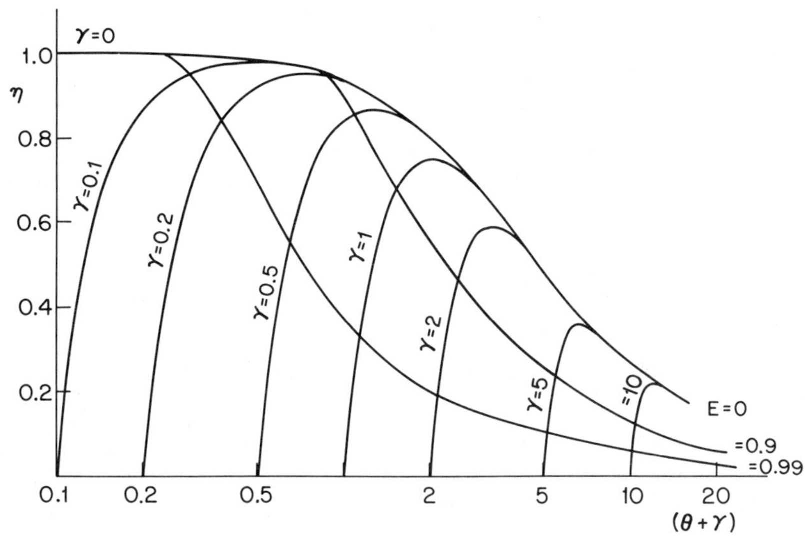

Figure 6 Particle effectiveness: first-order kinetics.

curves for the two extreme cases (S >> K and S << K) are very similar, at least within the accuracy with which film thicknesses can be controlled in practice. We can therefore draw conclusions from this graph for Monod kinetics with any value of S/K.

3. The Thick-Film, Flat-Plate Approximation:
$\theta > \xi$ and $(\theta + \gamma) >> \xi$

We are interested only in the region $\bar{r} << \theta + \gamma$, so Equation (7) can be rewritten

$$\frac{d^2C}{dr^2} = G \frac{dG}{dC} = Kf\left(\frac{S}{K}\right) \frac{CH(C_l)}{K + C} \tag{17}$$

Where G is defined as $dC/d\bar{r}$. The boundary conditions are $\bar{r} = \xi$, C = ES, and G = 0; $\bar{r} = 0$, C = S.

Integrating once with these conditions permits evaluation of the uptake (or production) rate per unit area of film:

$$N = D \frac{dC}{dr}\bigg|_{R+L} = \left(\frac{\bar{\mu}D}{YKf(S/K)}\right)^{1/2} G_0 = \left\{\frac{2\bar{\mu}KD}{Y}\left[\frac{S}{K}(1 - E) - \ln\frac{K + S}{K + ES}\right]\right\}^{1/2} \tag{18}$$

This form of the result is useful for thick films growing on nonspherical support particles, such as pall rings. For spherical particles it can be expressed as the effectiveness factor:

$$\eta = \frac{3NY}{(R + L)\mu(S)} = \frac{3}{\theta + \gamma} \left\{ 1 - E\left[\frac{1 - (K/ES) \ln (1 + ES/K)}{1 - (K/S) \ln (1 + S/K)} \right] \right\}^{1/2} \tag{19}$$

Figure 5 shows (for $E = 0$) how the other solutions approach this at large $\theta + \gamma$.

C. Implications for Reactor Design

With the definition of effectiveness factor given in Equation (8), the local reactor productivity (the rate of production or consumption of a component in unit volume of reactor) equals $\varepsilon_S \eta \mu(S_j)/Y$. So the overall productivity is a maximum if the designer:

1. Maximizes the average solids holdup in the reactor.
2. Maximizes the average value of $\mu(S_j)$. The inlet and outlet values of S are fixed design criteria. Mixing in the liquid phase reduces the average S value so a plug flow reactor is usually best.
3. Minimizes Y. Y can be written Y_X/X, where Y_X is the normal cell yield (gram cells per gram component) and X is the cell concentration in the particle (gram cells per cubic centimeter). The $\gamma = 0$ lines in Figures 5 and 6 apply to gel-immobilized cells and macroporous supports as well as to "natural" flocs and pellets of biomass. However, the cell concentration X in particles produced by flocculent organisms are usually higher than that in synthetic immobilized cell particles, so the former are preferred if a strain that produces a convenient size of floc is available. For a porous support in which the pores are completely filled with biomass, X is less than the "natural floc" value by a factor equal to the internal porosity.
4. Maximizes the average effectiveness factor. Figures (5) and (6) give some general rules about achieving this.

Particles of pure biomass set the upper limit on effectiveness. This is reasonable because support particles, being biologically inactive, represent wasted space in the reactor, although Andrews and Przezdzcieki (1986) showed that this is not necessarily true when the liquid-phase mass transfer resistance is significant. Figure 6 shows that this upper limit depends on the amount of the limiting component present, that is, on the value of E. For example, if $E = 0.9$ all the region above this line must be ignored.

If the biomass particles are too large, then resistance to mass transfer inside the particle reduces its effectiveness. Figures 5 and 6 show that keeping the dimensionless radius θ in the range 1-1.5 will keep the effectiveness factor above 0.9 except when the limiting component is almost exhausted (E = 0.99).

Following a line of constant γ on the graph shows the effect of adding biomass to a solid support particle. One layer of cells corresponds to approximately $\theta = 0.01$, so that solid particles with only a monolayer of cells have a very low effectiveness. (Gel immobilization is a preferable technique for nonflocculent cells.) Adding more layers of cells initially increases the effectiveness factor considerably. However, a film thickness is eventually reached at which the limiting component concentration at the base of the film reaches zero (the film becomes "thick": Fig. 4b). Any additional biomass beyond this point is redundant and only decreases the effectiveness. Between these two regions there is a film thickness that maximizes the effectiveness factor. Such an optimum must exist for any support particle and any form of the kinetics, and one goal of the designer should be to achieve this optimum thickness in the reactor.

The maximum possible effectiveness factor decreases as the support particle size is increased. On the other hand, it is easier to work with larger particles than with very small ones. A reasonable compromise is $\gamma = 1$, $\theta = 1$, which gives an effectiveness factor near 0.8 (Fig. 5) and is used as the basis for design studies. It also illustrates the superiority of fluidized beds over packed beds. The volume of biomass per unit volume of support particle x is related to θ and γ by

$$1 + x = \left(1 + \frac{\theta}{\gamma}\right)^3 \tag{20}$$

So $\theta = \gamma = 1$ corresponds to a biofilm volume seven times the volume of the solid supports. This amount of biomass can never grow in a packed bed; there is not enough interstitial volume between the particles. The rocks in a typical trickling filter correspond to a γ value of several hundred, which gives a very low effectiveness factor and illustrates that there is considerable room for improvement in bioreactor design.

An interesting exception to some of the rules given here is provided by substrate inhibition kinetics. This cannot be treated quantitatively with the equations given here, but qualitatively it is clear that, since μ is a decreasing function of S, the effectiveness factor can be larger than 1. Also, liquid-phase mixing, which reduces the average S in the reactor, may increase rather than decrease reactor productivity.

D. The Liquid-Fluidized Bed

When a liquid is pumped up through a bed of uniform particles, the variation in pressure drop and liquid holdup (equal to the bed porosity when no gas is present) with superficial velocity is as shown in Figure 7. At low velocities the particles form a packed bed with constant porosity and a pressure drop that is directly proportional to the liquid velocity. This can continue only until the pressure drop across the bed equals the buoyant weight of the bed per unit area. At this point, known as the minimum fluidization velocity u_{mf}, the particles are pushed apart from each other, their weight being supported by the flow of liquid. Further increases in velocity cause expansion of the bed but no change in pressure drop. In a fermenter the pressure drop is a function of both the amount of biomass and the amount of gas in the bed so it can provide useful operating information. In terms of head loss it is

$$\Delta h = H\left[\bar{\varepsilon}_s\left(\frac{\rho_p}{\rho} - 1\right) - \bar{\varepsilon}_g\right]$$

$$= V_s\left(\frac{\rho_s}{\rho} - 1\right)(1 + B\bar{x}) - H\bar{\varepsilon}_g \qquad \text{for supported films} \qquad (21)$$

The fluidized bed can exist until the superficial velocity reaches the settling velocity of the particles, at which point the particles are washed out of the bed ($\varepsilon = 1$). Expansion of the bed with increasing velocity is smooth and continuous. The distinction that is sometimes made between "expanded" beds and "fluidized" beds at some arbitrary level of bed expansion has no basis in the phenomena. If wall effects are negligible, which requires a bed diameter to particle diameter ratio of at least 100 (a criteria often not met in laboratory columns), the porosity of the bed is given by

$$\varepsilon^n = \frac{u}{u_t} \qquad (22)$$

Many correlations have been suggested for the constant n. Al-Djibouni and Garside (1979) give the simplest:

$$\frac{5.1 - n}{n - 2.7} = \frac{Re_t^{0.9}}{10} \qquad (23)$$

There are two complications in dealing with fermenters. The first is the presence of gas bubbles. The mechanics of three-phase fluidized beds have been reviewed by Muroyama and Fan (1985), but there is no information on beds in which the gas flow increases

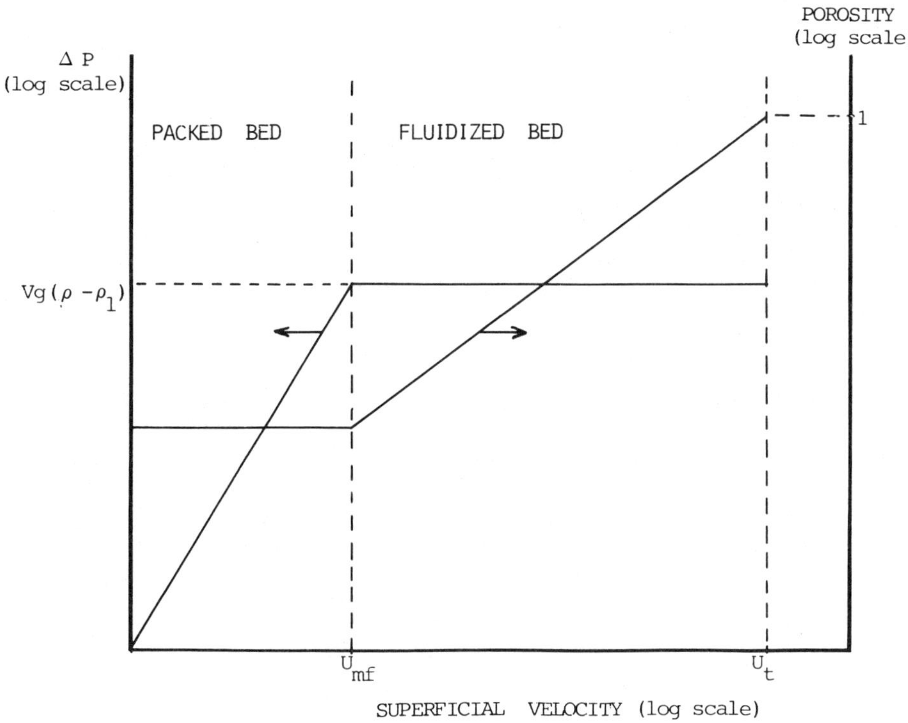

Figure 7 Behavior of a liquid fluidized bed.

up the bed as it does in a fermenter. The second complication is the changes in particle settling velocity produced by changes in the liquid density and viscosity as the fermentation progresses. Fermenting a 10% glucose solution to ethanol changes the liquid specific gravity from 1.02 to 0.99, which increases the settling velocity of a floc particle of specific gravity 1.03 by a factor of 4. When combined with stratification, this explains the observation made by Royston (1966) that the cell concentration has a maximum near the middle of a tower fermenter producing beer. This effect is far less significant for beds of denser particles or for wastewater treatment in which the change in liquid properties is negligible.

A bed fluidized by a liquid, sometimes called particulate fluidization, expands smoothly with none of the violent "bubbling" and mixing associated with gas-fluidized beds (so-called aggregative fluidization). It is natural to assume that the liquid phase is in plug flow, at least for tall thin beds with no gas. This is confirmed by both theoretical studies using accepted correlations for the liquid-phase axial dispersion coefficient (Andrews, 1982) and measurements showing definite gradients of product and substrate concentrations, pH, and so on, throughout the bed. The presence of a gas phase increases liquid mixing (Muroyama and Fan, 1985). The UASB reactor probably has the best-mixed liquid phase owing to its low aspect ratio (height-diameter) and large gas production. Bolle et al. (1986) found that their tracer data for the liquid phase fit a model consisting of two stirred tanks with short-circuiting caused by the gas flow. A wastewater treatment reactor with a high liquid recycle ratio can be considered a stirred tank reactor whatever the inherent mixing characteristics of the fluidized bed.

A major difference between packed and fluidized beds is that particles in a fluidized bed are free to move around. The mechanics of liquid-fluidized beds are not sufficiently studied to permit us to predict the exact solids mixing condition. However, it must fall somewhere between two limiting cases. If the tendency of the particles to move about is high, then the solids will be completely mixed, a disordered condition in which each particle spends an equal amount of time in all parts of the bed. A sample of particles drawn from any height would be a representative sample of all the particles in the bed. If, on the other hand, the tendency to move is small, the solids adopt the other extreme of complete stratification. This is a highly ordered situation in which each particle has a set position, with all those particles with a higher settling velocity below it and all those with a lower settling velocity above it. A sample of particles from a fixed height would all have the same settling velocity.

Factors that encourage solids mixing are as follows:

1. Low variance in particle settling velocity: obviously if this variance is zero there is no basis for stratification so the particles mix. This is particularly important in fermenters because the growth of biomass changes the settling velocity. Floc beds stratify stably because the largest particles have the highest settling velocity and go to the base of the bed, where, being exposed to the highest substrate concentrations, they grow fastest. For dense solid support particles, the growth of biofilm decreases the settling velocity (Fig. 3; $B = 0$) so heavily coated particles move up the bed to where the substrate concentration is lowest. A stable stratification with monosized support particles is possible only if particles are removed from the top of the bed, stripped of excess biomass, and returned to the bed. Unfortunately, this produces a steady-state distribution of film thicknesses that is the exact opposite of the optimum situation. Thinly coated particles go to the base of the bed where the substrate con-

centration is highest, and vice versa. Maintaining the constant, optimum value of θ given by Figure 5 or 6 requires thick films where the substrate concentration is high.

2. Large, heavy particles: the mixing tendency increases with the Re_t value of the particles. In fact, large lead shot fluidized by water behaves more like a gas-fluidized aggregative bed, with violent "bubbling" and mixing.
3. Bed porosity close to 0.7: Al-Djibouni and Garside (1979) found a maximum in the solids axial dispersion coefficient at this value.
4. A gas phase: gas bubbles can transport particles through the bed as well as increasing mixing by agitating the liquid phase (Muroyama and Fan, 1985).
5. Large bed diameter: care is always required in scaleup because wall effects in laboratory columns inhibit mixing.
6. Low aspect ratio: mixing is more significant in short wide beds than in tall thin beds, partly because the flow distribution at the bed inlet is inevitably somewhat uneven (depending on how well the flow distributor is designed). The mixing induced by these flow imperfections affects a much greater fraction of a low-aspect-ratio bed.
7. Inclination from the vertical: inclination of as little as 1° can cause a large-scale recirculation of the bed solids.

Which limiting case is closer to the situation in real fermenters? In fermenters both bed porosity and Re_t are usually small. In this situation, evidence from bimodal fluidized beds (beds containing only two types of particles) suggests complete stratification exists if the particle settling velocities vary by a factor of 2 (Al-Djibouni and Garside 1979). Andrews and Fonta (1988) observed stratification in a bed of monosized support particles even for the ethanol fermentation, where considerable gas is generated. The best evidence for stratification comes from UASB reactors (Bolle et al., 1986). These are large-scale, low-aspect-ratio beds with gas generation, but they do stratify into a lower "sludge bed" of larger, mainly granulated particles (which may not be fluidized) and an upper "sludge blanket" in which the biomass concentration decreases with height. It is therefore reasonable to take "complete stratification" as an operating assumption for fermenter design.

The factors that increase solids mixing also, as a general rule increase liquid mixing. This is fortunate because the worst case would be a combination of solids mixing and liquid plug flow. Cells normally function best when they can acclimate themselves to their environment, and cells on a particle moving through the gradients of concentration, pH, and so on, found in a plug-flow reactor would experience a constantly changing environment with no chance to acclimate. This situation, which can appear with monosized, dense support particles and no excess biofilm removal, should be avoided.

This is particularly true for anaerobic wastewater treatment because, even when the substrate is a simple sugar, this is a two-step process involving two different classes of bacteria working sequentially:

$$\text{Sugar} \xrightarrow[\text{bacteria}]{\text{acetogenic}} \text{acids} + H_2 \xrightarrow[\text{bacteria}]{\text{methanogenic}} CO_2 + CH_4$$

If the biofilm does not contain balanced numbers of acetogens and methanogens such that sugars diffuse in and CH_4 and CO_2 diffuse out, then there may be gradients up the bed, not just of substrate concentration but also of substrate composition. This is the situation in packed bed filters (Young and Dehab, 1982a). Acetogenic bacteria dominate near the reactor entrance, where their substrate is plentiful, generating excess acetic acid that diffuses into the liquid to be consumed by methanogens farther along the bed. Solids mixing could produce the worst possible case with some methanogen-rich films near the bed entrance and some acetogen-rich films near the exit, both exposed to inappropriate substrates. In common fluidized bed practice this problem is avoided by heavy liquid recycle, which eliminates gradients in the liquid phase.

III. DESIGN STUDIES

With the present state of knowledge it is not possible to design fluidized bed bioreactors purely from first principles. Considerable experimentation is always required at the laboratory and pilot scales, especially for wastewater treatment because of the variation among different wastewaters. Nevertheless, theoretical studies are useful to demonstrate the effects of the different variables and to set up ideal reactors against which the performance of real reactors can be judged. This comparison can identify areas of the reactor where there is room for improvement, as well as areas where more research is needed.

Since the main area of uncertainty is the variation of solids holdup in the bed due to gas generation and stratification, we take as an independent variable not the distance up the bed z but the volume of solids per unit area below a point v. These are related by

$$\frac{dv}{dz} = \varepsilon_S \qquad (24)$$

Assuming plug flow of the liquid, there is a conservation equation for each component in the bed:

$$-u \frac{dS}{dv} = \eta \frac{\mu(S_j)}{Y} \tag{25}$$

where $S = S_0$ at $v = 0$.

A. Ethanol Production from Glucose with *Zymomonas mobilis* in a Porous Support

The objective is 99% conversion of a 150 g/L glucose feed stream, using supports with an internal porosity $\delta = 0.4$ filled with biomass. The parameter values are given in Table 2 (Andrews, 1986). The $\bar{p}$ value is high, first because it represents a best fit of the linear inhibition model to data that are not exactly linear and, second, because this value is strictly the concentration that stops ethanol production, which is always higher than the concentration that stops cell growth [the difference is due to non-growth-associated activity (Lee and Rogers, 1983)].

Results from the previous section must be used to determine which are the limiting and rate-controlling components. Assume that N, P, S, and so on are in adequate supply. To choose between glucose and ethanol, note that S_p is defined as $\bar{p}$ - the actual product concentration and that the actual product concentration at a point is given by $Y(S_0 - S)/Y_p$. It follows that

$$\frac{DYS}{(DYS)_p} = \frac{D}{D_p} \frac{S}{(\bar{p}\ Y_p/Y - S_0) + S} \tag{26}$$

This is always less than 1 since $D/D_p = 0.54$ (taking the ratio as the same as that in water), and $S_0 < \bar{p}\ Y_p/Y$ or the product concentration would reach $\bar{p}$ before S reached zero and there would necessarily be unused substrate in the outflow, which is uneconomical. So glucose is the limiting component throughout the reactor. However, ethanol is rate controlling if (Sec. II.A)

$$DY(S - 4K) > D_pY_p(S_p - 0.8\bar{p}) \tag{27}$$

Eliminating S_p as above, it can be shown that this is true throughout the reactor as long as

$$S > 4K + 0.2 \frac{Y_p\bar{p}D_p}{YD} \tag{28}$$

Table 2 Parameter Values for Design Studies

Ethanol production by *Zymomonas mobilis*
Inlet glucose concentration: 150 g/L
Outlet glucose concentration: 1.5 g/L
Kinetics:

$$\frac{\mu}{Y_p} = \frac{\bar{\mu}}{Y_p}\left(1 - \frac{p}{\bar{p}}\right)$$

$$\frac{\bar{\mu}}{Y_p} = 0.60\delta \ \ \text{g/cm}^3 \text{ per h} \qquad \bar{p} = 140 \text{ g/L}$$

Yield of ethanol from glucose: $Y/Y_p = 0.48$
Diffusivity of ethanol in biomass = 1/2 value in water - 0.023 cm^2/h
Effective diffusivity in particle: $D_p = 0.023\ \delta\ cm^2/h$
Internal porosity of supports: $\delta = 0.40$

Anaerobic wastewater treatment
Inlet concentration: $S_0 = 1$ g COD per L
Outlet concentration: $S = 0.1$ g COD per L
Flow rate: $F = 40\ m^3/h$
Kinetics: Monod; $\bar{\mu}/Y = 0.002$ g COD per cm^3 per h; $K = 0.2$ g COD per L
Diffusivity in biomass = 1/2 of average value for organics in water = 0.016 cm^2/h

Thus ethanol is rate controlling and follows first-order (linear inhibition) kinetics, and glucose is limiting. This is a common situation for product-inhibited fermentations and is the basis of Figure 6. The value of E calculated from Equation (26) is 0.72 at the inlet and 0.99 at the outlet. Suppose we wish to maintain an effectiveness factor of 0.8 throughout the reactor. Figure 6 shows that this requires a dimensionless particle radius $\theta = 1.5$ at the inlet and $\theta = 0.4$ at the outlet. With $(\bar{\mu}/D_p Y_p \bar{p})^{1/2} = 10\ cm^{-1}$, this means particle diameters of 3 and 0.8 mm, respectively. Note that if we supply this range of particle sizes, bed stratification automatically gives the correct particle distribution: large at the bottom and small at the top.

The superficial liquid velocity is fixed by the need to keep the particles at the bottom of the bed fluidized to an appropriate porosity, say $\varepsilon = 55\%$. Suppose the support is a plastic of density 1.1 g/cm^3. With a biomass density of 1.03 g/cm^3, the density of the particle is $(0.4 \times 1.03) + (0.6 \times 1.1) = 1.072\ g/cm^3$. The

150 g/L glucose solution at the inlet has a density of 1.058 g/cm^3 and viscosity 1.57 cp. so U_t = 1.65 cm/s. Equation (22), which can be applied because there is no gas at the base of the bed, then gives u = 0.214 cm/s.

This calculation illustrates one reason for using support particles despite the fact that they reduce the biomass concentration in the bed. A flocculent strain of *Z. mobilis* that produces flocs of exactly the right size range exists and has been used in a laboratory-scale reactor with great success (Scott, 1983). However, the floc density is only 1.03 g/cm^3, less than that of the inlet liquid. So the preceding calculation would give a negative U_t, the particles cannot strictly form a fluidized bed, the superficial velocity must be very low, and control of the bed is difficult.

If η is assumed constant through the bed, Equation (25) can be integrated for ethanol concentration S_p using linear inhibition kinetics ($\mu / Y = \bar{\mu} S_p / Y_p \bar{p}$):

$$\ln \left(\frac{\bar{p}}{S_{pout}} \right) = \frac{\bar{\mu} \eta V}{u Y_p \bar{p}} \tag{29}$$

With η = 0.8 this gives V = 4 m. It is difficult to convert this to a bed height owing to the variation in solids holdup up the bed. In the absence of gas the 0.8 mm particles at the top of the bed (when liquid density = 0.986 g/cm^3 and viscosity = 1.33 cp) would have U_t = 1.39 cm/s and with u = 2.14 cm/s would form a bed with ε_s = 0.36 [Eq. (22) with $\varepsilon_s = 1 - \varepsilon$]. In practice the large volume of gas would reduce this at least an order of magnitude (neglecting what dissolves in the liquid, the gas superficial velocity would be 8.24 cm/s at normal temperature and pressure), which suggests an average solids holdup in the bed of the order of 0.2, a bed height of approximately 4/0.2 = 20 m, and a hydraulic residence time of 2.5 h. The corresponding volumetric productivity would be 27 g/L per h, slightly lower than values reported for beds of floc (Scott, 1983), principally because of the space occupied by the support particles in the reactor.

The value of this type of calculation is that it immediately eliminates some possibilities and suggests others. Repeating it with porous glass supports, for example, results in an impossibly tall bed of over 500 m. [At the other extreme, Andrews (1986) showed that a bed of flocs for aerobic wastewater treatment would be less than 1 mm tall.] The main limitation on the reactor productivity is obviously the low and unpredictable solids holdup near the top of the bed. A tapered bed is the best solution. Another would be to use more nearly monosized particles of, say, 2 mm diameter, hoping that the lower effectiveness factor at the top of the bed is balanced by the higher solids holdup. However, the solids mixing in a bed of monosized particles would move individual cells rapidly from low-ethanol to high-ethanol environments, which may degrade their performance. On the other hand, the biomass may grow into a continuous film outside the support particle, in which case the bed would stratify like a tower fermenter. These questions can only be resolved by experiment with the particular feed and microbial strain.

B. Anaerobic Wastewater Treatment with Solid Supports

The ethanol fermenters in the previous section could be characterized as single-pass, completely stratified fluidized beds. There is no reason that an anaerobic treatment unit cannot be designed (approximately) in a similar fashion, despite the fact that the differences between individual wastewaters and the complexity of the microbiology involved have produced a wide variation among published parameter values. Suppose the problem is as stated in Table 2. The parameter values follow the recommendations of Henze and Harremoes (1982) and Droste and Kennedy (1986). The $\bar{\mu}/Y$ value corresponds to a specific uptake rate of 1 g COD per g VSS per day and a film density of 40 g VSS per L. There is uncertainty in the D value, but this is less critical since only its square root appears in the calculations.

The bed would be designed to handle the maximum flow in a single pass, liquid recycle being used only to keep the bed fluidized properly when the wastewater flow is reduced. Figure 5 shows that a reasonable effectiveness factor ($\eta > 0.75$) can be achieved with $\gamma = \theta = 1$. This corresponds to support particle diameters of 2.4 mm near the bed entrance where $S/K = 5$ and 1 mm at the exit where $S/K = 1/2$. This variation should be sufficient to maintain stratification. This is essential because with large gradients of substrate concentration and composition in the liquid phase, the biomass on a particle moving up and down the bed would be exposed to changing environments much faster than it could acclimate to them.

In fact, supported film reactors for anaerobic wastewater treatment are not usually designed this way, but with heavy liquid recycle and monosized particles giving complete mixing of both the liquid and solid phases. The reasons are instructive. Basic reactor theory predicts that a plug flow reactor has a large advantage in productivity over a completely mixed reactor for the first-order (linear inhibition) kinetics that approximate the behavior of the ethanol fermentation. However, the advantage is far less for wastewater treatment where the kinetics are closer to zero order. Also, a mixed liquid phase can dilute the shock loads of toxic solutes that plague some wastewater treatment units, and the absence of gradients (particularly of pH) in the liquid phase simplifies control of the reactor.

Design of the mixed reactor proceeds as follows. Since all the particles are exposed to the outlet conditions, $\gamma = 1$ means they should be monosized with 1 mm diameter. The required film thickness ($\theta = 1$ or $L = 1$ mm) can be achieved by correctly selecting the particle density. High-density particles have a high shear stress at the surface, which tends to strip off biomass and therefore produces a thin steady-state biofilm. Thicker films are expected on less dense particles. The correct density must be determined experimentally for each particular wastewater. Andrews and Tien (1979) showed that as a first approximation the expansion of a fluidized bed as biomass grows in it is equal to x, the ratio of biomass volume to support particle volume. Since $\theta = \gamma = 1$ corresponds to $x = 7$, the correct particle is one that gives a sevenfold increase in bed volume between the bed with no biomass and the bed containing a steady-state biofilm thickness.

Note that all the particles are exposed to the same conditions so they all develop the same film thickness. There is no basis for bed stratification, and the particles mix. Note also that with heavy liquid recycle the amount of gas generated in a single pass through the reactor is small. Consequently, the solids holdup remains almost constant throughout the bed. This is the greatest advantage of this reactor over the plug flow, stratified bed type, in which the stratification and large gas holdup creates low solids holdups near the top of the reactor.

The conservation equation for COD in the completely mixed reactor is

$$F(S_0 - S) = V \frac{\bar{\mu}}{Y} \eta \frac{S}{K + S} \qquad (30)$$

This gives the solids volume in the bed as $V = 72\ m^3$, which gives a support particle volume (for $x = 7$) of 9 m^3, a bed volume (for average solids holdup of 45%) $V_b = 160\ m^3$, and an HRT for the system (based on bed volume) of 4 h. The final advantage of this type of reactor is that we are free to choose a convenient bed height (within limits) by adjusting the recycle ratio. Suppose for the purposes of illustration that the optimum support particle is found to be an activated carbon with wet density 1.4 g/cm^3. The settling velocity when this particle is coated with biofilm to $x = 7$ is 3.5 cm/s [Eq. (1)]. The superficial velocity required to fluidize these particles to a porosity of 55% is $u = 0.52$ cm/s [Eq.

(22)]. Since the flow through the bed is $F(1 + R_r)$, the height and diameter of the bed can be found from

$$\frac{\pi d^2}{4} = \frac{V_b}{H} = \frac{F}{u}(1 + R_r) \tag{31}$$

The bed dimensions calculated from this equation vary from 74 × 1.7 m (H × d) with $R_r = 0$ to 0.74 × 17 m with $R_r = 100$. Both are impractical. A reasonable compromise is $R_r = 7$, which is just sufficient to approximate complete mixing and gives a bed 9.3 × 4.7 m.

IV. CONCLUSIONS

Fluidized bed bioreactors are superior to packed beds for processes involving flocculent biomass because they can contain a higher concentration of active biomass without problems of clogging and gas binding in the bed. However, the solids in them tend to stratify based on either the support particle size or the biomass thickness, and together with gas production, this leads to a continuous decline in solids holdup up the bed. Tapered fluidized beds and mixed beds with liquid recycle are possible solutions to this problem.

Particles of pure biomass give the highest effectiveness factors, but the reactor designer has no control over the size and density to which they grow. Beds of these particles, tower fermenters, and UASB reactors are therefore more difficult to operate even with microbial strains like *Z. mobilis* that produce relatively uniform flocs. They stratify strongly so the base may become a packed mass, and small flocs are carried out of the top of the reactor by gas bubbles, requiring a well-designed disengagement section above the reactor. Periodic removal of excess biomass from the base of the bed may alleviate these difficulties.

Designing a reactor with support particles, either porous or solid, reduces to the problem of choosing the size, size distribution, and density of the supports. The sizes are fixed by the need to maintain a high effectiveness factor, the density by the need to maintain a reasonable bed height. Dense particles, such as sand, require a high liquid velocity to fluidize them and thus a very tall bed (or high liquid recycle rate) to achieve the desired contact time.

Dense support particles also produce a higher shear stress at the particle surface, which tends to strip off the biomass, producing a thinner steady-state biofilm thickness. It may be possible to choose the density of a solid support particle such that the steady-state film thickness is close to the optimum thickness. When the film grows past its optimum thickness even with very dense supports, particles must be removed from the top of the bed so that excess biofilm can be washed off. To ensure that all the particles move through the washing device in turn, the support particles must be monosized and they must be dense so that the addition of biofilm reduces their settling velocity. These beds tend to stratify the wrong way around, with thin biofilms near the base of the bed where the substrate concentration is highest and vice versa. This makes the biofilm thickness on the particles returned to the bed an important parameter that should be controlled as closely as possible.

V. GLOSSARY

Component A collective term for nutrients consumed by the biomass and products formed by it.

Flocculent Microorganisms that can secrete extracellular polysaccharides and thus stick to each other to form large clumps or flocs. This is taken here as the same as the ability to form a thick film of organisms, although this strictly requires that the cells can stick not only to each other but also to the support surface provided.

Fluidized bed In the sanitary engineering literature slightly expanded beds are called expanded, and only very expanded beds (more than twice their settled volume) are called fluidized. All beds expanded by the upflow of a liquid or gas (or both) are here called fluidized.

Hydraulic residence time (HRT) Unless otherwise specified this is the single-pass empty bed residence time (reactor volume over liquid flow rate through bed or reactor height over superficial velocity). The *bed HRT* is smaller owing to the liquid volumes in the reactor above and below the bed. When liquid recycle is used the *system HRT* is $(1 + R_r) \times$ HRT.

Limiting component Defined here as the component whose concentration reaches zero first inside a large bioparticle (for inhibitory products this implies the concentration reaches $\bar{p}$).

Mean cell residence time (MCRT) The mass of cells in the reactor divided by the cell outflow rate from the system.

Rate-controlling component Defined here as the component whose concentration is the first to reach the value at which it limits μ to $0.8\bar{\mu}$.

Reactor productivity The mass of product formed, or substrate consumed in the case of wastewater treatment, per unit volume of reactor per hour. For wastewater treatment it equals loading times fractional removal.

Solids holdup The volume of solids per unit volume of bed.

VI. NOMENCLATURE

B	density ratio (biofilm-liquid)/(support-liquid)
C	component concentration in the particle
d	bed diameter
D	effective diffusivity in the particle
e	exponent in drag coefficient correlation
E	$1 - (DYS)_1/(DYS)$
f(S/K)	function defined by Equation (6)
F	liquid flow rate
g	gravitational acceleration
G	$dC/d\bar{r}$
H	bed height
$H(C_1)$	Heavyside function
Δh	head loss across the bed
K	Monod half-velocity constant
L	biofilm thickness
n	exponent in bed porosity correlation
N	component consumption (or production) rate per unit area
$\bar{p}$	product concentration that stops product formation
r	radial position
$\bar{r}$	$(R + L - r)[\bar{\mu}/DYKf(S/K)]^{1/2}$
R	support particle radius
R_r	recycle ratio
Re_t	Reynolds number based on U_t and particle diameter
S	component concentration in the liquid
U	superficial liquid velocity
U_t	settling velocity
v	volume of solids per unit area below a point in the bed
V	volume of solids per unit area in the bed
x	volume of biofilm over volume of support particle
Y	yield coefficient; cm^3 biomass per g component
z	distance up the bed
γ	$R[\bar{\mu}/DYKf(S/K)]^{1/2}$
δ	internal porosity of support particle
ε	bed porosity
ε_s	local solids holdup in reactor
$\bar{\varepsilon}_s$	average solids holdup in reactor

$\bar{\varepsilon}_g$	average gas holdup in reactor
η	particle effectiveness factor
μ	specific growth rate of biomass
$\bar{\mu}$	maximum value of μ
ξ	dimensionless active depth
θ	$L[\bar{\mu}/DYKf(S/K)]^{1/2}$
ρ	liquid density
ρ_p	particle density

Subscripts

p	product (ethanol)
s	support particle
l	limiting component

REFERENCES

Al-Djibouni, M. R., and Garside, J. (1979). Particle mixing and classification in liquid fluidized beds. *Trans. Inst. Chem. Eng.* *57*:94-103.

Andrews, G. F. (1982). Fluidized-bed fermentors: A steady state analysis. *Biotechnol. Bioeng.* *24*:2013-2030.

Andrews, G. F. (1986). Selecting particles for fluidized-bed bioreactors with flocculant biomass. *Biotechnol. Prog.* *2*:16-22.

Andrews, G. F., and Fonta, J. P. (1986). A novel adsorbing bioreactor. *Biotechnol. Bioeng. Symp. Ser.* *17*:317-325.

Andrews, G. F., and Fonta, J. P. (1988). Biofilms on adsorbent particles. In *Bioreactor Immobilized Enzymes and Cells*, M. Moo-Young (Ed.). Elsevier, New York, pp. 111-124.

Andrews, G. F., and Przezdziecki, J. (1986). The design of fluidized bed fermentors. *Biotechnol Bioeng.* *28*:802-810.

Andrews, G. F., and Tien, C. (1979). The expansion of a fluidized bed containing biomass. *AIChE J.* *25*:720-723.

Atkinson, B., Black, G. M., and Pinches, A. (1981). The characteristics of solid supports and biomass support particles when used in fluidized beds. In *Biological Fluidized Bed Treatment of Water and Wastewater*, P. Cooper and B. Atkinson (Eds.). Wiley, New York, Chap. 5.

Behie, L. A., and Gaucher, C. M. (1988). The application of continuous three phase fluidized bed bioreactors to the production of pharmaceuticals. In *Bioreactor Immobilized Enzymes and Cells*, M. Moo-Young (Ed.). Elsevier, New York, pp. 197-212.

Bischoff, K. B. (1965). Effectiveness factors for general reaction rate forms. *AIChE J.* *11*:351-357.

Bland, R. R., Chen, H. C., Jewell, W. J., Bellamy, W. D., and Zall, R. R. (1982). Continuous high rate production of ethanol by *Zymomonas mobilis* in an attached film expanded bed fermentor. *Biotechnol. Lett. 4*:323-327.

Bolle, W. L., Van Bruegel, J., van Eybergen, G. C., Kossen, N. W. F., and Zoetemeyer, R. J. (1986). Modelling the liquid flow in UASB reactors. *Biotechnol. Bioeng. 28*:1615-1623.

Cooper, P. F., and Atkinson, B. (Eds.) (1981). *Biological Fluidized Bed Treatment of Water and Wastewater*. Wiley, New York.

Coulter, J. B., Soneda, S., and Ettinger, M. B. (1957). Anaerobic contact process for sewage disposal. *Sewage Indust. Waste 29*:468-477.

Cross, W. H., Chain, E. S. K., Pohland, F. G., Harper, S., Kharkar, S., Cheng, S. S., and Lu, F. (1982). Anaerobic biological treatment of coal gasifier effluent. *Biotechnol. Bioeng. Symp. Ser. 12*:349-358.

Dean, R. C., Karkare, S. B., Ray, N. G., Runstadler, P. W., and Venkatsubramanian, K. (1988). Large scale culture of hybridoma and mammalian cells in fluidized bed bioreactors. In *Bioreactor Immobilized Enzymes and Cells*, M. Moo Young (Ed.). Elsevier, New York, pp. 125-142.

Droste, R. L., and Kennedy, K. F. (1986). Sequential substrate utilization and effectiveness factors in fixed biofilms. *Biotechnol. Bioeng. 28*:1713-1721.

Fan, L. S., Fijie, K., and Long, T. (1986). Characteristics of a draft tube gas-liquid-solid fluidized bed bioreactor for phenol degradation. *Biotechnol. Bioeng. 30*:498-504.

Fernandes, X. A., Cantwell, A. D., and Mosey, F. E. (1985). Anaerobic biological treatment of sewage. *Water Pollut. Control 84*:99-110.

Francis, C. W., and Hancher, C. W. (1981). Biological denitrification of high nitrate wastes generated by the nuclear industry. In *Biological Fluidized Bed Treatment of Water and Wastewater*, P. Cooper and B. Atkinson (Eds.). Wiley, New York, Chap. 14.

Genung, R. K., Hancher, C. W., Rivera, A. L., and Harris, M. T. (1982). Energy conservation and methane production in municipal wastewater treatment using fixed-film anaerobic bioreactors. *Biotechnol. Bioeng. Symp. Ser. 12*:365-380.

Greenshields, R. N., and Smith, E. L. (1971). Tower fermentation systems and their applications. *Chem. Eng. 249*:182-190.

Henze, M., and Harremoes, P. (1982). Anaerobic treatment of wastewater in fixed-film reactors. In *Anaerobic Treatment of Wastewater in Fixed-Film Reactors*, M. Henze (Ed.). Pergamon, New York, pp. 1-102.

Jones, S. T., Korus, R. A., Admassu, W., and Heimsch, R. C. (1984). Ethanol fermentation in a continuous tower fermentor. *Biotechnol. Bioeng. 26*:742-747.

Klapwijk, A. H., Smit, A., and Morre, H. (1981). Denitrification of domestic wastewater in an upflow sludge blanket reactor. In *Biological Fluidized Bed Treatment of Water and Wastewater*, P. F. Cooper and B. Atkinson (Eds.). Wiley, New York.

Lee, K. J., and Rogers, P. L. (1983). The fermentation kinetics of ethanol production by *Zymomonas mobilis. Chem. Eng. J. 27*: B31-35.

Lettinga, G., van Velsen, A. F. M., Hobma, S. W., de Zeeuw, W., and Klapwijk, A. (1980). Use of the upflow sludge blanket reactor concept for wastewater treatment. *Biotechnol. Bioeng. 22*:699-734.

Lettinga, G., Hobma, S. W., Hulshoff, L. W., Pol, W. De Zeeuv, DeJong, P., Grin, P., and Roersma, R. (1982). Design operation and economy of anaerobic treatment. In *Anaerobic Treatment of Wastewater in Fixed Film Reactors*, M. Henze (Ed.). Pergamon, New York.

Lettinga, G., Hulshoff, L. W., Korster, I. W., Wiegant, W. M., DeZeeuw, W. J., Rinzema, A., Roersma, R. E., and Hobma, S. W. (1984). High rate anaerobic wastewater treatment using the UASB reactor under a wide range of temperature conditions. *Biotechnol. Genet. Eng. Rev. 2*:253-284.

Margaritis, A., and Wallace, J. B. (1982). The use of immobilized cells of *Zymomonas mobilis* in a novel fluidized bioreactor. *Biotechnol. Bioeng. Symp. Ser. 12*:147-153.

Minier, M., and Goma, G. (1982). Ethanol production by extractive fermentation. *Biotechnol. Bioeng. 24*:1565-1580.

Muroyama, K., and Fan, L. S. (1985). Fundamentals of gas liquid solid fluidization. *AIChE J. 31*:1-22.

Prince, I. G., and Barford, J. P. (1982). Tower fermentation of cane sugar juice. *Biotechnol. Lett. 4*:469-474.

Royston, M. G. (1966). Tower fermentation of beer. *Proc. Biochem. 1*:215-221.

Scott, C. D. (1983). Fluidized-bed bioreactors using flocculent *Z. mobilis. Ann. N.Y. Acad. Sci. 413*:449-455.

Shieh, W. K., Mulcahy, L. T., and LaMotta, E. J. (1981). Effectiveness factors for spherical particles coated with biofilm. *Trans. Inst. Chem. Eng. 59*:121-126.

Smith, E. L., and Greenshields, R. N. (1974). Tower fermentation systems and their application to aerobic processes. *Chem. Eng. 281*:28-34.

Stephanopoulos, G., San, K. Y., and Davison, B. H. (1985). A novel bioreactor-cell precipitator combination for high cell density fermentations. *Biotechnol. Prog. 1*:250-259.

Strandburg, G. W., Donaldson, T. L., and Arcuri, E. J. (1982). Continuous ethanol production by a flocculant strain of *Z. mobilis*. *Biotechnol. Lett. 4*:347-352.

Sutton, O. C., and Gaddy, J. L. (1980). Ethanol production in an immobilized cell reactor. *Biotechnol. Bioeng. 22*:1735-1748.

Tyagi, R. D., and Ghose, T. K. (1982). Studies on immobilized *Saccharomyces cerevisiae*. *Biotechnol. Bioeng. 24*:781-816.

Wiegant, W. M., and de Man, W. A. (1986). Granulation of biomass in thermophilic UASB reactors treating acidified wastewaters. *Biotechnol. Bioeng. 28*:718-727.

Wheatley, A. D., and Cassel, L. (1985). Effluent treatment by anaerobic biofiltration. *Water Pollut. Control 84*:10-22.

Young, J. C., and Dahab, M. F. (1982a). Effect of media design on performance. In *Anaerobic Treatment of Wastewater in Fixed-Film Reactors*, M. Henze (Ed.). Pergamon, New York, pp. 369-383.

Young, J. C., and Dahab, M. F. (1982b). Operational characteristics of anaerobic packed-bed reactors. *Biotechnol. Bioeng. Symp. Ser. 12*:303-316.

Young, J. C., and McCarty, P. L. (1969). The anaerobic filter for waste treatment. *J. Water Pollut. Control Fed. 41*:R160-173.

23

Start-up, Dynamics, and Control of Anaerobic Digesters

JOHN PATRICK BARFORD / *University of Sydney, Sydney, New South Wales, Australia*

I. INTRODUCTION

Over the past 15 years, substantial advances have been made in the technology and industrial application of anaerobic digestion. In the most advanced systems available, start-up is based on significant laboratory-scale (and sometimes pilot-scale) assessment of the particular waste to be treated. Such variables as the final design loading rate (kg COD per m^3 per day), expected conversion (percentage COD removed), and biomass and biogas yields (kg biomass formed per kg COD converted and m^3 methane per kg COD converted), where COD is chemical oxygen demand (kg/m^3), and nutrient requirements have been established at laboratory and/or pilot scale. In addition, the final plant is highly instrumented, making such occurrences as shock overloads and significant changes in pH and temperature highly unlikely, with all the important variables contributing to the final operation of the complete anaerobic digestion system subject to continuous monitoring and with alarms signaling deviation from the desired set points and defined control strategies. Such systems are consequently reliable during the start-up phase and stable in operation.

The level of stability in start-up and the operational control of these advanced anaerobic digestion systems are significantly different from the generally perceived (until quite recently) low level of reliability and stability of traditional anaerobic systems.

There are three points of importance that follow from this. First, if the very extensive strategies for waste evaluation, laboratory- and/or pilot-scale assessment, and advanced engineering design (including a high level of instrumentation and control) are undertaken, then a stable start-up and digester operation can be assured. Second, it must be appreciated that only recently has our knowledge of many basic aspects of anaerobic digestion (e.g., microbiology, nutritional requirements, toxins and inhibitor effects, and flocculation mechanisms) been sufficient to enable an analytic approach to be outlined that ensures rational design leading to stable start-up and operation. Finally, it is important to appreciate that many factors may contribute to the eventual inability, in the practical application of anaerobic digestion technology, to undertake these very extensive strategies for waste evaluation, laboratory-pilot-scale assessment, and advanced engineering design. This may be due to practical considerations, for example, time or budget restrictions, inappropriate level of technology for the particular application, inavailability of technical expertise, and component equipment.

It can be seen from the preceding discussion that it is difficult to generalize with respect to start-up and stability since a very large number of external factors contribute to the final design and its likelihood to result in a stable start-up and operation. Lettinga et al. (1983a) attempted to characterize some of these external factors:

1. The characteristics of the wastewater, e.g., the temperature and temperature variations; the nature, type, and concentration of the pollutants (complexity, composition, and dissolved fraction); and daily, weekly, and seasonal variations in the composition and in the COD, pH, salt content, and bicarbonate alkalinity
2. Local factors, such as cost of land, cost of labor, construction materials, and need for a foundation
3. Environmental restrictions set for the discharge of effluents to receiving waters, restrictions set for the emission of malodorous compounds, and the costs of disposal of effluents into the sewer system
4. Desired degree of sophistication of the treatment plant with respect to automation, control, construction, and so on
5. The use that will be made of the biogas and the extent to which future expansions of the production capacity of the industry should be taken into account
6. Availability of waste heat, costs of energy, costs of chemicals, and so on

All designs are thus an economic compromise between the ideal and the extent to which external factors impose their presence to compromise the ideal design strategy. Consequently, it is difficult to generalize with respect to the practical design of an anaerobic digester for a particular waste in a particular environmental and legislative situation since each design is a compromise resulting from the relative weights given to a variety of external factors.

In this chapter, it is not intended to discuss in detail the many factors that contribute to practical digester design. However, it is important to note that the final design decided upon may have a significant effect on the ease of start-up and operational control achieved. In this chapter, an ideal design strategy is detailed that incorporates an extensive assessment of the particular waste, selection of a suitable sludge, and the steps involved in the development of an industrial-scale anaerobic digester from laboratory to pilot to full scale. This is followed by a detailed discussion of dynamics and control of anaerobic digesters, including identification of factors affecting digestion control, control strategies commonly employed at industrial scale, and, finally, monitoring and instrumentation. The level of control and instrumentation presented is that of the most advanced industrial anaerobic digesters. Hence in both cases the discussion approaches the ideal strategies for optimum design. It is implied, and in fact demonstrable, that significant deviations from these strategies result in a digester design that is not optimal and is subject to nonquantified but significant operational difficulties both in start-up and during sustained operation.

II. STRATEGY FOR THE DEVELOPMENT OF AN INDUSTRIAL-SCALE ANAEROBIC DIGESTER FROM LABORATORY TO PILOT TO FULL SCALE

An essential aspect of successful start-up on a full scale is the availability of extensive experimental data from the laboratory and/or pilot scale. In this section, the range and extent of experimental data necessary are identified and the methods used to assist in the initial design of a full-scale digester are outlined. Finally, the strategy for start-up of an industrial-scale anaerobic digester (based on the laboratory-pilot-scale experimental data and the use of this data in full-scale design) is illustrated with respect to such factors as sludge origin and acclimatization, feeding strategies, and process monitoring.

A. Experimental Data from Laboratory and/or Pilot

A schematic diagram of the essential steps involved in undertaking a vigorous assessment of the suitability of anaerobic digestion waste treatment for a particular waste is outlined in Figure 1.

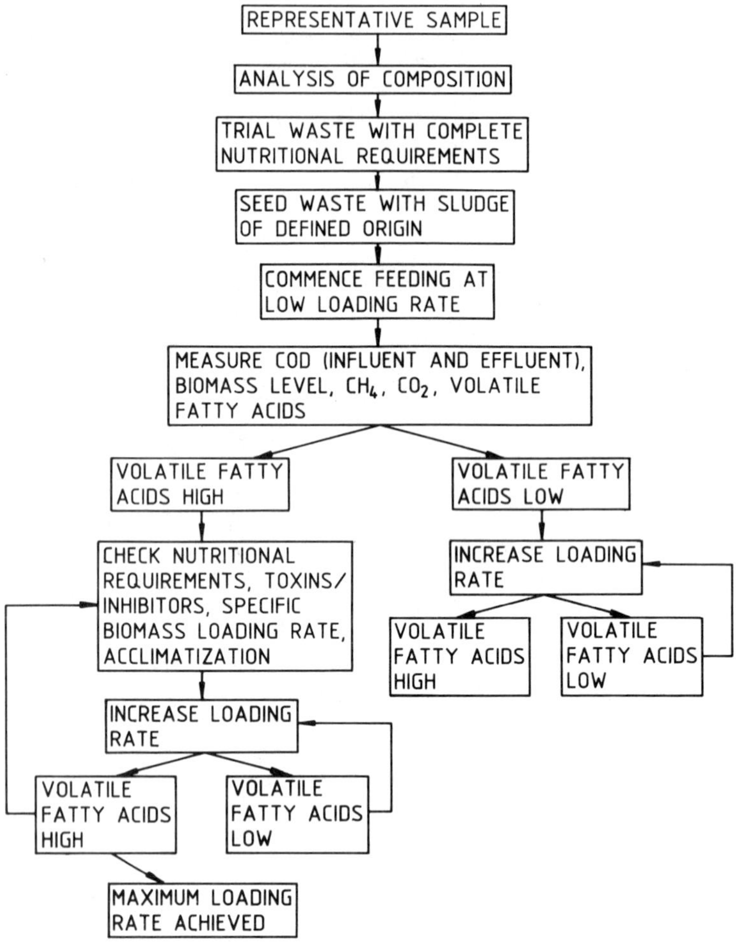

Figure 1 Strategy for development of an industrial-scale anaerobic digester from laboratory to pilot to full scale.

The process commences with a representative sample being obtained. Depending on the industry being studied and on other factors (such as a waste stream having a variable input in either composition or feed strength), this may be a straightforward task or may present an initial design choice. If the waste is relatively constant in composition and feed strength, representative sampling is straightforward. However, feed composition and strength often vary significantly. Generally, an equalization tank is placed before any full-scale anaerobic digester with the aim of reducing significant variations in waste strength. Most commonly, feed strength to digesters is maintained in the region of 5000-10,000 mg/L COD. Wastes of higher strengths are usually diluted (either by the addition of water or, more usually, by effluent recycle). This is certainly recommended for wastes of any complexity, although direct feeding of high-strength wastes for readily degradable wastes is possible (Barford et al., 1986). However, such feeding strategies are not generally recommended on the industrial scale, where they cause problems with effective feed distribution design necessary to avoid localized acid formation and where low volumetric flow may accentuate any aspect of bioreactor nonideality, such as bypassing. These aspects are discussed further in Section II.B.

Having obtained the best representative sample at an appropriate waste strength, it is then necessary for the composition of the waste to be comprehensively assessed. The extent to which the waste is analyzed is open to some question, although it is generally accepted that as a minimum the following should be analyzed: chemical oxygen demand or biologic oxygen demand (BOD), total Kjeldahl nitrogen, total suspended solids (TSS), and volatile suspended solids (VSS). These are generally, and desirably, estimated by standard methods in accordance with the American Public Health Association (1975). Modern instrumentation allows a much larger range of analyses to be routinely undertaken on various wastes to more completely quantify their composition. The elemental composition of a waste may be analyzed, after digestion by aqua regia, using inductively coupled plasma (ICP) emission spectrophotometry. We have routinely used this method in our laboratory to measure the composition of a number of wastes including stillages (Cail and Barford, 1985a-c), palm oil effluent (Cail and Barford, 1985d, e), and wool scouring effluent (Cail et al., 1986). In addition high-performance liquid chromatographic (HPLC) methods may be used to give the composition of the waste with respect to individual sugars (Prince and Barford, 1982), individual amino acids, individual vitamins, and so on.

These latter measurements, usually not undertaken as a routine, assume more importance when it is appreciated that it is a requirement for a rational laboratory-pilot-scale evaluation of a waste to provide a waste for trial that is nutritionally complete. This is essential since any deficiency seriously affects the performance of the laboratory trial upon which the ultimate full-scale design is based.

Current understanding of the nutritional requirements of anaerobic digesters has made significant advances in the past 10 years. However there does not yet exist a generally accepted unequivocal nutritionally complete medium. In laboratory-scale studies, the media used is one of the following two:

1. A synthetic, nondefined media (including such components as yeast extract, trace element solutions, and tryptic soy) (Isa et al., 1986; Hulshoff Pol et al., 1982)
2. An industrial waste supplemented with (generally) nitrogen, phosphorous, and various metals (Lettinga et al., 1984; Cail and Barford, 1985a-e; Cail et al., 1986; Lane, 1986; Gwatkin et al., 1987)

Adding nutrients to make an industrial waste nutritionally complete provides a significant challenge. Techniques (described above) used on a laboratory scale employing synthetic, nondefined media, although successful on this scale, do not generally provide an economically viable alternative at full scale. When the second option is employed, namely, supplementation of an industrial waste with defined chemical addition, the problem transfers to identifying the minimum requirement of a large range of nutrients, some possibly unknown as yet (e.g., growth factors) and often having interactions (e.g., metal ions and sulfides) (Callander and Barford, 1983b, c). Our knowledge of nutritional requirements and such interactions has dramatically improved over the past 10 years. The standard work for nutritional requirements in anaerobic digesters is that of Speece and McCarty (1964). However, recently, more detailed knowledge of the nutritional requirements of anaerobes has become available (Williams et al., 1986; Thauer and Morris, 1984; Diekert et al., 1985; Thauer et al., 1977), as well as significant understanding of their basic metabolism (Mah, 1983):

1. Importance of trace metals, such as iron (Hoban and van den Berg, 1979), nickel and cobalt (Williams et al., 1986; Thauer and Morris, 1984), and potassium, magnesium, and iron (Speece and McCarty, 1964).
2. Effect of sulfates and sulfides (Isa et al., 1986) and metal-sulfide interaction (Callander and Barford, 1983b, c).
3. Effect of toxins or inhibitors, such as detergents and accidental oil spillages (Gwatkin et al., 1987), disinfectants (Poels et al., 1984a), antibiotics (Poels et al., 1984a), petrochemical organics (Speece and Parkin, 1983), ammonium ions (Parkin et al., 1983), cyanide and chloroform (Yang et al., 1980), and formaldehyde (de Becker et al., 1983), among others.

Despite these advances, it is not possible either to define the full extent of the nutritional requirements of a particular waste from first principles or to identify the presence of toxins or inhibitors or predict their effect from first principles. These factors must be evaluated individually for each industrial waste. As shown in Figure 1, an industrial waste is generally supplemented with nitrogen, phosphorous (according to Speece and McCarty, 1964), and metals (using the increasingly available suggested strategies; e.g., Cail and Barford, 1985a). From this point, any further attention to nutritional aspects must be based on empirical studies. These are based on the results obtained when the waste is subjected to anaerobic digestion by a defined seed sludge.

After attempting to develop an industrial waste stream with complete nutritional requirements, the next step in the strategy for developing an industrial-scale anaerobic digester from laboratory to pilot to full scale is the seeding of this waste stream with a sludge of defined origin. It has been shown by practical experience in many laboratories that a sludge with an origin associated with the anaerobic digestion of the industrial waste in question, or of one of similar composition, affords the most desirable starting point. In our own laboratories, sludges taken from the settled material of an anaerobic lagoon treating cane juice stillage (Cail and Barford, 1985a), palm oil effluent (Cail and Barford, 1985d, e), and wool scouring effluent (Cail et al., 1986) have resulted in rapid and efficient initial digestion rates of these wastes. From this desirable starting point, the steady increase in loading rates until the maximum loading is reached (the next step in the strategy) can be performed with few operational problems. Such a seed sludge origin has many further advantages. It is acclimatized not only to the components of the waste that contribute to COD but also to any toxins or inhibitors that may be present. This latter factor generally makes a separate assessment of toxins and inhibitors less likely.

Despite the desirability of such a sludge origin, often, in a practical application, such a sludge may not be readily available either by virtue of existence or quantity (more likely a problem during full-scale start-up). In such cases it is common to use either a sludge of waste origin of similar composition and characteristics or, commonly, sewage digester sludge. The former strategy is suggested and confirmed experimentally by the ability of an acclimatized digester sludge to switch from one waste to another of similar characteristics without operational problems (Cail and Barford, 1985b). The latter has been discussed in detail by Lettinga et al. (1985) with particular reference to the upflow anaerobic sludge blanket (UASB) system in which sludge settleability is also an important consideration. Using such a sludge inevitably leads to a less rapid start-up of the laboratory reactor with sludge acclimatization necessary. The duration of this process is variable and poorly documented. Although Speece and Parkin (1983) report rapid acclimitization to various toxins, this does not guarantee that maximal feed loadings are achievable in such short acclimatization times. Obtaining maximal loadings on a laboratory scale is essential to an optimal full-scale design.

As a consequence, it is usual to seed a digester at a reasonable biomass concentration (1-5 kg VSS per m^3) and set a low initial loading rate in the region of 0.5-1 kg COD per m^3 per day.

From this starting point, the following strategy is employed, as illustrated in Figure 1. The following variables are measured: chemical oxygen demand of influent and effluent (mg/L), total suspended solids of digester contents (mg/L), volatile suspended solids (VSS) of digester contents (mg/L), volatile fatty acids (VFA), namely, acetic acid and propionic acid concentrations (mg/L), and methane and carbon dioxide contents of the gas formed (%). On the basis of these measurements, a continuing strategy is developed to obtain the maximum stable performance possible. At this point two factors are of importance. First, how does one know when optimal and then maximal digester performances are being achieved? Second, the digester type chosen and the characteristics of the waste have a significant effect on the ultimate maximal performance achievable and the strategy undertaken to achieve this. The maximum loading rate achievable is strongly dependent on waste and reactor type. Although it is difficult to generalize, Table 1 gives approximate figures for maximum loading rates at a conversion > 80% reduction of soluble COD for different reactor types (Callander and Barford, 1983a). Higher loading rates are generally possible for soluble wastes, albeit at the expense of treatment efficiency. Conversely, for wastes with considerable solids or insoluble components, an overall treatment efficiency of >80% would generally result in a considerably reduced loading rate.

Consequently, the strategy employed after the initial seeding will be similar for any reactor type and waste, but the maximum performance expected will vary from waste to waste and from reactor type to reactor type. This has been described in detail by Cail and Barford (1985a). A low initial loading rate (approximately 0.5-1 kg COD per m^3 per day) is used. Volatile fatty acid concentrations are monitored daily to check digester stability. The total VFA is not allowed to rise above 800-1000 mg/L. The space loading is only increased when VFA concentrations have fallen below 100 mg/L and then only in steps of 0.4-0.6 kg COD per m^3 per day. Some difference in detail exists between the system developed in our laboratory and that employed by others. For example, in the UASB system, higher loading rates and volatile fatty acid levels are tolerated to allow poorly settling sludge to wash out of the system (Lettinga et al., 1980, 1985). A comparison between this waste

Table 1 Performance Data of Modern Anaerobic Digestion Systems

Reactor type	Loading rates achievable (kg COD per m^3 per day)	Biomass level (g VSS per L)	Specific biomass loading rate (kg COD per kg VSS per day)
Contact process	5–10	20–30	0.17–0.5
UASB	10–15	30–40	0.25–0.5
Filter	10–20	10–25	0.4–2.0
Fluidized bed	20–30	30–40	0.5–1.0
Fixed film	10–20	12–16	0.8–1.2

treatment system and a similar system developed in our laboratories has been described by Cail and Barford (1985a, c). Similarly, somewhat faster initial loading rates may be used in anaerobic downflow stationary fixed film reactors to enhance biomass retention and growth rate (Kennedy and van den Berg, 1985).

Loading rates are consequently increased stepwise until maximal loading rates are obtained. As long as volatile fatty acid levels remain low, loading rates may be increased. However, volatile fatty acid levels may increase above that level defined as satisfactory. A number of factors may contribute to such a case, especially the following:

1. Maximum loading rate has been achieved.
2. Loading rate increase has outpaced the biomass concentration increase, leading to a level of specific biomass loading (kg COD per kg VSS per day) that cannot be tolerated.
3. A nutrient limitation has been reached.
4. A toxin or inhibitor has accumulated and is affecting biomass performance.

Although this is not an exhaustive list, it provides the most commonly observed problems in such waste evaluations. Strategies exist for considering each of these possibilities and modifying the strategies employed to overcome the immediate problems and allow a further increase in the maximum loading rate achievable. Briefly considering each of these factors:

1. If maximum loading rate has been achieved, the final maximum design loading rate has been obtained by definition.
2. If the loading rate increase has resulted in the specific biomass loading rates (kg COD per kg VSS per day) increasing to an excessive level, the solution will depend on the reactor type. If a CSTR with fixed biomass yield is employed, this factor will set the maximum loading rate possible since the biomass concentration may not be increased. For systems in which biomass concentration increases are possible, the problem will decrease with further biomass growth and accumulation. A reduction in the loading rate, in the short term, may assist the speed at which the reactor recovers by reducing the volatile fatty acid levels while biomass concentration is increasing.
3. The initial assessment is that nutrient requirements and waste supplementation has been inadequate or that toxins or inhibitors may have accumulated faster than the biomass may be able to acclimatize to them. The traditional method of assessing the extent to which these factors are affecting the ultimate performance of the digester is by the use of an assay procedure referred to as serum bottles (Owen et al., 1979). This method consists of a stoppered 50 mL conical flask, mounted in a water bath, and fitted with a 3 mm tube for gas removal and a 6 mm septum for gas purging and addition and removal of liquids using a 1 mL syringe. Hence, in principal, nutrient deficiencies may be identified and resolved by nutrient addition to a sample taken from the main laboratory digester and placed in this system. Volatile fatty acid levels would decrease, ideally, upon nutrient addition. With respect to the presence of toxins or inhibitors, these may be identified by the removal of toxicity or inhibition by digester supernatant dilution. Once identified, two solutions are available at present, either dilution of the feed or further long-term adaptation of the culture to the toxin or inhibitor.

Given these procedures, a full laboratory-scale evaluation may be obtained when the following data are available: loading rate (kg COD per m^3 per day), specific biomass loading (kg COD per kg VSS), treatment efficiency (% COD removed), total suspended solids (mg/L), volatile suspended solids (mg/L), and volatile fatty acids (mg/L).

Other factors that must be considered are generally a function of the reactor type used. For example, the UASB process (Lettinga et al., 1980, 1985; Hulshoff Pol et al., 1982) and upflow floc digester (Cail and Barford, 1985a-e; Barford et al., 1986; Cail et al., 1986) rely on the promotion of flocculation and granulation; the fluidized bed process relies on effective biomass attachment (Mulder, 1986; Borghans et al., 1986; Heijnen et al., 1986a, b; Enger et al., 1986; Salkineje-Salonea et al., 1983). Anaerobic downflow stationary fixed film reactors also rely on effective biomass attachment (van den Berg et al., 1985; Kennedy and Droste, 1985, 1986; Samson and Kennedy, 1985; Samson et al., 1985). Two-phase digesters rely on effective phase separation as well as biomass retention by flocculation, granulation, entrapment, or attachment (Pipyn et al., 1979; Isa et al., 1986; Huysman et al., 1983; Poels et al., 1984b). In general, it could be said that these factors play a small role in laboratory-scale anaerobic digesters and the reference works give a detailed description of how this is achieved. Information from full-scale digesters is not generally published since it is generally proprietary information, although experience at this level assumes equal if not greater importance.

B. Use of Laboratory and/or Pilot Experimental Data in Full-Scale Digester Design

Historically, with the very successful development of the UASB process, process development proceeded through laboratory to pilot to full scale (Lettinga et al., 1980, 1985). However, it is clear that major designers of full-scale anaerobic digesters now rely largely upon individual laboratory-scale waste evaluation and accrued experience with a number of wastes. In doing so there has been a general agreement that substantial loading rates and treatment efficiencies may be achieved by a number of reactor designs provided attention to engineering detail is paid on the full scale, adequate monitoring and instrumentation are undertaken at full scale, and a relatively conservative design loading criterion (perhaps 0.5-0.67 of the maximum design loadings obtained on laboratory scale) is used. Monitoring and instrumentation are the subject of a separate discussion in Section IV. Attention to engineering detail covers a range of factors. Given a relatively conservative choice of full-scale loading rate, most engineering effort has been placed in the following areas:

1. Maintaining active biomass at a desired concentration (kg VSS per m^3) and optimal specific sludge loading rate (kg COD per kg VSS per day).
2. Distribution of digester feed to avoid the localized formation of "acid spots" and any bypassing and dead regions.
3. Characterization of large-scale digesters using traditional nonideal reactor methodologies

Optimal specific sludge loading rate may be achieved by a well-managed sludge wastage strategy. Very few works have described the changing characteristics (i.e., sludge age) of biomass in long-term anaerobic digestion. Experience from our laboratories was presented by Barford et al. (1986) for a high-strength soluble waste. It is still quite difficult to accurately measure the activity of particular components of anaerobic digester flora, making selective wastage not possible at this stage.

In most cases, precautions are taken on a large scale to maintain the biomass concentration at a desired concentration. Much engineering design has consequently focused on an effective settling region at the digester exit (Lettinga et al., 1980, 1983a, b; 1985). This may be assisted by recycling of any significant biomass but has most recently led to the development of hybrid reactors in which advanced systems, such as the UASB, are followed by an aerobic filter of porous or nonporous material (e.g., Guiot and van den Berg, 1984b; Lettinga et al., 1983b; and commercialized by ACEC, Belgium).

Significant attention has been focused on the distribution of digester feed and the characterization of large-scale digesters using traditional nonideal reactor methodologies. Feed distribution, although a problem in all high-rate anaerobic digesters, has been discussed in detail only for the UASB system (Lettinga et al., 1983a, b). As a rough guideline, nozzle spacing of feed distributor points is recommended as <1 m^2 per nozzle at loading rates of 1-2 kg COD per m^3 per day for both dense flocculant (>40 kg dry solids per m^3) and granular sludge and 5 m^3 per nozzle at loading rates greater than 3 kg COD per m^3 per day for thin flocculant sludge (<40 kg dry solids per m^3).

Attempts to avoid, or at least quantify, the effect of dead spots and bypassing are generally performed on an operating digester by traditional nonideal reactor methodologies (Heertjes and van der Meer, 1978; Heertjes et al., 1982; Buijs et al., 1982; Samson and Kennedy, 1985; Samson et al., 1985; Kennedy and Guiot, 1986; Bolle et al., 1986a, b). In these studies, design factors important for the practical operation of digesters were obtained. With respect to the UASB digesters, for example, the work of Heertjes et al. (1982) and Buijs et al. (1982) indicated that the size of the sludge bed bypassing stream was influenced by the height of the sludge bed. In the case of a 200 m^3 reactor with a sludge bed only 0.4 m high, 53% of the influent bypassed the bed. The authors recommended that a sludge bed 1.5-2.5 m high would minimize this bypassing effect. They also suggested an optimum reactor height of 4-6 m to minimize dead zones while providing good mixing. Further studies that lead to quantitative data for short-circuiting and settler design have been undertaken by Bolle et al. (1986a, b).

It is clear from the preceding discussion that the design of full-scale digesters is based on laboratory-scale work modified to include significant input from engineering design experience on a full scale.

C. Start-up Strategies for Full-Scale Anaerobic Digesters

Very limited reported data exist on the start-up strategies of full-scale plants. By far the most complete description is that of Szendrey (1983), which describes the construction and start-up of a 3,500,000 (U.S.) gal downflow stationary fixed film reactor used by the Bacardi Corporation for the treatment of rum stillage. From this description, a good overall picture is given of many of the practical considerations that assume considerable importance in the construction and start-up of a reactor of this size. This case in particular is a good example of the forward planning required for such large systems, since in this case no signficiant quantities of anaerobic seed sludge were available. The seed sludge was grown up from a 3000 gal pilot filter, then transferred to a 10,000 gal NaOH storage tank temporarily converted for the purpose, then to a 750,000 gal holding tank as it was built. By the time the actual construction was finished 600,000 gal of active seed sludge had been produced. When the costs of the packing ($1,014,000) and total tankage ($3,121,400) are given, the significant scale of the project is clear. Although this description is reasonably complete, there are many proprietary aspects to the start-up as there are with other major suppliers of turnkey full-scale anaerobic digestion plants.

In general, it may be stated that start-up for a full-scale plant follows the same principles of start-up used in laboratory-scale digesters since the objectives are identical, namely, reliable and stable progression to a maximal loading rate that operates in a stable and controlled manner. However, there are also a large range of practical considerations that are as important, if not more so, when starting up on such a large scale. The necessity to consider sludge provision, as in the Bacardi case, is not uncommon. When starting up a UASB, either a large quantity of previously adapted granular sludge is used (which greatly assists rapid start-up) or the granular sludge must be developed from sewage sludge. This process has been outlined in detail. (Lettinga et al., 1980, 1983a, b; 1985). The start-up of any form of advanced digester may be further complicated when sewage sludge is not easily accessible. In fact, in some countries, sewage sludge is not available.

In an attempt to obtain some information on full-scale start-up procedures, the author wrote to all of the major companies involved in the design and commissioning of full-scale high-rate anaerobic digesters. As expected, little detailed information was forwarded with respect to digester start-up on the full scale. However, the publications received identified the major factors involved and some data relating to digester dynamics and stability were obtained. In the following section these are discussed.

III. DIGESTER STABILITY, DYNAMICS, AND CONTROL

As discussed in Section II.C, it may be considered that start-up of full-scale digesters is similar to that of laboratory- and/or pilot-scale digesters with significant engineering design components superimposed. In this section, the major factors involved in full-scale digester start-up are briefly outlined. Following this, the available experimental data concerning digester stability and dynamics (predominantly from laboratory and/or pilot scale) are discussed. Finally, the theoretical and practical basis for digester control is presented.

A. Major Factors Involved in Full-Scale Start-up

The following major high-rate anaerobic digestion systems are discussed to illustrate the range of factors involved: the upflow anaerobic sludge blanket, including modified UASB reactors; downflow stationary fixed film reactors and anaerobic filters; fluidized bed reactors; and two-phase digestion system.

1. *Upflow Anaerobic Sludge Blanket*

The major factors involved in the start-up of a UASB reactor are

1. Feed distribution
2. Solids-liquid-gas separation
3. Development of pelletization

The first two of these factors were discussed in Section II.C. The development of a well-pelletized (or granular) sludge with rapid settling properties is crucial to the optimum performance of the UASB. The factors contributing to this development are not well understood. Experiments on VFA-fed cultures have shown that three phases can be distinguished in the development of granules (Huishoff Pol et al., 1982). In phase 1 (low space loads), the sludge bed expands as gas production increases. The growth of filamentous organisms causes a deterioration in sludge settleability. In phase 2, these poorly settling fractions begin to be washed out of the reactor and the heavier particles begin to concentrate in the bottom of the reactor. Phase 3 occurs as granular growth begins to exceed washout and a net accumulation of biomass occurs. In this stage, loading rates can be increased more rapidly than in the first two stages. Furthermore, the experience gained by Lettinga and coworkers as a result of their extensive studies on a variety of wastes has enabled them to characterize the start-up and acclimatization of unadapted or poorly adapted seed sludge so that a granular sludge is developed:

1. The initial loading rate should be increased only when the volatile fatty acids (VFA) are well degraded. The increases should be relatively small to avoid overloading the biomass present.
2. Granulation is very waste dependent. It occurs most readily in easily acidified wastes, such as sugar beet and potato processing effluents, but is somewhat slower with render waste, distillery waste, and cornstarch waste.
3. The presence of Ca^{2+} appears to improve the pelletization process, but excess concentrations (200-300 mg/L) can cause problems by precipitating phosphorus.
4. High levels of suspended solids inhibit the formation of pellets.
5. It appears that for good pelletization to occur, the sludge loading must be above 0.6 kg COD per kg VSS per day.
6. High levels of NH_4^+ inhibit pelletization.
7. Two types of granules can form, consisting of predominantly either filamentous type or rod type organisms. The reasons for the difference are not clear.

In conclusion, it appears that although granulation can occur quite readily in some wastes (namely, 1-2 months), in others this process may take 6-9 months or granulation may not form at all.

Modified UASB reactors may be considered to be either digesters, such as the upflow floc digester (Cail and Barford, 1985a-e; Barford et al., 1986; Cail et al., 1986) in which flocculants are used to accelerate floc formation, or UASB reactors, which have a further anaerobic filter section attached to the top of the digester. The upflow floc digester has been directly compared to an UASB system with respect to start-up (on a laboratory or pilot scale) by Cail and Barford (1985a). They demonstrated in the laboratory that, compared with a control reactor operated along the lines of a UASB, start-up from a poorly adapted seed sludge was more rapid in the upflow floc digester. As expected this was largely due to improved biomass retention as a result of flocculant addition. Ultimately, after about 9 months, both systems developed a granular sludge and had similar performance.

When an UASB is modified to include an anaerobic filter, biomass retention within porous and nonporous packing has a significant effect on digester start-up. Factors affecting the colonization of nonporous and porous packing materials have been published (Huysman et al., 1983; Poels et al., 1984a). Although their precise rate of colonization and the specific contribution and requirements during start-up of a full-scale digester have not been presented, the process has been commercialized by ACEC (Belgium).

2. Downflow Stationary Fixed Film Reactors (DSFF) and Anaerobic Filters

The rate of start-up of such systems has been found to depend on the type of inoculum, the type and strength of waste, the level of volatile acids maintained, and the characteristics of the support material used and surface-volume rates (van den Berg et al., 1985; Kennedy and van den Berg, 1985). The full extent of the effect of these variables on start-up is discussed by these authors for laboratory- and/or pilot-scale reactors.

Operation in the downflow mode enables solids and undigested materials to be eliminated from the reactor itself. In common with all filters, there is no need for a gas-solid-liquid separator, and the evolution of gas in the channels results in a vigorous mixing action at the top of the bed where the feed is introduced. As a result of this vigorous mixing action at the top of the reactor, the design of the feed distributor is not critical; provided that there is a reasonable spread of entry points, the feed will be well mixed in the liquid head above the packing.

Once again the use of effluent recycle has been shown to have a highly beneficial effect on treatment rates and efficiencies in the DSFF reactor. Investigations by Duff and Kennedy (1983) showed that although recirculation had little effect on start-up, the maximum loading rate that could be applied to the reactor was higher at the higher rates of effluent recycle. They used recycle rates of up to 500:1.

Owing to the need to design the packing with channels sufficiently large to prevent clogging with suspended materials or surface biofilm, the specific surface areas in such reactors is limited to about 100 m^2/m^3. As a result, such reactors are not particularly well suited to the treatment of very dilute wastes (<2000 mg COD per L) owing to the short contact times involved with such effluents.

3. Fluidized Bed Reactors (FBR)

Two types of fluidized bed reactors are described in the literature. These are the anaerobic attached film expanded bed (AAFEB), which is operated in a partially fluidized state at about 10-20% bed expansion, and the fully fluidized bed reactors, which are operated at much higher (up to 88%) bed expansions (Jeris et al., 1974, 1977). The main factors affecting fluidized bed reactor design are

1. Fluidizing liquid distributor design
2. Particle, type, density, and size
3. Particle colonization by bacteria
4. Control of biomass thickness

The colonization of surfaces by microorganisms has widespread ramifications for both this reactor type and the previously described downflow attached film reactor. Considerable research has been undertaken on biofilm development. Ellwood et al. (1979), Characklis (1982), and Savage and Fletcher (1985) have published comprehensive reviews on this topic. The processes that contribute to biofilm development on a support phase are as follows.

1. The adsorption of organics. It has been found that in a microbial culture, an organic monolayer will rapidly deposit onto a clean surface. It has been suggested that this layer is a prerequisite for biologic attachment.
2. The rate of transport of microorganisms to the surface. This depends on the fluid flow characteristics.
3. Attachment to the surface. This is a complex topic involving a mixture of both physical and chemical forces. However, a common factor in the attachment process appears to be the production of polysaccharide binding material.
4. Microbial growth. This is affected by substrate diffusion into the biofilm matrix.
5. Re-entrainment of biofilm into the bulk liquid. This can occur as a result of fluid shear or sloughing of the whole film owing to nutrient starvation.

As expected, the colonization process is strongly affected by the nature of the support surface. Huysman et al. (1983) have published an interesting paper comparing the colonization of porous and nonporous packing materials. The materials tested were sepiolite (clay), zeolite (aluminosilicate), argex (fire-expanded clay), glass beads, activated carbon, sponge, unreticulated polyurethane foam, and reticulated polyurethane foam. Of the nonporous materials, the colonization velocity was greatest on sepiolite, decreasing through argex and zeolite, with no colonization occurring on the glass beads after 30 days. Reticulated polyurethane foam was the best of the porous materials. Thus from their results it appeared that the most important factors for rapid colonization were surface roughness for nonporous materials, and for porous materials the size of the pores and degree of porosity had the greatest influence. However, it should be borne in mind that in highly porous material the bacteria may face diffusional limitations if the length of the diffusional path between the cells and the free interspace solution is too great.

Van den Berg and Kennedy (1981) compared the colonization of needle-punched polyester construction material, red clay drain tiles, cylinders of grey potter's clay, and polyvinyl chloride (PVC) sheeting in downflow stationary fixed film reactors. Needle-punched polyester and red clay drain tiles were the most effective in developing an active biomass film in 40-50 days. PVC sheet gave the second worst performance, and a glass surface showed very little colonization at all. The effectiveness of needle-punched polyester was most likely due to the fine structure of the feltlike material, which would easily entrap the bacteria.

Two main types of anaerobic reactors depending on this process can be recognized. One, which consists of a tank randomly packed with support material, such as that described above, and operated in an upflow manner, and the other, the downflow stationary fixed film reactor in which the support surfaces are structured to allow relatively unimpeded flow down the surface and that, as the name implies, is operated in a downflow manner.

One of the most important factors in achieving uniform fluidization of the reactor is the design of the liquid distributor at the base of the reactor. Despite this, or perhaps because of it, there is little published information on different designs. Lettinga et al. (1983a) suggest that for good fluidization to be achieved there would need to be at least one influent entry point every square meter. Other papers have described the use of a graded bed of "pea" gravel or anthracite near the base of the reactor to assist in the uniform distribution of the fluid flow. If a reasonably even fluid flow is not achieved there is the serious possibility of channeling and short-circuiting of the liquid through the packed bed.

Most FBR reactors are designed as straight wall reactors of relatively high height-diameter ratio; however, a number of investigators point to some potential disadvantages with such systems. They have suggested that in addition to the difficulty of designing feed distributors that give even flow distribution throughout the reactor, such reactors at high bed expansions have a relatively narrow range of optimum operating conditions and thus low stability. To circumvent these problems, Scott and Hancher (1976) describe the development of a tapered column design that has the appearance of an inverted cone in which feed is introduced at the apex and flows upward. They suggested that owing to the gradual expansion in cross section, the flow should be more stable throughout the reactor than in conventional designs. Boening and Larsen (1982) compared the performance of a straight wall FBR and a tapered FBR reactor treating whey. Although it was not made entirely clear by the authors, the performance of the tapered reactor was said to be better.

With respect to support phase selection, a particular medium must be selected that requires minimal energy for fluidization (light and spherical) but that does not wash out of the reactor and can withstand the abrasive forces in the reactor. The optimum superficial velocity in an FBR is that which just suspends the largest particle but does not wash out the smallest; at this point bed porosity (or voidage) is minimal.

A variety of support materials have been used in fluidized bed reactors. These have included bonemeal, coal, aluminium oxide pellets, spent ion-exchange resins, and sand. Although the use of light particles is desirable from the viewpoint of reducing pumping costs, many of these materials are expensive and do not stand up particularly well to the attrition and shear forces experienced in the biologic film or particle separators. Sand appears to be one of the cheapest and toughest materials for use in a FBR.

4. *Two-Phase Digestion Systems*

These systems generally involve similar start-up problems as those associated with the UASB design or modified UASB designs, since most two-phase systems incorporate the same principles as UASB design, merely separating the phases (Cohen, 1983).

B. Digester Stability and Dynamics: Experimental Data

Significant experimental data exist establishing the operational stability and dynamic response of laboratory- or pilot-scale digesters to process pertubations for most of the modern digester types, including UASB (Gwatkin et al., 1987; Lettinga, 1980; Hulshoff Pol et al., 1982; Pipyn et al., 1979) and modified UASB reactors (Kennedy and Guiot, 1986; Guiot and van den Berg, 1984a, b), upflow floc digester (Cail and Barford, 1985b); anaerobic fixed film reactors (Kennedy et al., 1985; van den Berg et al., 1985; Kennedy and van den Berg, 1985), anaerobic filters (Szendrey, 1983), and fluidized beds (Mulder, 1986).

In general, it is clear from these works that modern anaerobic digesters are inherently stable to shock overloads. This is inevitably due to the larger biomass level existing in the digesters and an inherent buffer capacity due to the biomass operating at the low end of the range of specific sludge loading rates (kg COD per kg

VSS per day). This is also demonstrated during start-up following an extensive shutdown.

The following evidence for digester stability and controlled dynamic response has been presented on a laboratory and pilot scale. These are only illustrative and not meant to be exhaustive.

Gwatkin et al. (1987) illustrated the inherent stability of a laboratory-scale UASB digester to temperature, buffer capacity, pH, and recycling short-term variations.

Lettinga (1980), Hulshoff Pol et al. (1982), and Pipyn et al. (1979) have demonstrated that sudden organic overloading has only limited consequences (with only a small short-term reduction in COD removal recorded).

Cail and Barford (1985b) illustrated the stability of the upflow floc digester to shock overload by giving a prolonged shock load averaging 83 kg COD per m^3 per day to a sweet sorghum stillage digester (2.5-3.0 times the steady-state load) for 24 h. This led to a temporary rise in volatile fatty acid levels (to 3454 mg/L acetic and 3697 mg/L propionic), which was quickly reduced to preshock levels (<200 mg/L total acids) by reducing the space load to 24 kg COD per m^3 per day for 1-2 days before returning it to the preshock level of 30 kg COD per m^3 per day.

Kennedy et al. (1985) have illustrated the stability of the anaerobic fixed film reactors to a shock organic concentration overload by giving a five to sevenfold increase in organic load at an approximately constant hydraulic retention time. Provided the waste was well buffered, the digester recovered to normal performance in a relatively short period of time (24-72 h), with the dynamic response characterized by elevated volatile acid concentrations, decreased reactor pH, and excess biofilm sloughing.

These reactors have also been shown to be stable to hydraulic overloading (van den Berg and Kennedy, 1981, 1982) and changing feed from one waste to another (van den Berg and Kennedy, 1982). During hydraulic shock overloading, reactors returned to normal operation 24-48 h after a 10-fold increase in loading rate to 94 kg COD per m^3 per day for 24 h at 35°C. Although thermophilic digesters did not appear to have the same stability to hydraulic overloading, the studies were not as complete as the mesophilic experiments.

Mulder (1986) described a shock 10-fold feed concentration overload to a laboratory- or pilot-scale fluidized bed reactor. It was claimed that this shock gave no operational problems except that the COD removal efficiency was low during and shortly after the period of overload.

From these observations, taken in conjunction with the discussion in Section II.A with respect to toxins and inhibitors, it may be concluded that all high-rate anaerobic digestion processes are inherently stable to shock overloads when sufficient buffering capacity and/or pH control is maintained. There is no reason to believe that those laboratory or pilot plant observations do not translate to the full scale, provided the engineering design factors of importance, which have been discussed previously, are adequately addressed.

C. Digester Control

There are a number of levels on which digester control may be considered. The initial choice is whether an attempt will be made to understand the complex interactions occurring within a digester or whether a number of control variables will be identified, the process considered as a black box, and control strategies simply related by process control to these identified major control variables. Additionally, the scale of operation has a significant effect on the control strategies chosen.

Considering the scale of operation, it could generally be said that at laboratory scale, little process control of any sophistication is used. Basically important control variables, usually pH and volatile fatty acids, are identified and controlled within certain boundary conditions, often manually to some extent if not fully. Generally pH, for example, is measured by sampling at this scale with the digester pH set to the desired range (generally 6.5-7.5) by trial-and-error adjustment of the feed. Similarly, volatile acid levels are measured, generally by extraction, and gas chromatography methods using the supernatant of a digester sample. Again, a control range is defined. Generally, it is not advisable to allow the total volatile fatty acid levels to exceed 2000-3000 mg/L, with the desirable range 0-500 mg/L. When the feed rate to the laboratory digester is reduced manually, by trial and error, the volatile fatty acid levels are reduced to the desired operating range.

The extent to which these two processes are replaced by more sophisticated control processes is determined to a significant extent by scale in addition to likely operating costs on a large scale. As discussed in detail in Section IV, modern sophisticated anaerobic digestion systems generally include continuous on-line pH measurement and automatic control by in-line acid or alkali addition using in-line mixing. Although an in-line process control scheme could be envisaged for VFA control (using high-performance liquid chromatography in-line measurement, for example, or automated gas chromatography; Duffy, 1986), this would normally not be considered economically viable.

It is reasonable to say that any attempt to obtain a detailed understanding of the interactions that occur in anaerobic digesters have been limited to fundamental studies. These have been used only in a limited sense on a laboratory scale (Rozzi and Labellarte, 1984; Rozzi and Passino, 1985) for control and not, at this time, on a large scale. However, many attempts have been made to understand the complex interactions of digesters with the ultimate aim of allowing sophisticated process control to be undertaken at a later date. These "models" are based on a variety of structures and concepts, such as bicarbonate monitoring (Rozzi and Labellarte, 1984; Rozzi et al., 1985), kinetics (McCarty, 1971; Mosey, 1983), hydrogen gas concentration (Heyes and Hall, 1981; McCarty, 1981), substrate inhibition (Andrews, 1969), pH (Graef and Andrews, 1973), and propionic acid formation (Sinechal et al., 1979). Despite their lack of application, models of increasing sophistication have been proposed. These have resulted in an approach that clearly differentiates between physicochemical parameters and the various microbial populations that contribute to the overall microbial response (Rozzi et al., 1985; Rozzi and Passino, 1985; Heyes, 1984). In the author's opinion, although these models have not been applied on a full scale to any extent, they greatly contribute to our understanding of the anaerobic digestion system by systematically structuring the increasing knowledge that is becoming available in such areas as the kinetics of individual microbial species. In addition, they are starting to play a more important role in digester design and process control.

Even though process control is limited on a laboratory scale and the use of mathematical models is limited on the industrial scale, it should not be concluded that sophisticated monitoring and instrumentation does not exist on industrial-scale anaerobic digesters. In fact, as previously discussed, modern full-scale anaerobic digestion plants are highly instrumented to ensure stable operation and control. This generally involves attention to engineering design detail to measure and control such variables as liquid levels, pH, temperature, and feed rates. This is detailed and discussed in the next section.

IV. MONITORING AND INSTRUMENTATION OF A FULL-SCALE ANAEROBIC DIGESTER

As discussed previously, although control strategies are simplistic in their basis, the level of instrumentation and monitoring in recent advanced anaerobic digester design is high. Much of the more reliable operation and control of recent advanced systems can be directly attributed to this monitoring and instrumentation. Generally, the details of monitoring and instrumentation are only provided by companies to potential clients through detailed flow charts that are proprietary and generally covered by secrecy agreements. As a consequence, these are not readily available in published form.

An exception is the work of Pipyn et al. (1979), who published, in reasonable detail, a flowchart for a two-phase anaerobic upflow reactor, including some basic instrumentation. Although the commercial application of this system by ACEC (Belgium) has more sophisticated monitoring and instrumentation, the basic detail of instrumentation described in this work is indicative of the important variables onto which process control must be exerted.

Figure 2 gives a generalized schematic diagram for anaerobic digestion plant instrumentation. This diagram is meant to give only an indication of the minimum level of monitoring and instrumentation necessary and is not necessarily indicative of particular supplier or an advanced anaerobic digestion treatment system. Each supplier varies in detail from this generalized schematic.

In Figure 2, it is assumed that the process of anaerobic digestion can be performed on a full scale with two main tanks. The first tank may be considered either an equalization tank or an acidification tank (first reactor). The level of monitoring and instrumentation may vary depending on the definition of this stage. If the first tank is strictly a bioreactor, then it may be preceded by a third tank. For the purpose of this generalization, two tanks are considered, with the first tank closely monitored and controlled. The first reactor is controlled for temperature, pH of incoming feed, and volumetric flow rate into the reactor.

This is achieved by temperature, pH, and level indicator-controllers with alarms. From this reactor, effluent is taken and, after pH adjustment with an in-line mixing system using a pH indicator-controller with alarm, fed to the second reactor from which final effluent is either recycled to dilute the initial feed (helping to equalize waste strength variations) or discharged. The level of sophistication of control (e.g., temperature or level) of this reactor may vary and depends on the overall performance of the process (e.g., heat losses).

By these simple control sequences, it may be ensured that the digesters are operated in a fashion that minimizes if not eliminates the major causes of instability, namely, such process variations as pH, temperature, hydraulic overload, and to a large extent waste strength variations.

The potential for using other control variables (e.g., volatile fatty acid concentrations, hydrogen gas concentration, and treatment efficiency) have not been applied on a full scale and are still very much in the development stage. They will require a more mechanistic understanding of the complex interactions occurring in anaerobic digesters. As previously discussed, recent developments in modeling

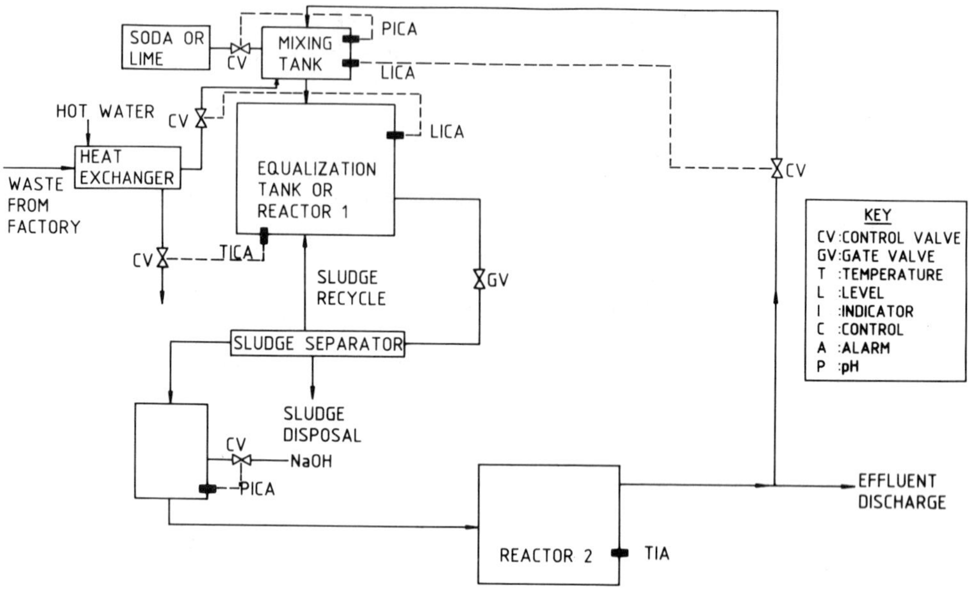

Figure 2 Generalized schematic diagram for anaerobic digestion plant instrumentation.

will assist this development. Other instrumentation also shows promise, for example the use of a microcalorimeter to detect, at an early stage, the presence of toxins (Heijen et al., 1986a).

Unlike the full scale plant, laboratory-scale anaerobic digestion systems are often manually controlled on a day-to-day basis (by measurement of VFA and treatment efficiency) and little precaution is taken to guard against hydraulic overload by erratic pump behavior by the use of level control. The lack of sophisticated control on this scale is deceptive in that it is totally inadequate for stable operation on a full scale.

V. CONCLUSIONS

When subject to a rigorous start-up procedure, modern anaerobic digestion plants at full scale can be relied upon for stable operation. Poor operational stability can generally be blamed on inadequate laboratory- and/or pilot-scale studies, poor start-up procedures, or insufficient attention to engineering design, including monitoring and instrumentation.

REFERENCES

American Public Health Association (1975). *Standard Methods for the Examination of Water and Wastewater*, 14th Ed. American Public Health Association, Washington, D.C.

Andrews, J. F. (1969). Dynamic model of the anaerobic digestion process. *J. Sanit. Eng. Div. Proc. ASCE SAI*:95-116.

Barford, J. P., Cail, R. G., Callander, I. J., and Floyd, E. J. (1986). Anaerobic digestion of high strength cheese whey utilizing semi-continuous digesters and chemical flocculant addition. *Biotechnol. Bioeng. 28*:1601-1607.

Boening, P. H., and Larsen, V. F. (1982). Anaerobic fluidised bed whey treatment. *Biotechnol. Bioeng. 14*:2539-2556.

Bolle, W. L., van Breugel, J., van Eybergen, G. C., Kossen, N. W. F., and Zoetemeyer, R. J. (1986a). Modelling the liquid flow in upflow anaerobic sludge blanket reactors. *Biotechnol. Bioeng. 28*: 1615-1620.

Bolle, W. L., van Breugel, J., van Eybergen, G. C., Kossen, N. W. F., and van Gils, W. (1986b). An integral dynamic model for the UASB reactor. *Biotechnol. Bioeng. 28*:1621-1636.

Borghans, A. J. M. L., van Gils, W. M. A., Hiusman, B. C., and Wind, E. (1986). 10 Years experience of the full-scale anaerobic treatment of sugar industry waste water in the Netherlands with the biothane process, poster and lecture presentation, EWPCA Conference, September, Amsterdam.

Buijs, C., Heertjes, P. M., and van der Meer, R. R. (1982). Distribution and behaviour of sludge in upflow reactors for anaerobic treatment of wastewater. *Biotechnol. Bioeng. 24*:1975-1989.

Cail, R. G., and Barford, J. P. (1985a). A comparison of an upflow floc (tower) digester and UASB system treating cane juice stillage. *Agric. Wastes 14*(4):291-300.

Cail, R. G., and Barford, J. P. (1985b). An evaluation of the performance of an upflow floc (tower) digester treating sugar beet and sweet sorghum stillages. *Biomass 6*:279-285.

Cail, R. G., and Barford, J. P. (1985c). The development of granulation in an upflow floc digester and an upflow anaerobic sludge blanket digester treating cane juice stillage. *Biotechnol. Lett. 7*(7):493-498.

Cail, R. G., and Barford, J. P. (1985d). Mesophilic semi-continuous anaerobic digestion of palm oil mill effluent. *Biomass 7*:287-295.

Cail, R. G., and Barford, J. P. (1985e). Thermophilic semi-continuous anaerobic digestion of palm-oil mill effluent. *Agric. Wastes 13*:295-304.

Cail, R. G., Barford, J. P., and Lichacz, R. (1986). Anaerobic digestion of wool scouring wastewater in a digester operated semi-continuously for biomass retention. *Agric. Wastes 18*:27-38.

Callander, I. J., and Barford, J. P. (1983a). Recent advances in anaerobic digestion technology. *Proc. Biochem. 18*(4):24-30.
Callander, I. J., and Barford, J. P. (1983b). Precipitation, chelation, and the availability of metals as nutrients in anaerobic digestion. 1. Methodology. *Biotechnol. Bioeng. 25*:1937-1958.
Callander, I. J., and Barford, J. P. (1983c). Precipitation, chelation, and the availability of metals as nutrients in anaerobic digestion. 2. Applications. *Biotechnol. Bioeng. 25*:1959-1972.
Callander, I. J., and Barford, J. P. (1984). Improved anaerobic digestion of pig manure using a tower fermenter. *Agric. Wastes 11*(1):1-24.
Characklis, W. G. (1982). Fouling biofilm development: A process analysis. *Biotechnol. Bioeng. 24*:1923-1960.
Cohen, A. (1982). Optimisation of anaerobic digestion of soluble carbohydrate containing waste waters by phase separation. Ph.D. thesis, University of Amsterdam.
Cohen, A. (1983). Two phase digestion of liquid and solid wastes. Proceedings of the Third International Symposium on Anaerobic Digestion, Boston, pp. 123-138.
de Becker, P., Jans, T., and Piscaer, P. (1983). Anaerobic treatment of formaldehyde-containing wastewater. In *Proc. Eur. Symp. AWWT, Noordwijkerhaut*, W. J. van den Brink (Ed.). TNO Corporate Communication Department, The Hague, The Netherlands, pp. 449-463.
Diekert, G., Fuchs, G., and Thauer, R. K. (1985). Properties and function of carbon monoxide dehydrogenase from anaerobic bacteria. In *Microbial Gas Metabolism: Mechanistic, Metabolic and Biotechnological Aspects*, Society for General Microbiology, Cambridge University Press, Cambridge.
Duff, S. J. B., and Kennedy, K. J. (1983). Effect of effluent reticulation on start-up and steady-state operation of the downflow stationary fixed film (DSFF) reactor. *Biotechnol. Lett. 5*:317-320.
Duffy, M. L. (1986). Advantages and disadvantages of wide-bore, thick-film capillary columns in GC. *Int. Lab. 1*(1):78-87.
Ellwood, D. C., Melling, J., and Rutter, P. (1979). *Adhesion of Microorganisms to Surfaces*. Academic Press, London.
Enger, W. A., van Gils, W. M. A., Heijen, J. J., and Koevoets, W. A. A. (1986). Full-scale performance of a fluidised bed in a two-stage anaerobic waste water treatment at Gist-Biocades, lecture presentation. EWPCA Conference, September, Amsterdam.
Graef, S. F., and Andrews, J. F. (1973). Mathematical modelling and control of anaerobic digestion. *14th Annual Meeting. AIChE Symp. Ser. 136*:101-131.
Guiot, S. R., and van den Berg, L. (1984a). Dynamic performance of an anaerobic reactor combining an upflow sludge blanket and

a filter for the treatment of sugar waste. 39th Industrial Waste Conference, Purdue University, pp. 704-717.

Guiot, S. R., and van den Berg, L. (1984b). Performance and biomass retention of an upflow anaerobic reactor combining a sludge blanket and a filter. *Biotechnol. Lett.* *6*(3):161-164.

Gwatkin, P. G., Dodds, J. G., and Lane, A. G. (1987). Pilot-scale anaerobic upflow sludge bed treatment of dried vine-fruit wash-water. *Food Technol. Aust.* *39*(8):373-382.

Heertjes, P. M., and van der Meer, R. R. (1978). Dynamics of liquid flow in an upflow reactor, used for anaerobic treatment of wastewaters. *Biotechnol. Bioeng.* *20*:1577-1592.

Heertjes, P. M., Kuijvenhoven, L. J., and van der Meer, R. R. (1982). Fluid flow pattern in upflow reactors for anaerobic treatment of wastewaters. *Biotechnol. Bioeng.* *24*:443-459.

Heijnen, J. J., Hoeks, F. W. J. M. M., and Mourik, J. (1986a). Application of a microcalorimeter as a fast toximeter for the protection of anaerobic treatment systems, poster presentation. EWPCA Conference, September, Amsterdam.

Heijnen, J. J., Hoeks, F., and Mourik, J. (1986a). Application of a microcalorimeter as a fast toximeter for the protection of anaerobic treatment systems, poster presentation. EWPCA September, Amsterdam.

Heyes, R. H. (1984). The kinetic modelling of anaerobic digester organisms, Ph.D. thesis. University of New South Wales, Kensington, Australia.

Heyes, R. H., and Hall, R. J. (1981). Anaerobic digestion modelling—the role of H_2. *Biotechnol. Lett.* *3*(8):431-436.

Hoban, D. J., and van den Berg, L. (1979). Effect of iron on conservation of acetic acid to methane during methanogenic fermentations. *J. Appl. Bacteriol.* *47*:153-159.

Hulshoff Pol, L. W., de Zeeuw, W. J., Velzeboer, C. T. M., and Lettinga, G. (1982). Granulation in UASB reactors. Proceedings IAWPR Specialised Seminar on Anaerobic Treatment of Wastewater in Fixed Film Reactor, Copenhagen, June 16-18, pp. 305-318.

Huysman, P., van Meenen, P., Van Assche, P. and Verstraete, W. (1983). Factors affecting the colonization of non porous and porous packing materials in model upflow methane reactors. *Biotechnol. Lett.* *5*(9):643-648.

Isa, Z., Grusenmeyer, S., and Verstraete, W. (1986). Sulfate reduction relative to methane production in high rate anaerobic digestion: Technical aspects. *Appl. Environ. Microbiol.* *51*(3):572-579.

Jeris, J. S., Beer, C., and Mueller, J. A. (1974). High rate biological denitrification using a granular fluidisation bed. *J. Water Pollut. Control Fed.* *46*(9):2118-2128.

Jeris, J. S., Owens, R. W., Hickey, R., and Flood, F. (1977). Biological fluidised bed treatment for BOD and nitrogen removal. *J. Water Polut. Control Fed. 49*:816-829.
Kennedy, K. J., and Droste, R. L. (1985). Startup of anaerobic downflow stationary fixed film (DSFF) reactors. *Biotechnol. Bioeng. 27*:1152-1165.
Kennedy, K. J., and Droste, R. L. (1986). Anaerobic fixed film reactors treating carbohydrate wastewater. *Water Res. 20*(6): 689-695.
Kennedy, K. J., and Guiot, S. R. (1986). Anaerobic upflow bed-filter-development and application. Proceedings of the IAWCPR Post Conference International Seminar on Anaerobic Treatment in Tropical Countries, São Paolo, Brazil, August.
Kennedy, K. J., and van den Berg, L. (1985). Anaerobic downflow stationary fixed film reactors. In *Comprehensive Biotechnology*, C. L. Cooney and A. E. Humphrey (Eds.). Pergamon Press, New York, pp. 1027-1049.
Kennedy, K. J., Muzar, M., and Copp, G. H. (1985). Stability and performance of mesophilic anaerobic fixed-film reactors during organic overloading. *Biotechnol. Bioeng. 27*:86-93.
Lane, A. G. (1986). Start-up, operating requirements and granule formation during upflow sludge bed treatment of a strong food processing effluent. *Environ. Technol. Lett. 7*(11):555-564.
Lettinga, G. (1980). Anaerobic treatment of low strength wastes. Inter University Course on Anaerobic Digestion, Dijon, December.
Lettinga, G., van Velsen, A. F. M., Hobma, S. W., De Zeeuw, W., and Klapwijk, A. (1980). Use of the upflow sludge blanket (USB) concept for biological wastewater treatment, especially for anaerobic treatment. *Biotechnol. Bioeng. 22*:699-734.
Lettinga, G., Hobma, S. W., Hulshoff Pol, L. W., de Zeeuw, W., de Jong, P., Grin, P., and Roersma, R. (1983a). Design operation and economy of anaerobic treatment. *Water Sci. Technol. 15*: 177-195.
Lettinga, G., Hulshoff Pol, L. W., Wiegant, W., de Zeeuw, W., Hobma, S. W., Grin, P., Roersma, R., Sayed, S., and van Velsen, A. F. M. (1983b). Upflow sludge blanket processes. Proceedings of the Third International Symposium on Anaerobic Digestion, Boston, August, pp. 139-156.
Lettinga, G., Hulshoff Pol, L. W. H., Koster, I. W., Wiegant, W. M., de Zeeuw, W. J., Rinzema, A., Grin, P. C., Roersma, R. E., and Hobma, S. W. (1984). High-rate anaerobic waste-water treatment using the UASB reactor under a wide range of temperature conditions. *Biotechnol. Genet. Eng. Rev. 2*:253-284.
Lettinga, G., de Zeeuw, W., Hulshoff Pol, L., Wiegant, W., and Rinzema, A. (1985). Anaerobic wastewater treatment based on

biomass retention with emphasis on the UASB-process. Proceedings of the International Symposium on Anaerobic Digestion. Guangzhou, China, pp. 279-301.
Mah, R. A. (1983). Interactions of methanogens and non-methanogens in microbial ecology. Proceedings of the Third Symposium on Anaerobic Digestion, Boston, August, pp. 11-22.
McCarty, P. L. (1971). Energetics and kinetics of anaerobic treatment. *Adv. Chem. Ser. 105*:91-107.
McCarty, P. L. (1981). One hundred years of anaerobic treatment. Proceedings of the Second International Symposium on Anaerobic Digestion, Travemunde, Germany, September, pp. 3-22.
Mosey, F. E. (1983). Kinetic descriptions of anaerobic digestion. Proceedings of the Third International Symposium on Anaerobic Digestion, Boston, August, pp. 37-52.
Mulder, A. (1986). Anaerobic purification of acidified yeast wastewater in laboratory fluidised bed reactors, poster presentation. EWPCA. Conference, September, Amsterdam.
Owen, W. F., Stuckey, D. C., Healy, F. B., Young, L. Y., and McCarty, P. L. (1979). Bioassay for monitoring biochemical methane potential and anaerobic toxicity. *Water Res. 13*:485-492.
Parkin, G. F., Speece, R. E., Yang, C. H. J., and Kocher, W. M. (1983). Response of methane fermentation systems to industrial toxicants. *J. Water Pollut. Control Fed. 55*:44-53.
Pipyn, P., Verstraete, W., and Ombregt, J. P. (1979). A pilot scale anaerobic upflow reactor treating distillery wastewaters. *Biotechnol. Lett. 1*:495-500.
Poels, J., Van Assche, P., and Verstraete, W. (1984a). Effects of disinfectants and antibiotics on the anaerobic digestion of piggery waste. *Agric. Wastes 9*:239-247.
Poels, J., van Assche, P., and Verstraete, W. (1984b). High rate anaerobic digestion of piggery manure with polyurethane sponges as a support material. *Biotechnol. Lett 6*(11):747-752.
Prince, I. G., and Barford, J. P. (1982). Tower fermentation of sugar cane juice. *Biotechnol. Lett. 4*:469-474.
Rozzi, A., and Labellarte, G. (1984). Direct bicarbonate determination in anaerobic digester liquors by measurement of the pressure of carbon dioxide. *Proc. Biochem. 19*(12):201-203.
Rozzi, A., and Passino, R. (1985). Mathematical models in anaerobic treatment processes. In *Mathematical Models in Biological Waste Water Treatment*, S. E. Jorgensen and M. J. Gromiec (Eds.). Elsevier, Amsterdam, pp. 637-690.
Rozzi, A., Di Pinto, A. C., and Brunetti, A. (1985a). Anaerobic process control by bicarbonate monitoring. *Environ. Technol. Lett. 6*:594-601.

Rozzi, A., Merlini, S., and Passino, R. (1985b). Development of a four population model of the anaerobic degradation of carbohydrates. *Environ. Technol. Lett.* *6*:610-619.

Salkineje-Salonea, M. S., Nyns, E. J., Sutton, P. M., van den Berg, L., and Wheatley, A. D. (1983). Starting-up of an anaerobic fixed film reactor. *Water Sci. Technol.* *15*:305-308.

Samson, R., van den Berg, R., and Kennedy, K. J. (1985). Mixing characteristics and startup of anaerobic downflow stationary fixed film (DSFF) reactors. *Biotechnol. Bioeng.* *27*:10-19.

Samson, S., and Kennedy, K. J. (1985). Effect of reactor height on mixing characteristics and performance of the anaerobic downflow stationary fixed-film (DSFF) reactor. *J. Biotechnol.* *2*:95-106.

Savage, D. W., and Fletcher, M. (Eds.) (1985). *Bacterial Adhesion.* Plenum Press, New York.

Scott, C. D., and Hancher, C. W. (1976). Use of a tapered fluidised bed as a continuous bioreactor. *Biotechnol. Bioeng.* *18*:1393-1403.

Sinechal, X. J., Installe, M. J., and Nyns, E. J. (1979). Differentiation between acetate and higher volatile acids in the modelling of the anaerobic biomethanisation process. *Biotechnol. Lett.* *1*: 309-314.

Speece, R. E., and McCarty, P. L. (1964). Nutrient requirements and biological solids accumulation in anaerobic digestion. Adv. Water Pollut. Res. Proceedings of the First International Conference on Water Pollution Research *2*:305-322.

Speece, R. E., and Parkin, G. F. (1983). The response of methane bacteria to toxicity. Proceedings of the Third International Symposium on Anaerobic Digestion, Boston, August, pp. 23-35.

Szendrey, L. M. (1983). Start up and operation of the Bicardi corporation anaerobic filter. Proceedings of the Third International Anaerobic Digestion Conference, Boston, August.

Thauer, R. K., and Morris, J. G. (1984). Metabolism of chemotropic anaerobes: Old views and new aspects. In *The Microbe 1984*, Part II, *Prokaryotes and Eukaryotes*, Society for General Microbiology Symposium 36, D. P. Kelly and N. G. Carr (Eds.). Society for General Microbiology, Cambridge University Press, Cambridge, pp. 123-168.

Thauer, R. K., Jungermann, K., and Decker, K. (1977). Energy and conservation in chemotrophic anaerobic bacteria. *Bacteriol. Rev.* *41*(1):100-180.

van den Berg, L., and Kennedy, K. J. (1981). Support materials for stationary fixed film reactors for high-rate methanogenic fermentations. *Biotechnol. Lett* *3*(4):165-170.

van den Berg, L., and Kennedy, K. J. (1982). Comparison between intermittent and continuous loading of stationary fixed film reactors for methane production from wastes. *J. Chem. Technol. Biotechnol.* *32*:427-432.

van den Berg, L., Kennedy, K. J., and Samson, R. (1985). Anaerobic downflow stationary fixed film reactors: Performance under steady state and non-steady state conditions. *Water Sci. Technol.* *17*:89-102.

Williams, C. M., Shih, J. C. H., and Spears, J. W. (1986). Effect of nickel on biological methane generation from a laboratory poultry waste digester. *Biotechnol. Bioeng.* *28*:1608-1610.

Yang, C. H. J., Speece, R. E., Parkin, G. F., Gosset, J., and Kocher, W. M. (1980). The response of methane fermentation to cyanide and chloroform. *Prog. Water Technol.* *12*:977-989.

Index